Quantum Transport
Introduction to Nanoscience

Quantum transport is a diverse field, sometimes combining seemingly contradicting concepts – quantum and classical, conducting and insulating – within a single nano-device. Quantum transport is an essential and challenging part of nanoscience, and understanding its concepts and methods is vital to the successful design of devices at the nano-scale.

This textbook is a comprehensive introduction to the rapidly developing field of quantum transport. The authors present the comprehensive theoretical background, and explore the groundbreaking experiments that laid the foundations of the field. Ideal for graduate students, each section contains control questions and exercises to check the reader's understanding of the topics covered. Its broad scope and in-depth analysis of selected topics will appeal to researchers and professionals working in nanoscience.

Yuli V. Nazarov is a theorist at the Kavli Institute of Nanoscience, Delft University of Technology. He obtained his Ph.D. from the Landau Institute for Theoretical Physics in 1985, and has worked in the field of quantum transport since the late 1980s.

Yaroslav M. Blanter is an Associate Professor in the Kavli Institute of Delft University of Technology. Previous to this Neuroscience, he was a Humboldt Fellow at the University of Karlsruhe and a Senior Assistant at the University of Geneva.

Quantum Transport

Introduction to Nanoscience

YULI V. NAZAROV
Delft University of Technology

YAROSLAV M. BLANTER
Delft University of Technology

CAMBRIDGE UNIVERSITY PRESS
Cambridge, New York, Melbourne, Madrid, Cape Town,
Singapore, São Paulo, Delhi, Mexico City

Cambridge University Press
The Edinburgh Building, Cambridge CB2 8RU, UK

Published in the United States of America by Cambridge University Press, New York

www.cambridge.org
Information on this title: www.cambridge.org/9780521832465

First published 2009

A catalogue record for this publication is available from the British Library

Library of Congress Cataloguing in Publication Data
Nazarov, Yuli V.
Quantum transport : introduction to nanoscience / Yuli V. Nazarov and
Yaroslav M. Blanter.
p. cm.
Includes bibliographical references and index.
ISBN 978-0-521-83246-5
1. Electron transport. 2. Quantum theory. 3. Nanoscience. I. Blanter,
Yaroslav M., 1967– II. Title.
QC176.8.E4N398 2009
620′.5–dc22
2008054041

ISBN 978-0-521-83246-5 Hardback

Contents

Preface *page* vii

Introduction 1

1 Scattering 7

1.1 Wave properties of electrons 7
1.2 Quantum contacts 17
1.3 Scattering matrix and the Landauer formula 29
1.4 Counting electrons 41
1.5 Multi-terminal circuits 49
1.6 Quantum interference 63
1.7 Time-dependent transport 81
1.8 Andreev scattering 98
1.9 Spin-dependent scattering 114

2 Classical and semiclassical transport 124

2.1 Disorder, averaging, and Ohm's law 125
2.2 Electron transport in solids 130
2.3 Semiclassical coherent transport 137
2.4 Current conservation and Kirchhoff rules 155
2.5 Reservoirs, nodes, and connectors 165
2.6 Ohm's law for transmission distribution 175
2.7 Spin transport 187
2.8 Circuit theory of superconductivity 193
2.9 Full counting statistics 205

3 Coulomb blockade 211

3.1 Charge quantization and charging energy 212
3.2 Single-electron transfers 223
3.3 Single-electron transport and manipulation 237
3.4 Co-tunneling 248
3.5 Macroscopic quantum mechanics 264
3.6 Josephson arrays 278
3.7 Superconducting islands beyond the Josephson limit 287

4 Randomness and interference **299**
4.1 Random matrices 299
4.2 Energy-level statistics 309
4.3 Statistics of transmission eigenvalues 324
4.4 Interference corrections 336
4.5 Strong localization 363

5 Qubits and quantum dots **374**
5.1 Quantum computers 375
5.2 Quantum goodies 386
5.3 Quantum manipulation 397
5.4 Quantum dots 406
5.5 Charge qubits 427
5.6 Phase and flux qubits 436
5.7 Spin qubits 445

6 Interaction, relaxation, and decoherence **457**
6.1 Quantization of electric excitations 458
6.2 Dissipative quantum mechanics 470
6.3 Tunneling in an electromagnetic environment 487
6.4 Electrons moving in an environment 499
6.5 Weak interaction 513
6.6 Fermionic environment 523
6.7 Relaxation and decoherence of qubits 538
6.8 Relaxation and dephasing of electrons 549

Appendix A Survival kit for advanced quantum mechanics 562
Appendix B Survival kit for superconductivity 566
Appendix C Unit conversion 569
References 570
Index 577

Preface

This book provides an introduction to the rapidly developing field of quantum transport. Quantum transport is an essential and intellectually challenging part of nanoscience; it comprises a major research and technological effort aimed at the control of matter and device fabrication at small spatial scales. The book is based on the master course that has been given by the authors at Delft University of Technology since 2002. Most of the material is at master student level (comparable to the first years of graduate studies in the USA). The book can be used as a textbook: it contains exercises and control questions. The program of the course, reading schemes, and education-related practical information can be found at our website www.hbar-transport.org.

We believe that the field is mature enough to have its concepts – the key principles that are equally important for theorists and for experimentalists – taught. We present at a comprehensive level a number of experiments that have laid the foundations of the field, skipping the details of the experimental techniques, however interesting and important they are. To draw an analogy with a modern course in electromagnetism, it will discuss the notions of electric and magnetic field rather than the techniques of coil winding and electric isolation.

We also intended to make the book useful for Ph.D. students and researchers, including experts in the field. We can liken the vast and diverse field of quantum transport to a mountain range with several high peaks, a number of smaller mountains in between, and many hills filling the space around the mountains. There are currently many good reviews concentrating on one mountain, a group of hills, or the face of a peak. There are several books giving a view of a couple of peaks visible from a particular point. With this book, we attempt to perform an overview of the whole mountain range. This comes at the expense of detail: our book is not at a monograph level and omits some tough derivations. The level of detail varies from topic to topic, mostly reflecting our tastes and experiences rather than the importance of the topic.

We provide a significant number of references to current research literature: more than a common textbook does. We do not give a representative bibliography of the field. Nor do the references given indicate scientific precedences, priorities, and relative importance of the contributions. The presence or absence of certain citations does not necessarily reflect our views on these precedences and their relative importance.

This book results from a collective effort of thousands of researchers and students involved in the field of quantum transport, and we are pleased to acknowledge them here. We are deeply and personally indebted to our Ph.D. supervisors and to distinguished senior colleagues who introduced us to quantum transport and guided and helped us, and to comrades-in-research working in universities and research institutions all over the world.

This book would never have got underway without fruitful interactions with our students. Parts of the book were written during our extended stays at Weizmann Institute of Science, Argonne National Laboratory, Aspen Center of Physics, and Institute of Advanced Studies, Oslo.

It is inevitable that, despite our efforts, this book contains typos, errors, and less comprehensive discourses. We would be happy to have your feedback, which can be submitted via the website www.hbar-transport.org. We hope that it will be possible thereby to provide some limited "technical" support.

Introduction

It is an interesting intellectual game to compress an essence of a science, or a given scientific field, to a single sentence. For natural sciences in general, this sentence would probably read: *Everything consists of atoms*. This idea seems evident to us. We tend to forget that the idea is rather old: it was put forward in Ancient Greece by Leucippus and Democritus, and developed by Epicurus, more than 2000 years ago. For most of this time, the idea remained a theoretical suggestion. It was experimentally confirmed and established as a common point of view only about 150 years ago.

Those 150 years of research in atoms have recently brought about the field of *nanoscience*, aiming at establishing control and making useful things at the *atomic scale*. It represents the common effort of researchers with backgrounds in physics, chemistry, biology, material science, and engineering, and contains a significant technological component. It is technology that allows us to work at small spatial scales. The ultimate goal of nanoscience is to find means to build up useful artificial devices – *nanostructures* – atom by atom. The benefits and great prospects of this goal would be obvious even to Democritus and Epicurus.

This book is devoted to *quantum transport*, which is a distinct field of science. It is also a part of nanoscience. However, it is a very unusual part. If we try to play the same game of putting the essence of quantum transport into one sentence, it would read: *It is not important whether a nanostructure consists of atoms*. The research in quantum transport focuses on the properties and behavior regimes of nanostructures, which do not immediately depend on the material and atomic composition of the structure, and which cannot be explained starting by classical (that is, non-quantum) physics. Most importantly, it has been experimentally demonstrated that these features do not even have to depend on the size of the nanostructure. For instance, the transport properties of quantum dots made of a handful of atoms may be almost identical to those of micrometer-size semiconductor devices that encompass billions of atoms.

The two most important scales of quantum transport are conductance and energy scale. The measure of conductance, G, is the conductance quantum $G_Q \equiv e^2/\pi\hbar$, the scale made of fundamental constants: electron charge e (most of quantum transport is the transport of electrons) and the Planck constant $\hbar$ (this indicates the role of quantum mechanics). The energy scale is determined by flexible experimental conditions: by the temperature, $k_B T$, and/or the bias voltage applied to a nanostructure, eV. The behavior regime is determined by the relation of this scale to internal energy scales of the nanostructure. Whereas physical principles, as stressed, do not depend on the size of the nanostructure, the internal scales do. In general, they are *bigger* for smaller nanostructures.

This implies that the important effects of quantum transport, which could have been seen at room temperature in atomic-scale devices, would require helium temperatures (4.2 K), or even sub-kelvin temperatures, to be seen in devices of micrometer scale. This is not a real problem, but rather a minor inconvenience both for research and potential applications. Refrigeration techniques are currently widely available. One can achieve kelvin temperatures in a desktop installation that is comparable in price to a computer. The cost of creating even lower temperatures can be paid off using innovative applications, such as quantum computers (see Chapter 5).

Research in quantum transport relies on the nanostructures fabricated using nanotechnologies. These nanostuctures can be of atomic scale, but also can be significantly bigger due to the aforementioned scale independence. The study of bigger devices that are relatively easy to fabricate and control helps to understand the quantum effects and their possible utilization before actually going to atomic scale. This is why quantum transport tells what can be achieved if the ultimate goal of nanoscience – shaping the world atom by atom – is realized. This is why quantum transport presents an indispensable "*Introduction to nanoscience*."

Historically, quantum transport inherits much from a field that emerged in the early 1980s known as *mesoscopic physics*. The main focus of this field was on quantum signatures in semiclassical transport (see, e.g., Refs. [1] and [2], and Chapter 4). The name *mesoscopic* came about to emphasize the importance of intermediate (meso) spatial scales that lie between micro-(atomic) and macroscales. The idea was that quantum mechanics reigns at microscales, whereas classical science does so at macroscale. The mesoscale would be a separate kingdom governed by separate laws that are neither purely quantum nor purely classical; rather, a synthesis of the two. The mesoscopic physics depends on the effective dimensionality of the system; the results in one, two, and three dimensions are different. The effective dimensionality may change upon changing the energy scale. In these terms, quantum transport mostly concentrates on a zero-dimensional situation where the whole nanostructure is regarded as a single object characterized by a handful of parameters; the geometry is not essential. Mesoscopics used to be a very popular term in the 1990s and used to be the name of the field reviewed in this book. However, intensive experimental activity in the late 1980s and 1990s did not reveal any sharp border between meso- and microscales. For instance, metallic contacts consisting of one or a few atoms were shown to exhibit the same transport properties and regimes as micron-scale contacts in semiconductor heterostructures. This is why the field is called now quantum transport, while the term *mesoscopic* is now most commonly used to refer to a cross-over regime between quantum and classical transport.

The objects, regimes, and phenomena of quantum transport are various and may seem unlinked. The book comprises six chapters that are devoted to essentially different physical situations. Before moving on to the main part of the book, let us present an overview of the whole field (see the two-dimensional map, Fig. 1). For the sake of presentation, this map is rather Procrustean: we had to squeeze and stretch things to fit them on the figure. For instance, it does not give important distinctions between normal and superconducting systems. Still, it suffices for the overview.

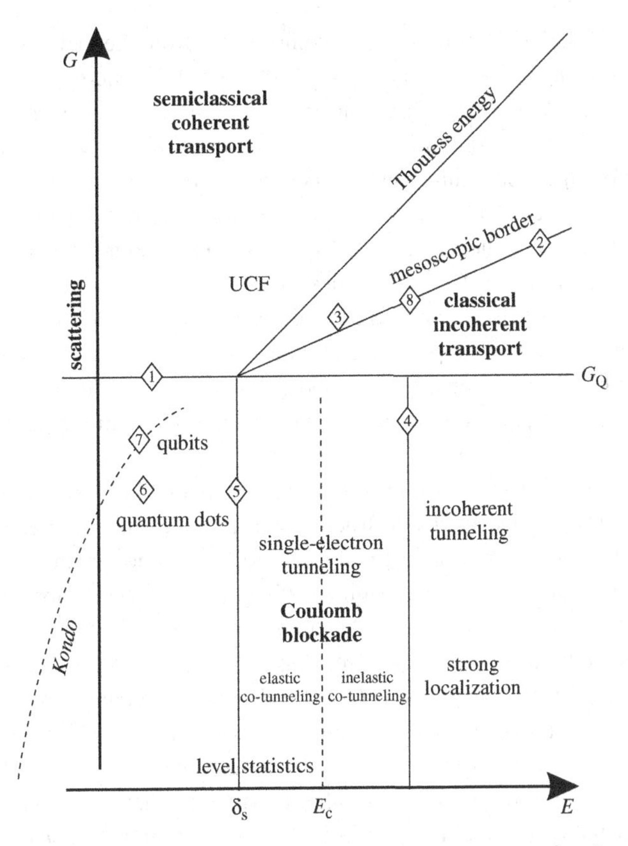

Fig. 1. **Map of quantum transport. Various important regimes are given here in a log–log plot. The numbered diamonds show the locations of some experiments described in the book (see the end of this Introduction for a list).**

The axes represent the conductance of a nanostructure and the energy scale at which the nanostructure is operated; i.e. that set by temperature and/or voltage. This is a log–log plot, and allows us to present in the same plot scales that differ by several orders of magnitude. There is a single universal measure for the conductance – the conductance quantum G_Q. If $G \gg G_Q$, the electron conductance is easy: many electrons traverse a nanostructure simultaneously and they can do this in many ways, known as *transport channels*. For $G \ll G_Q$, the transport takes place in rare discrete events: electrons tunnel one-by-one. The regions around the cross-over line $G \simeq G_Q$ attract the most experimental interest and are usually difficult to comprehend theoretically.

There are several internal energy scales characterizing the nanostructure. To understand them, let us consider an example nanostructure that is of the same (by order of magnitude) size in all three dimensions and is connected to two leads that are much bigger than the nanostructure proper. If we isolated the nanostructure from the leads, the electron energies become discrete, as we know from quantum mechanics. Precise positions of the energy levels would depend on the details of the nanostructure. The energy measure of such quantum discreteness is the *mean level spacing* δ_S – a typical energy distance between the adjacent levels. Another energy scale comes about from the fact that electrons are charged particles carrying an elementary change e. It costs finite energy – the *charging energy* E_C – to add an extra electron to the nanostructure. This charging energy characterizes the interactions of electrons. At atomic scale, $\delta_S \simeq 1\,\text{eV}$ and $E_C \simeq 10\,\text{eV}$. These internal scales are smaller for bigger structures, and E_C is typically much bigger than δ_S.

As seen in Fig. 1, these scales separate different regimes at low conductance $G \ll G_Q$. At high conductance, $G \gg G_Q$, the electrons do not stay in the nanostructure long enough to feel E_C or δ_S. New scales emerge. The time the electron spends in the nanostructure gives rise to an energy scale: the *Thouless energy*, E_{Th}. This is due to the quantum uncertainty principle, which relates any time scale to any energy scale by $(\Delta E)(\Delta t) \sim \hbar$. The Thouless energy is proportional to the conductance of the nanostructure, $E_{Th} \simeq \delta_S G/G_Q$, and this is why the corresponding line in the figure is at an angle in the log–log plot.

Another slanted line in the upper part of Fig. 1 is due to the electron–electron interaction, which works destructively. It provides intensive energy relaxation of the electron distribution in a nanostructure and/or limits the quantum-mechanical *coherence*. On the right of the line, the quantum effects in transport disappear: we are dealing with classical incoherent transport. At the line, the inelastic time, τ_{in}, equals the time the electron spends in the nanostructure, that is, $\hbar/\tau_{in} \simeq E_{Th}$. The corresponding energy scale can be estimated as $\simeq \delta_S(G/G_Q)^2 \gg E_{Th}$. In the context of mesoscopics, Thouless has suggested that extended conductors are best understood by subdividing a big conductor into smaller nanostructures. The size of such nanostructure is chosen to satisfy the condition $\hbar/\tau_{in} \simeq E_{Th}$. This is why all experiments where mesoscopic effects are addressed are actually located in the vicinity of the line; we call it the *mesoscopic border*.

Once we have drawn the borders, we position the material contained in each chapters on the map. Chapter 1 is devoted to the *scattering* approach to electron transport. It is an important concept of the field that at sufficiently low energies any nanostructure can be regarded as a (huge) scatterer for electron waves coming from the leads. At $G \gg G_Q$, the validity of the scattering approach extends to the mesoscopic border. At energies exceeding the Thouless energy, the energy dependence of the scattering matrix becomes important. In Chapter 1, we explain how the scattering approach works in various circumstances, including a discussion of superconductors and time-dependent and spin-dependent phenomena. We relate the transport properties to the set of transmission eigenvalues of a nanostructure – its "pin-code." The basics explained in Chapter 1 relate, in one way or another, to all chapters.

If we move up along the conductance axis, $G \gg G_Q$, the scattering theory becomes progressively impractical owing to a large number of transport channels resulting in a bigger scattering matrix. Fortunately, there is an alternative way to comprehend this *semi-classical coherent regime* outlined in Chapter 2. We show that the properties of nanostructures are determined by *self-averaging* over the quantum phases of the scattering matrix elements. Because of this, the laws governing this regime, being essentially quantum, are similar to the laws of transport in classical electric circuits. We explain the machinery necessary to apply these laws – quantum *circuit theory*. The quantum effects are frequently concealed in this regime; for instance, the conductance is given by the classical Ohm's law. Their manifestations are most remarkable in superconductivity, the statistics of electron transfers, and spin transport. Remarkably, there is no limitation to quantum mechanics at high conductances as soon as one remains above the mesoscopic border.

Chapter 3 brings us to the lower part of the map – to conductances much lower than G_Q. There, the charging energy scale E_C becomes relevant, manifesting a strong interaction between the electrons (the *Coulomb blockade*). This is why we concentrate on the energies of the order of E_C, disregarding the mean level spacing δ_S. Transport in this *single-electron tunneling* regime proceeds via incoherent transfer of single electrons. However, the transfers are strongly correlated and can be precisely controlled – one can manipulate electrons one-by-one. The quantum correction to single-electron transport is *co-tunneling*, i.e. cooperative tunneling of two electrons. The energy scale $\sqrt{E_C\delta_S}$ separates inelastic and elastic co-tunneling. In the elastic co-tunneling regime, the nanostructure can be regarded as a scatterer in accordance with the general principles outlined in Chapter 1. The combination of the Coulomb blockade and superconductivity restores the quantum coherence of elementary electron transfers and provides the opportunity to build quantum devices of almost macroscopic size.

The material discussed in Chapter 4 is spread over several areas of the map. In this chapter, we address the statistics of persistent fluctuations of transport properties. We start with the statistics of discrete electron levels – this is the domain of low conductances, $G \ll G_Q$, and low energies, of the order of the mean level spacing. Then we go to the different corner, to $G \gg G_Q$ and the energies on the left from the mesoscopic border, to discuss fluctuations of transmission eigenvalues – the *universal conductance fluctuations* (UCF) – and the interference correction to transport, *weak localization*. The closing section of Chapter 4 is devoted to strong localization in disordered media, where electron hopping is the dominant mechanism of conduction. This implies $G \ll G_Q$ and high energies.

A fascinating development of the field is the use of nanostructures for quantum information purposes. Here, we do not need a flow of quantum electrons, but rather a flow of quantum information. Chapter 5 presents qubits and quantum dots, perhaps the most popular devices of quantum transport. For both devices, the discrete nature of energy levels is essential. This is why they occupy the energy area left of the level spacing δ_S on the map. We also present in Chapter 5 a comprehensive introduction to quantum information and manipulation.

In Chapter 6 we discuss interaction effects that do not fit into the simple framework of the Coulomb blockade. Such phenomena are found in various areas of the map. We start this chapter with a discussion of the underlying theory, called *dissipative quantum mechanics*. We study the effects of an electromagnetic environment on electron tunneling, remaining in the area of the Coulomb blockade. We go up in conductance to understand the fate of the Coulomb blockade at $G \gtrsim G_Q$ and the role of interaction effects at higher conductances. The electrons in the leads provide a specific (fermionic) environment responsible for the *Kondo effect* in quantum dots. The Kondo energy scale depends exponentially on the conductance and is given by the curve on the left side of the map. Finally, we discuss energy dissipation and dephasing separately for qubits and electrons. In the latter case, we are at the mesoscopic border.

At high energies one leaves the field of quantum transport: transport proceeds as commonly taught in courses of solid-state physics.

We have not yet mentioned the numbered diamonds in the map. These denote the location of several experiments presented in various chapters of the book.

(1) Discovery of conductance quantization (Section 1.2);
(2) interference nature of the weak localization (Section 1.6);
(3) universal conductance fluctuations (Section 1.6);
(4) single-electron transistor (Section 3.2);
(5) discrete states in quantum dots (Section 5.4);
(6) early qubit (Section 5.5);
(7) Kondo effect in quantum dots (Section 6.6);
(8) energy relaxation in diffusive wires (Section 6.8).

1 Scattering

1.1 Wave properties of electrons

Quantum mechanics teaches us that each and every particle also exists as a wave. Wave properties of macroscopic particles, such as brickstones, sand grains, and even DNA molecules, are hardly noticeable to us; we deal with them at a spatial scale much bigger than their wavelength. Electrons are remarkable exceptions. Their wavelength is a fraction of a nanometer in metals and can reach a fraction of a micrometer in semiconductors. We cannot ignore the wave properties of electrons in nanostructures of this size. This is the central issue in quantum transport, and we start the book with a short summary of elementary results concerning electron waves.

A quantum electron is characterized by its *wave function*, $\Psi(\boldsymbol{r}, t)$. The squared absolute value, $|\Psi(\boldsymbol{r}, t)|^2$, gives the probability of finding the electron at a given point $\boldsymbol{r}$ at time t. Quantum states available for an electron in a vacuum are those with a certain wave vector $\boldsymbol{k}$. The wave function of this state is a *plane wave*,

$$\Psi_{\boldsymbol{k}}(\boldsymbol{r}, t) = \frac{1}{\sqrt{\mathcal{V}}} \exp\left(\mathrm{i}\boldsymbol{k} \cdot \boldsymbol{r} - \mathrm{i}E(\boldsymbol{k})t/\hbar\right), \tag{1.1}$$

$E(\boldsymbol{k}) = \hbar^2\boldsymbol{k}^2/2m$ being the corresponding energy. The electron in this state is spread over the whole space of a very big volume $\mathcal{V}$; the squared absolute value of Ψ does not depend on coordinates. The prefactor in Eq. (1.1) ensures that there is precisely one electron in this big volume. There are *many* electrons in nanostructures. Electrons are spin $1/2$ fermions, and the Pauli principle ensures that each one-particle state is either empty or filled with one fermion. Let us consider a cube in k-space centered around $\boldsymbol{k}$ with the sides $\mathrm{d}k_x, \mathrm{d}k_y, \mathrm{d}k_z \ll |\boldsymbol{k}|$. The number of available states in this cube is $2_\mathrm{s}\mathcal{V}\ \mathrm{d}k_x\ \mathrm{d}k_y\ \mathrm{d}k_z/(2\pi)^3$. The factor of 2_s comes from the fact that there are two possible spin directions. The fraction of states filled in this cube is called an electron *filling factor*, $f(\boldsymbol{k})$. The particle density n, energy density $\mathcal{E}$, and density of electric current $\boldsymbol{j}$ are contributed to by all electrons and are given by

$$\begin{bmatrix} n \\ \mathcal{E} \\ \boldsymbol{j} \end{bmatrix} = \int 2_\mathrm{s} \frac{\mathrm{d}^3\boldsymbol{k}}{(2\pi)^3} \begin{bmatrix} 1 \\ E(\boldsymbol{k}) \\ e\boldsymbol{v}(\boldsymbol{k}) \end{bmatrix} f(\boldsymbol{k}). \tag{1.2}$$

Here we introduce the electron charge e and the velocity $\boldsymbol{v}(\boldsymbol{k}) = \hbar\boldsymbol{k}/m$. Quantum mechanics puts no restriction on $f(\boldsymbol{k})$. However, the filling factor of electrons in an *equilibrium* state at a given electrochemical potential μ and temperature T is set by Fermi–Dirac statistics:

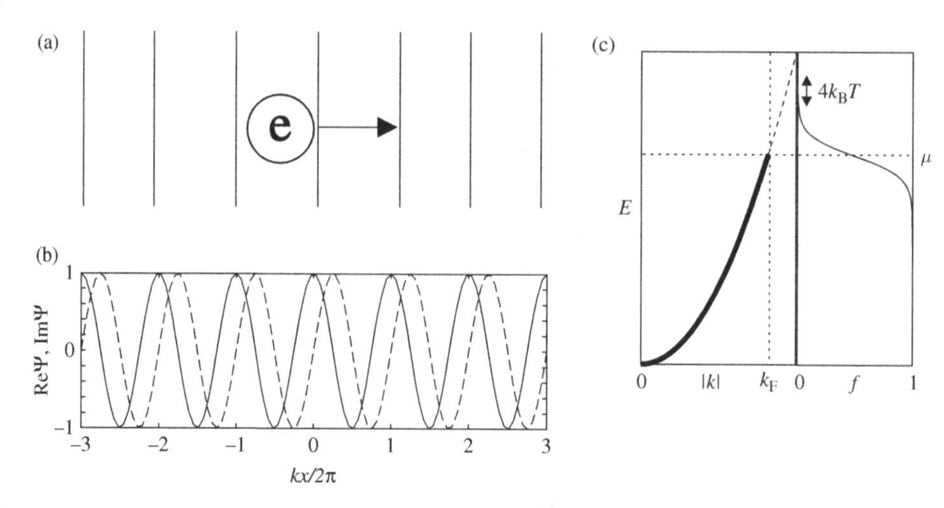

Fig. 1.1. **Electrons as waves. (a) An electron in a vacuum is in the plane wave state with the wave vector $\boldsymbol{k}$. (b) The profile of its wave function Ψ. (c) At zero temperature, the electrons fill the states with energies below the chemical potential μ ($|k| < k_F$). At a given temperature, the filling factor f is a smoothed step-like function of energy.**

$$f_{\text{eq}}(\boldsymbol{k}) = f_{\text{F}}(E(\boldsymbol{k}) - \mu) \equiv \frac{1}{1 + \exp((E - \mu)/k_B T)}. \tag{1.3}$$

The chemical potential at zero temperature is known as the Fermi energy, E_{F}.

Control question 1.1. What is the limit of $f_F(E)$ at $T \to 0$? Hint: see Fig. 1.1.

Next, we consider electrons in the field of electrostatic potential, $U(\boldsymbol{r}, t)/e$. The wave function $\Psi(\boldsymbol{r}, t)$ of an electron is no longer a plane wave. Instead, it obeys the time-dependent Schrödinger equation, given by

$$\mathrm{i}\hbar \frac{\partial \Psi(\boldsymbol{r}, t)}{\partial t} = \hat{H} \Psi(\boldsymbol{r}, t); \quad \hat{H} \equiv -\frac{\hbar^2}{2m} \nabla^2 + U(\boldsymbol{r}, t). \tag{1.4}$$

This is an evolutionary equation: it determines Ψ in the future given its instant value. The evolution operator $\hat{H}$ is called the *Hamiltonian*. For the time being, we concentrate on the stationary potential, $U(\boldsymbol{r}, t) \equiv U(\boldsymbol{r})$. The wave functions become stationary, with their time dependence given by the energy

$$\Psi(\boldsymbol{r}, t) = \exp(-\mathrm{i}Et/\hbar)\psi_E(\boldsymbol{r}).$$

The Schrödinger equation reduces to

$$E\psi_E(\boldsymbol{r}) = \hat{H}\psi_E(\boldsymbol{r}) = \left[-\frac{\hbar^2}{2m} \nabla^2 + U(\boldsymbol{r}) \right] \psi_E(\boldsymbol{r}). \tag{1.5}$$

The Hamiltonian becomes the operator of energy, while the equation becomes a linear algebra relation defining the eigenvalues E and the corresponding eigenfunctions Ψ_E of this operator. These eigenfunctions form a *basis* in the Hilbert space of all possible wave functions, so that an arbitrary wave function can be expanded, or represented, in this basis. The first (gradient) term in the Hamiltonian describes the kinetic energy; the second term, $U(\boldsymbol{r})$, represents the potential energy.

A substantial part of quantum mechanics deals with the above equation. It cannot be readily solved for an arbitrary potential, and our qualitative understanding of quantum mechanics is built upon several simple cases when this solution can be obtained explicitly. Following many good textbooks, we will concentrate on the one-dimensional motion, in which the potential and the wave functions depend on a single coordinate x. However, we pause to introduce a key concept that makes this one-dimensional motion more physical.

1.1.1 Transmission and reflection

Let us confine electrons in a tube – a *waveguide* – of rectangular cross-section that is infinitely long in the x direction. We can do this by setting the potential U to zero for $|y| < a/2$, $|z| < b/2$ and to $+\infty$ otherwise. We thus create walls that are impenetrable to the electron and are perpendicular to the y and z axes. We expect a wave to be reflected from these walls, changing the sign of the corresponding component of the wave vector, $k_y \to -k_y$ or $k_z \to -k_z$. This suggests that the solution of the Schrödinger equation is a superposition of incident and reflected waves of the following kind:

$$\psi(x,y,z) = \exp(ik_x x) \sum_{s_y,s_z=+,-} C_{s_y s_z} \exp(s_y i k_y y) \exp(s_z i k_z z). \tag{1.6}$$

Since the infinite potential repels the electron efficiently, the wave function must vanish at the walls, $\psi(x, y = \pm a/2, z) = \psi(x, y, z = \pm b/2) = 0$. This gives a linear relation between $C_{s_y s_z}$ that determines these superposition coefficients. To put it simply, the walls have to be in the nodes of a standing wave in both y and z directions. This can only happen if $k_{y,z}$ assume quantized values $k_y^n = \pi n_y/a, k_z^n = \pi n_z/b$, with integers $n_y, n_z > 0$ corresponding to the number of half-wavelengths that fit between the walls. The notation we use throughout the book here we introduce for the compound index $n = (n_y, n_z)$. The wave function reads as follows:

$$\begin{aligned}
\psi_{k_x,n}(x,y,z) &= \psi_{k_x}(x)\Phi_n(y,z); \\
\psi_{k_x}(x) &= \exp(ik_x x); \\
\Phi_n(y,z) &= \frac{2}{\sqrt{ab}} \sin(k_y^n(y - a/2)) \sin(k_z^n(z - b/2)).
\end{aligned} \tag{1.7}$$

The transverse motion of the electron is thus *quantized*. The electron in a state with the given n (these states are called *modes* in wave theory and *transport channels* in nanophysics) has only one degree of freedom corresponding to one-dimensional motion. The energy spectrum consists of one-dimensional branches shifted by a channel-dependent energy E_n (see Fig. 1.2), given by

$$E_n(k_x) = \frac{(\hbar k_x)^2}{2m} + E_n; \quad E_n = \frac{\pi^2 \hbar^2}{2m}\left(\frac{n_y^2}{a^2} + \frac{n_z^2}{b^2}\right). \tag{1.8}$$

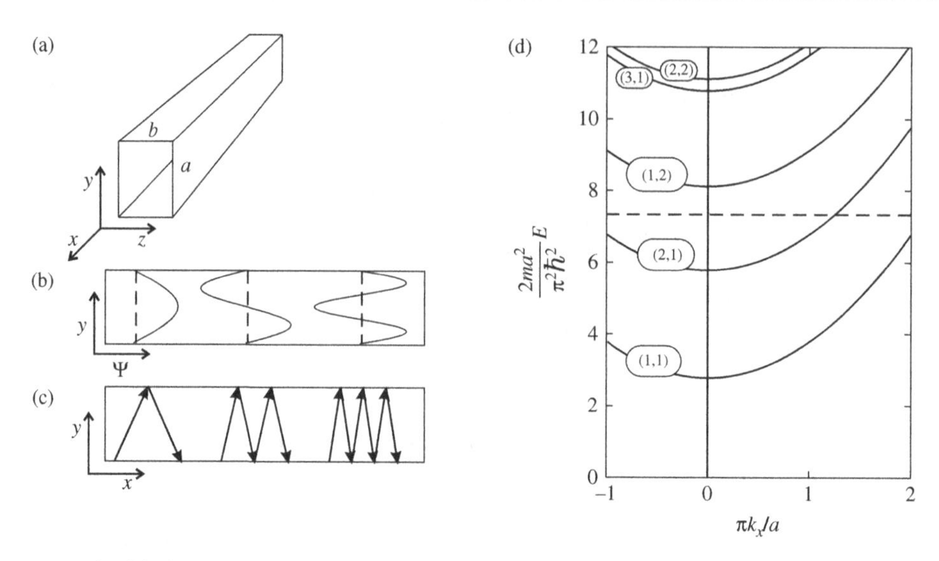

Fig. 1.2. Waveguide. (a) Electrons are confined in a long tube of rectangular cross-section. (b) Wave function profiles of the modes (1,1), (2,1), and (3,1). (c) Corresponding trajectories of a classical particle reflecting from the waveguide walls. (d) Energy spectrum of electron states in the waveguide ($b/a = 0.7$). At the chemical potential shown by the dashed line, the electrons are present only in the modes (1,1) and (2,1).

Let us add some more design to our waveguide. We cross it with a *potential barrier* of simple form,

$$U(x) = \begin{cases} U_0, & 0 < x < d \\ 0, & \text{otherwise.} \end{cases} \tag{1.9}$$

The possible solutions outside the barrier for a given n and energy are plane waves of the form of Eqs. (1.7). It is important to note that there are *two* possible solutions with $k_x = \pm k = \pm\sqrt{2m(E - E_n)}/\hbar$, corresponding to the waves propagating to the right (positive sign) or to the left. A wave sent from the left is scattered at the barrier, part of it being reflected back, another part being transmitted. We have

$$\psi(x) = \begin{cases} \exp(ikx) + r\exp(-ikx), & x < 0 \\ B\exp(i\kappa x) + C\exp(-i\kappa x), & 0 < x < d \\ t\exp(ikx), & x > d, \end{cases} \tag{1.10}$$

where $\kappa = \sqrt{2m(E - E_n - U_0)}/\hbar = \sqrt{k^2 - 2mU_0/\hbar^2}$. The wave function and its x-derivative must be continuous at $x = 0$ and $x = d$. These four conditions give four linear equations for the unknown coefficients r, B, C, and t. The most important for us are the transmission amplitude t and the reflection amplitude r. The *transmission coefficient*, $T(E) = |t|^2$, determines which fraction of the wave is transmitted through the obstacle. The *reflection coefficient*, $R(E) = |r|^2 = 1 - T(E)$, determines the fraction reflected back. We find

$$T(E) = \frac{4k^2\kappa^2}{(k^2 - \kappa^2)^2 \sin^2 \kappa d + 4k^2\kappa^2}. \tag{1.11}$$

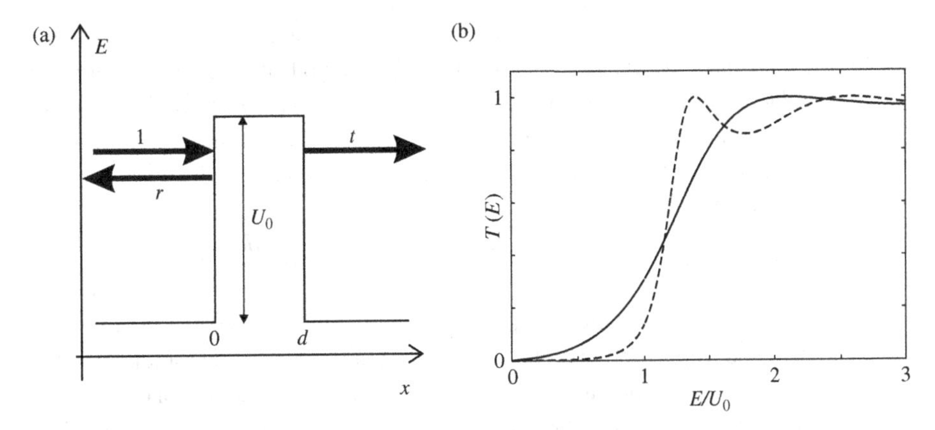

Fig. 1.3. Potential barrier. (a) Scattering of an electron wave at a rectangular potential barrier. (b) Transmission coefficient (see Eq. (1.11)) of the barrier for two different thicknesses, $d\sqrt{2mU_0}/\hbar = 3$ (solid) and 5 (dashed). For the thicker barrier the transmission coefficient is close to the classical one, $T(E) = 1$ at $E > U_0$.

Control question 1.2. Find the coefficients r, t, B, and C in terms of κ, k, and d.

In classical physics, particles with energies below the barrier ($E < U_0$) would be totally reflected ($T = 0$), while particles with energies above the barrier would be fully transmitted ($T = 1$). Quantum mechanics changes this: electrons are transmitted and reflected at any energy (Fig. 1.3). Even an electron with an energy well below the barrier (corresponding to imaginary κ) has a finite, albeit an exponentially small, chance of being transmitted, $T(E) \propto \exp(-2d\sqrt{2m(U_0 + E_n - E)}/\hbar) \ll 1$. This is called *tunneling*.

The above consideration is not limited to barriers localized within a certain interval of x. For any barrier, the solution very far to the left, $x \to -\infty$, can be regarded as a superposition of incoming and reflected waves, $\psi = \exp(ikx) + r\exp(-ikx)$. Very far to the right, $x \to \infty$, the solution is a transmitted wave, $\psi = t\exp(ikx)$. To calculate t and r, we have to solve the Schrödinger equation everywhere and match these two asymptotic solutions.

1.1.2 Electrons in solids

The above discussion concerns electrons in a vacuum. The electrons in nanostructures are not in a vacuum, rather they are in a solid state medium such as a metal or a semiconductor. What does this change? Surprisingly, not much. A crystalline lattice of a solid state medium provides a periodic potential relief. The solutions of the Schrödinger equation for such a potential are no longer plane waves as in Eq. (1.1), but rather are *Bloch waves*,

$$\psi_{\boldsymbol{k},p}(\boldsymbol{r}) = \exp(i\boldsymbol{k}\boldsymbol{r})\, u_{\boldsymbol{k},p}(\boldsymbol{r}), \tag{1.12}$$

where $u_{\boldsymbol{k},P}$ is a periodic function with the same periods as the lattice. The vector of *quasimomentum*, $\hbar\boldsymbol{k}$, is defined up to a period of a reciprocal lattice, and the index P labels different energy bands. The energy $E_P(\boldsymbol{k})$ is a *periodic* function of quasimomentum. This implies that it is bounded. Therefore, the spectrum at a given $\boldsymbol{k}$ consists of discrete values corresponding to energy bands. The electron velocity in the given state $(\boldsymbol{k}, P)$ is given by

$$\boldsymbol{v}_P(\boldsymbol{k}) = \frac{1}{\hbar}\frac{\partial E_P(\boldsymbol{k})}{\partial \boldsymbol{k}}.$$

With these notations, Eq. (1.2) remains valid. The integration over $\mathrm{d}^3\boldsymbol{k}$ must be replaced by the summation over the energy band index P and integration over the quasimomentum within the reciprocal lattice unit cell (or the first Brillouin zone).

This summarizes the differences between the descriptions of an electron in a vacuum and in a crystalline lattice.

Note that the above discussion disregards the interaction between electrons. However, there are many electrons in a solid state medium, they are charged, and they interact with each other. One would have to deal with the Schrödinger equation for a many-body wave function that depends on coordinates of all electrons in the nanostructure, which is a formidable task. What makes the above discussion relevant?

This was a Nobel Prize question (awarded to Lev Landau in 1962). The above description is relevant because we "cheat." We do not describe the real interacting electrons. Indeed, we cannot, nor do we have to. Rather, we implicitly consider the quantum transport of *quasi*electrons (or *quasi*particles), elementary charged excitations above the ground state of all the electrons present in the solid state. The interaction between these excitations is weak and in many instances can be safely disregarded.

Let us give a short summary of the arguments that justify this implicit substitution for the important case of a metal. By definition, a metal is a material that can be charged with no energy cost. This means that the energy required to add some charge Q into a piece of metal is $\mu Q/e$, where μ is the chemical potential.

We now describe this quantum mechanically. Before the charge was added, the piece of metal was in its ground state. Let us add one elementary charge. This drives the system to an excited state, which corresponds to creating precisely one *quasi*electron. By symmetry consideration, this state should have a certain quasimomentum and spin $1/2$. To conform to the definition of the metal, the energy of this state has to be equal to μ, $E(\boldsymbol{k}) = \mu$. This condition defines a surface in three-dimensional space of quasimomentum, the *Fermi surface*. Fermi surfaces can look rather complicated. For example, the Fermi surface of gallium looks like the fossil of a dinosaur – to this end, a very symmetric dinosaur. The Fermi surface of free electrons is a sphere: noble metals provide a good approximation to it (see Fig. 1.4). In the following, we count the energy of quasiparticles from the Fermi level μ.

Let us concentrate on the situation when the temperature and applied voltage are much smaller than μ. This sets the energy scale $\Delta E \simeq \max(eV, k_{\mathrm{B}}T) \ll \mu$ available for quasiparticles, which are therefore all located close to the Fermi surface. The important parameter is the density of states ν at the Fermi surface, defined as the number of states per energy interval in a unit volume. The density of the quasiparticles is therefore $\nu\Delta E$, much

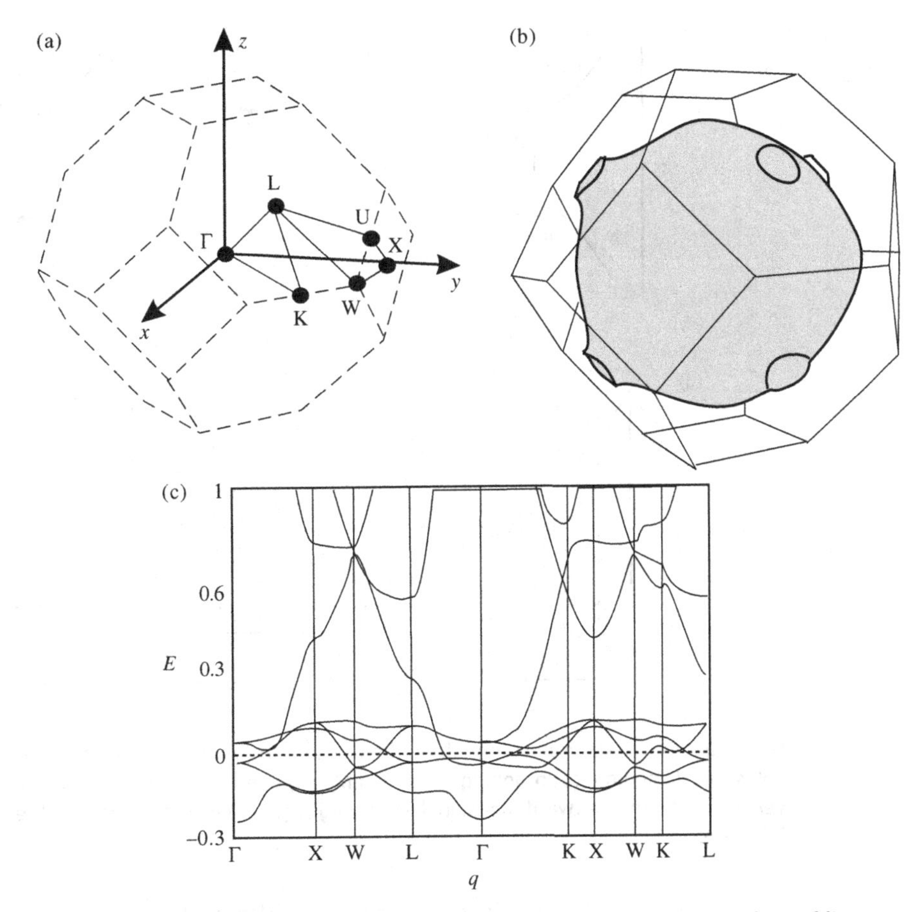

Fig. 1.4. **A realistic metal: silver. (a) Brillouin zone with symmetry points Γ, X, W, L, and K and lines. (b) Fermi surface. (c) Energy bands plotted along the symmetry lines.**

smaller than the density of the original electrons in the metal. The smaller ΔE is, the bigger the distance between the quasiparticles. This explains why the interaction is negligible: the quasiparticles just do not come together to interact.

The original electrons interact according to Coulomb's law. The quasiparticles are not original electrons, and the residual interaction between them is strongly modified. First of all, the electric field around each quasiparticle is *screened* by electrons forming the ground state since they redistribute to compensate the quasiparticle charge. This quenches the long-range repulsion between the quasiparticles. The interaction may be mediated by phonons (vibrations of the crystalline lattice) and is not even always repulsive. This may drive the metal to a superconducting state.

The above arguments allow us to start the discussion of quantum transport with the notion of non-interacting (quasi)electrons. We will see that the interactions may not always be disregarded in the context of quantum transport. The above arguments do not work if interaction occurs at mesoscopic rather than at microscopic scales.

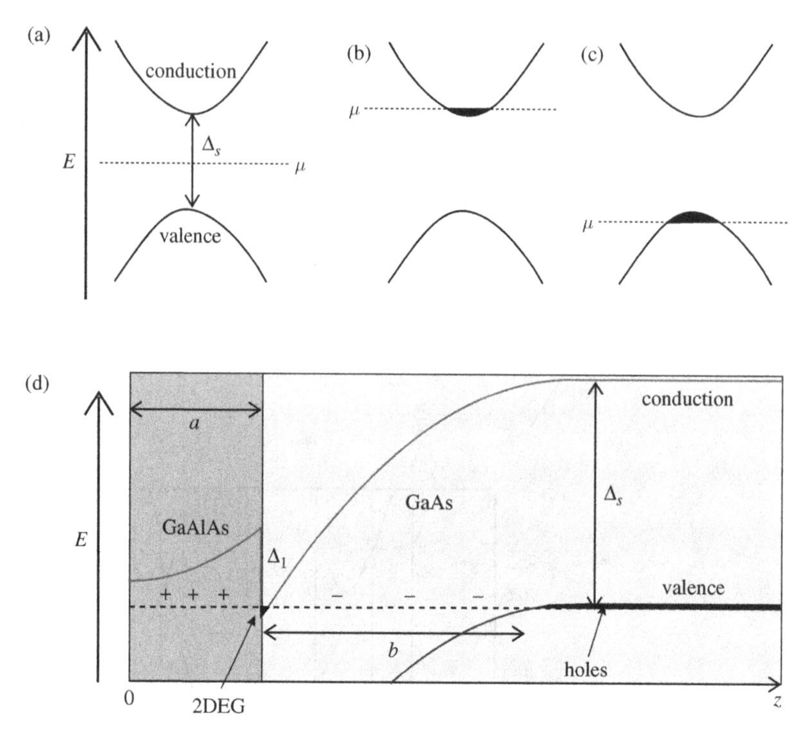

Fig. 1.5. **Energy bands in a semiconductor. Black-filled regions in (b) and (c) indicate carriers: electrons or holes. (a) No doping; (b) n-doping; (c) p-doping. (d) Band edges in GaAlAs–GaAs heterostructure versus the depth *z*. A two-dimensional electron gas (2DEG) is formed close to the GaAlAs–GaAs interface.**

1.1.3 Two-dimensional electron gas

There is a long way to go from metal solids to practical nanostructures, and this way has been found during the technological developments of the second half of the twentieth century. It started with semiconductors: insulators with a relatively small gap separating conduction (empty) and valence (occupied) bands. Of all the rich variety of semiconductor applications, one is of particular importance for quantum transport, and that is the making of an artificial and easily controllable metal from a semiconductor. This is achieved by a process called *doping*, in which a small controllable number of impurities are added to a chemically pure semiconductor. Depending on the chemical valence of the impurity atom, it either gives an electron to the semiconductor (the atom works as an n-dopant) or extracts one, leaving a hole in the semiconductor (p-dopant). Even a small density of the dopants (say, 10^{-4} per atom) brings the chemical potential either to the edge of the conduction band (n-type semiconductor) or to the edge of the valence band (p-type semiconductor); see Fig. 1.5. In both cases, the semiconductor becomes a metal with a small carrier concentration. A rather simple trick of doping different areas of a semiconductor with p- and n-type dopants creates p-n junctions, transistors (for which W. Shockley, J. Bardeen, and W. H. Brattain received the Nobel Prize in 1956), and most of the power of semiconductor electronics.

A disadvantage of the resulting metal is that it is rather dirty. Indeed, it is made by impurities, so that the number of scattering centers approximately equals the number of carriers. It is advantageous to separate spatially the dopants and the carriers induced. In the course of these attempts, the two-dimensional electron gas (2DEG) has been put into practice.

The most convenient way to make a 2DEG involves a selectively doped GaAlAs–GaAs heterostructure, a layer of n-doped GaAlAs on the surface of a p-doped GaAs crystal. The lattice constants of the two materials match, providing a clean, defect-free interface between them. The semiconducting energy gap in GaAlAs is bigger than in GaAs, and the expectation is that the electrons from n-dopants in GaAlAs eventually reach the GaAs. Why would these carriers stay near the surface? To understand this, let us consider the electrostatics of the whole structure (see Fig. 1.5). In the one-dimensional (1d) geometry given, the potential energy of the electrons is $U(z) = e\Phi(z)$. The electrostatic potential $\Phi(z)$ and charge density $\rho(z)$ are related by the Poisson equation:

$$\frac{d^2\Phi(z)}{dz^2} = 4\pi\rho(z)/\epsilon,$$

where we assume the same dielectric constant ϵ in both materials. If no carriers are present in GaAlAs, the dopants with volume density n_1 make a parabolic potential profile in the material, $U(z) = U(0) + (2\pi e^2/\epsilon)n_1 z^2$, $0 < z < a$, a being thickness of the layer.[1] If we cross the interface, there is a drop in potential energy that equals the energy mismatch $\Delta_1 \approx 0.2$ eV between the conduction bands of the materials. Let us assume that the electrons are concentrated close to the interface at the GaAs side and figure out the conditions at which it actually happens. If the *surface* density of the electrons equals n_0, the electric field in the z direction jumps at the interface, i.e.

$$\left(\frac{d\Phi}{dz}\right)_{z=a+0} - \left(\frac{d\Phi}{dz}\right)_{z=a-0} = -(4\pi e/\epsilon)n_0.$$

The bulk GaAs is p-doped, so there are supposed to be holes. However, the holes are separated from the interface and the electrons by a *depletion* layer of thickness b. The negatively charged dopants (with volume density n_2) in this layer form an inverse parabolic profile, $U(z) = U(a+0) + (dU(z=a+0)/dz)(z-a) - (2\pi e^2/\epsilon)n_2(z-a)^2$, $a < z < b$.

This allows us to determine conditions for the stability of this charge distribution. Since electrons at the interface and holes in the valence band share the same chemical potential, the difference of the potential energies just equals the semiconducting gap, $\Delta_s = 1.42$ eV in GaAs, $U(a+b) - U(a+0) = \Delta_s$.[2] Further, the holes are in equilibrium, so the electrostatic force $-dU(z)/dz$ vanishes at the edge of the depletion layer, $z = a + b$. To ensure that there are no carriers in the GaAlAs layer, one requires $U(0) > U(a+b)$. Solving for everything, we obtain

$$n_0 = n_1 a - \sqrt{\frac{\Delta_s n_2 \epsilon}{2\pi e^2}}.$$

[1] Typically, $a = 50$ nm. A simple technique to reduce the disorder is not to dope GaAlAs in a *spacer* layer adjacent to the interface.

[2] To write this, we disregard the kinetic energy of both electrons and holes in comparison with Δ_s.

Control question 1.3. What is the thickness of the depletion layer in terms of n_2 and Δ_s?

The carriers in GaAlAs are absent only if the dopant density is sufficiently low, $n_1 < \Delta_1(\epsilon/2\pi a^2 e^2)$. Since $n_0 > 0$, the desired charge distribution occurs in a certain interval of the dopant density: $\Delta_1(\epsilon/2\pi a^2 e^2) > n_1 > \sqrt{\Delta_s n_2 \epsilon/2\pi a^2 e^2}$. The 2DEG structure is also stable in the limit of vanishing p-dopant density, $n_2 \to 0$. In this case, the surface density of the 2DEG just equals that of the n-dopants, $n_0 = n_1 a$.

Now we turn to the details of electron wave functions and spectrum in the 2DEG. The electrons concentrated near the interface experience the potential that depends only on z. As for waveguides, we can separate the electron motion in the z-direction from that in the xy plane. The motion in the x, y directions is free, whereas it is finite in the z direction: The potential $U(z)$ takes the form of a triangular-shaped well. So, the motion in the z direction is quantized, giving rise to a series of discrete energy levels E_n with corresponding wave functions $\Phi_n(z)$ of the localized states (Fig. 1.6). The wave functions of the electron states are plane waves in the x, y directions and can be presented as $\psi_{k_x,k_y,n}(x, y, z) \propto e^{i(k_x x+k_y y)}\Phi_n(z)$. Each n thus gives rise to a subband of two-dimensional states with energies given by ($\boldsymbol{k} \equiv (k_x, k_y)$)

$$E_n(\boldsymbol{k}) = E_n + \frac{\hbar^2 k^2}{2m}.$$

Here, m is the effective mass of electrons in GaAs, equal to 0.067 of the electron mass in a vacuum. If we count energy from the bottom of the infinitely deep triangular well, the energy levels read $E_n = c_n((U')^2\hbar^2/(2m))^{1/3}$, where $U' \equiv dU(z = a + 0)/dz$, and $c_1 = 2.338$ and $c_2 = 4.082$.

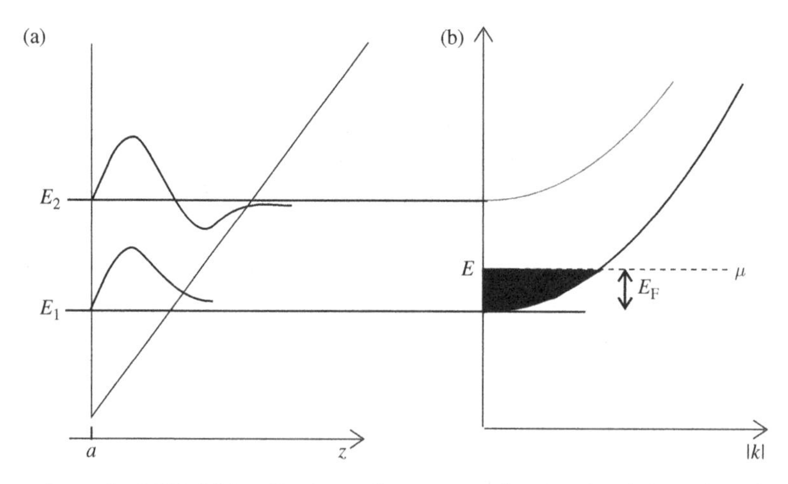

Fig. 1.6. Energy spectrum of a 2DEG. (a) Localized wave functions and energy levels in a triangular well potential near the interface. (b) All electrons are accommodated in the lowest subband.

Control question 1.4. What is U' in terms of n_0, n_1, and a? Why can E_0 not exceed Δ_1?

Let us put electrons into the levels. We note that the density of states in each subband does not depend on energy and equals

$$\nu = 2_s \int \mathrm{d}\boldsymbol{k}\, \delta E - E_n(\boldsymbol{k}) = \frac{2_s m}{2\pi \hbar^2} \Theta(E - E_n).$$

Taking GaAs mass, we obtain $\nu = 2.8 \times 10^{-4}\,\mathrm{meV}^{-1}\,\mathrm{nm}^2$. The surface densities n_0 in a 2DEG are in the range 1–$4 \times 10^{-4}\,\mathrm{nm}^{-2}$. In this density range, all electrons are accommodated in the lowest subband. This implies that the corresponding Fermi energy, E_F, counted from the subband edge, $E_\mathrm{F} = n_0/\nu$, does not reach the edge of the second subband, $E_\mathrm{F} < E_2 - E_1$. This Fermi energy is smaller by two orders of magnitude than a typical Fermi energy in metals. The Fermi wavelength, $\lambda = \sqrt{2\pi/n_0}$, ranges from 40 to 80 nm and also exceeds the typical Fermi wavelength in metals by two orders of magnitude. This means that the quantum effects in a 2DEG can be seen at much larger space scales than in common metals.

1.2 Quantum contacts

A common nanostructure does not even remotely resemble an infinitely long waveguide. However, the physics of quantum transport is surprisingly similar to that of a waveguide. The recognition of this fact and its experimental verification was, and still is, one of the main events in the history of the field. We introduce this important idea in two steps. In this section, we consider in detail the quantum point contact (QPC) – a system *without* potential barriers – and show that it is equivalent to a waveguide *with* a potential barrier. In Section 1.3, we turn to the more complicated case of a generic nanostructure.

1.2.1 Adiabatic quantum transport

We start by looking at a waveguide of variable cross-section. The waveguide is extended along the x axis, bounded by impenetrable potential walls, and has a rectangular cross-section, $|y| < a(x)/2$, $|z| < b(x)/2$, with the dimensions varying as one moves along the contact. Far to the right and the left, $x \to \pm\infty$, these dimensions assume constant values a_∞ and b_∞. In the middle, the walls come closer, forming a constriction (Fig. 1.7). The solutions, Eqs. (1.7), found for the ideal waveguide do not apply to this case, and solving the Schrödinger equation is cumbersome. (The variables in the three-dimensional Schrödinger equation do not separate and the motion does not become one-dimensional.)

We obtain, however, a general understanding of the quantum waves in the system by looking at an *adiabatic* waveguide [3]. Its dimensions are assumed to vary smoothly so that the length scale at which they change is much longer than the dimensions themselves:

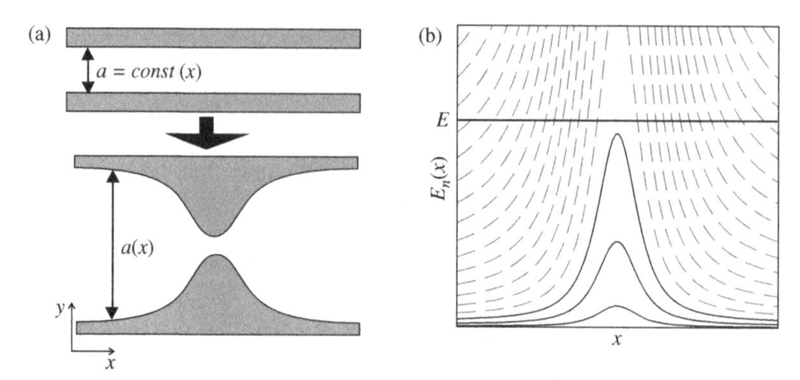

Fig. 1.7. (a) From a waveguide of constant cross-section to an adiabatic waveguide. (b) Effective potential energy for the transport channels. At the given energy *E*, only three transport channels (solid curves) are open.

$|a'(x)|, |b'(x)| \ll 1$, $a(x)|a''(x)|$, $b(x)|b''(x)| \ll 1$. Under these conditions, the walls are *locally* flat and parallel, so that locally the wave functions can be approximated by those of the ideal waveguide (Eqs. (1.7)). The variables are locally separated, so that

$$\psi_n(x, y, z) = \psi(x)\Phi_n(a(x), b(x), y, z), \tag{1.13}$$

where the transverse wave functions $\Phi(a, b, y, z)$ are given by Eqs. (1.7). The function $\psi(x)$ corresponds to one-dimensional motion and satisfies

$$\left(-\frac{\hbar^2}{2m}\frac{\partial^2}{\partial x^2} + E_n(x)\right)\psi(x) = E\psi(x). \tag{1.14}$$

Here, E_n presents a channel-dependent energy introduced by Eq. (1.8). This energy depends on x via the waveguide dimensions $a(x), b(x)$:

$$E_n(x) = \frac{\pi^2\hbar^2}{2m}\left(\frac{n_y^2}{a^2(x)} + \frac{n_z^2}{b^2(x)}\right). \tag{1.15}$$

We note that this term plays the role of *potential* energy for one-dimensional motion. Strangely, this potential energy depends on the channel index. Let us plot these energies versus x (see Fig. 1.7). For each channel we see a potential barrier forming in the narrowest part of the constriction. The bigger the numbers n_y, n_z, the higher the barrier.

Let us concentrate on a given energy E. In a given channel, we compare it with the maximal barrier height assuming an impenetrable barrier. If E exceeds the height, the electrons coming to the constriction traverse it; otherwise, they are reflected back. Since the barrier height increases with the channel index, there are only a finite number of *open* channels where electrons can pass the constriction. All other channels are *closed.*

Therefore, the adiabatic waveguide of variable cross-section without a potential barrier appears to be the same as for an ideal waveguide with a potential barrier, already considered in Section 1.1. For each channel, we define transmission and reflection amplitudes in the usual way (Fig. 1.3) and we end up with the channel-dependent transmission coefficient $T_n(E)$. It appears that the adiabaticity also implies an almost classical potential barrier, so that $T = 1$ for open channels and $T = 0$ for closed ones. The only exception is a narrow energy interval where the energy almost aligns with the top of the barrier. Remarkably, the electrons in the closed channels are almost perfectly reflected in spite of the absence of potential barriers in the system.

Control question 1.5. Assume the waveguide dimensions depend on x as follows: $a(x) = b(x) = a_\infty - a_0/(1 + (x/\xi)^2)$, $a_0 < a_\infty$. At which energy does the first open channel appear? At which energy are there three open channels?

Now we are ready to turn to our core business: the quantum transport. First, we determine the electric current in the constriction. The first step is to adapt the expression for the current given by Eq. (1.2) to the case of quantized transverse motion. We do this by replacing the integration over k_y and k_z by the summation over their discrete quantized values k_y^n and k_z^n, as follows:

$$\int \frac{dk_x}{2\pi}\frac{dk_y}{2\pi}\frac{dk_z}{2\pi}(\cdots) \rightarrow \int \frac{dk_x}{2\pi}\, ab \sum_n (\cdots). \tag{1.16}$$

This procedure describes an ideal waveguide, in which case k_x stands for the wave vector. In our case, the waveguide is not ideal and the wave vector depends on x. However, we have chosen the shape in such a way that for $x \rightarrow \pm\infty$ the waveguide *is* ideal, and we can use Eq. (1.16) to evaluate the full current I via the cross-section located infinitely far to the left from the constriction. Note that we are free to choose the cross-section in an arbitrary way since the charge conservation law implies that the stationary full current flowing through any cross-section is the same. We get the full current by multiplying the current density j_x by the cross-section area $a_\infty b_\infty$, thus absorbing the factors in Eq. (1.16),

$$I = 2_s e \sum_n \int_{-\infty}^{\infty} \frac{dk_x}{2\pi} v_x(k_x) f_n(k_x), \tag{1.17}$$

the velocity being $v_x = \hbar k_x/m$. Let us concentrate on the filling factors $f_n(k_x)$, which are different for open ($T = 1$) and closed ($T = 0$) channels (Fig. 1.8). If the channel is closed, all electrons passing the cross-section from the left are reflected from the barrier and subsequently pass the same cross-section from the right. Therefore, in a closed channel there is the same amount of right- and left-going electrons, and the filling factors are the same for the two momentum directions, $f_n(k_x) = f_n(-k_x)$. Since their velocities are opposite, the contribution of the closed channels to the net current vanishes. Thus, we concentrate on open channels.

For open channels, the filling factors for the two momentum directions are not the same. To realize this fact, we have to understand how the electrons get to the waveguide. This leads us to the concept of a *reservoir*. Any nanostructure taking part in quantum transport is part of an electric circuit. This means that it is connected to large, *macroscopic* electric pads each kept at a certain voltage (electrochemical potential). These pads contain a large number of electrons at thermal equilibrium. These electrons are characterized by the filling factor, Eq. (1.3), which depends only on the energy and the chemical potential of the corresponding reservoir. In our setup, the waveguide is connected to two such reservoirs:

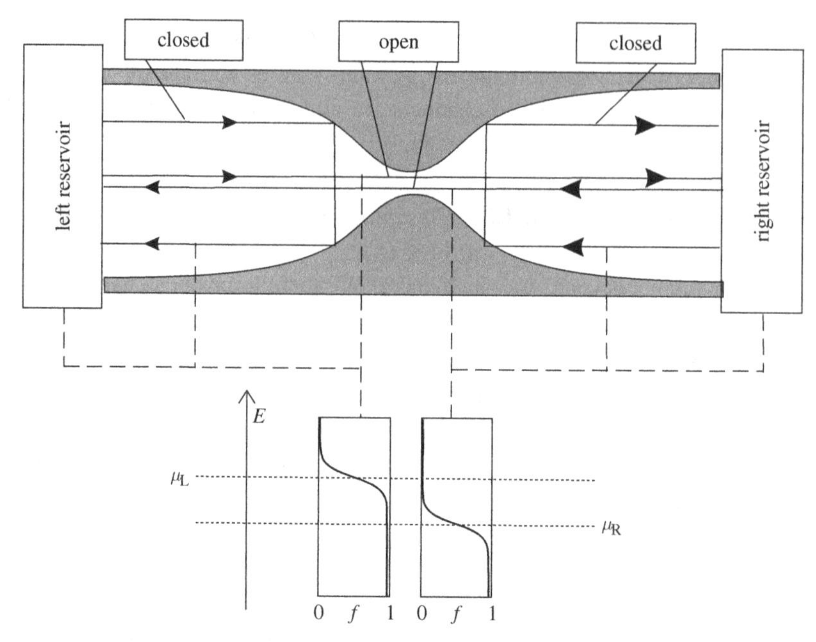

Fig. 1.8. **Filling factors in a quantum paint contact for open and closed channels.**

left ($x \to -\infty$) and right ($x \to \infty$). Electrons with $k_x > 0$ come from the left reservoir and have the filling factor $f_L(E) \equiv f_F(E - \mu_L)$. Electrons with $k_x < 0$ come to the cross-section having passed the constriction. Therefore, they carry the filling factor of the right reservoir, $f_R(E) \equiv f_F(E - \mu_R)$.

Since the filling factors depend only on the energy, it is natural to replace k_x in favor of the total energy E for each momentum direction. Since the velocity is $v_x = \hbar^{-1}(\partial E/\partial k_x)$, we have $\mathrm{d}E = \hbar v(k_x)\mathrm{d}k_x$, and this cancels the velocity in Eq. (1.17). Thus, we end up with the remarkably simple expression,

$$I = \frac{2_s e}{2\pi\hbar} \sum_{n:\mathrm{open}} \int \mathrm{d}E \left[f_L(E) - f_R(E)\right]$$
$$\equiv \frac{2_s e}{2\pi\hbar} N_{\mathrm{open}}(\mu_L - \mu_R) \equiv G_Q N_{\mathrm{open}} V. \tag{1.18}$$

We have integrated over energy; this yields a factor $\mu_L - \mu_R$. The simplest way to integrate is to assume vanishing temperature. Then the filling factors $f_{L,R} = \Theta(\mu_{L,R} - E)$ differ only within an energy strip $\min(\mu_{L,R}) < E < \max(\mu_{L,R})$ and are constant within the strip. The width of the strip is given by $|\mu_L - \mu_R|$.

Exercise 1.1. (a) Making use of the explicit form of the Fermi distribution f_F, show that the integral

$$\int \mathrm{d}E \left[f_L(E) - f_R(E)\right]$$

equals $\mu_L - \mu_R$ at any temperature.

(b) Prove that the integral retains the same value for any function $f(E)$ expressing the filling factors, $f_{L,R} = f(E - \mu_{L,R})$, provided $f \to 0$ at $E \to \infty$ and $f \to 1$ at $E \to -\infty$.

The difference in the chemical potentials corresponds to the voltage difference applied, $V = (\mu_L - \mu_R)/e$. The voltage difference drives the current; there is no current at $V = 0$ since that corresponds to the state of thermodynamic equilibrium. The factor $\mu_L - \mu_R$ is the same for all open channels. Therefore the current is proportional to the number of open channels, N_{open}, and the voltage. The proportionality coefficient is called the *conductance quantum* and conventionally defined as[3] $G_Q = 2_s e^2/(2\pi\hbar)$. The conductance of the system, I/V, appears to be quantized in units of G_Q. This factor is made up from fundamental constants. The conductance quantum does not depend on material properties, nanostructure size, and geometry, or from a concrete theoretical model used to evaluate the transport properties.

Equation (1.18) is a specific case of the celebrated Landauer formula. We have derived the formula for the case when $T(E)$ can be either zero or one. The general case is treated in Section 1.3.

Let us remark that we have obtained this very general relation in the framework of a specific model of a constricted adiabatic waveguide with impenetrable walls and rectangular cross-section. Now we discuss why this result holds in a far more general setup. First, let us get rid of the assumption of impenetrable walls and rectangular cross-section, and introduce a waveguide with an arbitrary confining potential $U_x(y, z)$. Provided the waveguide is adiabatic, we can still separate the variables in the Schrödinger equation and write the solution in the form of Eq. (1.13), where the transverse wave functions now obey the following equation:

$$\left[-\frac{\hbar^2}{2m}\left(\frac{\partial^2}{\partial y^2} + \frac{\partial^2}{\partial z^2}\right) + U_x(y, z)\right]\Phi_n(x; y, z) = E_n(x)\Phi_n(x; y, z), \tag{1.19}$$

where the discrete index n labels the transverse states, $E_n(x)$ being the channel-dependent potential energy (see Eq. (1.15)). This is the only change compared with the previous model.

Next, let us note that the number of open channels, and, consequently, the conductance of our system, are determined only by the narrowest part of the waveguide. Therefore we can change the shape of the waveguide without changing its transport properties, provided the narrowest cross-section stays the same. Let us see what happens if we change it in the way shown in Fig. 1.9, sending a_∞ and b_∞ to infinity. The structure we end up with – a *quantum point contact* (QPC) – is not a waveguide at all. In particular, in a waveguide we have a finite number of channels at each energy and the spectrum consists of discrete energy branches (Fig. 1.2). In contrast to this, the number of transport channels approaching the QPC is infinite, and the energy spectrum is continuous. Of all these channels, only a *finite* number are transmitted through the constriction.

[3] Eventually, it could be more logical to incorporate the number of spin directions 2_s into the number of open channels, so that one has two transport channels for each n.

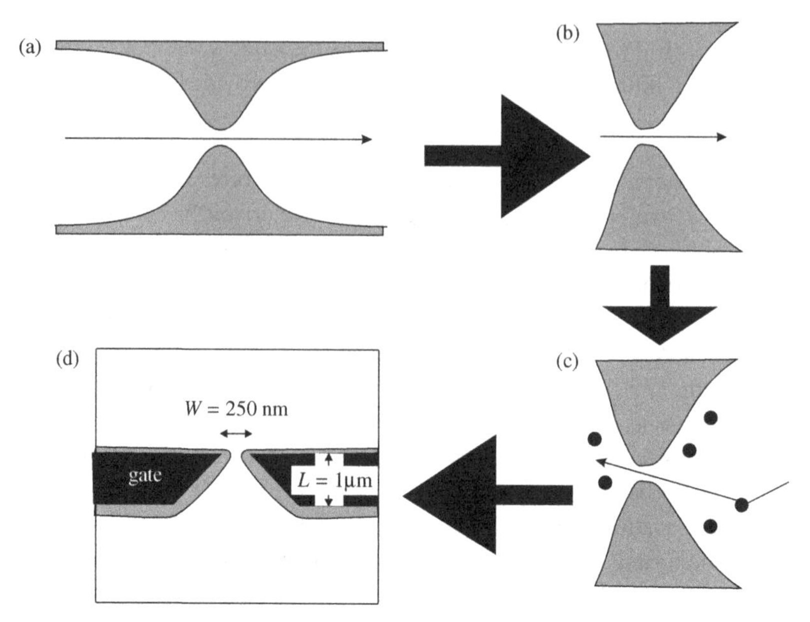

Fig. 1.9. From a model to real life. (a) Adiabatic waveguide with a finite number of transport channels. (b) QPC: infinite number of transport channels, finite number of open channels. (c) Scattering around the QPC is negligible provided the resistance of scattering region $R_{\rm QPC}$ is much smaller. (d) Experimental realization of QPC in a 2DEG with a split gate (adapted from Ref. [4]).

1.2.2 Experimental evidence of conductance quantization

Quantization of conductance was first observed in GaAlAs–GaAs semiconductor heterostructures [4, 5]. In these heterostructures, the electrons are confined near the surface forming a 2D electron gas (2DEG); see Section 1.1. In terms of our waveguide, this means that one of the dimensions $b \to 0$. Then only the lowest subband ($n_z = 1$) is relevant. In addition, two gate electrodes were imposed on the top of the heterostructure. These electrodes were electrically isolated from the 2DEG. However, they were used to shape this 2DEG; the potential applied to these electrodes repels the electrons from them (Fig. 1.9), creating surrounding impenetrable walls. The constriction is formed by the walls in the gap between the electrodes, the width corresponding to the dimension a of our model waveguide. Increasingly negative voltage created greater repulsion, shifting the walls outwards and therefore making the constriction narrower. The minimum width $a_{\rm min}$ is thus controlled by the gate voltage.

The number of open channels, in its turn, is determined by the minimum width. Let us write down this dependence explicitly for a model that disregards the potential inside the 2DEG and assumes infinite potential outside (a more realistic model is treated later in this section). A new channel with index $n = (n_y, 1)$ opens when the energy position of the top of the barrier, W_n, passes the Fermi energy as we change $a_{\rm min}$:

$$W_n \equiv \frac{\hbar^2\pi^2}{2a_{\rm min}^2 m} n_y^2 = E_{\rm F} = \frac{\hbar^2 k_{\rm F}^2}{2m}, \tag{1.20}$$

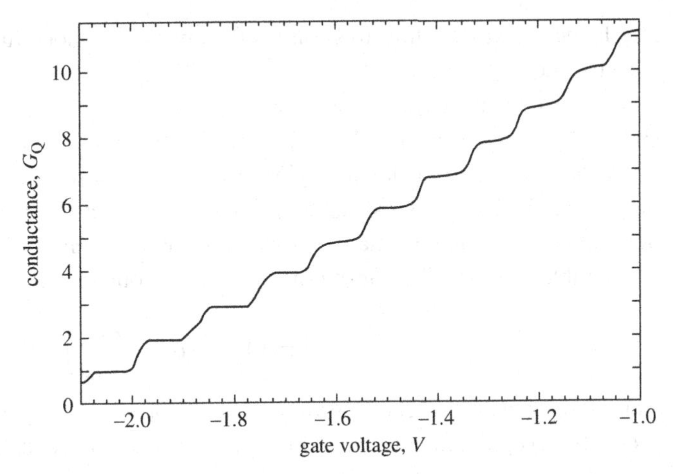

Fig. 1.10. **Experimental evidence of conductance quantization: discrete transport channels are visible. Adapted from Ref. [4].**

and thus

$$N_{\text{open}} = \left[k_F a_{\min}/\pi\right], \qquad (1.21)$$

where the brackets denote the integer part. Thus, by varying the gate voltage, one changes the number of open channels. Therefore, the gate voltage dependence of the conductance is expected to look like a set of stairs with step height G_Q, and this is what was measured in the experiment discussed in Ref. [4] (see Fig. 1.10).

Control question 1.6. Comparing Figs. 1.9 and 1.10, estimate the upper bound for k_F in the 2DEG used in the experiment discussed in Ref. [4].

At the time of the experiment, it was expected that such quantization would be observed for an ideal waveguide, but it was a complete surprise to observe it in a relatively short constriction in a far from ideal system. We cite Ref. [4]:

> *We propose an explanation of the observed quantization of the conductance, based on the assumption of quantized transverse momentum in the contact constriction. In principle this assumption requires a constriction much longer than wide, but presumably the quantization is conserved in the short and narrow constriction of the experiment.*

This quote, written a year before a theoretical understanding of quantum point contact was achieved, shows the essence of this section: discrete transport channels do not need waveguides to persist.

The conductance steps observed in the experiment are not very sharp. In reality, the transmission coefficient of a given channel does not change abruptly from zero (closed) to one (open). This change is only abrupt if the reflection from the barrier is classical. As we have learned in the example of a simple barrier considered in Section 1.1, quantum mechanics makes the transmission coefficient a continuous function of energy. This will also be true for a QPC.

To illustrate this, let us concentrate on the energy E that lies close to the top of the barrier W_n at certain n_y, $E = W_n + \delta E$, $|\delta E| \ll W_n$. In this case, the potential close to the top of the barrier can be expanded in Taylor series, $E_n(x) = W_n - |E_n''(0)|x^2/2$ (we assume that the top of the barrier is located at $x = 0$). Thus, electrons with $E \approx W_n$ are transmitted through a *parabolic* potential. This well known quantum-mechanical problem was solved by Kemble in 1935. The transmission coefficient obtained from the solution,

$$T_n(E) = \left[1 + \exp\left(-\frac{\delta E}{\hbar\Omega_n}\right)\right]^{-1}, \tag{1.22}$$

smoothly joins the two classical values, $T_n = 1$ at $\delta E \gg \hbar\Omega_n$ and $T_n = 0$ for $|\delta E| \gg \hbar\Omega_n$, $\delta E < 0$, thus providing the smearing of the steps at energy scale $\hbar\Omega_n$. This energy scale is given by

$$\hbar\Omega_n = \frac{\hbar}{2\pi}\sqrt{\frac{|E_n''(0)|}{m}} = \frac{\hbar^2 n_y}{2m}\sqrt{\frac{a''}{a^3}} = W_n\frac{\sqrt{a''a}}{\pi^2 n_y},$$

with a'', a being taken at $x = 0$. We see that this energy scale is much smaller than W_n provided the adiabaticity condition $a''a \ll 1$ is met. Moreover, this smearing scale is much smaller than the energy distance between the opening of consecutive channels, $W_{n+1} - W_n$. This makes the cross-over sharp, even for moderately adiabatic constrictions ($a''a \sim 0.3$ for the experiment mentioned).

To evaluate the conductance, we cannot use Eq. (1.18), since it only applies for $T = 0$ or $T = 1$. We need the full Landauer formula (to be derived in Section 1.3):

$$I = \frac{G_Q}{e}\sum_n \int dE\; T_n(E)\left[f_L(E) - f_R(E)\right].$$

At temperatures and voltages much smaller than $\hbar\Omega_n$, the integration over energy is essentially multiplication of the integrand with eV, with the substitution $E \to \mu$. The opening of the nth channel occurs at $|\mu - W_n| \simeq \hbar\Omega_n$ and is described by

$$\frac{G}{G_Q} = n - 1 + T_n(\mu) = n - 1 + \left[1 + \exp\left(-\frac{\mu - W_n}{\hbar\Omega_n}\right)\right]^{-1}. \tag{1.23}$$

Provided the electrochemical potential μ does not fit this narrow interval, we return to the classical situation: n channels are fully opened at $W_n < \mu < W_{n+1}$.

1.2.3 Electrostatic shaping of 2DEG

The QPC in the experiment considered was shaped electrostatically with the gate electrodes. We have understood that the conductance is determined by the number of open channels, this number being changed by the gate voltage. Let us now explore the details of the electrostatics of the shaping.

In the following we present a piece of classical rather than quantum physics; we discuss it here because of practical importance and because it is a piece of interesting physics. We

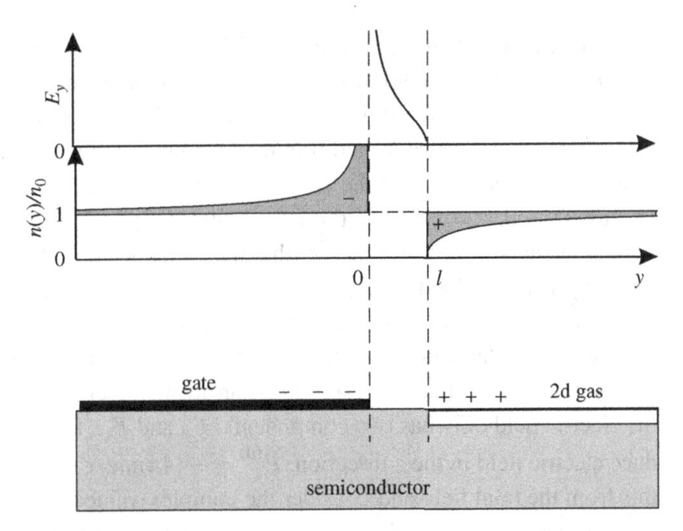

Fig. 1.11. Depletion of 2d gas by a side gate.

loosely follow Refs. [7] and [8] in this discussion. It turns out that plain classical electrostatics is more important for the 2DEG shaping than any quantum effects related to the motion of electrons in the 2DEG. To demonstrate this, let us first make a simple estimation of the electrostatic potential involved. In the heterostructures used, electrons are confined to a plane. Let us look at one semi-infinite gate electrode repelling a semi-infinite 2DEG (Fig. 1.11), both lying in the plane $z = 0$ and separated by distance l. The 2DEG density is defined by the fixed density n_0 of donors located beneath the heterostructure surface. This means that far from the gate ($y \to \infty$) the electron density equals n_0 to compensate for this positive-charge background. The same compensation takes place in the gate far from the 2DEG ($y \to -\infty$). The charge density is not compensated for at distances of the order of l. This resembles a capacitor with the metal plates wrenched to be in the same plane. The voltage across the capacitor can be still estimated using the formula for a planar capacitor, $V_g = (4\pi l/\epsilon)(Q/S)$. The electric field is concentrated in the semiconductor rather than in the vacuum, due to the large dielectric constant $\epsilon \gg 1$. The charge density Q/S is thus of the order of en_0, so that $V_g \sim eln_0/\epsilon$. Let us compare this with the typical kinetic energy of electrons in the 2DEG, $E_F = (\hbar k_F)^2/(2m) \sim \hbar^2 n_0/m$. We see that the potential energy dominates provided $l \gg a_B \equiv \hbar^2\epsilon/(e^2 m)$, a_B being the Bohr radius in the semiconductor (a_B is 10 nm for GaAs). This condition is thus fulfilled for l in the 100 nm range. In this parameter range, the 2DEG can be considered as an ideal conductor that screens the electric field, very much like a metallic gate electrode.

A surprising result of this approximation is that at $V_g \to 0$, $l \to 0$, and it looks like the 2DEG is depleted from under the gate. This is because, in the above reasoning, we have disregarded the distance d from the top metallic gate to the 2DEG proper in comparison with l. If we take d into account, the 2DEG stays under the gate at zero gate voltage, but is depleted at relatively small gate voltages $V_g = 4\pi en_0 d/\epsilon$. We will use this voltage as a reference for further discussion.

In what follows, we consider electrostatics to quantify the gate voltage dependence of the 2DEG shape.

Although the charge density in the 2DEG and the gate electrode is concentrated in the $z = 0$ plane, the distribution of electrostatic potential experienced and created by electrons is essentially three-dimensional; i.e., one has to consider the potential in the whole three-dimensional space to find it in the (x, y) plane in which the electrons are located. In principle, one has to solve the electrostatic problem separately in two half-spaces: in the vacuum ($z > 0, \epsilon = 1$) and in the semiconductor ($z < 0, \epsilon \gg 1$). There is, however, a simplification stemming from the fact that the charges are at the interface only: one replaces the setup with an effective medium with the effective dielectric constant $(\epsilon + 1)/2 \approx \epsilon/2$. This brings about extra symmetry with respect to $z \to -z$. To simplify it further, we do not consider x dependence of the potential. This implies the layout extended along the x axis, the geometry that corresponds to an ideal waveguide considered previously. Thus, the electric field only has two components, E_y and E_z. The positively charged donors produce electric field in the z direction, $E_z^{(\mathrm{d})} = -(4\pi n_0 e/\epsilon)\mathrm{sign}\, z$. It is convenient to subtract this from the total field and consider the complex-valued field $\mathcal{E} \equiv E_y + \mathrm{i}E_z$ produced by all other charges. This allows us to incorporate a common trick that enables us to solve a variety of electrostatic problems: if $\mathcal{E}$ is an analytic function of the complex coordinate $u \equiv y + \mathrm{i}z$, it automatically satisfies the Poisson equation for $z \neq 0$.

This solution must obey the boundary conditions that we now describe. Both the metallic gate ($y < 0$) and the 2DEG ($y > l$) are ideal conductors, so the potential is constant along each conductor. The in-plane component of the field, E_y, thus vanishes at $y < 0$ and $y > l$. The component E_z should experience a jump at $z = 0$ proportional to the charge density (with the density of donors subtracted). Symmetry requires $E_z(y + \mathrm{i}0) = -E_z(y - \mathrm{i}0)$. We conclude that $E_z(0) = 0$ for $0 < y < l$, whereas far from the capacitor, $y \to \pm\infty$,

$$E_z(y + \mathrm{i}0) = -E_z(y - \mathrm{i}0) = \frac{4\pi n_0 e}{\epsilon}.$$

In addition, the 2DEG must be in mechanical equilibrium; no net force is acting on electrons close to its boundary. This implies that $E_y(l) = 0$. This does not apply to the gate electrode; the electrostatic force acting on the charge in there may be compensated for by elastic forces keeping the gate at the substrate so that $E_y(0) \neq 0$.

The above conditions are easy to reformulate in terms of the complex function $\mathcal{E}$; it is real and single-valued at $0 < y < l$ and $z = 0$, it is imaginary, and has branch cuts elsewhere at $z = 0$, $\mathcal{E}(y + \mathrm{i}0) = -\mathcal{E}(y - \mathrm{i}0)$. This is the characteristic property of the square root function! We conjecture that $\mathcal{E}$ is a square root of a product or a ratio of two polynomials. The simplest guess satisfying the boundary conditions at $y \to \infty$, $y < 0$, and $y = l$ reads

$$\mathcal{E} = \frac{4\pi n_0 e}{\epsilon}\sqrt{\frac{l - u}{u}}. \tag{1.24}$$

There is an important theorem in electrostatics that guarantees the uniqueness of this solution, so we do not have to consider this further.

Control question 1.7. Check that all boundary conditions are indeed satisfied. To this end, explicitly evaluate the imaginary and real parts of the above expression (E_z, E_y).

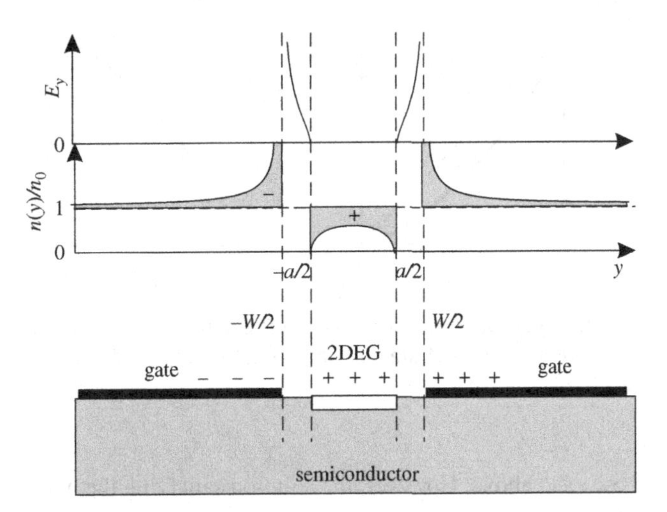

Fig. 1.12. **Split gate: electrostatic shaping of a 2DEG.**

Now we relate the distance l to the gate voltage:

$$V_g = \int_0^l dy\, E_y = \frac{4\pi n_0 e}{\epsilon} \int_0^l dy \sqrt{\frac{l-y}{y}} = \frac{2\pi^2 n_0 e l}{\epsilon}, \tag{1.25}$$

where $l = \epsilon V_g / 2\pi^2 n_0 e$. This quantifies our initial estimate. The distance l is commonly called the *depletion length*.

Exercise 1.2. Consider a thin long gate hanging at a distance a from the 2DEG plane. If the width of the gate is much smaller than a, it can be viewed as an infinitely thin wire with uniform charge density $\tilde{q}$ per unit length. Upon increasing $\tilde{q}$, the electron density underneath the gate decreases. Evaluate the critical value of $\tilde{q}$ at which the 2DEG is completely depleted at a certain position, i.e. its density approaches zero there.

Let us return to the shaping of the QPC. We now have two gates with the same potential V_g separated by distance W. A strip of 2DEG of width a is formed in between (see Fig. 1.12). We begin with a naive and straightforward model. Assume that the 2DEG is depleted by both gates independently. Then $a = W - 2l$. At a certain pinch-off voltage $V_p = \pi^2 n_0 e W/\epsilon$, we have $a = 0$, and the strip disappears. In addition, we assume that the potential is zero in the 2DEG and increases steeply beyond. This allows us to use Eq. (1.21) for the number of open channels, yielding

$$N_{\text{open}} = \left[\frac{k_F a}{\pi}\right] = \left[\frac{k_F W}{\pi} \frac{V_p - V_g}{V_p}\right]. \tag{1.26}$$

Thus, the conductance decreases linearly with the gate voltage.

Let us try to improve on the model. First, let us improve our consideration of electrostatics. There are two gates, left ($y < -W/2$) and right ($y > W/2$), and the 2DEG is confined in the area $|y| < a/2$, with a to be determined. The boundary conditions are similar to

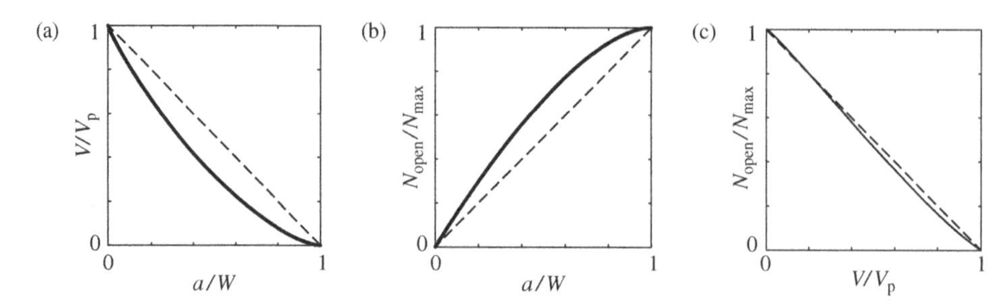

Fig. 1.13. Split gate. (a) Voltage versus a/W. (b) The number of open channels versus a/W. (c) The resulting voltage dependence of the number of channels. Dashed lines show estimates based on Eq. (1.26).

those used above. However, the solution satisfying the new boundary conditions is now as follows:

$$\mathcal{E} = \frac{4\pi n_0 e}{\epsilon}\sqrt{\frac{u^2 - (a/2)^2}{(W/2)^2 - u^2}}; \tag{1.27}$$

this is distinct from Eq. (1.24). We can express the gate voltage in terms of a via the following relation:

$$V_g = \int_{-W/2}^{-a/2} \mathrm{d}y\, E_y = \frac{4\pi n_0 e}{\epsilon}\int_{-W/2}^{-a/2} \mathrm{d}y \sqrt{\frac{y^2 - (a/2)^2}{(W/2)^2 - y^2}}.$$

The integral can be explicitly evaluated at $a = 0$ to yield the pinch-off voltage $V_p = 2\pi n_0 e W/\epsilon$. It differs from our naive estimate by the factor of $2/\pi$. Also, the dependence of the strip width on the gate voltage is not linear (see Fig. 1.13).

Next, we improve on the relation between the number of open channels and the width of the strip a. Naively, we have assumed that all potential in the 2DEG is screened. This is indeed true for a large potential of the order of V_g. In fact, there is a residual potential in the 2DEG which is of the order $E_F/e \ll V_g$. One can see this from the fact that the electron density in the 2DEG differs from the bare value of n_0, i.e.

$$n(y) = \frac{\epsilon\mathcal{E}(y + \mathrm{i}0)}{4\pi e} = n_0\sqrt{\frac{(a/2)^2 - y^2}{(W/2)^2 - y^2}}. \tag{1.28}$$

We can easily evaluate the residual potential while assuming that the density varies at a scale much longer than the Fermi wavelength. In this case, the relation between the density, the Fermi wave vector, and the residual potential $U_r(y)/e$ is local and is that of a uniform 2DEG. The relation between the Fermi wave vector, k_F, and the electron density, $n = 2_s k_F^2/(2\pi)^2$, was derived in Section 1.1. The sum of the potential and the kinetic energy of an electron equals E_F, thus $E_F = U_r + \hbar^2 k_F^2/2m$. We can therefore express the residual potential as follows:

$$\frac{U_r(y)}{E_F} = 1 - \frac{n(y)}{n_0}, \tag{1.29}$$

and we can study the electron states in the corresponding Schrödinger equation.

There is a shortcut, however. In the ideal waveguide, we saw that transverse wave functions obey the condition that an integer number of half-wavelengths is contained between the walls. In other words, the total phase change of the wave function between the walls should be πn_y, n_y being a positive integer. The maximum n_y allowed at a given energy yields the number of open channels at this energy. If the wave vector varies smoothly with y, the phase change equals $\int dy\ k_y(y)$. The maximum n_y at energy E_F is given by the substitution $k_y(y) = k_F(y)$, and the number of open channels is thus given by

$$N_{\text{open}} = \frac{1}{\pi}\int_{-a/2}^{a/2} dy\ k_F(y) = \frac{k_F}{\pi}\int_{-a/2}^{a/2} dy \sqrt{\frac{n(y)}{n_0}}, \tag{1.30}$$

with k_F evaluated at the density n_0. The number of channels calculated from Eqs. (1.30) and (1.28) is shown in Fig. 1.13. It deviates substantially from the assumed linear dependence on a.

Let us now complete the work and plot the number of open channels as a function of the gate voltage normalized to V_p. What a sad irony! The function evaluated hardly differs from the naive linear estimate of Eq. (1.26). This is intrinsic for all detailed studies of quantum transport: harder work that is intended to include all possible parameters characterizing a nanostructure yields very little improvement in comparison with the "naive" reasoning, provided the reasoning is *correctly* based on general laws of quantum transport.

1.3 Scattering matrix and the Landauer formula

In Sections 1.1 and 1.2, we studied electron transport in idealized waveguides with or without a potential barrier. These waveguides not only illustrate the concepts of quantum transport, but also model concrete experimental situations. A waveguide with no potential barrier models a QPC, a constriction created by gates in a 2DEG. A waveguide with a potential barrier models electron propagation through an insulating layer between two metals.

Real nanostructures can be made in a variety of ways, and can be more complicated. Modern fabrication technology allows for sophisticated semiconductor heterostructures, combining and shaping different metals, using nanotubes, molecules, and even single atoms as elements of an electron transport circuit. Various means can be used to control the transport properties of a fabricated nanostructure. It is only possible to describe all this in a single book because all these systems obey the general laws of quantum transport that we formulate in this section.

There is a common feature of all fabrication methods: two nanostructures that are intended to be identical, that is, are made with the same design and technology, are *never* identical (see Fig. 1.14). Beside the artificial features brought by design, there is also *disorder* originating from defects of different kind inevitably present in the structure. The position of and/or potential created by such defects is random, and in most cases can be neither controlled nor measured. It is unlikely that this situation will change with further technological developments; even if one achieves a perfect control of every atom in

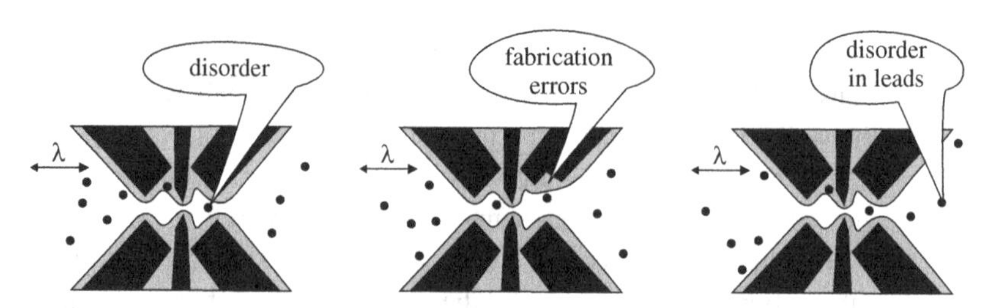

Fig. 1.14. **Nanostructures of an identical design are never identical.**

a nanostructure, one would not be able to control all the atoms in the macroscopic contact leads, which cannot be separated from the nanostructure. The defects scatter electrons, affecting the transport properties. Conductance of the structure is thus random, depending on a specific realization of disorder in the structure and in the leads; this means there is a formidable number of uncontrollable parameters.

Fortunately, the transport properties of any nanostructure can be expressed through a smaller set of parameters. The condition for this is that electrons traverse the structure without energy loss, so they experience only elastic scattering. These conditions for a given structure are always achieved at sufficiently low temperature and applied voltage. The scattering is characterized by a *scattering matrix* that contains information about electron wave functions far from the structure. The transport is described by a set of *transmission eigenvalues* derived from this scattering matrix. A great deal of literature on quantum transport, and a great deal of this book, is in fact devoted to evaluation of the transmission eigenvalues and establishing their general properties. In this section, we derive the relation between conductance and the transmission eigenvalues and thus demonstrate that understanding the transmission properties of a system automatically means understanding its transport properties.

1.3.1 Scattering matrix

We have mentioned in Section 1.2 that any nanostructure taking part in quantum transport is part of an electric circuit. It is connected to several *reservoirs*, which are in thermal equilibrium and are characterized by a fixed voltage. In this section, we only consider the case when there are two reservoirs (referred to as left and right). Generalization to many reservoirs is given in Section 1.5. Between the reservoirs is the *scattering region* – the nanostructure proper. Let us start with a feature borrowed from the QPC model of Section 1.2: ideal waveguides connect the reservoirs and the scattering region (Fig. 1.15). This is convenient since the scattering only takes place in a finite region, the reservoirs being far from this region. The wave functions may have very complicated forms in the scattering region, but in the waveguides they are always combinations of plane waves. The left and right waveguides do not have to have the same axis and the same cross-section. This is why it is convenient to introduce the separate coordinates $x_L < 0, y_L, z_L$ and $x_R > 0, y_R, z_R$ for the left and right waveguides, respectively. Generally, a wave function at fixed energy E can be presented as a linear combination of the plane waves

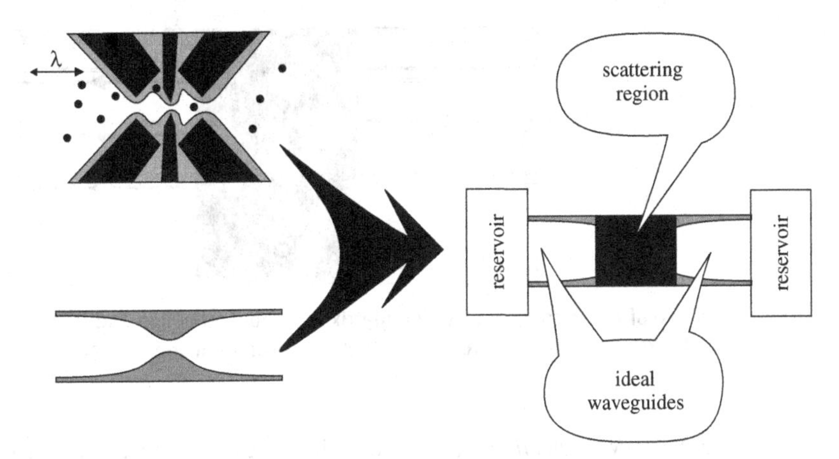

Fig. 1.15. **Scattering approach to quantum transport. Ideal waveguides and reservoirs from QPC plus scattering in between form an adequate model of transport in any nanostructure.**

$$\psi(x_\mathrm{L}, y_\mathrm{L}, z_\mathrm{L}) = \sum_n \frac{1}{\sqrt{2\pi\hbar v_n}} \Phi_n(y_\mathrm{L}, z_\mathrm{L}) \left[a_{\mathrm{L}n} \mathrm{e}^{\mathrm{i}k_x^{(n)} x_\mathrm{L}} + b_{\mathrm{L}n} \mathrm{e}^{-\mathrm{i}k_x^{(n)} x_\mathrm{L}} \right] \tag{1.31}$$

and

$$\psi(x_\mathrm{R}, y_\mathrm{R}, z_\mathrm{R}) = \sum_m \frac{1}{\sqrt{2\pi\hbar v_m}} \Phi_m(y_\mathrm{R}, z_\mathrm{R}) \left[a_{\mathrm{R}m} \mathrm{e}^{-\mathrm{i}k_x^{(m)} x_\mathrm{R}} + b_{\mathrm{R}m} \mathrm{e}^{\mathrm{i}k_x^{(m)} x_\mathrm{R}} \right]. \tag{1.32}$$

Here we label the transport channels in the left and right waveguides by the indices n and m, respectively. The corresponding transverse wave functions are Φ_n and Φ_m, and the energies of the transverse motion are E_n, E_m. For any transport channel n or m, be it in the left or right waveguide, the energy E fixes the value of the wave vector $k_x^{(n)} = \sqrt{2m(E - E_n)}/\hbar$. Transport is due to propagating, not evanescent, waves, and $k_x^{(n)}$ has to be real. Then, only a finite number of open channels, N_L to the left and N_R to the right, exist at a fixed energy E. We explicitly write the square roots of velocities v_n in each channel. This is to ensure that the current density does not contain these factors and is expressed in terms of $a_{\mathrm{L}n}$, $b_{\mathrm{L}n}$ or $a_{\mathrm{R}m}$, $b_{\mathrm{R}m}$ only.

In Eqs. (1.31) and (1.32) the coefficients $a_{\mathrm{L}n}$, $a_{\mathrm{R}m}$ are the amplitudes of the waves coming from the reservoirs, and $b_{\mathrm{L}n}$, $b_{\mathrm{R}m}$ are the amplitudes of the waves transmitted through or reflected back from the scattering region. These coefficients are therefore not independent: the amplitude of the wave reflected from the obstacle linearly depends on the amplitudes of incoming waves in all the channels,

$$b_{\alpha l} = \sum_{\beta=\mathrm{L,R}} \sum_{l'} s_{\alpha l, \beta l'} a_{\beta l'}, \quad \beta = \mathrm{L,R}, \quad l = n, m. \tag{1.33}$$

The proportionality coefficients are combined into a $(N_\mathrm{L} + N_\mathrm{R}) \times (N_\mathrm{L} + N_\mathrm{R})$ *scattering matrix* $\hat{s}$. It has the following block structure:

$$\hat{s} = \begin{pmatrix} \hat{s}_\mathrm{LL} & \hat{s}_\mathrm{LR} \\ \hat{s}_\mathrm{RL} & \hat{s}_\mathrm{RR} \end{pmatrix} \equiv \begin{pmatrix} \hat{r} & \hat{t}' \\ \hat{t} & \hat{r}' \end{pmatrix}. \tag{1.34}$$

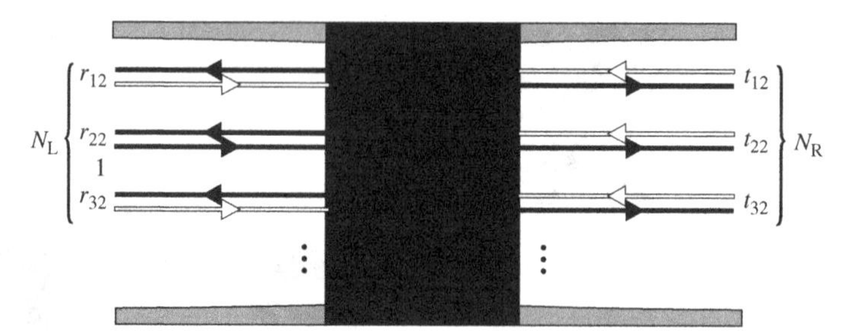

Fig. 1.16. Structure of two-terminal scattering matrix. We show reflection and transmission amplitudes of the electron wave coming from the left in the second transport channel, $n' = 2$.

The $N_L \times N_L$ *reflection matrix* $\hat{r}$ describes the reflection of the waves coming from the left. Thus, $r_{nn'}$ is the amplitude of the following process: the electron coming from the left in the transverse channel n' is reflected to the channel n. Consequently, $|r_{nn'}|^2$ is the probability of this process. The $N_R \times N_R$ reflection matrix $\hat{r}'$ describes reflection of particles coming from the right. Finally, the $N_R \times N_L$ *transmission matrix* $\hat{t}$ is responsible for the transmission through the scattering region (see Fig. 1.16).

Control question 1.8. Explain the block structure of the scattering matrix in the following cases: (i) electrons in all transport channels are reflected from the nanostructure with no transmission remaining in the same channel; (ii) electrons in all transport channels are reflected from the nanostructure with no transmission, but do not remain in the same channel; (iii) electrons in all channels are transmitted without any reflection and do not have to remain in the same channel.

An important condition on the scattering matrix is imposed by symmetry with respect to time reversal. If this symmetry holds, the scattering matrix is symmetric, $\hat{s} = \hat{s}^T$. So, the reflection matrices are symmetric, and $\hat{t}' = \hat{t}^T$. The applied magnetic field B changes sign upon time reversal. In this case, the time-reversal symmetry relates the elements of the scattering matrix at opposite values of magnetic field, $r_{nn'}(B) = r_{n'n}(-B)$, $r'_{mm'}(B) = r'_{m'm}(-B)$, $t_{mn}(B) = t'_{nm}(-B)$.

Any scattering matrix satisfies the unitarity condition, $\hat{s}^\dagger\hat{s} = \hat{1}$. The diagonal element of $\hat{s}^\dagger\hat{s}$ is given by

$$\left(\hat{s}^\dagger\hat{s}\right)_{nn} = \sum_{n'} |r_{nn'}|^2 + \sum_m |t_{mn}|^2 = 1, \tag{1.35}$$

since it represents the total probability of an electron in channel n being either reflected or transmitted to any channel.

Exercise 1.3. The above unitarity condition is best expressed using reflection–transmission block structure, Eq. (1.34), and provides several important details of the scattering approach. To see this: (i) write down the condition $\hat{s}^\dagger\hat{s} = \hat{1}$ explicitly in block notation and show that it gives rise to three independent conditions on the matrices $\hat{r}, \hat{r}', \hat{t}$, and $\hat{t}'$; (ii) do the same with the condition $\hat{s}\hat{s}^\dagger = \hat{1}$ and show that it gives rise to three extra conditions; (iii) use the derived conditions to demonstrate that $\hat{t}'\hat{t}'^{\dagger} = \hat{r}\hat{t}^{\dagger}\hat{t}\hat{r}^{-1}$; (iv) derive the corresponding condition for $\hat{t}'^{\dagger}\hat{t}'$; (v) making use of the above results, prove that the matrices $\hat{t}\hat{t}^{\dagger}, \hat{t}^{\dagger}\hat{t}, \hat{t}'\hat{t}'^{\dagger}$, and $\hat{t}'^{\dagger}\hat{t}'$ all have the same set of non-zero eigenvalues $\{T_{\mathrm{p}}\}$ – transmission eigenvalues; (vi) show that $\hat{r}\hat{r}^{\dagger}, \hat{r}^{\dagger}\hat{r}, \hat{r}'\hat{r}'^{\dagger}$, and $\hat{r}'^{\dagger}\hat{r}'$ all have the same set of eigenvalues $\{R_{\mathrm{p}}\}$ different from 1 and that $R_{\mathrm{p}} = 1 - T_{\mathrm{p}}$.

1.3.2 Transmission eigenvalues

We now turn to the calculation of the current, using Eq. (1.17) as the starting point. Let us calculate the current through a cross-section located in the left waveguide. The electrons with $k_x > 0$ originate from the left reservoir, and their filling factor is therefore $f_{\mathrm{L}}(E)$. Now, the electrons with $k_x < 0$ in a given channel n are coming from the scattering region. A fraction of these electrons originate from the left reservoir and are reflected; they carry the filling factor $f_{\mathrm{L}}(E)$. This fraction is determined by the probability of being reflected to channel n from all possible starting channels n', $R_n(E) = \sum_{n'} |r_{nn'}|^2$. Other electrons are transmitted through the scattering region, their filling factor being $f_{\mathrm{R}}(E)$. The resulting filling factor for $k_x < 0$ is therefore $R_n f_{\mathrm{L}}(E) + (1 - R_n) f_{\mathrm{R}}(E)$. For the current we write

$$\begin{aligned} I &= 2_s e \sum_n \left\{ \int_0^{\infty} \frac{\mathrm{d}k_x}{2\pi} v_x(k_x) f_{\mathrm{L}}(E) \right. \\ &\quad \left. + \int_{-\infty}^{0} \frac{\mathrm{d}k_x}{2\pi} v_x(k_x) \left[R_n(E) f_{\mathrm{L}}(E) + (1 - R_n(E)) f_{\mathrm{R}}(E) \right] \right\} \\ &= 2_s e \sum_n \int_0^{\infty} \frac{\mathrm{d}k_x}{2\pi} v_x(k_x)(1 - R_n(E)) \left[f_{\mathrm{L}}(E) - f_{\mathrm{R}}(E) \right]. \end{aligned} \tag{1.36}$$

To derive the final equation line, we have changed k_x to $-k_x$ in the second integral in Eq. (1.36). We use the unitarity relation, Eq. (1.35), to prove that

$$1 - R_n = \sum_m |t_{mn}|^2 = (\hat{t}^{\dagger}\hat{t})_{nn}.$$

Now we repeat the trick of the previous section, changing variables from k_x to E, and we arrive at the following expression:

$$I = \frac{2_s e}{2\pi} \int_0^{\infty} \mathrm{d}E \; \mathrm{Tr}\left[\hat{t}^{\dagger}\hat{t}\right] \left[f_{\mathrm{L}}(E) - f_{\mathrm{R}}(E) \right], \tag{1.37}$$

where we have used the short-hand notation

$$\mathrm{Tr}\left[\hat{t}^{\dagger}\hat{t}\right] = \sum_n \left(\hat{t}^{\dagger}\hat{t}\right)_{nn}.$$

Alternatively, the trace can be presented as a sum of eigenvalues T_p of the Hermitian matrix $\hat{t}^{\dagger}\hat{t}$, the *transmission eigenvalues*. Because of the unitarity of the scattering matrix, T_p are real numbers between zero and one.

The transmission eigenvalues depend on energy. However, in the linear regime, when the applied voltage is much smaller than the typical energy scale of this dependence, they can be evaluated at the Fermi surface, and we obtain the following expression for the conductance:

$$G = G_Q \sum_p T_p(\mu). \tag{1.38}$$

Calculation of the current in the right waveguide gives the same result: current is conserved.

Equation (1.38) is known as the (two-terminal) *Landauer formula*. Rolf Landauer [8] pioneered the scattering approach to electrical conduction many years ago. At the time, he was met with distrust since the common view on conduction was based on the semiclassical approach outlined in Chapter 2. Views changed drastically after the publication of Ref. [4].

Exercise 1.4. Let us evaluate corrections coming from a smooth energy dependence of transmission coefficients near the Fermi energy. Let us assume that $G(E) = G_Q \sum_p T_p$ can be expanded near E_F, $G(E) = G_0 + G_1(E - E_F) + G_2(E - E_F)^2 + \cdots$. We keep the first three terms in the expansion. Calculate the corresponding contributions to the current, Eq. (1.37), assuming $eV, k_BT \ll E_F$, $\mu_L = E_F$.

Hint:

$$\int \frac{x^2 \, dx}{\cosh^2 x} = \frac{\pi^2}{6}.$$

We have derived Eq. (1.38) assuming the nanostructure to be connected to ideal waveguides that support N_L and N_R transport channels. Now we can get rid of this unrealistic assumption by repeating the reasoning we used for the QPC. Let us unfold the waveguides so that their cross-sections become infinite; it should not change the transport properties of the nanostructure. The number of transport channels becomes infinite, $N_L, N_R \to \infty$. This means that there are infinitely many transmission eigenvalues. This also means that the total number of transport channels $N_{L,R}$ is an "unphysical" quantity: it characterizes an auxiliary model rather than the nanostructure, and no transport property of a nanostructure would eventually depend on $N_{L,R}$.[4]

How do we reconcile the finite conductance given by Eq. (1.38) with the infinite number of transmission eigenvalues? The implication is that infinitely many transmission eigenvalues are concentrated very close to zero transmission, so that they contribute neither to conductance nor to any other transport property.

To evaluate the transmission eigenvalues of a given nanostructure, one solves the Schrödinger equation in the scattering region and matches the two asymptotics, Eqs. (1.31) and (1.32), extracting the scattering matrix. The solutions depend on all the details, such as the location of the gates defining the nanostructure design and the given configuration of the disorder. Even for relatively simple systems, this is a time-consuming task, without

[4] Confusingly, this "number of transport channels" is commonly used in the literature to characterize the area of the (narrowest) cross-section of a nanostructure.

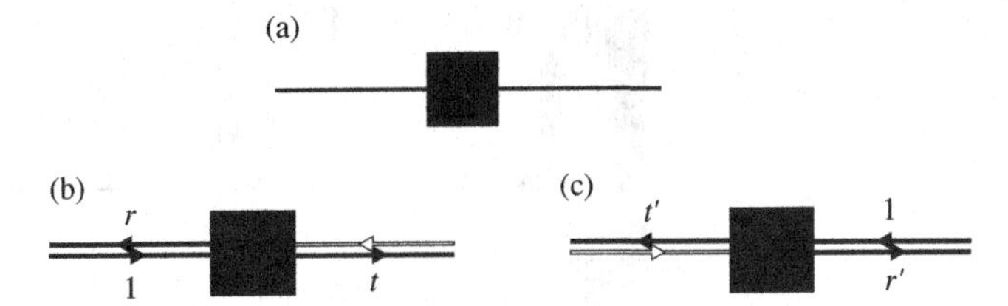

Fig. 1.17. One-channel scatterer. (a) Notation; (b) transmission and reflection amplitudes of the waves coming from the left; (c) same for the waves coming from the right.

much intellectual impact. A calculation for a given system gives us no idea what the result would be if we were to change a design detail or the disorder configuration.

This makes it important to comprehend the general properties of transmission eigenvalues, those depending on the system design rather than on the details.

One channel

Let us start with a simple example: a scatterer that can transmit only one transport channel (for a given energy). All but one of the transmission eigenvalues are zero. The structure is thus characterized by a single transmission eigenvalue, T. This is precisely the transmission coefficient we have discussed for the potential barrier in Section 1.1; $R = 1 - T$ is the reflection coefficient. The scattering matrix is a 2×2 matrix and contains more parameters, since in Eq. (1.34) r, r', and t are complex numbers, constrained by the conditions of unitarity (see Fig. 1.17). There are three independent parameters T, θ, and η:

$$\hat{s} = \begin{pmatrix} \sqrt{R}e^{i\theta} & \sqrt{T}e^{i\eta} \\ \sqrt{T}e^{i\eta} & -\sqrt{R}e^{i(2\eta-\theta)} \end{pmatrix}. \tag{1.39}$$

The phases θ and η do not manifest themselves in the transport in a single nanostructure of this type. As we show in Section 1.6, these phases are relevant if we combine two structures producing quantum interference effects.

Control question 1.9. How many independent parameters characterize a general scattering matrix for two channels? Assume time-reversal symmetry.

For the ideal systems considered previously – a rectangular potential barrier and a QPC – the scattering does not mix different transport channels. An electron in channel n can either be reflected back and stay in the same channel or be transmitted through the barrier to end up in an identical channel at the other side. Therefore the matrix of such an ideal system is *block-diagonal* – the matrices r, r', t, and, importantly, $t^\dagger t$ are diagonal. Thus, the transmission eigenvalues for these systems are just the transmission coefficients in the channels.

1.3.3 Distribution of transmission eigenvalues

The transmission eigenvalues T_p depend on disorder configuration and therefore are random (Fig. 1.18). We need a quantity that characterizes the *design* of a nanostructure rather

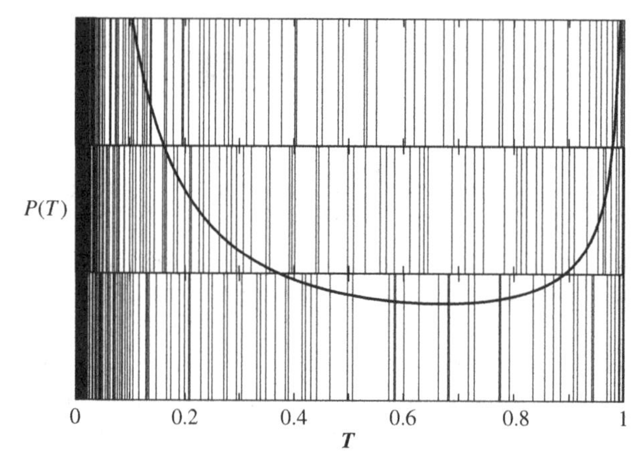

Fig. 1.18. **Transmission eigenvalues. We show the transmission eigenvalues for three disorder realizations of a diffusive conductor with a nominal resistance of 350 Ω. The transmission distribution (thick solid line) is given by Eq. (1.43).**

than a concrete disorder configuration. This is provided by the distribution function of transmission eigenvalues (transmission distribution) $P(T)$. Suppose we make an ensemble of nanostructures sharing an identical design and differing in disorder configurations. Each nanostructure provides a set of transmission eigenvalues. Let us concentrate on a narrow interval of transmissions from T to $T + \mathrm{d}T$; count the number of transmission eigenvalues that fall into this interval, and divide this by the total number of nanostructures. In the limit of a large ensemble, the result converges to $P(T)\mathrm{d}T$. Mathematically, the transmission distribution is thus defined as follows:

$$P(T) = \left\langle \sum_p \delta\left(T - T_p(E)\right) \right\rangle. \tag{1.40}$$

The angular brackets in Eq. (1.40) mean the ensemble average, that is, the average over all formally identical nanostructures in the ensemble. The function $P(T)$ facilitates evaluating other averages. The average of an arbitrary function of the transmission eigenvalues becomes

$$\left\langle \sum_p f(T_p) \right\rangle = \int_0^1 \mathrm{d}T\ f(T)P(T). \tag{1.41}$$

In particular, one integrates $G_Q T P(T)$ to obtain the average conductance $\langle G \rangle$.

What is the use of the above relation? If the average conductance of a nanostructure much exceeds the conductance quantum, $\langle G \rangle \gg G_Q$, the transmission eigenvalues are dense, the typical spacing between the eigenvalues being much less than one. This means that the sums over the transmission eigenvalues can be replaced by the integrals according to Eq. (1.41). The transport properties are thus *self-averaged* in this limit, their fluctuations being much smaller than the average values. The transport properties appear to be almost insensitive to a specific disorder configuration. The fluctuations of transport properties may become significant if $\langle G \rangle \simeq G_Q$ and the transmission eigenvalues are sparse.

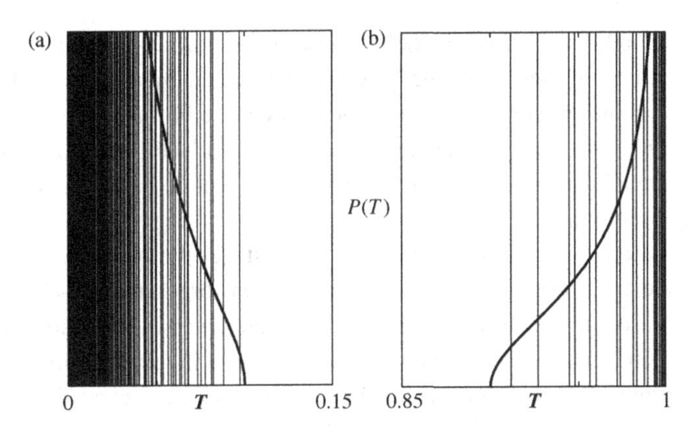

Fig. 1.19. **Examples of transmission distribution. (a) Tunnel junction in series with a diffusive conductor of a small resistance. (b) QPC with 20 open channels in series with a diffusive conductor of a small resistance.**

A fair part of this book either quantifies the transmission distribution or makes use of it. In the rest of this section, we provide examples of $P(T)$ without quantifying it.

Let us start with a QPC and consider the energy at which a finite number N_{open} of transport channels are open ($T = 1$). An infinite number of channels are closed ($T = 0$). The corresponding transmission distribution consists of two delta-functional peaks,

$$P(T) = N_{\text{open}}\delta(1 - T) + \infty\delta(T).$$

Closed channels do not play any role in the transport, and the part with the open channels leads to the expression for the conductance, $G = G_Q N_{\text{open}}$ that we have already seen. We will ignore the part proportional to $\delta(T)$ and write, for a clean QPC,

$$P_{\text{QPC}}(T) = N_{\text{open}}\delta(1 - T). \tag{1.42}$$

The transmission eigenvalues in a clean QPC are highly degenerate. If we add a small number of defects to the QPC, this degeneracy is lifted. The scattering at the defects mixes the channels: an incident electron in open channel n, which without scattering would pass the constriction, can now be reflected to any channel n', or be transmitted to an arbitrary channel m. These processes modify the transmission matrix and, consequently, the transmission eigenvalues. If such channel mixing is weak, so that the probabilities of scattering from open channels are small, we expect that all transmission eigenvalues remain close to one. The role of disorder is thus to lift the degeneracy (see Fig. 1.19). This regime is realized when the contribution of disorder to the total conductance of the system is sufficiently small, the resistance R due to defects being much smaller than the resistance of the QPC. As resistance R is increased further to values of the order of $1/G_{\text{QPC}}$, the transmission eigenvalues are spread over the whole interval $0 < T < 1$.

A complementary example is a *tunnel junction*. Let us take a sufficiently wide ideal potential barrier at an energy much below the top energy of the barrier. All the transmission coefficients are guaranteed to be small, $T \ll 1$. If the channels do not mix, the transmission eigenvalues are just these coefficients and the transmission distribution concentrates near

$T = 0$. If we add some defects next to the barrier, the channels mix. Some electrons after being reflected from the barrier are reflected by defects back to the barrier. They get a "second chance" to tunnel through. Because of this, some transmission eigenvalues grow with increasing defect resistance R. Similarly to QPC, the transmission eigenvalues are spread over the whole interval $0 < T < 1$ if R is comparable with the resistance of the tunnel junction.

Let us add more defects. At some stage, the resistance due to the defects dominates the total resistance. At this point, we can forget about a QPC or a tunnel junction being present in the structure. The electron that traverses the scattering region experiences many scattering events at the defects. Its motion is highly random. This corresponds to *diffusion* provided the conductance of the structure still much exceeds the conductance quantum. The transmission distribution in a diffusive structure appears to be universal – not depending on the details of the structure design (Fig. 1.18),

$$\rho_{\mathrm{D}} = \frac{\langle G \rangle}{2G_{\mathrm{Q}}} \frac{1}{T\sqrt{1-T}}. \tag{1.43}$$

The integral of the transmission distribution over T gives the total number of transport channels. This integral diverges for the bimodal distribution, Eq. (1.43) indicating an infinite number of channels that may take part in diffusive transport.

Control question 1.10. How would one prove using the Landauer formula that the averaged conductance corresponding to Eq. (1.43) is indeed $\langle G \rangle$? What is the total number of transport channels? Explain the result.

1.3.4 Scattering in operator formalism

It is customary and convenient to treat electrons in solids with the aid of creation and annihilation operators (see Appendix A). Here we present the operator formalism for the scattering approach and re-derive the Landauer formula. We follow the derivation of Ref. [9].

An arbitrary wave function in the left waveguide is represented as a sum of plane waves Eq. (1.31). These plane waves, however, do not form a basis, since they only represent asymptotic expressions of wave functions, which have a complicated form in the scattering region and do not have to be orthogonal. It is convenient to use *scattering states* – the states which originate from the reservoirs as plane waves and then are partially transmitted through the scattering region and partially reflected back. The scattering state originating from the left reservoir has the form

$$\begin{aligned}\psi_{\mathrm{L}n}(x_{\mathrm{L}}, y_{\mathrm{L}}, z_{\mathrm{L}}) = {} & \frac{1}{\sqrt{2\pi\hbar v_n(E)}} \Phi_n(y_{\mathrm{L}}, z_{\mathrm{L}}) \mathrm{e}^{\mathrm{i}k_x^{(n)} x_{\mathrm{L}}} \\ & + \sum_{n'} \frac{1}{\sqrt{2\pi\hbar v_{n'}(E)}} r_{n'n}(E) \Phi_{n'}(y_{\mathrm{L}}, z_{\mathrm{L}}) \mathrm{e}^{-\mathrm{i}k_x^{(n')} x_{\mathrm{L}}},\end{aligned} \tag{1.44}$$

in the left waveguide, and

$$\psi_{Ln}(x_R, y_R, z_R) = \sum_m \frac{1}{\sqrt{2\pi\hbar v_m(E)}} t_{mn}(E)\Phi_m(y_R, z_R)e^{-ik_x^{(m)}x_R} \tag{1.45}$$

in the right waveguide. Analogously, there are scattering states ψ_{Rm} originating from the right reservoir.

For each of these states, we can introduce creation and annihilation operators. Let us introduce the creation operators $\hat{a}^\dagger_{Ln}(E)$ and $\hat{a}^\dagger_{Rm}$, which create electrons in the scattering states with energy E, originating from the left reservoir in transport channel n and from the right reservoir in transport channel m, respectively. The conjugated operators $\hat{a}_{Ln}(E)$ and $\hat{a}_{Rm}$ annihilate particles in the same states. The operators $\hat{a}^\dagger$ and $\hat{a}$ are sufficient for the quantum-mechanical description of the system.

For convenience, we introduce another set of operators. The operator $\hat{b}^\dagger_{Ln\sigma}(E)$ creates an electron with energy E and spin projection σ in transport channel n in the left waveguide moving to the left. A similar creation operator for right-movers in the right waveguide is $\hat{b}^\dagger_{Rm\sigma}(E)$, and the annihilation operators are $\hat{b}_{Ln\sigma}(E)$ and $\hat{b}_{Rm\sigma}(E)$, respectively. These operators are linearly related to the set $\hat{a}$ via the scattering matrix,

$$\begin{aligned} \hat{b}_{\alpha l\sigma}(E) &= \sum_{\beta=L,R} \sum_{l'} s_{\alpha l,\beta l'}(E)\hat{a}_{\beta l'\sigma}(E); \\ \hat{b}^\dagger_{\alpha l\sigma}(E) &= \sum_{\beta=L,R} \sum_{l'} s_{\beta l',\alpha l}(E)\hat{a}^\dagger_{\beta l'\sigma}(E), \quad \alpha = L, R, \quad l = n, m. \end{aligned} \tag{1.46}$$

Since electrons are fermions, the operators $\hat{a}$ obey anticommutation relations:

$$\begin{aligned} &\hat{a}^\dagger_{\alpha l\sigma}(E)\hat{a}_{\beta l'\sigma'}(E') + \hat{a}_{\beta l'\sigma'}(E')\hat{a}^\dagger_{\alpha l\sigma}(E) = \delta_{\alpha\beta}\delta_{ll'}\delta_{\sigma\sigma'}\delta(E - E'); \\ &\hat{a}_{\alpha l\sigma}(E)\hat{a}_{\beta l'\sigma'}(E') + \hat{a}_{\beta l'\sigma'}(E')\hat{a}_{\alpha l\sigma}(E) = 0; \\ &\hat{a}^\dagger_{\alpha l\sigma}(E)\hat{a}^\dagger_{\beta l'\sigma'}(E') + \hat{a}^\dagger_{\beta l'\sigma'}(E')\hat{a}^\dagger_{\alpha l\sigma}(E) = 0. \end{aligned} \tag{1.47}$$

Equations (1.47) occur since the operators $\hat{a}$ form a basis. In the same way, the operators describing left-moving electrons in the left waveguide and right-moving electrons in the right waveguide also form a basis, and similar relations hold between the operators $\hat{b}$ and $\hat{b}^\dagger$. However, the operators $\hat{a}$ and $\hat{b}$ do not obey such relations, as is evident from Eqs. (1.46).

Exercise 1.5. Making use of Eqs. (1.46) proves that the operators $\hat{b}$ and $\hat{b}^\dagger$ satisfy the same anticommutation relations Eqs. (1.47) as the operators $\hat{a}$ and $\hat{a}^\dagger$.

Now we consider the quantum-mechanical averages of the products of creation and annihilation operators. Since the right-moving particles in the left waveguide originate from the left reservoir, we have

$$\left\langle \hat{a}^\dagger_{\alpha l\sigma}(E)\hat{a}_{\beta l'\sigma'}(E') \right\rangle = \delta_{\alpha\beta}\delta_{ll'}\delta_{\sigma\sigma'}\delta(E - E')f_\alpha(E), \quad \alpha = L, R. \tag{1.48}$$

The average product of two creation or two annihilation operators is always zero.

Let us proceed by writing down the field operators $\hat{\Psi}_\sigma(\boldsymbol{r},t)$ and $\hat{\Psi}^\dagger_\sigma(\boldsymbol{r},t)$, which annihilate and create the electron with the given spin projection at a given point and time moment. In the left waveguide, we have

$$\hat{\Psi}_\sigma(\boldsymbol{r},t) = \int \mathrm{d}E\ \mathrm{e}^{-\mathrm{i}Et/\hbar} \sum_n \frac{\Phi_n(y_\mathrm{L}, z_\mathrm{L})}{\sqrt{2\pi\hbar v_n(E)}} \left[\hat{a}_{\mathrm{L}n\sigma} \mathrm{e}^{\mathrm{i}k_x^{(n)} x_\mathrm{L}} + \hat{b}_{\mathrm{L}n\sigma} \mathrm{e}^{-\mathrm{i}k_x^{(n)} x_\mathrm{L}}\right];$$
$$\hat{\Psi}^\dagger_\sigma(\boldsymbol{r},t) = \int \mathrm{d}E\ \mathrm{e}^{\mathrm{i}Et/\hbar} \sum_n \frac{\Phi^*_n(y_\mathrm{L}, z_\mathrm{L})}{\sqrt{2\pi\hbar v_n(E)}} \left[\hat{a}^\dagger_{\mathrm{L}n\sigma} \mathrm{e}^{-\mathrm{i}k_x^{(n)} x_\mathrm{L}} + \hat{b}^\dagger_{\mathrm{L}n\sigma} \mathrm{e}^{\mathrm{i}k_x^{(n)} x_\mathrm{L}}\right].$$

The basic course of quantum mechanics teaches us that if we know the wave function of the system, we can write down the expression for the current density. Now, the formulae for the field operators enables us to write the operator of current in the left waveguide,

$$\hat{I}(x_\mathrm{L},t) = \frac{\hbar e}{2\mathrm{i}m} \sum_\sigma \int \mathrm{d}y_\mathrm{L}\,\mathrm{d}z_\mathrm{L} \left[\hat{\Psi}^\dagger_\sigma \frac{\partial}{\partial x_\mathrm{L}} \hat{\Psi}_\sigma - \left(\frac{\partial}{\partial x_\mathrm{L}} \hat{\Psi}^\dagger_\sigma\right) \hat{\Psi}_\sigma\right]. \quad (1.49)$$

To calculate the average current, we only need to know the *time-averaged* current operator. To avoid dealing with ill defined delta-functions, we perform the following trick. Imagine that all the quantities are periodic in time with the period $\mathcal{T} \to \infty$. The allowed values of energy are then found from the condition that the exponents of the type $\exp(\mathrm{i}Et)$ are also periodic, hence $E = 2\pi q\hbar/\mathcal{T}$ with an integer q. Consequently, we replace $\int \mathrm{d}E$ by $2\pi\hbar/\mathcal{T}\sum_n$. We use

$$\left\langle \mathrm{e}^{\mathrm{i}(E-E')t} \right\rangle_t = \delta_{qq'},$$

where the angular brackets here denote the time average. This means that in the expression for the current Eq. (1.49) both field operators must be evaluated at the same energy. We obtain

$$\left\langle \hat{I} \right\rangle_t = \frac{G_\mathrm{Q}}{e} \left(\frac{2\pi\hbar}{\mathcal{T}}\right)^2 \sum_{n\sigma} \sum_E \left[\hat{a}^\dagger_{\mathrm{L}n\sigma}(E)\hat{a}_{\mathrm{L}n\sigma}(E) - \hat{b}^\dagger_{\mathrm{L}n\sigma}(E)\hat{b}_{\mathrm{L}n\sigma}(E)\right]. \quad (1.50)$$

Equation (1.50) has an easy interpretation: the current in the left waveguide is the number of particles moving to the right (represented by $\hat{a}^\dagger\hat{a}$) minus the number of particles moving to the left ($\hat{b}^\dagger\hat{b}$), summed over all channels and energies.

Eliminating $\hat{b}$ in favor of $\hat{a}$, we write

$$\left\langle \hat{I} \right\rangle_t = \frac{G_\mathrm{Q}}{e} \left(\frac{2\pi\hbar}{\mathcal{T}}\right)^2 \sum_{n\sigma} \sum_{\alpha\beta,ll'} \sum_E \hat{a}^\dagger_{\alpha l\sigma}(E)\hat{a}_{\beta l'\sigma}(E)$$
$$\times \left[\delta_{\alpha\mathrm{L}}\delta_{\beta\mathrm{L}}\delta_{nl}\delta_{nl'} - s^*_{\alpha l,\mathrm{L}n}(E)s_{\mathrm{L}n,\beta l'}(E)\right]. \quad (1.51)$$

The last step is to perform the quantum-mechanical average of Eq. (1.51) and to find the average current. At first glance, this makes no sense, since according to Eq. (1.48) the average of the product of two operators at the same energy is infinite. However, for the discretized energies we have to replace the delta-function with the Kronecker delta-symbol,

$$\delta(E - E') \to \frac{\mathcal{T}}{2\pi\hbar}\delta_{qq'}.$$

This cancels one factor of $\mathcal{T}$. Now we can take the limit $\mathcal{T} \to \infty$ and from the discrete sum come back to the integral over energies. Taking into account that the averaging procedure yields $\alpha = \beta$, $l = l'$, and using the unitarity condition Eq. (1.35), we safely arrive at the Landauer formula, Eq. (1.37).

1.4 Counting electrons

Any experimental measurement is in fact a result of the average of many readings of a measuring device. This is because the readings differ, or fluctuate, even if the parameters controlling the physical situation do not change. Each individual reading is random, being a result of interplay between many factors beyond our control. Some of them come from the measuring device, reflecting its imperfectness; some are intrinsic for the system being measured; some cannot be controlled at all owing to quantum mechanics. An example from quantum transport is the measurement of the electric current in a nanostructure at a given voltage. The electron transfer is a stochastic process, and the number of electrons traversing the nanostructure during a given time interval Δt is random. Even if we measure the current with an ideal ammeter, the readings would differ.

Given the situation, there are two possible courses of action. First, and most common, is to get rid of the fluctuations by averaging over a large number of readings. The result of such an approach is the average current, the quantity we have studied so far. Alternatively, one can study the statistics of these fluctuations by trying to measure a probability of a certain current read-out. This would generally require more measurements, but rewards us with more information. As we show in this section, the statistics of the electron transfer reveals information about a nanostructure that cannot be readily accessed by means of average current measurements.

Let us recall some general concepts of probability theory. Suppose we make a measurement counting some random events during a certain time interval Δt. This can be, for instance, the number of babies born in Nashville, Tennessee, during a week; the number of birds crossing the Continental Divide in an easterly direction during an hour; or the number of electrons passing from one to another reservoir via a nanostructure during one nanosecond. The number of events N measured during the time interval is a *random number*. On repeating the measurement many times, we obtain different results. Summing them up and dividing by the number of measurements, we get the *average* number, $\langle N \rangle$. If the conditions remain the same during the measurement, and do not change from one measurement to another (for instance, if birds are only counted at the same time of day and during the same season), this average is proportional to Δt.

Besides the average, we may want to know the *distribution* of the results – the probability, P_N, that precisely N events will be observed in a measurement. To quantify this probability, one repeats identical measurements M_{tot} times and counts the number of measurements M_N that give the count N (Fig. 1.20). The ratio M_N/M_{tot} yields the probability P_N in the limit $M_N \gg 1$. This probability distribution is normalized, $\sum_N P_N = 1$. Once we know it, we can estimate not only the average,

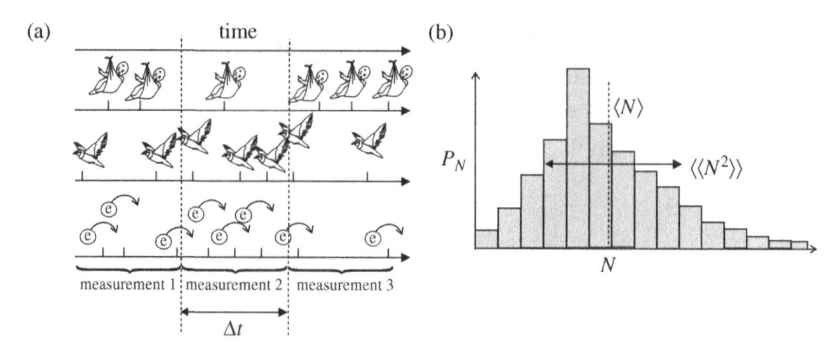

Fig. 1.20. **Counting statistics. (a) Examples of countable events. (b) Counting statistics is characterized by a probability distribution of counts.**

$$\langle N \rangle = \sum_N N P_N,$$

but also, for instance, the variance, or second cumulant, of N, which measures the degree of the deviation from the average,

$$\langle\langle N^2 \rangle\rangle = \left\langle (N - \langle N \rangle)^2 \right\rangle = \sum_N N^2 P_N - \left(\sum_N N P_N \right)^2.$$

The description of the statistics using the distribution function P_N is not always the most convenient one. In the example with the birds, let us suppose that we know the probability distributions P^{s} (of the birds that cross the Divide south of Independence Pass) and P^{n} (for those which cross north of this point). These events are statistically independent, since the birds hardly coordinate their itineraries. The total distribution is given by a convolution of the two,

$$P_N^{\mathrm{tot}} = \sum_{M=0}^{N} P_M^{\mathrm{s}} P_{N-M}^{\mathrm{n}}.$$

Most conveniently, this is expressed in terms of *characteristic function* of a probability distribution,

$$\Lambda(\chi) = \left\langle \mathrm{e}^{\mathrm{i}\chi N} \right\rangle = \sum_N P_N \mathrm{e}^{\mathrm{i}\chi N}.$$

For independent events, the characteristic function of the total distribution is just a product of characteristic functions of each type of events, $\Lambda^{\mathrm{tot}}(\chi) = \Lambda^{\mathrm{s}}(\chi)\Lambda^{\mathrm{n}}(\chi)$. This is a handy property if the outcome of the measurement is contributed to by various sources.

Differentiating the function $\ln \Lambda(\chi)$ k times with respect to $\mathrm{i}\chi$ and setting $\chi = 0$ subsequently, we generate the kth cumulant of the distribution. Thus, the first derivative produces the average N, and the second derivative reproduces the variance. How do the cumulants depend on Δt? Let us divide the interval Δt into two shorter intervals Δt_1 and Δt_2 (but still long enough so that many events occur during each of them). Events occurring during each of these intervals are statistically independent provided the intervals are long enough. Therefore the characteristic function is a product of the characteristic functions describing the intervals. Its logarithm is the sum of corresponding logarithms, $\ln \Lambda(\chi, \Delta t) = \ln \Lambda(\chi, \Delta t_1) + \ln \Lambda(\chi, \Delta t_2)$. Since $\Delta t = \Delta t_1 + \Delta t_2$, we conclude that $\ln \Lambda(\chi, \Delta t)$, and therefore all the cumulants are proportional to the time of measurement Δt.

1.4.1 Statistics of electron transfers

Now we return to electrons in nanostructures. An event is a transfer of an electron from one reservoir to another. The quantity to count in this case is the charge Q passed from left to right during the time Δt. We assume that this measurement time is long enough, so that $Q \gg e$ and the laws of statistics apply. On average, $\langle Q \rangle = \langle I \rangle \Delta t$, and we have already spent quite some time calculating the (average) current $\langle I \rangle$. We now make a step further and describe the *statistical properties* of a random variable Q. This challenging task can be accomplished within the scattering approach outlined in the preceding text. The result in the form of a compact Levitov formula is given in this section. The derivation of this formula is not elementary and can be found in Section 2.9 and in Ref. [10].

We discuss two simple limiting cases that will prepare the reader before revealing such an important piece of information.

Let us first assume that electrons are transferred only in one direction and that transfers are uncorrelated. To calculate the characteristic function, we divide the interval Δt into very short intervals dt. The probability of transferring one electron during this short interval is given by $\Gamma\, dt \ll 1$, Γ being the transfer rate; the probability of transferring no electrons is, consequently, $1 - \Gamma\, dt$. We neglect the probability of transferring more than one electron since this probability is proportional to $(dt)^2$ and is therefore much less than $\Gamma\, dt$. The characteristic function for a short interval is thus given by

$$\Lambda_{dt}(\chi) = \left\langle e^{i\chi Q/e} \right\rangle = (1 - \Gamma\, dt) + (\Gamma\, dt)e^{i\chi}.$$

Since the electrons pass independently, the characteristic function for the whole interval is just a product, i.e.

$$\Lambda_{\Delta t}(\chi) = (\Lambda_{dt}(\chi))^{\Delta t/dt} = \exp\left(\Gamma \Delta t(e^{i\chi} - 1)\right) = \exp\left(\tilde{N}(e^{i\chi} - 1)\right). \quad (1.52)$$

Here $\tilde{N} \equiv \Gamma \Delta t$ is the average number of electrons transferred, $\tilde{N} = \langle Q \rangle / e$. Taking the inverse Fourier transform, we find for the probability P_N for N particles to be transferred during the time Δt,

$$\begin{aligned} P_N &= \int_0^{2\pi} \frac{d\chi}{2\pi} \Lambda(\chi) e^{-iN\chi} \approx \int_0^{2\pi} \frac{d\chi}{2\pi} e^{-i\chi N + \tilde{N}(e^{i\chi} - 1)} \\ &= \frac{\tilde{N}^N}{N!} e^{-\tilde{N}\Delta t}. \end{aligned} \quad (1.53)$$

Equation (1.53) is recognized as the *Poisson distribution*. As we will see, this situation of uncorrelated electron transfer occurs in tunnel junctions, where all transmission eigenvalues are small. In this case, the currents are small, implying that the time intervals between successive transfers are large. Therefore it is easy to understand why they do not correlate.

Control question 1.11. Calculate the average number of particles $\langle N \rangle$, and the second $\langle N^2 \rangle - \langle N \rangle^2$ and third $\langle N^3 \rangle - 3\langle N^2 \rangle \langle N \rangle + 2\langle N \rangle^3$ cumulants for the Poisson distribution in Eq. (1.52).

An opposite example is an ideally transmitting channel at zero temperature. In this case, the electrons are in ideal wave states and their momentum in the transport direction is a well defined quantum number, which does not fluctuate. Since the total current is just a sum over the momenta of individual electrons, it does not fluctuate either. The distribution $P_N = \delta(\tilde{N} - N)$ provides the characteristic function $\Lambda(\chi) = \exp(\mathrm{i}\chi(N - \tilde{N}))$. The transfers are thus correlated ideally.

For intermediate transmissions $0 < T_p < 1$, the transmitted electrons are correlated, but not ideally. The many-channel, finite-temperature result for the characteristic function is given by the *Levitov formula*,

$$\ln \Lambda(\chi) = 2_s \Delta t \int \frac{\mathrm{d}E}{2\pi\hbar} \sum_p \ln \Big\{ 1 + T_p \left(\mathrm{e}^{\mathrm{i}\chi} - 1 \right) f_\mathrm{L}(E) \left[1 - f_\mathrm{R}(E) \right] + T_p \left(\mathrm{e}^{-\mathrm{i}\chi} - 1 \right) f_\mathrm{R}(E) \left[1 - f_\mathrm{L}(E) \right] \Big\}. \tag{1.54}$$

The logarithm of the characteristic function is a sum over transport channels; this suggests that electron transfers in different channels are independent. Also, the logarithm is an integral over the energy, suggesting that electrons are transferred independently in each energy interval. Importantly, the electron transfers from the left to the right and from the right to the left do correlate. To stress this, let us consider an energy strip where $f_\mathrm{L} = f_\mathrm{R} = 1$ so that the electron states are filled in both reservoirs. The net current is zero. If the transfers were uncorrelated, they would give rise to current fluctuations. However, the formula gives no events in this case: transfers to the left are blocked by electrons filling states in the left reservoir, and the same is true for transfers to the right.

To comprehend the formula, let us consider the limit of negligible temperature, $eV \gg k_\mathrm{B}T$. In this case, the integration over energy is confined to the energy strip $\min(\mu_{\mathrm{L,R}}) < E < \max(\mu_{\mathrm{L,R}})$ and the integrand does not depend on energy. Recalling that $\mu_\mathrm{L} - \mu_\mathrm{R} = eV$, we obtain

$$\ln \Lambda(\chi) = \pm \frac{2_s eV \Delta t}{2\pi\hbar} \sum_p \ln \left[1 + T_p \left(\mathrm{e}^{\pm\mathrm{i}\chi} - 1 \right) \right], \tag{1.55}$$

where the upper and lower signs refer to the case of positive and negative voltages, respectively. Let us, for simplicity, consider $V > 0$. We define $N_\mathrm{at} = 2_s \Delta t eV / 2\pi\hbar$ and assume it to be integer. The characteristic function becomes

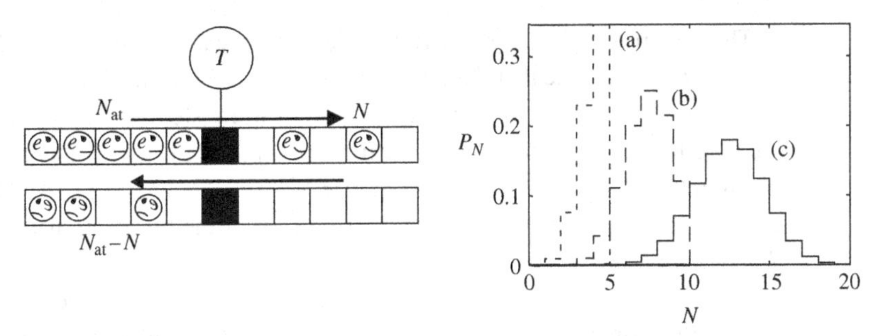

Fig. 1.21. **Electrons gambling. Binomial distribution of transmitted electrons for the winning chance (transmission eigenvalue) $T = 0.7$ and for different numbers of attempts: $N_{at} = 5$ (a); 10 (b); 20 (c).**

$$\Lambda(\chi) = \prod_p \Lambda_p(\chi); \tag{1.56}$$

$$\Lambda_p = \left((1-T_p) + \left(T_p e^{i\chi}\right)\right)^{N_{at}} = \sum_{N=0}^{N_{at}} \binom{N_{at}}{N} T_p^N (1-T_p)^{N_{at}-N} e^{iN\chi}.$$

We made use of Newton's binomial theorem in the preceding expression. Let us concentrate on one channel p and take the inverse Fourier transform of Λ_P. We obtain the *binomial distribution* of the number of electrons transferred,

$$P_N^{(p)} = \binom{N_{at}}{N} T_p^N (1-T_p)^{N_{at}-N}, \tag{1.57}$$

which allows for a frivolous interpretation. The point is that the binomial distribution is known from theory of gambling: For a given winning chance T_p and number of game slots (*number of attempts*) N_{at} it yields the probability to win N times (Fig. 1.21).

At zero temperature and positive voltage, all the electrons are incident from the left reservoir trying to get to the right. The interpretation suggests that the stream of incident electrons is very regular: the time interval between the arrivals of two adjacent electrons is the same, $\Delta t/N_{at} = e/G_Q V$. Each of them either passes through the scatterer (with probability T_p) or is reflected back (with probability $R_p = 1 - T_p$). The average number of those which pass is $N_{at}T_p$, conforming to the Landauer formula. The distribution P_N given by Eq. (1.57) is the probability that, out of N_{at} electrons arriving at the scatterer, N pass through and $N_{at} - N$ are reflected back.

For more than one channel, the binomial distribution no longer holds. However, we obtain a convolution of binomial distributions corresponding to each channel.

Exercise 1.6. (i) Derive from Eq. (1.56) the expression for the distribution P_N for the case of two channels and check that it is indeed a convolution of two binomial distributions. (ii) Show that if the transmission eigenvalues in the two channels are identical, $T_1 = T_2 = T$, it reduces to the binomial distribution Eq. (1.57) with N_{at} replaced with $2N_{at}$. Explain the result.

The electrons appear on the right side of the scatterer in an irregular fashion. If T_p is small, we can assume that the intervals between those which have passed are long, random and *independent* – the two subsequent electron transfers are thus uncorrelated. Indeed, if we take the Levitov formula in the limit of $T_p \ll 1$, it yields the characteristic function given in Eq. (1.52) with $\tilde{N}/\Delta t = G_Q V/e \sum_p T_p = GV/e = \langle I \rangle /e$. The Poisson distribution given in Eq. (1.53) is thus the limiting case of the binomial distribution Eq. (1.57) for $T \ll 1$ and $N \ll N_{\mathrm{at}}$.

If the transmission eigenvalues do not depend on energy, the integral over energy in Eq. (1.54) can be taken explicitly at an arbitrary relation between eV and $k_B T$. The characteristic function then becomes

$$\ln \Lambda(\chi) = \frac{2_s k_B T \Delta t}{2\pi\hbar} \sum_p \left\{ \operatorname{arccosh}^2 \left[T_p \cosh\left(\frac{eV}{2k_B T} + i\chi \right) + (1 - T_p) \cosh\left(\frac{eV}{2k_B T} \right) \right] - \left(\frac{eV}{2k_B T} \right)^2 \right\}. \tag{1.58}$$

1.4.2 Noise and third cumulant

We obtain the cumulants of transferred charge by differentiating Eqs. (1.54), (1.55), and (1.58) with respect to χ at $\chi = 0$.

We start with the most general, Eq. (1.54), to calculate the first cumulant of the transmitted change – the average charge $\langle Q \rangle$:

$$\langle Q \rangle = e \left. \frac{\partial \ln \Lambda}{\partial (i\chi)} \right|_{\chi=0} = \frac{2_s e \Delta t}{2\pi\hbar} \sum_p \int dE \; T_p(E) \left[f_L(E) - f_R(E) \right]. \tag{1.59}$$

A comparison with Eq. (1.37) shows that, as expected, $\langle Q \rangle = \langle I \rangle \Delta t$, and Eq. (1.54) is in full agreement with the Landauer formula.

Now we look at the second cumulant of the transmitted charge, $\langle\langle Q^2 \rangle\rangle \equiv \langle Q^2 \rangle - \langle Q \rangle^2$. Differentiating Eq. (1.54) twice with respect to $i\chi$ and setting $\chi = 0$, we obtain

$$\langle\langle Q^2 \rangle\rangle = \frac{2_s e^2 \Delta t}{2\pi\hbar} \sum_p \int dE \left\{ T_p \left[f_L(1 - f_L) + f_R(1 - f_R) \right] + T_p \left(1 - T_p \right) (f_L - f_R)^2 \right\}. \tag{1.60}$$

The first feature to note is that, in contrast to $\langle Q \rangle$, the second cumulant of the transmitted charge does not vanish at equilibrium. Indeed, for $V = 0$ ($f_L = f_R$), only the first term in Eq. (1.60) contributes, and we obtain

$$\langle\langle Q^2 \rangle\rangle_{\mathrm{eq}} = \frac{2_s e^2 k_B T \Delta t}{\pi\hbar} \sum_p T_p = 2G_Q k_B T \Delta t. \tag{1.61}$$

To understand the meaning of this expression, we define the correlation function of current fluctuations (usually known as *current noise power*),

$$S(\omega) = \left\langle \hat{I}(t)\hat{I}(t') + \hat{I}(t')\hat{I}(t) - 2\left\langle \hat{I}(t) \right\rangle \left\langle \hat{I}(t') \right\rangle \right\rangle_\omega . \tag{1.62}$$

If the measurement time Δt is long enough, the second cumulant of the transmitted charge is expressed via zero-frequency noise, $\langle\langle Q^2\rangle\rangle = \Delta t S(0)/2$. On the other hand, at equilibrium the correlation functions like Eq. (1.62) obey the fluctuation–dissipation theorem, $S(0) = 4k_B TG$, which agrees with Eq. (1.61). These equilibrium current fluctuations are known as *Nyquist–Johnson noise* and are present in any system, independent of the nature of this current.

Let us now turn to the opposite limit $eV \gg k_B T$, the so-called *shot noise* limit. In this case, only the second term in Eq. (1.60) survives. For energy-independent transmission eigenvalues, this gives [12, 13]

$$\langle\langle Q^2\rangle\rangle = \Delta t S(0)/2; \quad S(0) = 2eG_Q V \sum_p T_p(1 - T_p). \tag{1.63}$$

Control question 1.12. Equation (1.63) states that neither open ($T_p = 1$) nor closed ($T_p = 0$) channels contribute to the noise. Explain why.

This non-equilibrium noise (*shot noise*) appears because of the discreteness of the electron charge – electrons arrive at the scatterer one-by-one. If we know only the average current in the system, it is impossible to guess what the shot noise would be – in other words, shot noise probes the transmission properties of the systems differently from those probed by the conductance (for a review, see Ref. [14]).

Equation (1.63) interpolates between the two examples considered at the beginning of this section. For the tunnel junction, $T_p \ll 1$, we have $S(0) = 2e\langle I\rangle$. This *Schottky formula* follows from the Poisson distribution, Eq. (1.53). The *Poisson value* of shot noise shows the maximal possible level of this type of noise. Shot noise is always suppressed compared with this value due to the factors $(1 - T_p)$. The ratio between these two, given by

$$F = \frac{S(0)}{2e\langle I\rangle} = \frac{\sum_p T_p(1 - T_p)}{\sum_p T_p}, \quad 0 \le F \le 1, \tag{1.64}$$

is known as the *Fano factor*. In a quantum point contact, all the channels are either fully open or fully closed, and thus there is no shot noise, $F = 0$.

The third system we often mention is a diffusive wire. The channels with $T \sim 1$ available in a diffusive wire noticeably suppress the shot noise. Using Eq. (1.43) to average Eq. (1.63), we find that the shot noise is given by $S(0) = 2e\langle I\rangle/3$, or, in other words, the Fano factor equals $1/3$ [15].

Exercise 1.7. (i) Verify that for diffusive conductors the Fano factor is $1/3$. (ii) For a symmetric chaotic cavity (Section 4.3) the distribution of transmission eigenvalues is $P(T) = 1/(\pi\sqrt{T(1-T)})$. Calculate the Fano factor.

We now illustrate this theoretical conclusion by considering the results of an experimental study of shot noise in diffusive wires described in Ref. [11]. The authors prepared several samples of gold wires, about 1 μm long, differing in their cross-sections. This resulted in a difference in conductance, which varied from about $75G_Q$ to $340G_Q$. For the

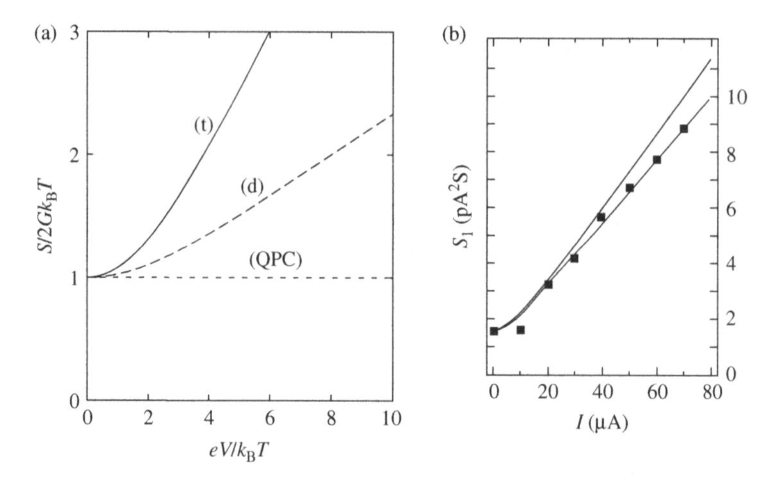

Fig. 1.22. **Noise. (a) Voltage dependence of noise for tunnel junction (t), diffusive conductor (d), and quantum point contact (QPC). (b) Experimental results adapted from Ref. [11].**

samples with low resistance, the measured noise clearly demonstrates 1/3-suppression (see the lower solid curve in Fig. 1.22). For the sample with the highest resistance, however, a noise enhancement in comparison with the 1/3-suppression was observed. This is because in such diffusive wires the effects of electron–electron interactions become important, and this increases noise.

For a general relation between eV and k_BT, one obtains a simple relation

$$S(0) = 2G\left(eVF\coth(eV/2k_BT) + 2k_BT(1-F)\right), \tag{1.65}$$

plotted in Fig. 1.22 for a tunnel junction, a diffusive wire and a QPC.

Third cumulant

The noise measurement is more complicated than the average current measurement since it requires the collection of more measurement results to achieve decent accuracy. The direct measurement of higher cumulants is even more challenging. However, the third cumulant has been measured recently [16]. At equilibrium ($V = 0$) the characteristic function of Eq. (1.54) becomes even in χ, and therefore all the odd cumulants disappear. We have seen this already for the average charge, or current. The same is true for the third cumulant: it is odd in voltage and absent at equilibrium. From Eq. (1.55) we obtain the third cumulant in the shot noise regime, as follows:

$$\langle\langle Q^3\rangle\rangle = e^3\frac{\partial^3}{\partial(\mathrm{i}\chi)^3}\ln\Lambda(\chi) = e^2VG_Q\Delta t\sum_p T_p(1-T_p)(1-2T_p). \tag{1.66}$$

Again, for the tunnel junction $T \ll 1$ we obtain $\langle\langle Q^3\rangle\rangle = e^2\langle I\rangle\Delta t$, which can be derived directly from the Poisson distribution. For a QPC, where the electron stream is regular, the

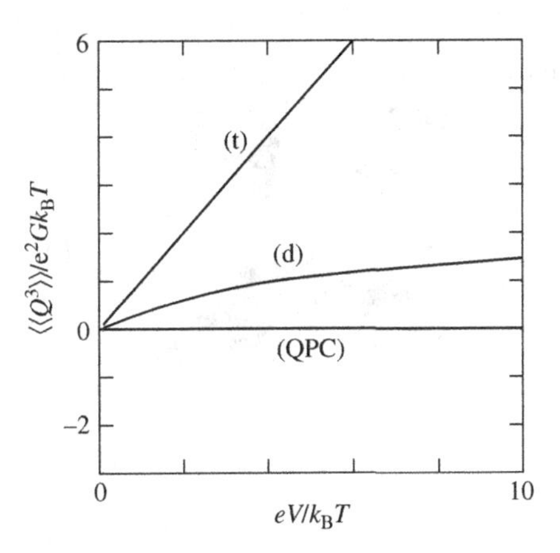

Fig. 1.23. Voltage dependence of the third cumulant of the transferred charge for (t) a tunnel junction, (d) a diffusive conductor, and (QPC) a quantum point contact.

$\langle\langle Q^3\rangle\rangle$ disappears as expected. In a diffusive wire, the calculation with transmission probability, Eq. (1.43), yields $\langle\langle Q^3\rangle\rangle = e^2\langle I\rangle\Delta t/15$. The third cumulant can be either positive or negative; the open channels with $T_p > 1/2$ favor a negative sign.

The full voltage dependence is given by

$$\langle\langle Q^3\rangle\rangle = e^2 G_Q \Delta t \sum_p T_p \Big[eV - 3T_p X(eVX - 2k_B T) + T_p^2(eV(3X^2-1) - 6k_B T X)\Big], \quad X \equiv \coth(eV/2k_B T). \tag{1.67}$$

This expression is plotted for several types of nanostructures in Fig. 1.23.

1.5 Multi-terminal circuits

A nanostructure is typically connected to several (more than two) electrodes. Some of them (gates) are used to form and/or control the nanostructure, and no electron transfer takes place between them and the nanostructure (see Section 1.7). Others (terminals) either pass the current through the system or are kept at zero current and serve to measure voltage. Terminals are electronic reservoirs. If we send electrons from one terminal, they are either reflected back, or, after spending some time inside the nanostructure, exit to any of the other terminals. In Sections 1.3 and 1.4, we discussed two-terminal nanostructures. In this section, we describe the transport properties of a *multi-terminal* nanostructure by generalizing the scattering approach developed. The model is very similar: a finite-size scattering region is connected to N equilibrium reservoirs (kept at fixed voltages V_α, $\alpha = 1, \ldots, N$) by ideal waveguides (Fig. 1.24). The wave functions in these waveguides are plane waves. Introducing for each waveguide α a set of local coordinates $x_\alpha > 0, y_\alpha, z_\alpha$ (the axis x_α is directed along the waveguide from the scattering region to the reservoir), we write these functions as follows:

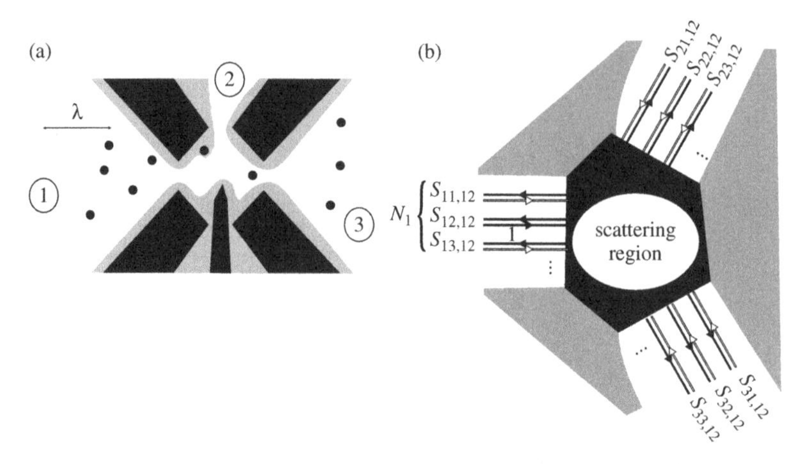

Fig. 1.24. Multi-terminal circuit. (a) There are eight electrodes connected to this nanostructure: five gates and three terminals. (b) Structure of multi-terminal scattering matrix. We show amplitudes of the wave coming in the second channel of the first terminal.

$$\psi(x_\alpha, y_\alpha, z_\alpha) = \sum_n \frac{1}{\sqrt{2\pi\hbar v_{\alpha n}}} \Phi_{\alpha n}(y_\alpha, z_\alpha) \left[a_{\alpha n} \mathrm{e}^{-\mathrm{i}k_x^{(\alpha n)} x_\alpha} + b_{\alpha n} \mathrm{e}^{\mathrm{i}k_x^{(\alpha n)} x_\alpha} \right]. \tag{1.68}$$

Here, and in the rest of this section, the first indices α, β label different terminals, and the second ones m, n refer to transport channels inside each terminal; $\Phi_{\alpha n}$ and $E_{\alpha n}$ are respectively the transverse wave function and the energy of the transverse motion in the transport channel n in the terminal α. The (real) wave vector in the same channel depends on the energy E, $k_x^{(\alpha n)} = \sqrt{2m(E - E_{\alpha n})}/\hbar$.

The scattering matrix $\hat{s}$ linearly relates the amplitudes of incoming $a_{\beta m}$ and outgoing $b_{\alpha n}$ waves, i.e.

$$b_{\alpha n} = \sum_{\beta m} s_{\alpha n, \beta m} a_{\beta m}. \tag{1.69}$$

The scattering matrix is unitary, $\hat{s}^\dagger \hat{s} = \hat{1}$; this provides the conservation of the number of electrons. Its diagonal blocks $\hat{s}_{\alpha\alpha}$ describe the reflection of the electrons incident from the reservoir α back to the same reservoir (possibly changing the transport channel). The off-diagonal blocks $\hat{s}_{\alpha\beta}$ are responsible for the transmission of electrons from terminal β to terminal α.

Similarly to the two-terminal case, the matrix satisfies the time reversibility relations,

$$s_{\alpha n, \beta m}(B) = s_{\beta m, \alpha n}(-B), \tag{1.70}$$

where the change of sign of magnetic field B indicates time-reversed situation.

1.5.1 Multi-terminal Landauer formula

Consider a current flowing through the cross-section of the waveguide α, in the direction from the scattering region (to the reservoir). The electrons with $k_x < 0$, originated from reservoir α, are described by the distribution function $f_\alpha(E)$. The electrons with $k_x > 0$ come from various reservoirs. The fraction of particles that are incident from the reservoir β in the transport channel m and that end up in the waveguide α in the transport channel n, is given by $|s_{\alpha n,\beta m}|^2$, their distribution function being $f_\beta(E)$. Thus, the filling factor for the particles with $k_x > 0$ is given by

$$\sum_{\beta m} \left|s_{\alpha n,\beta m}\right|^2 f_\beta(E),$$

and we write for the current, in terminal α,

$$\begin{aligned} I_\alpha &= 2_s e \sum_n \left\{ \int_{-\infty}^{0} \frac{\mathrm{d}k_x}{2\pi} v_x(k_x) f_\alpha(E) \right. \\ &\left. + \int_0^\infty \frac{\mathrm{d}k_x}{2\pi} v_x(k_x) \sum_{\beta m} \left|s_{\alpha n,\beta m}\right|^2 f_\beta(E) \right\} \\ &= 2_s e \sum_n \int_0^\infty \frac{\mathrm{d}k_x}{2\pi} v_x(k_x) \sum_{\beta m} \left\{ \left|s_{\alpha n,\beta m}(E)\right|^2 - \delta_{\alpha\beta}\delta_{mn} \right\} f_\beta(E). \end{aligned} \tag{1.71}$$

Changing variables from k_x to E, we arrive at the following expression:

$$\begin{aligned} I_\alpha &= -\frac{G_Q}{e} \int_0^\infty \mathrm{d}E \sum_{\beta mn} \left\{ \delta_{\alpha\beta}\delta_{mn} - \left|s_{\alpha n,\beta m}\right|^2 \right\} f_\beta(E) \\ &= -\frac{G_Q}{e} \int_0^\infty \mathrm{d}E \sum_\beta \mathrm{Tr} \left\{ \delta_{\alpha\beta} - \hat{s}^\dagger_{\alpha\beta}\hat{s}_{\alpha\beta} \right\} f_\beta(E), \end{aligned} \tag{1.72}$$

where the trace is taken over the transport channels n, and the matrix $\hat{s}^\dagger_{\alpha\beta}$ is the conjugate of $\hat{s}_{\alpha\beta}$. In its turn, $\hat{s}_{\alpha\beta}$ is a block of the matrix $\hat{s}$, which describes the transmission of electrons from terminal β to terminal α (for $\alpha \neq \beta$) or their reflection back to α (for $\alpha = \beta$).

We first note that, due to the unitarity condition, the currents in all terminals add up to zero, $\sum_\alpha I_\alpha = 0$. This *current conservation* holds in both linear and non-linear regimes.

Consider now the linear regime. Let us keep all chemical potentials equal to E_F, except for the chemical potential at one terminal γ, $\mu_\gamma = E_\mathrm{F} + eV_\gamma$. This means that the voltage V_γ is applied between terminal γ and all other terminals. This voltage induces the current in all terminals. Let us calculate the current through terminal α induced by this voltage. We first note that if, in Eq. (1.72), all distribution functions are the same (equal, for example, to $f(E) = (\exp(E - E_\mathrm{F})/k_\mathrm{B}T + 1)^{-1}$), the current I_α vanishes due to the unitarity of the scattering matrix – no current is induced at equilibrium. This means that instead of $f_\beta(E)$ we can write $f_\beta(E) - f(E)$, without affecting the current. The only surviving term in the sum is then for $\beta = \gamma$. Assuming finally that the scattering matrix depends on the energy on a scale much larger than the applied voltage eV_γ, we obtain $I_\alpha = G_{\alpha\gamma}V_\gamma$, with

$$G_{\alpha\gamma} = -G_Q\, \mathrm{Tr}\left[\delta_{\alpha\gamma} - \hat{s}^\dagger_{\alpha\gamma}\hat{s}_{\alpha\gamma}\right], \tag{1.73}$$

where the scattering matrix is evaluated at $E = E_\mathrm{F}$. Equation (1.73) is the generalization of the Landauer formula, Eq. (1.38), to the multi-terminal case.

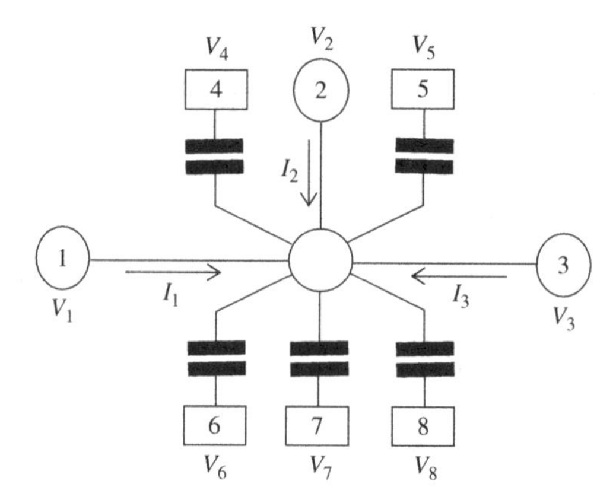

Fig. 1.25. Electric circuit corresponding to the nanostructure in Fig. 1.24.

Repeating the same arguments as in the two-terminal case (Section 1.3), we conclude that the assumption that the number of channels supported by the waveguides is finite is not necessary, and that the Landauer formula holds also for an infinite number of transport channels.

In the linear regime, the contributions to the current from different terminals add up. If voltages V_β are applied to terminals β, the current induced in terminal α is found to be the sum, $I_\alpha = \sum_\beta G_{\alpha\beta} V_\beta$. The coefficients $G_{\alpha\beta}$ are the elements of the conductance matrix. Current conservation requires that $\sum_\alpha G_{\alpha\beta} = 0$. Because of the time-reversibility properties of the scattering matrix, the conductance, given in Eq. (1.73), has the following symmetry:

$$G_{\alpha\beta}(B) = G_{\beta\alpha}(-B),$$

which is a particular case of the Onsager symmetry relations. Thus, in the absence of magnetic field, $B = 0$, an N-terminal nanostructure possesses $N(N-1)/2$ independent conductances, $(N-1)$ diagonal and $[(N-1)^2 - (N-1)]/2$ off-diagonal ones; others are fixed from the symmetry and current conservation. In particular, for a two-terminal system, $N = 2$, we have only one independent conductance, $G = -G_{\text{LL}} = -G_{\text{RR}} = G_{\text{LR}} = G_{\text{LR}}$, which we calculated in Section 1.3.

We can also introduce the resistance matrix $R_{\alpha\beta}$, $V_\alpha = \sum_\beta R_{\alpha\beta} I_\beta$, which is the inverse of the conductance matrix. Obviously, it has the symmetry properties $R_{\alpha\beta}(B) = R_{\beta\alpha}(-B)$.

1.5.2 Voltage probes and two-terminal measurement

If one needs to characterize, investigate, or just simply test a macroscopic electric circuit, a voltage probe is an indispensable tool. One contacts by the probe different points of the circuit and reads the voltage. An ideal voltage probe has an infinite resistance and therefore is not invasive; it does not perturb the distribution of the currents and voltages in the circuit. In this way one can, for instance, compare the resistances of different elements of the circuit by comparing the voltage drops across the elements.

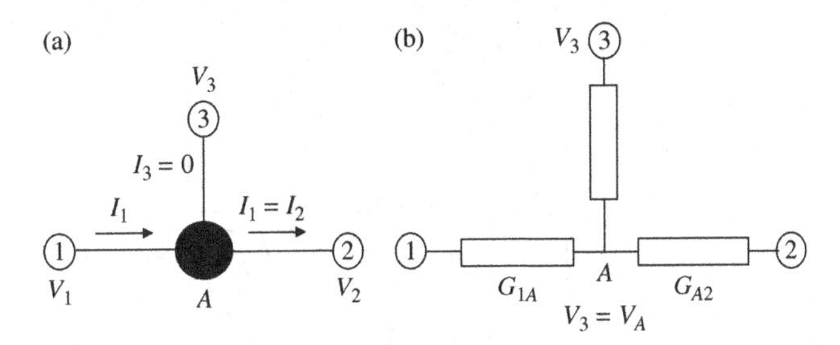

Fig. 1.26. **(a) Three-terminal circuit with terminal 3 as a voltage probe. (b) A classical circuit could be presented in this way, and conductances of the elements G_{1A}, G_{A2} can be determined from the voltage V_3 measured.**

How do we implement the idea of a voltage probe for a nanostructure? It may seem that the main problem is to make a small, nano-size, probe. In fact, this appears to be an achievable task: a classical example is provided by scanning tunneling microscopy (STM), in which the tunnel contact of the STM tip can be freely positioned and provides nano-scale resolution of a non-invasive electric measurement. Therefore one can measure the local voltage in the nanostructures. The main problem arises in interpreting the results of such a measurement. Quantum transport does not obey the laws of classical circuitry. As we see in what follows, the results can be very surprising.

The voltage probe can be easily considered within the multi-terminal scattering approach. Let us start with the simplest three-terminal geometry (Fig. 1.26) and make the third terminal a voltage probe. In the linear regime, the current to terminal 3 is given by

$$I_3 = G_{31}(V_1 - V_3) + G_{32}(V_2 - V_3),$$

where we have used $G_{33} = -G_{31} - G_{32}$. Since we are connecting this terminal to an ideal voltmeter, this current has to be zero. This happens when the voltage applied to the third terminal is given by

$$V_3 = \frac{G_{31}V_1 + G_{32}V_2}{G_{31} + G_{32}} = \frac{V_1 \operatorname{Tr} \hat{s}_{31}^{\dagger}\hat{s}_{31} + V_2 \operatorname{Tr} \hat{s}_{32}^{\dagger}\hat{s}_{32}}{\operatorname{Tr} \hat{s}_{31}^{\dagger}\hat{s}_{31} + \operatorname{Tr} \hat{s}_{32}^{\dagger}\hat{s}_{32}}. \tag{1.74}$$

This is the voltage read by the voltmeter. To make the measurement non-invasive, we have to ensure that $G_{32}, G_{31} \ll G_{12}$, or, equivalently, that $\hat{s}_{31}, \hat{s}_{32} \to 0$. The voltage remains finite in this limit.

One gets the same result in a classical electric circuit in which two conductances G_{31} and G_{32} are connected in series; the voltage V_3 is the potential at point A between the conductances. For a classical circuit, we could separate the circuit into two elements with conductances G_{1A} and G_{A2}. From elementary circuit theory rules we immediately obtain

$$G_{1A} = G_{12}\frac{V_1 - V_2}{V_1 - V_3}; \quad G_{A2} = G_{12}\frac{V_2 - V_1}{V_2 - V_3}. \tag{1.75}$$

Let us see if this separation works for an elementary example of quantum transport: a single-channel conductor with transmission coefficient T (Fig. 1.27) connecting terminals 1 and 2. We attach the voltage probe 3 to the left waveguide, between reservoir 1 and the

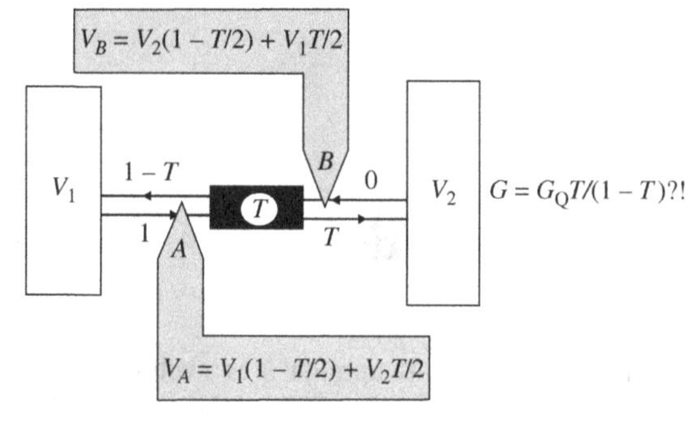

Fig. 1.27. The "wrong" Landauer formula illustrates the non-local nature of conductance in nanostructures.

scattering region, to measure the potential V_A left of the scattering region. To calculate this voltage, we have to know the conductances G_{31} and G_{32}, which are proportional to the probability of an electron being transferred from 1 and 2, respectively, to 3. Assume that the probability of tunneling into the voltage probe is $w \ll 1$. The electron incident from 1 has two possible routes into 3: either directly (probability w), or after first being reflected from the scattering region (probability $(1 - w)(1 - T)w \approx (1 - T)w$). Up to the terms proportional to w^2, these two processes are independent, and the total probability is given by the sum of the two. Thus, $G_{31} = G_Q w(2 - T)$. Similarly, the probability of going from 2 to 3 is the product of the probabilities of passing the scattering region (T) and of tunneling to 3 (w). Thus, $G_{32} = G_Q wT$. It boils down to $V_A = V_1(1 - T/2)$ $+V_2T/2$.

This is somewhat unexpected. In fact, our general picture implies no resistance between point A and the left reservoir, so we would like to have $V_A = V_1$. Let us check the previous result for V_A with a more rigorous argument. Suppose the current from point A to 3 in each energy strip $(E, E + \mathrm{d}E)$ is proportional to the probability of finding an electron at point A with energy E, the *local* filling factor $f_A(E)$; whereas the current from 3 to point A is proportional to $f_3(E)$. The net current in terminal 3 is thus given by

$$I_3 = (G_Q/e) \int \mathrm{d}E \; w(E)(f_A(E) - f_3(E)). \tag{1.76}$$

What is f_A? In fact, we have already evaluated the filling factors in the waveguide (see Eq. (1.36)): $(1 - T)f_1(E) + Tf_2(E)$ for left-going and $f_1(E)$ for right-going electrons. Since the probe is equally coupled to both left-going and right-going electrons, the resulting filling factor is just an average of the two, $f_A(E) = (1 - T/2)f_1(E) + (T/2)f_2(E)$. We integrate this over energy, and find the voltage V_3 at which I_3 vanishes. Provided the tunneling probability $w(E)$ does not depend on energy, we reproduce the previous result, $V_A = V_1(1 - T/2) + V_2T/2$.

Let us move voltage probe 3 to point B on the other side of the scattering region. Repeating the same arguments, we find $V_B = V_2(1 - T/2) + V_1T/2$. It looks like we manage to separate the one-channel scatterer into three parts, the conductances being given by

$$G_{1A} = G_{12}\frac{V_1 - V_2}{V_1 - V_A} = 2G_Q;$$
$$G_{AB} = G_{12}\frac{V_1 - V_2}{V_A - V_B} = G_Q\frac{T}{1-T}; \quad (1.77)$$
$$G_{B2} = G_{12}\frac{V_1 - V_2}{V_B - V_2} = 2G_Q.$$

This reasoning was originally provided by Landauer [8], who assigned the resistance $1/G_{AB}$ to the scatterer. This differs from the two-terminal Landauer formula, Eq. (1.38), by the factor $1 - T$. Indeed, for a quantum point contact ($T = 1$) the two-terminal Landauer formula gives the resistance R_Q, whereas Eqs. (1.77) yield zero resistance R_{AB} – the voltage does not drop across the QPC. This sounds very intuitive: since the scattering is absent, it does not provide any resistance. The voltage in this case drops between 1 and A as well as between B and 2. The "elements" with conductances $G_{1A} = G_{B2} = G_Q/2$ were called "contact resistances." Their resistances add with R_{AB} to provide the correct answer for a two-terminal circuit.

The persistence of Rolf Landauer in attracting the attention of the scientific community to these questions, and the fascination he managed to convey, have laid the foundations of modern quantum transport. However, nobody can *apply* a voltage difference to a scatterer in such a way that its conductance equals G_{AB}. The voltage in quantum transport can only be applied to the reservoirs. And point A, where only a single transport channel is present, is too small for a reservoir. However, as we have just shown, the voltage difference can be *measured* between any points of the nanostructure. All this proves that quantum transport is very non-local; in general, a nanostructure cannot be separated into elements having definite resistance. As we see in Chapter 2, this property is partially restored for nanostructures that encompass many open transport channels so that the typical conductance is much greater than G_Q.

If we really want to apply the voltage difference to a scatterer in such a way that Eqs. (1.77) are reproduced, we need at least a *four-terminal* circuit. The difference between the two Landauer formulas, Eq. (1.38) and Eqs. (1.77), in our opinion, best illustrates the distinction between two-terminal and multi-terminal systems.

Control question 1.13. Why does the "wrong" Landauer formula give the correct result for a tunnel junction, $T \ll 1$?

From an experimental point of view, it is frequently convenient to measure the I–V characteristics of a structure by the *two-probe method.* In this method, two terminals (1 and 2) are used to pass the current I through the structure and two extra terminals (3 and 4) measure the voltage drop across the sample (Fig. 1.28). This is convenient way of getting rid of a series resistance in leads 1 and 2 compared with the two-terminal setup. Intuitively, the result of the two-probe measurement, $\mathcal{R} \equiv (V_3 - V_4)/I$, should coincide with the result of the two-terminal measurement, R_{AB}. Is this really so?

Generally, it is not. We can access the result using the scattering formalism. Assuming for simplicity that $G_{43}, G_{34} \ll G_{41}, G_{42}$ and $G_{31}, G_{32} \ll G_{12}, G_{21}$ (non-invasive voltage probes), we obtain

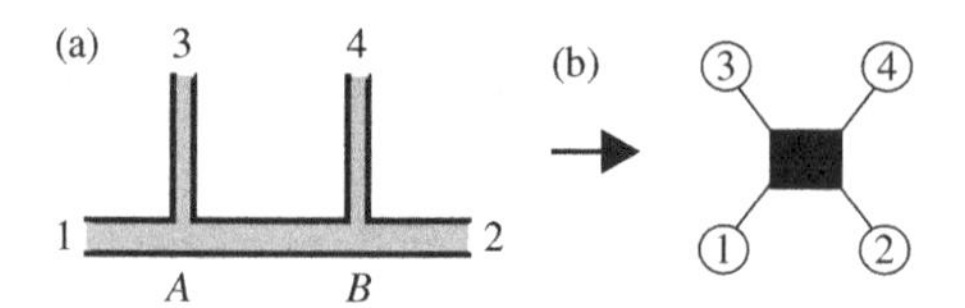

Fig. 1.28. A common scheme for two-probe measurement of the resistance R_{AB}, (a) does not generally work in quantum transport. The reason is that the equivalent four-terminal nanostructure (b) cannot be generally separated into elements of definite resistance.

$$V_3 - V_4 = (V_1 - V_2)\left(\frac{G_{31}}{G_{31}+G_{32}} - \frac{G_{41}}{G_{41}+G_{42}}\right). \tag{1.78}$$

Taking into account that the current under this assumption is given by $I = G_{21}(V_1 - V_2)$, we obtain the result of the two-probe measurement:

$$\mathcal{R} = \frac{1}{G_{21}}\left(\frac{G_{31}}{G_{31}+G_{32}} - \frac{G_{41}}{G_{41}+G_{42}}\right). \tag{1.79}$$

The observation to make is that this resistance does not display any symmetry with respect to time reversal, $\mathcal{R}(B) \neq \mathcal{R}(-B)$ [17]. This is in contrast to the time-reversal symmetry of two-terminal resistance R_{AB}, $R_{AB}(B) = R_{AB}(-B)$. The measurements of magnetoresistance [18] that seemingly disobeyed the Onsager relations have inspired the current interest in quantum transport.

1.5.3 Beam splitters

A basic element of a two-terminal nanocircuit is a scattering region – a region where an incoming electron makes a choice between transmission and reflection, characterized by a scattering matrix. The simplest single-channel example of such a scattering matrix is presented in Fig. 1.17. More sophisticated two-terminal nanostructures can be made by combining these basic elements in series. We discuss in Section 1.6 how to find the scattering matrix of the resulting nanostructure from those of the elements. However, all such nanostructures remain two-terminal ones. To do this in a multi-terminal arrangement, we need to introduce a new basic element – a *beam splitter*, where electrons can be reflected back or transmitted to several transport channels that end up in different terminals. The name beam splitter comes from optics, where it is really about splitting light beams. The simplest beam splitter mixes three transport channels (Fig. 1.29).

The scattering matrix for a general three-channel beam splitter is a 3×3 symmetric (in the absence of magnetic field) matrix, constrained by the conditions of unitarity. Such a scattering matrix is parameterized by five real numbers (apart from the insignificant overall phase factor). This is too much for a simple model, and we restrict ourselves to two cases when the scattering matrix depends on a single parameter only. In Section 1.6 we use these scattering matrices to model quantum interference.

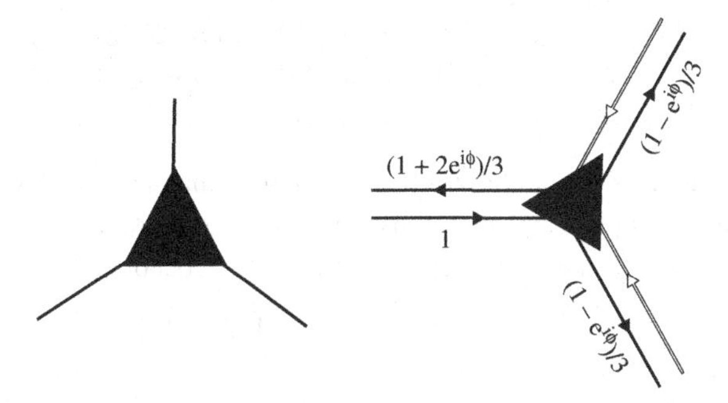

Fig. 1.29. Fully symmetric three-channel beam splitter. We show the scattering amplitudes of an incoming wave.

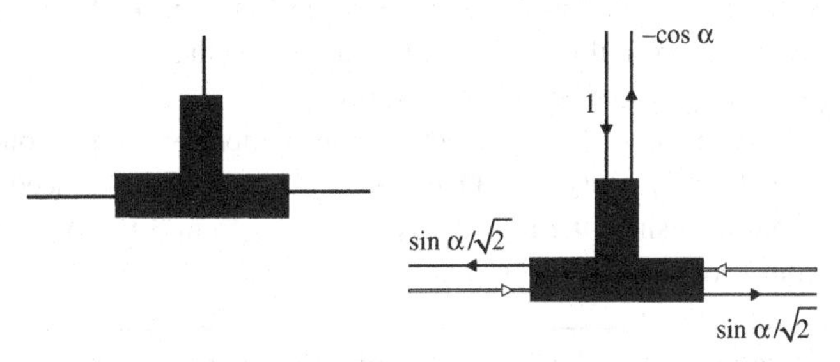

Fig. 1.30. T-beam splitter. If $\alpha = 0$, electrons traverse between 1 and 2 without reflection. If $\alpha = \pi/2$, electrons from 3 get to either 1 or 2 without reflection.

If all three channels are identical (fully symmetric beam splitter), all diagonal elements of the scattering matrix must be the same, and all off-diagonal elements must be the same. The unitary matrix satisfying this condition can be parameterized as follows:

$$\hat{s} = \frac{1}{3}\begin{pmatrix} 1+2e^{i\phi} & 1-e^{i\phi} & 1-e^{i\phi} \\ 1-e^{i\phi} & 1+2e^{i\phi} & 1-e^{i\phi} \\ 1-e^{i\phi} & 1-e^{i\phi} & 1+2e^{i\phi} \end{pmatrix}. \tag{1.80}$$

The phase ϕ is responsible for the coupling of the leads to the beam splitter. Indeed, the probability of reflection to the same lead is the absolute value of the diagonal element squared, and equals $R = [5 + 4\cos\phi]/9$. It varies between $R = 1$ ($\phi = 0$, full reflection) and $R = 1/3$ ($\phi = \pi$, the incoming stream is equally divided between all three leads).

Exercise 1.8. Regarding the beam splitter described by the scattering matrix in Eq. (1.80) as a three-terminal system, calculate the corresponding conductance matrix.

This restriction is no good for modeling purposes. Let us consider a T-beam splitter, which is symmetric with respect to the exchange of channels 1 and 2, and also its scattering matrix (Fig. 1.30). We choose the matrix elements to be real. One of the two possible realizations is given by

$$\hat{s} = \begin{pmatrix} -\sin^2(\alpha/2) & \cos^2(\alpha/2) & \sin(\alpha)/\sqrt{2} \\ \cos^2(\alpha/2) & -\sin^2(\alpha/2) & \sin(\alpha)/\sqrt{2} \\ \sin(\alpha)/\sqrt{2} & \sin(\alpha)/\sqrt{2} & -\cos(\alpha) \end{pmatrix}. \quad (1.81)$$

The angle α parameterizes the coupling of channel 3 to channels $1-2$. For $\alpha = 0$, channel 3 is uncoupled and electrons go from 1 to 2, or in the opposite direction without any scattering. If $\alpha = \pi/2$, the scattering matrix becomes

$$\hat{s} = \begin{pmatrix} -1/2 & 1/2 & 1\sqrt{2} \\ 1/2 & -1/2 & 1\sqrt{2} \\ 1\sqrt{2} & 1\sqrt{2} & 0 \end{pmatrix} \quad (1.82)$$

and describes the ideal beam splitting: the electrons coming from lead 3 are equally distributed between 1 and 2 and are not reflected back. An electron coming from 1 has, in this case, probability $1/2$ of being transmitted to 3, probability $1/4$ of proceeding to 2, and probability $1/4$ of being reflected back.

One might want to use a better beam spitter, for instance one where an electron coming from 1 always gets to 3. One would want too much. Indeed, time-reversibility relations forbid this: since the probability of getting from 3 to 1 is $1/2$, the probability of getting from 1 to 3 must be the same.

Exercise 1.9. Show that the only beam splitter where the electron coming from 1 always gets to 3 has only three non-zero elements, $s_{13} = s_{31}$, and s_{22}, with the absolute value of all these elements equal to one.

1.5.4 Counting statistics and noise

The transmission properties of a multi-terminal nanostructure are fully described by the distribution function of transmitted charge $P(Q_1, Q_2, \ldots, Q_N)$, which is the probability that the charge Q_α passed through the terminal α (from the scattering region to the reservoir) during the time Δt. Current conservation requires that this function is proportional to $\delta(Q_1 + \cdots + Q_N)$. It is more convenient to introduce the characteristic function $\Lambda(\{\chi_\alpha\})$, which is defined as a Fourier transform:

$$\Lambda(\{\chi_\alpha\}) = \sum_{Q_1,\ldots,Q_N} P(\{N_\alpha\}) \exp\left[\mathrm{i}(\chi_1 Q_1 + \cdots + \chi_N Q_N)/e\right]. \quad (1.83)$$

As a consequence of the current conservation, it only depends on the differences $\chi_\alpha - \chi_\beta$.

In the multi-terminal case, the characteristic function can be brought into a compact form analogous to the Levitov formula, Eq. (1.54). For the characteristic function, we have

$$\ln \Lambda(\{\chi_\alpha\}) = 2_s \Delta t \int \frac{\mathrm{d}E}{2\pi\hbar} \, \mathrm{Tr} \, \ln\left\{1 + \hat{f} + \hat{f}\hat{s}^\dagger\hat{\tilde{s}}\right\}, \quad (1.84)$$

where the trace is taken not only over the transport channels, but *also* over the terminals. In Eq. (1.84), $\hat{f}$ is the diagonal matrix with the matrix elements $f_\alpha(E)$ for all transport channels in the terminal α, and

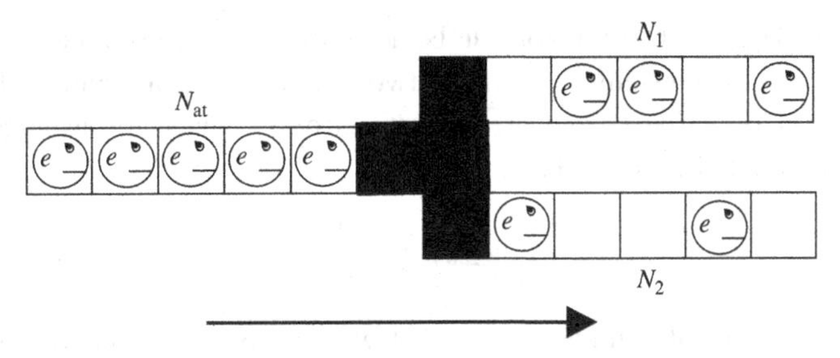

Fig. 1.31. Counting statistics of a reflectionless T-beam splitter.

$$\hat{\tilde{s}}_{\alpha\beta} = \hat{s}_{\alpha\beta} e^{i(\chi_\alpha - \chi_\beta)}.$$

To start with, let us create a link to the Levitov formula, Eq. (1.54), for two terminals. We show this explicitly for the simplest case of a single-channel scatterer. In this case, the argument of the logarithm is a 2×2 matrix. Using the parameterization given in Eq. (1.39), we write this matrix explicitly as follows:

$$\begin{pmatrix} 1 - f_L + f_L\left(R + Te^{i\chi}\right) & f_L RTe^{i(\eta-\theta)}\left(-1 + e^{-i\chi}\right) \\ f_R RTe^{i(\theta-\eta)}\left(1 - e^{i\chi}\right) & 1 - f_R + f_R\left(R + Te^{-i\chi}\right) \end{pmatrix}, \chi \equiv \chi_R - \chi_L.$$

The trace of a logarithm of a matrix equals the logarithm of the determinant of this matrix. Calculating the determinant and taking into account that $T + R = 1$, we reproduce Eq. (1.54).

Let us discuss a simple multi-terminal example. We take a reflectionless T-beam splitter, restrict ourselves to zero temperature limit, and choose voltages $V_1 = V_2 = 0$, $V_3 = V > 0$. The contribution to the integral in Eq. (1.84) comes from the energy strip $\mu_1 = \mu_2 < E < \mu_3 = \mu_2 + eV$. The integrand does not depend on the energy within the strip. We create a 3×3 matrix by substituting $f_1 = f_2 = 0$, $f_3 = 1$, and using Eq. (1.82) for $\hat{s}$. We calculate the determinant of the matrix:

$$\Lambda(\chi_1, \chi_2, \chi_3) = \left(\frac{\exp(i\chi_2)}{2} + \frac{\exp(i\chi_1)}{2}\right)^{N_{at}} \exp(-iN_{at}\chi_3),$$

$N_{at} = \Delta t G_Q V/e$ being the number of electrons coming from reservoir 3. The resulting distribution of the transmitted charges $Q_i = eN_i$ is binomial,

$$P_{N_1,N_2,N_3} = \delta(N_1 + N_2 - N_3)\delta(N_3 - N_{at})\binom{N_{at}}{N_1}\left(\frac{1}{2}\right)^{N_1}\left(\frac{1}{2}\right)^{N_2}. \tag{1.85}$$

What does this mean? The first delta-function is just the charge conservation; i.e. the total number of electrons going to 1 and 2 is the same as the number of electrons coming from 3. The second delta-function shows that the current coming from 3 does not fluctuate; we expected this from the fact that there is no reflection back to 3. The binomial shows that the electrons coming to the splitter with regular intervals reach either 1 or 2 with equal probabilities (Fig. 1.31). Thus, the currents to 1 and 2 do fluctuate. However, their fluctuations are strictly opposite: each extra electron that gets to 1 implies the lack of an electron getting to 2.

The general situation can be accessed for the *noise*, the second-order cumulants of the charges transmitted. As in the two-terminal case, all cumulants of the transmitted charge can be derived from the derivatives of Λ. Thus, the average charge $\langle Q_\alpha \rangle$ transmitted through the terminal α is given by

$$\langle Q_\alpha \rangle = e \left. \frac{\partial \ln \Lambda}{\partial (i\chi_\alpha)} \right|_{\chi_\beta = 0}, \quad \beta = 1, \dots, N.$$

A technical complication is that we are now differentiating the trace of a logarithm of a matrix. This is done using the relation $(\mathrm{Tr}\ \ln \hat{A})' = \mathrm{Tr}\ \hat{A}^{-1}\hat{A}'$. The argument of the logarithm turns to the unit matrix due to the unitarity constraint after we set all the counting fields χ_β to zero; differentiating the matrix over χ_α and calculating the trace, we arrive at $\langle Q_\alpha \rangle = I_\alpha \Delta t$, where the average current I_α is given by Eq. (1.72).

Let us now calculate the second cumulant of the charge,

$$\langle\langle Q_\alpha Q_\beta \rangle\rangle = e^2 \left. \frac{\partial^2 \ln \Lambda}{\partial (i\chi_\alpha) \partial (i\chi_\beta)} \right|_{\chi_\gamma = 0}, \quad \gamma = 1, \dots, N.$$

Treating the trace of logarithm in the same manner, after some manipulations with the unitarity condition we arrive at the following expression:

$$\begin{aligned} \langle\langle Q_\alpha Q_\beta \rangle\rangle &= \Delta t S_{\alpha\beta}(0)/2; \\ S_{\alpha\beta}(0) &= G_Q \int \mathrm{d}E \sum_{\gamma\delta} \mathrm{Tr} \left\{ \left[\delta_{\alpha\gamma}\delta_{\alpha\delta} - \hat{s}^\dagger_{\alpha\gamma}(E)\hat{s}_{\alpha\delta}(E) \right] \right. \\ &\quad \times \left. \left[\delta_{\beta\delta}\delta_{\beta\gamma} - \hat{s}^\dagger_{\beta\delta}(E)\hat{s}_{\beta\gamma}(E) \right] \right\} \\ &\quad \times \left\{ f_\gamma(E) \left[1 - f_\delta(E) \right] + f_\delta(E) \left[1 - f_\gamma(E) \right] \right\}, \end{aligned} \qquad (1.86)$$

where the trace is again taken over the transport channels. The matrix $S_{\alpha\beta}(\omega)$ is the current noise at the frequency ω, defined as

$$S_{\alpha\beta}(\omega) = \left\langle \hat{I}_\alpha(t)\hat{I}_\beta(t') + \hat{I}_\beta(t')\hat{I}_\alpha(t) - 2\left\langle \hat{I}_\alpha(t) \right\rangle \left\langle \hat{I}_\beta(t') \right\rangle \right\rangle_\omega, \qquad (1.87)$$

where the Fourier component is taken with respect to $t - t'$.

Exercise 1.10. Calculate the noise matrix $S_{\alpha\beta}$ for the beam splitter described by Eq. (1.80).

We now discuss the properties of the multi-terminal noise formula, Eq. (1.86). If there are only two terminals, we have $S_{\mathrm{LL}} = S_{\mathrm{RR}} = -S_{\mathrm{LR}} = -S_{\mathrm{RL}}$, and Eq. (1.86) reproduces the expressions of Section 1.3.

Then we consider the equilibrium, when the distribution functions $f_\alpha(E)$ in all the reservoirs are the same. In the linear regime, we assume that the scattering matrices are evaluated at the Fermi energy, and the integration of the Fermi functions over E provides the factor $k_{\mathrm{B}}T$. Using the unitarity of the scattering matrix, we write

$$\sum_{\gamma\delta} \mathrm{Tr} \left\{ \hat{s}^\dagger_{\alpha\gamma}\hat{s}_{\alpha\delta}\hat{s}^\dagger_{\beta\delta}\hat{s}_{\beta\gamma} \right\} = \mathrm{Tr}\ \delta_{\alpha\beta},$$

which provides the expression for the equilibrium (Nyquist–Johnson) noise,

$$S^{eq}_{\alpha\beta}(0) = -2k_{\rm B}T\left(G_{\alpha\beta} + G_{\beta\alpha}\right), \tag{1.88}$$

in accordance with the fluctuation-dissipation theorem.

At zero temperature, noise takes a simpler form,

$$\begin{aligned} S_{\alpha\beta}(0) = G_{\rm Q} \int {\rm d}E \sum_{\gamma\neq\delta} {\rm Tr}\left[\hat{s}^{\dagger}_{\alpha\gamma}(E)\hat{s}_{\alpha\delta}(E)s^{\dagger}_{\beta\delta}(E)\hat{s}_{\beta\gamma}(E)\right] \\ \times\left\{f_{\gamma}(E)\left[1 - f_{\delta}(E)\right] + f_{\delta}(E)\left[1 - f_{\gamma}(E)\right]\right\}. \end{aligned} \tag{1.89}$$

In particular, for $\alpha \neq \beta$ this can be rewritten as

$$S_{\alpha\beta}(0) = -2G_{\rm Q} \int {\rm d}E\ {\rm Tr}\left[\left(\sum_{\gamma} \hat{s}^{\dagger}_{\alpha\gamma}\hat{s}_{\beta\gamma} f_{\gamma}\right)\left(\sum_{\delta} \hat{s}^{\dagger}_{\beta\delta}\hat{s}_{\alpha\delta} f_{\delta}\right)\right],$$

and is obviously negatively defined, since it includes a product of a matrix with its conjugate. Thus, the current correlations at different terminals are always negative at zero frequency. We have already seen this for the beam splitter. The proof of this fact only uses that electrons are fermions. As a matter of fact, it turns out that this statement is not correct for bosons – the zero-frequency cross-correlations of bosons (for example photons in the different arms of an interferometer) can very well be positive.

1.5.5 Multi-terminal scattering in operator formalism

We now extend the operator formalism, developed in Section 1.3, to multi-terminal systems, and show how it can be used to calculate noise. First we introduce scattering states. The state $\psi_{\alpha n}$, which originates from reservoir α in transport channel n, is given by its asymptotic expressions,

$$\begin{aligned} \psi_{\alpha n}(x_{\alpha}, y_{\alpha}, z_{\alpha}) = & \frac{1}{\sqrt{2\pi\hbar v_{\alpha n}(E)}}\Phi_{\alpha n}(y_{\alpha}, z_{\alpha}){\rm e}^{-{\rm i}k_x^{(\alpha n)}x_{\alpha}} \\ & + \sum_m \frac{1}{\sqrt{2\pi\hbar v_{\alpha m}(E)}} s_{\alpha m,\alpha n}(E)\Phi_{\alpha m}(y_{\alpha}, z_{\alpha}){\rm e}^{{\rm i}k_x^{(\alpha m)}x_{\alpha}} \end{aligned} \tag{1.90}$$

in terminal α and

$$\psi_{\alpha n}(x_{\beta}, y_{\beta}, z_{\beta}) = \sum_m \frac{1}{\sqrt{2\pi\hbar v_{\beta m}(E)}} s_{\beta m,\alpha n}(E)\Phi_{\beta m}(y_{\beta}, z_{\beta}){\rm e}^{{\rm i}k_x^{(\beta m)}x_{\beta}} \tag{1.91}$$

in terminal $\beta \neq \alpha$.

Next, we proceed with the creation $\hat{a}^{\dagger}_{\alpha n\sigma}(E)$ and annihilation $\hat{a}_{\alpha n\sigma}(E)$ operators for the scattering states. Another set of operators, $\hat{b}^{\dagger}_{\alpha n\sigma}(E)$ and $\hat{b}_{\alpha n\sigma}(E)$, describe electrons moving along waveguide α in transport channel n from the scattering region. These two sets are related via the scattering matrix:

$$\hat{b}_{\alpha n\sigma}(E) = \sum_{\beta m} s_{\alpha n,\beta m}(E)\hat{a}_{\beta m\sigma}(E); \tag{1.92}$$

$$\hat{b}^{\dagger}_{\alpha n\sigma}(E) = \sum_{\beta m} s^{*}_{\beta m,\alpha n}(E)\hat{a}^{\dagger}_{\beta m\sigma}(E). \tag{1.93}$$

The operators $\hat{a}$ obey the anticommutation relations, Eqs. (1.47), which are the same as for two-terminal systems. The average product of a creation and an annihilation operator is also the same:

$$\left\langle \hat{a}^{\dagger}_{\alpha n\sigma}(E)\hat{a}_{\beta m\sigma'}(E')\right\rangle = \delta_{\alpha\beta}\delta_{nm}\delta_{\sigma\sigma'}\delta(E-E')f_{\alpha}(E). \tag{1.94}$$

Writing down the field operators in lead α,

$$\hat{\Psi}_{\sigma}(\boldsymbol{r}_{\alpha},t) = \int \mathrm{d}E\ \mathrm{e}^{-\mathrm{i}Et\hbar}\sum_{n}\frac{\Phi_{\alpha n}(y_{\alpha},z_{\alpha})}{\sqrt{2\pi\hbar v_{\alpha n}(E)}}\left[\hat{a}_{\alpha n\sigma}\mathrm{e}^{-\mathrm{i}k_x^{(\alpha n)}x_{\alpha}} + \hat{b}_{\alpha n\sigma}\mathrm{e}^{\mathrm{i}k_x^{(\alpha n)}x_{\alpha}}\right]$$

and

$$\hat{\Psi}^{\dagger}_{\sigma}(\boldsymbol{r}_{\alpha},t) = \int \mathrm{d}E\ \mathrm{e}^{\mathrm{i}Et\hbar}\sum_{n}\frac{\Phi^{*}_{\alpha n}(y_{\alpha},z_{\alpha})}{\sqrt{2\pi\hbar v_{\alpha n}(E)}}\left[\hat{a}^{\dagger}_{\alpha n\sigma}\mathrm{e}^{\mathrm{i}k_x^{(\alpha n)}x_{\alpha}} + \hat{b}^{\dagger}_{\alpha n\sigma}\mathrm{e}^{-\mathrm{i}k_x^{(\alpha n)}x_{\alpha}}\right],$$

we construct the operator of current in the terminal α,

$$\hat{I}_{\alpha}(x_{\alpha},t) = \frac{\hbar e}{2\mathrm{i}m}\sum_{\sigma}\int \mathrm{d}y_{\alpha}\ \mathrm{d}z_{\alpha}\left[\hat{\Psi}^{\dagger}_{\sigma}\frac{\partial}{\partial x_{\alpha}}\hat{\Psi}_{\sigma} - \left(\frac{\partial}{\partial x_{\alpha}}\hat{\Psi}^{\dagger}_{\sigma}\right)\hat{\Psi}_{\sigma}\right]. \tag{1.95}$$

Making the time periodic, with period $\mathcal{T}$ (and discrete energies $E = 2\pi q\hbar/\mathcal{T}$), similarly to how it was done in Section 1.3, we obtain the following expression for the *time-averaged* current operator in terminal α,

$$\begin{aligned}\left\langle \hat{I}_{\alpha}\right\rangle_t &= -\frac{e}{2\pi\hbar}\left(\frac{2\pi\hbar}{\mathcal{T}}\right)^2\sum_{n\sigma}\sum_{E}\left[\hat{a}^{\dagger}_{\alpha n\sigma}(E)\hat{a}_{\alpha n\sigma}(E) - \hat{b}^{\dagger}_{\alpha n\sigma}(E)\hat{b}_{\alpha n\sigma}(E)\right]\\ &= -\frac{e}{2\pi\hbar}\left(\frac{2\pi\hbar}{\mathcal{T}}\right)^2\sum_{n\sigma}\sum_{\beta\gamma,ll'}\sum_{E}\hat{a}^{\dagger}_{\beta l\sigma}(E)\hat{a}_{\gamma l'\sigma}(E)\\ &\quad\times\left[\delta_{\alpha\beta}\delta_{\alpha\gamma}\delta_{nl}\delta_{nl'} - s^{*}_{\beta l,\alpha n}(E)s_{\alpha n,\gamma l'}(E)\right].\end{aligned} \tag{1.96}$$

We have not yet taken the spin variables into consideration, and therefore the operator in Eq. (1.96) describes the current of particles with a given spin projection (which is the same for both projections). Using Eqs. (1.94), $\delta(0) \to \mathcal{T}/(2\pi\hbar)$, we reproduce Eq. (1.72) for the average current.

Our next task is to calculate the zero-frequency noise, Eq. (1.87). At zero frequency, the Fourier transform means integration over $t-t'$. It is convenient to discretize the time again, then the integration over time means time-averaging multiplied by the period $\mathcal{T}$. We also want to calculate the time-averaged (with respect to $(t+t')/2$) noise, and thus we have to time-average all the current operators in Eq. (1.87) and multiply the result by $\mathcal{T}$. In other words, for the calculation of the zero-frequency noise, it is enough to know only the time-averaged current operators. Substituting Eq. (1.96), we arrive at a *very* cumbersome expression that contains the quantum-mechanical average of four creation and annihilation operators. In order to proceed, we have to learn how to deal with averages such as

$$\langle \hat{a}_1^\dagger \hat{a}_2 \hat{a}_3^\dagger \hat{a}_4 \rangle,$$

where the subscript indices label the scattering states (they include the energy, the terminal, and the transport channel). This average is easy to calculate since we consider non-interacting electrons. We apply *Wick's theorem*, which states that the average of a product of an even number of creation and annihilation operators equals the sum of possible products of averages of pairs of such operators. Wick's theorem relies on the fact that the free-electron Hamiltonian is quadratic in creation and annihilation operators and does not hold, for instance, for interacting electrons. In our case, the average of $\langle \hat{a}_2 \hat{a}_4 \rangle$ is zero, and we are left with only two possible pairings: (i) 1 with 2 and 3 with 4, and (ii) 1 with 4 and 2 with 3,

$$\langle \hat{a}_1^\dagger \hat{a}_2 \hat{a}_3^\dagger \hat{a}_4 \rangle = \langle \hat{a}_1^\dagger \hat{a}_2 \rangle \langle \hat{a}_3^\dagger \hat{a}_4 \rangle + \langle \hat{a}_1^\dagger \hat{a}_4 \rangle \langle \hat{a}_2 \hat{a}_3^\dagger \rangle. \tag{1.97}$$

In the final term we have the average of $\hat{a}\hat{a}^\dagger$, which is calculated using the commutation relation,

$$\langle \hat{a}_2 \hat{a}_3^\dagger \rangle = \delta_{23} - \langle \hat{a}_3^\dagger \hat{a}_2 \rangle.$$

After all these manipulations, taking care of delta-functions of zero argument, and finally returning to the continuous energy variable, we arrive at Eq. (1.86).

1.6 Quantum interference

Our intuition is based on our everyday experience with classical physics. It is easier for us to understand the scattering of classical particles rather than quantum ones. Fortunately enough, some aspects of quantum transport can be readily understood in terms of scattering of *classical* particles. This scattering is characterized by probabilities rather than amplitudes. In this section, we concentrate on the effects that cannot be understood in such terms – the effects of *quantum interference*.

From the early days of quantum mechanics, it was traditional to illustrate the difference between classical and quantum mechanics using the *two-slit experiment*. Consider a quantum particle that can propagate from an initial point to a final point in two different ways: trajectories that go via one of the two slits in the screen (Fig. 1.32). The corresponding quantum amplitudes are A_1 and A_2, and the propagation probability for each trajectory is given by the absolute value of the amplitude squared, $P_{1,2} = |A_{1,2}|^2$. In classical physics, the total probability is just the sum of the two,

$$P_{\mathrm{cl}} = P_1 + P_2. \tag{1.98}$$

In quantum mechanics, amplitudes are added rather than probabilities. The total probability is the absolute value of the total amplitude squared, i.e.

$$P_{\mathrm{qm}} = |A_1 + A_2|^2 = |A_1|^2 + |A_2|^2 + A_1 A_2^* + A_1^* A_2 = P_{\mathrm{cl}} + 2\mathrm{Re}\, A_1 A_2^*. \tag{1.99}$$

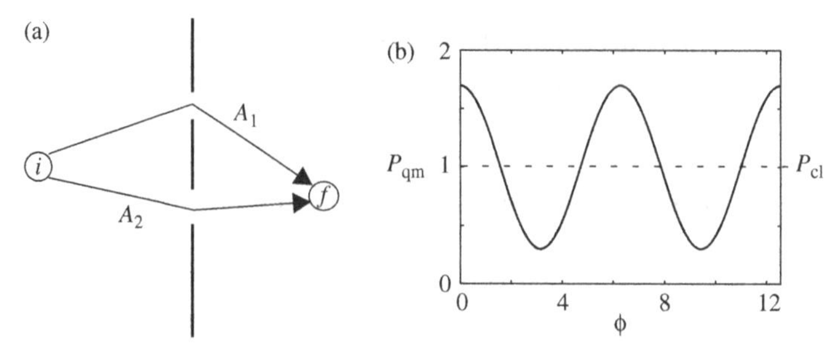

Fig. 1.32. Two-slit experiment. Quantum interference between two trajectories (a) results in oscillatory dependence of the propagation probability on the phase shift between two amplitudes ((b), plotted for $P_1/P_2 = 6$).

The final term in Eq. (1.99) represents the effect of the interference between the waves propagating along the two trajectories and cannot be accounted for in the classical theory. This is what is lost when we try to describe quantum transport classically.

To observe and refine the interference effect, one affects the relative phase of two amplitudes not changing their absolute values related to classical probabilities. For an actual two-slit experiment, this is achieved by a small displacement of the initial and/or the final point with respect to the slits. In terms of this *phase shift* ϕ, the probability reads.

$$P_{qm} = P_{cl} + 2\sqrt{P_1 P_2} \cos\phi. \tag{1.100}$$

The oscillatory dependence on the phase shift thus signals the quantum interference and is used to identify it experimentally. The interference may be constructive when $P_{qm} > P_{cl}$ ($\cos\phi > 0$) or destructive in the opposite case. If $P_1 = P_2$, one can tune the phase shift to suppress completely the quantum-mechanical probability.

Control question 1.14. Consider a non-ideal experiment in which the phase difference fluctuates during the measurement time over a typical scale much bigger than 2π. What will be the value of P_{qm} measured in this experiment?

The quantum interference and Eq. (1.99) are not just about the two-slit experiment; this occurs whenever quantum particles propagate. Throughout the rest of this section, we analyze the basic examples of interference effects in transport and describe their experimental manifestations.

1.6.1 Phase shifts

To understand phase shifts, we start with one-dimensional motion, for example, in a certain transport channel (Fig. 1.7). We assume that the effective one-dimensional potential $E_0(x)$ is sufficiently smooth so no scattering occurs. Nothing would happen to a classical particle in this case. However, a quantum electron traveling in this potential acquires a *phase*. To evaluate this phase, let us consider the wave function of an electron moving from the left to the right. The absence of scattering implies that this function can be considered in the semiclassical approximation:

$$\psi(x) = \exp(\mathrm{i}\phi(x)); \frac{\mathrm{d}\phi}{\mathrm{d}x} = k(x) \equiv \sqrt{2m(E - E_0(x))}/\hbar, \tag{1.101}$$

$k(x)$ being the "local" wave vector at point x. The electron moving from point x_1 to point x_2 collects the phase $\phi = \phi(x_1) - \phi(x_2)$. If the potential along the channel does not vary, it is just kL, L being the distance between the points.

The absolute value of the phase shift is usually of no interest. Besides, it is difficult to control. The relative change of the phase shift is more interesting. First, let us note that ϕ depends on energy. We take the derivative with respect to energy and note that $\mathrm{d}k(x)/\mathrm{d}E = 1/\hbar v(x)$, where $v(x)$ is the velocity of the electron. This yields

$$\frac{\mathrm{d}\phi}{\mathrm{d}E} = \int_{x_1}^{x_2} \frac{\mathrm{d}x}{\hbar v(x)} = \frac{\tau}{\hbar}, \tag{1.102}$$

τ being the time of flight between the points at a given energy. Second, one can shift the phase by modifying the potential within the channel, for instance with the help of gate electrodes. The shift is given by

$$\Delta\phi = \int eV(x)\frac{\mathrm{d}x}{\hbar v(x)} \simeq eV\tau/\hbar. \tag{1.103}$$

The phase shifts can be defined in a similar fashion for an electron that is not confined to a transport channel but moves in 3D space along a certain classical trajectory $\boldsymbol{x}(t)$. It is convenient to integrate over t in the above formulas, as follows:

$$\frac{\mathrm{d}\phi}{\mathrm{d}E} = \int_{t_1}^{t_2} \frac{\mathrm{d}t}{\hbar} = \frac{\tau}{\hbar}; \; \Delta\phi = \int eV(\boldsymbol{x}(t))\frac{\mathrm{d}t}{\hbar}.$$

The phase shift due to energy or potential is called the *dynamical* phase. It has an important property: if an electron takes a time-reversed path, so that it moves from point x_2 to x_1, the phase shift acquired is precisely the same.

As noticed by Aharonov and Bohm about fifty years ago, the phase shifts due to the magnetic field are more complicated and interesting. The magnetic phase accumulated along the trajectory explicitly depends on the vector potential $\boldsymbol{A}(x)$,

$$\phi_{\mathrm{mag}} = \frac{e}{\hbar c}\int_{t_1}^{t_2} \boldsymbol{A}\cdot\boldsymbol{v}(t)\mathrm{d}t = \frac{e}{\hbar c}\int_{x_1}^{x_2} \boldsymbol{A}\cdot\mathrm{d}\boldsymbol{x},$$

and is *opposite* for the time-reversed path. Indeed, one can describe precisely the same physical situation in a different gauge, shifting the vector potential by an arbitrary gradient field, $\boldsymbol{A} \to \boldsymbol{A} + \nabla\chi(x)$. This phase shift explicitly depends on $\chi(x)$, which makes it "unphysical" and unobservable.

The gauge-invariant, and thus observable, quantity is the magnetic phase accumulated along the *closed* path where the electron returns to the same point. It is proportional to the magnetic flux Φ enclosed by this closed trajectory,

$$\phi_{\mathrm{mag}} = \frac{e}{\hbar c}\oint \boldsymbol{B}\cdot\mathrm{d}\boldsymbol{S} = \frac{\pi\Phi}{\Phi_0},$$

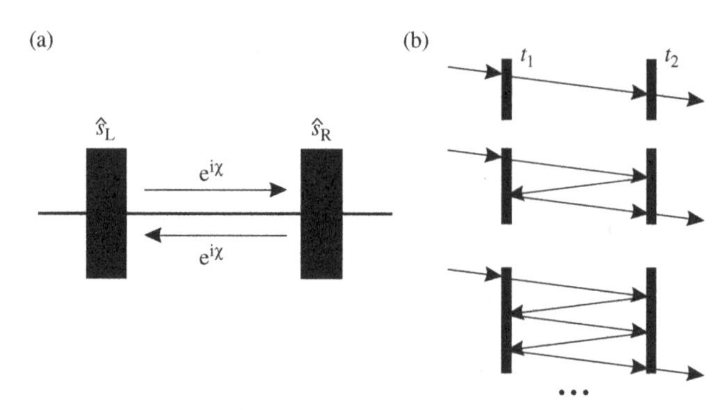

Fig. 1.33. (a) Double-junction nanostructure consists of two scatterers in series. The electron acquires the phase shift χ when traveling between the scatterers. (b) The transmission results from the interference of all trajectories shown, and thus depends on the phase shift.

where the magnetic flux quantum $\Phi_0 \equiv \pi\hbar c/e$.[5] The difference in the phase shifts along this trajectory and the time-reversed one does not contain the dynamical phase and thus does not depend on a concrete shape of the trajectory: it is $2\pi\Phi/\Phi_0$ precisely.

Generally speaking, any periodic dependence of a physical quantity on Φ/Φ_0 is called an Aharonov–Bohm (AB) effect. There may be an AB effect observed in the two-slit experiment described above, although the particle never makes a closed trajectory. Magnetic phase shifts along each trajectory, $\phi_{1,2}$, are not gauge-invariant and cannot be separately observed. However, the probability depends only on their difference $\phi_1 - \phi_2$, which is the gauge-invariant phase shift that would be acquired along a closed path made of trajectory 1 and *time-reversed* trajectory 2.

The magnetic phase shift is not the only effect of the magnetic field – it can also change the shape of the electron wave function and its spin state. However, for low magnetic fields the interference AB effect is the most important one. We note that the effect exists even if the magnetic flux is enclosed in a finite area and the electrons moving along the trajectories do not in fact feel the magnetic field! What matters for the interference effect is the total flux enclosed by the two trajectories, not the distribution of the magnetic field.

1.6.2 Double junction

The simplest quantum transport setup that demonstrates quantum interference effects consists of two scatterers in series (Fig. 1.33).

To keep it simple, we begin the discussion with only one transport channel. The scatterers are characterized by corresponding 2×2 scattering matrices $\hat{s}_{L,R}$, or, equivalently, by transmission $t_{L,R}$, $t'_{L,R}$ and reflection $r_{L,R}$, $r'_{L,R}$ amplitudes. Importantly, the electron

[5] A word of warning: 20% of publications in the field of quantum transport use a definition of Φ_0 that differs by a factor of 2.

acquires phase shift χ when traveling between the scatterers. We consider the dynamical phase only, so that this phase shift does not depend on the direction of propagation.

Let us now consider the amplitude of transmission through both scatterers. An electron can make it to the right of the two scatterers in a number of ways that differ by the number of attempts to penetrate the right scatterer (Fig. 1.33). We will call these ways "processes" (they can be viewed as classical processes of transmission, reflection, or propagation) or "trajectories" (like in the two-slit experiment, they can be associated with motion in space).

The simplest process is to get through the right scatterer at the first attempt. Its amplitude is a product of the three amplitudes of successive elementary processes: transmission through the left scatterer t_L, propagation between the scatterers $\exp(i\chi)$, and transmission through the right scatterer t_R, $A_1 = t_L \exp(i\chi) t_R$. If the first attempt is not successful, the electron is reflected back (r_R), propagates back to the left scatterer ($\exp(i\chi)$), is reflected to the right (r'_L), propagates again ($\exp(i\chi)$), and, if lucky, is transmitted through the right scatterer (t_R). The resulting amplitude of such a process is thus $A_1 = t_L r'_L r_R t_R \exp(3i\chi)$. More complex trajectories involve multiple trips of electrons back and forth between the two scatterers. The amplitude of a trajectory with m attempts is given by

$$A_m = t_L t_R \left(r'_L r_R\right)^{m-1} e^{i(2m-1)\chi}. \tag{1.104}$$

The total quantum-mechanical amplitude of propagation to the right is a sum of the amplitudes of all the processes, i.e.

$$t = \sum_{m=1}^{\infty} A_m = t_L t_R e^{i\chi} \sum_{m=0}^{\infty} \left(r'_L r_R e^{2i\chi}\right)^m = \frac{t_L t_R e^{i\chi}}{1 - r'_L r_R e^{2i\chi}}. \tag{1.105}$$

Control question 1.15. The amplitude diverges if the denominator equals zero. When does this happen?

The squared absolute value of the amplitude yields the transmission coefficient,

$$T = \left| \frac{t_L t_R e^{i\chi}}{1 - r'_L r_R e^{2i\chi}} \right|^2 = \frac{T_L T_R}{1 + R_L R_R - 2\sqrt{R_L R_R} \cos 2\chi}. \tag{1.106}$$

Here $T_{L,R} = |t_{L,R}|^2$ and $R_{L,R} = 1 - T_{L,R}$ are transmission and reflection coefficients for the individual scatterers. In the final relation, we conveniently include the phases of reflection amplitudes in χ, $2\chi \to 2\chi + \arg(r'_L r_R)$ is now the phase collected during the round trip experienced by an electron traveling from the left scatterer to the right one and back.

The transmission coefficient depends periodically on the phase χ. We stress that the phase depends on energy by virtue of Eq. (1.102) and this implies periodic dependence of transmission on energy, with period $2\pi\hbar/\tau$, τ being the round-trip time. The minimum value of the transmission coefficient is achieved for $\chi = \pi$ and the maximal value is achieved for $\chi = 0$:

$$T_{min} = \frac{T_L T_R}{(1 + \sqrt{R_L R_R})^2} < T(\chi) < \frac{T_L T_R}{(1 - \sqrt{R_L R_R})^2} = T_{max}.$$

The difference is best seen when both scatterers have low transparency, $T_{L,R} \ll 1$. In this case, $T_{min} \approx T_L T_R \ll 1$, which is not surprising. To get to the right, an electron has to pass

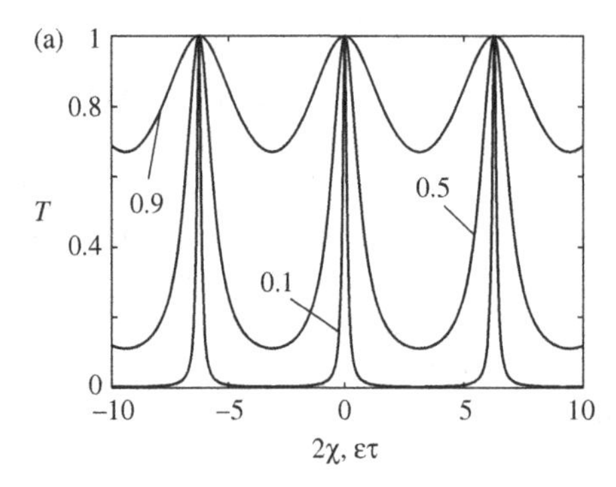

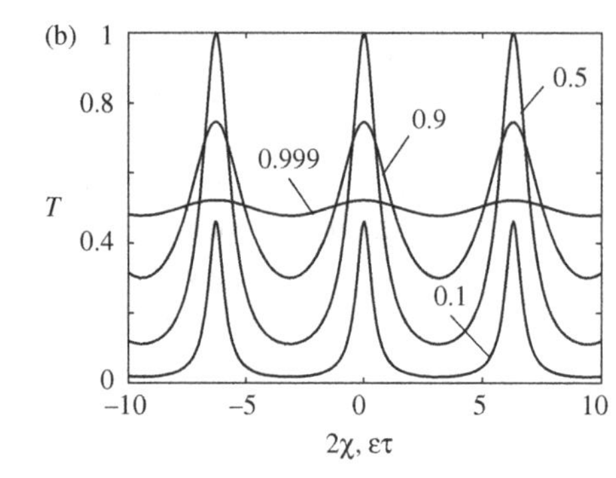

Fig. 1.34. **Transmission via double junction versus phase shift or energy. (a) Symmetric scatterers with $T_1 = T_2$ corresponding to the curve labels. The maximum transmission is always 1 in this case. (b) Non-symmetric scatterers, $T_1 = 0.5$, and T_2 corresponds to the curve labels. The narrow transmission resonances for $T_{1,2} \ll 1$ indicate formation of discrete energy levels at corresponding energies.**

two scatterers; hence, the total probability is the product of the probabilities of passing an individual scatterer. Let us now look at $T_{\max}$. Expanding the denominator up to linear terms in $T_{L,R}$, we find, surprisingly, that

$$T_{\max} = \frac{4T_L T_R}{(T_L + T_R)^2}, \tag{1.107}$$

which is not at all small. For example, if the two scatterers are identical, $T_L = T_R$, the maximal value of the transmission coefficient may become one – the system of two identical scatterers of very low transparency at certain values of energy becomes fully transparent. In this case, the energy dependence of the transmission coefficient has a resonant structure – $T(E)$ is very small for all energies except in the close vicinity of the values $\chi = \pi n$, at which it peaks (Fig. 1.34). This phenomenon is known as *resonant tunneling* or *Fabry–Perot resonances*. The resonances indicate the formation of discrete energy levels at the corresponding energies. Indeed, in the limit of vanishing transmission, the space between the scatterers is totally isolated from the leads, the motion is confined to this space, and the energy spectrum becomes discrete.

Let us concentrate on a single transmission resonance at $\chi = 0$ that occurs at $E = E_0$ and expand χ in the vicinity of this point, $\chi = (E - E_0)/2W$, W being of the order of energy distance between the resonances. The transmission assumes a Lorentzian shape,

$$T(E) = \frac{T_L T_R}{((T_L + T_R)/2)^2 + ((E - E_0)/W)^2}. \tag{1.108}$$

The energy width of the Lorentzian is given by $w = W(T_L + T_R)$. This allows for a somewhat unexpected interpretation in terms of decay rates. The particle in the resonance is confined, flying back and forth, between the junctions. There is a finite probability per unit time – a rate – that it will tunnel away, either through the left (Γ_L) or the right (Γ_R) junctions. By virtue of quantum uncertainty, the width is associated with the total decay rate, $w = \hbar(\Gamma_R + \Gamma_L)$; this yields the *Breit–Wigner* formula generic for resonances,

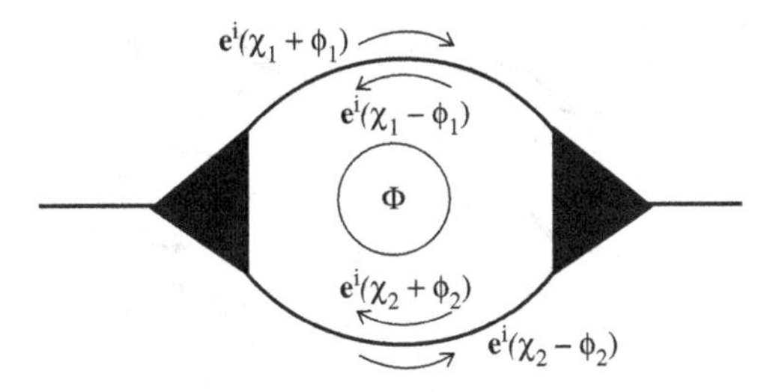

Fig. 1.35. Aharonov–Bohm ring made of two beam splitters penetrated by the magnetic flux Φ. Phase shifts in both arms are different for the opposite directions of propagation, $\phi_1 + \phi_2 = \pi\Phi/\Phi_0$.

$$T(E) = \frac{\Gamma_L \Gamma_R}{((\Gamma_L + \Gamma_R)/2)^2 + ((E - E_0)/\hbar)^2}, \tag{1.109}$$

where the rates are given by $\hbar\Gamma_{L,R} = WT_{L,R}$.

To zoom in on the difference between quantum and classical transport, let us calculate the classical probability of the propagation to the right. It is readily given by the sum of probabilities, $|A_m|^2$, of all processes as follows:

$$T_{cl} = \sum_{m=0}^{\infty} |A_m|^2 = T_L T_R \sum_{m=0}^{\infty} (R_L R_R)^m = \frac{T_L T_R}{1 - R_L R_R}. \tag{1.110}$$

This does not depend on the phase shift χ. The classical transmission through a double junction is thus phase- and energy-independent. This means that resonant tunneling is a purely quantum effect. Furthermore, if both scatterers have low transparency ($T_{L,R} \ll 1$), we can write the total classical conductance in the following form:

$$G_{cl} = G_Q T_{cl} = \frac{G_L G_R}{G_L + G_R} \text{ or } \frac{1}{G_{cl}} = \frac{1}{G_L} + \frac{1}{G_R},$$

which is easily recognized as Ohm's law for two resistors $1/G_L$ and $1/G_R$ in series. Everything that goes beyond Ohm's law physics (including resonant tunneling) cannot be described in classical terms and results from the quantum interference. We will discuss this in more detail in Section 2.1.

1.6.3 Aharonov–Bohm ring

Let us now deal with magnetic phases and devise the simplest model in which the Aharonov–Bohm effect is manifest in quantum transport [19]. We consider transmission through a ring connected to two reservoirs (Fig. 1.35). The ring consists of two arms, each supporting one transport channel. We treat dynamical ($\chi_{1,2}$) and magnetic ($\phi_{1,2}$) phases. An electron moving clockwise along the upper arm of the ring collects the phase $\phi_1 + \chi_1$, while an electron moving anticlockwise collects the phase $-\phi_1 + \chi_1$. An electron moving along the lower arm collects the phase $\phi_2 + \chi_2$ (clockwise) and $-\phi_2 + \chi_2$ (anticlockwise). As discussed, we expect that the transmission square depends on the gauge-invariant combination of the phases $\phi_1 + \phi_2 \equiv \phi_{AB} = \pi\Phi/\Phi_0$, which corresponds to the phase collected by the trajectory encircling the ring. The arms and the reservoirs are connected by

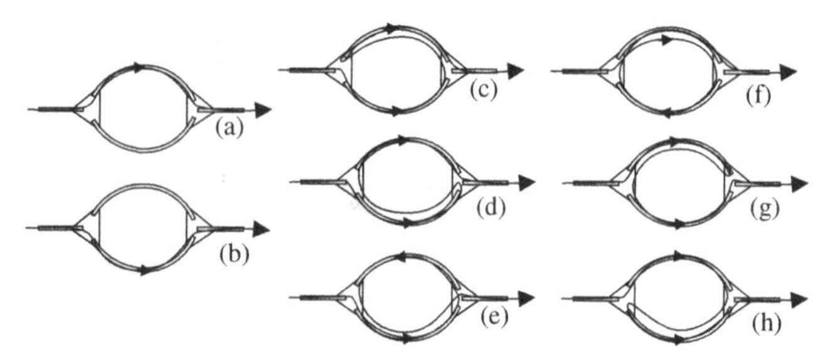

Fig. 1.36. Several example trajectories interfering in an AB ring. Interference of (a) and (b) provides a Φ_0 periodic contribution that depends on dynamical phase (*universal conductance fluctuations*). Interference in pairs (d), (e) and (f), (g) leads to a Φ_0 periodic contribution that survives averaging over dynamical phase (*weak localization*). Interference in pairs (c), (h) and (g), (d) yields a contribution time-reversed to that of pair (a), (b).

two beam splitters described by the scattering matrix in Eq. (1.82). Our goal is to find the transmission probability through the ring. As before, we proceed by identifying all the possible processes and summing up their amplitudes.

We immediately note that the number of possible processes is much larger than for the double junction, since, after each scattering, the electron can take either the upper or lower arm. For a given number of attempts m, this yields 2^{2m-1} possible processes. We plot in Fig. 1.36 two trajectories with $m = 1$ and some with $m = 2$.

Let us consider the amplitude of trajectory (a). An electron first enters the ring picking up the factor $1/\sqrt{2}$ from the scattering matrix of the beam splitter (Eq. (1.82)), then travels inside the ring acquiring the factor $e^{i(\chi_1+\phi_1)}$ and exits via the right beam splitter (factor $1/\sqrt{2}$ again), so that

$$t_{(a)} = \frac{1}{\sqrt{2}} e^{i(\chi_1+\phi_1)} \frac{1}{\sqrt{2}} = \frac{1}{2} e^{i(\chi_1+\phi_1)}.$$

As for the amplitude of process (h), the two first factors are the same. Then the electron goes through the beam splitter to the other arm (factor $1/2$), passes the lower arm clockwise ($e^{i(\chi_2+\phi_2)}$), is reflected from the beam splitter ($-1/2$), passes the lower arm counterclockwise ($e^{i(\chi_2-\phi_2)}$) and finally exits ($1/\sqrt{2}$). The amplitude is thus given by

$$t_{(h)} = \frac{1}{\sqrt{2}} e^{i(\chi_1+\phi_1)} \frac{1}{2} e^{i(\chi_2+\phi_2)} \left(-\frac{1}{2}\right) e^{i(\chi_2+\phi_2)} \frac{1}{\sqrt{2}} = -\frac{1}{8} e^{i(\chi_1+\phi_1+2\chi_2)}.$$

In this way we can determine the amplitude of any given trajectory.

Before summing up, let us look at the interference contributions of selected pairs of trajectories. Interference of the simplest trajectories (a) and (b) already gives the AB effect. It oscillates as a function of flux with period $2\Phi_0$, as follows:

$$P_{ab}^{int} = 2\text{Re}\, t_{(a)} t_{(b)}^* \propto \cos(\chi_1 - \chi_2 + \phi_{AB}). \tag{1.111}$$

Importantly, this contribution depends on dynamical phases and would disappear if one averages over these phase shifts. Contributions of this type are called *universal conductance fluctuations*. Why fluctuations? For nominally identical nanostructures, phase shifts are random. So, this contribution is individual for each nanostructure and will disappear if we average over a large ensemble of nominally identical ones.

Let us look at the interference of (d) and (e). This contribution does not depend at all on the dynamical phase and oscillates with a *twice shorter* period Φ_0,

$$P_{\text{de}}^{\text{int}} \propto \cos(2\phi_{\text{AB}}).$$

This is because of the very special relationship between these trajectories. Trajectories (d) and (e) comprise trajectory (b) with an extra closed loop orbiting the ring. The only difference is the direction of the loop: the electron orbits counterclockwise along (d) and counterclockwise along (e). As we mentioned above, the difference of phase shifts in this case does not depend on dynamical phase. The contribution of this type is called *weak localization correction.* Since it does not depend on the dynamical phase, it survives the ensemble averaging. We stress the generality of the effect: whatever the nanostructure, for any trajectory that contains a loop one finds a counterpart, the trajectory that differs by the direction of orbit only. The interference between pairs of such trajectories determines the weak localization correction in any nanostructure.

The AB ring under consideration is a two-terminal system. As we have learned from Section 1.5, this implies that its conductance and transmission is even in a magnetic field, $T(\phi_{\text{AB}}) = T(-\phi_{\text{AB}})$. Since $P_{\text{ab}}^{\text{int}}$ is neither even nor odd in flux, there must be time-reversed counter-terms proportional to $\cos(\chi_1 - \chi_2 - \phi_{\text{AB}})$ that ensure the symmetry. Such terms arise, for example, from the interference in pairs (c), (h) and (g), (d). Taken together, these terms provide

$$P^{\text{int}} \propto \cos(\chi_1 - \chi_2)\cos(\phi_{\text{AB}}).$$

Therefore, the extrema of universal conductance fluctuations are pinned to integer values of flux, $\phi_{\text{AB}} = \pi\Phi_0$. With changing dynamical phases, the minima can change to maxima and back.

To sum up the amplitudes of all processes, let us note that the total amplitude may be presented as follows:

$$t = \frac{1}{\sqrt{2}}(t_{\text{u}} + t_{\text{d}})\frac{1}{\sqrt{2}},$$

the square root factors corresponding to the entrance to and exit from the ring. The amplitude t_{u} describes all the processes when the electron first enters the upper arm of the ring; it does not matter what it does later. In the same way, t_{d} describes the processes when the electron first enters the lower arm of the ring.

What are the trajectories contributing to t_{u}? The simplest is when the electron enters the upper arm and then exits through the right beam splitter (Fig. 1.36(a)). A more sophisticated trajectory follows the upper arm clockwise (factor $e^{i(\phi_1+\chi_1)}$), is reflected back (factor $-1/2$), follows the upper arm counterclockwise (factor $e^{i(-\phi_1+\chi_1)}$), is reflected again ($-1/2$), and starts in the upper arm. Afterwards, it can do a lot of things: it can exit, or make more turns. It is important, however, that *the sum* of all these options is again t_{u} – the sum of all trajectories starting in the upper arm. Then, instead of being reflected into the upper arm, the electron can make it to the lower arm (factor $1/2$ rather than $-1/2$), and proceed in the lower arm (amplitude t_{d}). Finally, the third variant: the electron follows the upper arm clockwise (factor $e^{i(\phi_1+\chi_1)}$), is transmitted into the lower arm (1/2), and propagates through the lower arm clockwise ($e^{i(\phi_2+\chi_2)}$). Then it can either go to the upper arm ($(1/2) \cdot t_u$) or to the lower arm ($(-1/2) \cdot t_d$). This exhausts the possibilities. Summing them up, we obtain the following equation:

$$t_u = e^{i(\phi_1+\chi_1)} + \frac{1}{4}\left(e^{2i\chi_1} + e^{i(\chi_1+\chi_2+\phi_{AB})}\right)(t_u - t_d)\,. \tag{1.112}$$

Examining in the same way the trajectories starting in the lower arm, we obtain the second equation,

$$t_d = e^{i(\chi_2-\phi_2)} + \frac{1}{4}\left(e^{2i\chi_2} + e^{i(\chi_1+\chi_2-\phi_{AB})}\right)(t_d - t_u)\,, \tag{1.113}$$

whence

$$\begin{aligned} t &= \frac{t_u + t_d}{2} \\ &= \left(e^{i(\chi_1+\phi_1)} + e^{i(\chi_2-\phi_2)} + e^{i(\chi_1+2\chi_2+\phi_1)} + e^{i(2\chi_1+\chi_2-\phi_2)}\right) \\ &\quad \times \left(1 - \frac{e^{2i\chi_1} + e^{2i\chi_2} + 2e^{i(\chi_1+\chi_2)}\cos\phi_{AB}}{4}\right)^{-1}, \end{aligned} \tag{1.114}$$

and the conductance is given by $G = G_Q T$, $T = |t|^2$. A compact expression for a symmetric ring, $\chi_1 = \chi_2 = \chi/2$, is given by

$$G = G_Q \frac{(1 - \cos\chi)(1 + \cos^2\phi_{AB})}{\sin^2\chi + [\cos\chi - (1 + \cos\phi_{AB})/2]^2}\,. \tag{1.115}$$

Exercise 1.11. Find the transmission amplitude when both beam splitters are fully symmetric, Eq. (1.80).

Let us plot the resulting conductance versus flux. Figure 1.37 presents examples of such $2\Phi_0$-periodic curves for several values of $\chi_{1,2}$. This is to be contrasted with the conductance averaged over dynamical phases. As expected, it is clearly Φ_0-periodic.

More dependence of dynamical phases is presented in Fig. 1.38. There, the phases χ_1 and χ_2 are swept in opposite directions. As we see from Eq. (1.103), this can be realized by applying opposite electrostatic potentials to the upper and lower arms (electric field in vertical direction) by means of the gate electrodes. As expected, the extrema of the conductance remain at the same positions but change from maxima to minima and back. By a particular choice of phases, the first harmonic with the period $2\Phi_0$ can be tuned to zero, where the conductance is almost flux-independent.

1.6.4 Experiments on quantum interference

We will now describe three pioneering experiments in which quantum interference was first observed in electron transport. To avoid any misunderstanding, we stress that the

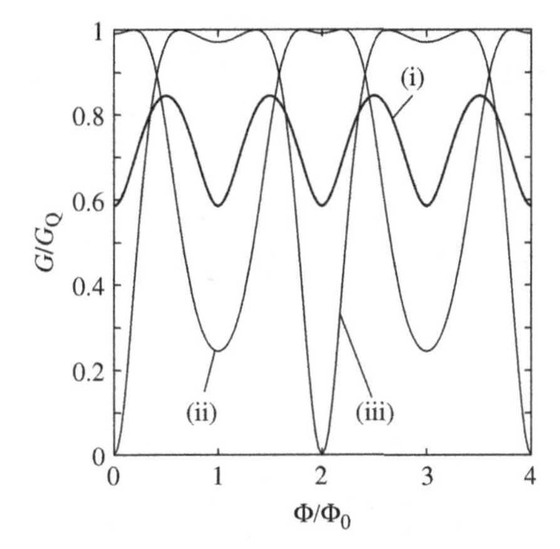

Fig. 1.37. Curve (i) presents the conductance of an AB ring averaged over dynamical phases $\chi_{1,2}$. It oscillates with period Φ_0. This is contrasted with non-averaged magnetoconductance curves ($\chi_1 = 4A$, $\chi_2 = 0.2$ for (ii) and $\chi_1 = -\chi_2 = 1$ for (iii)) which oscillate with period $2\Phi_0$.

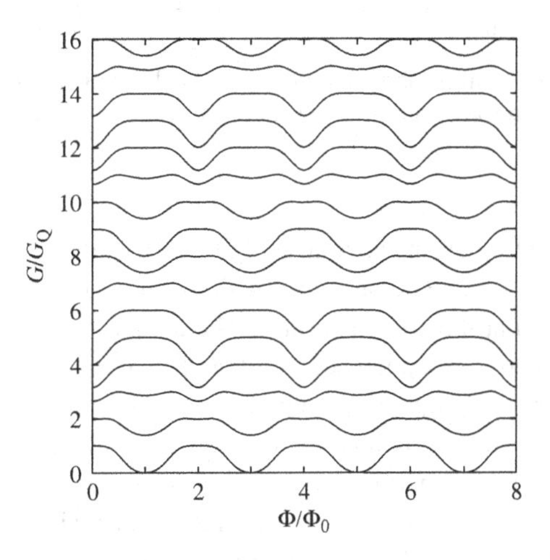

Fig. 1.38. Flux-dependent conductance of an AB ring at $\chi_{1,2} = 1.3 \mp \zeta$. Different curves correspond to ζ changing from 0 to 2π with step $\pi/8$. The curves are offset by conductance quantum for clarity. They display a characteristic pattern of extrema fixed at integer values of Φ/Φ_0 that change from minima to maxima and back when sweeping the dynamical phase.

simple one-channel models described above have very little to do with the actual experimental situation. All experiments have been performed for structures whose geometrical size exceeded the electron wavelength by many orders of magnitude. The conduction involved at least several thousand transport channels. At least in the first experiment, the electrons did not keep the quantum coherence throughout the length of the structure; inelastic processes forbade that. At this point, we cannot provide an adequate model for these experiments: the quantitative theory will be elaborated in subsequent chapters.

The interference in quantum transport is so wonderfully universal, however, that a qualitative understanding can be achieved with the models in hand. Let us see how it works.

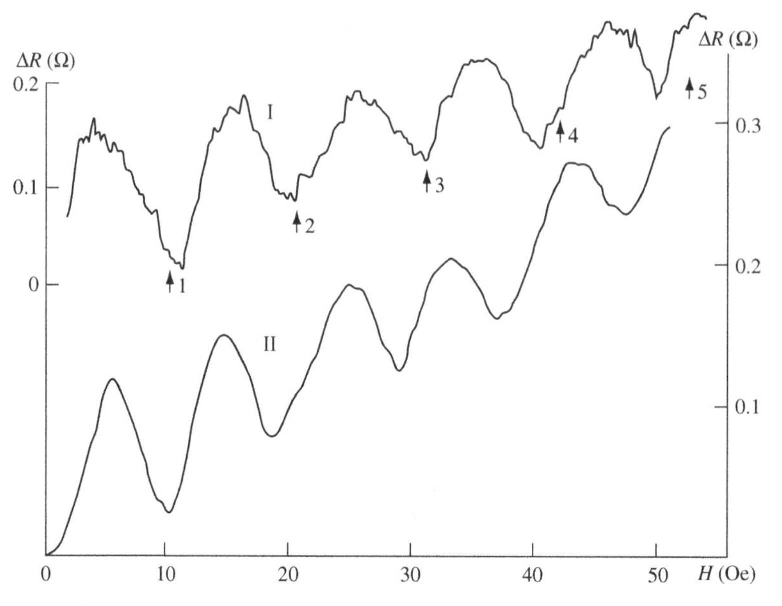

Fig. 1.39. The experiment by Sharvin and Sharvin [20] has proven the interference nature of the weak localization correction. The resistance of a metal cylinder was found to oscillate with period Φ_0 (see curve (i) in Fig. 1.37).

In 1981, Sharvin and Sharvin [20] fabricated thin metal cylinders by condensing magnesium vapor on a silicon thread with a diameter of 1500 nm. Magnesium formed a 100 nm thick film at the silicon surface. The authors estimated that electrons at low temperature could keep their coherence at a length scale L_ϕ of the order of the circumference of the cylinder. The length of the cylinder, $L = 1$ cm, was much bigger than this coherence length. The whole sample in this case can be viewed as $L/L_\phi \approx 300$ coherent conductors in series.

Sharvin and Sharvin applied a magnetic field parallel to the axis and found oscillatory dependence of the resistance (Fig. 1.39). The main period of this dependence corresponded to the flux Φ_0 via the cylinder cross-section.

The resistance of each coherent conductor contained an interference contribution that fluctuates depending on the dynamical phases. However, these contributions were added in series and the fluctuations were averaged out. The resistance change observed thus presented the *averaged* conductance change, called the weak localization correction.

In 1985, Webb and co-authors [21] fabricated a device with spatial dimensions smaller than L_ϕ: the first coherent conductor (Fig. 1.40). This simple device was made of 40 nm thick gold wires and contained a ring of diameter approximately 800 nm. The authors observed resistance oscillations with a twice bigger period $2\Phi_0$. This identifies the oscillations as "universal conductance fluctuations". They are different for nominally identical devices depending on dynamical phases.

The oscillations were not ideally periodic due to the finite thickness of the wires. The trajectory loops inside the wires had slightly different areas and thus were penetrated by slightly different flux. This produced the uncertainty of the magnetic field period.

In several years, the same group investigated an electrostatic AB effect [22]. They fabricated a similar device comprising antimony wires that form a square loop 820 nm on a

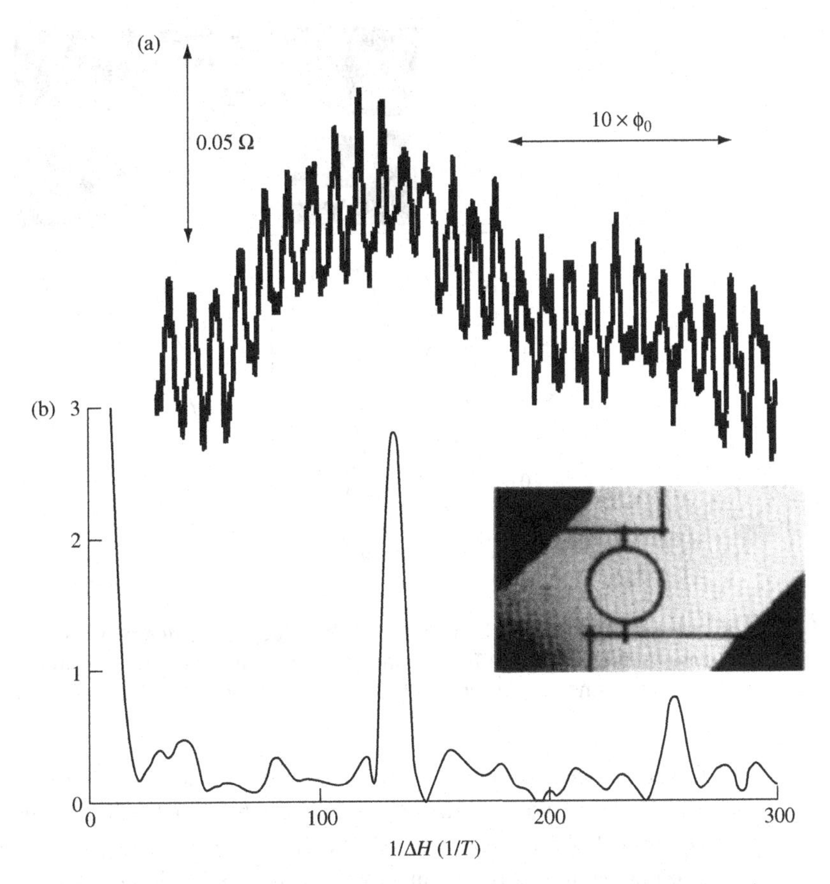

Fig. 1.40. **First observation of "universal conductance fluctuations" [21] in an AB device (inset). (a) Magnetoresistance oscillates with period $2\Phi_0$, while its average value would oscillate with a twice shorter period. (b) Fourier power spectrum of the oscillations contains peaks corresponding to both periods.**

side (Fig. 1.41). Gate electrodes were set close to two sides. They observed magnetoresistance oscillations with period $2\Phi_0$. They demonstrated that the gate voltage can be used to tune the positions of these oscillations: the phase is changed by π so that the positions of the minima correspond to former positions of maxima. The explanation is that the gate voltage provides the shift of dynamical phase in the arms of the loop. This was discussed in our model of the AB ring and presented by the curves plotted in Fig. 1.38. In fact, the effect of the gate voltage is strongly suppressed in this setup. The gate voltage would be screened out in an ideal conductor, and only a small fraction of it eventually reaches the electrons interfering in the loop. This is why a relatively high voltage difference is required for the phase shift.

Control question 1.16. What is the size of the square loop in Fig. 1.41(a) (experiment in Ref. [22]). Hint: see Fig. 1.41(b).

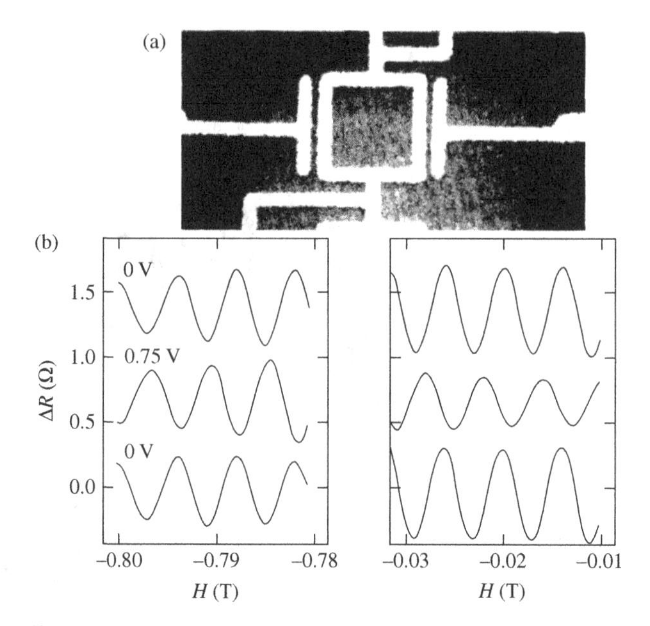

Fig. 1.41. Electrostatic AB effect [22]. (a) The conducting loop was gated with two capacitance probes (T-wires in the photo). (b) The minima of AB oscillations are changed to maxima by applying the gate voltage, since the latter affects dynamical phase of the electrons in the ring (see Fig. 1.38).

To summarize, the qualitative features of the experiments can be understood using our simple model of the one-channel AB ring, which includes periodicity and dependence on dynamical phases. Interestingly, the one-channel model even gives a correct estimation of the magnitude of the interference effect: a fraction of G_Q per coherent conductor. Experimentally, this estimation holds for much less resistive conductors, with $G \gg G_Q$ and consequently a large number of transport channels. The reasons for this are explained in Chapters 2 and 4.

1.6.5 Interference and combining scattering matrices

Real systems are more complicated than the simple models described above – they contain more transport channels and more scatterers. A scattering matrix of such a complex system is a combination of scattering matrices of individual scatterers and phase shifts acquired by an electron traveling between the scatterers (these phase shifts are different for different channels and form a special scattering matrix that is diagonal in the channel index). The general principle, which we illustrate with a number of examples in the following, is that output channels of one scatterer serve as input channels of others.

To characterize the transport, we may proceed as we did before: consider all possible transmission processes in which an electron starts in a reservoir in a certain channel, scatters to another channel at a scatterer, acquires a phase factor when getting to next scatterer, and repeats this as many times as required to reach the same or another reservoir. The amplitude of each process is again the product of the partial amplitudes, and we have to

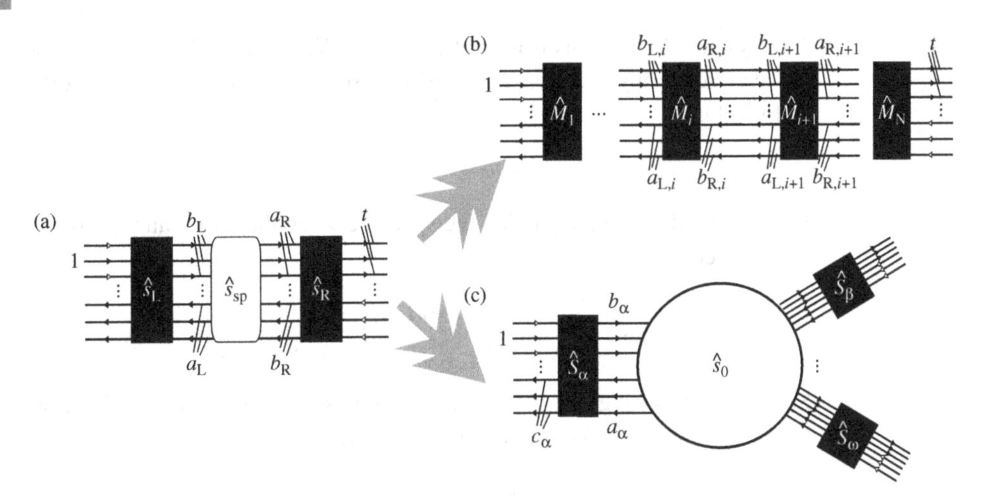

Fig. 1.42. **Interference results from combining scattering matrices. Three examples of combining considered in this section: (a) general double-junction structure; (b) stack of scatterers considered with transfer matrices; (c) a node allowing for multi-terminal geometry.**

sum up all processes that begin in a certain channel α and end in a certain channel β. In this way, we obtain the element $s_{\alpha\beta}$ of the whole scattering matrix. However, this is hardly practical since the number of possible processes to consider increases exponentially with the increase of different scattering possibilities. This makes the bookkeeping difficult. To overcome these difficulties, we return to the scattering approach and learn how to combine scattering matrices of individual scatterers into the scattering matrix of the whole structure.

The general solution of this problem is too involved and hardly instructive, so we consider below only three specific examples. First, we consider the general multi-channel double-junction setup. One can complicate this setup in two ways (Fig. 1.42). One can stack more scatterers in series, or one can add more terminals and scatterers to the space between the scatterers, making it a *node*. These three examples will reappear many times in this book.

Let us start with the general double junction. Now each of the scatterers is characterized by a scattering matrix with the block structure given in Eq. (1.34). Consider an electron incident from the left of the left scatterer in channel n; the amplitude is 1 in this channel and 0 in all other channels. Our goal is to determine the amplitudes t_m in the outgoing channels on the right side (see Fig. 1.42(a)). We consider the auxiliary amplitudes: those in the channels leaving the left scatterer to the right ($b_{\mathrm{L}m}$) or leaving the right scatterer to the left ($b_{\mathrm{R}m}$) and those in the channels coming to the left scatterer from the right ($a_{\mathrm{L}m}$) or to the right scatterer from the left ($a_{\mathrm{R}m}$).

These amplitudes are related via scattering matrices of the scatterers, i.e.

$$
\begin{aligned}
b_{\mathrm{L}m} &= t^{\mathrm{L}}_{mn} + \sum_l r'^{\mathrm{L}}_{ml} a_{\mathrm{L}l}; \\
b_{\mathrm{R}m} &= \sum_l r^{\mathrm{R}}_{ml} a_{\mathrm{R}l}; \quad t_m = \sum_l t^{\mathrm{R}}_{ml} a_{\mathrm{R}l}.
\end{aligned}
\tag{1.116}
$$

This system of equations is not yet complete: We need to relate the amplitudes a_m and b_m. Since no scattering is assumed between the scatterers, these amplitudes differ by the phase shifts only:

$$a_{\mathrm{R}m} = \mathrm{e}^{\mathrm{i}\chi_m} b_{\mathrm{L}m},\ a_{\mathrm{L}m} = \mathrm{e}^{\mathrm{i}\chi_m} b_{\mathrm{R}m}. \tag{1.117}$$

In other words, we can say that the space between the two scatterers is also described by the "scattering matrix,"

$$\hat{s}_{\mathrm{sp}} = \begin{pmatrix} 0 & \exp(\mathrm{i}\hat{\chi}) \\ \exp(\mathrm{i}\hat{\chi}) & 0 \end{pmatrix}, \quad \left(\mathrm{e}^{\mathrm{i}\chi}\right)_{mn} = \mathrm{e}^{\mathrm{i}\chi_m}\delta_{mn}. \tag{1.118}$$

Solving the resulting system of linear equations, we find t_m. Identifying it with the element t_{nm} of the overall transmission matrix, we write

$$\hat{t} = \hat{t}_{\mathrm{R}}\mathrm{e}^{\mathrm{i}\hat{\chi}}\left(\hat{1} - \hat{r}'_{\mathrm{L}}\mathrm{e}^{\mathrm{i}\hat{\chi}}\hat{r}_{\mathrm{R}}\mathrm{e}^{\mathrm{i}\hat{\chi}}\right)^{-1}\hat{t}_{\mathrm{L}}. \tag{1.119}$$

For one channel, Eq. (1.119) reduces to Eq. (1.105).

We learn from this result that the rules for combining scattering matrices are not simple: the scattering matrix of a compound object is neither a product nor a sum of the scattering matrices of its components; rather, it is a cumbersome combination. If we increase the number of constituents, the complexity of the resulting expression increases beyond any reasonable level. This provides a strong motivation for developing a semiclassical approach to combining the multi-channel scattering matrices; as we show in the following chapters, it is simpler and more intuitive.

Transfer matrices

One can add more scatterers to a double-junction system, stacking them in series. There is a convenient trick we can use to find the resulting scattering matrix of the stack: the *transfer matrix* technique.

Let us concentrate on one scatterer in the stack. It is fully described by either the scattering or the transfer matrix. While its scattering matrix relates the amplitudes of outgoing waves to the amplitudes of incoming waves (see Eq. (1.33)), its transfer matrix relates the amplitudes of the waves right of the scatterer to the amplitudes of the waves left of the scatterer – it "transfers" an electron across the scatterer from the left to the right.

To define it formally, we introduce four vectors of amplitudes: those incoming from the left (right) $\boldsymbol{a}_{\mathrm{L(R)}}$ and those going out to the left (right) $\boldsymbol{b}_{\mathrm{L(R)}}$. In these notations, the definition of the scattering matrix $\hat{s}$, Eq. (1.33), takes the following form:

$$\begin{pmatrix} \boldsymbol{b}_{\mathrm{L}} \\ \boldsymbol{b}_{\mathrm{R}} \end{pmatrix} = \hat{s}\begin{pmatrix} \boldsymbol{a}_{\mathrm{L}} \\ \boldsymbol{a}_{\mathrm{R}} \end{pmatrix}, \tag{1.120}$$

and the transfer matrix $\hat{M}$ is given by

$$\begin{pmatrix} \boldsymbol{b}_{\mathrm{R}} \\ \boldsymbol{a}_{\mathrm{R}} \end{pmatrix} = \hat{M}\begin{pmatrix} \boldsymbol{a}_{\mathrm{L}} \\ \boldsymbol{b}_{\mathrm{L}} \end{pmatrix}. \tag{1.121}$$

To find the relation between the matrices, we introduce the block structure in $\hat{M}$, similar to those in $\hat{s}$:

$$\hat{M} = \begin{pmatrix} \hat{m}_1 & \hat{m}_2 \\ \hat{m}_3 & \hat{m}_4 \end{pmatrix}. \tag{1.122}$$

In Eq. (1.120) we re-express $\boldsymbol{a}_\mathrm{R}, \boldsymbol{b}_\mathrm{R}$ in terms of $\boldsymbol{a}_\mathrm{L}, \boldsymbol{b}_\mathrm{L}$ to obtain

$$\begin{aligned} \hat{m}_1 &= \hat{t} - \hat{r}'\hat{t}'^{-1}\hat{r} = \left(\hat{t}^\dagger\right)^{-1}; \quad \hat{m}_2 = \hat{r}'\hat{t}'^{-1}; \\ \hat{m}_3 &= -\hat{t}'^{-1}\hat{r}'; \quad \hat{m}_4 = \hat{t}'^{-1}. \end{aligned} \tag{1.123}$$

Reversed relations express $\hat{t}, \hat{t}', \hat{r}$, and $\hat{r}'$ in terms of $\hat{m}_{1-4}$. We no longer need their explicit form.

Let us look at the next scatterer to the right. The incoming (outgoing) waves on its left are, in fact, outgoing (incoming) waves on the right of the previous scatterer, $\boldsymbol{b}_\mathrm{R} \to \boldsymbol{a}_\mathrm{L}$, $\boldsymbol{a}_\mathrm{R} \to \boldsymbol{b}_\mathrm{L}$. Then it follows directly from Eq. (1.121) that the matrix product of transfer matrices of these two scatterers transfers the electron from the left to the right of the two. This is valid in general – the transfer matrix of several objects in series is just a product of their transfer matrices. For example, the transfer matrix of a double junction is a product of the transfer matrices of the left scatterer, of the space between the scatterers (to be obtained from the matrix $\hat{s}_\mathrm{sp}$), and of the right scatterer. This simple combining property makes the transfer matrices indispensable.

The scattering matrix is unitary as a consequence of the conservation of the number of particles. The transfer matrix is not unitary, but it obeys a constraint that expresses the conservation of flux. On the left of a scatterer, the flux I is given by

$$I = \begin{pmatrix} \boldsymbol{a}_\mathrm{L}^* \\ \boldsymbol{b}_\mathrm{L}^* \end{pmatrix} \hat{\sigma}_z \begin{pmatrix} \boldsymbol{a}_\mathrm{L} \\ \boldsymbol{a}_\mathrm{L} \end{pmatrix}, \quad \hat{\sigma}_z = \begin{pmatrix} 1 & 0 \\ 0 & -1 \end{pmatrix}, \tag{1.124}$$

where $\hat{\sigma}_z$ is the Pauli matrix that labels left-going amplitudes with "1" and right-going amplitudes with "-1." The flux must be the same on the right of the scatterer, where the amplitudes are transformed with $\hat{M}$. This implies that

$$\hat{M}^\dagger \hat{\sigma}_z \hat{M} = \hat{M} \hat{\sigma}_z \hat{M}^\dagger = \hat{\sigma}_z. \tag{1.125}$$

We have learned in previous sections that the transport properties can be easily expressed in terms of T_p, eigenvalues of transmission matrix square $\hat{t}^\dagger\hat{t}$. It is advantageous to find these eigenvalues directly from the transfer matrix without calculating $\hat{t}$ explicitly. For this purpose, consider the Hermitian matrix $\hat{M}^\dagger\hat{M}$. Its eigenvalues M_p are real, positive numbers. It follows from the conservation law, Eq. (1.125), that if M_p is an eigenvalue, $1/M_p$ is also an eigenvalue. To prove this, assume that $\boldsymbol{u}_p$ is an eigenvector corresponding to the eigenvalue M_p, $\hat{M}^\dagger\hat{M}\boldsymbol{u}_p = M_p\boldsymbol{u}_n$. From Eq. (1.125), for any $\boldsymbol{u}_p$ we have $\hat{M}^\dagger\hat{M}\sigma_z\boldsymbol{u}_p = \sigma_z(\hat{M}^\dagger\hat{M})^{-1}\boldsymbol{u}_p$. Since $\boldsymbol{u}_p$ is an eigenvector of $\hat{M}^\dagger\hat{M}$, it is also an eigenvector of the matrix $(\hat{M}^\dagger\hat{M})^{-1}$ with the eigenvalue $1/M_p$. We thus conclude that the vector $\sigma_z\boldsymbol{u}_p$ is an eigenvector of the matrix $\hat{M}^\dagger\hat{M}$ with the corresponding eigenvalue $1/M_p$.

Exercise 1.12. Derive Eq. (1.119) using the transfer matrix approach and find the transmission eigenvalues T_p.

Furthermore, using Eqs. (1.123), we can prove the following relation between squares of transmission and transfer matrices:

$$\left(\hat{M}^\dagger\hat{M}+\left(\hat{M}^\dagger\hat{M}\right)^{-1}+2\right)^{-1}=\frac{1}{4}\begin{pmatrix}\hat{t}^\dagger\hat{t} & 0\\ 0 & \hat{t}'\hat{t}'^\dagger\end{pmatrix}. \tag{1.126}$$

This implies the following relation between the eigenvalues:

$$T_p=\frac{4}{M_p+2+1/M_p}. \tag{1.127}$$

Scatterers connected to a node

Another way to sophisticate the double-junction setup is to add more terminals to the space between the scatterers. This is the only way to proceed if we wish to model multi-terminal nanostructures. The space between the scatterers becomes a common node connected to all terminals. We will assume that each terminal is separated from the node by a corresponding scatterer. Let us label the terminals by Greek indices. Each scatterer is described by its own scattering matrix $\hat{s}^\alpha$. The node – the space between the scatterers – is described by the scattering matrix $\hat{s}_0$. It has the same function as the matrix $\hat{s}_{\text{sp}}$ for the double-junction system. An important distinction is that we cannot find a natural diagonal form for $\hat{s}_0$. This was possible for $\hat{s}_{\text{sp}}$ because we can ensure that each channel that starts from the left scatterer ends at the right one. Now each channel that starts from terminal α can end at any other terminal or get back. This situation can only be described with a general unitary matrix, $\hat{s}_0$.

Our goal is to calculate the scattering matrix of the compound system: the node with the scatterers.

We proceed very much in the same way as for a double junction. Consider an electron incident in lead α in transport channel n. For each lead β and each transport channel m, we look at three amplitudes: that in the channel reflected outside, $c_{\beta m}$, that coming from inside, $a_{\beta m}$, and that going inside, $b_{\beta m}$. From the definition of the scattering matrix, we have

$$\begin{aligned} c_{\beta m} &= \sum_l t'^{\beta}_{ml}a_{\beta l}+\delta_{\alpha\beta}r^{\beta}_{mn};\\ b_{\beta m} &= \sum_l r'^{\beta}_{ml}a_{\beta l}+\delta_{\alpha\beta}t^{\beta}_{mn}. \end{aligned} \tag{1.128}$$

To complete the system, we need to relate $a_{\beta m}$ and $b_{\beta m}$. This is provided by the scattering matrix of the node,

$$a_{\beta m}=\sum_{\gamma l} s^0_{\beta m,\gamma l}b_{\gamma l}. \tag{1.129}$$

Now, the matrices $\hat{r}^\alpha$ and $\hat{t}^\alpha$ are defined in the space of transport channels of the contact α, whereas the matrix $\hat{s}_0$ describes *all* transport channels in *all reservoirs*. In order to deal with the matrices of the same dimensionality, we define the "big" matrices

$$\begin{aligned} r_{\alpha n,\beta m} &= \delta_{\alpha\beta} r^{\alpha}_{mn}; \quad r'_{\alpha n,\beta m} = \delta_{\alpha\beta} r'^{\alpha}_{mn}; \\ t_{\alpha n,\beta m} &= \delta_{\alpha\beta} t^{\alpha}_{mn}; \quad t'_{\alpha n,\beta m} = \delta_{\alpha\beta} t'^{\alpha}_{mn}, \end{aligned} \tag{1.130}$$

and solve linear equations (1.128) and (1.129) to find

$$c_{\beta m} = r_{\beta m,\alpha n} + \left[\hat{t}'\hat{s}_0\left(1 - \hat{r}'\hat{s}_0\right)^{-1}\hat{t}\right]_{\beta m,\alpha n}.$$

Identifying this with the corresponding elements of the scattering matrix, we find finally that

$$\hat{s} = \hat{r} + \hat{t}'\hat{s}_0\left(1 - \hat{r}'\hat{s}_0\right)^{-1}\hat{t}. \tag{1.131}$$

Exercise 1.13. Consider a beam splitter, Eq. (1.80), connected by three leads intercepted by identical one-channel junctions with the scattering matrix in Eq. (1.39). (i) Using Eq. (1.131), find the full scattering matrix of the nanostructure; (ii) sketch the energy dependence of the transmission probability $|t_{12}(E)|$; (iii) using the multi-terminal Landauer formula, calculate the conductance matrix.

1.7 Time-dependent transport

We have studied in detail the dc electron transport in nanostructures and have understood that it is determined by the voltages applied to the leads and the scattering matrix of the nanostructure. If there is no possibility of changing the scattering matrix without making a completely new nanostructure, quantum transport would be an extremely boring field, at least from an experimental point of view.

Fortunately, such possibilities exist, and most frequently they are realized with *gates* – bulk metallic electrodes that are electrically disconnected from the scattering region as well as from the leads. The electrostatic potential of the gates can thus be varied independently of that of the leads. We have already seen in Section 1.2 that a dc voltage applied to the gates may change the width of a quantum point contact, thus affecting the number of open channels. Generally, the gates affect the scattering matrix of a nanostructure. It is very handy that the gates are electrically disconnected: there are no dc currents to the gates, only to the leads. If there is no voltage difference between the leads, applying the voltage difference to the gate does not drive the nanostructure out of equilibrium.

While a gate is not coupled to the scattering region electrically, it is always coupled *capacitively*. The voltage on the gate induces some stationary charge distribution, both in the nanostructure and the leads. If the leads and the scattering region were ideal conductors, the charge would be accumulated in infinitesimally thin layer at the surface. This layer would screen the electric field of the gate inside the conductors and there would be no effect on the scattering properties. In reality, the charge induced is spread over finite width,

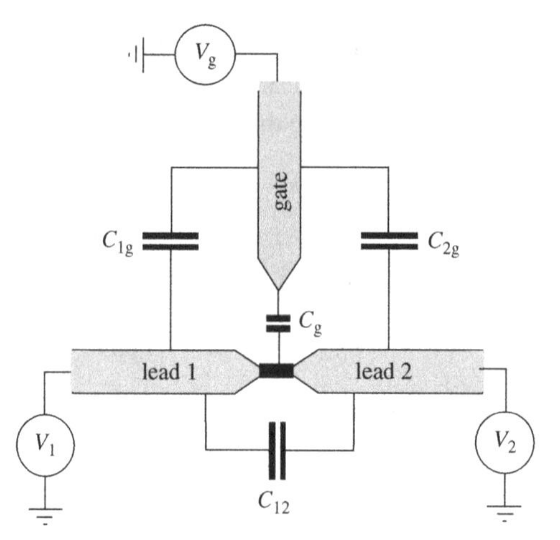

Fig. 1.43. Nanostructure with two leads and a gate. The large external capacitances C_{12}, C_{1g}, and C_{2g} provide paths for ac current, shunting the nanostructure at sufficiently high frequencies.

and the screening is not perfect. For bulk leads, this is irrelevant. For a sufficiently small nanostructure, a penetrated electric field changes the scattering matrix. A convenient model is that of a double junction. From all the charge induced by the gate, only a small part q accumulated between the scatterers can affect the transport. This charge is proportional to the gate potential V_g with respect to the ground, $q = -C_g V_g$, assuming grounded leads. The coefficient C_g is the *capacitance* between the nanostructure and the gate. Since the leads and the gate are bulk and the nanostructure is small, the charge induced in the leads and the corresponding capacitances usually greatly exceed C_g (Fig. 1.43).

One tends to under-appreciate the ultimate convenience provided by a dc electric measurement in the investigation of nanostructures. The nanostructure size is negligible in comparison with the dimensions of the gates and leads. Nevertheless, we are sure that the dc current goes through the nanostructure and is completely defined by its scattering properties. Full appreciation is only achieved when we understand that it is no longer convenient if, instead of dc voltage, we apply ac voltage and try to measure ac current.

The point is that one cannot measure an ac *particle* current; rather, one measures electric current. At finite frequency all capacitors become conductors. A *displacement* current $I = \dot{Q} = C\dot{V}$ flows between the plates of each capacitor. If we look at Fig. 1.43, we see that the *external* capacitances – those between the leads, and between the leads and the gate – are large, matching the large size of these conductors. At a sufficiently high frequency, the displacement currents dominate, and the current goes through the capacitors, completely bypassing the nanostructure. Given the dc conductance G of the nanostructure, the frequency is estimated by comparison of displacement and particle current, $GV \simeq \omega CV$, $\omega_{ext} \simeq G/C$. To make this practical, we take $C \simeq 10^{10}$ F corresponding to the electrodes of a meter scale and $1/G \equiv 1\,\mathrm{k}\Omega$ to estimate $\omega_{\mathrm{ext}} \simeq 10^7$ Hz. This frequency is low at quantum transport scale: the conductance of the nanostructure stays the same as at zero frequency. It would be interesting to access the frequency dependence of the nanostructure conductance – this would supply more detailed information about the quantum transport. However, the above reasoning shows that this is not possible. The measured frequency dependence is determined by the inter-lead capacitances.

1.7.1 The ac current response

The macroscopic capacitances mentioned above can, in principle, be characterized accurately. This allows us, in principle, to single out the small frequency-dependent response of the nanostructure proper. Can one evaluate this response using the scattering matrix only? A short answer is that the scattering approach only works if the frequency scale is small in comparison with the scale E_i that sets the energy dependence of the scattering matrix. For few-channel nanostructures, this energy scale can be readily associated with a typical time that the electrons spend traversing the nanostructure. Indeed, if we recall our favorite model of two scatterers in series (Section 1.6), the energy dependence scale is set by the inverse time of flight between the scatterers. At frequencies below this scale, the response does not depend on frequency and corresponds to the conductance given by the (multi-terminal) Landauer formula. It is intuitively clear that if the frequency of the electric field that drives the electrons through the nanostructure is higher than the inverse time of flight, the driven electrons do not traverse the nanostructure; instead they oscillate inside without getting to the leads.

One might think that it is enough to solve the time-dependent Schrödinger equation inside the nanostructure with the time-dependent potential incorporating the high-frequency driving field, and consequently find the high-frequency particle current. It is far more complicated than this, however. At a low frequency, we have frequently been helped by the fact that the particle currents through any cross-section of the nanostructure are the same and equal the electric currents. This is not automatically guaranteed at a finite frequency. The electrons could have been accumulated between two cross-sections. The conservation of the number of electrons, the continuity equation,

$$\dot{\rho}(\boldsymbol{r},t) + \operatorname{div}\boldsymbol{j}(\boldsymbol{r},t) = 0,$$

implies that $J_2(\omega) - J_1(\omega) = -\mathrm{i}\omega N(\omega)$, where $J_{1,2}$ are Fourier components of the particle currents through the cross-sections 1,2 and $N(\omega)$ is the Fourier component of the number of electrons accumulated between the cross-sections. This is not the only problem. At a low frequency, the particle current did not depend on the details of the voltage distribution across the nanostructure. Instead, it was completely defined by the overall voltage drop. At a finite frequency, different voltage distributions with the same overall drop will result in different particle currents.

Both problems are solved by careful consideration of *capacitive* response in the nanostructure [23]. The accumulation of particles is the accumulation of charge; this accumulated charge produces the electric field that tries to suppress the particle/charge accumulation. This brings about yet another frequency scale, the inverse RC-time of the structure, $\tau_{RC} \simeq C_g/G$. We stress that $C_g \ll C$, and thus the defined frequency scale greatly exceeds ω_{ext}.

Control question 1.17. What is this scale for a nanostructure of micrometer size and $1/G = 1\,\mathrm{k}\Omega$?

At frequencies much less than $1/\tau_{RC}$, the charge accumulation is strongly suppressed, and the particle current is the same in all cross-sections and equals the electric current. The distribution of voltage in the nanostructure is adjusted to ensure the absence of charge accumulation. If the frequencies are still smaller than the energy scale E_i, the resulting current response conforms to the scattering matrix result. Otherwise, the scattering approach fails in general. At frequencies much greater than $1/\tau_{RC}$, the capacitive response dominates: the electric currents mainly go via capacitors between the different parts of the nanostructure as well as via capacitors to the gates and leads. Substantial current goes to the gate electrode(s).

The frequency dependence of particle current at $\omega \ll 1/\tau_{RC}$ can be commonly regarded as an *inductive* response. To illustrate this, let us consider free electrons subject to an external uniform electric field $\boldsymbol{E}(t)$. The field accelerates the electrons so that their velocities are $\dot{\boldsymbol{v}} = e\boldsymbol{E}/m$, resulting in a current proportional to $\boldsymbol{E}/\mathrm{i}\omega$. This is kinetic inductance, to be distinguished from geometric inductance due to the time dependence of the magnetic field generated by the current. We can readily apply this to a quantum point contact with N_{open} open channels to obtain $I(\omega) = G_{\mathrm{Q}} N_{\mathrm{open}} v_{\mathrm{F}}/\mathrm{i}\omega L$, L being the constriction length. We see that the inductive response approaches the order of conductance at $\omega \simeq v_{\mathrm{F}}/L$; this frequency scale is indeed related to the typical traversal time L/v_{F}. The above estimation is valid for quasiballistic nanostructures, in which electrons experience few scatterings. The estimation is modified if there are many scatterings: $G_{\mathrm{Q}} N_{\mathrm{open}}$ is replaced by the nanostructure conductance G, and L is replaced by a typical distance l traveled by the electron between two scatterings (the mean free path). In this case, the inductive response only becomes important at a frequency scale corresponding to the inverse of scattering time v_{F}/l. We stress that in this case the scattering time is much shorter than the traversal time through the whole structure. Thus the frequency response of nanostructures with many scatterings, in particular diffusive ones, exhibit no features at the inverse traversal time. This is in line with the absence of the energy dependence of the scattering matrix at the corresponding energy scale.

The above reasoning can be summarized in a simple model describing the nanostructure made of resistances, inductances, and capacitances to the gate and the leads (Fig. 1.44). The ac voltage drops over the resistors, and inductors model the voltage distribution over the nanostructure. An interesting model situation that we make use of is when the voltage difference between the points of the nanostructure and one of the leads is negligible, so the voltage drops between the other lead and the last scatterers of the nanostructure. As an example, let us take a two-junction one-channel nanostructure and choose $T_{\mathrm{L}} \ll T_{\mathrm{R}} \ll 1$: the voltage drops mainly at the left scatterer, while the voltage in the middle approximately equals V_{R}. The interesting aspect is that in this case the nanostructure can be treated within the scattering approach. Since there are no voltage drops inside, the electrons traversing the nanostructure do not experience an ac field and thus propagate at the same energy. However, the scattering matrix may exhibit a significant energy dependence.

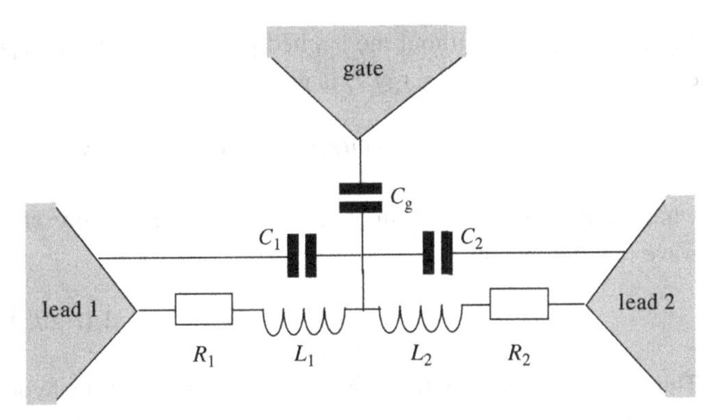

Fig. 1.44. **Simple circuit illustrating ac responses of a nanostructure: resistive, inductive, and capacitive.**

The reasoning of this subsection eventually discourages the detailed investigation of ac current response in quantum transport. It is difficult to quantify this response, theoretically as well as experimentally. If the difficulty is overcome, the result is most likely to be expressed by means of a simple circuit, as in Fig. 1.44, with no interesting physics.

It is much more interesting to study a dc transport in the presence of time-dependent voltages and/or frequency-dependent noise. In both cases, the electric measurement is performed at low frequency so that it takes no effort to single out the nanostructure response. However, these phenomena are essentially non-linear since, in linear approximation, the ac drive causes an ac response only. The phenomena and situations that occur in time-dependent quantum transport are too many and diverse to be described within this book. We concentrate in what follows on three concepts that, on one hand, are often encountered in various circumstances, and, on the other hand, are of general importance.

1.7.2 Tien–Gordon effect

In 1963, Tien and Gordon [24] put forward a simple, intuitive, and unusually practical description of quantum transport that is valid for a variety of two-terminal nanostructures biased simultaneously by dc and ac voltages. This description relates the dc current in the presence of ac voltage modulation with frequency Ω to I–V curves $I(V)$ of the same nanostructure in the *absence* of ac modulation:

$$I_{\mathrm{dc}}(V) = \sum_l p_l I\left(V + \hbar\Omega l/e\right). \tag{1.132}$$

Here the coefficients p_l depend on the amplitude and shape of the modulation. For a simple harmonic signal $V_{\mathrm{ac}}(t) = \tilde{V}\sin\Omega t$, these read $p_l = J_l^2(e\tilde{V}/\hbar\Omega)$. The I–V curve with modulation is thus a linear superposition of dc voltage I–V curves shifted by quantized voltages $\hbar\Omega l/e$ with coefficients depending on the ac power.

Let us concentrate first on one lead only (for instance, the left one) and forget about electron transfers to another lead. Let us apply a time-dependent potential $V(t)$ to the lead. If we neglect all other electrodes, the potential is *spatially uniform*, i.e. the same in all points of the lead. Without the applied potential, the time-dependent wave function of an electron state with the energy E is given by

$$\Psi(\boldsymbol{r}, t) = \exp\left(-\frac{\mathrm{i}}{\hbar}Et\right)\psi_E(\boldsymbol{r}), \tag{1.133}$$

whatever the complicated function ψ_E may be. In the presence of the potential, the same wave function becomes

$$\Psi(\boldsymbol{r}, t) = \exp\left(-\frac{\mathrm{i}}{\hbar}Et - \frac{\mathrm{i}e}{\hbar}\int^t V(t')\mathrm{d}t'\right)\psi_E(\boldsymbol{r}). \tag{1.134}$$

Thus, the uniform potential does not modify the coordinate dependence of any state, only adding the extra *phase* to the usual term $Et/\hbar$ in its wave function.

We stress that this extra phase is the same for all electron states in the lead and thereby it produces strictly no physical effect if we forget about electron transfers. Since the phase is not a gauge-invariant quantity, we could easily produce such shifts by choosing a different gauge. This is not surprising: a potential that is constant in space cannot affect the electron motion.

Let us now assume that the time-dependent potential is periodic and let us expand the resulting wave function in Fourier series. As an example, we take $V(t) = V + \tilde{V}\sin\Omega t$, a constant plus a single harmonic. Using

$$\exp\left(\frac{\mathrm{i}eW}{\hbar\Omega}\cos\Omega t\right) = \sum_{l=-\infty}^{\infty} a_l \exp(-\mathrm{i}l\Omega t), \quad a_l = J_l\left(\frac{e\tilde{V}}{\hbar\Omega}\right),$$

where J_l is the Bessel function of lth order, we find

$$\Psi(\boldsymbol{r}, t) = \sum_{l=-\infty}^{\infty} a_l \exp\left(-\frac{\mathrm{i}}{\hbar}(E + eV - \hbar\Omega l)t\right)\psi_E(\boldsymbol{r}). \tag{1.135}$$

The wave function given by Eq. (1.135) is made up of components with discrete energies $\varepsilon = E + eV - \hbar l\Omega$.

The intensities of the components are given by the squares of Fourier amplitudes, and the energy distribution is given by

$$P_E(\varepsilon) = \sum_l p_l \delta(\varepsilon - E - eV + \hbar l\Omega), \tag{1.136}$$

to be compared (Fig. 1.45) with $P_E(\varepsilon) = \delta(\varepsilon - E)$ without the potential. The electron can thus be found in a set of discrete energy bands labeled l; these are called *side bands* provided $l \neq 0$. The probabilities $p_l = |a_l|^2$, given by the square of the Bessel function, are even in l and normalized:

$$\sum_{l=-\infty}^{\infty} p_l = 1. \tag{1.137}$$

Note that Eqs. (1.135) and (1.137) are not restricted to the case of harmonic potential $V(t)$: they retain the same form for any periodic potential, with the only difference that the probabilities p_l are more complicated functions of the parameters characterizing the potential.

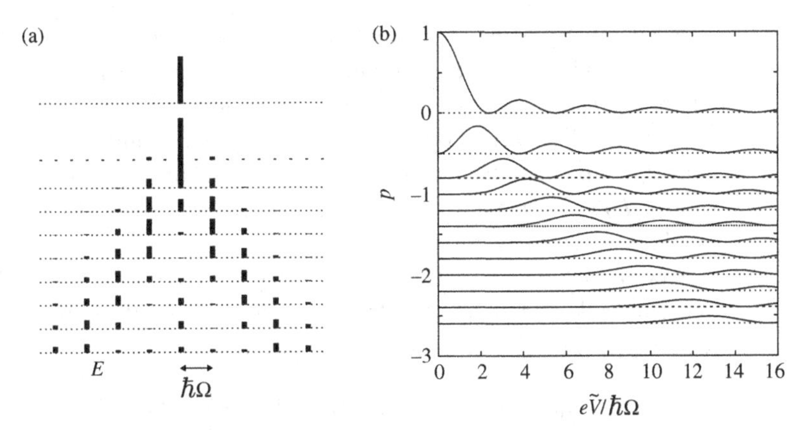

Fig. 1.45. **Development of side bands in the presence of ac voltage. (a) Side band weights p_l for $e\tilde{V}/\hbar\Omega$ ranging from 0 to 4.5 with step 0.5. (b) First 12 p_l plotted versus $e\tilde{V}/\hbar\Omega$; oscillations are clearly visible. The curves are offset for clarity.**

These side bands do not give rise to any physical effect unless we include the electron transfers to the right lead not subjected to any potential. We now turn to electron tunneling thus assuming $T_p \ll 1$. In this case, the electron transfers in either direction, channel, and energy interval are rare independent events. As in Section 1.4, the corresponding statistics in the low-frequency limit are generated by the characteristic function $\Lambda(\chi)$,

$$\ln \Lambda_{\Delta t}(\chi) = \Delta t \left\{ \left(\mathrm{e}^{\mathrm{i}\chi} - 1 \right) \Gamma_{\mathrm{LR}} + \left(\mathrm{e}^{-\mathrm{i}\chi} - 1 \right) \Gamma_{\mathrm{RL}} \right\}. \tag{1.138}$$

Here $\Gamma_{\mathrm{LR(RL)}}$ are the tunneling rates from the left to the right (from the right to the left) given, for dc voltage bias, by

$$\Gamma_{\mathrm{LR}} = 2_{\mathrm{s}} \sum_p \int T_p(E) f_{\mathrm{L}}(E)(1 - f_{\mathrm{R}}(E)) \frac{\mathrm{d}E}{2\pi\hbar};$$

$$\Gamma_{\mathrm{RL}} = 2_{\mathrm{s}} \sum_p \int T_p(E) f_{\mathrm{R}}(E)(1 - f_{\mathrm{L}}(E)) \frac{\mathrm{d}E}{2\pi\hbar}.$$

The filling factor is given by the equilibrium Fermi function in the right lead, $f_{\mathrm{R}}(E) = f_{\mathrm{F}}(E)$, and is shifted by the voltage bias in the left lead $f_{\mathrm{L}}(E) = f_{\mathrm{F}}(E - eV)$. Here and in the following we make a rather specific assumption concerning the transmissions: if they depend on energy, this energy is always counted from the chemical potential in the right electrode. This means that the contact is electrostatically coupled to the right lead only, so that the voltage applied to the left lead does not change the shape of the scattering potential in the contact.

What changes if we add an ac component to the applied voltage? Consider Γ_{RL} first. Electrons are transferred from the filled states on the right to the empty states on the left. The overall transfer probability at a given energy in channel p is thus a product of transmission probability $T_p(E)$, the fraction of filled states on the right ($f_{\mathrm{R}}(E)$), and the fraction of empty states on the left. The latter would be just $1 - f_{\mathrm{L}}(E)$ in the dc case. Now it is more complicated. On the left, the electron can end up in any discrete energy band with index l. By doing so, it would absorb ($l < 0$) or emit ($l > 0$) l energy quanta $\hbar\Omega$: photons forming the ac electric field. This is why the Tien–Gordon effect is sometimes referred as *photon-assisted tunneling*. The chance of reaching a certain band is given by p_l, and the filling factor corresponding to this band is shifted by energy $eV - \hbar\Omega l$. Summing up all probabilities, we find that the fraction of empty states is given by $1 - \tilde{f}_{\rm L}(E)$ with $\tilde{f}_{\rm L}(E) = \sum_l p_l f_{\rm F}(E - eV + \hbar\Omega l)$. The same reasoning for $\Gamma_{\rm LR}$ also gives that $f_{\rm L}(E)$ should be replaced by $\tilde{f}_{\rm L}(E)$.

Now we use the characteristic function given by Eq. (1.138) with the rates modified to calculate dc current and low-frequency noise. Since the only effect of ac voltage is to replace $f_{\rm L}(E)$ by $\tilde{f}_{\rm L}(E)$, the average dc current becomes

$$\begin{aligned} I_{\rm dc} &= \frac{G_{\rm Q}}{e}\sum_p \int {\rm d}E\; T_p(E)\left[\tilde{f}_{\rm L}(E) - f_{\rm R}(E)\right] \\ &= \frac{G_{\rm Q}}{e}\sum_{lp} p_l \int dE\; T_p(E)\left[f_{\rm F}(E - eV - \hbar\Omega l) - f_{\rm F}(E)\right] \\ &= \sum_l p_l I\,(V + \hbar\Omega l/e)\,; \end{aligned} \tag{1.139}$$

this proves Eq. (1.132). To derive the final equality, we note that each term corresponding to a given band l is, by virtue of Landauer formula, the dc current produced by voltage $V_l = V + \hbar\Omega l/e$. In the same situation, it is easy to derive a similar formula for zero-frequency current noise, $S = 2\langle\langle Q^2\rangle\rangle/\Delta t$,

$$S_{\rm dc} = \sum_l p_l S\,(V + \hbar\Omega l/e)\,, \tag{1.140}$$

where $S(V)$ is likewise the noise at dc bias given by Eq. (1.60).

Exercise 1.14. (i) Using Eq. (1.139), show that $I_{\rm dc}(V)$ for a double junction with $T_{\rm L}, T_{\rm R} \ll 1$ is a staircase-like curve. (ii) Calculate positions and heights of the steps, considering separately the cases when $\omega = \Omega$, where $\hbar\omega$ is the distance between the transmission resonances, and $\omega \neq \Omega$.

We have still assumed very small transmissions. It is remarkable that Eq. (1.139) and Eq. (1.140) are in fact valid in the much more general situation of arbitrary transmission. This can be verified by a rather tedious calculation. A straightforward hypothesis would be that the whole counting statistics can be obtained with the same trick of shifting and adding counting statistics at dc voltage. But this is not true: the counting statistics of ac transport is much more complicated. To understand the reason for this, consider the contributions in the characteristic function from a correlated transfer of two electrons, those arising in the second order in $T_p \ll 1$. In the dc case, the correlation takes place if the electrons have the same energy. In the presence of ac bias, the electrons starting in the right lead at energies E and $E + \hbar\Omega n$ can end up on the left with the same energy (although in different side bands) and thus correlate. So, in general, the electrons are correlated even if they are at different energies. Therefore, the transfers within different energy intervals are not independent. The latter implies that Eq. (1.138) cannot be valid generally. These processes are negligible in

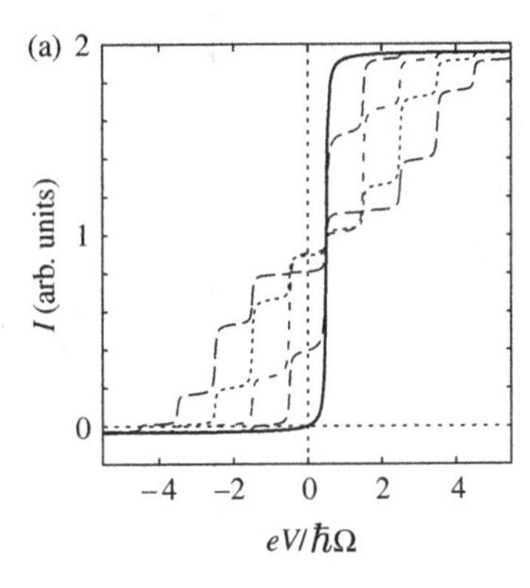

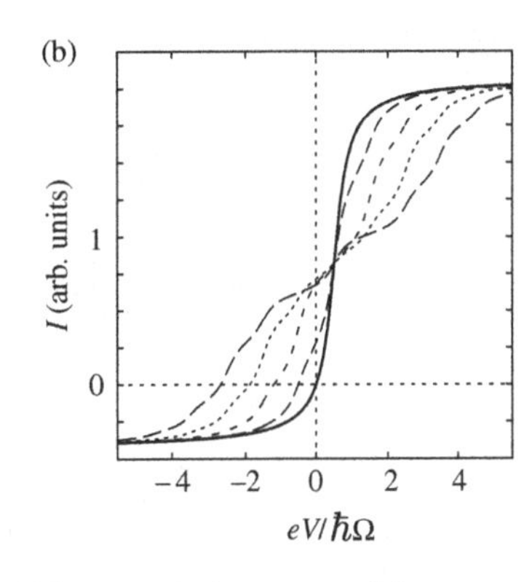

Fig. 1.46. **Tien–Gordon modification of dc I–V curves (thick solid lines). (a) Sharp step of dc I–V curve at $V_0 = 0.5\hbar\Omega$ is multiplied in the presence of ac voltage to be seen at any $V = V_0 + \hbar\Omega l/e$. Subsequent curves correspond to $e\tilde{V}/\hbar\Omega$, ranging from 0 to 4 with step 1. (b) The same curves for a less sharp step. The peculiarities are barely seen and dc voltage mainly results in overall smoothing of the curve.**

the tunneling limit $T_p \ll 1$, where the full statistics still corresponds to the Poissonian expression in Eq. (1.138) with ac-modified rates.

Let us now discuss the general properties of the Tien–Gordon formulas Eqs. (1.132) and (1.140). The resulting I–V curve is a sum of shifted dc I–V curves. This means that if the dc curve has any isolated peculiarity, i.e. peak, step, cusp, etc., at voltage V_0, there are multiple peculiarities of the same sort at voltage positions $V_0 + \hbar\Omega l/e$. The magnitude of the original peculiarity (for instance, height of a step) is spread between the new ones with the weights p_l. As a result, the overall curve eventually becomes smoother (Fig. 1.46). If the dc current $I(V)$ is linear in voltage (which is always the case if the transmission eigenvalues do not depend on energy at the scale of eV), the sum rule given by Eq. (1.137) ensures no effect of ac voltage on dc current, $I_{\mathrm{dc}}(V) = I(V) = GV$.

The Tien–Gordon effect is essentially quantum-mechanical: $\hbar$ is involved in the conversion between frequency and voltage or energy. The effect sets a quantum scale of ac voltage, $\tilde{V}_{\mathrm{scale}} \equiv \hbar\Omega/e$. If $\tilde{V} \ll \tilde{V}_{\mathrm{scale}}$, few photons are emitted or absorbed in the course of electron transfer. In the opposite limit, $\tilde{V} \gg \tilde{V}_{\mathrm{scale}}$, the ac modulation is too slow to provide quantum effects and is in fact adiabatic. The Tien–Gordon formula in this limit becomes evidently simple, as follows:

$$I_{\mathrm{dc}} = \int_0^{2\pi/\Omega} I(V + \tilde{V}\sin\Omega t)\frac{\Omega\, \mathrm{d}t}{2\pi}, \tag{1.141}$$

the time-dependent current adiabatically follows $V(t)$, and the dc current is the average of the time-dependent current over the period.

> **Control question 1.18.** Assume that the current $I(V)$ is a sharp step function like that shown in Fig. 1.46. What does the current I_{dc} look like for $\tilde{V} \gg \tilde{V}_{\mathrm{scale}}$? For $\tilde{V} \ll \tilde{V}_{\mathrm{scale}}$?

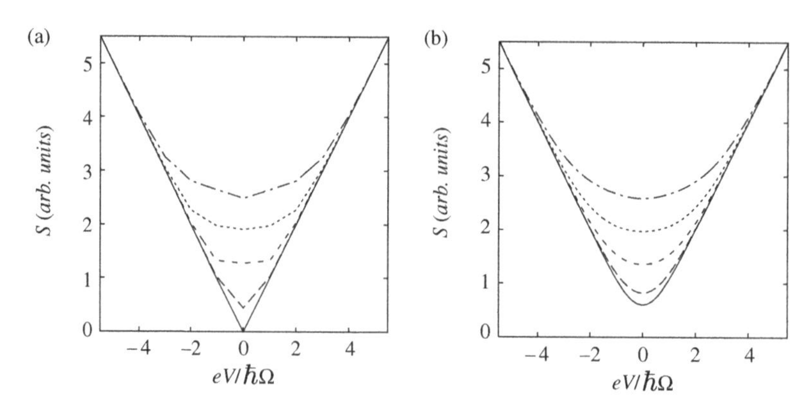

Fig. 1.47. **Tien–Gordon modification of voltage-dependent noise. (a) For $k_BT \to 0$ the noise curve has a cusp at V=0. This cusp is multiplied in the presence of ac voltage. (b) If the cusp is smoothed by the temperature ($k_BT = \hbar\Omega$ is taken), the ac voltage just smoothes it further. This results in increased noise at zero dc voltage. The values of $e\tilde{V}/\hbar\Omega$ corresponding to different curves are the same as in Fig. 1.46.**

At dc voltage bias, the current is always absent if the voltage is zero. Moreover, the current is positive at positive bias and negative at negative bias, so that the positive power, IV, is dissipated in the course of electron transport. This power is supplied by the voltage source. These properties generally do not hold in the presence of an ac voltage. There can be a net dc current at zero dc bias provided the dc I–V curve is not odd in voltage: the ac current is *rectified*, as in a common diode. Consequently, dissipated power $I_{\mathrm{dc}}V$ does not have to be positive. This implies that the dc voltage source may gain energy in the process of electron transfer. This energy comes from the absorption of an ac electromagnetic field.

The same picture of superimposed curves applies to the low-frequency noise, Eq. (1.140). We note that even for energy-independent transmission eigenvalues the noise is modified by ac voltage. At vanishing temperatures, $S(V)$ has a cusp at $V = 0$. The corresponding peculiarities in $S_{\mathrm{dc}}(V)$ are found at voltages $\hbar l\Omega/e$ (Fig. 1.47).

1.7.3 Quantum noise

The current through a nanostructure is, in principle, time-dependent even at dc voltage bias since it consists of transfers of individual electrons. This is manifested in frequency-dependent noise, so we address this topic when discussing time-dependent transport.

We have discussed low-frequency noise thoroughly in Section 1.4. As discussed above, the measurements at low and high frequencies are essentially different. At low frequencies, noise can be readily measured as slow current fluctuations. The discussion implied that the noise persists at high frequency as well, at least up to frequencies of the order of the attempt frequency estimated for one channel. How do we measure high-frequency noise?

This question has a fundamental aspect. If the current were a classical fluctuating variable $I(t)$, the finite-frequency noise would be defined as

$$S_{\mathrm{class}}(\omega) = 2\int_{-\infty}^{\infty} \mathrm{d}t\, \mathrm{e}^{-\mathrm{i}\omega t}\, \langle I(\tau)I(\tau+t)\rangle\,.$$

For dc voltages, the correlator of currents at two different moments of time $\langle I(\tau)I(t+\tau)\rangle$ only depends on the difference t and does not depend on τ. It is convenient to set $\tau = 0$. The classical variables commute, so $\langle I(0)I(t)\rangle = \langle I(t)I(0)\rangle = \langle I(0)I(-t)\rangle$, which makes classical noise real and even in frequency.

In reality, the currents are quantum operators, and noise is defined in terms of the average product of these operators:

$$S(\omega) = 2\int_{-\infty}^{\infty} dt\, e^{-i\omega t}\left\langle \hat{I}(\tau)\hat{I}(t+\tau)\right\rangle.$$

Unlike classical variables, the operators do not commute. This implies that, for a quantum noise, $S(\omega) \neq S(-\omega)$.

How do we measure this noise? A classical detector sensitive to a certain frequency would measure the square of the Fourier component (of the current) at this frequency and is not able to distinguish between positive and negative frequencies. To measure quantum noise, we need a *quantum* detector able to make such a distinction [25, 26].

Let us consider the transitions between the quantum states of a detector and concentrate on two states $|a\rangle$ and $|b\rangle$. We assume that this transition is induced by the weak time-dependent perturbation proportional to the current, $\hat{H} = \alpha|b\rangle\langle a|\hat{I}(t) + \text{h.c.}$ In this situation, one uses the Fermi golden rule to find the transition rate:

$$\Gamma_{a\to b} = \frac{|\alpha|^2}{2\hbar^2} S\left(\frac{E_b - E_a}{\hbar}\right). \tag{1.142}$$

We see that the transition rate is proportional to the noise at the frequency $\hbar\omega = E_b - E_a$ and thereby one can measure noise by measuring the transition rate. In its turn, the transition rate is measured at low frequency rather than at frequency ω; one may say that the current fluctuations are thereby rectified. This is, for instance, how a photodiode works: the high-frequency signal – light – causes transitions of electrons, which are detected as a dc output current. Remarkably, this detector distinguishes between positive and negative frequencies. If $E_b > E_a$, the detector absorbs energy from noise and senses positive frequency. Otherwise, it emits energy and senses negative frequency. The classical noise is even in frequency and thus causes equal transition rates in both directions: $|a\rangle \to |b\rangle$ as well as $|b\rangle \to |a\rangle$.

As a warm-up, let us look at quantum noise of the nanostructure in equilibrium at temperature T. This noise tries to bring the detector to the same temperature so that the probability of finding the detector in a state with energy E would obey the Boltzmann distribution, $p(E) \propto \exp(-E/k_B T)$. For two states $|a\rangle$ and $|b\rangle$, this implies $p_a/p_b = \exp(-(E_a - E_b)/k_B T)$. However, by virtue of detailed balance, the number of transitions from $|a\rangle$ to $|b\rangle$ should be the same as that from $|b\rangle$ to $|a\rangle$, i.e. $\Gamma_{a\to b}p_a = \Gamma_{b\to a}p_b$. Recalling Eq. (1.142), we see that

$$\frac{S(\omega)}{S(-\omega)} = e^{-\hbar\omega/k_B T},$$

irrespective of the nature of the nanostructure.

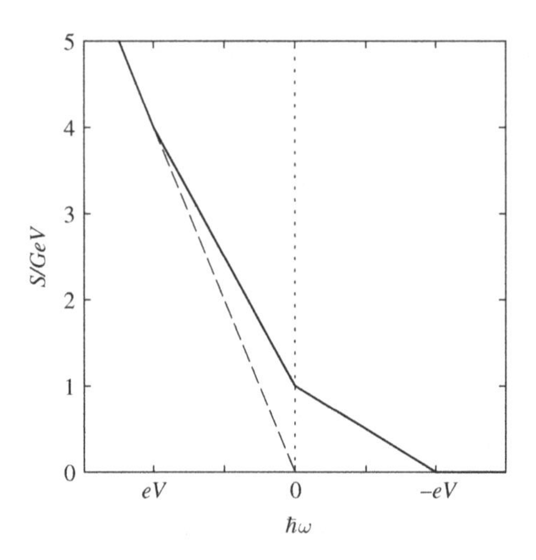

Fig. 1.48. Frequency dependence of the non-equilibrium noise (Fano factor $F = 0.5$). Dashed line: corresponding equilibrium noise ($V = 0$).

Another relation between $S(\omega)$ and $S(-\omega)$ is not restricted to equilibrium and stems from the Kubo theory of linear response,

$$S(\omega) - S(-\omega) = 4\hbar\omega \,\mathrm{Re}G(\omega),$$

where $G(\omega)$ is the finite-frequency conductance expressing the current response of the nanostructure on the ac voltage applied. These two relations determine the noise at equilibrium as follows:

$$S(\omega) = \frac{4\hbar\omega \,\mathrm{Re}G(\omega)}{\exp(\hbar\omega/k_B T) - 1}. \tag{1.143}$$

At zero temperature, S is only non-zero at negative frequencies, $S = -4\hbar\omega \,\mathrm{Re}G(\omega)\times \theta(-\omega)$ (Fig. 1.48). Indeed, an equilibrium nanostructure at zero temperature is in its lowest energy state and therefore cannot give any energy to the detector – it can only absorb energy from the detector to undergo a transition to an excited state.

Now we turn to non-equilibrium quantum noise. We first recall the zero-frequency expression, Eq. (1.60), and present it in the following form:

$$S(0) = 2eG_Q \int \mathrm{d}E \sum_p \{T_p(E)\left[f_R(E)(1 - f_L(E)) + f_L(E)(1 - f_R(E))\right] - T_p^2(E)\left[f_L(E) - f_R(E)\right]^2\}. \tag{1.144}$$

An actual calculation of the finite-frequency noise yields a similar expression, with some energies shifted by $\hbar\omega$:

$$\begin{aligned} S(\omega) = 2eG_Q \int \mathrm{d}E \sum_p \{&T_p(E)\left[f_R(E)(1 - f_L(E - \hbar\omega)) + f_L(E + \hbar\omega)(1 - f_R(E))\right] - T_p(E)T_p(E + \hbar\omega)\left[f_L(E) - f_R(E)\right] \\ &\times \left[f_L(E + \hbar\omega) - f_R(E + \hbar\omega)\right]\}. \end{aligned} \tag{1.145}$$

The first group of terms in square brackets describes one-electron transfers corresponding to the rates $\Gamma_{\mathrm{RL,LR}}$ discussed earlier in this section. It allows for an easy interpretation in terms of photon-assisted tunneling. The first term in the group depicts the transmission of an electron from the right with energy E. This event produces noise – an opportunity to give out the energy $\hbar\omega$. Therefore, the energy in the left lead is shifted by this amount. Similarly, the second term gives the transmission of the electron from the left, with energy $E + \hbar\omega$, that ends up in the right lead with energy E. In the tunneling limit, the noise is determined by these terms. Generally, the second group with T_p^2 also contributes. These terms correspond to the correlated transfer of two electrons through the scattering region, the energies of the two differing by $\hbar\omega$. Their contribution is symmetric in frequency and disappears in equilibrium.

Let us concentrate on the zero-temperature case and assume that the transmission eigenvalues as functions of energy do not change on the scale of frequency and voltage. Then we have a simple expression for quantum noise (Fig. 1.48),

$$S(\omega) = 2G \begin{cases} -2\hbar\omega, & \hbar\omega < -e|V| \\ (e|V| - \hbar\omega) - (1 - F)(e|V| + \hbar\omega), & -e|V| < \hbar\omega < 0 \\ F(e|V| - \hbar\omega), & 0 < \hbar\omega < e|V| \\ 0, & e|V| < \hbar\omega, \end{cases}$$

where all transmission eigenvalues are incorporated into the Fano factor $F \equiv \sum_p T_p(1 - T_p)/\sum_p T_p$ and the conductance G. The noise at positive frequencies is restricted by $e|V|/\hbar$: The maximum energy quantum the detector can take from the noise is the work from the voltage source upon transfer of a single electron through the nanostructure. There is piecewise linear dependence on both frequency and voltage. The noise increases at large negative frequencies, indicating the increasing ability of the nanostructure to absorb bigger energy quanta.

Exercise 1.15. (i) Find an argument to explain that Eq. (1.145), for finite temperature and a one-channel quantum point contact with $T = 1$, reproduces Eq. (1.143). (ii) Deduce the equilibrium noise given by corrections to the quantum noise expression assuming a small reflection coefficient, $R = 1 - T \ll 1$.

1.7.4 Adiabatic pumping

There are numerous examples of pumping and pumps in technology and nature, ranging from bicycle pumps to human hearts. To give a definition, a pump produces a net flow (of water, air, charge, etc.) in the absence of constant drive (water level or air pressure difference) by cyclic action. For electric transport, this means a dc current induced by an oscillatory perturbation in the absence of dc voltage. As we saw at the beginning of this section, this perturbation can be an applied ac voltage: in this case, the ac voltage may be rectified to produce a dc current.

This is not the only way to realize pumping in quantum transport. Instead of applying the voltage to the terminals, one may change the scattering matrix of the nanostructure by applying the ac voltage to the gates. We consider *adiabatic pumping*, implying that the frequency of this perturbation is the smallest of all possible frequency scales. In fact, this frequency must be smaller than the inverse time for an electron to traverse the nanostructure.

Our starting point is the relation between the charge transferred to a certain transport channel j upon a slow variation of the scattering matrix, $\hat{s} \to \hat{s} + \delta\hat{s}$,

$$\delta Q_j = -\frac{\mathrm{i}e}{2\pi}\left(\delta\hat{s}\,\hat{s}^\dagger\right)_{jj}. \tag{1.146}$$

This is a short-hand relation for the case where we can disregard the energy dependence of the scattering matrix. If this is not the case, one integrates this expression over energies, with the weight given by the energy derivative of the filling factor, $\partial f(E)/\partial E$. The charge depends only on the variation of the scattering matrix and does not depend on how fast this matrix has been charged: this expresses the adiabaticity.

To evaluate the current flowing to a certain terminal α, we take the time derivative and sum over all channels that belong to this terminal. Thereby we relate the current to the time derivative of the scattering matrix [27],

$$I_\alpha = \frac{\mathrm{i}e}{2\pi}\int \mathrm{d}E\,\frac{\partial f(E)}{\partial E}\sum_\gamma \mathrm{Tr}\,\dot{\hat{s}}_{\alpha\gamma}(E)\hat{s}^\dagger_{\alpha\gamma}(E), \tag{1.147}$$

where, as in Section 1.5, $\hat{s}_{\alpha\gamma}(E)$ is a block of the scattering matrix that describes transmission of electrons from terminal γ to terminal α.

To understand these relations better, let us inspect two specific cases. First, let us consider only one terminal supporting one channel; in this case, electrons are reflected ideally from the scattering region. The scattering matrix is just a single number presenting the reflection amplitude: $\hat{s} = r = \exp(\mathrm{i}\theta)$. The charge in the channel at a certain length L is given by

$$Q = Le\int_{-\infty}^{\infty}\frac{\mathrm{d}k_x}{2\pi}f(k_x) = ek_\mathrm{F}L/\pi. \tag{1.148}$$

The change of the reflection phase θ is equivalent to the x-shift of the scatterer, resulting in effective elongation δL of the channel. To see this, we recall that the wave function reads $\psi(x) \propto \exp(\mathrm{i}k_\mathrm{F}x) + \exp(-\mathrm{i}k_\mathrm{F}x)\exp(\mathrm{i}\theta)$. Thus, $\delta\theta = 2k_\mathrm{F}\delta L$ and

$$\delta Q = \frac{e}{2\pi}\delta\theta = -\frac{\mathrm{i}e}{2\pi}\delta s\, s^\dagger, \tag{1.149}$$

in full accordance with Eq. (1.146). The relation between the number of particles and the reflection phase at Fermi energy is known as the *Friedel sum rule* in solid state physics.

Secondly, let us recover the Landauer formula from Eq. (1.147). Let us apply voltages V_α to the terminals. As we have seen at the beginning of the section, the wave functions in each lead acquire phase factors $\exp(\mathrm{i}\phi_\alpha)$, $\phi_\alpha \equiv -(e/\hbar)\int^t V_\alpha(t')\mathrm{d}t'$. We consider vanishing voltages so that the phases change adiabatically and can be treated as time-independent ones. The scattering matrix relates the wave amplitudes in incoming and outgoing channels. Since each amplitude has acquired the terminal-dependent phase factor, the scattering matrix changes as follows:

$$\hat{s}_{\alpha\beta} \to \mathrm{e}^{\mathrm{i}(\phi_\alpha-\phi_\beta)}\hat{s}_{\alpha\beta}. \tag{1.150}$$

For the time derivative of the scattering matrix we therefore obtain

$$\dot{\hat{s}}_{\alpha\beta} = \frac{ie}{\hbar}(V_\beta - V_\alpha)\hat{s}_{\alpha\beta}.$$

Substitution into Eq. (1.147) immediately yields

$$I_\alpha = \sum_\beta G_{\alpha\beta} V_\beta,$$

with conductances given by the Landauer relation, Eq. (1.73). These two checks convince us of the validity of Eq. (1.146)

Let us now consider pumping in a two-terminal one-channel conductor described by the scattering matrix given in Eq. (1.39). If we vary adiabatically the three independent parameters $T = 1 - R, \theta$, and ϕ, the charge transferred to the left (right) terminal becomes

$$\delta Q_\mathrm{L}(t) = -\delta Q_\mathrm{R}(t) = \frac{e}{2\pi}\left\{R\delta\theta(t) + T\delta\phi(t)\right\}, \tag{1.151}$$

where $\delta\theta$ and $\delta\phi$ are shifts of the corresponding parameters. As we have seen, the shift of θ arises from the displacement of the scatterer in the channel ($\delta\theta = 2k_\mathrm{F}\delta L$); the first term simply describes this effect. If $R = 1$, the scatterer does not leak and the left side of the channel becomes effectively longer to accumulate extra charge δQ_L, while the right side becomes shorter, losing the same charge. If $R \neq 1$, the scatterer leaks, producing a counterflow of charge; this reduces the accumulated charge. As we see from Eq. (1.150), the variation of ϕ corresponds to the voltage applied across the scatterer. The second term in Eq. (1.151) therefore describes the current produced by this voltage. We have already studied this effect at the beginning of the section so we assume that ϕ is not varied. We are ready to pump. Let us vary R and θ in a cyclic fashion. The instant values of these two specify a point in the parameter space of the scattering matrix, which, in our case, is a two-dimensional manifold (R, θ). The point moves upon varying the parameters. All points swept in the course of cyclic motion make a closed contour ∂C in this space. To visualize this, let us plot it in convenient coordinates $X = \sqrt{R}\cos\theta, Y = \sqrt{R}\sin\theta$. Since $R \leq 1$, all possible scattering matrices fall into the unit circle $X^2 + Y^2 \leq 1$. The net charge pumped over a period is given by

$$\begin{aligned} Q &= \frac{e}{2\pi}\int_0^{\text{period}} \mathrm{d}t\, R(t)\frac{\mathrm{d}\delta\theta}{\mathrm{d}t} = \frac{e}{2\pi}\oint_{\partial C} R\,\mathrm{d}\theta \\ &= \frac{e}{\pi}\int_C \mathrm{d}X\,\mathrm{d}Y = \frac{e}{\pi}\mathcal{A}_C, \end{aligned} \tag{1.152}$$

where $\mathcal{A}_C$ is $\pm$area enclosed by the contour ∂C, where the plus (minus) sign is for a contour swept counterclockwise (clockwise). The net charge pumped thus has a straightforward geometrical meaning.

Control question 1.19. Can you see from the derivation why there is no charge transferred if only T is varied?

There are contours and contours. The contour that comes to mind first is simply a circle of radius $\sqrt{R}$. Its area is given by $\mathcal{A}_C = \pi R$, and the charge eR is pumped over a period.

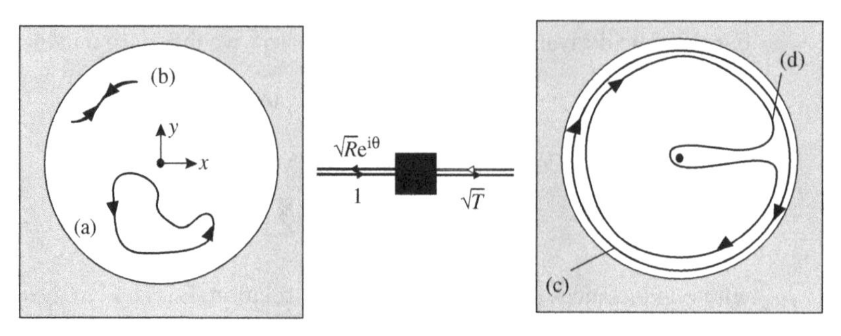

Fig. 1.49. Different pumping cycles of a one-channel scatterer plotted in convenient coordinates (see the text). The net charge pumped over the period is given by the area enclosed by the closed contour (a) presenting the pumping cycle. The pumping requires variation of two independent parameters: contour (b), corresponding to a one-parameter cycle, encloses no area. Cycle (c) is forbidden because the corresponding contour encloses the origin (the cycle would give rise to a net shift of the scatterer). Cycle (d), which is very similar to cycle (c), is allowed.

This looks suspicious: if $R = 1$, no charge is ever transferred via the scatterer. The point is that this simplest contour is not legitimate. Indeed, variation of the phase θ means that the scatterer moves. If $\oint d\theta \neq 0$, as for the contour under consideration, there is *net* displacement of the scatterer over the period and it does not return to the same position. The action is thus not cyclic: this is not a pumping! For a true cycle, there is no net phase shift, $\oint d\theta = 0$. In other words, a legitimate contour may not encircle the origin. We stress that the charge can only be pumped if at least two parameters of the scattering matrix – for example, R and θ, or any two independent combinations of those – are varied. Otherwise, the scattering matrix will pass each point in parameter space twice during the cycle (going back and forth). The contour presenting the cycle (contour (b) in Fig. 1.49) would be just a curve enclosing no area – no charge is pumped.

This is also true in the many-channel case. Consider a multi-channel multi-terminal scattering matrix that depends on time via two parameters $X_{1,2}(t)$ (they can be, for instance, potentials of the gate electrodes). The charge transferred to channel j upon variation of these parameters is (Eq. (1.146))

$$\delta Q_j = -\frac{ie}{2\pi}\left[\left(\frac{\partial \hat{s}}{\partial X_1}\hat{s}^\dagger\right)_{jj}\delta X_1(t) + \left(\frac{\partial \hat{s}}{\partial X_2}\hat{s}^\dagger\right)_{jj}\delta X_2(t)\right]. \tag{1.153}$$

As above, let us consider a contour ∂C in two-dimensional space (X_1, X_2) drawn in the course of the cyclic motion of the parameters. By virtue of Green's formula of vector analysis,

$$\oint_{\partial C} \boldsymbol{F}\, d\boldsymbol{l} = \int_C dX_1\, dX_2 \left(\frac{\partial F_2}{\partial X_1} - \frac{\partial F_1}{\partial X_2}\right),$$

which is valid for an arbitrary vector field $\boldsymbol{F} \equiv (F_1, F_2)$. The charge pumped over the cycle is again given by the integral over the area C enclosed, i.e.

$$\delta Q_j = \frac{e}{\pi} \int_C K_j(X_1, X_2) \mathrm{d}X_1 \, \mathrm{d}X_2;$$
$$K_j(X_1, X_2) \equiv \mathrm{Im} \left(\frac{\partial \hat{s}}{\partial X_1} \frac{\partial \hat{s}^\dagger}{\partial X_2} \right)_{jj}. \tag{1.154}$$

Thus, the charge pumped also has a geometric meaning for the multi-channel case. The detailed structure of the scattering matrix is incorporated into the functions $K_j(X_1, X_2)$, which are called curvatures.

Exercise 1.16. Consider a beam splitter that is symmetric with respect to the exchange of leads 1 and 2. It is characterized by the following scattering matrix:

$$\hat{s} = \begin{pmatrix} \sqrt{1-2T} & \sqrt{T}e^{i\theta} & \sqrt{T}e^{i\theta'} \\ \sqrt{T}e^{i\theta} & \sqrt{1-2T} & \sqrt{T}e^{i\theta'} \\ \sqrt{T}e^{i\theta'} & \sqrt{T}e^{i\theta'} & \sqrt{1-2T}e^{i\chi} \end{pmatrix},$$

which is unitary provided the constraints $2\sqrt{T(1-2T)}\cos\theta + T = 0$ and $\sqrt{T}\exp(i\theta) + \sqrt{1-2T}(1+\exp(2i\theta' - i\chi)) = 0$ are fulfilled. This leaves only two independent parameters, T and χ. Consider the pumping cycle, which is a small circle in the T–χ plane: $(T - T_0)^2 + \chi^2 = (\delta T)^2$, $\delta T < T_0$. The circle surrounds the point $T = T_0$, $\chi = 0$. Calculate the charge pumped into all three leads during the cycle.

In the course of adiabatic pumping the parameters vary in a smooth, continuous fashion. Intuitively, one expects that the response to this variation varies continuously as well. However, the response in our case is the charge transferred – a discrete quantity. This is why the adiabatic pumping is always noisy. At non-zero temperature, this is obvious: the noise arises from thermally activated electrons going back and forth. In addition, the pumping itself produces noise that persists even at zero temperature. The low-frequency counting statistics of pumping can be conveniently expressed in terms of the probabilities p_n of transfering n elementary charges per cycle [28]. We illustrate this with small cycles where the change of the scattering matrix is small. In this case, the electron transfers are rare, $p_0 \approx 1$, $p_{\pm 1} \ll 1$, and we have a bidirectional Poissonian statistics given by Eq. (1.138) with $\Gamma_{\mathrm{LR,RL}} = p_{\pm 1}/\mathrm{period}$. Let us specify a small cycle for our one-channel model by $R(t) = R + r\cos(\Omega t)$, $\theta(t) = \eta\,\cos(\Omega t + \Phi)$, $r, \eta \ll 1$. The probabilities are given by

$$p_{\pm 1} = T \left(\frac{r^2}{R} + \eta^2 R \mp 2r\eta \sin\Phi \right). \tag{1.155}$$

The current and the noise are given by $I = e(p_1 - p_{-1})/\mathrm{period}$ and $S = 2e(p_1 + p_{-1})/\mathrm{period}$, respectively; they are plotted in Fig. 1.50. Whereas the current essentially depends on the phase shift, the noise does not. If R and θ vary in phase or in anti-phase, the current vanishes. Interestingly, the pumping can be optimized by tuning the parameters to $r = \eta R$ and $\Phi = -\pi/2$. In this case, $p_{-1} = 0$ and all electrons are transferred in the same direction.

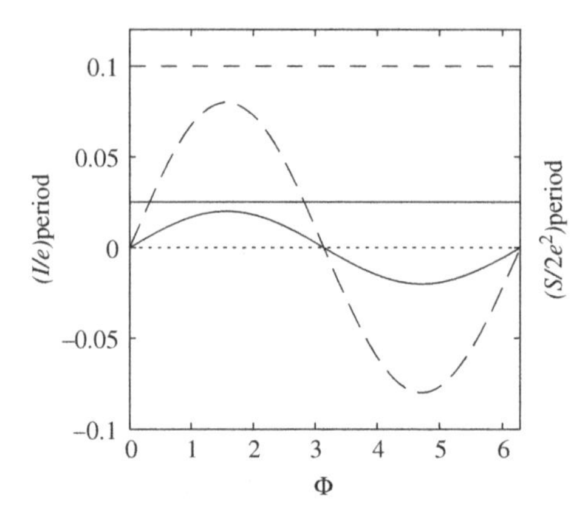

Fig. 1.50. Example: current and noise for a small pumping cycle described in the text. The horizontal lines represent noise, which does not depend on Φ. The curves denote the current, which is an oscillating function of Φ. The parameter η differs for dashed ($\eta = 0.05$) and solid ($\eta = 0.2$) curves, respectively. All other parameters are the same: $T = R = 0.5$, $r = 0.1$.

1.8 Andreev scattering

In this section, we consider electron transport in nanostructures which are connected not only to the reservoirs in the normal state, but also to one or several reservoirs that are in the superconducting state. Electron properties of superconductors differ from those of normal metals, as explained in Appendix B. The energies of the quasiparticle states are separated from the Fermi energy by the superconducting gap Δ. Let us count energy measured from the Fermi level. If a piece of a normal metal is brought into a contact with a superconductor, an electron with an energy above Δ can enter the superconductor, where it will be converted into a quasiparticle of the same energy. This, however, does not work at $E < \Delta$ since there are no quasiparticles. Therefore, for voltages and temperatures below Δ, no current may flow to the superconductor according to the scattering approach considered in the preceding sections.

1.8.1 Andreev reflection

Charge transfer may proceed, however, by a different mechanism: an electron coming from a normal metal to a superconductor can be reflected back as a hole. While this process conserves energy, it does not conserve charge in the normal metal: since the charges of an electron and a hole are opposite, a charge deficit of $2e$ arises. This implies that a Cooper pair with charge $2e$ has been added on the superconducting side. This transfers the charge from the normal metal into the superconductor. Let us note that the momentum of the hole $\hbar k_{\mathrm{h}}$ is almost equal to that of the electron, $\hbar k_{\mathrm{h}} = \hbar k_{\mathrm{e}} - 2E/v_{\mathrm{F}}$ (Fig. 1.51). Since $|E| \ll E_{\mathrm{F}}$, $k_{\mathrm{h}} \approx k_{\mathrm{e}} \approx k_{\mathrm{F}}$. However, the velocity of the holes, $v_{\mathrm{h}} = \hbar^{-1}(\partial E/\partial k_{\mathrm{h}})$ is opposite to that of electrons; holes with $k_{\mathrm{h}} > 0$ actually move away from the superconductor. This process is called *Andreev reflection* [29].

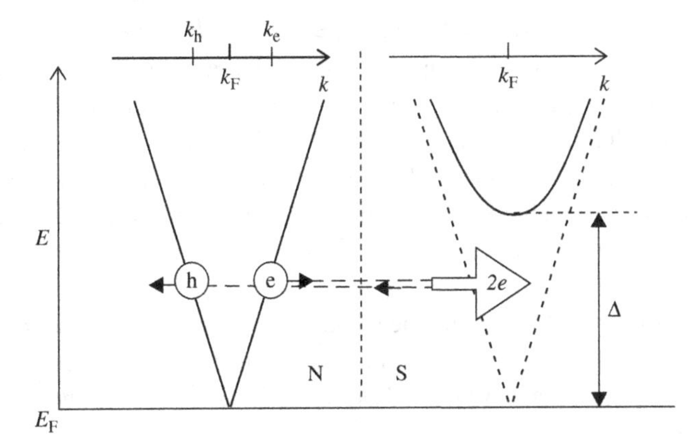

Fig. 1.51. Andreev reflection: an electron coming from a normal (N) metal to a superconductor (S) is reflected as a hole with the same energy and approximately the same momentum.

Let us elaborate on the quantitative description of Andreev reflection. In the presence of superconductivity, an excitation in a metal is conveniently represented by a *two-component* wave function, the components describing electrons ($\psi_e(\boldsymbol{r})$) and holes ($\psi_h(\boldsymbol{r})$). The wave function obeys the *Bogoliubov – de Gennes* (BdG) *equation*, which is a generalization of the Schrödinger equation:

$$\begin{pmatrix} \hat{H} & \Delta e^{i\varphi} \\ \Delta e^{-i\varphi} & -\hat{H}^* \end{pmatrix} \begin{pmatrix} \psi_e(\boldsymbol{r}) \\ \psi_h(\boldsymbol{r}) \end{pmatrix} = E \begin{pmatrix} \psi_e(\boldsymbol{r}) \\ \psi_h(\boldsymbol{r}) \end{pmatrix}. \tag{1.156}$$

Here the energy is counted from the Fermi level, so that $\hat{H} = \hat{H}_0 - E_F$, $\hat{H}_0 = -(\hbar^2/2m)(\nabla + ie\boldsymbol{A}(\boldsymbol{r})/\hbar c)^2 + U(\boldsymbol{r})$ being the Hamiltonian for electrons in the absence of any superconductors. As explained in Appendix B, the superconductivity mixes electrons and holes. With these cross-terms, the Hamiltonian becomes a 2×2 matrix. The cross-terms are off-diagonal elements of the matrix and are complex numbers with modulus Δ and phase φ. These values are position-dependent and vanish in the normal part of the nanostructure. It is enough for our purposes to assume that Δ and φ are constant in the superconducting reservoir, with Δ being equal to the superconducting energy gap far in the reservoir.[6]

To understand the meaning of Eq. (1.156), let us first consider a normal metal, in which the Hamiltonian is diagonal and the equations for the electron and hole components separate. The solutions are plane waves $\psi_{e,h}(\boldsymbol{r}) \propto \exp(i\boldsymbol{k}\boldsymbol{r})$. Substituting this into Eq. (1.156), and considering only excitations close to the Fermi surface, $|E| \ll E_F$, we find $k = k_F \pm E/\hbar v_F$, where $\pm$ represents electron and hole components, respectively. Note that the momenta of both electron-like and hole-like solutions can be either above or below k_F. This is in conflict with the conventional definition of quasiparticles in a normal metal,

[6] Strictly speaking, the values of $\Delta(\boldsymbol{r})$ and $\varphi(\boldsymbol{r})$ actually depend on the solutions of BdG equations at all energies. Consequently, the superconducting pair amplitude Δ is suppressed in the region adjacent to the normal reservoir. However, the suppression does not play an important role and can be disregarded for model purposes [30].

where the quasiparticles with $k > k_F$ ($k < k_F$) are called electrons (holes). We can easily sort out this problem for the normal metal. Indeed, BdG equations allow for solutions with positive energies, $E = |\xi|$, where we have defined $\xi = \hbar v_F(k - k_F)$, as well as for the solutions with negative energies, $E = -|\xi|$. The latter are not independent from the former; they are obtained from each other by a flip of components. Thus, BdG equations contain a double set of solutions. The solutions with negative energies would represent electrons with $k < k_F$ and holes with $k > k_F$ – contradicting the conventional definition of a quasiparticle. To conform to the conventional definition of electrons and holes in a normal metal, we retain the solutions with positive energies only.

Let us now look at the solutions of Eq. (1.156) in a superconductor. Substituting $\psi_{e,h}$ in the form of plane waves and assuming $\Delta, E \ll E_F$, we find that the corresponding energies are given by

$$E = \sqrt{\xi^2 + \Delta^2}, \quad \xi = \hbar v_F(k - k_F). \tag{1.157}$$

For $E > \Delta$, quasiparticles can freely propagate in a superconductor and have an energy spectrum given by Eq. (1.157) rather than $E = |\xi|$. For $E < \Delta$, quasiparticles in a bulk superconductor do not exist.

We consider next an *ideal* (no scattering) contact between a normal metal ($x < 0$) and a superconductor ($x > 0$). Since the transport channels are not mixed, it suffices to consider one transport channel n (this channel index will be suppressed where it does not lead to the confusion). Let us look at the solutions of the form $\psi_{e,h}(x) \propto \tilde{\psi}_{e,h}(x) \exp(ik_F^{(n)} x)$ that correspond to an electron propagating to the right and a hole moving in the opposite direction. The envelope function $\tilde{\psi}(x)$ varies at a space scale that is much bigger than the electron wavelength and satisfies the following BdG equation:

$$\begin{pmatrix} -i\hbar v_F \, d/dx & \Delta(x)e^{i\varphi} \\ \Delta(x)e^{-i\varphi} & i\hbar v_F \, d/dx \end{pmatrix} \begin{pmatrix} \tilde{\psi}_e(x) \\ \tilde{\psi}_h(x) \end{pmatrix} = E \begin{pmatrix} \tilde{\psi}_e(x) \\ \tilde{\psi}_h(x) \end{pmatrix}. \tag{1.158}$$

In the normal metal, we take the wave function in the form

$$\tilde{\psi}(x < 0) = \begin{pmatrix} e^{ixE/\hbar v_F} \\ r_A e^{-ixE/\hbar v_F} \end{pmatrix}, \tag{1.159}$$

which describes the incoming electron and the outgoing Andreev-reflected hole. The hole amplitude acquires an extra factor r_A: the amplitude of Andreev reflection.

For $E < \Delta$, there are no solutions extending to the bulk of the superconductor. There is, however, an *evanescent* solution falling off away from the normal reservoir. This is given by

$$\tilde{\psi}(x > 0) = C \begin{pmatrix} f_e \\ f_h \end{pmatrix} e^{-x\sqrt{\Delta^2 - E^2}/\hbar v_F}, \tag{1.160}$$

where C is an arbitrary constant and the coefficients $f_{e,h}$ are to be found from Eq. (1.158) (the BdG equation) and the normalization condition $|f_e|^2 + |f_h|^2 = 1$.

Control question 1.20. What are the explicit forms of $f_{e,h}$?

The typical scale of penetration into the superconductor – the superconducting correlation length – is of the order $\hbar v_F/\Delta \gg \lambda_F$ and diverges at the threshold energy $E = \Delta$.

Now let us find the amplitudes r_A and C matching both solutions at $x = 0$. The derivatives do not have to be matched since the effective BdG equation contains the first derivatives only. The amplitude of Andreev reflection is given by

$$r_A(E) = e^{i\chi} = e^{-i\varphi}\left(\frac{E}{\Delta} - i\frac{\sqrt{\Delta^2 - E^2}}{\Delta}\right), \quad \chi = -\arccos\left(\frac{E}{\Delta}\right) - \varphi. \qquad (1.161)$$

As expected, the electron is fully Andreev reflected ($|r_A|^2 = 1$). The phase of the outgoing hole is shifted by χ with respect to the phase of the incoming electron. The phase shift between the amplitudes of an incoming hole and an outgoing electron, calculated similarly, equals $\tilde{\chi} = -\arccos(E/\Delta) + \varphi$.

Exercise 1.17. (i) Write down the solutions of the BdG equation, Eq. (1.158), for energies above the threshold, $E > \Delta$. (ii) Matching these solutions with the solutions in the normal metal, show that the amplitude of Andreev reflection is given by

$$r_A = e^{-i\varphi}\left(\frac{E}{\Delta} - \frac{\sqrt{E^2 - \Delta^2}}{\Delta}\right). \qquad (1.162)$$

(iii) Note that $|r_A|^2 < 1$ and describe the corresponding scattering process. (iv) Find the asymptotic expression for the probability of Andreev reflection for $E \gg \Delta$.

1.8.2 Andreev conductance

We now consider a more general situation in which a nanostructure is placed between the normal and superconducting reservoirs. The nanostructure in the normal state is described by a scattering matrix $\hat{s}(E)$ that generally depends on energy. Quite amusingly, the same scattering matrix determines the properties of Andreev reflection, which is now combined with the common "normal" reflection of electrons or holes coming to the nanostructure from either side. The scattering theory for Andreev reflection was first put forward by Blonder, Tinkham, and Klapwijk [31].

To start with, we have to find the scattering matrix for electrons and holes. For electrons at energy $E > 0$, this is obviously $\hat{s}_e(E) = \hat{s}(E)$. The holes at the same energy involve states below the Fermi level, and their scattering is related to $\hat{s}(-E)$. However, as we have seen, an electron and a hole at the same momentum have opposite velocities, so the incoming electrons correspond to outgoing holes and vice versa. To account for this, one replaces $\hat{s}$ by $\hat{s}^{-1}$. In addition, the holes obey a time-reversed Hamiltonian: to account for this, the scattering matrix must be transposed. Therefore,

$$\hat{s}_h(E) = (\hat{s}(-E)^{-1})^T = \hat{s}^*(-E). \qquad (1.163)$$

Control question 1.21. Which property of the scattering matrix guarantees Eq. (1.163)?

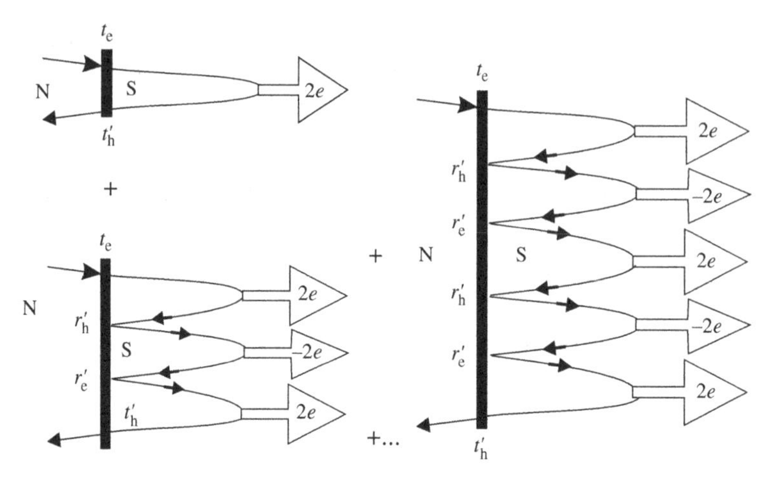

Fig. 1.52. Andreev conductance. The amplitude of the Andreev equation to the normal (N) lead from the nanostructure adjacent to a superconductor (S) is contributed by the processes that differ in the number of electron trips between the nanostructure and superconductor.

To simplify the notation, we consider a one-channel scatterer. For this setup, $r_e = r(E)$, $r_h = r^*(-E)$, $t_e = t(E)$, and $t_h = t^*(-E)$.

Let us calculate the amplitude of the Andreev reflection assuming $E < \Delta$. We use the same approach as in Section 1.6 to sum the amplitudes of various processes that convert the incoming electron to the outgoing hole (Fig. 1.52). The simplest process involves the electron transmission through the nanostructure (amplitude t_e), Andreev reflection from the superconductor (the phase factor $\exp(i\chi)$, see Eq. (1.161)) and transmission through the nanostructure in the backward direction as a hole (t'_h). Thus, the total amplitude of this process is given by $r_{a0} = t_e t'_h \exp(i\chi)$. The next process (Fig. 1.52) involves the reflection of the hole (amplitude r'_h). The hole is Andreev-reflected ($\exp(i\tilde{\chi})$), the resulted electron is reflected again (r'_e), and is converted back into a hole in the superconductor ($\exp(i\chi)$). Finally, the hole transmits through the nanostructure. The extra steps mentioned result in $r_{a1} = r_{a1} r'_h e^{i\tilde{\chi}} r'_e e^{i\chi}$. More complicated processes differ in the number of electron trips between the nanostructure and the superconductor, so that $r_{an} = r_{a0}(r'_h e^{i\tilde{\chi}} r'_e e^{i\chi})^n$. Summing them up, we obtain the total amplitude of Andreev reflection:

$$r_A = \sum_{n=0}^{\infty} r_{an} = \frac{t_e t'_h e^{i\chi}}{1 - r'_e r'_h e^{i(\chi+\tilde{\chi})}}. \tag{1.164}$$

Let us disregard the energy dependence of the scattering matrix: as we have seen, this is plausible if the energy scale associated with the dwell time in the nanostructure exceeds the energies involved, i.e. eV or Δ. In this case, the scattering matrices for electrons and holes are complex-conjugate.

To simplify, let us assume a low voltage $eV \ll \Delta$. We note that $\chi + \tilde{\chi} = -2\arccos(E/\Delta)$; since the relevant energies E are of the order of eV, one can approximate

$\chi + \tilde{\chi} = -\pi$. The Andreev reflection coefficient, $R_A = |r_A|^2$, is given by

$$R_A = \frac{T^2}{(2-T)^2}, \quad (1.165)$$

and is unambiguously determined by T, the transmission eigenvalue of the corresponding transport channel for the nanostructure in the normal state. The normal reflection coefficient is $R_N = 1 - R_A$. For ideal contact ($T = 1$), we recover the earlier result $R_A = 1$, $R_N = 0$. Note that the result in Eq. (1.165) is a consequence of quantum interference. A "classical" calculation (summing up the probabilities $|r_{an}|^2$ rather than the amplitudes) would yield a wrong result.

To calculate the conductance, we note the analogy with normal scattering. Indeed, the fraction R_N of incoming electrons is normally reflected; these electrons do not contribute to the current. The Andreev reflection process (probability R_A) results in the charge transfer of $2e$ (rather than e in the normal case). Thus, the *Andreev conductance* becomes $G_A = 2G_Q R_A$. The same reasoning actually reproduces the whole counting statistics of Andreev transport: it is given by the Levitov formula with $e \to 2e$ and $T \to R_A$.

Exercise 1.18. Determine the noise in Andreev transport. Express the result in terms of the Fano factor (see Eq. (1.64)). What is the upper boundary for the Fano factor?

For many transport channels, one obtains a sum over the channels:

$$G_A = 2G_Q \sum_p (R_A)_p = 2G_Q \sum_p \frac{T_p^2}{(2-T_p)^2}. \quad (1.166)$$

Now we can analyze this formula, employing the notion of the distribution of transmission eigenvalues.

Exercise 1.19. Assume that the nanostructure is diffusive so that the distribution of transmission eigenvalues is given by Eq. (1.43). Express the Andreev conductance in terms of the conductance in the normal state.

If the nanostructure is of a tunnel type, $T_p \ll 1$, we end up with $G_A = G_Q \sum_p T_p^2/2$. Andreev conductance is thus proportional to the second power of the transmission eigenvalues. This reflects the fact that Andreev reflection requires two transmission events, one of an electron and one of a hole. Note that in the normal state the conductance is much higher, being proportional to the first power of T_p. For an ideal contact ($T_p = 1$), the situation is reversed: $G_A = 2G$, the factor of 2 reflecting the fact that Andreev reflection transfers double charge.

1.8.3 Andreev bound states

Consider now a superconducting junction: a nanostructure placed between two superconductors that have the same superconducting gap Δ but differ in their phases. We assume that the nanostructure is sufficiently short, not manifesting the energy dependence of its scattering matrix at energy scale Δ. Under these conditions, it is not important whether the nanostructure is made of a normal metal or a superconductor, or even an insulator. The absence of energy dependence implies that the electrons spend a very short time τ_d in the nanostructure; by virtue of the Heisenberg uncertainty principle, this time is too short to allow a response to the superconductivity inside the nanostructure, $\tau_d\Delta \ll \hbar$. The scattering matrix of the nanostructure is thus its scattering matrix in the normal state.

Let us consider an electron in the nanostructure at sufficiently low energy. It will experience Andreev reflections trying to get to either superconductor. The resulting hole experiences the same problem: it cannot escape the nanostructure and is converted back to an electron in the course of the escape attempt. We conclude that an electron/hole in the nanostructure must perform a so-called finite motion. Quantum mechanics teaches us that any finite motion of a particle gives rise to discrete energy levels. Indeed, a nanostructure between the superconducting reservoirs kept at different phases gives rise to a set of bound states for quasiparticles – *Andreev bound states*. Let us calculate the energies of these states.

First consider again a single channel. The scattering matrix of the nanostructure relates the amplitudes of outgoing and incoming states with respect to the nanostructure,

$$\begin{pmatrix} \boldsymbol{b}_{\mathrm{e}} \\ \boldsymbol{b}_{\mathrm{h}} \end{pmatrix} = \begin{pmatrix} \hat{s} & 0 \\ 0 & \hat{s}^* \end{pmatrix} \begin{pmatrix} \boldsymbol{a}_{\mathrm{e}} \\ \boldsymbol{a}_{\mathrm{h}} \end{pmatrix}, \tag{1.167}$$

where the two components of the amplitude vectors correspond to the left and right side of the nanostructure, respectively,

$$\boldsymbol{b}_{\mathrm{e}} = \begin{pmatrix} b_{\mathrm{Le}} \\ b_{\mathrm{Re}} \end{pmatrix}; \quad \boldsymbol{b}_{\mathrm{h}} = \begin{pmatrix} b_{\mathrm{Lh}} \\ b_{\mathrm{Rh}} \end{pmatrix},$$

and similarly for the incoming amplitudes $\boldsymbol{a}_{\mathrm{e}}$, $\boldsymbol{a}_{\mathrm{h}}$. The scattering of holes, as mentioned, is given by the complex-conjugate matrix $\hat{s}^*$. The nanostructure does not convert electrons to holes; this is why the matrix appearing in Eq. (1.167) is block-diagonal.

Andreev reflection from the superconductors converts electrons to holes and vice versa, yielding the following complementary relation between $\boldsymbol{a}$ and $\boldsymbol{b}$:

$$\begin{pmatrix} \boldsymbol{a}_{\mathrm{e}} \\ \boldsymbol{a}_{\mathrm{h}} \end{pmatrix} = \begin{pmatrix} 0 & \hat{s}_{\mathrm{eh}} \\ \hat{s}_{\mathrm{he}} & 0 \end{pmatrix} \begin{pmatrix} \boldsymbol{b}_{\mathrm{e}} \\ \boldsymbol{b}_{\mathrm{h}} \end{pmatrix}, \tag{1.168}$$

with

$$\hat{s}_{\mathrm{eh}} = \begin{pmatrix} \mathrm{e}^{\mathrm{i}\tilde{\chi}_{\mathrm{L}}} & 0 \\ 0 & \mathrm{e}^{\mathrm{i}\tilde{\chi}_{\mathrm{R}}} \end{pmatrix}; \quad \hat{s}_{\mathrm{he}} = \begin{pmatrix} \mathrm{e}^{\mathrm{i}\chi_{\mathrm{L}}} & 0 \\ 0 & \mathrm{e}^{\mathrm{i}\chi_{\mathrm{R}}} \end{pmatrix}.$$

The Andreev reflection phases are given by Eq. (1.161): $\chi_{\mathrm{L,R}} = -\varphi_{\mathrm{L,R}} - \arccos(E/\Delta)$, $\tilde{\chi}_{\mathrm{L,R}} = \varphi_{\mathrm{L,R}} - \arccos(E/\Delta)$, $\varphi_{\mathrm{L,R}}$ being the superconducting phases of the left and right reservoirs.

Control question 1.22. Explain the structure of the matrix in Eq. (1.168).

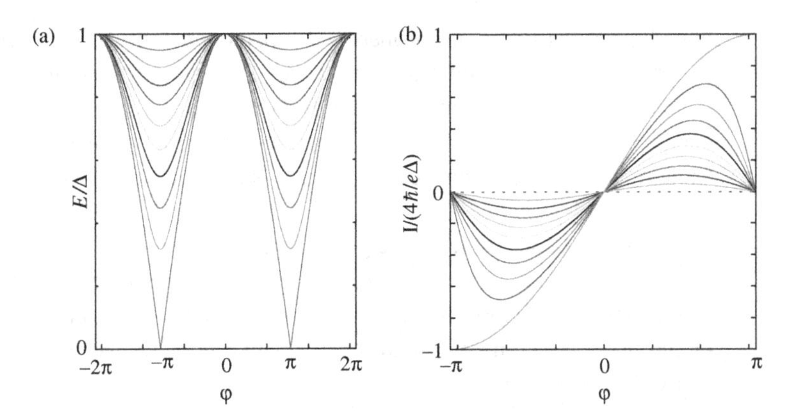

Fig. 1.53. (a) The energies of Andreev bound states versus φ for T_p, ranging from 0.1 (upper curve) to 1 (lowest curve) with step 0.1. (b) Corresponding superconducting currents (the upper curve at positive φ corresponds to $T_p = 1$).

Two systems of linear equations, Eqs. (1.167) and (1.168), have non-zero solutions only if the product $\hat{\Pi}$ of the 4×4 matrices in these relations has an eigenvalue 1. Indeed, if one excludes $\boldsymbol{a}$ from the equations, the equation for $\boldsymbol{b}$ reduces to the eigenvalue problem $\hat{\Pi}\boldsymbol{b} = \boldsymbol{b}$, and solutions exist only if $\det(\hat{\Pi} - \hat{1}) = 0$. Transforming this condition, one obtains the energy of the bound state:

$$E = \Delta\sqrt{1 - T\sin^2(\varphi/2)}, \tag{1.169}$$

where T is the transmission eigenvalue corresponding to the scattering matrix $\hat{s}$, and $\varphi = \varphi_{\mathrm{L}} - \varphi_{\mathrm{R}}$ is the phase difference across the junction [32].

Control question 1.23. Can you trace how Eq. (1.169) emerges from the conditions imposed on the Andreev phases $\chi_{\mathrm{L,R}}$?

For many channels, an Andreev bound state appears in each channel with the energy given by

$$E_p = \Delta\sqrt{1 - T_p\sin^2(\varphi/2)}. \tag{1.170}$$

The energy is modulated by the phase difference φ. For $\varphi = 0$, $E_p = \Delta$ for all channels. In this case, the states are not really bound: they are at the edge of a continuous quasiparticle spectrum in the superconductor. The minimum value $\Delta\sqrt{1 - T_p}$ is achieved at $\varphi = \pi$ (Fig. 1.53).

1.8.4 Josephson effect

So far we have considered the bound states for excitations. For example, an excitation can be a quasiparticle cooled down in the vicinity of the nanostructure: it will be trapped in the bound state. An important property of superconductivity is the correspondence between the properties of the excitations and those of the ground state of the superconductor. This is manifested in the symmetry of the BdG equation with respect to positive and negative energies. The solutions at negative energies can be associated with the filled levels contributing to the ground-state energy, which is the sum of single-particle excitation energies E_n, $E_g = -\sum E_n$.[7]

Let us now concentrate on the ground-state energy of the system. It is contributed to by all excitation energies: those corresponding to propagating quasiparticles above the superconducting gap and those of the bound Andreev states. Only the latter contributions depend on the superconducting phase difference between the reservoirs φ. We concentrate on this phase-dependent part:

$$E(\varphi) = \sum_p E_p(\varphi) = \Delta \sum_p \sqrt{1 - T_p \sin^2(\varphi/2)}. \tag{1.171}$$

We will see now that the phase-dependent energy gives rise to a persistent current in the ground state – a *supercurrent*. Let us slowly vary the phase difference. The energy shift per unit time is given by

$$\frac{\mathrm{d}E}{\mathrm{d}t} = \frac{\partial E(\varphi)}{\partial \varphi} \frac{\mathrm{d}\varphi}{\mathrm{d}t}.$$

The global gauge invariance (see Appendix B) dictates that the time derivative of the superconducting phase is simply the potential of the corresponding superconductor, $\dot{\varphi} = 2eV/\hbar$. The energy change per unit time is the power dissipated at the junction. On the other hand, this power is the product of current and voltage. We conclude that the current in the junction is given by

$$I(\varphi) = -\frac{2e}{\hbar} \sum_p \frac{\partial E_p}{\partial \varphi} = \frac{e\Delta}{2\hbar} \sum_p \frac{T_p \sin\varphi}{\sqrt{1 - T_p \sin^2(\varphi/2)}}. \tag{1.172}$$

The supercurrent – or *Josephson* current – is an odd periodic function of the phase difference, and vanishes at $\varphi = 0$. In particular, for a tunnel junction $T_p \ll 1$, the supercurrent reads $I(\varphi) = I_c \sin\varphi$, where the amplitude $I_c = (e\Delta/2\hbar)\sum_p T_p = (\pi\Delta/2e)G_N$, where G_N is the conductance of the junction in the normal state. Here, I_c is the maximum possible supercurrent achieved at $\varphi = \pi/2$. Historically, the first superconducting junctions were tunnel ones. Usually, the term *Josephson junction* implies the above relation, between the current and phase, that corresponds to the Josephson energy $E_J(\varphi) = -E_J \cos\varphi$, $E_J = \hbar I_c/2e$. In principle, any nanostructure can serve as a Josephson junction; the current–phase characteristics essentially depends on the transmission eigenvalues. For example, a quantum point contact ($T_p = 1$) gives $I(\varphi) = I_c \sin(\varphi/2)$, $I_c = (\pi\Delta/e)G_N$ and maximum current is achieved at $\varphi = \pi$ (Fig. 1.53).

[7] Intuitively, one would include the spin degeneracy in the state count. This would be wrong: the sum is over orbital states. The reason is that the BdG equations provide a double set of solutions, as mentioned in Section 1.8.1.

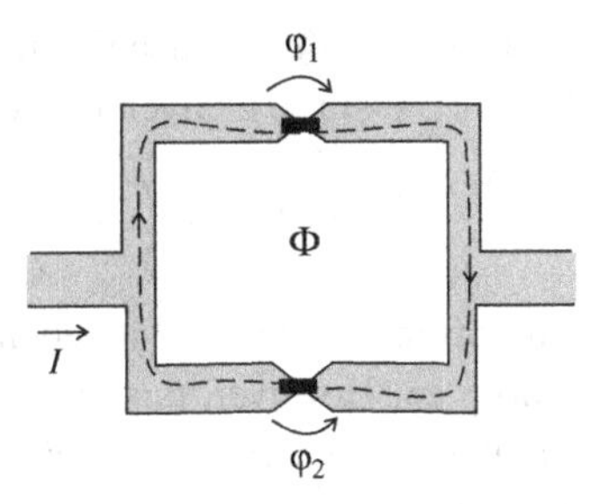

Fig. 1.54. The dc SQUID layout: two Josephson junctions with phase differences $\varphi_{1,2}$. The difference, $\varphi_1 - \varphi_2$, is determined by the flux Φ through the SQUID loop.

Exercise 1.20. Find the maximum supercurrent for a one-channel Josephson junction with transmission coefficient T. At which value of the phase is this current achieved?

The Josephson effect is essentially quantum-mechanical. We have already seen some examples of phenomena in which quantum mechanics plays an important role, such as the transmission through a double junction or the Andreev reflection from a single interface; in those cases, we were always able to present a classical analog of the effect to give them some meaning. In contrast, the Josephson effect is formulated in a way that cannot be interpreted classically: in classical physics, the phase of the wave function does not exist, and therefore a supercurrent cannot occur. The Josephson effect is one of the best illustrations of the concepts of quantum mechanics.

Josephson junctions are applied in many areas where a sensitive measurement of magnetic fields is an issue. Such a measurement is performed with a superconducting quantum interference device (SQUID). In the conceptually simplest version (dc SQUID), the device is a large superconducting loop with two arms intercepted by Josephson junctions (Fig. 1.54). The current through the device is the sum of the currents through both junctions, $I = I_1(\varphi_1) + I_2(\varphi_2)$, where we denote the phase drops at the junctions by φ_1 and φ_2, respectively.

Let us consider a magnetic field B applied perpendicular to the plane of the SQUID. The magnetic field modifies the phase drops at the junctions making them unequal. Indeed, the global gauge invariance requires that the phase and the vector potential always come in the combination $\nabla\varphi - (2e/\hbar c)\boldsymbol{A}$. Let us integrate this combination over the SQUID loop (the integration contour is given by the dashed curve in Fig. 1.54). In doing so, we can neglect the phase gradients in the bulk superconductors, but we must keep the phase drops at the Josephson junctions. This yields

$$\oint \mathrm{d}\boldsymbol{r}\,(\nabla\varphi - (2e/\hbar c)\boldsymbol{A}(\boldsymbol{r})) = \varphi_1 - \varphi_2 + 2\pi\Phi/\Phi_0,$$

where we have used the Stokes theorem transforming the contour integral of the vector potential to the area integral of the field – the magnetic flux Φ through the SQUID loop, $\Phi_0 = \pi\hbar c/e$ being the flux quantum (see Section 1.6). The phase shift along the closed contour is zero: $\varphi_1 - \varphi_2 = -2\pi\Phi/\Phi_0$.

Now we calculate the Josephson current in the SQUID, assuming equal tunnel Josephson junctions, $I = I_c(\sin\varphi_1 + \sin\varphi_2)$. Given the relation between the phase drops, we obtain

$$I = 2I_c \cos\left(\frac{\pi\Phi}{\Phi_0}\right) \sin\left(\varphi_2 + \frac{\pi\Phi}{\Phi_0}\right). \tag{1.173}$$

If we fix the total current I, the only independent parameter that can adjust to the current is the phase drop φ_2. However, this is only possible if the current I does not exceed $I_{max} = 2I_c|\cos(\pi\Phi/\Phi_0)|$. Otherwise, the current is not a supercurrent and a finite voltage is measured on the SQUID. The area of the loop can be quite large, even of meter scale. Since the SQUID measures Φ, the total flux through the whole area of the loop, it is sensitive to astonishingly small magnetic fields.

Control question 1.24. Which magnetic field significantly changes the critical current of a SQUID with dimensions $1\text{ m} \times 1\text{ m}$?

Exercise 1.21. Consider a SQUID in which one of the junctions is of tunnel origin, $I_b(\varphi) = I_b \sin\varphi$ and another is characterized by $I_s(\varphi)$ to be determined from the measurement of the SQUID critical current. Assume $I_b \gg I_s$. (i) Show that, in zeroth order in I_s/I_b, the critical current does not depend on the flux. At which value of φ_b is this critical current achieved? (ii) Compute the critical current in the next order and explain how to recover $I_s(\varphi)$ from the measurement.

1.8.5 Superconducting junction at constant voltage bias

If a constant voltage V is applied across a superconducting junction, the phase difference increases linearly with time, $\varphi = (2eV/\hbar)t$. The junction between the two superconductors reacts to the constant voltage in a manner very different from junctions in the normal state: it produces an ac current (*ac Josephson effect*) that oscillates at the Josephson frequency $2eV/\hbar$.

If the voltage is low, $eV \ll \Delta$, the origin of the ac current is easy to comprehend. The linear sweep of the phase gives rise to the oscillating motion of the energies of the Andreev bound states. The ac current is given by Eq. (1.172), with $\varphi = (2eV/\hbar)t$. There is no dc current. The latter would imply the energy dissipation. However, the energies of all bound states return to the same position over a time period $\pi\hbar/eV$.

If the voltage eV is comparable with the value of the superconducting gap Δ, the situation is more complicated. Since the Josephson frequency is comparable with the energies of the states, we cannot expect that the energies adiabatically follow the time-dependent phase. Besides, there is a dissipation: the junction creates quasiparticles above the gap that leave the junction region carrying the energy away. Thus, we expect to observe a dc current.

To see how this works, let us first consider an open channel between two superconducting electrodes biased at finite constant voltage V. As already noted, the current does not actually depend on the spatial distribution of the voltage. We can assume that the voltage

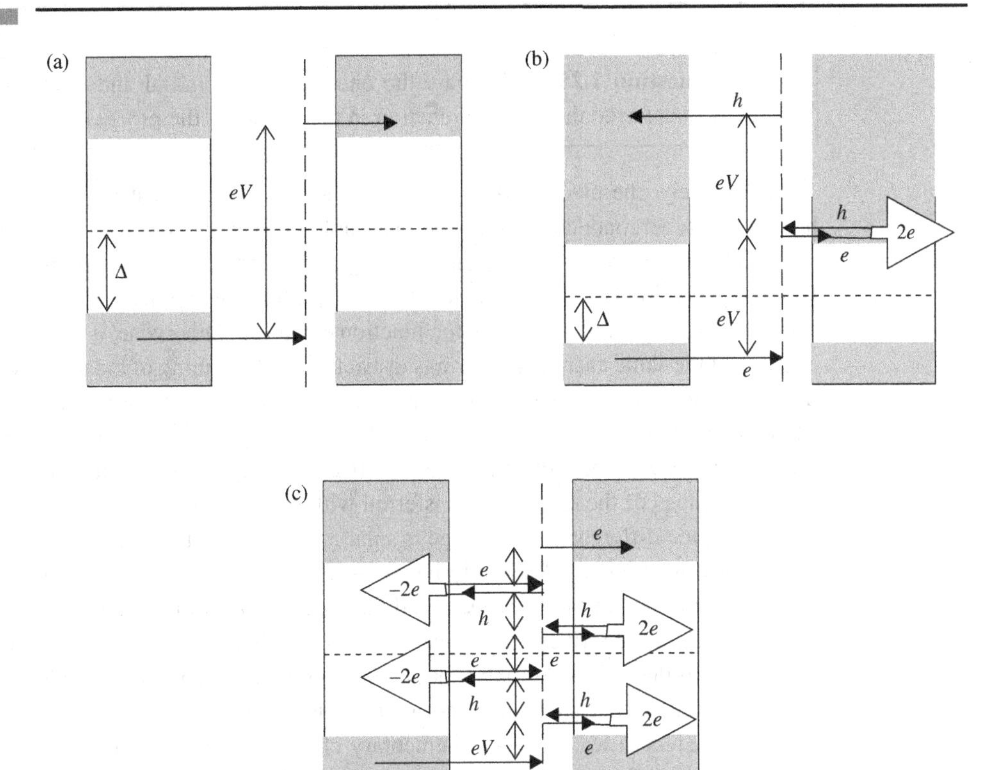

Fig. 1.55. **Elementary scattering processes in a voltage-biased open channel between two superconductors. Electrons (holes) acquire energy eV when crossing the dashed line from the left (right). Quasiparticle states are available in the shaded regions. (a) If $eV > 2\Delta$, a quasiparticle can be transferred from the left to the right in one shot. (b) Alternatively, it may be Andreev-reflected and get to the left at a higher energy. (c) Multiple Andreev reflections are required for such processes at $eV \ll \Delta$. The process shown transfers five elementary charges and is enabled at $5eV > 2\Delta$.**

drops over an arbitrary point of the channel (given by the dashed line in Fig. 1.55). When an electron (hole) crosses the point from the left to the right it increases (decreases) its energy by eV. It decreases (increases) the energy by the same amount while crossing from the right to the left. Since the energies are changing in the process of transmission, we cannot separate positive and negative energies in BdG equations as we did before and we consider quasiparticle energies of both signs. Then, in both left and right electrodes only quasiparticle states with $|E| > \Delta$ are available (Fig. 1.55). Let us consider an electron coming from the left superconducting electrode with an energy E slightly below $-\Delta$. It crosses the point where the voltage drops so it arrives at the right electrode with energy $E + eV$. If $E + eV > \Delta$, it may leave the junction, reaching the quasiparticle states available at this energy. This requires $eV > 2\Delta$. Thereby, the electron from negative energies has been transformed into a quasiparticle at positive energies. This can be seen as the generation of two quasiparticles of two *positive* energies: one with energy $-E > \Delta$ and another with $E + eV > \Delta$ (Fig. 1.55 (a)).

Control question 1.25. Compare the energies of the initial and final states. What charge is transferred through the junction in the course of the process?

Alternatively, the electron does not leave the junction but instead is converted into a hole at the superconducting electrode. The hole crosses the voltage drop from the right to the left, increasing its energy and arriving at the left electrode with energy $E + 2eV$. If it escapes to the left electrode, we have quasiparticles with energies $-E$ and $E + 2eV$, and the charge transferred through the junction equals $2e$. Otherwise, it is converted into an electron of the same energy. This brings us back to the beginning of the process: an electron incoming from the left. We conclude that Andreev reflections can help a process to result in any number of charges transferred, although only two quasiparticles are created. Since the probabilities of Andreev reflections (Eq. (1.162)) quickly decrease with increasing energy, the probabilities of the processes transferred with multiple charge are small.

This is quite different if the voltage is small, $eV \ll \Delta$. Let us again consider an incoming electron with energy E slightly below $-\Delta$. If $E + eV > -\Delta$, there are no available quasiparticle states in the right lead, and the electron has to turn into a hole by Andreev reflection. The hole arrives at the left lead with the slightly larger energy $E + 2eV < \Delta$, so Andreev reflection is the only option. The process of subsequent Andreev reflections continues until the energy of an electron or a hole exceeds Δ. We conclude that each such process transfers at least $2\Delta/eV$ elementary charges. Generally, at $eV \simeq \Delta$ we find the processes that differ in the charge transferred, en. The n-process involves $n - 1$ Andreev reflections and starts at threshold voltage $eV_n > 2\Delta/n$. The onset of each process produces a singularity in I–V curves at the corresponding voltage. These singularities – subgap structure at I–V curves – form an experimental signature of these *multiple Andreev reflections*.

The quantitative theory should include the scattering between the superconducting electrodes. We sketch the general approach below [33, 34].

The setup is the same as the one used to calculate Andreev bound states. The important difference is that the amplitudes of the electrons and holes are not at the same energy: instead, the time-dependent amplitude is a superposition of all energies separated by eV, i.e.

$$\psi(t) = \sum_n \psi_n \mathrm{e}^{-\mathrm{i}(E+neV)t/\hbar}.$$

Each of the amplitudes ψ_n can describe incoming or outgoing electrons or holes. These amplitudes are related by the scattering matrix of the nanostructure,

$$\begin{pmatrix} \boldsymbol{b}_{\mathrm{e}} \\ \boldsymbol{b}_{\mathrm{h}} \end{pmatrix} = \begin{pmatrix} \hat{s} & 0 \\ 0 & \hat{s}^* \end{pmatrix} \begin{pmatrix} \boldsymbol{a}_{\mathrm{e}} \\ \boldsymbol{a}_{\mathrm{h}} \end{pmatrix}, \tag{1.174}$$

where, in distinction from Eq. (1.167), the amplitudes are shifted in energy by eV to incorporate the voltage drop,

$$\boldsymbol{b}_{\mathrm{e}} = \begin{pmatrix} b_{\mathrm{Le},n} \\ b_{\mathrm{Re},n+1} \end{pmatrix}; \quad \boldsymbol{b}_{\mathrm{h}} = \begin{pmatrix} b_{\mathrm{Lh},n} \\ b_{\mathrm{Rh},n-1} \end{pmatrix},$$

and similarly for $\boldsymbol{a}$.

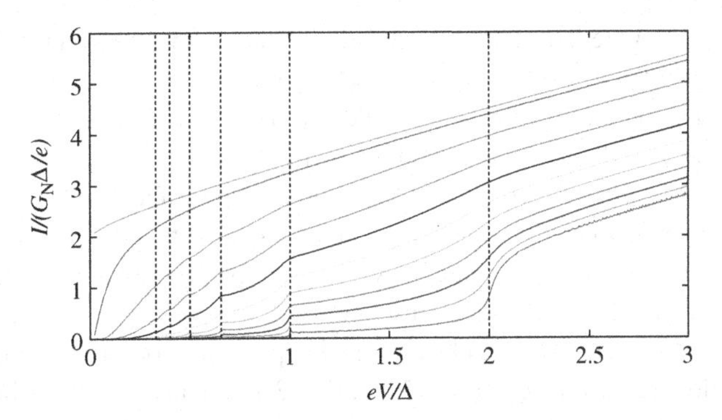

Fig. 1.56. *I–V* curves of a single-channel superconducting junction. The transmission eigenvalue T_p increases from 0.1 (lowest curve) to 1 (upper curve) with step 0.1 except for the curve below the upper curve, for which $T_p = 0.98$. Vertical dotted lines indicate threshold voltages V_1–V_6.

Let us consider a scattering state for the case when a quasiparticle with energy $E < -\Delta$ comes from the left superconducting electrode. At the left superconductor, the energy of electrons and holes stays the same. The corresponding complementary relation for $\boldsymbol{a}$, $\boldsymbol{b}$ does not mix the amplitudes of different energies (different n),

$$\begin{pmatrix} a_{\mathrm{Le},n} \\ a_{\mathrm{Lh},n} \end{pmatrix} = \begin{pmatrix} 0 & r_{\mathrm{A}}^{(n)} \\ r_{\mathrm{A}}^{(n)} & 0 \end{pmatrix} \begin{pmatrix} b_{\mathrm{Le},n} \\ b_{\mathrm{Lh},n} \end{pmatrix} + \begin{pmatrix} u(E) \\ v(E) \end{pmatrix} \delta_{n0}, \tag{1.175}$$

where the Andreev reflection amplitudes $r_{\mathrm{A}}^{(n)}$ are taken at corresponding energies $E_n = E + eVn$ and are given by Eqs. (1.161) and (1.162). We disregard the dependence of the amplitudes on the superconducting phase since the superconducting phase difference is already taken into account by the voltage drop between the electrodes. We can thus conveniently set both $\varphi_{\mathrm{R,L}}$ to 0. The second term in Eq. (1.175) accounts for the incoming quasiparticle at energy E (hence $n = 0$). We learn from Appendix B that a quasiparticle excitation is a superposition of an electron and a hole, with $u, v = ((1 \pm \sqrt{1 - (\Delta/E)^2})/2)^{1/2}$ being the superposition coefficients. The amplitude of the incoming quasiparticle enters the equations for a_n and b_n as a free term. Similar relations hold at the right superconductor. Since no quasiparticle comes from the right, one has simply

$$\begin{pmatrix} a_{\mathrm{Re},n} \\ a_{\mathrm{Rh},n} \end{pmatrix} = \begin{pmatrix} 0 & r_{\mathrm{A}}^{(n)} \\ r_{\mathrm{A}}^{(n)} & 0 \end{pmatrix} \begin{pmatrix} b_{\mathrm{Re},n} \\ b_{\mathrm{Rh},n} \end{pmatrix}. \tag{1.176}$$

The scattering state is found by solving the resulting (in principle, infinite) system of equations for a_n, b_n. The solution is cumbersome and has to be analyzed numerically. Each scattering state, with quasiparticles coming either from the left or from the right at all negative energies, provides a contribution to the current that is obtained by integration over all energies. The result is a function of the ratio eV/Δ and of the transmission coefficient T_p in the channel, $I_p = (G_Q\Delta/e)\mathcal{I}(eV/\Delta, T_p)$ (Fig. 1.56). The total current is the sum over all transmission channels.

We note that $\mathcal{I}$ is strongly suppressed below the voltage $2\Delta/e$ if $T_p \ll 1$. Indeed, the charge transfer below this threshold requires at least one Andreev reflection. To perform this, the electron or hole should traverse the scattering region one more time. The probability of this is suppressed by a factor T_p. Similarly, the charge transfer below V_n requires n Andreev reflections and is suppressed by a factor T_p^n. At large voltages $eV \gg \Delta$, the superconductivity is not important for charge transfer, and current approaches its value in normal metal, $\mathcal{I} \simeq T_p(eV/\Delta)$. In principle, there are singularities in I–V curves at each threshold voltage V_n corresponding to the onset of a process with charge en transferred. The singularities are clearly visible up to $T_p \simeq 0.7$. In the tunneling regime, the singularities are steps (see Section 3.7.2, Eq. (3.102)) that are increasingly rounded upon increasing T_p. The actual singularities that survive the rounding, even at $T_p \to 1$, are jumps of the second derivative of the current and are not visible with the naked eye.

Beside the dc current, the scattering approach allows us to find the harmonics I_m of the ac current at multiples of the Josephson frequency, $I(t) = \sum_m I_m \exp(2eVtm/\hbar)$ [34]. Indeed, each scattering state contributes to the time-dependent current, given by

$$I(t) \propto \left(|\psi_{\mathrm{e}}(\boldsymbol{r}, t)|^2 - |\psi_{\mathrm{h}}(\boldsymbol{r}, t)|^2 \right),$$

where $\psi_{\mathrm{e,h}}$ are electron and hole components of the amplitude. Substituting the time-dependent amplitudes, we see that

$$I_m \propto \sum_n \left(b^*_{\mathrm{Le},n} b_{\mathrm{Le},n+m} - b^*_{\mathrm{Lh},n} b_{\mathrm{Lh},n+m} - a^*_{\mathrm{Le},n} a_{\mathrm{Le},n+m} + a^*_{\mathrm{Lh},n} a_{\mathrm{Lh},n+m} \right).$$

The harmonics also exhibit subgap singularities at $V = V_n$.

It is clear that the processes involving a transfer of multiple charges should lead to interesting and non-trivial full counting statistics; this has been analyzed in Ref. [35] in detail. As we have seen, at low voltages and transmissions, an elementary process of charge transfer involves the transfer of many ($\simeq 2\Delta/eV$) elementary charges. To make an analogy, the electrons do not traverse the nanostructure as separate independent vehicles; rather, they are organized in long trains of $\simeq 2\Delta/eV$ coaches. This enhances the Fano factor, and accounts for its large values. This enhancement has been confirmed experimentally [36] for nanostructures with a well controlled set of the transmissions T_p.

1.8.6 Nanostructure pin-code: experimental

We have already mentioned many times that the scattering and transport properties of a nanostructure are completely determined by the full set of transmission eigenvalues, known as the "pin-code." It is very difficult to crack the pin-code in the course of measurement in the normal state since the Landauer conductance gives only the sum of all T_p. On the contrary, the I–V curves of a superconducting junction are reasonably sensitive to individual eigenvalues T_p. The brilliant experiment described in Ref. [37] demonstrates how one can extract all the relevant transmission eigenvalues just by measuring the I–V curves.

The experiments were performed with superconducting break junctions. In the break junction technique [38], a long and narrow wire is deposited on an elastic substrate. In the

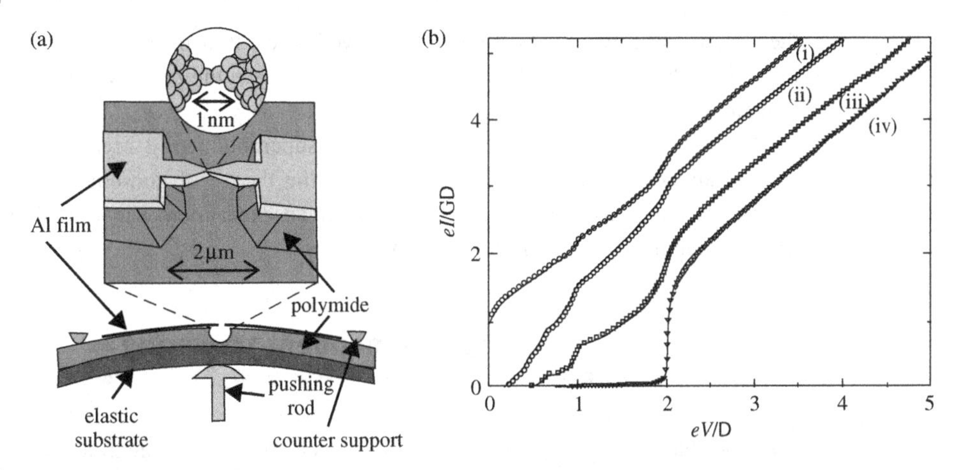

Fig. 1.57. **Experimental determination of nanostructure pin-code [37] (a) Break junction: experimental layout. (b) A fit of I–V curves in the superconducting state reveals the individual transmission eigenvalues. (i) $T_1 = 0.997, T_2 = 0.46, T_3 = 0.29$; (ii) $T_1 = 0.74, T_2 = 0.11$; (iii) $T_1 = 0.46$, $T_2 = 0.35, T_3 = 0.007$; (iv) $T_1 = 0.0025$.**

course of the experiment, the substrate is bent so that the wire stretches and eventually breaks; hence the name "break junctions." The substrate bending can be controlled with high precision, so that it is possible to stabilize the system immediately before the wire breaks. At this moment, the narrowest place of the wire is only several atoms wide, and the voltage drops at narrowest place. Thereby one creates an atomic-size nanostructure with a few open transport channels. One monitors the conductance during the experiment and tunes the nanostructure to any desired value of G.

The samples used in Ref. [37] were suspended aluminum microbridges (Fig. 1.57), 2 μm long and 100 nm thick, constricted in the middle to approximately 100 nm. This is still too wide for a few-channel junction, and further narrowing of the constriction has been achieved with the break junction technique. From both sides, the bridge opens to large (dozens of microns long) pads glued to an elastic organic (polyamide) substrate. The substrate was mounted on a bending mechanism, which was adjusted in such a way that a micron-long displacement of the mechanism resulted in a well controlled change in the distance between the clamping points of the bridge of only 0.2 nm. The samples were first broken and then brought back into contact to form a nanostructure with a few open transport channels. The experiment was performed at ultra-low temperature (~ 1 mK, well below the temperature of the superconducting transition). In the course of the experiment, the clamping points were slowly pushed apart. This diminishes, and effectively reduces the number of, transmission eigenvalues: the conductance goes down. The setup was stable enough that the deformation could be stopped at any point (corresponding to a particular set of transmission eigenvalues) and the dc current versus the applied voltage could be measured at this point.

Fitting the I–V curves using a sum of contributions of individual transport channels, one can very precisely determine all the relevant transmission eigenvalues. The number of transmissions T_p taken into consideration is determined by the accuracy of the fit, and the authors were able to resolve up to the five biggest transmission eigenvalues. Examples of similar fits are presented in Fig. 1.57, where we see the precision of the fits and of the transmission eigenvalues extracted.

The importance of these experiments on superconducting break junctions goes far beyond an experimental check of validity of the theory of non-equilibrium transport in Josephson junctions. The experiments provide an experimental justification of the basics of scattering theory of quantum transport.

1.9 Spin-dependent scattering

Electrons have spin 1/2. This implies that the electron wave function is a two-component quantity – a *spinor*, given by

$$\psi(\boldsymbol{r}) = \begin{pmatrix} \psi_{\uparrow}(\boldsymbol{r}) \\ \psi_{\downarrow}(\boldsymbol{r}) \end{pmatrix},$$

where $\psi_{\uparrow(\downarrow)}$ correspond to the states with spin "up" ("down") with respect to a given axis. Spin is a physical quantity, very much like electric charge or momentum. It has three components x, y, and z, making a pseudovector. The corresponding operator is expressed in terms of the pseudovector of 2×2 *Pauli matrices* $\hat{\boldsymbol{\sigma}}$, $\hat{\boldsymbol{S}} = \hbar\hat{\boldsymbol{\sigma}}/2$ that act on spinors. Frequently, electron spin can be disregarded, as we have been doing so far. In the absence of interactions that explicitly depend on spin, the wave functions $\psi_{\uparrow}$ and $\psi_{\downarrow}$ are identical. The only fact to take into account is that the number of electron states is twice that without spin. In quantum transport, this only leads to the factor 2_{s} in the conductance quantum. In this section, we consider circumstances in which the spin-dependent interactions cannot be disregarded. This may happen due to three factors: spin-splitting in a magnetic field, interaction with an exchange field in ferromagnets, and spin-orbit interaction. All these factors can be incorporated into the scattering matrix, making it spin-dependent.

Zeeman splitting

A magnetic field $\boldsymbol{B}$ does many things to electrons: it produces phase shifts (described in Section 1.6), and it also tries to bend electron trajectories into Larmor circles. These *orbital* effects will be disregarded in this section. The magnetic field also interacts with electron spin, so that the spin-dependent Hamiltonian reads $\hat{H} = g\mu_{\mathrm{B}}\boldsymbol{B} \cdot \hat{\boldsymbol{\sigma}}/2$. Here the combination of fundamental constants $\mu_{\mathrm{B}} = e\hbar/2mc$ is the Bohr magneton, and $g = 2$ for electrons in a vacuum. In semiconductor heterostructures, the value of this g-factor may be significantly modified and even change sign; for example, $g = -0.44$ for electrons in bulk GaAs. Thus, the energy of the state with the spin projection parallel (antiparallel) to the magnetic field is shifted up (down) by $g\mu_{\mathrm{B}}B/2$. This is known as *Zeeman splitting*. To understand the effect of this splitting on quantum transport, let us recall the model of an adiabatic wave guide (Section 1.2) that describes a quantum point contact. Within each transport channel n, the electrons with the spin projection $\pm\hbar/2$ (spin up and spin down) feel different effective potential energies, given by

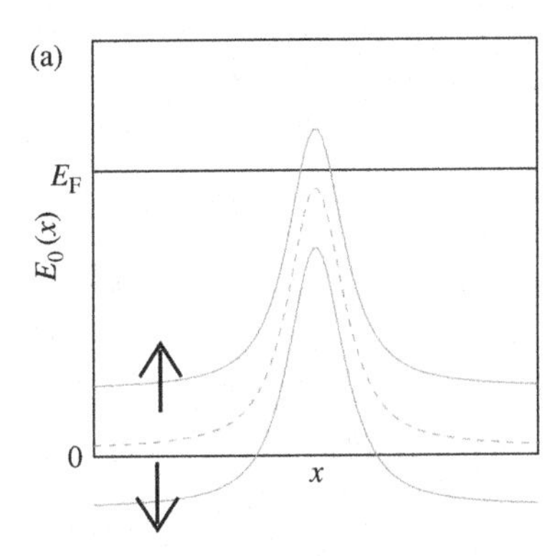

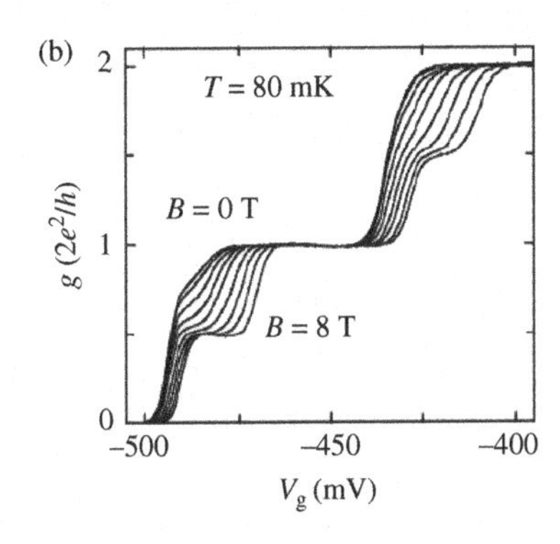

Fig. 1.58. QPC as a spin filter. (a) Electrons see the constriction as a spin-dependent potential barrier (only E_0 is shown). Dotted line: E_0 at zero magnetic field. (b) QPC conductance quantization upon increasing magnetic field. Adapted from Ref. [39].

$$E_{\uparrow,\downarrow} = E_n(x) \pm \frac{g\mu_B B}{2}.$$

At first, it seems that the Zeeman splitting does not affect transport since in a uniform magnetic field the final term does not depend on the coordinate. However, if we concentrate on a given total energy E, electrons with different spin projections see potential barriers of different heights. It is easy to find the energy window in which a certain transport channel is open for spin-down electrons but closed for spin-up electrons. If the chemical potentials of the leads exist in this window, the quantum point contact (QPC) works as a *spin filter* – it only lets through the electrons with spin down (see Fig. 1.58). If, under these circumstances, one changes the number of open channels with gate voltage, one sees extra plateaus arising at half-integer values of the conductance $G_Q/2, 3G_Q/2, \ldots$ This is the simplest illustration of spin-dependent scattering.

Exercise 1.22. Find the quantitative form of the QPC linear conductance versus gate voltage in the presence of a spin-polarizing magnetic field. Use Eq. (1.23), assuming a linear dependence of the barrier height W_1 on the gate voltage, $W_1 - \mu = \alpha(V_g^{(0)} - V_g)$ ($V_g^{(0)}$ is the gate voltage corresponding to the middle of the first conductance step in the absence of the magnetic field). How big should the magnetic field be to enable the resolution of $G_Q/2$ plateau?

Ferromagnets

A uniform magnetic field favors a certain projection of spin, for example spin down in all parts of the sample – in all reservoirs as well as the scattering region. It is much more interesting if different projections of the spin are favorable in different regions. This could be achieved with a magnetic field that is directed differently in different regions. However, such a magnetic field is difficult to create, and besides, the Zeeman energy for magnetic

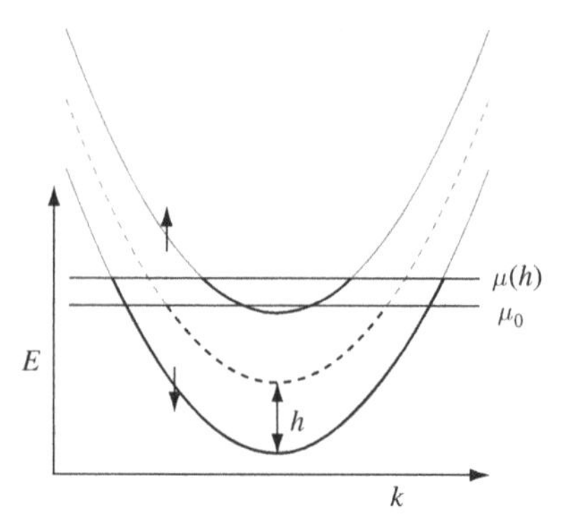

Fig. 1.59. **Energy bands in a ferromagnet. Spin-splitting results in different numbers of spin-down and spin-up electrons. Thicker lines: occupied states; dotted line: $h = 0$.**

fields available remains numerically small, not exceeding 10 meV. Instead, one can use *ferromagnets* (more precisely, ferromagnetic metals). In ferromagnets, there is a preferential direction for electron spins – that of the magnetization of the ferromagnet. This results from the spontaneous breaking of time-reversal symmetry. Indeed, a magnet reversed in time has the opposite magnetization. Electrons with different spin projections on the magnetization axis have different energies corresponding to the same quasimomentum, $E_{\uparrow}(\boldsymbol{k}) \neq E_{\downarrow}(\boldsymbol{k})$. There is a spin-splitting similar to Zeeman splitting. The energy difference is called the *exchange field* since it arises from an exchange interaction between electron spins, and its $\boldsymbol{k}$-dependence is frequently disregarded. The exchange field may be as high as 1 eV, several orders of magnitude greater than the Zeeman splitting due to the magnetic field in the magnet.

The exchange field $\boldsymbol{h}$ is a pseudovector quantity directed along the magnetization. Combining magnets with different orientations of the magnetization and/or normal metals with no exchange field, one achieves the situation where favorable spin directions are different across the nanostructure.

Exercise 1.23. Let us consider a simplified model of a ferromagnetic metal with a parabolic spectrum and $\boldsymbol{k}$-independent exchange field.

$$E_{\uparrow,\downarrow}(\boldsymbol{k}) = \frac{\hbar^2 k^2}{2m} \pm |h|$$

(see Fig. 1.59). (a) Calculate the density of electrons of different spin directions, $n_{\uparrow,\downarrow}$, at given μ making use of Eq. (1.2). (b) The total density of electrons remains fixed, independent of the exchange field. Express the chemical potential in terms of the density and the exchange field. (c) At which exchange field are all electrons polarized in the same direction? Express this field in terms of μ_0, the chemical potential at zero exchange field.

Spin-orbit interaction

There is another spin-dependent interaction which, in contrast to Zeeman splitting and ferromagnetism, does not require breaking of time-reversal symmetry, and therefore in principle persists in all materials. This interaction takes the following form:

$$\hat{H}_{\mathrm{SO}} = \frac{\mu_{\mathrm{B}}}{2c}(\hat{\boldsymbol{\sigma}} \cdot \boldsymbol{E} \times \hat{\boldsymbol{v}}), \tag{1.177}$$

where $\boldsymbol{v} \equiv -(\mathrm{i}\hbar/m)\partial/\partial\boldsymbol{r}$ is the velocity operator and $\boldsymbol{E}$ is the electric field originating from either the crystal lattice potential or various defects in the nanostructure. This interaction has a simple interpretation: special relativity implies that an electron moving with velocity $v \ll c$ in the electric field "sees" an extra magnetic field $\boldsymbol{B} = (\boldsymbol{v} \times \boldsymbol{E})/c$ that provides Zeeman splitting of its spin states. Note, however, that this simple interpretation produces a wrong factor of 2 in the above relation.

Without spin-orbit interaction, electron states in a crystal solid can be regarded as plane waves (Bloch states) with a certain wave vector $\boldsymbol{k}$. Surprisingly, the spin-orbit interaction brings almost no change to this picture. The reason for this is time-reversal symmetry, which requires $E_{\downarrow}(\boldsymbol{k}) = E_{\uparrow}(-\boldsymbol{k})$. Thus, if the crystal lattice possesses inversion symmetry, the energy does not change upon inversion of $\boldsymbol{k}$, $E(\boldsymbol{k}) = E(-\boldsymbol{k})$. From this it follows that $E_{\downarrow}(\boldsymbol{k}) = E_{\uparrow}(\boldsymbol{k})$, so that spin-orbit interaction does not lead to any spin-splitting and the electron states remain double-degenerate. If we fill a spin-up state with an electron, the state remains occupied – spin is conserved despite the presence of spin-orbit interaction.

This drastically changes if spin-orbit interaction is combined with scattering. In this case, the scattered electron not only changes its momentum, but also can change – flip – its spin. The probability of the flip is proportional to the strength of the spin-orbit interaction. We see that spin-orbit interaction is a relativistic effect that must be weak for electrons moving with typical velocities $v \lesssim 10^{-2}c$. The probability of the flip contains a small factor $(v/c)^4$. This small factor is compensated by Z^4, where Z is the atomic number of the solid constituents. We see that the flip probability is negligible for solids made up of from light atoms, approaches $1/2$ for heavier atoms, and is about 10^{-4} for elements from the middle of the periodic table.

Thus spin-flip scattering does not conserve spin, so that electrons in bulk materials keep their spin only for a typical time τ_{sf}. Since the flip probability is small, this time is much longer than the typical scattering time τ it takes an electron to forget the direction of its momentum.

We should note that spin-orbit interaction is not the only source of spin-flip scattering. Electrons can flip spin while scattering on localized magnetic moments, which we discuss in detail in Section 6.6.2.

Exercise 1.24. For a GaAs-based two-dimensional electron gas, the spin-dependent part of the Hamiltonian is given by ($z \perp$ the 2DEG plane)

$$\hat{H}_{\mathrm{SO}} = \alpha(\hat{\sigma}_x k_y - \hat{\sigma}_y k_x) + \beta(\hat{\sigma}_x k_x - \hat{\sigma}_y k_y).$$

The first term is called the Rashba interaction and originates from the asymmetry of the electron confinement in the z direction. The second term, the Dresselhaus interaction, originates from the absence of inversion symmetry in the bulk GaAs. Show that at given $k_{x,y}$ this Hamiltonian amounts to an effective exchange field. Find the direction of the field and associated spin-splitting energy.

1.9.1 Scattering matrix with spin

In the presence of a spin-dependent interaction, the elements of the scattering matrix depend not only on the incoming and outgoing transport channels p, p', but also on the spin projections $\alpha, \beta = \uparrow, \downarrow$ of incoming and outgoing states: $s_{p',\beta;p,\alpha}$. If we disregard spin-dependent interactions, as we did in previous sections, $s_{p',\beta;p,\alpha} = s_{p';p}\delta_{\alpha\beta}$; the scattering conserves the spin projection and does not depend on it. The scattering matrix can be presented with the aid of Pauli matrices,

$$s_{p',\beta;p,\alpha} = s^{(0)}_{p',p}\delta_{\alpha\beta} + \boldsymbol{s}_{p',p} \cdot \boldsymbol{\sigma}_{\beta\alpha}, \tag{1.178}$$

where $\hat{s}^{(0)}$ and the pseudovector $\hat{\boldsymbol{s}}$ are matrices in the space of transport channels.

Unitarity and time-reversal symmetry impose several conditions on the spin dependence of the scattering matrix. Let us start with the spin-orbit interaction when the time reversibility holds, so that, for any permitted process, the time-reversed process is also permitted and has the same amplitude. A normal wave function (for example, of spinless particles) is complex-conjugated under time-reversal. It is more complicated with spinors: under time reversal a spinor ψ is transformed to ψ^{tr}:

$$\psi^{tr} = \hat{g}\psi^*, \quad \hat{g} = \begin{pmatrix} 0 & -1 \\ 1 & 0 \end{pmatrix}. \tag{1.179}$$

This is the only transformation that assures that the spin operator flips sign under time reversal, $\boldsymbol{S} \to -\boldsymbol{S}$. Consider now an initial state characterized by the spinor $\psi_{\rm i}$, which is transformed to the final one, $\psi_{\rm f}$, by means of the scattering matrix $\hat{s}$, $\psi_{\rm f} = \hat{s}\psi_{\rm i}$. The time-reversed process is characterized by the same amplitude, which means $\psi_{\rm i}^{\rm tr} = \hat{s}\psi_{\rm f}^{\rm tr}$. Using Eq. (1.179) we obtain

$$\psi_{\rm i}^* = -\hat{g}\hat{s}\hat{g}\psi_{\rm f}^* \Rightarrow \psi_{\rm i} = -\hat{g}\hat{s}^*\hat{g}\psi_{\rm f}. \tag{1.180}$$

Taking into account the unitarity of $\hat{s}$, we come to the symmetry relation

$$\hat{s}^{\rm T} = -\hat{g}\hat{s}\hat{g}. \tag{1.181}$$

In Pauli notation, this reads as follows:

$$s^{(0)}_{p',p} = s^{(0)}_{p,p'}; \ \boldsymbol{s}_{p',p} = -\boldsymbol{s}_{p',p}. \tag{1.182}$$

Note that if the scattering matrix does not have any spin structure ($\hat{\boldsymbol{s}} = 0$), Eq. (1.181) simply states that it is symmetric – the condition we have used throughout the preceding sections. From these symmetries, one proves that transmission eigenvalues – eigenvalues of $\hat{t}^\dagger\hat{t}$ – remain doubly degenerate in the presence of a spin-orbit interaction. This is analogous to the well known Kramer theorem in quantum mechanics, which states that the levels – the eigenvalues of a Hamiltonian – remain doubly degenerate with respect to spin provided time reversibility holds.

If the spin-dependent part of the scattering matrix arises due to either a magnetic or an exchange field, its time-reversal properties are exactly opposite, since $\hat{\boldsymbol{s}}$ is proportional to this field and the field changes sign upon time reversal. If spin-dependent scattering is weak, as is usually the case, $\hat{s}^{(0)} \gg \hat{\boldsymbol{s}}$, and the whole spin-dependent part of the scattering matrix can be presented as a sum of two independent contributions $\hat{\boldsymbol{s}} = \hat{\boldsymbol{s}}^{(\mathrm{SO})} + \hat{\boldsymbol{s}}^{(\mathrm{M})}$, where, by definition,

$$\boldsymbol{s}^{(\mathrm{M})}_{p',p} = \boldsymbol{s}^{(\mathrm{M})}_{p',p};\ \boldsymbol{s}^{(\mathrm{SO})}_{p',p} = -\boldsymbol{s}^{(\mathrm{SO})}_{p',p}. \quad (1.183)$$

Another set of conditions arises from the unitarity of the scattering matrix. If $\hat{s}^{(0)} \gg \hat{\boldsymbol{s}}$, the spin-independent part $\hat{s}^{(0)}$ is approximately unitary. As for the spin-dependent part, in this limit it satisfies

$$\hat{s}^{(0)\dagger}\hat{\boldsymbol{s}} + \hat{\boldsymbol{s}}^{\dagger}\hat{s}^{(0)} = 0$$

for all cases: spin-orbit interaction, exchange field, and magnetic field. This yields a relation between the real and imaginary parts of $\hat{\boldsymbol{s}}$, generally speaking, not a simple one. To make it simple, one presents the (unitary) scattering matrix as a matrix exponent of a Hermitian matrix, $\hat{s} = \exp(\mathrm{i}\hat{H})$. If we separate $\hat{H}$ into a spin-dependent part $\hat{H}^{(0)}$ and a spin-independent part $\hat{\boldsymbol{H}}$, the unitarity requires that $\hat{\boldsymbol{H}} = \hat{\boldsymbol{H}}^{\dagger}$, and time reversibility yields

$$\boldsymbol{H}^{(\mathrm{M})}_{p',p} = \boldsymbol{H}^{(\mathrm{M})}_{p',p};\ \boldsymbol{H}^{(\mathrm{SO})}_{p',p} = -\boldsymbol{H}^{(\mathrm{SO})}_{p',p}. \quad (1.184)$$

We illustrate the above relations using a one-channel scatterer. Without spin, the scattering matrix is a symmetric 2×2 matrix. With spin, it becomes a 4×4 matrix, or, equivalently, is characterized by four 2×2 matrices $\hat{s}^{(0)}, \hat{\boldsymbol{s}}$. We concentrate on the case of weak spin-dependent scattering, with $\hat{s}^{(0)}$ given by

$$\hat{s}^{(0)} = \begin{pmatrix} r & t \\ t & r' \end{pmatrix}.$$

The spin-orbit part can be presented as follows:

$$\hat{\boldsymbol{s}}^{(\mathrm{SO})} = \boldsymbol{\xi} t \begin{pmatrix} 0 & \mathrm{i} \\ -\mathrm{i} & 0 \end{pmatrix},$$

with real vector $\boldsymbol{\xi} \ll 1$. Note that the reflection matrix does not depend on the spin, so the electrons can only be reflected with the same spin. For two or more transport channels, electrons can flip the spin only if they are reflected to a different channel. The probability of spin-flip occurring during transmission depends on the polarization of the incoming electron, given by $p = (\boldsymbol{\xi} \times \boldsymbol{n})^2$ for an electron with spin in the direction of unit vector $\boldsymbol{n}$.

Control question 1.26. What is the average (over initial spin directions) spin-flip probability?

The spin-flip probability vanishes for $\boldsymbol{\xi} \parallel \boldsymbol{n}$. If there are more channels, the vectors $\boldsymbol{\xi}$ in different channels generally point out in different directions, so that the spin-flip probability never vanishes.

We restrict the discussion of the spin effects of exchange and magnetic field to the case when the vectors $\boldsymbol{s}^{(\mathrm{M})}_{p',p}$ are all parallel to the same fixed unit vector $\boldsymbol{m}$. This is the case when there is only one favored spin direction in the nanostructure, for example when the magnetization directions of all ferromagnetic reservoirs are collinear, or there is only one ferromagnetic reservoir, or there is a uniform external magnetic field. In this case, we have two separate unitary scattering matrices for spins up and down with respect to $\boldsymbol{m}$, i.e.

$$\hat{s} = \hat{s}_\uparrow \frac{1 + \boldsymbol{m} \cdot \boldsymbol{\sigma}}{2} + \hat{s}_\downarrow \frac{1 - \boldsymbol{m} \cdot \boldsymbol{\sigma}}{2}, \tag{1.185}$$

where, for a one-channel conductor, one has

$$\hat{s}_\uparrow = \begin{pmatrix} r_\uparrow & t_\uparrow \\ t_\uparrow & r'_\uparrow \end{pmatrix}; \quad \hat{s}_\downarrow = \begin{pmatrix} r_\downarrow & t_\downarrow \\ t_\downarrow & r'_\downarrow \end{pmatrix}. \tag{1.186}$$

There are two distinct transmission eigenvalues $T_\uparrow$ and $T_\downarrow$. If the incoming electrons are spin-polarized in the direction of $\boldsymbol{m}$, they do not experience any spin-flip. It looks like there are two species of electrons, with spins up and down, that are scattered differently, and that the number of electrons of each species is conserved. We stress that this does not happen if the electrons are polarized in any other direction. The probabilities of spin-flip reflection and transmission are given by $(\boldsymbol{n} \times \boldsymbol{m})^2 |r_\uparrow - r_\downarrow|^2/4$ and $(\boldsymbol{n} \times \boldsymbol{m})^2 |t_\uparrow - t_\downarrow|^2/4$, respectively. Spin-flip may occur even for closed channels with $|r_\uparrow|^2 = |r_\downarrow|^2 = 1$.

1.9.2 Spin currents

Spin is a quantity that can be transported with electron flow, very much like charge. Let us consider a flow of electrons with spin up and a flow of electrons with spin down in the opposite direction. In this situation, there is no net electric current. However, there is *spin current* – the spin component in the direction of the quantization axis is transferred by the flow. In contrast to the electric current, spin current is a current of a pseudovector quantity. The spin current density operator has nine components and is given by

$$\hat{j}^{(\mathrm{S})}_{\alpha;\beta} = \frac{\hbar}{2} \hat{\sigma}_\alpha \hat{v}_\beta,$$

$\hat{\sigma}_\alpha$ and $\hat{v}_\beta$ being Cartesian components of the Pauli matrix vector and velocity operator, respectively. The total spin current in a nanostructure is determined as the integral of the current density over a cross-section and has three components corresponding to three Pauli matrices.

It would be nice if spin current were a conserving quantity, and in the absence of spin-dependent scattering this is indeed the case. However, without spin-dependent interaction the spin current can be neither excited nor measured, at least by electric means.

As we mentioned above, spin and spin currents are conserved at the time scale τ_{sf} and, consequently, at the length scale L_{sf} covered by moving electrons during this time. In practice, this scale rarely exceeds several micrometers. Transfer of spin leads to *spin accumulation* at this length scale, resulting in non-equilibrium spin-dependent chemical potentials $\mu_{\uparrow,\downarrow}$. It is convenient to regard the parts of the leads adjacent to the nanostructure at space scale L_{sf} as "spin-reservoirs" where spin and spin current are conserved. This does not correspond to the current definition of a reservoir, and in principle the "spin-reservoirs" can be regarded as parts of the nanostructure, so that spin current is both excited and detected within the nanostructure. We will discuss these issues extensively in Section 2.7; at the moment we just make use of the concept of "spin-reservoir." We note that there are two types of "spin-reservoirs": normal, where all three components of spin current are conserved and all three components of spin can be accumulated, and ferromagnetic. In a ferromagnet, only one component of spin – that parallel to the magnetization – is conserved, and only this component can be accumulated. The reason for this is that scattering in a ferromagnetic "spin-reservoir" is different for different spin projections. As we have seen, this causes spin-flip for spin components perpendicular to the magnetization axis and they are not conserved.

Let us illustrate spin and electric currents with a single-channel scatterer, described by Eq. (1.186), placed between ferromagnetic (left) and normal (right) "spin-reservoirs." First, let us concentrate on the spin components in the direction of magnetization (z axis). Regarding electrons as "species" and applying the Landauer formula to the particle currents $J_{\uparrow,\downarrow}$ of each species, we immediately obtain

$$\begin{pmatrix} J_\uparrow \\ J_\downarrow \end{pmatrix} = \frac{G_Q}{2e}\begin{pmatrix} T_\uparrow[\mu_{L\uparrow} - \mu_{R\uparrow}] \\ T_\downarrow[\mu_{L\downarrow} - \mu_{R\downarrow}] \end{pmatrix}, \tag{1.187}$$

where the different chemical potentials $\mu_{\uparrow,\downarrow}$ account for spin accumulation in the "spin-reservoirs." We define the spin accumulation as $W_z^{L,R} \equiv (\mu_{L,R\uparrow} - \mu_{L,R\downarrow})/\hbar$.

Since electric current and the z component of spin current are given by $e(J_\uparrow + J_\downarrow)$ and $\hbar(J_\uparrow - J_\downarrow)/2$, respectively, we obtain

$$\begin{aligned} I &= GV + \frac{\hbar G_P}{2e}(W_z^L - W_z^R), \\ I_z^{(S)} &= \frac{\hbar G_P}{2e}V + \frac{\hbar^2 G}{4e^2}(W_z^L - W_z^R), \end{aligned} \tag{1.188}$$

where V is the voltage difference between left and right, G is the Landauer conductance of the nanostructure, and G_P characterizes the spin-polarization properties of the contact,

$$G, G_P = G_\uparrow \pm G_\downarrow = G_Q \frac{T_\uparrow \pm T_\downarrow}{2}.$$

We see from this that spin current is generated by applying a voltage difference across the contact, provided $T_\uparrow \neq T_\downarrow$. In addition, spin accumulation, which can arise in the course of spin transport, generates electric current and can be detected. Note the remarkable symmetry: both effects are characterized by the same coefficient G_P. This is an example of Onsager relations for generalized fluxes and forces.

What about other components of spin current? The situation is rather unusual. As we have mentioned, the ferromagnet supports x, y components of neither spin current nor spin accumulation. Thus, $I_{x,y}^{(S)} = 0$ on the left of the contact. Normally, one concludes that currents on the right are also absent. Such a conclusion, however, relies on current conservation in the scattering region, and this does not hold for x, y components of the spin current! Let us assume that some spin accumulation $W^{\mathrm{R}}_{x,y}$ has been created in the normal "spin-reservoir" and evaluate the resulting spin current.

This can be achieved in a way similar to the derivation of the Landauer formula in Section 1.3. The difference is that a given state coming to the scatterer from the right is a spinor, $\psi_{\mathrm{i}} \exp(-ik_x x)$, whereas the corresponding outgoing state is $\psi_{\mathrm{f}} \exp(ik_x x)$, ψ_{i} and ψ_{f} being related by the scattering matrix. Both states contribute to the density of spin current,

$$j^{(\mathrm{S})}_{\alpha} = \frac{\hbar}{2} v_x \left(\psi_{\mathrm{i}}^* \hat{\sigma}_{\alpha} \psi_{\mathrm{i}} - \psi_{\mathrm{f}}^* \hat{\sigma}_{\alpha} \psi_{\mathrm{f}}\right); \; \psi_{\mathrm{f}} = \begin{pmatrix} r_{\uparrow} & 0 \\ 0 & r_{\downarrow} \end{pmatrix} \psi_{\mathrm{i}}.$$

Substituting spinors polarized in the x, y directions, and integrating over k_x or energy, we obtain the spin currents:

$$\begin{aligned} I^{(\mathrm{S})}_x &= \frac{\hbar^2}{4e^2}\left(\mathrm{Re}\, G_{\uparrow\downarrow} W^{\mathrm{R}}_x + \mathrm{Im}\, G_{\uparrow\downarrow} W^{\mathrm{R}}_y\right), \\ I^{(S)}_y &= \frac{\hbar^2}{4e^2}\left(\mathrm{Re}\, G_{\uparrow\downarrow} W^{\mathrm{R}}_y + \mathrm{Im}\, G_{\uparrow\downarrow} W^{\mathrm{R}}_x\right), \end{aligned} \tag{1.189}$$

where the complex conductance $G_{\uparrow\downarrow}$ is given by

$$G_{\uparrow\downarrow} = G_{\mathrm{Q}}(1 - r_{\uparrow} r^*_{\downarrow}).$$

This coefficient is called the *mixing* conductance since it highlights the fact that electrons cannot really be considered as two independent species with spin up and spin down, and the scattering eventually mixes the up and down components of the spinors. We see that there is some mixing conductance even if the usual conductance is absent ($T_{\uparrow} = T_{\downarrow} = 0$), for example, at the boundary between a metal and a ferromagnetic insulator. There, the spin current flows even in the absence of actual electron transfer. It arises from spin precession of the electrons that hit the boundary, feeling the exchange field in the insulator.

For a non-polarizing contact ($r_{\uparrow} = r_{\downarrow}$), the mixing conductance coincides with the usual one, along with the spin-dependent conductances: $G_{\uparrow\downarrow} = G_{\uparrow} = G_{\downarrow} = G$. The resulting relations can readily be generalized for multi-channel conductors. The spin-dependent conductances, similar to the Landauer conductances, are simply the sums of the corresponding transmission eigenvalues, i.e.

$$G^{\alpha} = \frac{G_Q}{2} \sum_{pq} |t^{\alpha}_{pq}|^2 = \frac{G_{\mathrm{Q}}}{2} \sum_{p} T^{\alpha}_p. \tag{1.190}$$

The complex mixing conductance cannot be expressed in the transmission eigenvalues since, as mentioned above, it is not entirely determined by transmission. However, it is related to the reflection matrices at the normal-metal side of the contact:

$$G^{\uparrow\downarrow} = G_{\mathrm{Q}} \sum_{pq} \left[\delta_{pq} - r^{\uparrow}_{pq} (r^{\downarrow}_{pq})^*\right]. \tag{1.191}$$

It follows that $\mathrm{Re}\, G^{\uparrow\downarrow} \geq (G^{\uparrow} + G^{\downarrow})/2$.

1.9.3 Spin and interference

It turns out that the spin-orbit interaction may strongly affect quantum interference of electron waves at length scales exceeding L_{sf}. We analyze the details of this effect in Section 4.4. Here we give a simple explanation stemming from the fundamentals of spin.

In Section 1.6, when discussing interference, we identified pairs of electron trajectories, the interference of which provides the so-called weak localization contribution to scattering probabilities. This contribution does not disappear upon averaging over random phase shifts and shows up in the Aharonov–Bohm effect.

These trajectories contain a loop and differ by the direction in which the loop is traversed. Propagation amplitudes in the loop for these two directions, $A_{1,2}$, correspond to the same scattering process with initial and finite states interchanged. If these amplitudes are regarded as matrices, they are thus transposed, $A_1 = A_2^T$. In the absence of AB flux, time reversibility implies that $A_1 = A_2$. Then, the probability is determined by $|A_1 + A_2|^2 = 4|A_1|^2$: the interference contribution $P^{\text{int}} = A_1^* A_2 + A_1 A_2^*$ is the same as the classical contribution $P^{\text{cl}} = |A_1|^2 + |A_2|^2$. There is *constructive interference* of the two trajectories. AB flux modulates interference, $P = P^{\text{int}} \cos(2\phi_{\text{AB}}) + P^{\text{cl}}$, suppressing the probability at small values of flux.

This changes in the presence of spin-orbit scattering. The amplitudes A_1 and A_2 now become 2×2 matrices in the spin space. The relation between these transmission amplitudes follows from Eq. (1.182). Therefore, if $\hat{A}_1 = A_0 \hat{1} + \boldsymbol{A} \cdot \hat{\boldsymbol{\sigma}}$, another amplitude reads $\hat{A}_2 = A_0 \hat{1} - \boldsymbol{A} \cdot \hat{\boldsymbol{\sigma}}$. The total probability is obtained by summing up over two initial and two final spin states, labeled α, β:

$$P = \sum_{\alpha\beta} |A_1^{\alpha\beta} e^{i\phi_{\text{AB}}} + A_2^{\alpha\beta} e^{-i\phi_{\text{AB}}}|^2 = P^{\text{cl}} + P^{\text{int}},$$

$$P^{\text{cl}} = \sum_{\alpha\beta} \left(|A_1^{\alpha\beta}|^2 + |A_2^{\alpha\beta}|^2 \right) = 4(|A_0|^2 + \boldsymbol{A} \cdot \boldsymbol{A}^*), \tag{1.192}$$

$$P^{\text{int}} = 4(|A_0|^2 - \boldsymbol{A} \cdot \boldsymbol{A}^*) \cos(2\phi_{\text{AB}}).$$

The total probability at zero flux, $P(0) = 8|A_0|^2$, is not sensitive to the spin-dependent part of scattering, but the interference contribution is. The sign of the interference contribution depends on the relation between spin-dependent and spin-independent scattering, and the latter depends on the length of the loop. For loops shorter than L_{sf}, the spin-independent scattering dominates, $|A_0| \gg |\boldsymbol{A}|$, and interference remains constructive, $P(0) = 2P^{\text{cl}} > P^{\text{cl}}$. For loops longer than L_{sf}, transmissions with and without spin-flip at any initial spin-polarization should have the same weight: the spin is forgotten in the course of scattering. This implies that $|A_0| = |A_x| = |A_y| = |A_z|$, and $P(0) = P^{\text{cl}}/2 < P^{\text{cl}}$ – i.e. that the interference is destructive. Thus, a sufficiently strong spin-flip scattering charges the sign of the interference correction; this is often referred to as *weak antilocalization*.

2 Classical and semiclassical transport

We devoted Chapter 1 to a purely quantum-mechanical approach to electron transport: the scattering approach. Electrons were treated as quantum waves that propagate between reservoirs – the contact pads of a nanostructure. The waves experience scattering, and the transport properties are determined by the scattering matrix of these waves. As we have seen, this approach becomes progressively impractical with the increasing number of transport channels, and can rarely be applied for $G \gg G_Q$, where G is the conductance of the system.

A different starting point is well known from general physics, or, more simply, from general life experience, which is rather classical. In this context, a nanostructure is regarded as an element of an electric circuit, which conducts electric currents. If one makes a more complicated circuit by combining these elements, one does not have to involve quantum mechanics to figure out the result. Rather, one uses Ohm's law or, generally, Kirchhoff rules. The number of parameters required for this description is fewer than in the quantum-mechanical scattering approach. For example, the phase shifts of the scattering matrix do not matter.

In this chapter, we will bridge the gap between these opposite starting points. The first bridge is rather obvious: it is important to understand that these two opposite approaches do not contradict each other. In Section 2.1, we illustrate the difference and the link between the approaches with a comprehensive example of a double-junction nanostructure. We derive the rules of common circuit theory, that is, a classical description of electron transport, *from* quantum mechanics, using conservation laws for number-of-particles, charge, heat, and balance equations that follow from these laws. In Section 2.2 we present the traditional description of electron transport in solids based on Boltzmann and drift-diffusion equations. Further, we see how it evolves into a finite-element approach of a circuit theory.

We will build another bridge by considering *semiclassical* transport. To this end, we introduce Green's functions, which allow us to take quantum coherence into account (Section 2.3). We will show that quantum mechanics eventually brings about some new conservation laws and balance equations that are absent in classical physics (Section 2.4). From these laws, we will derive a template for a quantum circuit theory (Section 2.5). Thereby the interference effects, which constitute the difference between classical and quantum transport, can be incorporated into a circuit theory.

After establishing the rules of the game, we can use the template by assigning the physical significance to its components. Thereby, we make quantum circuit theories suitable for specific tasks. We consider a number of specific circuit theories describing distribution of transmission eigenvalues (Section 2.6), spin transport (Section 2.7), superconductivity (Section 2.8), and full counting statistics (Section 2.9).

2.1 Disorder, averaging, and Ohm's law

2.1.1 Double-junction paradox

To appreciate the points we are to make in this chapter, we first consider an example of a double tunnel junction, studied in detail in Section 1.6. Let us present it in a somewhat paradoxical way by confronting two naive estimates of the conductance of this system. Consider the situation when both scatterers are tunnel junctions, so that $t_{L,R} \ll 1$.

One way to look at it is to observe that an electron should hop over both junctions in order to go from the left to the right. According to the rules of quantum mechanics, the amplitude A of such a process should be proportional to the product of partial amplitudes, $A \propto t_L t_R$. The amplitudes of more complicated processes that involve reflections from the junctions should be proportional to the higher powers of $t_{L,R}$ and thus can be disregarded. The Landauer formula states that the conductance of each junction is proportional to the absolute value of the corresponding amplitude squared, $G_{L,R} \propto |t_{L,R}|^2$, and the overall conductance is proportional to $|A|^2$. We come to the conclusion that the overall conductance scales as a product of conductances of the two tunnel junctions,

$$G \sim G_L G_R / G_Q.$$

Now, let us employ our common sense. Each tunnel junction is a resistor. Resistances of the junctions are inverse conductances, $1/G_{L,R}$. Ohm's law states that the overall resistance of two resistors in series is a sum of the two resistances:

$$G = \frac{1}{1/G_L + 1/G_R} = \frac{G_L G_R}{G_L + G_R}.$$

There is an apparent contradiction between these two estimations. To make it even more obvious, let us reduce the conductance of each junction by a factor of 2. The "quantum" formula yields that overall conductance is decreased by a factor of 4, whereas Ohm's law predicts the factor of 2. This is the *double-junction paradox* (see Fig. 2.1). Which estimation is the correct one?

Fortunately, we know the full quantum-mechanical solution from Section 1.6. Neither of the estimates, "classical" nor "quantum," is correct. The total conductance of a double junction cannot be expressed only in terms of the conductances of the two junctions. Instead, it shows a resonant structure as a function of the phase shift χ acquired by an electron when traveling between the junctions (see Eq. (1.106)):

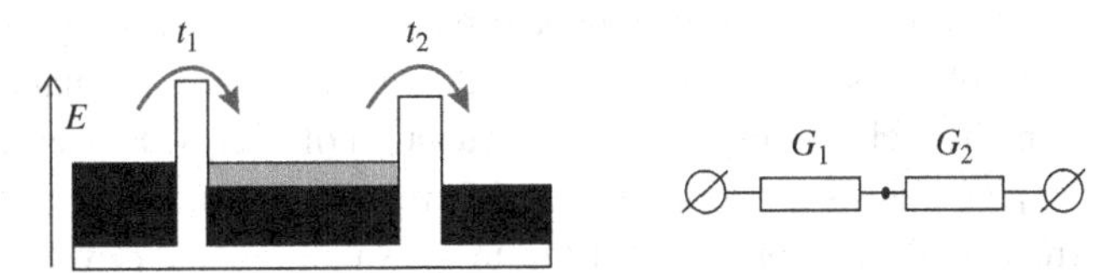

Fig. 2.1. Double tunnel junction. A quick quantum estimate of its conductance and Ohm's law seem to contradict each other.

$$G = G_Q \frac{T_L T_R}{1 + R_L R_R + 2\sqrt{R_L R_R} \cos \chi}, \quad R_{L,R} = 1 - T_{L,R}.$$

This is a very important and general statement (it is not limited to the double junction): although transport properties of a single nanostructure depend on transmission eigenvalues only, the transport properties of a combined nanostructure are not uniquely determined by the transmission eigenvalues of its constituents. They also depend on phase shifts.

In our example, the "quantum" estimate works if we are far from the resonance, $\cos \chi \neq -1$. In this case, the conductance scales as $G_L G_R$. At the resonance, $\cos \chi = -1$, the conductance does not scale at all; for $G_L = G_R$ it always stays equal to G_Q, for example.

There is, however, an easy way to obtain Ohm's law from the quantum-mechanical result. In fact, the classical expression for the double-junction conductance, Eq. (1.110), can be obtained by averaging $T(\chi)$ over the phase shift, $T_{cl} = \langle T(\chi) \rangle_\chi$. This proves that the averaging over χ is equivalent to summing the probabilities rather than amplitudes – precisely the approach of classical physics. It is important to note that this procedure of averaging has an immediate physical sense. As we have discussed in Section 1.6, the acquired phase is proportional to energy, $d\chi/dE \propto \tau/\hbar$, τ being a typical time of electron propagation between the junctions. Therefore the averaging over χ is equivalent to the averaging of the transmission eigenvalue over a wide energy interval, as the Landauer formula suggests.

Exercise 2.1. Average Eq. (1.106) over the phase shift χ and recover Ohm's law in the limit $T_{L,R} \ll 1$.

The analysis of a single-channel double junction brings us to two important ideas. First, Ohm's law is reproduced only if we average over phase shifts, thus disregarding quantum interference. Secondly, this averaging makes physical sense: it corresponds to an averaging over wide energy bands.

Let us turn to a multi-channel double junction. To start with, consider a simple model of *independent* channels, where an incoming electron may go back and forth between the junctions but always remains in the same channel. For this simple model, the conductance is just a sum over channels,

$$G = \sum_p \frac{T_{L,p} T_{R,p}}{1 + R_{L,p} R_{R,p} + 2\sqrt{R_{L,p} R_{R,p}} \cos \chi_p}, \tag{2.1}$$

with transmission coefficients and phase factors depending on the channel index p. The result of the summation is sketched in Fig. 2.2 for six channels. The contribution from each channel is a regular periodic function of energy. However, the periods and offsets of the phase shifts are different for different channels. This is why the result exhibits small irregular fluctuations around the mean value; i.e., *self-averaging* takes place. The more channels that take part in the sum, the smaller the relative fluctuations. In the limit of a large number of channels, we expect that the conductance is close to its average value.

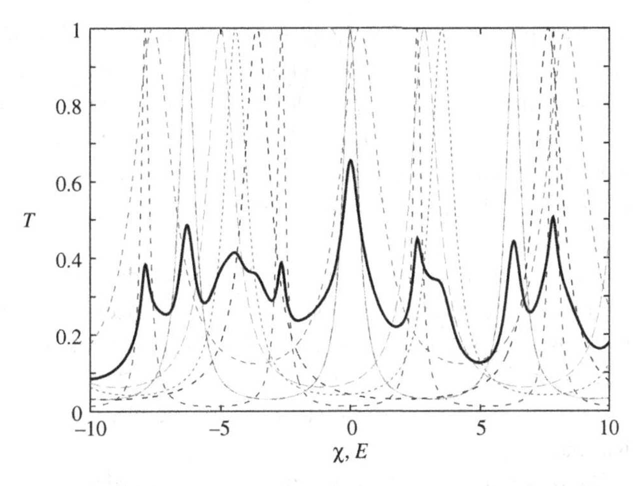

Fig. 2.2. **A large number of channels ($G \gg G_Q$) in a nanostructure gives rise to the self-averaging of transport properties. Plot: transmission eigenvalues of six independent channels with random phase shifts. Thick line: transmission averaged over the channels.**

There is, however, a detail that requires some extra discussion. The average conductance for independent channels is not quite the same as predicted by Ohm's law:

$$G = G_Q \sum_p \frac{T_{L,p} T_{R,p}}{T_{L,p} + T_{R,p}} \neq G_Q \frac{\sum_p T_{L,p} \sum_p T_{R,p}}{\sum_p T_{L,p} + \sum_p T_{R,p}} \equiv G_{\mathrm{Ohm}}. \tag{2.2}$$

It turns out that the simple model misses an important feature: the channels are generally mixed, whereas electrons are scattered between the barriers. This scattering is described by a general (unitary) scattering matrix $\hat{s}$ – a *unitary mixer*, as discussed in Section 1.6.5. The assumption of the independent-channel model is that this matrix is diagonal. In this case, the number of phase shifts equals the number of channels. Yet generally $\hat{s}$ does not have to be diagonal. Moreover, it is natural to assume that the non-diagonal elements are equally important and are of the same order of magnitude. Thus, for a multi-channel system we have more complicated phase shifts. Averaging over all possible phase shifts, we recover Ohm's law in the limit of a large number of transport channels.

2.1.2 Random phase shifts

The example with the double junction illustrates the generic situation in quantum transport, which we encountered in Section 1.6: transport properties depend on phase shifts. We know already (from Section 1.3) that these phase shifts are impossible to control due to the disorder. Furthermore, the phase shifts are large provided the size of a nanostructure L exceeds the Fermi wavelength by at least several times, since the phase of the wave function is of the order of $k_F L$. A misplaced atom or an alien atom provides a phase shift of the order of π, at least for one transport channel, and scrambles the whole picture.

We know that formally identical nanostructures are random, meaning that there exists random configuration of disorder. The bigger the nanostructure, the bigger the number of unknown parameters. If we knew the precise configuration of disorder, we could figure out what the phase shifts are. We do not know. But do we really need to know?

Let as draw an analogy to statistical physics. Even the smallest volume of gas consists of zillions of molecules. Although the gas is characterized by the enormous number of the coordinates and velocities of the molecules, quite fortunately we do not have to know them all to deal with the gas. What we need is a handful of numbers: pressure, temperature, volume, etc.

The same occurs in quantum transport: the randomness of the phase shifts and the related complexity eventually lead to simplicity. We expect that in the limit of a large number of channels the conductance, along with all the other transport properties, is *self-averaged* and has a negligible dependence on phase shifts. The nanostructure then can be characterized using a handful of parameters: most generally, those that determine the transmission distribution.

The difference with the gas is that a given disorder configuration and given phase shifts are "frozen": they do not change in time. A fluctuation – a deviation from the average value – in a gas lives for a very short time, whereas fluctuation of conductance in a nanostructure persists at given control parameters. If one does not trust self-averaging and wishes to check if it is good, one in principle should fabricate many formally identical nanostructures and measure the conductance. In practice, this is inconvenient. Instead, one changes control parameters, such as gate voltages or magnetic field. The phase shifts strongly depend on these parameters, and the averaging over a sufficiently wide region of their variation is equivalent to averaging over phase shifts.

When do we expect self-averaging? The effective number of transport channels, which is given by the conductance of the structure, $N \simeq G/G_Q$, must be large.[1] This requires $G \gg G_Q$. In this case, the conductance fluctuations are small in comparison with the averaged conductance $\langle G \rangle$; as we will see in Chapter 4, they are of the order of G_Q. If $G \simeq G_Q$, the fluctuations are of the order of the mean value. An extra condition is the efficient mixing of the channels. Indeed, we have seen that the averaging is incomplete for the example of independent channels. Fortunately, any realistic nanostructure contains enough disorder to provide the mixing. So the condition rather forbids some especially degenerate theoretical models.

Let us illustrate this general discussion with an explicit calculation for a multi-terminal node connected to a number of scatterers (Section 1.6). Quantum mechanically, this system is described by the scattering matrix, Eq. (1.131). We now perform the averaging of the conductance over the *unitary mixer* $\hat{s}_0$. Consider the probability that an electron incident in lead α in the transport channel n ends up in lead β, transport channel m. Using the notation of Section 1.6, this probability is given by

$$|c_{\beta m}|^2 = \left| \sum_{m'} t^{\beta}_{mm'} a_{\beta m'} \right|^2,$$

[1] This holds for coherent electron transport described by scattering theory. Large incoherent conductors can self-average at $G \ll G_Q$.

provided $\beta \neq \alpha$. If the scattering matrix $\hat{s}_0$ of the node is randomly distributed, the averaged probability of ending up in a channel does not depend on the channel. If the number of the channels is sufficiently large, one can disregard the fluctuations of this probability. This is why we can assume that the amplitudes $a_{\beta m'}$ of electrons leaving the node have the same magnitude $|a|^2$ but different random phases. Averaging over these phases (i.e. throwing out all terms containing phase factors) and summing over m, we obtain the element of conductance matrix $G_{\beta\alpha}$,

$$G_{\beta\alpha} = G_Q \sum_m |c_{\beta m}|^2 = G_Q |a|^2 \sum_{mm'} |t^{\beta}_{mm'}|^2 = G_\beta |a|^2,$$

where G_β is the conductance of the barrier separating the node from the reservoir β. To complete the calculation, we need to find $|a|^2$. To this end, we involve the current conservation: the number of particles entering the node equals the number of particles exiting the node,

$$\sum_{\beta m} |b_{\beta m}|^2 = \sum_{\beta m} |a|^2.$$

Expressing the amplitudes b in terms of the amplitudes a (special care has to be taken with lead α), taking their absolute values squared and averaging again over random phases of a, we obtain

$$|a|^2 = \frac{G_\alpha}{\sum_\gamma G_\gamma},$$

which yields Ohm's law,

$$G_{\beta\alpha} = \frac{G_\alpha G_\beta}{\sum_\gamma G_\gamma}, \quad \alpha \neq \beta. \tag{2.3}$$

A similar calculation can be performed for the component $G_{\alpha\alpha}$, to yield

$$G_{\alpha\alpha} = -G_\alpha + \frac{G_\alpha^2}{\sum_\gamma G_\gamma}.$$

As it should be, the total current is conserved: $\sum_\beta G_{\alpha\beta} = 0$.

To summarize this section, we have shown that the conductance of a nanostructure in the classical limit $G \gg G_Q$ obeys Ohm's law and that this fact is in full agreement with the quantum-mechanical treatment.

Exercise 2.2. Consider a ring (Section 1.6) connected to the reservoirs by two identical beam splitters (see Eq. (1.81)). (i) Explain why the use of Ohm's law for the setup is less obvious than for a double junction. (ii) Argue that the only way to implement Ohm's law is to ascribe the voltages to the arms of the ring. (iii) Given the potentials of the reservoirs, compute the currents in the beam splitters. (iv) Show that the current conservation in each arm suffices to determine the voltages in the arms. (v) Calculate the total conductance of the ring.

2.2 Electron transport in solids

As we have just seen, the main difficulty with the true quantum-mechanical approach to transport is that the scattering matrix becomes very complicated depending on barely controllable phase shifts. It is worth noting that the same problem was encountered about 80 years ago when scientists recognized that electrons in solids are waves rather than particles. In an ideal crystal solid, electron wave functions are Bloch waves and are reasonably similar to plane waves. However, the impurities in solids considerably complicate the wave functions. Each impurity acts as a scattering center, re-emitting incoming waves, and each wave function becomes a superposition of Bloch waves scattered by each impurity in the sample. The true quantum picture thus becomes hopelessly complicated.

The way out was quickly found and is described in (for example) Refs. [40] and [41]. One uses common sense instead of quantum mechanics and describes quantum electrons as a statistical ensemble of particles, or "balls." The classical particles are characterized by their coordinates $\boldsymbol{r}$ and (quasi)momenta $\boldsymbol{p}$, points in six-dimensional phase space. However, one does not need information concerning the coordinates of all the particles. Instead, the statistical ensemble is described by a distribution function in six-dimensional space that shows how many particles are there in an element of the space $\mathrm{d}\boldsymbol{p}\,\mathrm{d}\boldsymbol{r}$. A quantum particle cannot be described by a coordinate and a momentum simultaneously. However, a statistical ensemble of quantum particles can be described by a function of $\boldsymbol{p}, \boldsymbol{r}$ that is similar to the distribution function. Qualitatively, the dependence of this function on momenta and coordinates is characterized by two scales δp, δr. For electrons in a metal, $\delta p \simeq \hbar k_{\mathrm{F}}$. By virtue of the Heisenberg uncertainty relation $\delta r\ \delta p \geq \hbar/2$, the classical limit is achieved if the function only slowly depends on the coordinates, $\delta r \gg k_{\mathrm{F}}^{-1}$. In this limit, the function can be unambiguously identified with the distribution function of the classical particles.

2.2.1 Boltzmann equation

To start with a statistical description, one introduces the non-equilibrium *filling factor* (or *distribution function*), which shows how many particles there are in a small volume of phase space around the point $(\boldsymbol{r}, \boldsymbol{p})$. In principle, one can express this filling factor quantum mechanically in terms of the wave functions ψ_n,

$$f(\boldsymbol{r}, \boldsymbol{p}) = \sum_{\text{filled}} \int \mathrm{d}\boldsymbol{y}\ \psi_n^*(\boldsymbol{r} + \boldsymbol{y}/2)\psi_n(\boldsymbol{r} - \boldsymbol{y}/2)\exp(-\mathrm{i}(\boldsymbol{p} \cdot \boldsymbol{y})/\hbar), \tag{2.4}$$

and try to implement the Schrödinger equation to find the equations for $f(\boldsymbol{r}, \boldsymbol{p})$. However, this is highly redundant. There is a much simpler way to obtain the equations from a reasoning based on particle balance. Let us consider the motion of particles concentrated in an element of the phase space near the point $(\boldsymbol{r}, \boldsymbol{p})$. The time derivatives of their coordinates and momenta are all the same within the element,

$$\dot{\boldsymbol{r}} = \boldsymbol{v}(\boldsymbol{p}), \text{ velocity;}$$
$$\dot{\boldsymbol{p}} = \boldsymbol{F}(\boldsymbol{r}), \text{ force ,}$$

since velocity and force are functions of the point in the phase space. In this case, we can say that the element shifts with all the particles it contains. The number of particles in this element obviously stays the same. Moreover, the six-dimensional volume of this element does not change either. This follows from the fact that the motion is governed by the Hamilton function $H(\boldsymbol{r}, \boldsymbol{p})$ and

$$\boldsymbol{v} = \frac{\partial H}{\partial \boldsymbol{p}}; \ \boldsymbol{F} = -\frac{\partial H}{\partial \boldsymbol{r}}.$$

The Hamilton function commonly used encompasses the parabolic electron spectrum and an external (non-periodic) potential given by

$$H(\boldsymbol{p}, \boldsymbol{r}) = \frac{p^2}{2m} + U(\boldsymbol{r}).$$

This disregards the effects of a periodic crystalline potential in solids. These can be incorporated by modifying the electron spectrum. We do not consider this modification to simplify the outlining the issue.

From this it follows that the total time derivative of the filling factor, $\mathrm{d}f(t, \boldsymbol{r}(t), \boldsymbol{p}(t))/\mathrm{d}t$, is simply zero. Rewriting this in terms of partial derivatives, we obtain the balance equation of the first stage:

$$0 = \frac{\mathrm{d}f}{\mathrm{d}t} = \frac{\partial f}{\partial t} + \boldsymbol{v}\frac{\partial f}{\partial \boldsymbol{r}} + \boldsymbol{F}\frac{\partial f}{\partial \boldsymbol{p}}. \tag{2.5}$$

This suffices to describe the filling factor if the Hamilton function varies smoothly at the space scale k_F^{-1}. However, this is usually not the case: one should not forget impurities and other defects that created the problem. The impurities give rise to a potential profile that is sharp at space scale k_F^{-1}. This potential causes scattering: the particle changes its momentum by a value of the order of $\hbar k_F$. To incorporate the scattering into the particle balance, we introduce *scattering rates* $W_{\boldsymbol{p},\boldsymbol{p}'}$: the probabilities per unit time of going from a state with momentum $\boldsymbol{p}$ to a state with momentum $\boldsymbol{p}'$. The number of particles leaving state $\boldsymbol{p}$ to all possible states $\boldsymbol{p}'$ is then given by $\int W_{\boldsymbol{p},\boldsymbol{p}'} f_{\boldsymbol{p}}\, \mathrm{d}\boldsymbol{p}'/(2\pi\hbar)^3$, whereas the number of particles coming to state $\boldsymbol{p}$ from all possible states $\boldsymbol{p}'$ is given by $\int W_{\boldsymbol{p}',\boldsymbol{p}} f_{\boldsymbol{p}'}\, \mathrm{d}\boldsymbol{p}'/(2\pi\hbar)^3$. Summing up everything, we arrive at the *Boltzmann* equation for the non-equilibrium filling factor:

$$\frac{\partial f_{\boldsymbol{p}}}{\partial t} = -\boldsymbol{v}\frac{\partial f_{\boldsymbol{p}}}{\partial \boldsymbol{r}} - \boldsymbol{F}\frac{\partial f_{\boldsymbol{p}}}{\partial \boldsymbol{p}} + \int \frac{\mathrm{d}\boldsymbol{p}'}{(2\pi\hbar)^3}(W_{\boldsymbol{p}'\boldsymbol{p}} f_{\boldsymbol{p}'} - W_{\boldsymbol{p}\boldsymbol{p}'} f_{\boldsymbol{p}}). \tag{2.6}$$

Note first that the scattering rates W themselves may depend on the filling factor. This is why the Boltzmann equation is generally non-linear in f. If the source of the scattering is particle–particle collisions, like in the original Boltzmann equation for a gas, this is indeed the case. Usually such collisions are accompanied by energy transfer, and are therefore inelastic.

At low energies, however, most scattering is *elastic* – electrons are scattered off the potential of impurities without energy change. Correspondingly, the rates contain a delta-function, $\delta(E(\boldsymbol{p}) - E(\boldsymbol{p}'))$, that accounts for the energy conservation. The rates also do not depend on the filling factors, so the Boltzmann equation for elastic scattering is a linear one. In this section, we will assume that the elastic scattering dominates. We will consider the details of inelastic processes in Chapter 6.

The elastic rates generally obey the symmetry relation $W_{\boldsymbol{p},\boldsymbol{p}'} = W_{-\boldsymbol{p}',-\boldsymbol{p}}$ owing to time-reversal symmetry. Indeed, upon time reversal the initial state interchanges with the final state and the momenta change in sign.

Control question 2.1. Show that if one takes the equilibrium Fermi distribution as $f_{\boldsymbol{p}}$, and assumes elastic scattering rates, the final term in Eq. (2.6) vanishes.

Since the Boltzmann equation, Eq. (2.6), is solely based on the classical balance reasoning, one can ask: is quantum mechanics needed there? Quantum mechanics is in fact required to evaluate the scattering probabilities $W_{\boldsymbol{p},\boldsymbol{p}'}$ using the electron wave functions without impurities – plane or Bloch waves – and concrete models of impurity potential. A customary model for the scattering rates is based on the Born approximation: the rates are evaluated using the Fermi golden rule, assuming that the impurity potential $\tilde{U}_{\text{imp}}$ is a perturbation. The rates are given by

$$W_{\boldsymbol{p},\boldsymbol{p}'} = \frac{2\pi}{\hbar} c_{\text{imp}} |\tilde{U}_{\text{imp}}((\boldsymbol{p} - \boldsymbol{p}')/\hbar)|^2 \delta(E(\boldsymbol{p}) - E(\boldsymbol{p}')), \tag{2.7}$$

where $\tilde{U}_{\text{imp}}(\boldsymbol{k})$ is a Fourier component of the potential of an impurity or a defect of a given sort and c_{imp} is the concentration of these impurities. Equation (2.7) is the result of the averaging over the positions of impurities required to suppress possible interference effects. This makes the scattering events on different impurities independent. If there are several sorts of impurities present in the system, the rates are the sum of contributions of each sort. The scattering is elastic; as seen explicitly from Eq. (2.7), the energy of the initial state is the same as the energy of the final state. Due to this energy conservation, the final term in Eq. (2.6), known as the *collision integral*, vanishes, provided the distribution function f depends only on the energy $E(\boldsymbol{p})$ and not on the direction of the momentum $\boldsymbol{p}$. In particular, the Fermi distribution function is the function of energy only. Therefore, it nullifies the collision integral and satisfies the Boltzmann equation in the absence of external forces. This is not surprising: the electrons are supposed to be at thermodynamic equilibrium under these conditions. It is the external force $\boldsymbol{F}$ that drives the system out of equilibrium and modifies the distribution function.

A common customization of the scattering rate is based on the convenient assumption that $\tilde{U}_{\text{imp}}(\boldsymbol{k})$ does not depend on $\boldsymbol{k}$. This assumption is physically justified when the size of an impurity is smaller than k_{F}^{-1}, so that $\tilde{U}_{\text{imp}}$ can be regarded as constant. This situation is sometimes called "white noise scattering" since it takes place for the delta-function correlations of the random scattering potential. In this case, the scattering is characterized by a single parameter: The momentum relaxation time is given by $\tau_p^{-1} = (2\pi \nu c_{\text{imp}}/\hbar)|U_{\text{imp}}|^2$. The Boltzmann equation now reads

$$\frac{\partial f}{\partial t} = -\boldsymbol{v}\frac{\partial f}{\partial \boldsymbol{r}} - \boldsymbol{F}\frac{\partial f}{\partial \boldsymbol{p}} + (\langle f \rangle - f)\,\tau_p^{-1}, \tag{2.8}$$

where the angular brackets denote the averaging of f over the angles – over all directions of $\boldsymbol{p}$ at a given energy E. It is customary to define the mean free path $l = v\tau_P$, which describes the typical distance covered by the electron between scattering events.

Since the scattering is elastic, the filling factors at different energies enter the scattering terms independently. Inelastic scattering, which may originate from either electron–electron or electron–photon interaction, mixes the filling factors at different energies.

2.2.2 Drift-diffusion equation

As discussed, the Boltzmann equation given in Eq. (2.8) is valid at spatial scales exceeding k_F^{-1}. We have just seen that the equation defines a new spatial scale: the mean free path l. At spatial scales much bigger than l, the electrons scatter many times, each time changing direction randomly. Thereby they forget the initial direction of their momentum. However, they do remember their initial energy since the scattering at impurities is elastic. Thus, at length scales greater than l, the filling factor is isotropic and depends on energy only:

$$f(\boldsymbol{r}, \boldsymbol{p}) \Rightarrow f(E) \text{ (isotropic)}.$$

This allows us to proceed to a simpler equation, known as the *drift-diffusion* equation. It holds separately at each energy provided there is no inelastic scattering. The derivation is worth giving here. Let us substitute into the Boltzmann equation the filling factor in the form $f(\boldsymbol{r}, \boldsymbol{p}) = f(E) + f^{(1)}(\boldsymbol{r}, \boldsymbol{p})$, $f^{(1)}$ being the small anisotropic part, $f^{(1)} \ll f(E)$, $\langle f^{(1)} \rangle = 0$, and the average is taken over the directions of $\boldsymbol{p}$. As a first step, we average the resulting equation over the angle. The scattering term is zero upon integration over the angle, irrespective of the specific form of f. The rest of the terms result in the following:

$$\frac{\partial f}{\partial t} + \nabla \cdot \boldsymbol{j} = 0; \; \boldsymbol{j} = \langle \boldsymbol{v} f^{(1)} \rangle.$$

Note that this represents a conservation law, or, in other terms, a continuity equation for the number of particles at a given energy. The notation $\boldsymbol{j}$ stands for the particle current density per element of momentum space. As a second step, we must obtain the relation between this current density and the isotropic part of f. To this end, we take the difference of the Boltzmann equation and its angular average. The scattering terms yield the contribution proportional to $f^{(1)}$, and we disregard the anisotropic part in comparison with the isotropic one in the rest of the terms. This expresses the anisotropic part in terms of the isotropic one, i.e.

$$-\frac{f^{(1)}}{\tau_P} = \boldsymbol{v}\frac{\partial f}{\partial \boldsymbol{r}} + (\boldsymbol{F} \cdot \boldsymbol{v})\frac{\partial f}{\partial E},$$

where we have used $\boldsymbol{v} \equiv \partial E/\partial \boldsymbol{p}$. The current density is now readily obtained, and we combine these two relations into the drift-diffusion equation:

$$\frac{\partial f}{\partial t} + \nabla \cdot \boldsymbol{j} = 0; \; \boldsymbol{j} = -D\boldsymbol{F}\frac{\partial f}{\partial E} - D\frac{\partial f}{\partial \boldsymbol{r}}. \tag{2.9}$$

In the second part of Eq. (2.9), the first term describes the drift: an external force $\boldsymbol{F}$ produces a flux of particles. The second term is the *diffusion*: the current density is proportional and opposite to the density gradient trying to compensate it. The coefficient of proportionality, $D(E) = \tau_P(E)v^2(E)/3$, is the diffusion coefficient.

Control question 2.2. Derive the expression for the diffusion coefficient D in one, two, and three dimensions.

The quantities ν, τ_P, and D depend on the energy at the scale of the Fermi energy. As we have already noted in Section 1.2, electron transport usually takes place in a narrow energy strip near the Fermi energy. Under these assumptions, one can disregard this energy dependence, replacing ν, τ_P, and D by their values at the Fermi energy. It is also convenient to count energy from the Fermi energy, as we frequently do. Another convenient notation we introduce now concerns the definitions of the current and current density. For circuit-theory applications, it is convenient to work with currents and current densities that have dimensions of conductance and conductivity, respectively. Such a current density is related to particle current density per element of the momentum space by the normalization $\boldsymbol{j} \to \boldsymbol{j}e^2\nu$. The new notation is not expected to mislead the reader since, from dimensional analysis, one always understands what quantity is considered. Using the new notation, the drift-diffusion equation is given by

$$e^2\nu\frac{\partial f}{\partial t} + \nabla\cdot\boldsymbol{j} = 0;\ \boldsymbol{j} = -\sigma\boldsymbol{F}\frac{\partial f}{\partial E} - \sigma\frac{\partial f}{\partial \boldsymbol{r}}. \tag{2.10}$$

The conductivity $\sigma = e^2 D\nu$ is thereby expressed in terms of the diffusion coefficient and the density of states. The relation between the conductivity and the diffusion coefficient is known as the *Einstein relation*.

Now we are ready for the next simplification step. We integrate the drift-diffusion equation over energy, assuming that ν and σ are energy-independent in a narrow region near the Fermi surface. Using the relations for the charge density ρ and the electric current density $\boldsymbol{I}$, we have

$$\begin{pmatrix}\rho\\ \boldsymbol{I}\end{pmatrix} = \int \mathrm{d}E \begin{pmatrix}\nu e f(E)\\ \boldsymbol{j}(E)/e\end{pmatrix}, \tag{2.11}$$

to obtain

$$\frac{\partial\rho}{\partial t} = -\nabla\cdot\boldsymbol{I};\ \boldsymbol{I} = \sigma\boldsymbol{E} - D\nabla\rho. \tag{2.12}$$

The first part of Eq. (2.12) is simply the charge conservation law; gives the electric current in terms of the electric field and the charge density gradient. We can employ yet another simplification: in most conductors there is no volume charge density, the charge being concentrated at the surface of the conductor, and thus $\rho = 0$. In the stationary case, $\boldsymbol{E} = -\nabla V$, $V(\boldsymbol{r})$ being the electrostatic potential. This brings us to the *Laplace equations*, the solutions of which define current and potential distribution in conductors characterized by (position-dependent) conductivity:

$$\nabla\cdot\boldsymbol{I} = 0;\ \boldsymbol{I} = -\sigma(\boldsymbol{r})\nabla V. \tag{2.13}$$

2.2.3 Distribution function in one dimension

To illustrate Eq. (2.10), we consider it in a one-dimensional geometry, where the distribution function depends only on one coordinate (x) over the length L. Such a situation is likely to occur if a uniform piece of metal of constant cross-section is placed between two massive contacts. One example is a sandwich: a metallic film of thickness L and other dimensions much larger than L placed between two semi-infinite electrodes. In this case, all quantities depend only on the coordinate x (with the possible exception of the regions near the film boundaries, which are disregarded). Another setup is a wire, which can be described by the drift-diffusion equation if its smallest width d is much greater than the mean free path l. If the wire is long ($d \ll L$), the distribution function depends only on the coordinate x along the wire and stays constant over the cross-section (with the possible exception of the wire ends, which are also disregarded). For definiteness, we will talk about a wire.

Equation (2.10) allows us to find the distribution function at any point of the wire. Indeed, since there is no field inside the wire, $\boldsymbol{F} = 0$, the equation reads $\partial^2 f/\partial x^2 = 0$. Thus f is a linear function of x. The wire is attached to the two reservoirs, left (at $x = 0$) and right ($x = L$), described by the distribution functions $f_L(E)$ and $f_R(E)$, respectively. This provides the boundary conditions at the ends of the wire: $f(0) = f_L, f(L) = f_R$. Taking this into account, we obtain

$$f(x, E) = \frac{L - x}{L} f_L(E) + \frac{x}{L} f_R(E). \tag{2.14}$$

The current in Eq. (2.11) is given by $I = (\sigma/eL) \int dE(f_L - f_R)$.

It might seem that the appearance of the distribution function, Eq. (2.14), implies that there is a finite density of charge in the system. Indeed, one can try to compute the difference of charge densities at points x_1, x_2 as follows:

$$\rho(x_1) - \rho(x_2) = e\nu \int dE \, (f(x_1, E) - f(x_2, E)) = \frac{x_2 - x_1}{L} eV\nu,$$

assuming a voltage difference V between the electrodes. This would be contradictory: the charge density would produce an electric field that should be absent in a metal. The point is that the bottom of the energy band E_0, which sets the lower limit of integration in the above expression, depends on the position x along the conductor. For instance, in the left reservoir it is higher than in the right reservoir by the amount equal to the applied voltage eV. The x-dependence of the bottom of the band is determined from the condition that the charge density in the conductor is homogeneous:

$$\begin{aligned}\rho(x_1) - \rho(x_2) = e\Bigg(&\int_{E_0(x_1)}^{\infty} \nu(E - E_0(x))dE \; f(x_1, E) \\ &- \int_{E_0(x_1)}^{\infty} \nu(E - E_0(x))dE \; f(x_2, E)\Bigg) = 0.\end{aligned}$$

Control question 2.3. How does $E_0(x)$ depend on x?

Consider first a vanishing temperature. Then, each of the functions $f_{L,R}$ is a step function: $f_L(E) = \theta(E_F + eV - E)$, $f_R(E) = \theta(E_F - E)$. The distribution function in the wire then takes a *double-step* shape: it equals zero for $E > eV$, unity for $E < 0$, and $1 - x/L$ for the energies between zero and eV. The position of an intermediate step depends on the point x. Such an arrangement obviously produces the current density $I = \sigma V/L$. Electron–electron interactions smear the double-step curve (see Section 6.8).

As a different example, let us assume that there is no voltage bias (the chemical potentials are the same on the left and on the right), but the temperatures are different. The distribution function in the wire, Eq. (2.14), is given by

$$f(x, E) = \frac{L - x}{L} f_F(E, T_L) + \frac{x}{L} f_F(E, T_R), \tag{2.15}$$

where the second argument of a Fermi distribution function indicates the temperature at which it is evaluated. Note that it does not have the form of a Fermi function, and thus the distribution inside the wire cannot be characterized by a certain temperature.

Let us calculate the thermal current generated by the temperature difference in the reservoirs. Similar to electric current, it is expressed in terms of the distribution function,

$$Q = \int \mathrm{d}E (j(E)\, E/e^2). \tag{2.16}$$

Substituting Eq. (2.15) and assuming again that the temperature difference is small, we obtain $Q = -K\Delta T$, with the *thermal conductance* $K = (\pi^2/3)(\sigma/Le^2)k_B^2 T$. Note that the electric and thermal conductance obey the relation $K/TG = \pi^2 k_B^2/3e^2$. This is actually more general (not specific to the one-dimensional case), and is well known from the theory of solids as the *Wiedemann–Franz law*. This law holds provided that the electron scattering is predominantly elastic and that all non-electronic contributions (for example those from phonons) can be disregarded.

Exercise 2.3. Derive the Wiedemann–Franz law from Eqs. (2.15) and (2.16).

The final remark we make is about the replacement of the energy-dependent density of states, or conductivity, by their values at the Fermi surface. We have already seen that in nanostructures, for example in the double junction, the transmission coefficient may be a rapidly changing function of energy, and such an approximation should be taken with a certain amount of caution. There are situations when this approximation does not suffice to obtain an effect known to persist from general reasoning. In solids, such an example is the *thermoelectric effect*, in which the current is generated by the temperature difference (with no voltage applied). Consider the distribution function given in Eq. (2.15). The temperature gradient is not a force, and thus, if there is no voltage, one has the current density, $j = -\sigma\, \partial f/\partial r$, and the total current, $I = \int \mathrm{d}E\ j(E)/e$. Using Eq. (2.15), we obtain

$$I = \frac{1}{eL} \int \mathrm{d}E\ \sigma\, (f_F(E, T_L) - f_F(E, T_R))\ .$$

We see that, if the conductivity σ is taken to be energy-independent, the integral vanishes. For example, if the temperature difference is small, $\Delta T \equiv T_L - T_R \ll T_{L,R}$, one uses the following approximation:

$$f_F(E, T_L) - f_F(E, T_R) \approx \frac{\partial f_F}{\partial T} \Delta T = -\frac{E - E_F}{k_B T^2} \frac{\partial f_F}{\partial E} \Delta T,$$

and since $\partial f_F / \partial E$ is an even function of $E - E_F$ (the delta-function at zero temperature), the integral is zero. To obtain the thermoelectric effect, we have to take into account that the conductivity is (weakly) energy-dependent and expand it around the Fermi energy, $\sigma(E) \approx \sigma(E_F) + (E - E_F)\sigma'$. Calculating the current, we obtain

$$I = \frac{\pi^2}{3k_B^2} \frac{\sigma' T}{eL} \Delta T.$$

Thus, the thermoelectric effect vanishes at zero temperature and is due only to the energy derivative of the conductivity. In this sense, the thermoelectric current is very small. Indeed, let us compare it with the electric current $I_{el} = \sigma V/L$. The energy dependence of σ occurs at the scale of E_F, thus $\sigma' \sim \sigma/E_F$. We then have $I/I_{el} \sim k_B^2 T \Delta T / eV E_F$. Typically, E_F is of the order of electron-volts. For the two currents to be of the same order, even for room temperature ($k_B T/E_F \sim 10^{-2}$), we would need to apply the temperature difference $k_B \Delta T \sim 10^2$ eV.

2.3 Semiclassical coherent transport

2.3.1 Green's functions

The Boltzmann and drift-diffusion equations disregard the coherence of electron waves from the very beginning, and thus are not very useful for a description of quantum transport in nanostructures. What we need is a rigorous formalism that, on the one hand, is semiclassical, while, on the other hand, at least partially preserves this coherence. Such a formalism is based on the *semiclassical Green's functions*, and is detailed in the following. We will see that the formalism is very similar to the traditional semiclassical description. We will find analogs of the Boltzmann equation as well as of the drift-diffusion equation. There is, however, an important difference: semiclassical Green's functions, in contrast to the distribution function, are able to retain information about the coherence. In the semiclassical approximation, only the relevant part of this information is retained: the coherence that survives isotropization of electrons by scattering. The information is stored in the *matrix structure* of the Green's functions.

Originally, the formalism was developed for the description of superconductors, but it proved to be much more broad. For the moment, we stick to the original formulation [42, 43, 44], and show later how the formalism should be adapted for a broader class of problems, including those not involving superconductivity.

As we discussed in Section 1.8, the starting point for describing electron states in superconductors are the BdG equations (see Eq. (1.156)) for the two-component wave function. For $\Delta = 0$ (no superconductivity), they are two uncoupled equations describing single particles in a metal, either as electrons with energies E or holes with energies $-E$. A finite

Δ couples electrons and holes, so that the actual single-particle state is a coherent superposition of an electron and a hole. Forty years ago it was recognized with fascination that this coherence survives isotropization by impurity scattering and persists at long spatial and time scales. It was also understood that the traditional particle-balance approach cannot account for this coherence and must be replaced by the quantum-mechanical Green's function approach.

A historical comment would be in order here. Even 15 years ago, Green's function methods were regarded as advanced knowledge available to only a selected few, a subject not to be taught to lay persons. Accordingly, the wider scientific community tended to regard Green's function users as math freaks not capable of thinking in physical terms or of communicating relevant results. Although such attitudes have not disappeared completely, there has been tremendous progress in popularizing these methods, as well as in the understanding of "the physics behind" them. This is why we dare to address the topic in this comprehensive text. We will not present any general introduction to quantum Green's functions here, however. They may be of various sorts and used for various purposes.

In this chapter, we use the Keldysh approach, outlined below. This is designed to describe open quantum systems out of thermodynamic equilibrium. This suits the goals of quantum transport. The starting point of the approach is to consider the evolution of the density matrix $\hat{\rho}$ of a quantum system subject to the many-body Hamiltonian $\hat{\mathcal{H}}(t)$ at the time interval (t_0, t_1). The density matrix can be presented as a linear combination of the products of some "bra" and "ket" wave functions,

$$\hat{\rho}(t) = \sum_{i,j} \rho_{ij} |\Psi_{\text{ket}}^{(i)}(t)\rangle\langle\Psi_{\text{bra}}^{(j)}(t)|,$$

and satisfies the Heisenberg equation,

$$\frac{\partial \hat{\rho}}{\partial t} = -\frac{\mathrm{i}}{\hbar}\left(\hat{\mathcal{H}}(t)\hat{\rho}(t) - \hat{\rho}(t)\hat{\mathcal{H}}(t)\right).$$

To find the evolution, one solves the above equation, introducing unitary operators $\hat{U}^{(\pm)}(t_0, t_1)$, where $\pm$ stands for "ket" ("bra"), so that

$$\rho(t_1) = \hat{U}^{(+)}(t_0, t_1)\rho(t_0)\hat{U}^{(-)}(t_0, t_1). \tag{2.17}$$

The evolution operator $\hat{U}^{(+)}(t_0, t_1)$ is a product of elementary operators $\hat{U}^{(+)}(t, t + \mathrm{d}t) = \exp\{-\mathrm{i}\hat{\mathcal{H}}(t)\mathrm{d}t/\hbar\}$ that describe the evolution within a short time interval $(t, t + \mathrm{d}t)$. These elementary operators are time-ordered so that $\hat{U}^{(+)}(t, t + \mathrm{d}t)$ stands on the left from $\hat{U}^{(+)}(t', t' + dt)$ provided $t > t'$, i.e.

$$\hat{U}^{(+)} = \mathrm{T}\exp\left\{-\frac{\mathrm{i}}{\hbar}\int_{t_0}^{t_1} \mathrm{d}t\ \hat{\mathcal{H}}(t)\right\}, \tag{2.18}$$

where T denotes the time ordering. The evolution operator for "bra" is the inverse of $\hat{U}^{(+)}(t', t' + \mathrm{d}t)$ and is consequently anti-time-ordered (notation $\tilde{T}$):

$$\hat{U}^{(-)} = \tilde{\mathrm{T}}\exp\left\{\frac{\mathrm{i}}{\hbar}\int_{t_0}^{t_1} \mathrm{d}t\ \hat{\mathcal{H}}(t)\right\} = \left(\hat{U}^{(+)}\right)^{-1}. \tag{2.19}$$

One speaks of the Keldysh contour that has a forward ("ket") part (for $\hat{U}^{(+)}$), going from t_0 to t_1, and a backward ("bra") part (for $\hat{U}^{(-)}$), going back in time from t_1 to

t_0. Then, all elementary evolution operators in $\hat{U}^{(\pm)}$ are ordered along the contour. The Keldysh diagram technique is, by construction, a perturbation theory for the evolution kernel, Eq. (2.17). Contributions to this quantity arise from perturbations to $\hat{\mathcal{H}}$ in both forward or backward contours. This is why all elements of the diagram technique and all physical quantities in the Keldysh approach have an extra *Keldysh index* $\pm$ that denotes the forward or backward parts of the contour.

Extended Keldysh formalism

An interesting recent advance is the modification of the Keldysh technique that makes it suitable for the problems of counting statistics. In this case, one operates with an "extended" Keldysh approach. There, the evolution of the "ket" and "bra" wave functions is governed by *different* Hamiltonians $\hat{\mathcal{H}}^{\pm}(t)$. One works with an "extended" density matrix $\hat{\tilde{\rho}}$, satisfying

$$\frac{\partial \hat{\tilde{\rho}}}{\partial t} = -\frac{\mathrm{i}}{\hbar}\left(\hat{\mathcal{H}}^{+}(t)\hat{\tilde{\rho}}(t) - \hat{\tilde{\rho}}(t)\hat{\mathcal{H}}^{-}(t)\right).$$

Therefore, the evolution is different in the forward and backward parts of the Keldysh contour and is given by

$$\hat{\tilde{\rho}}(t_1) = \hat{U}^{(+)}(t_0,t_1)\rho(t_0)\hat{U}^{(-)}(t_0,t_1), \tag{2.20}$$

$$\hat{U}^{(+)} = \mathrm{T}\exp\left\{-\frac{\mathrm{i}}{\hbar}\int_{t_0}^{t_1} \mathrm{d}t\ \hat{\mathcal{H}}^{(+)}(t)\right\}, \tag{2.21}$$

$$\hat{U}^{(-)} = \tilde{\mathrm{T}}\exp\left\{\frac{\mathrm{i}}{\hbar}\int_{t_0}^{t_1} \mathrm{d}t\ \hat{\mathcal{H}}^{-}(t)\right\} \neq (\hat{U}^{(+)})^{-1}, \tag{2.22}$$

$$\hat{\mathcal{H}}^{(\pm)} = \hat{\mathcal{H}} \pm \hat{\mathcal{I}}\chi. \tag{2.23}$$

Since the Hamiltonians are different, the $\tilde{\rho}$ evolved in this way is not a density matrix. For example, its trace $\mathrm{Tr}[\hat{\tilde{\rho}}] \neq 1$, whereas it should equal unity for any density matrix. Nevertheless, this trace is not useless – it gives the characteristic function of counting statistics of the variable $\hat{\mathcal{Q}} = \int_{t_0}^{t_1} \mathrm{d}t\ \hat{\mathcal{I}}(t)$,

$$\Lambda(\chi; t_1 - t_0) = \mathrm{Tr}\left(\hat{\tilde{\rho}}(t_1)\right). \tag{2.24}$$

It is customary to define Green's functions as averages of pairs of electron creation/annihilation operators that are time-ordered along the Keldysh contour. However, in this chapter we concentrate on non-interacting electrons in the semiclassical limit. In this case, things become much simpler. We state that the Green's functions are *matrices* that depend on two coordinates $\boldsymbol{r}, \boldsymbol{r}'$ and two time moments t, t': $G^{\alpha\beta}(\boldsymbol{r}, t; \boldsymbol{r}', t')$. For Keldysh Green functions in superconductivity, each index α, β is a composite index containing three indices, each taking two values: Keldysh (corresponding to either the forward or backward contour), spin (either "up" or "down"), and the so-called *Nambu index* that distinguishes electrons and holes. Thus, the Green function is eventually an 8×8 matrix. To operate with these indices, it is convenient to use Pauli matrices. We introduce the matrices $\hat{\tau}_{1,2,3}$, acting in Keldysh space, as follows:

$$\hat{\tau}_1 = \begin{bmatrix} 0 & 1 \\ 1 & 0 \end{bmatrix}; \; \hat{\tau}_2 = \begin{bmatrix} 0 & -\mathrm{i} \\ \mathrm{i} & 0 \end{bmatrix}; \; \hat{\tau}_3 = \begin{bmatrix} 1 & 0 \\ 0 & -1 \end{bmatrix}.$$

We also introduce similar matrices $\sigma_{1,2,3}$ and $\eta_{1,2,3}$, acting on spin and Nambu indices, respectively. In the following, we denote the Green's functions with the matrix structure with a check, $\check{G}$, to distinguish them from "hats" of the operators. It is also common to use the following parameterization of the Green's function in Nambu space:

$$\check{G} = \begin{pmatrix} \hat{G} & \hat{F} \\ \hat{F}^{\dagger} & -\hat{G} \end{pmatrix}. \tag{2.25}$$

Here, the Green's function $\hat{G}$ (which may still be a matrix in spin and Keldysh spaces) describes electrons and holes, whereas $\hat{F}$ and $\hat{F}^{\dagger}$ are the components related to the superconductivity; they describe, for example, the mixing of electrons and holes and the creation of Cooper pairs.

In stationary conditions, Green's functions depend only on the time $t - t'$, so one can work in an energy representation, $\check{G}(t - t') \to \check{G}(\epsilon)$. The Green's functions obey the following equation:

$$(\check{E} - \hat{H}_{\boldsymbol{r}})\check{G}(\boldsymbol{r}, \boldsymbol{r}'; \epsilon) = \delta(\boldsymbol{r} - \boldsymbol{r}'); \tag{2.26}$$
$$\check{E} \equiv \epsilon\hat{\eta}_3 + \frac{1}{2}\Delta(\boldsymbol{r})(\mathrm{i}\hat{\eta}_2 + \hat{\eta}_1) + \frac{1}{2}\Delta^*(\boldsymbol{r})(\mathrm{i}\hat{\eta}_2 - \hat{\eta}_1),$$

where $\Delta(\boldsymbol{r})$ is a complex quantity, denoted as $\Delta e^{\mathrm{i}\varphi}$ in Eq. (1.156).

Control question 2.4. What is the matrix structure of the Green's function if there is no superconductivity and spin-dependent scattering is absent?

If we set the right-hand side of Eq. (2.26) to zero, it becomes equivalent to the Schrödinger equation (or the BdG equation if superconductivity is present). Its solution can be obtained in terms of solutions of the Schrödinger (BdG) equations – the scattered electron waves studied in Chapter 1. The right-hand side ensures proper normalization of these solutions, since the Schrödinger equation defines the wave functions only up to a factor. Equation (2.26) does not contain information about the filling of electron states. To take this information into account, and thereby find a unique solution for the Green's functions, one has to require that the Green's functions "at infinity", that is, far from the nanostructure, assume their equilibrium values. In the realm of quantum transport, the "infinities" correspond to the reservoirs or leads. They are in thermodynamic equilibrium, and are characterized by chemical potentials and temperatures. Once the solution for the Green's functions is obtained, one can express physical observables in terms of these functions. In quantum transport, we are mostly interested in the charge density and the electric current density. For superconducting Green's functions, they are expressed as follows:

$$\rho(\boldsymbol{r}) = \frac{e}{4} \int \frac{d\epsilon}{2\pi} \, \mathrm{Tr}\{\hat{\tau}_3 \check{G}(\boldsymbol{r}, \boldsymbol{r}; \epsilon)\}; \tag{2.27}$$

$$\boldsymbol{I}(\boldsymbol{r}) = \int \frac{d\epsilon}{2\pi} \lim_{\boldsymbol{r}' \to \boldsymbol{r}} \frac{-ie}{4m} \, \mathrm{Tr}\{\hat{\tau}_3 \hat{\eta}_3 \, (\nabla_{\boldsymbol{r}} - \nabla_{\boldsymbol{r}'}) \, \check{G}(\boldsymbol{r}, \boldsymbol{r}'; \epsilon)\}. \tag{2.28}$$

Let us note that Eq. (2.26) is in fact a rewriting of the Schrödinger (BdG) equation: its solutions can be obtained from the wave functions satisfying the equations. From this we conclude that, at this stage, the Green's function formalism based on Eq. (2.26) is completely equivalent to the scattering approach outlined in Chapter 1, although it does not explicitly introduce the scattering matrix. For simple scattering that encompasses several channels the "blunt" scattering approach is definitely simpler and provides more intuition.

The situation changes for $G \gg G_Q$: scattering matrices become too complicated to handle, and the Green's function method becomes more advantageous. The power of the Green's function method relies on opportunities of *semiclassical approximation*. For Green's functions, one can derive the semiclassical approach straightforwardly and rigorously from quantum mechanics rather than the "balance" reasoning of Section 2.2. We outline the derivation below. The main point we would like to make is that there are Green's function counterparts to Boltzmann and drift-diffusion equations. However, these counterparts bring about an important new element absent in Section 2.2 – the "check" matrix structure of Green's functions that accounts for quantum coherence.

2.3.2 Eilenberger equation

Note first that Eq. (2.26) can be equivalently presented in the conjugated form, where the Hamiltonian acts on $\boldsymbol{r}'$ rather than on $\boldsymbol{r}$, and the matrix $\check{E}$ is on the right of $\check{G}$:

$$\check{G}(\boldsymbol{r}, \boldsymbol{r}'; \epsilon)(\check{E} - \hat{H}_{\boldsymbol{r}'}) = \delta(\boldsymbol{r} - \boldsymbol{r}'). \tag{2.29}$$

This equation can be obtained from Eq. (2.26) if one treats $\check{G}(\boldsymbol{r}, \boldsymbol{r}')$ as an operator in $\boldsymbol{r}$-space and multiplies Eq. (2.26) by $(\check{E} - \hat{H})^{-1}$ from the left and by $(\check{E} - \hat{H})$ from the right.

The first step in the semiclassical approach is to take the difference of the direct and conjugated Eq. (2.26) and (2.29) to obtain

$$\left[\check{E}, \check{G}\right] - \left(\frac{\hbar^2 (\nabla^2_{\boldsymbol{r}'} - \nabla^2_{\boldsymbol{r}})}{m} + U(\boldsymbol{r}) - U(\boldsymbol{r}') \right) \check{G}(\boldsymbol{r}, \boldsymbol{r}'; \epsilon) = 0. \tag{2.30}$$

This is still an exact equation. To make it semiclassical, let us introduce the so-called Wigner representation of the Green's function:

$$\check{G}(\boldsymbol{r}, \boldsymbol{p}\,; \epsilon) \equiv \int d\boldsymbol{y} \, e^{-i\boldsymbol{p}\cdot\boldsymbol{y}/\hbar} \check{G}(\boldsymbol{r} + \boldsymbol{y}/2, \boldsymbol{r}' - \boldsymbol{y}/2; \epsilon). \tag{2.31}$$

Let us note the similarity with Eq. (2.4) for the "quantum" non-equilibrium filling factor. This establishes an analogy between a Green's function and a distribution function. We now substitute the Wigner representation into Eq. (2.30), assuming that the dependence of $\check{G}$ on $\boldsymbol{r}$ at the scale of the wavelength λ_{F} is smooth, $\partial/\partial\boldsymbol{r} \ll p/\hbar$. We thus obtain the Green's function analog of the first-stage balance equation, Eq. (2.5):

$$\frac{\mathrm{i}\left[\check{E},\check{G}\right]}{\hbar}+\boldsymbol{v}\frac{\partial\check{G}}{\partial\boldsymbol{r}}+\boldsymbol{F}\frac{\partial\check{G}}{\partial\boldsymbol{p}}=0. \tag{2.32}$$

The difference from Eq. (2.5) is the commutator that appears due to the matrix structure of the Green's functions. This structure is absent for the distribution function in the Boltzmann equation. Also, the equation is for a quantity that depends on a greater number of variables: whereas $f \equiv f(\boldsymbol{r},\boldsymbol{p})$, the Green's function depends also on the energy parameter ϵ. In fact, this dependence is redundant. The Green's function has a sharp singularity at $\epsilon = \pm E(p)$, for electrons and holes, respectively. To remove the redundancy and to obtain simpler Green's functions, an important step is taken in the course of the derivation. Traditionally, it is called, rather mathematically, "Integration over ξ." It follows from the observation that we have already made: that quantum transport takes place at energies close to the Fermi surface. Here we also note that, in Eq. (2.32) (i) important values of $\boldsymbol{p}$ are close to the Fermi surface, $p \simeq \hbar k_{\mathrm{F}}$, and (ii) the coefficients at this scale barely depend on the distance to the Fermi surface. For instance, the velocity $\boldsymbol{p}/m$ can be replaced by $\boldsymbol{v} = v_{\mathrm{F}}\boldsymbol{n}$, where $\boldsymbol{n}$ (the unit vector in the direction of momentum) parameterizes the Fermi surface. This is why the equation holds if we *integrate* it over $\xi \equiv \boldsymbol{p}^2/2m - E_{\mathrm{F}}$. This corresponds to the averaging over wide energy intervals mentioned in Section 2.1. From now on, we will work with "ξ-integrated", or *semiclassical*, Green's functions $\check{G}(\boldsymbol{r},\boldsymbol{n};\epsilon)$, defined as follows:

$$\check{G}(\boldsymbol{r},\boldsymbol{n};\epsilon)=\frac{\mathrm{i}}{\pi}\int \mathrm{d}\xi\ \check{G}(\boldsymbol{r},\boldsymbol{p}\ ;\epsilon).$$

The Green's function defined in this way is conveniently dimensionless and satisfies Eq. (2.32).

Let us now outline how to take the scattering by impurities into account. Since the Keldysh technique is essentially a perturbation theory, one can proceed with perturbations in terms of impurity potential and derive equations for Green's functions *averaged* over different realizations of this potential – over positions of impurities and defects. The effect of impurities can be incorporated into Eq. (2.26) by adding impurity "self-energy" $\check{\Sigma}(\boldsymbol{r},\boldsymbol{r}')$, to the Hamiltonian, $\hat{H} \to \hat{H} + \check{\Sigma}(\boldsymbol{r},\boldsymbol{r}')$. The function $\check{\Sigma}$, in contrast to the impurity potential, is a smooth function of coordinates. Repeating the steps that led us to Eq. (2.32), we obtain an analog of the Boltzmann equation, Eq. (2.6), known as the *Eilenberger equation* (we skip the terms containing force $\boldsymbol{F}$):

$$\frac{\mathrm{i}\left[\check{E},\check{G}\right]}{\hbar}+\boldsymbol{v}\frac{\partial\check{G}}{\partial\boldsymbol{r}}=\frac{\mathrm{i}\left[\check{\Sigma},\check{G}\right]}{\hbar}. \tag{2.33}$$

The commutator of $\check{\Sigma}$ and $\check{G}$ plays the role of the scattering terms in the Boltzmann equation, and $\check{\Sigma}$ is expressed in terms of the scattering rates $W_{\boldsymbol{pp}'}$ as a *check*:

$$\check{\Sigma}(\boldsymbol{p})=\int\frac{\mathrm{d}\boldsymbol{p}'}{(2\pi\hbar)^3}W_{\boldsymbol{pp}'}\check{G}(\boldsymbol{p}')=\int\frac{\mathrm{i}\nu}{\pi}\frac{\mathrm{d}\boldsymbol{n}'}{4\pi}W_{\boldsymbol{nn}'}\check{G}(\boldsymbol{n}'). \tag{2.34}$$

The simplest model assumption of the "white noise" potential and isotropic scattering that led us to Eq. (2.8) can be implemented here also; this yields

$$\check{\Sigma} = \frac{\mathrm{i}}{2\tau_P}\langle \check{G}(\boldsymbol{n})\rangle. \tag{2.35}$$

The Eilenberger equation depends on the same parameters as the Boltzmann equation and has the same range of validity. There are two important distinctions arising from the matrix structure. First, in contrast to all the semiclassical relations we studied, the Eilenberger equation is non-linear, since $\check{\Sigma} \propto \check{G}$. The non-linearity thus arises from averaging over impurities. Secondly, the Eilenberger equation has an important integral of motion, $\partial_{\boldsymbol{r}}(\check{G}(\boldsymbol{n},\boldsymbol{r}))^2 = 0$. At "infinity" (far from all interfaces) the Green's functions assume equilibrium values. This brings us to an important conclusion, i.e.

$$\check{G}^2 = \check{1} \tag{2.36}$$

everywhere in the nanostructure. Surprisingly, although this relation, sometimes called the "normalization condition," plays an important role in practical calculations, and is very simple to write, we are not able to offer a simple explanation.

Let us recall the extended Keldysh formalism at this point. In quantum transport, its main application is the counting statistics of electron transfers. In this case, the operators $\hat{\mathcal{I}}$ and $\hat{\mathcal{Q}}$ in Eq. (2.23) are, respectively, the many-body operators of current and transferred charge. For non-interacting electrons, the Green's functions are 2×2 matrices in Keldysh indexes, satisfying

$$(E - \hat{H} - \tau_3\chi\hat{I})\check{G}(\boldsymbol{r},\boldsymbol{r}')\check{G} = 0, \tag{2.37}$$

$\hat{I}$ being the single-particle operator of the full current through a certain cross-section. Since the current is conserved, this cross-section can be chosen arbitrarily. It is convenient to choose the cross-section to be far from the nanostructure so that it bisects an adjacent reservoir. In this case, the effect of the counting field χ can be incorporated into the boundary condition in the corresponding reservoir by a modification of the corresponding Green's function,

$$\check{G}(\chi) = \exp(-\mathrm{i}\chi\tau_3/2)\check{G}(\chi = 0)\exp(\mathrm{i}\chi\tau_3/2), \tag{2.38}$$

$\check{G}(\chi = 0)$ being the usual equilibrium Keldysh Green's function of the corresponding reservoir.

All steps and equations of the semiclassical approach remain precisely the same for the extended χ-dependent Green's functions. This enables evaluation of full counting statistics for a variety of nanostructures. The approach can be extended to superconducting nanostructures with minor effort and has been successfully used for that.

Control question 2.5. (i) Write the Eilenberger equation, Eq. (2.33), in components in Nambu space. (ii) Find its solutions in a normal metal ($\Delta = 0$) and without scattering ($\Sigma = 0$).

2.3.3 Usadel equation

We proceed now to the analog of the drift-diffusion equation. As in the balance approach, the equation is valid at scales exceeding the mean free path l. To derive it from the Eilenberger equation, one eventually performs the same steps as for Eq. (2.9). The Green's function in this limiting case is almost isotropic and can be represented as $\check{G}(\boldsymbol{r},\boldsymbol{n}) = \check{G}(\boldsymbol{r}) + \check{G}^{(1)}(\boldsymbol{r},\boldsymbol{n})$, $\check{G}^{(1)}$ being the small anisotropic term, $\check{G}^{(1)} \ll \check{G}$, $\langle \check{G}^{(1)} \rangle = 0$. Due to the normalization condition $\check{G}^2 = \check{1}$, the matrices $\check{G}^{(1)}$ and $\check{G}$ anticommute, $\check{G}^{(1)}\check{G} = -\check{G}\check{G}^{(1)}$. As the first step, we substitute this representation into the Eilenberger equation and average the resulting equation over the angle. The scattering term drops out, yielding

$$\frac{ie^2\nu}{\hbar}[\check{E},\check{G}] + \frac{\partial \check{\boldsymbol{j}}}{\partial \boldsymbol{r}} = 0; \; \check{\boldsymbol{j}} = \langle \boldsymbol{v}\check{G}^{(1)} \rangle. \tag{2.39}$$

This is the first time we encounter the *matrix* current $\check{\boldsymbol{j}}$, a quantity that will play an important role in quantum circuit theory. The coefficient $e^2\nu$ before the commutator arises from the fact that the so-defined current has the dimension of conductance, as introduced previously (Section 2.2). Apart from the commutator term, the current $\check{\boldsymbol{j}}$ is conserved. To find the relation between the matrix current and the isotropic part of the Green's function $\check{G}$, we subtract the angle-averaged Eilenberger equation from the original one. In this way, we obtain (assuming $1/\tau_P \gg \epsilon, \Delta$)

$$\frac{1}{2\tau_P}[\check{G}^{(1)},\check{G}] = \frac{\check{G}^{(1)}\check{G}}{\tau_P} = \boldsymbol{v}\frac{\partial \check{G}}{\partial \boldsymbol{r}},$$

which immediately gives us the *Usadel equation*,

$$\frac{ie^2\nu}{\hbar}[\check{E},\check{G}] + \frac{\partial \check{\boldsymbol{j}}}{\partial \boldsymbol{r}} = 0; \; \check{\boldsymbol{j}} = -\sigma(\boldsymbol{r})\check{G}\nabla\check{G}. \tag{2.40}$$

This describes coherent transport in the diffusion regime, when $\check{G}(\boldsymbol{r})$ are continuous functions of coordinates. If there are tunnel barriers in the structure, this is not the case, and $\check{G}(\boldsymbol{r})$ are generally different on the two sides of the barrier. To describe the situation, one has to implement a boundary condition on the barrier. This condition relates a component of the matrix current normal to the barrier to the Green's functions at both sides $\check{G}_1$ and $\check{G}_2$,

$$\check{j}_\perp = \frac{g_\mathrm{T}}{2}\left[\check{G}_1, \check{G}_2\right], \tag{2.41}$$

where g_T is the conductance of the barrier per unit area. One can understand this condition better if one considers its counterpart for a non-matrix current, as we did in Section 2.2. In this case, it relates the normal component of the current to the local drop of filling factors across the barrier:

$$j_\perp = g_\mathrm{T}\left[f_1(E) - f_2(E)\right].$$

What one actually measures is the current density per unit energy $\boldsymbol{j}$. It is expressed via diagonal components of the matrix current (see Eq. (2.28)):

$$\boldsymbol{j} = \frac{e}{4}\,\mathrm{Tr}\left(\tau_3\eta_3\check{\boldsymbol{j}}\right). \tag{2.42}$$

Exercise 2.4. Consider Eq. (2.40) for a normal metal with homogeneous conductivity $\sigma(\boldsymbol{r}) = \sigma$. (i) Write the Usadel equation, Eq. (2.40), in components. (ii) Assuming 1d geometry, solve the Usadel equation with the boundary conditions $\hat{G}|_{x=\pm\infty} = \hat{\tau}_3 - \hat{\tau}_1 + i\hat{\tau}_2 - 2f_{\mathrm{L,R}}(E)(\hat{\tau}_3 + i\hat{\tau}_2)$ (see Eq. (2.83) below). (iii) Using Eq. (2.42), calculate the conductivity and show that it indeed equals σ.

2.3.4 Semiclassical approach and coherence

The semiclassical Green's function approach outlined above is valid provided the mean free path is much longer than the Fermi wavelength, $k_{\mathrm{F}}l \gg 1$. There are corrections to the conductance, which are of order G_{Q} and have essentially quantum origin: they reveal the coherence of electron waves since they depend on the interference pattern of the waves created by the defects. In one dimension, they become so strong that the whole semiclassical approach fails (see Section 4.5). These corrections can be of the two types – weak localization and universal conductance fluctuations. We have already seen how they arise in the example of an Aharonov–Bohm ring (see Section 1.6), and we will consider them in more detail in Chapter 4. We only mention now that the quantum corrections are obviously affected by both weak magnetic field and (gate) voltage, which allows their experimental observation. If the measurement is performed with a coherent conductor, with a dimension smaller than the coherence length determined by inelastic scattering, its conductance exhibits irregular dependence on the magnetic field on the scale of G_{Q}, the conductance fluctuations. These fluctuations are determined by random phase shifts unique to this conductor. A conductor much longer than the coherence length can be seen as a classical circuit composed of many coherent conductors with dimensions determined by this coherent length. Their universal conductance fluctuations are independent, and average out when the conductance of the whole sample is measured. This is how we access weak localization correction: the G_{Q} correction to the conductance of each coherent conductor *averaged* over many of them.

It is important to note that coherent effects in conductors with $G \gg G_{\mathrm{Q}}$ are not limited to G_{Q} corrections. They can be large if one measures anything *but* the conductance, for example, noise. The semiclassical Green's function approach can be slightly modified to include coherence and to evaluate the transmission distribution for any nanostructure with $G \gg G_{\mathrm{Q}}$. For this purpose, one introduces a "fake" matrix structure that accounts for coherence. As a matter of fact, the Eilenberger and Usadel equations do not depend on the concrete matrix structure and can be applied to systems that have nothing to do with superconductivity. With this "fake" matrix structure, the commutators disappear from the Usadel equation, and the matrix current is conserved exactly:

$$\frac{\partial \check{\boldsymbol{j}}}{\partial \boldsymbol{r}} = 0; \ \check{\boldsymbol{j}} = -\sigma(\boldsymbol{r})\check{G}\nabla\check{G}. \tag{2.43}$$

In this situation, one can derive from the Usadel equation an analog of the Laplace equation. Let us describe a two-terminal nanostructure. If the Green's functions in the terminals are $\check{G}_1$, $\check{G}_2$, the Green's function everywhere in the nanostructure can be expressed as follows:

$$\check{G}(\boldsymbol{r}) = \exp\{u(\boldsymbol{r})\check{M}\}\check{G}_1, \tag{2.44}$$

where $\check{M}$ is a constant matrix that anticommutes with $\check{G}_1$. This yields

$$\check{\boldsymbol{j}} = -\check{M}\nabla u(\boldsymbol{r}),$$

and the Usadel equation yields the linear equation

$$\nabla(\sigma(\boldsymbol{r})\nabla u(\boldsymbol{r})) = 0,$$

which is precisely the Laplace equation, Eq. (2.13), for the voltage distribution in a conductor of an arbitrary shape. Suppose we know this voltage distribution with the boundary conditions $u \to 0$ (1) in the left (right) reservoir, respectively. Then one immediately restores the solution of the Usadel equation: the constant matrix $\check{M}$ is found from the condition given by

$$\check{G}_2 = \exp\{\check{M}\}\check{G}_1 \Rightarrow \check{M} = \ln\{\check{G}_1\check{G}_2\}.$$

It is clear from Eq. (2.13) that the full current – the current density integrated over a cross-section – is proportional to the full electric current at the voltage drop $u_{\rm R} - u_{\rm L} = 1$, which is just $G_{\rm D}$, the conductance of this arbitrary-shaped conductor:

$$\check{I} = G_{\rm D}\ln\{\check{G}_1\check{G}_2\}. \tag{2.45}$$

We show in Section 2.6 that, based on this relation, the transmission distribution for a diffusive conductor is universal and does not depend on its shape, geometry, and other parameters. The only relevant quantity is the conductance $G \gg G_{\rm Q}$.

The semiclassical Green's function method outlined in this section has many advantages: it can be rigorously derived from exact quantum mechanics, and microscopic details and material parameters can be taken into account with any precision desired. However, the method may become quite disastrous in concrete applications: one has to solve a non-linear differential equation with complicated boundary conditions. This is why in the next section we turn to quantum circuit theory, which is a finite-element approximation to the semiclassical Green's function method. As we will see, quantum circuit theory can be very easy to apply.

Exercise 2.5. Take the Green's functions $\check{G}_1$ and $\check{G}_2$ in the form described in Exercise 2.4 (Eq. (2.83)) and use Eq. (2.45) to reproduce the linear conductance.

2.3.5 Supercurrent from the Eilenberger and Usadel equations

In the following, we give concrete examples of typical work flows in the course of application of the Eilenberger and Usadel equations in order to illustrate the corresponding techniques.

Both examples concern the evaluation of the equilibrium supercurrent. Under equilibrium conditions, the Keldysh structure of the Green's function can be expressed in terms of the equilibrium Fermi distribution function $f(\epsilon) \equiv (1 - \tanh(\epsilon/2k_B T))$:

$$\check{G}(\epsilon) = \begin{bmatrix} \check{R}\tilde{f} + \check{A}f & (\check{A} - \check{R})f \\ (\check{A} - \check{R})\tilde{f} & \check{R}f + \check{A}\tilde{f} \end{bmatrix}, \tag{2.46}$$

where $\tilde{f} \equiv (1 - f) \equiv f(-\epsilon)$. The 2×2 matrices $\check{R}$ ($\check{A}$) are retarded (advanced) Green's functions, respectively. They are analytic functions of complex ϵ at Im $\epsilon > 0$ (< 0). We thus need to evaluate these energy-dependent matrices in any point of the structure. In the case of the Eilenberger equation, we also need to specify their dependence on the momentum direction $\boldsymbol{n}$. In fact, both matrices satisfy the same equation and are related to each other by $\check{R}(\epsilon, \boldsymbol{n}) = -\check{\eta}_3 \check{A}(-\epsilon, -\boldsymbol{n})\check{\eta}_3$. So we need to solve for one of these: we choose a retarded one.

We consider a 1d geometry where two massive superconductors are separated by a layer of a normal metal of the width d. The axis z is perpendicular to the layer plane. Absolute values of the order parameter Δ in the superconductors are the same. The phases of the order parameter are φ_1 and φ_2, respectively, for the left and right superconductor. In other words,

$$\Delta(\boldsymbol{r}) = \begin{cases} \Delta e^{i\varphi_1} & z < -d/2 \\ 0 & |z| < d/2 \\ \Delta e^{i\varphi_2} & z > d/2. \end{cases}$$

The current is in the z direction, and its density at a given point is expressed in terms of $\check{R}$ and $\check{A}$ (see Eq. (2.42)) as follows:

$$I_z(z) = \frac{1}{4}\mathrm{Tr}\int d\epsilon\ \hat{\tau}_z \check{\eta}_3 \check{I} = \frac{e\nu v_F}{4}\int d\boldsymbol{n}\ d\epsilon\ \tanh\frac{\epsilon}{2T}\ n_z \mathrm{Tr}\left(\check{\eta}_3(\check{R} - \check{A})\right).$$

To start with, we assume that the thickness of the normal layer d, as well as the superconducting coherence length $\xi = \hbar v_F/\Delta$, are much smaller than the mean free path. We can therefore disregard the impurity scattering. This allows us to stick to the Eilenberger equation and to the simplest version of it, where $\check{\Sigma} = 0$. The resulting equation is *linear* in the Green's function:

$$-i[\check{E}, \check{R}] = \hbar \boldsymbol{v} \frac{\partial \check{R}}{\partial \boldsymbol{r}}.$$

It is interesting to note that, in this limit, the Eilenberger equation splits into separate equations for separate electron trajectories that go straight through the structure. We parameterize a trajectory with parameter τ that has the dimension of time, $\boldsymbol{r}(\tau) = \boldsymbol{r}(0) + \boldsymbol{v}\tau$. At a given trajectory, $\hbar\partial\check{R}(\tau)/\partial\tau = -i[\check{E}, \check{R}]$.[2] A convenient choice is $z(0) = 0$. In this case, each trajectory remains in the normal layer provided $|\tau| < \tau_0/2 \equiv d/2|v_z| \equiv d/2v_F|n_z|$. Let us concentrate first on the trajectories that go from the left to the right superconductor ($v_z > 0$).

[2] This does not imply a time dependence of $\check{R}$: τ gives a *position* at the trajectory.

It is clear that the solutions of the above linear equation are many in number. We have to choose a proper one from the unitary condition $\check{R} = 1$ and the behavior of the solution at $z \to \pm\infty$. For the Eilenberger equation, this choice is not evident and requires some discussion. Let us note first that for an interval where $\check{E} = \mathrm{const}(\tau)$ it is not a problem to provide a general solution for $\check{R}$ in the special basis – let us call it the local basis – where $\check{E}$ is diagonal:

$$\check{E} \to \begin{pmatrix} E & 0 \\ 0 & -E \end{pmatrix}; \; \check{R} = \begin{pmatrix} r & r_+ \mathrm{e}^{-\mathrm{i}2E\tau/\hbar} \\ r_- \mathrm{e}^{\mathrm{i}2E\tau/\hbar} & -r \end{pmatrix}.$$

Here we choose Im $E > 0$, and $r, r_\pm$ are piecewise constants taking generally different values in different intervals. In our setup, there are three such intervals: the normal metal, the right, and the left superconductor. By virtue of the unitarity condition, $r^2 + r_+ r_- = 1$. Let us note that the exponent coming with r_+ grows without bound at $z \to +\infty$, while that with r_- grows without bound at $z \to -\infty$. Since unbounded solutions must not occur, we conclude that $r_- = 0$ in the left superconductor and $r_+ = 0$ in the right superconductor. Since $\check{R}$ is continuous, this establishes two good boundary conditions at $\tau = \pm\tau_0$ for the Green's function in the normal metal. Unfortunately, they are in local bases of the corresponding superconductors and not in the local basis of the normal metal. To proceed, we have to present these conditions in an arbitrary basis where $\check{E}$ is not diagonal. The best way to do this is to use the projection matrices. Let us note that in the local bases of the left and right superconductors the equilibrium Green's functions are diagonal, $\check{R}_{\mathrm{eq}} = \eta_3 = \check{E}/E$. Let us consider two matrices $P_\pm = (1 \pm \check{R}_{\mathrm{eq}})/2$. It is evident from the explicit form of these matrices in the local basis,

$$P_+ = \begin{pmatrix} 1 & 0 \\ 0 & 0 \end{pmatrix}; \; P_- = \begin{pmatrix} 0 & 0 \\ 0 & 1 \end{pmatrix},$$

that they can be used to cut – or project – matrix elements of other matrices. For instance, for any $\check{M}$,

$$\check{P}_+ \check{M} \check{P}_+ = \begin{pmatrix} M_{11} & 0 \\ 0 & 0 \end{pmatrix}, \; \check{P}_+ \check{M} \check{P}_- = \begin{pmatrix} 0 & 0 \\ M_{12} & 0 \end{pmatrix}.$$

This enables us to write the conditions in the arbitrary basis, for example

$$r_+ = 0 \Leftrightarrow (1 + \check{R}_{\mathrm{eq}})\check{R}(1 - \check{R}_{\mathrm{eq}}) = 0.$$

Thereby we come to two boundary conditions imposed on $\check{R}$ in the normal metal at the boundaries with both superconductors:

$$(1 - \check{R}_1)\check{R}(\tau = -\tau_0/2)(1 + \check{R}_1) = 0;$$
$$(1 + \check{R}_2)\check{R}(\tau = \tau_0/2)(1 - \check{R}_2) = 0,$$

where $\check{R}_{1(2)}$ stand for the equilibrium functions in the left (right) superconductor. We stress the general nature of the boundary conditions presented. We have derived them for 2×2 matrices, but in fact they are valid for any matrix structure and are indispensable in applications of the Eilenberger equation. Let us also note the difference between the conditions on the left and on the right: these conditions depend on the direction of electron propagation.

Control question 2.6. Give the conditions for a trajectory going from the right to the left.

Exercise 2.6. Consider the limit $d \to 0$ so that $\check{R}(-\tau_0/2) = \check{R}(\tau_0/2) \equiv \check{R}$. Show that the matrix $\check{R} = (2 - \check{R}_1 + \check{R}_2)(\check{R}_1 + \check{R}_2)^{-1}$ satisfies both boundary conditions and the unitarity condition. Use only the fact that $\check{R}_{1,2}^2 = 1$.

Let us give the explicit form of $\check{R}_{1,2}$:

$$\check{R}_{1,2} = \frac{1}{E}\begin{pmatrix} \epsilon & \Delta e^{i\varphi_1} \\ -\Delta e^{-i\varphi_1} & -\epsilon \end{pmatrix}, \qquad E = \sqrt{(\epsilon + i0)^2 - \Delta^2}. \tag{2.47}$$

In the superconducting gap, $|\epsilon| < \Delta$, E is purely imaginary, $E = i\sqrt{\Delta^2 - \epsilon^2}$. Beyond the gap, E seems to be purely real. The small imaginary part i0 that distinguishes between retarded and advanced Green's functions is, however, important here and must not be omitted. An ambiguous sign of the square root has to be chosen regarding i0 and should satisfy $\operatorname{Im} E > 0$. This suggests that beyond the gap E is *odd* in ϵ, $E = \operatorname{sign}(\epsilon)\sqrt{\epsilon^2 - \Delta^2}$. In the normal metal, $E = \epsilon$.

With all this taken into account, we bring the boundary conditions to the explicit form:

$$e^{i(\epsilon\tau_0/\hbar - \varphi_1)}(E - \epsilon)r_+ + 2\Delta r + e^{-i(\epsilon\tau_0/\hbar - \varphi_1)}(E + \epsilon)r_- = 0;$$
$$e^{i(-\epsilon\tau_0/\hbar - \varphi_2)}(E + \epsilon)r_+ - 2\Delta r + e^{i(\epsilon\tau_0/\hbar + \phi_2)}(E - \epsilon)r_- = 0.$$

We then make use of the unitarity condition[3] and solve for the most important part of the Green's function, $\operatorname{Tr}(\check{\eta}_3\check{R})/2 = r$:

$$r = -i\tan\left(\frac{\epsilon\tau_0}{\hbar} - \frac{\varphi}{2} + \arcsin\frac{\epsilon}{\Delta} + i0\right). \tag{2.48}$$

Here, φ is the phase difference between the superconductors, $\varphi = \varphi_1 - \varphi_2$. The expression in the above form is most comprehensive at $|\epsilon| < \Delta$. Beyond the gap region the same expression can be cast to the following form:

$$r = i\cot\left(\frac{\epsilon\tau_0}{\hbar} - \frac{\varphi}{2} + i\,\operatorname{arccosh}\frac{|\epsilon|}{\Delta}\right). \tag{2.49}$$

Let us note that in the gap region the Green's function diverges in a set of energy points where the argument of tan approaches $\pi/2 + n\pi$. This gives the discrete energies of Andreev bound states developed along the trajectory. Both positive and negative ϵ give rise to positive bound state energies E_A that satisfy $\arcsin(E_A/\Delta) + E_A\tau_0 = \pm\varphi + n\pi$. If the trajectory is short, $\tau_0 \ll \Delta/\hbar$, there is a single bound state $E_A = \Delta\cos(\varphi/2)$. We have already seen this in Section 1.8: this is the Andreev bound state for a short (Eq. (1.170)) and transparent ($T_p = 1$) junction. With increasing length of the trajectory, more and more Andreev bound states appear in the gap region emerging from the continuum of delocalized quasiparticle states beyond the gap region (Fig. 2.3(a)). At sufficiently large τ_0, there

[3] Strictly speaking, the conduction determines $\check{R}$ upon a ± 1 factor. However, the correct sign is readily fixed by comparing the resulting expression with its known limits, for example, at $\epsilon \gg \Delta$.

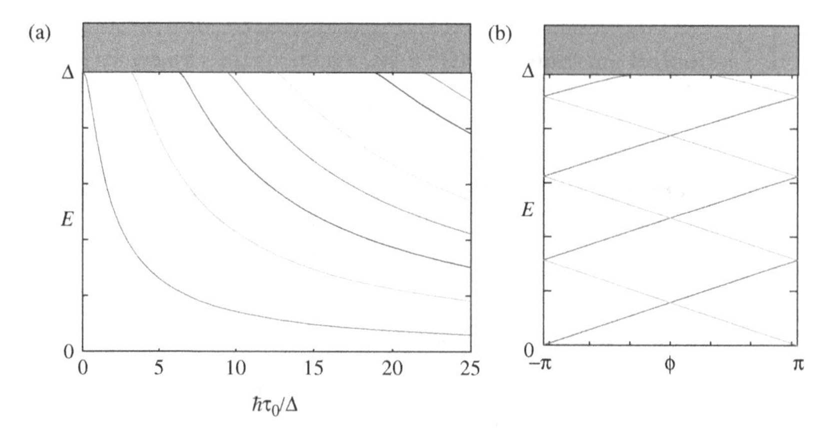

Fig. 2.3. **Discrete Andreev bound states developed along a trajectory. The shaded rectangles denote the continuum of quasiparticle states above Δ. (a) Energies of the states at $\varphi = 0$ versus the trajectory length $v_F\tau_0$. (b) Shoe lacing: phase dependence of the energies at $\tau_0 = 9\Delta/\hbar$.**

are $\approx 2\tau_0\Delta/\pi\hbar$ bound states in the gap region. The phase dependence of all these levels resembles a criss-cross shoe lacing (Fig. 2.3(b)).

We are almost ready to compute the current. We have the retarded Green's function, and can express the advanced function by making use of the relation $\check{A}(\boldsymbol{n}, \epsilon) = -\check{\eta}_3\check{R}(-\boldsymbol{n}, \epsilon)\check{\eta}_3$. Note that flipping $\boldsymbol{n}$ flips the trajectory direction: trajectories going from the left to the right change to those going from the right to the left. This does not change τ_0, but does flip the superconducting phase difference: $\varphi \to -\varphi$. Therefore, the current density reads $I_z = e\nu v_F/2 \int \mathrm{d}\epsilon\ \mathrm{d}\boldsymbol{n}\ \tanh(\epsilon/2k_BT)|n_z|\mathrm{Re}(r(\epsilon, \varphi) + r(-\epsilon, -\varphi))$. In the gap region, the real part of the Green's function consists of δ peaks corresponding to the discrete Andreev states:

$$\mathrm{Re}\, r = \pi \sum_n \delta\left(\frac{\epsilon\tau_0}{\hbar} - \frac{\varphi}{2} + \frac{\epsilon}{\Delta} + \pi n\right).$$

Integrating over energies, we recover the contribution of the Andreev bound states to the current: it resembles a familiar relation (Eq. (1.172)):

$$I_z = e\pi\nu v_F \int \mathrm{d}\boldsymbol{n}\ |n_z| \sum_n \frac{\partial E_{A,n}(\varphi)}{\partial\varphi}. \tag{2.50}$$

Comparing this with Eq. (1.172), we find the density of transport channels per area and per element of direction $\mathrm{d}\boldsymbol{n}$ as follows:

$$\mathrm{d}N_{\mathrm{tr}} = \frac{\pi\hbar\nu v_F}{2}\, \mathrm{d}A\ \mathrm{d}\boldsymbol{n}\ |n_z|.$$

There is also a contribution to the current from the energy region beyond the gap. Generally, this contribution cannot be disregarded. However, in two opposite limits of long and short trajectories it is negligible, and the current is given by Eq. (2.50). The limit of short trajectories and vanishing temperatures has been already discussed in Section 1.8.4: the supercurrent is given by $I(\phi) = (\pi\Delta/e)G_N \sin(\varphi/2)$, where in our case the normal-state conductance of the structure $G_N = e^2 A\nu v_F/4$.

Let us investigate the opposite limit of long trajectories assuming vanishing temperature. In this limit, the Andreev bound states are linear functions of the phase difference $\partial E_{A,n}/\partial\varphi = \pm\hbar/2\tau_0$. Looking at Fig. 2.3, we note that the sign of the derivatives alternates from level to level. Therefore, the contributions of the levels cancel each other, except at the last "lace," which oscillates with τ_0. Averaging over these oscillations, we obtain $\sum_n \langle\partial E_{A,n}/\partial\varphi\rangle = \hbar\varphi/2\pi\tau_0$. Integrating over $\boldsymbol{n}$, we recover the "sawtooth" (piecewise-linear) current–phase relation: $I_z = (evv_F^2/6d)\varphi$ at $|\varphi| < \pi$.

Such a simple form of the answer suggests that there exists some simple explanation. Indeed, the current in this limit can eventually be evaluated without going to microscopic details. Since the Andreev bound states in the limit are dense in the gap region, the electrons in the normal metal layer are fully involved in the motion of the superconducting condensate. This motion, as mentioned in Appendix B, is characterized by the superfluid velocity $v_s = (\hbar/2m)(\nabla\varphi - (2e/c)\boldsymbol{A})$. In our case, the phase gradient $\nabla_z\varphi = \varphi/d$. The current density is then given by $I_z = env_s$, where n is the total electron density. This coincides with the above relation derived with the Eilenberger equation.

Let us turn to the Usadel equation. We now assume that the structure contains a sufficient number of impurities such that the thickness of the normal metal layer along with the superconducting coherence length, $\xi = \sqrt{D\hbar/\Delta}$, are bigger than the mean free path; this enables the use of the equation. The unitarity condition $\check{R}^2 = 1$ imposed on the retarded Green's function in Nambu space is readily resolved by the following parameterization in terms of two variables, θ and χ:

$$\check{R} = \begin{pmatrix} \cosh\theta & \sinh\theta e^{i\chi} \\ \sinh\theta e^{-i\chi} & -\cosh\theta \end{pmatrix}. \tag{2.51}$$

This parameterization is equally applicable in the global basis where η_3 is diagonal and in a local basis where $\check{E}$ is diagonal. The variables θ and χ are generally complex.

Let us first discuss general solutions of the Usadel equation in a local basis. Substituting $\check{R}$ in the above parameterization into Eq. (2.40), we obtain ($' \equiv \partial_z$)

$$\left(\chi' \sinh^2\theta\right)' = 0; \tag{2.52}$$

$$\theta'' = -\frac{2iE}{\hbar D}\sinh\theta + \frac{\chi'^2}{2}\sinh 2\theta. \tag{2.53}$$

These equations have two (complex) integrals of motion:

$$\chi' \sinh^2\theta = J; \tag{2.54}$$

$$\frac{(\theta')^2}{2} + \frac{2iE}{\hbar D}(\cosh\theta - 1) + \frac{J^2}{2\sinh^2\theta} = K. \tag{2.55}$$

The first integral of motion manifests the conservation of a component of the matrix current in the Usadel equation, i.e. that commuting with $\check{E}$.

Let us first analyze the solutions in the superconductors. At $z \to \pm\infty$, the Green's functions should approach $\check{R}$. For the parameterization in use, this implies $\theta \to 0$ at $z \pm \infty$. The only solution satisfying this corresponds to $K = J = 0$, and we obtain

$$\chi' = 0; \tag{2.56}$$

$$\theta' = \pm 2^{3/2} e^{i\pi/4} \sqrt{\frac{E}{\hbar D}} \sinh \frac{\theta}{2}, \tag{2.57}$$

where the $+$ $(-)$ sign refers to the left (right) superconductor. The latter equations can be easily integrated to obtain θ, χ inside the superconductors in terms of their boundary values. For instance, in the right superconductor,

$$\chi(z) = \chi(d/2);$$

$$\frac{\tanh(\theta(z)/4)}{\tanh(\theta(d/2)/4)} = \exp\left(-e^{i\pi/4}\sqrt{\frac{2E}{\hbar D}}(z - d/2)\right). \tag{2.58}$$

At the same time, Eqs. (2.56) and (2.57) give the boundary conditions at the interfaces $z = \pm d/2$, thereby enabling us to solve for the Green's function in a normal metal. These conditions relate the derivatives of χ, θ to their values at the corresponding interface. Equations (2.56) and (2.57) give them in local bases of the superconductor: they have to be brought to an arbitrary basis with projection matrices $\check{P}_\pm = (1 \pm \check{R}_{eq})/2$, as we did for the Eilenberger equation. We do not do this here, turning instead to simpler boundary conditions.

These simpler boundary conditions hold whenever we can neglect the change in the Green's functions in the superconductors compared with that in the normal metal. To see how this works, consider small energies $|\epsilon| \ll \Delta$. The Green's function changes at a typical length $\sqrt{E/\hbar D}$. Since in the normal metal $E = \epsilon$ and in the superconductor $|E| = \Delta$, this length is much bigger in the normal metal. The derivatives of the Green's functions are the same at the interface; therefore, the change of the Green's function in the superconductor is much smaller and can be neglected. Generally, we need to compare the resistance of the superconductor layer of thickness ξ to the resistance of the normal metal, either at thickness $\sqrt{\epsilon/\hbar D}$ or at full thickness d: the change is roughly proportional to the resistance, and is negligible if the resistance is small.

If we may neglect the change of the Green's function in a superconductor, the Green's functions at the edges of a normal metal must match the equilibrium Green's functions in the corresponding superconductors, $\check{R}(\mp d/2) = \check{R}_{1,2}$. From now on, we use the θ–χ parameterization in the local basis of the normal metal. Recalling Eq. (2.47), we see that the boundary conditions become

$$\theta(d/2) = \theta(-d/2) = \theta_S; \ \cosh(\theta_S) \equiv \frac{\epsilon}{\sqrt{\Delta^2 - (\epsilon + i0)^2}};$$

$$\chi(-d/2) = \phi_1; \ \chi(d/2) = \phi_2,$$

and no longer contain derivatives. Owing to the symmetry between left and right, θ is an even and $\chi - (\varphi_1 + \varphi_2)$ is an odd function of the coordinate z. So we have two differential equations, Eqs. (2.52) and (2.53), supplemented by the boundary conditions: a computer-ready problem.

An alternative solution strategy relies on the integrals of motion, Eqs. (2.54) and (2.55). With these, one expresses derivatives in terms of θ. For example, it follows from Eq. (2.55) that

$$\theta' = \sqrt{2(K - J^2 \sinh^{-2}\theta/2 + \sinh^2(\theta/2)(\mathrm{i}\epsilon/\hbar D))} \equiv F(\theta, K, J).$$

This allows us to avoid the explicit computation of θ, χ at all z. One expresses the layer thickness and the phase drop in terms of these derivatives to arrive at

$$\frac{d}{2} = \int_{\theta(0)}^{\theta_S} \frac{\mathrm{d}\theta}{\theta'} = \int_{\theta(0)}^{\theta_S} \frac{\mathrm{d}\theta}{F(\theta)} \equiv F_1(K, J, \theta(0), \theta_S); \tag{2.59}$$

$$\phi_2 - \phi_1 = \int_{-d/2}^{d/2} \mathrm{d}z\, \chi' = J^2 \int_{-d/2}^{d/2} \frac{\mathrm{d}z}{\sinh^2\theta}$$

$$= 2\int_{\theta(0)}^{\theta_S} \frac{\mathrm{d}\theta}{F(\theta)\sinh^2\theta} \equiv F_2(K, J, \theta(0), \theta_S)\,, \tag{2.60}$$

where $F_{1,2}$ are expressed in terms of incomplete elliptic integrals. Since θ is even in z, $\theta'(0) = 0$ and $K = J^2 \sinh^{-2}\theta(0) + (4\mathrm{i}\epsilon/\hbar D)\sinh^2\theta(0)$. This leaves two ordinary complex equations to solve and find $J, \chi(0)$ at given ϵ. It is precisely this first integral of motion that we need in order to compute the current density. Indeed, in the Usadel formalism, the current density per energy interval reads (Eq. (2.40)) $\check{j} = -\sigma\check{G}\check{G}'$. We recall Eq. (2.46) to express $\check{G}$ via the retarded and advanced Green's functions, to make use of the parameterization Eq. (2.51), and to compute the trace with $\check{\tau}_3\check{\eta}_3/4$ (Eq. (2.42)) to arrive at

$$I_z = -\sigma/e \int \mathrm{d}\epsilon\, \tanh\frac{\epsilon}{2k_B T}\, \mathrm{Im}\, J(\epsilon),$$

where we take into account that $J_R(\epsilon) = J_A(-\epsilon)$. Therefore, the fact that J is an integral of motion is the manifestation of current conservation.

The resulting equations generally prohibit analytical solutions. These can be obtained in several limiting cases, and we analyze one of them below. We restrict our consideration to energies that are higher than the Thouless energy of the junction, but still much lower than the gap, $E_{Th} = \hbar D/d^2 \ll |\epsilon| \ll \Delta$. Naturally, this makes sense only if $E_{Th} \ll \Delta$, or, equivalently, $d \gg \xi$: the normal metal layer must be sufficiently thick. In this limit, the Green's function in the normal metal essentially deviates from its equilibrium value $\theta = 0$ only near the interfaces, at distances $\simeq (\epsilon/\hbar D)^{-1/2} \ll d$. Therefore, we do not expect a large supercurrent: let us see how small it is.

Since $\epsilon \ll \Delta$, $\theta(\pm d/2) = \theta_S \approx \mathrm{i}\pi/2$, and $\chi = \phi_{1,2}$ in the left and right superconductor, respectively. Let us first investigate the Green's function near the left interface. In doing so, we can disregard the mere existence of the right interface and assume $\theta \to 0$ at $z \to \infty$. We have been dealing with solutions of this type in the superconductors, and the solution sought is obtained by minor adaptation of Eq. (2.58):

$$\tanh\frac{\theta(z)}{4} = \mathrm{i}\tan\left(\frac{\pi}{8}\right)\exp\left(-\frac{z - d/2}{L(\epsilon)}\right),$$

$$L(\epsilon) \equiv \mathrm{e}^{\mathrm{i}\pi/4}\sqrt{\frac{2\epsilon}{\hbar D}};\, \chi(z) = \varphi_1.$$

This falls off exponentially with increasing z. Sufficiently far from the interfaces, $\theta \approx 0$, so we can expand as follows:

$$(\theta e^{\pm i\chi})_1 = \kappa \exp(-z/L(\epsilon))e^{\pm\varphi_1},$$

where κ is an exponentially small number. Now let us turn to the right interface. Repeating the steps, we find another approximate solution that is accurate closer to the right interface:

$$(\theta e^{\pm i\chi})_2 = \kappa \exp(z/L(\epsilon))e^{\pm\varphi_2}; \qquad \kappa \equiv i4\tan\left(\frac{\pi}{8}\right)\exp\left(-\frac{d}{L(\epsilon/2)}\right)$$

far from the interfaces.

The crucial observation is that, in most of the normal metal layer, $\theta \approx 0$, and the Usadel equation is *linear* in terms of the non-diagonal matrix elements. Because of this, the true solution is just a linear superposition of the two approximate solutions:

$$\theta e^{\pm i\chi} = (\theta e^{\pm i\chi})_1 + (\theta e^{\pm i\chi})_2.$$

Now we can evaluate the integral of motion, given by

$$J(\epsilon) = \chi' \sinh^2\theta \approx \chi'\theta^0 = 2\kappa^2/L(\epsilon)\sin\varphi.$$

It does not depend on z as it should by virtue of current conservation.

The remaining task to complete is the integration over ϵ. It does not make sense to do this at vanishing temperature $k_B T \ll E_{Th}$ since, in this case, the main contribution to the current comes from the energies not covered by the present consideration. So we do it at a higher temperature that still does not exceed the superconducting gap, $E_{Th} \ll k_B T \ll \Delta$. Let us note that, at real energy, $L(\epsilon)$ has both real and imaginary parts. This causes $J(\epsilon)$ to oscillate quickly. To evaluate the integral, we shift the integration contour in the plane of complex ϵ to the imaginary axis where the integrand does not oscillate. The thermal factor $\tanh(\epsilon/2k_B T)$ has poles at the imaginary axis, and the main contribution to the integral comes from a pair of closest poles at $\epsilon = \pm\pi k_B T$. Finally, we obtain (using $\tan(\pi/8) = \sqrt{2} - 1$)

$$I = 64(\sqrt{2}-1)^2 \frac{\sigma}{ed} k_B T \sqrt{\left(\frac{2\pi k_B T}{E_{Th}}\right)} \exp\left(-\sqrt{\frac{2\pi k_B T}{E_{Th}}}\right) \sin\varphi. \qquad (2.61)$$

We see that the supercurrent in this regime is always proportional to $\sin\varphi$. As we have seen in Section 1.8, this I–φ characteristic is typical for tunnel junctions. However, the electrons do not tunnel; rather, they diffuse through the metal layer. What tunnels are the Cooper pairs: the $\sin\varphi$ term indicates that the Cooper pair transfers are independent. This is in line with the exponentially suppressed supercurrent: the Cooper pairs really see the normal metal layer as a kind of almost impenetrable tunnel barrier. Qualitatively, this suppression can be understood in terms of scattering theory. Energies $\ll E_{Th}$ bring essential energy dependence to transmission amplitudes. An amplitude of a Cooper pair transfer involves two transmission amplitudes at $\pm\epsilon$ and is therefore sensitive to energy-dependent and random phases of the transmission amplitudes. The self-averaging over these random phases efficiently suppress the resulting Cooper pair amplitude. This also explains the temperature dependence of the supercurrent.

The following remark concerns both examples given in the preceding text. We have evaluated the current density in a normal metal and found it constant in z, as dictated by current conservation. If we evaluate it beyond the normal layer, we would be surprised: the computed current density vanishes inside the superconductors at a rather short length scale $\sim\xi$! It is clear what happens: at length scale $\sim\xi$, the current coming from the normal layer is converted to the bulk supercurrent, the motion of the superconducting condensate. This motion gives rise to a phase gradient in the superconductor, and therefore the superconducting order parameter should depend on z. Our calculation has failed to encompass this effect since we forced the constant superconducting order parameter at $|z| > d/2$. The accurate solution of superconductivity equations requires *self-consistency*. The superconducting order parameter, $\Delta(\boldsymbol{r})$, should be determined from the solution for the Green's functions. Such a solution conforms to current conservation. In practice, the accurate account of self-consistency is seldom required for nanostructures. This is in contrast to bulk superconductors. Another simplifying assumption made was to disregard the dependence of the Green's functions on transverse coordinates. However natural this assumption might seem, it is not compatible with the Meissner effect in superconductors: they cannot support a uniform current density. The magnetic fields produced by currents repel them from the superconducting bulk to a thin surface layer. So, in fact, our consideration is restricted to the structures of small transverse dimensions.

In this technical subsection, we have exemplified a typical application of the Eilenberger and the Usadel equations in their differential form to a superconducting heterostructure. Even for the simplest geometry considered, the technicalities are rather difficult. Besides, the simplest geometry cannot account for self-consistency and the Meissner effect. This calls for a more comprehensive approach, and provides the extra motivation to turn to the finite-element techniques considered in the rest of this chapter.

2.4 Current conservation and Kirchhoff rules

Laplace equations (see Eqs. (2.13)) for continuous distribution of voltages and currents are physically obvious and therefore simple. The position-dependent conductivity allows us to model conductors of any shape and design, and solving Laplace equations will give currents and voltages. The problem is that the equations are partial differential ones – this means that they are not easy to solve either numerically or analytically except for several primitive geometries. We know that for electric circuits there is a fortunate way out – instead of solving differential Laplace equations, one splits the circuit into finite elements and solves several algebraic equations.

Any circuit consists of *terminals* and *nodes* denoted by a Latin index, say k, and characterized by voltages V_k. The voltages at the terminals are fixed, while the voltages in the nodes need to be found. Terminals and nodes are connected by *connectors* (Fig. 2.4). The current through a connector between the nodes (terminals) i and k is proportional to the voltage drop at the connector:

$$I_{ik} = G_{ik}(V_i - V_k). \tag{2.62}$$

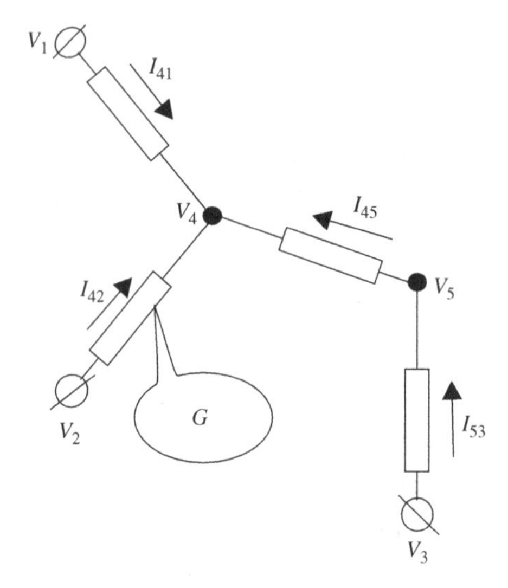

Fig. 2.4. A common electric circuit consists of terminals (1–3), nodes (4,5), and connectors. Each connector is characterized by its conductance G. The voltages $V_{1,2,3}$ are fixed in the reservoirs. The voltages at nodes $V_{4,5}$ are determined from the balance of currents in the circuit, and the current in each connector is finally determined from these voltages.

How do we calculate the voltages at the nodes? The rules of circuit theory, the Kirchhoff rules, are in fact conservation laws, or balance equations. Current is conserved. This implies that, for each node k, the sum of the currents coming from all connectors equals zero, i.e.

$$\sum_k I_{ik} = 0. \tag{2.63}$$

This yields a set of equations for voltages at the nodes. The solution of these equations allows one to find the current in any connector and between the terminals.

The solution of the Laplace equations for an arbitrary geometry with any desired accuracy is, in principle, obtained by the finite-element approach in the limit of a large number of circuit elements. In such a scheme, the volume of the conductor is separated into elementary volumes, for example small cubes. The elementary volumes then become the nodes of the circuit. Making elementary volumes smaller and smaller boosts accuracy, simultaneously increasing the number of nodes.

It is good that this number does not have to be high for the purposes of design or any other practical activity; a few elements suffice. If a highly accurate solution of the Laplace equations is desired, it is found by a computer program that eventually implements a finite-element, circuit-theory approach with thousands of elements.

In the following, we consider the rules of *quantum circuit theory*, which is more powerful than the standard circuit theory based on Ohm's laws. The principle is essentially the same – the system is split into finite elements, and these elements are combined using balance equations. The difference now is that (i) properties of these finite elements are found from quantum-mechanical laws; (ii) quantum mechanics adds new conservation laws generally absent in classical circuit theory.

2.4.1 Double junction revisited

Before starting this complicated business, we would like to demonstrate some general ideas of the finite-element approach and the importance of balance equations using a particular simple example. We turn again to the double junction setup, which we have discussed thoroughly in Section 2.1. The finite-element approximation of this setup is straightforward: the space between the junctions is a node, and the two junctions are Kirchhoff connectors between the node and the left and right terminals.

We will discuss the energy dependence of the distribution function of electrons in the node in three different situations that differ in the role played by inelastic scattering in the node. To understand this role, let us first analyze an infinite system that does not know about the reservoirs. If the scattering in the system is elastic only, the energy dependence is undefined. Any isotropic function of energy would satisfy the balance equations for the filling factor, even if does not have the form of an equilibrium Fermi distribution. This changes if inelastic scattering is present in the node. The electrons can either lose or gain the energy. If the inelastic processes come from electron–electron scattering, the electrons can only exchange energy, in which case the total energy and number of electrons are conserved. In this case, the equilibrium distribution function is a Fermi distribution function. However, two parameters of this function – temperature and chemical potential – remain undefined. Finally, if electrons experience inelastic scattering on phonons (the vibrations of crystal lattice), the electron system can lose or gain energy from the lattice and the electron temperature is that of the crystal. The chemical potential is still undefined since the total number of electrons is conserved.

Let us now consider coupling to the reservoirs and see how the balance equations set these undefined quantities.

First, we assume that no inelastic scattering takes place when electrons traverse the double junction. This implies conservation of current *at each energy*. The reservoirs (terminals) are characterized by filling factors $f_{L,R}(E)$. The assumption is that the node is characterized by the filling factor $f_N(E)$. By virtue of Landauer formula, the currents through each junction *to* the node are proportional to the corresponding filling factor differences:

$$j_1 = G_1(f_L(E) - f_N(E)); \ j_2 = G_2(f_R(E) - f_N(E)).$$

The conservation law gives simply $j_1 + j_2 = 0$. From this we readily calculate the filling factor and current at each energy:

$$f_N = \frac{G_1 f_L + G_2 f_R}{G_1 + G_2}; \ j = j_1 = -j_2 = \frac{G_1 G_2}{G_1 + G_2}(f_L - f_R).$$

At each energy, f_N is thus in between f_L and f_R. It does not have the form of a Fermi function; instead, at low temperatures it develops two steps as a function of energy, which correspond to steps in f_L and f_R (see Eq. (2.14)). The full electric current is obtained by integration over the energy,

$$I = \frac{1}{e}\int dE\ j(E) = \frac{G_1 G_2}{G_1 + G_2}V, \tag{2.64}$$

and satisfies Ohm's law provided the transmission eigenvalues of the tunnel junctions do not depend on energy.

Let us turn to the second situation. If one increases the spacing between the barriers, more and more inelastic electron–electron collisions take place. Electrons exchange energy, and this results in their thermalization. The filling factor has the form of the Fermi distribution,

$$f_{\mathrm{N}}(E) = \frac{1}{1 + \exp(E - \mu)/k_{\mathrm{B}}T^*}, \tag{2.65}$$

and is a smooth function of energy. This function depends on two unknown parameters: the chemical potential μ and the temperature T^* of the node. To determine these two parameters, we apply two conservation laws: charge and energy conservation. The charge and heat flows through junctions 1 and 2 are given by

$$eI_1 = G_1 \int dE(f_{\mathrm{L}}(E) - f_{\mathrm{N}}(E)),$$
$$eI_2 = G_2 \int \mathrm{d}E(f_{\mathrm{R}}(E) - f_{\mathrm{N}}(E))$$

and

$$e^2 q_1 = G_1 \int \mathrm{d}E\,E(f_{\mathrm{L}}(E) - f_{\mathrm{N}}(E)),$$
$$e^2 q_2 = G_2 \int \mathrm{d}E\;E(f_{\mathrm{R}}(E) - f_{\mathrm{N}}(E)).$$

The balance of the charge flows,

$$I_1 + I_2 = 0,$$

yields

$$\mu = \frac{G_1\mu_{\mathrm{L}} + G_2\mu_{\mathrm{R}}}{G_1 + G_2}.$$

On computing the current, we recover the same relation as for the previous situation of negligible inelastic scattering, Eq. (2.64).

To calculate the temperature of the node, we employ the balance of heat flows, $q_1 + q_2 = 0$. At equilibrium (zero voltage), the temperature of the node obviously equals the temperature of both terminals. When voltage is applied and both terminals are kept at zero temperature, T^* is determined by the voltage:

$$k_{\mathrm{B}}T^* = \frac{\sqrt{3}}{\pi}\frac{\sqrt{G_1 G_2}}{G_1 + G_2} eV.$$

Generally, it is of the order of the applied voltage and temperature in the reservoirs, whichever is greater. The result for several ratios G_1/G_2 is plotted in Fig. 2.5.

Exercise 2.7. Evaluate the effective temperature T^* for the finite temperature T of both terminals given the voltage difference eV. Hint: the expression for heat flow through each terminal consists of separate terms that depend either on two chemical potentials or on two temperatures.

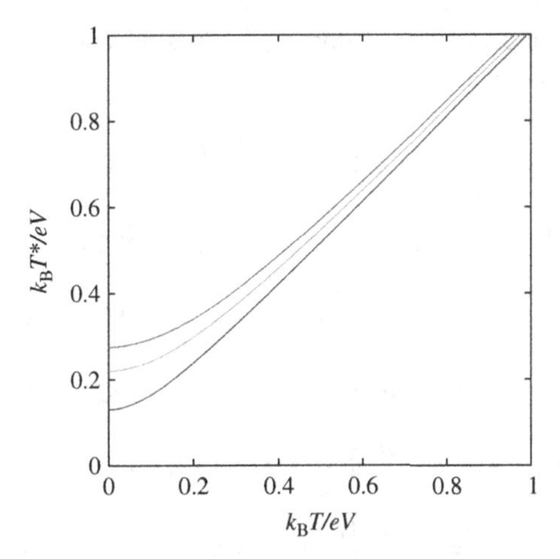

Fig. 2.5. **Effective temperature of the node in the double-junction system versus temperature at fixed voltage V; $G_1 \geq G_2$, and G_1/G_2 takes values 1, 2 and 4 from the top curve down.**

Finally, we present the third case. Upon further increase of the size of the node, the latter can efficiently exchange heat with external world. The heat conservation law does not apply, and the effective temperature is that of the external environment, $T^* \simeq 0$. The charge conservation yields the same equations as before, so the current is again given by Ohm's law.

The three cases we have considered demonstrate the power and wide applicability of the classical conservation laws, balance equations, and circuit theory.

It is important to understand that the same Ohm's law holds in all three cases under quite different conditions. In the first situation, the electrons retain their quantum coherence while traversing the nanostructure. It can be regarded as a single scatterer, and is characterized by transmission eigenvalues. Under these conditions, one can use the quantum circuit theory described below, for example to find the distribution of the transmission eigenvalues. In the two latter cases, the electron coherence is destroyed by inelastic processes, and therefore only classical balance applies.

2.4.2 Matrix currents and leakage currents

We have already encountered the conservation of a matrix current when discussing the Usadel equation. Here we comprehend matrix currents at a more general level. It turns out that quantum mechanics gives rise to conservation laws which are absent in classical mechanics, or at least are not associated with any quantities of obvious physical significance. This sounds strange at least. Since we will base the whole of quantum circuit theory on these laws, we should avoid any possible misunderstanding and distrust. This is why we revise the concept step by step.

Let us recall first how the conservation of probability current is proven in quantum mechanics. The proof can be found in the first few pages of any good introductory textbook in quantum mechanics and goes as follows. Consider a wave function $\psi(\boldsymbol{r})$ that obeys the Schrödinger equation:

$$E\psi = \hat{H}\psi = \left(-\frac{\hbar^2}{2m}\nabla^2 + U(\boldsymbol{r})\right)\psi(\boldsymbol{r}).$$

Let us try the following expression for the vector density of the probability current:

$$\boldsymbol{j}(\boldsymbol{r}) = -\frac{\mathrm{i}\hbar}{2m}(\psi^*\nabla\psi - \psi\nabla\psi^*). \tag{2.66}$$

We show that the divergence of this current is zero, i.e.

$$\begin{aligned}\nabla\cdot\boldsymbol{j}(\boldsymbol{r}) &= -\frac{\mathrm{i}\hbar}{2m}\nabla\cdot(\psi^*\nabla\psi - \psi\nabla\psi^*)\\ &= -\frac{\mathrm{i}\hbar}{2m}\left\{\left[\nabla\psi^*\cdot\nabla\psi - \nabla\psi\cdot\nabla\psi^*\right] + \left[\psi^*\nabla^2\psi - \psi\nabla^2\psi^*\right]\right\}\\ &= \frac{\mathrm{i}}{\hbar}\left[\psi^*\left[E - U(x)\right]\psi - \psi\left[E - U(x)\right]\psi^*\right] = 0,\end{aligned}$$

where we have assumed that the first term in the second line vanishes identically, and we have transformed the second term using the Schrödinger equation.

Thus, we have proved the current conservation irrespective of any concrete realization of the potential $U(\boldsymbol{r})$, $\nabla\cdot\boldsymbol{j}(\boldsymbol{r}) = 0$. The quantum expression given in Eq. (2.66) has an obvious classical analog – the particle current. The introductory textbook stops here, and we proceed further.

Let us look at two *different* solutions χ, ψ of the *same* Schrödinger equation,

$$E\psi = \hat{H}\psi = \left(-\frac{\hbar^2}{2m}\nabla^2 + U(\boldsymbol{r})\right)\psi(\boldsymbol{r})$$

and

$$E'\chi = \hat{H}\chi = \left(-\frac{\hbar^2}{2m}\nabla^2 + U(\boldsymbol{r})\right)\chi(\boldsymbol{r}),$$

and try the following combination:

$$\boldsymbol{j}(\boldsymbol{r}) = -\mathrm{i}\frac{\hbar}{2m}\left(\chi\nabla\psi - \psi\nabla\chi\right).$$

The same calculation gives us

$$\nabla\cdot\boldsymbol{j}(\boldsymbol{r}) = \frac{\mathrm{i}}{\hbar}(E - E')\chi\psi. \tag{2.67}$$

For $E = E'$, we obtain the conserving current. In general, this current does not have an obvious classical analog, at least for $\chi \neq \psi^*$. Thus, quantum mechanics can give rise to new conservation laws absent in classical mechanics.

Let us illustrate this statement with two examples. First, we address spin currents (see Section 1.9). The electron wave function is a spinor with two components corresponding to two possible values of electron spin, $s = \pm\hbar/2$. Provided the Hamiltonian is spin-independent and time-reversible, the two components, $\psi_\uparrow$ and $\psi_\downarrow$, and their complex conjugates all satisfy the same Schrödinger equation. To implement the general scheme outlined above, we consider the 2×2 matrix current, given by

$$j_{\alpha\beta}(\boldsymbol{r}) = -\frac{\mathrm{i}\hbar}{2m}(\psi_\alpha^* \nabla \psi_\beta - \psi_\beta \nabla \psi_\alpha^*),$$

where each of the indices α, β can assume the values $\uparrow$, $\downarrow$. For the time-dependent Schrödinger equation, we can prove along the same lines that

$$\frac{\partial \rho_{\alpha\beta}}{\partial t} + \nabla \cdot j_{\alpha\beta}(\boldsymbol{r}) = 0,$$

where $\rho_{\alpha\beta}$ is the electron density matrix in spin indices. We note now that any 2×2 matrix can be expanded in the basis containing the diagonal unity matrix $\hat{1}$ and three Pauli matrices. If we expand the above equation, the part proportional to $\hat{1}$ gives the conservation law for particle density, whereas the parts proportional to the Pauli matrices give conservation laws for the x, y, z components of spin density.

One may argue that the spin density can be regarded as a pseudovector classical quantity, and hence that these conservation laws are still of classical nature. However, we can unleash our imagination and consider particles of higher spin, say $S = 5/2$. The spin density is still a pseudovector, so we expect four conserving currents again. How many quantum laws are there? The electron wave functions in this case are six-component spinors. Thus the matrix current is a 6×6 matrix, and we count 36 conservation laws.

The second example we provide is in the context of Andreev scattering, where wave functions obey the Bogoliubov–de Gennes equation (see Section 1.8). In this case, a wave function also has two components, a superposition of electron and hole components. If the Hamiltonian is time-reversible, the electron ψ_e and hole ψ_h components in the normal metal (where $\Delta = 0$) satisfy the same Schrödinger equation with energies equal to $\pm E$. Therefore, the matrix current, given by

$$j_{\alpha\beta}(\boldsymbol{r}) = -\mathrm{i}\frac{\hbar}{2m}\left(\psi_\alpha^* \nabla \psi_\beta - \psi_\beta \nabla \psi_\alpha^*\right),$$

where Nambu indices α and β mark electrons or holes, is conserved at $E = 0$, exactly at the Fermi surface.

Since the Green's functions comprise two wave functions, the Green's function formalism is able to utilize these conservation laws. Indeed, we have already seen that the Usadel equation (Eq. (2.40)) can be cast into a conservation law at zero energy.

Leakage matrix currents

The same Usadel equation, however, indicates that matrix currents are not always precisely conserved: at non-zero energy, the matrix current $\check{j}$ is obviously not conserved, i.e. $\nabla \cdot \check{j} = -(\mathrm{i}e^2\nu/\hbar)[\check{E}, \check{G}] \neq 0$. Formally, this prohibits the use of conservation laws. Practically, it is easy to get around this prohibition. One can always *redefine* matrix currents in such a way that they *are* conserved. This is achieved by introducing additional "*leakage*" matrix currents.

To understand this concept, let us draw an analogy with a leaking electrical device. Consider a thin metallic film of thickness d with conductivity σ, which is separated by a resistive tunnel barrier (conductance per unit area given by g_T) from a bulk grounded electrode. The current density $\boldsymbol{j}$ in the film is almost uniform and is a vector in two directions in the plane of the film. The voltage distribution in the film satisfies

$$\nabla \cdot \boldsymbol{j} + g_T V = 0; \ \boldsymbol{j} = -\sigma d \nabla V, \tag{2.68}$$

a relation similar to the Usadel equation. Indeed, the charge in the film is not conserved: it leaks to the ground. However, the relation can be presented in current-balance form. We simply introduce the surface density of the leakage current, $j_{\rm lc} = g_T V$, so that

$$\nabla \cdot \boldsymbol{j} + j_{\rm lc} = 0.$$

Exercise 2.8. Assume a semi-infinite film in the half-plane $x > 0$. Its left end ($x = 0$) is connected to a terminal at fixed voltage V_0. Find the voltage and current distributions in the film. Find the resistance between the terminal and the ground.

Thus encouraged, we return to the Usadel equation and introduce the (volume) density of the matrix leakage current as follows:

$$\check{j}_{\rm lc} = \frac{{\rm i}e^2\nu}{\hbar}\left[\check{E}, \check{G}\right].$$

The full matrix current flowing to each elementary volume of the structure is now redefined to include the leakage. In this case, it is conserved in each volume, i.e.

$$\check{I}_{\rm full} = \int {\rm d}S(\check{j} \cdot \boldsymbol{N}) + \int {\rm d}V\ \check{j}_{\rm lc} = \check{I} + \check{I}_{\rm lc} = 0.$$

The first integration is over the surface of the volume, $\boldsymbol{N}$ being the normal to the surface, and the second one is over the volume.

If one stays with partial differential equations, the leakage current introduced in this way is an unnecessary trick: one can (try to) solve equations without it. However, leakage currents become very helpful when constructing a quantum finite-element approach.

2.4.3 Finite-element approximation

We have just learned that quantum mechanics provides extra conservation laws that account for the coherence of electron waves, and that conserved currents acquire a matrix structure. Now we are ready to construct a finite-element theory – a quantum circuit theory – based on the conservation of matrix current. This theory greatly resembles common electric circuit theory. The latter works with three elements: terminals, connectors, and nodes. The voltages at the terminals are fixed, and the voltages at the nodes are determined from the current conservation in each node. Finally, these voltages determine the currents in each conductor and, most importantly, to and from the terminals. The same elements are also there in quantum circuit theory (see Fig. 2.6). We will use the term "reservoirs" instead of terminals, since these elements are just the reservoirs described in Chapter 1 – large contact pads adjacent to the nanostructure.

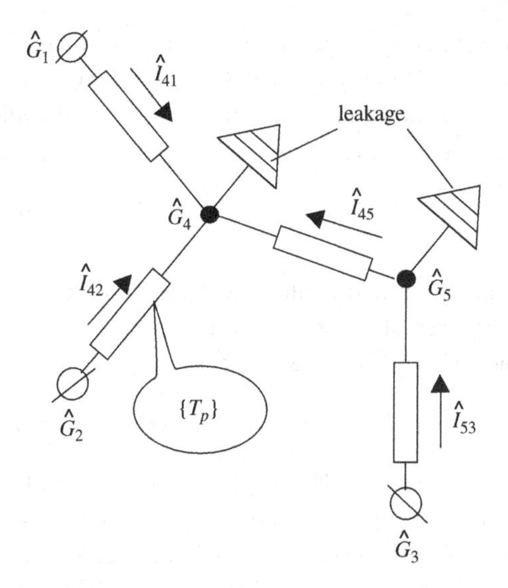

Fig. 2.6. A quantum circuit consists of reservoirs (1–3), nodes (4,5), and connectors. Each connector is characterized by the set of its transmission eigenvalues $\{T_p\}$ or, equivalently, by their distribution. Matrix voltage $\check{G}_{1,2,3}$ is fixed in the reservoirs. Matrix voltages $\check{G}_{4,5}$ are determined from the balance of matrix currents that also include "leakage" currents.

In common circuit theory, the counterpart of the current is the voltage (drop). Is the counterpart of the matrix current the *matrix voltage*? Obviously, it must be a matrix. Less obviously, this matrix $\hat{G}$ can always be chosen to satisfy the following properties:

$$\mathrm{Tr}\,\hat{G} = 0, \quad \hat{G}^2 = \hat{1}. \tag{2.69}$$

These relations come from the Green's function approach outlined in Section 2.3, and the matrix voltage in fact corresponds to the ξ-integrated isotropic Green's function discussed there. The above properties thus follow from the microscopic theory of disordered nanostructures based on the Green's function approach. However, the applications of quantum circuit theory do not require any knowledge of Green's functions and can be presented following circuit-theory rules and the simple examples outlined in the following. We note several properties of such matrices that will be used frequently in the text. The fact that $\check{G}^2 = 1$ implies that eigenvalues of the matrix $\check{G}$ can only be ± 1. Since $\mathrm{Tr}\,\check{G} = 0$, there must be an equal number of positive and negative eigenvalues. For two such matrices, $\check{G}_1$ and $\check{G}_2$, one checks by direct multiplication that $\check{G}_1\check{G}_2 = (\check{G}_2\check{G}_1)^{-1}$. This implies that $\check{G}_1\check{G}_2$ and $\check{G}_2\check{G}_1$ always *commute* and therefore can be diagonalized simultaneously. The sum and the difference of $\check{G}_{1,2}$ *anticommute*.

Control question 2.7. Can you prove that $\check{G}_1\check{G}_2 = \left(\check{G}_2\check{G}_1\right)^{-1}$ explicitly?

If we consider the difference in the limit $\check{G}_1 \to \check{G}_2$, we see that any variation of $\check{G}$ anticommutes with $\check{G}$.

As in common circuit theory, the matrix voltage is fixed in the terminals (reservoirs). Each node is also characterized by a matrix voltage. To solve a circuit means finding the matrix voltage at the nodes as a function of matrix voltages in the reservoirs. To do so, one implements the conservation of matrix current in each node,

$$\sum_{\text{connectors}} \check{I} = 0, \tag{2.70}$$

the quantum Kirchhoff rule. The matrix current in each connector is expressed via two matrix voltages at its ends, just as electric current via a connector is expressed via the difference of the voltages at its ends; i.e.

$$\check{I} = F(\check{G}_1, \check{G}_2) \leftrightarrow I = G(V_1 - V_2). \tag{2.71}$$

Whereas it takes a single parameter, conductance, to characterize a connector in common circuit theory, a connector in quantum circuit theory is characterized by its "pin-code" – a set of its transmission eigenvalues T_p. For connectors with $G \gg G_Q$, we will not operate with T_p but rather with their distribution $\rho(T)$. For instance, a quantum point contact and a tunnel junction of the same conductance are indistinguishable in the framework of common circuit theory. In quantum circuit theory, these connectors are very different since they have different transmission distributions. We will show below (Section 2.6) that there is a universal relation that gives the matrix current in terms of $G_{1,2}$ and T_p irrespective of the concrete physics that gives rise to the matrix structure.

Similar to common circuit theory, the goal of quantum circuit theory is to find the matrix currents flowing to each reservoir as functions of the matrix voltages fixed in the terminals; this becomes an easy task once the matrix voltages in the nodes are known. As we will see from the examples, the physical information is finally extracted from these matrix currents. "Leakage" currents require some more reasoning. To incorporate them into the same template, one introduces fictitious reservoirs and fictitious connectors to these reservoirs from each node. These elements, unlike real terminals, cannot be positioned in space, even at an intuitive level, since a quantum leakage current only describes the loss of coherence and does not leak anywhere in space. However, the practical advantage is overwhelming, since currents from the fictitious terminals enter the current balance in precisely the same fashion as from the real ones and can be treated in the same manner.

Control question 2.8. Separate the nanostructure given in Fig. 1.14 into finite elements.

Variational principle

Many, if not all, physical laws can be formulated as variational principles. Newton's equations of motions, the thermodynamic laws, and the Schrödinger equation can all be derived by minimizing the corresponding (energy) functionals. The same applies to quantum circuit theory, in which Kirchhoff rules – balance equations for the matrix current – can be most consistently derived from a variational principle.

To formulate a variational principle for quantum circuit theory, we introduce an action that depends on matrix voltages in all nodes and reservoirs. The actual voltages at the nodes are determined from the condition that this action has a minimum with respect to variations of $\check{G}$ at the nodes, $\delta\mathcal{S} = 0$.

This action in circuit theory can be presented as a sum of contributions of each connector:

$$\mathcal{S} = \sum_c \mathcal{S}_c(\check{G}_{1c}, \check{G}_{2c}), \tag{2.72}$$

where $1c$ and $2c$ are the nodes at the ends of connector c. Variations of $\check{G}$ are not arbitrary since they must obey the normalization condition $\check{G}^2 = \check{1}$, which implies $(\delta\check{G})\check{G} + \check{G}(\delta\check{G}) = 0$. It is advantageous to parameterize such variations by an infinitesimally small matrix $\delta\check{w}$ at which no constraints are imposed, i.e.

$$\delta\check{G} = (\delta\check{w})\check{G} - \check{G}(\delta\check{w}), \tag{2.73}$$

so that $\delta\check{G}$ automatically anticommutes with $\check{G}$. The current through the connector to node $1c$ is then given by

$$\check{I}_{1c} = G_Q \frac{\delta\mathcal{S}}{\delta\check{w}_{1c}} = G_Q \left[\frac{\delta\mathcal{S}}{\delta\check{G}_{1c}}, \check{G}_{1c} \right]. \tag{2.74}$$

If we vary the matrix voltage at node i, the variation of the total action is given by

$$\frac{\delta\mathcal{S}}{\delta\check{w}_i} = \frac{1}{G_Q} \sum_j \check{I}_{ij}, \tag{2.75}$$

where the summation runs over the connectors that end at node i. The extremum condition $\delta\mathcal{S} = 0$ thus assures that the Kirchhoff rule given in Eq. (2.70) is fulfilled.

There is another relation worth mentioning. Suppose we have found the matrix voltages at the nodes that provide the extremum of the action. The resulting action depends now only on the matrix voltages of the reservoirs. If we vary the result with respect to these voltages, we obtain matrix currents that are flowing to the corresponding reservoirs,

$$\check{I}_r = G_Q \frac{\delta\mathcal{S}}{\delta\check{w}_r}.$$

We will use the variational approach several times in this book, and we always provide the concrete expressions for the action.

2.5 Reservoirs, nodes, and connectors

2.5.1 Separation into finite elements

First, we need to address two important questions: (1) how do we present a real nanostructure in terms of quantum circuit theory, and (2) how do we subdivide it into finite elements? The situation is once again similar to that of common circuit theory: although there is no general algorithm we can employ to make a subdivision such as this for an arbitrary structure, a suitable subdivision can be found in any realistic situation. Nanostructures are often devised as a combination of elements such as tunnel junctions, quantum point contacts, or diffusive layers. In this case, the subdivision is already made at the nanostructure design stage. The electric properties of a structure serve as a good guide for its subdivision into elements. If there is a voltage drop somewhere in a nanostructure, the place where it occurs is regarded as a connector. If there is a place where voltage hardly changes, and it is not adjacent to the contact pads, it has to be treated as a node. Since any voltage distribution can be presented as a sequence of drops and plateaus with any fixed accuracy, such subdivision can always be carried out, as long as the number of elements increases with increasing accuracy.

In principle, this is enough to proceed. However, it is a good idea to provide a specific example of finite-element separation to introduce the concepts involved. For this purpose, we consider a finite-element presentation of a diffusive conductor described by the Usadel equation, Eq. (2.40), as a combination of tunnel junctions in series.

Instead of continuous space $\boldsymbol{r}$, we take a connected discrete set of $\boldsymbol{r}_i$ such that the Green's functions $\check{G}(\boldsymbol{r}_i)$ at neighboring points of the set are close to each other. We associate a connector with each nearest-neighbor connection in the set in such a way that it simulates continuous conductivity of the system. Let us demonstrate how to choose such connectors for a cubic lattice of $\boldsymbol{r}_i$ with periods a along the x, y and z axes. Let us expand $\check{G}$ in the vicinity of node i: $\check{G}(\boldsymbol{r}) = \check{G}(\boldsymbol{r}_i) + \check{\Xi} \cdot (\boldsymbol{r} - \boldsymbol{r}_i)$, $\check{\Xi} a \ll \check{G}$. Since $\check{\Xi}$ is proportional to a variation of $\check{G}$, $\check{\Xi}$ and $\check{G}$ anticommute.

From Eq. (2.40) we obtain the continuous matrix current density:

$$\check{j} = \sigma \check{G}(\boldsymbol{r}_i)\check{\Xi}. \tag{2.76}$$

In the network, the α component of the current density ($\alpha = x, y, z$) is given by the matrix current via a connector in the α direction divided by the area a^{-2} of the corresponding face of the cube that bounds the node i; i.e.,

$$\check{j}^{\alpha} = a^{-2}\check{I}_{\alpha}. \tag{2.77}$$

We see that if we choose

$$I_{ik} = \frac{g_{ik}}{2}[\check{G}_i, \check{G}_k], \tag{2.78}$$

k being the node neighboring i, with $g_{ik} = \sigma a$, we reproduce Eq. (2.76). To prove this we make use of the fact that $\check{\Xi}\check{G} + \check{G}\check{\Xi} = 0$. Later in this section, we demonstrate that the matrix current in the form given in Eq. (2.78) corresponds to the notion of a tunnel junction. Thus, indeed, the discretization of a diffusive conductor means representing it using a set of tunnel junctions.

Now we can rewrite Eq. (2.40) as a Kirchhoff rule:

$$\sum_k \check{I}_{ik} + \check{I}_{\mathrm{lc}} = 0. \tag{2.79}$$

We have already mentioned that, since the Usadel equation does not conserve current, the matrix leakage current $\check{I}_{\mathrm{lc}}$ has to be introduced at each node. This is the same as attaching a fictitious terminal to each node via a connector that supports the leakage current. Indeed, the expression for the leakage current has a similar form to that of Eq. (2.78):

$$\check{I}_{\text{lc}} = \frac{g_{if}}{2}\left[\check{G}_i, \check{G}_{\text{f}}\right], \tag{2.80}$$

with the matrix voltage $\check{G}_{\text{f}} = \check{E}/\sqrt{\epsilon^2 - \Delta^2(\boldsymbol{r})}$ in the fictitious terminal, the matrix $\check{E}$ defined in Eq. (2.26), and the corresponding "conductance"

$$g_{if} = \text{i}2e^2\nu V_i\sqrt{\epsilon^2 - \Delta^2(\boldsymbol{r})}/\hbar$$

being proportional to the volume $V_i = a^3$ associated with node i. The matrix current in Eq. (2.80) describes two processes that may be viewed as a "leakage of coherence." Namely, the terms proportional to ϵ describe the decoherence between electrons and holes, that is they describe the fact that the electrons and holes at the same energy difference ϵ from the Fermi surface have slightly mismatching wave vectors. The terms proportional to Δ are responsible for the conversion between quasiparticles and Cooper pairs that form the superconducting condensate. This explains the leakage of quasiparticles.

An important practical issue to address is the choice of the lattice parameter a that sets the fineness of the discretization. There is no limitation on a from *below* – we can discretize the conductor down to the mean free path, the scale at which Usadel equation ceases to hold. However, a finer mesh increases computational efforts. From the practical point of view, we obviously want to keep the number of elements as low as possible. What are the limitations on a from *above*?

We can gain some intuition about this by looking at a leaking electric device described by Eq. (2.68). The voltage distribution in one-dimensional geometry is given by $V(x) = V(0)\exp(-\xi x)$, $\xi = \sqrt{\sigma d/g_{\text{T}}}$, where $V(0)$ is the voltage at the edge of the film. To reproduce this distribution, the distance between the nodes should be smaller than ξ in the region of several ξ from the edge. At larger distances from the edge, one node would suffice. An optimal mesh is therefore non-uniform. Similarly, the nodes for the Usadel setup should be closer than the coherence length $\xi = \sqrt{D\max(\Delta, \epsilon)}$ at the scale of several ξ from the edge. However, it makes no sense to keep the mesh this fine through the whole sample; at larger distances from the edge, one or two nodes suffice.

Let us now present the above Kirchhoff rules in the form of an action. We start with a single conductor. Looking at Eq. (2.74) it is easy to guess that the simplest form, Eq. (2.78), of the current through connector c between nodes i and j corresponds to the simplest action:

$$\check{I} \propto [\check{G}_i, \check{G}_j] \Leftarrow \mathcal{S}_{ij} \propto \text{Tr}\{\check{G}_i\check{G}_j\}. \tag{2.81}$$

Control question 2.9. Can you restore the proportionality coefficients in Eq. (2.81)?

The contribution of an individual connector to the action is thus given by $\mathcal{S}_c = (g_{ik}/2G_{\text{Q}})\,\text{Tr}\{\check{G}_i\check{G}_j\}$, both for real and leakage currents. The total action is therefore the sum over all connectors, real or fictitious:

$$\mathcal{S} = \sum_c \mathcal{S}_c.$$

We will see in Section 2.5.2 that this form of the action is characteristic of a tunnel junction. Indeed, this is consistent with the form of the matrix current via a tunnel boundary (see Eq. (2.41)).

Let us now go back and obtain the continuous limit ($a \to 0$) of the action. First we note that we can add an arbitrary constant to the action. For the moment, we choose it in a convenient way such that, for any conductor c, $\mathcal{S} = 0$ if the Green's functions are the same at both ends. Then we prove that, for conductor c,

$$\mathcal{S}_c = \frac{g_{ik}}{2G_Q} \mathrm{Tr}\{\check{G}_i \check{G}_k - 1\} = \frac{g_{ik}}{4G_Q} \mathrm{Tr}\{\check{G}_i \check{G}_k + \check{G}_k \check{G}_i - 2\}$$
$$= -\frac{g_{ik}}{4G_Q} \mathrm{Tr}\{(\check{G}_i - \check{G}_k)^2\},$$

the action depends only on the difference of the Green's function at the ends. To derive this, we use the cyclic permittivity of matrices under the trace function, $\mathrm{Tr}(AB) = \mathrm{Tr}(BA)$, and the normalization condition. The difference becomes a spatial derivative in the continuous limit, and we obtain

$$\mathcal{S} = -G_Q^{-1} \int \mathrm{d}\boldsymbol{r}\, \sigma(\boldsymbol{r})\, \mathrm{Tr}\{(\nabla \check{G})^2\} + \mathrm{i}\pi\nu \int \mathrm{d}\boldsymbol{r}\, \mathrm{Tr}\{\check{E}\check{G}(\boldsymbol{r})\}, \tag{2.82}$$

where the first term describes the usual connectors and the second one describes leakage currents. The variation of this action yields the Usadel equation.

2.5.2 Reservoirs and nodes

A reservoir in quantum circuit theory is very much the same as that discussed in the scattering approach (see Chapter 1). It presents an "infinity," a bulk lead in contact with the nanostructure containing many electron states in local equilibrium. In the scattering approach, a reservoir is characterized by an energy-dependent filling factor, $f(E)$, which is, in principle, a Fermi distribution characterized by a certain temperature and chemical potential.

In the Green's function formalism, and in quantum circuit theory, the reservoir is characterized by an energy-dependent matrix voltage $\check{G}$. Different choices of $\check{G}$ correspond to different physical theories and situations. For example, if we are interested only in average currents inside the nanostructure and in those going to/coming from the reservoirs, and therefore we are after Landauer formulas, each reservoir is characterized by an energy-dependent 2×2 matrix in Keldysh space,

$$\check{G}_{\mathrm{usual}} = \begin{bmatrix} 1 - 2f(E) & -2f(E) \\ -2 + 2f(E) & 2f(E) - 1 \end{bmatrix} \equiv \hat{\tau}_3 - \hat{\tau}_1 + \mathrm{i}\hat{\tau}_2 - 2f(E)(\hat{\tau}_3 + \mathrm{i}\hat{\tau}_2), \tag{2.83}$$

corresponding to usual (not extended) Keldysh Green's functions. If one wishes to address, for example, the full counting statistics of electron transfers (Levitov formula), one does this with slightly more general matrices,

$$\check{G} = \mathrm{e}^{-\mathrm{i}\chi\hat{\tau}_3/2} \check{G}_{\mathrm{usual}} \mathrm{e}^{\mathrm{i}\chi\hat{\tau}_3/2} = \begin{bmatrix} 1 - 2f(E) & -2f\mathrm{e}^{\mathrm{i}\chi} \\ (-2 + 2f(E))\mathrm{e}^{-\mathrm{i}\chi} & 2f(E) - 1 \end{bmatrix}, \tag{2.84}$$

corresponding to the extended Keldysh Green's functions. The parameter χ is a counting field that counts electrons going to the reservoir. Solving the quantum circuit theory equations yields $\Lambda(\{\chi_\alpha\})$, the characteristic function of charge transfers.

The next example is for nanostructures where spin-injection is of importance. As discussed in Section 1.9, we can ascribe a non-equilibrium spin accumulation to a reservoir. Correspondingly, the filling factor acquires spin structure and is, by itself, a 2×2 matrix $\check{\rho}$ in spin indices. The corresponding matrix voltage $\check{G}$ is then a 4×4 matrix made of 2×2 spin blocks as follows:

$$\check{G} = \begin{bmatrix} 1 - 2\check{\rho} & -2\check{\rho} \\ -2 + 2\check{\rho} & 2\check{\rho} - 1 \end{bmatrix}. \tag{2.85}$$

Superconductivity introduces the coherence between electrons and holes, and a superconducting reservoir in equilibrium is generally characterized by a 4×4 matrix in Nambu (electron–hole) and Keldysh indices. It is convenient to present it in the form of Keldysh blocks in the following way:

$$\check{G}_{\text{sup}} = \frac{1}{2}\begin{bmatrix} \check{R} + \check{K} + \check{A} & -\check{R} + \check{K} + \check{A} \\ -\check{R} - \check{K} + \check{A} & \check{R} - \check{K} + \check{A} \end{bmatrix}, \tag{2.86}$$

where $\check{R}$, $\check{A}$, and $\check{K}$ are 2×2 matrices in Nambu space. The advantage of this representation is that the circuit-theory relations for $\check{R}$ and $\check{A}$ involve only these two matrices, so that one can first evaluate them in the network, and then proceed with $\check{K}$. The matrices $\check{R}$ and $\check{A}$ are associated with retarded and advanced Green's functions, respectively, and are therefore analytical in the upper (lower) half-plane of the complex variable ϵ. They also satisfy separately the normalization condition $\check{R}^2 = \check{A}^2 = 1$. In an equilibrium superconducting reservoir, they are given by

$$\begin{pmatrix} \check{R} \\ \check{A} \end{pmatrix} = \frac{\pm 1}{\sqrt{(\epsilon \pm \mathrm{i}\delta)^2 - |\Delta|^2}} \begin{bmatrix} \epsilon & \Delta^* \\ -\Delta & -\epsilon \end{bmatrix}, \tag{2.87}$$

with $\delta \to +0$ and $\check{K} = (\check{R} - \check{A}) \tanh \epsilon/2T$.

Control question 2.10. At what energies are the Green's functions given by Eq. (2.87) real? At what energies are they imaginary? Hint: $\sqrt{z} = \sqrt{|z|}\exp(\mathrm{i}\varphi/2)$.

If the setup contains both normal and superconducting reservoirs, we need to express the Green's function in the normal reservoir with Nambu matrices. The normal reservoir may be biased at voltage V with respect to the superconductors. The resulting 4×4 matrix is obviously diagonal in the Nambu space and is given by

$$\hat{R} = -\hat{A} = \hat{\eta}_3; \;\; \hat{K} = \begin{bmatrix} \tanh \frac{\epsilon + eV}{2T} & 0 \\ 0 & -\tanh \frac{\epsilon - eV}{2T} \end{bmatrix}. \tag{2.88}$$

We have given examples of reservoirs for the problems that we will address later in this chapter. They by no means exhaust all the possibilities. Note first of all that the size and form of the matrix voltage – the Green's function – in a reservoir depends not only on the reservoir itself, but also on what other reservoirs are present in the circuit. (In the above example, when some reservoirs are superconducting, and some are normal, we need to keep an extra dimension in the Nambu space, even for the normal reservoirs.) Moreover, it depends on the problem in hand – the Green's functions in use are different for the calculation of average current and full counting statistics in the same system. More complicated situations can be envisaged where the matrix structure is more complex, combining simpler structures. For example, to describe the full counting statistics of a stationary, non-equilibrium superconducting system with spin-injection, one makes use of 8×8 energy-dependent matrices with Keldysh, Nambu, and spin indices that also include counting fields.

Exercise 2.9. Write down the matrix Green's function in a normal reservoir for the above problem of full counting statistics of a stationary non-equilibrium superconducting system with spin injection.

In the absence of time-dependent drives, the quantum circuit theory equations apply separately at each energy. An interesting and important twist in the matrix structure comes about in a time-dependent situation, for example if the voltages applied to the reservoirs are *time-dependent*. In this case, the Green's functions depend on both time arguments, $\check{G} = \check{G}(t_1, t_2)$, rather than on the time difference only. Their Fourier components correspondingly depend on two energies, $\check{G}(\epsilon_1, \epsilon_2)$. The point is that a time (energy) coordinate can be regarded as an extra matrix index. The time-dependent Green's function satisfies the same normalization condition, which is now expressed in terms of integration over time:

$$\check{G}^2 = \check{1} \Leftarrow \int \mathrm{d}t \; \check{G}(t_1, t)\check{G}(t, t_2) = \delta(t_1 - t_2).$$

Control question 2.11. How may this condition be rewritten in an energy representation?

In fact, the matrix structure can also be very simple; as we will show in the next section, the transmission distribution can be evaluated with a matrix voltage as simple as

$$\check{G} = \begin{bmatrix} 0 & \mathrm{e}^{\mathrm{i}\phi} \\ \mathrm{e}^{-\mathrm{i}\phi} & 0 \end{bmatrix}. \tag{2.89}$$

We now turn to the *nodes*. In common circuit theory, a node is similar to a terminal: its state is described by the same parameter, voltage. The same is true in quantum circuit theory: the state of a node is characterized by a matrix voltage $\check{G}$. This is, in fact, the basic approximation of the quantum circuit theory: a node is about the same as a reservoir. The only difference is that the Green's function of the node is determined from the balance equations. It will have the same matrix structure as ascribed to the reservoirs, but does not have to assume any specific form characteristic of a reservoir. This is in line with the classical balance reasoning of Section 2.1, where we revisit the double junction. In that case, the node between two tunnel junctions is characterized by its own filling factor; its energy dependence does not have to be that of a Fermi distribution function.

We stress again that this simple correspondence "node = reservoir" is an approximation of the full scattering approach of Chapter 1. Indeed, in the scattering approach a node has a different function: it is a unitary mixer being described by an energy-dependent scattering matrix that relates electron waves going in and out of the node. The matrix reflects the random configuration of disorder in the node. Thus, the node is characterized by a multitude of energy-dependent functions, components of the scattering matrix. Only the self-averaging over disorder configurations, which occurs in the limit $G \gg G_Q$, enables the approximation "node = reservoir" and allows one to ascribe a certain matrix voltage to a node.

In microscopic terms, the "node = reservoir" equation implies that Green's functions must be isotropic over a part of the nanostructure assigned to a node. This required isotropization certainly takes place in the above example. There, the discretization led to an elementary cube. The size of this cube was still greater than the mean free path. This presumes that the cube contains many impurities. The impurities account for diffusive scattering and thus for isotropization on this spatial scale. Surprisingly, the nodes can be associated with parts of the nanostructure where there are no impurities so the transport is formally ballistic. Let us look, for example, at a model nanostructure that is a bounded region of metal connected to the reservoirs through ballistic quantum point contacts with $G \gg G_Q$. There are no impurities in the metal, so all electron scattering – reflection – takes place at the boundaries only. Such a model structure is called a "cavity" or a "quantum cavity," since it resembles the cavity resonators used in the microwave technique. If the size of the cavity exceeds by far the Fermi wavelength, the electron transport is almost classical: electrons enter and leave the nanostructure following a certain classical trajectory. If the cavity size exceeds the cross-sections of the point contacts, the trajectory involves several reflections from the boundaries. It is the chaotic character of this classical motion that produces isotropization: after several reflections, the memory of the initial direction of momentum is lost. This is why, although the transport in the cavity is ballistic, one can actually regard the whole cavity as a single node connected to the reservoirs by means of the ballistic point contacts. This isotropization requirement can be reformulated as follows: the conductance of the contacts must be much less than the estimation of the system conductance based on its geometric size, for example, $G_Q k_F R$ for a two-dimensional nanostructure of spatial dimensions R. The leakage current from the nanostructure is still given by Eq. (2.80) for a diffusive system; it is not sensitive to the concrete mechanism that has provided the isotropization.

2.5.3 Connectors

A connector element of a circuit theory is completely described by a relation between the current in the connector and the states of the nodes at the ends of the connector. For classical (linear) electric circuits, it is just Ohm's law: $I = G(V_1 - V_2)$. It is characterized by a single parameter, conductance G. In our quantum circuits, a connector element describes an *elastic* scattering between the electron states of the nodes, and is characterized by the corresponding scattering matrix. As discussed, in quantum transport we usually do not need all elements of the scattering matrix: all transport properties are determined by the "pin-code," the set of transmission eigenvalues T_p. Moreover, in the limit $G \gg G_Q$ there are many transmission eigenvalues, and their distribution $\rho(T)$ is an adequate characteristic. Therefore, a connector is characterized by the distribution of its transmission eigenvalues. In all other respects, a connector is a black box: we do not (have to) know (and do not want to know) what is inside.

This is not to promote ignorance: one has to know inside details to decide whether a given part of a nanostructure (or the whole nanostructure) can be regarded either as a single connector or represented as a set of connectors and nodes. The decision may be different depending on the circumstances, even for the same structure. Let us take the example of a long diffusive wire. If we consider transport of normal electrons, and we are sure that no inelastic scattering occurs in the wire, it is acceptable to regard it as a single connector with diffusive distribution of transmission eigenvalues. If the reservoirs are superconducting, one may need to incorporate the leakage currents from the Usadel equation by splitting the wire into shorter connectors and nodes. Estimating the relative importance of two terms in the Usadel equation, we conclude that it becomes necessary if the wire length exceeds the energy-dependent coherence length ξ (see Section 2.5.1). Inelastic scattering will also become important at a certain length L_{in} (see Section 6.8). If the wire length exceeds L_{in} one would again split the wire into connectors and nodes where the inelastic relaxation takes place, similar to the double junction in Section 2.4.

Let us turn to the expression for the matrix current in a connector in terms of the matrix voltages $\check{G}_{1,2}$ of the corresponding nodes or reservoirs at the two ends of the connector. The expression is eventually the same for all possible physical situations that the matrix structure may represent. The current is given by

$$\hat{I} = G_Q \sum_p \frac{T_p(\check{G}_1\check{G}_2 - \check{G}_2\check{G}_1)}{2 + (T_p/2)(\check{G}_1\check{G}_2 + \check{G}_2\check{G}_1 - 2)}. \tag{2.90}$$

This corresponds to the following connector action:

$$\mathcal{S} = \frac{1}{2}\sum_p \mathrm{Tr}\left\{\ln\left[1 + \frac{T_p}{4}(\check{G}_1\check{G}_2 + \check{G}_1\check{G}_2 - 2)\right]\right\}. \tag{2.91}$$

To prevent any possible misunderstanding, we recall that the matrices $\check{G}_1\check{G}_2$ and $\check{G}_1\check{G}_2$ commute, so the exact order of factors in the above expressions is of no importance. A detailed derivation of the expressions is given in Ref. [45].

Exercise 2.10. Show that Eqs. (2.90) and (2.91) are in fact equivalent and follow from each other.

The simplest example of a quantum circuit is a nanostructure without nodes (just a connector between two reservoirs), which has been studied in detail in Chapter 1. Amazingly, all possible physical situations concerning such a nanostructure and all possible transport relations are contained in a very compact form in Eqs. (2.90) and (2.91). Now we derive several fundamental relations of quantum transport starting from Eq. (2.90) for the Landauer matrix current.

Naturally enough, we start with the Landauer formula. Take a connector between two reservoirs in the normal state. Substituting the matrix voltages $\check{G}_1, \check{G}_2$ in the form given by Eq. (2.83), we find that, in this case, $\check{G}_1\check{G}_2 + \check{G}_2\check{G}_1 = \check{2}$, and the matrix current (Eq. 2.90)) is given by

$$\hat{I} = \frac{G_Q}{2} \sum_p T_p \left[\check{G}_1, \check{G}_2\right] = 2G_Q \sum_p T_p (f_1 - f_2) \begin{bmatrix} 1 & 1 \\ -1 & -1 \end{bmatrix}.$$

To obtain the current through the connector in each energy interval, we take the trace of the matrix current, similar to Eq. (2.42), in the context of the Usadel equation. For normal transport, the relation is given by

$$j = \frac{1}{4}\mathrm{Tr}\left(\hat{\tau}_3 \hat{I}\right). \tag{2.92}$$

The trace yields the familiar relation $j = G_Q \sum_p T_p(f_1 - f_2)$. Integrating it over energy yields the Landauer formula, Eq. (1.38).

Let us now turn to Levitov formula. It is easier to derive it from the connector action than from the expression for the current. In the context of full counting statistics, we have to work with matrices, as in Eq. (2.84). Since the result depends only on the difference of counting fields across the connector, we can attach the counting field χ only to the second reservoir. Taking the logarithm of the matrix in this case is trivial, since the anticommutator of two traceless 2×2 matrices is proportional to the unity matrix. In this particular case, direct matrix multiplication yields

$$\check{G}_1\check{G}_2 + \check{G}_2\check{G}_1 = \left[2 + 4\left(\mathrm{e}^{\mathrm{i}\chi} - 1\right) f_1 (1 - f_2) + 4\left(\mathrm{e}^{-\mathrm{i}\chi} - 1\right) f_2 (1 - f_1)\right]\check{1}$$

and

$$\mathcal{S} = \ln\left\{1 + T_p\left[\left(\mathrm{e}^{\mathrm{i}\chi} - 1\right) f_1 (1 - f_2) + \left(\mathrm{e}^{-\mathrm{i}\chi} - 1\right) f_2 (1 - f_1)\right]\right\}.$$

Integration over energy yields the Levitov formula, Eq. (1.54).

Exercise 2.11. Derive the Andreev conductance in Eq. (1.166) starting from Eq. (2.90) and using the superconducting matrix voltages given by Eqs. (2.86), (2.87), and (2.88).

Thus, the Landauer matrix current relations given by Eqs. (2.90) and (2.91) provide a uniform and compact presentation of all these quantum transport formulas that we have previously obtained by different methods and in different contexts. In the limit $G \gg G_Q$, the summation over the channels in these expressions can be straightforwardly replaced by the integration over T with the weight given by the transmission distribution $\rho(T)$. There is, however, an alternative equivalent representation of a Landauer connector, which is frequently more convenient. Since $\check{G}_1\check{G}_2 = (\check{G}_1\check{G}_2)^{-1}$, the relations can be written in terms of one-parametric functions of a matrix, $\mathcal{I}$ and $\bar{S}$, as follows:

$$\check{I} = \mathrm{i}\mathcal{I}(-\mathrm{i}\ln(\check{G}_1\check{G}_2)); \quad \mathcal{S} = \mathrm{Tr}\, \overline{S}(-\mathrm{i}\ln(\check{G}_1\check{G}_2)); \tag{2.93}$$

$$\mathcal{I}(\phi) = G_{\mathrm{Q}} \sum_p \frac{T_p \sin\phi}{1 - T_p \sin^2(\phi/2)}; \tag{2.94}$$

$$\bar{S}(\phi) = \frac{1}{2}\sum_p \ln(1 - T_p \sin^2(\phi/2)); \quad -4G_{\mathrm{Q}}\frac{\partial \bar{S}}{\partial \phi} = \mathcal{I}(\phi). \tag{2.95}$$

Many formulas of quantum transport become more elegant if expressed in terms of $\mathcal{I}$ and $\bar{S}$, rather than in terms of $\rho(T)$. In this way we obtain the following:

$$\text{Landauer conductance: } G = \dot{\mathcal{I}}(0), \tag{2.96}$$

$$\text{Andreev conductance: } G_{\mathrm{A}} = \dot{\mathcal{I}}\left(\frac{\pi}{2}\right), \tag{2.97}$$

$$\text{Fano factor: } F = \frac{1}{3}\frac{1 - \dddot{\mathcal{I}}(0)}{\dot{\mathcal{I}}(0)}, \tag{2.98}$$

and the full counting statistics in the shot noise limit $eV \gg k_{\mathrm{B}}T$ is given by

$$\ln \Lambda(\chi) = \bar{S}(2\,\mathrm{arcsinh}^2\sqrt{\mathrm{e}^{\mathrm{i}\chi} - 1}), \tag{2.99}$$

with $\dot{\mathcal{I}} \equiv \partial\mathcal{I}/\partial\phi$.

Exercise 2.12. Derive Eq. (2.99) from the Levitov formula in the corresponding limit.

In the subsequent sections of this chapter we will make use of three basic types of connectors – tunnel junctions, quantum point contacts, and diffusive conductors. Each is characterized by the distribution of transmission eigenvalues discussed in Section 1.3. Now we specify the matrix current and the connector action for these three main types of connectors. They are easily obtained from Eqs. (2.94) and (2.95). Thus, for a tunnel junction all transmission eigenvalues are very small, $T_p \ll 1$. Expanding these expressions in the first power of T_p, we obtain

$$\mathcal{I} = G_{\mathrm{T}} \sin\phi; \quad \bar{S} = -\frac{G_{\mathrm{T}}}{2G_{\mathrm{Q}}}\sin^2(\phi/2), \tag{2.100}$$

where $G_{\mathrm{T}} = G_{\mathrm{Q}}\sum T_p$ is the tunnel conductance. Furthermore, for a ballistic quantum point contact all transmission eigenvalues equal either zero or one. Hence we have

$$\mathcal{I} = 2G_{\mathrm{B}}\tan(\phi/2); \quad \bar{S} = \frac{G_{\mathrm{B}}}{G_{\mathrm{Q}}}\ln\cos(\phi/2), \tag{2.101}$$

with the total conductance $G_{\mathrm{B}} = G_{\mathrm{Q}}N$, N being the number of eigenvalues with $T_p = 1$. For a diffusive connector, the expressions for the matrix current and the connector action can be obtained directly from the Usadel equation, Eq. (2.45). We give the result here,

$$\mathcal{I} = G_{\mathrm{D}}\phi; \quad \bar{S} = \frac{G_{\mathrm{D}}}{8G_{\mathrm{Q}}}\phi^2, \tag{2.102}$$

and explain it using simple arguments in Section 2.6.

Exercise 2.13. Derive Eq. (2.102) from Eq. (2.45).

2.5.4 Summary of the circuit-theory rules

We have not as yet fully developed any concrete quantum circuit theory suitable for a concrete physical situation; rather, we have discussed its general features. Thereby, we have prepared a *template* that we will adapt for several examples. Before doing this, let us summarize the template rules.

(i) Define the circuit. This includes a proper subdivision of the nanostructure into reservoirs, connectors, and nodes, specifying the type of each conductor in terms of its transmission distribution – usually tunnel junction, ballistic contact, or diffusive conductor.
(ii) Define the matrix structure of the theory: what are the matrix voltages characterizing the reservoirs?
(iii) Write down the Kirchhoff rules (see Eq. (2.70)) for each node with leakage currents taken into account. Information about the types of connectors is used at this point.
(iv) For given values of matrix voltages find, from the Kirchhoff rules, the matrix voltages $\check{G}$ at each node.
(v) Find the matrix currents in the circuit using Eq. (2.90).
(vi) If necessary, repeat steps (iv) and (v) at each energy. Extract physical values from the matrix currents.

An almost equivalent approach is to use the variational principle. In this case, at step (v) one finds the action for each connector and sums the individual contributions to obtain the action for the whole nanostructure.

2.6 Ohm's law for transmission distribution

Let us start with the examples of specific quantum circuit theories. We know from Chapter 1 that the "pin-code" of T_p and the distribution $\rho(T)$ of these transmission eigenvalues play a central role in quantum transport. In this section, we address the transmission distribution of a compound nanostructure comprising elements with known transmission distributions. For this purpose, we develop a specific quantum circuit theory based on the conservation law for a matrix current. The resulting scheme is surprisingly similar to common circuit theory based on Ohm's law and is applicable at $G \gg G_Q$ where the law applies. This is why it is natural to call it a generalized Ohm's law.

As discussed in Chapter 1, the distribution of transmission eigenvalues of a nanostructure is defined as follows:

$$\rho(T) = \sum_p \left\langle \delta(T - T_p) \right\rangle ,$$

where summation is over all transport channels of a nanostructure, and the average is over all possible phase shifts, that is, over an ensemble of formally identical nanostructures. We work assuming $G \gg G_Q$. In this limit, the transmission eigenvalues are dense in the interval (0, 1), and self-averaging takes place.

To derive the scheme, we first relate the transmission distribution to a matrix structure. Let us consider a connector between reservoirs (nodes) 1 and 2. For the matrix voltages of these reservoirs (nodes) we make the simplest choice possible, i.e.

$$\check{G}(\phi_{1,2}) = \begin{bmatrix} 0 & \exp(\mathrm{i}\phi_{1,2}) \\ \exp(-\mathrm{i}\phi_{1,2}) & 0 \end{bmatrix}.$$

All possible Green's functions thus belong to the same manifold, which is parameterized with a single parameter ϕ, referred to below as a *phase*. The normalization condition $\check{G}^2 = 1$ is automatically satisfied on this manifold. This phase should not be confused with the phase of the electron waves or with the superconducting phase: it is an auxiliary parameter to help us with our task, and does not have an obvious physical meaning. Let us substitute these matrices into the expression for the matrix current, Eq. (2.90). We obtain

$$\check{I} = \begin{bmatrix} \mathrm{i} & 0 \\ 0 & -\mathrm{i} \end{bmatrix} \mathcal{I}(\phi_1 - \phi_2);\ \mathcal{I}(\phi) = G_Q \sum_p \frac{T_p \sin\phi}{1 - T_p \sin^2(\phi/2)}.$$

There are two important features to note. First, the matrix current through the connector for this choice of matrix voltages depends only on the phase difference across the connector. In a sense, this is similar to a common electric conductor: the current depends only on the voltage difference across it. Secondly, the matrix structure involved is a trivial one: for any situation, the current is proportional to the same matrix, $\hat{I} \propto \mathrm{i}\sigma_z$. Therefore, the matrix current in this particular circuit theory can be conveniently regarded as a scalar. The one-parameter function $\mathcal{I}(\phi)$ characterizing the type of connector has already been introduced by Eq. (2.92). In terms of transmission distribution, it can be rewritten as follows:

$$\mathcal{I}(\phi) = G_Q \int \mathrm{d}T\ \rho(T) \frac{T \sin\phi}{1 - T \sin^2(\phi/2)}. \tag{2.103}$$

If we know the transmission distribution, we can evaluate the phase dependence of the current $I(\phi)$: the I–ϕ characteristic of a connector. We have already exploited this idea in Section 2.5 to give these phase dependences for the basic types of connectors.

Control question 2.12. What are the current–phase relations for hypothetical conductors with the following distributions of transmission eigenvalues (normalize these distributions to the total conductance): (i) $\delta(T - 1/2)$; (ii) $\delta(T - 1)$; (iii) uniform.

It works the other way round too: the transmission distribution can be retrieved from the phase dependence of the current. To this end, one uses the theory of the functions of a complex variable and considers $\mathcal{I}(\phi)$ in the plane of the complex phase. This allows one

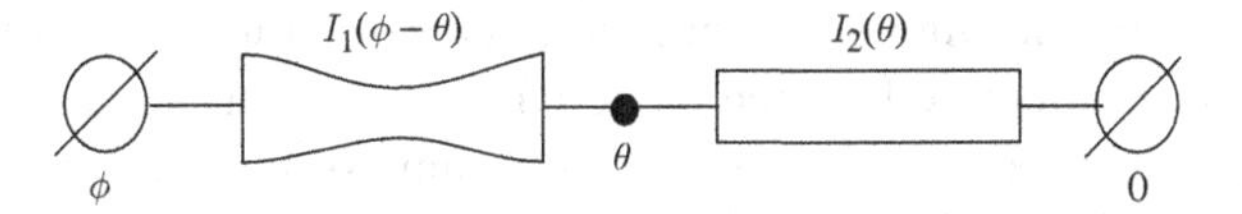

Fig. 2.7. **Phase drops and matrix currents in the simplest compound nanostructure.**

to invert Eq. (2.103) and directly express the transmission distribution in terms of $\mathcal{I}(\phi)$, as follows:

$$\rho(T) = \frac{\rho_D(T)}{\pi G} \operatorname{Re}\left(\mathcal{I}\left(\pi - 0 + 2i \operatorname{arccosh}(T^{-1/2})\right)\right),$$
$$\rho_D(T) \equiv \frac{G}{2G_Q T\sqrt{1-T}}. \tag{2.104}$$

The "0" in Eq. (2.104), as is common in the theory of the functions of a complex variable, stands for an infinitesimally small positive number. This is important since $\mathcal{I}$, as a function of a complex variable, may have cuts at Re $\phi = \pi + 2\pi n$, and "0" determines at which side of the cut the function is to be evaluated. The function ρ_D is already familiar to us; this is the transmission distribution of a diffusive conductor (see Eq. (1.43)).

Exercise 2.14. Prove Eq. (2.104) starting from Eq. (2.103). For this, assume the following parameterization of the complex phase: $\phi = \pi - 0 + i\mu$, and make use of the Cauchy relation $\text{Im}\{(A + i0)^{-1}\} = -i\pi\delta(A)$.

A rather unexpected use of Eqs. (2.103) and (2.104) is to evaluate the transmission distribution of a quantum circuit made of connectors and nodes. To see this, let us construct the simplest two-terminal quantum circuit from two connectors with known transmission distributions. It is presented in Fig. 2.7. For each element, the transmission distribution defines the I–ϕ characteristics – the relation between the current and the phase drop at the element. We set the phase to ϕ and 0 for the two terminals, respectively. Let us ascribe an (as yet unknown) phase θ to the node. Then the current in the left connector is given by $\mathcal{I}_1(\phi - \theta)$, and that in the second one is given by $\mathcal{I}_2(\theta)$, with $\mathcal{I}_1$ and $\mathcal{I}_2$ being properties of the connectors. Next, we use the current conservation in the node,

$$\mathcal{I}_2(\theta) = \mathcal{I}_1(\phi - \theta) = \mathcal{I}(\phi),$$

to find the phase, θ. From this we find the resulting current $\mathcal{I}(\phi)$, which can be evaluated in either connector due to the current conservation. This is the I–ϕ characteristic of the whole compound circuit. We know, however, that, for any connector, including the compound one, the characteristic unambiguously determines the transmission distribution (Eq. (2.104)). So we have found that one!

This reasoning is readily generalized to an arbitrary circuit with two terminals. We ascribe the phases to the terminals, and find, using the Kirchhoff rules, the resulting phase in each node of the circuit. Then the total current, being expressed as a function of the phase difference between the terminals, ϕ, provides the transmission distribution for the whole setup.

Before moving on to concrete examples, let us consider the "linear regime" where the phase difference between the terminals $\phi \ll 1$. This also implies that the phase drops at each connector are much less than unity, so that $\mathcal{I}_c \approx G_c(\phi_{1c} - \phi_{2c})$. By virtue of Eq. (2.93), G_c is just the Landauer conductance of this connector. This makes the solution of quantum Kirchhoff equations both easy and trivial. We end up with a *linear* circuit, which is equivalent to the corresponding circuit of ohmic resistors. The I–ϕ characteristics becomes $\mathcal{I}(\phi) = G\phi$, G being given by Ohm's law. Non-trivial information about the transmission distribution is obtained beyond the linear regime.

Let us apply this general technique to a number of concrete examples.

2.6.1 Double tunnel junction

Let us start with an example to which we have already paid some attention in this book – two tunnel junctions in series. The space between the junctions is now a node. Under current assumptions, the character of the electron motion between the barriers does not really matter. It can be diffusive, ballistic, or even tunnel, provided the resistance of the region between the junctions is negligibly small in comparison with the resistance of each junction.

The current through each connector is given by Eq. (2.100). We write down the condition of current conservation in the node,

$$I(\phi) = G_1 \sin\,\theta = G_2 \sin(\phi - \theta),$$

where $G_{1,2}$ are the tunnel conductances of the junctions, and θ is the phase in the node. Solving for θ, we obtain

$$I(\phi) = \frac{G_1 G_2 \sin\phi}{\sqrt{G_1^2 + G_2^2 + 2G_1 G_2 \cos\phi}}\,. \tag{2.105}$$

Performing analytical continuation to complex phase and using Eq. (2.104), we readily find the transmission distribution (see Fig. 2.8),

$$\rho_{\mathrm{DJ}}(T) = \frac{G}{\pi G_{\mathrm{Q}}} \frac{1}{\sqrt{T^3(T_c - T)}},\ T < T_c, \qquad T_c \equiv \frac{4G_1G_2}{(G_1+G_2)^2}\,, \tag{2.106}$$

while the total conductance of the system, $G = G_1G_2/(G_1 + G_2)$, is given by Ohm's law.

This transmission distribution differs drastically from the one for a single tunnel barrier: the transmission eigenvalues are not small and concentrated near $T = 0$; instead, typically, $T \simeq 1$, especially if the junctions are of comparable conductances. Since we have already explained the double-junction "paradox" in detail (see Section 2.1.1), this should be of no surprise to the reader. This is yet another manifestation of resonant transmission via electron states localized between the junctions. Another prominent feature of the distribution is that it reaches $T \simeq 1$ only if the conductances of the junctions are the same. Otherwise, there is a gap in the distribution at $1 > T > T_c$. This is also in agreement with the quantum results of Section 2.1.1. If the transmission amplitudes via two junctions are not the same, the highest possible transmission coefficient is less than unity. Only if the probabilities of

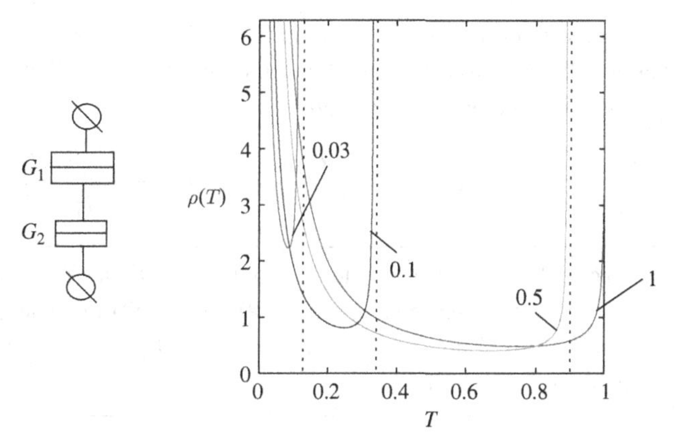

Fig. 2.8. **Transmission distribution of two tunnel junctions in series is bimodal at "resonance" $G_1 = G_2$ and is restricted from the side of high transmissions for unequal junctions. The curve labels denote the parameter G_1/G_2.**

escaping from the node to both sides are equal is the transmission in the resonant state ideal. Note, however, that the prediction of the gap under the current assumption $G \ll G_Q$ does not imply a strict prohibition of transmission eigenvalues in the gap region. There can be always a quirk: a peculiar realization of random phase factors that brings a transmission eigenvalue to the "forbidden" interval. The prediction of the gap just implies that such quirks are highly improbable. More sophisticated techniques [46] based on the evaluation of a non-trivial extremum of a circuit-theory action estimates this probability as $\ln p \propto -G/G_Q$. The probability is thus exponentially small in the limit $G \gg G_Q$.

It is interesting to provide here a phenomenological derivation of the double-junction transmission distribution based solely on transmission resonances. We will assume $G_1 = G_2$ for simplicity, and look at a symmetric resonance where maximum transmission is ideal. In the vicinity of the transmission peak, the transmission assumes Lorentzian shape (see Eq. (1.106)):

$$T = \frac{1}{1 + (\chi/w)^2} \Rightarrow \frac{\mathrm{d}T}{\mathrm{d}\chi} = \frac{2}{w} \frac{\chi/w}{(1 + (\chi/w)^2)^2} = \pm \frac{2}{w} T^2 \sqrt{\frac{1}{T} - 1}, \tag{2.107}$$

χ being a small deviation of the phase shift from its value at resonance position, and $w \ll \pi$ gives the width of the resonance. Let us make a natural assumption: that the phase shifts are random and are distributed uniformly. This reproduces the distribution,

$$\rho(T)\mathrm{d}T = \text{const.} \cdot \mathrm{d}\chi \Rightarrow \rho_{\mathrm{DJ}}(T) = \text{const.} \left| \frac{\mathrm{d}\chi}{\mathrm{d}T} \right| = \frac{G}{\pi G_Q} \frac{1}{\sqrt{T^3(1-T)}}, \tag{2.108}$$

where the constant is determined from the normalization condition, $G_Q \int T\rho(T)\mathrm{d}T = G$. Similar reasoning reproduces the transmission distribution for unequal junctions as well. There are obvious logical caveats in this derivation. For instance, we assume that *all* resonances reach the ideal transmission. For many channels and random scattering between them, this assumption is not justified. The "derivation" therefore cannot be taken seriously: its value is mnemonic rather than scientific.

The distribution function given in Eq. (2.108) and/or the I–ϕ characteristic Eq. (2.105) can now be used to calculate physical quantities. Thus, for the Fano factor and Andreev conductance one obtains, respectively,

$$F = 1 - \frac{G_1 G_2}{(G_1 + G_2)^2}, \quad G_A = \frac{G_1^2 G_2^2}{(G_1^2 + G_2^2)^{3/2}}. \tag{2.109}$$

Exercise 2.15. Evaluate the full counting statistics for a double junction making use of the general equations Eqs. (2.109) and (2.99) and the concrete Eq. (2.105). Why might the full counting statistics expression be understood in terms of the transfer of half-integer charges *e/2*?

It is interesting to note that the double-junction transmission distribution unexpectedly reappears in rather different setups not related to any tunneling. For instance, the same transmission distribution holds for a very disordered interface [47]. As we will see in Section 6.5, the renormalization by electron–electron interactions causes the transmission distribution of most conductors to converge to Eq. (2.108).

2.6.2 Two ballistic contacts

As the next example, we consider the transmission through two point contacts in series. We repeat the same procedures using the I–ϕ characteristic of point contacts, Eq. (2.101). The conductances of the contacts, $G_{1,2}$, are determined by the numbers of open channels supported by the contacts. From the current balance in the node,

$$I(\phi) = 2G_1 \tan(\theta/2) = 2G_2 \tan((\phi - \theta)/2),$$

we find

$$I(\phi) = (G_1 + G_2)\cot(\phi/2)\left\{\left[1 + \frac{4G_1 G_2}{(G_1 + G_2)^2}\tan^2(\phi/2)\right]^{1/2} - 1\right\}, \tag{2.110}$$

and eventually the following transmission distribution [48]:

$$\rho(T) = \frac{G_1 + G_2}{\pi G_Q}\frac{1}{T}\sqrt{\frac{T - T_c}{1 - T}}, \quad T > T_c, \quad T_c \equiv \left(\frac{G_1 - G_2}{G_1 + G_2}\right)^2, \tag{2.111}$$

and $\rho(T) = 0$ otherwise (see Fig. 2.9).

All properties of the distribution are almost exactly reversed compared with those of the double-junction transmission distribution. Generically, the distribution is concentrated at open channels with $T \simeq 1$ as opposite to almost closed channels $T \ll 1$. It also has a gap at low transmissions $0 < T < T_c$ rather than at high ones. The gap is closed only if the two quantum point contact conductances are the same, in which case the distribution becomes symmetric with the minimum at $T = 1/2$. If one of the conductances (say G_1) is much larger than another one, the distribution shrinks to $T = 1$, approaching a distribution of a single quantum point contact with the conductance G_2. Let us give here the limiting form

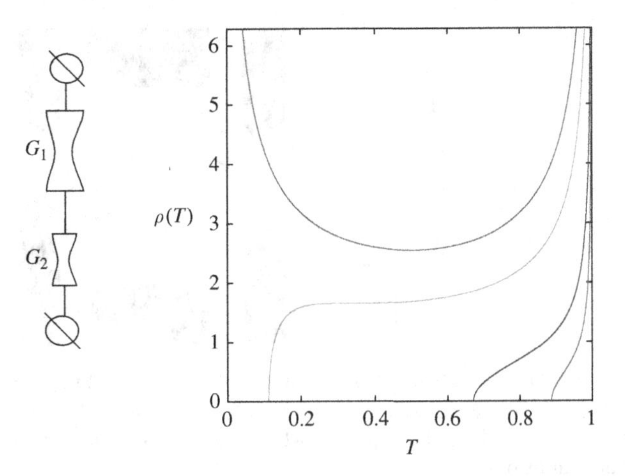

Fig. 2.9. **Transmission distribution of two ballistic contacts in series is bimodal at $G_1 = G_2$. For unequal conductances, the transmission eigenvalues avoid lower values, so that the distribution is concentrated near 1 if $G_1 \ll G_2$. The parameter G_1/G_2 takes the values 1, 0.5, 0.1, 0.03 from the upper curve down.**

of this distribution, which is more convenient to regard as the distribution of reflection eigenvalues $R = 1 - T$, $R \ll 1$:

$$\rho(R) = \frac{4G_2}{\pi R_c}\Theta(R_c - R)\sqrt{\frac{R_c}{R} - 1}; \; R_c \equiv \frac{4G_2}{G_1} \ll 1. \tag{2.112}$$

The noise suppression and the Andreev conductance for two ballistic junctions in series are as follows (see Eqs. (2.98) and (2.97)):

$$F = \frac{G_1 G_2}{(G_1 + G_2)^2}, \; G_{\mathrm{A}} = (G_1 + G_2)\left(1 - \frac{G_1 + G_2}{\sqrt{G_1^2 + G_2^2 + 6G_1G_2}}\right). \tag{2.113}$$

Ballistic cavity paradox

Two quantum point contacts in series give the circuit-theory description of a ballistic cavity discussed in Section 2.5.2: a model system where electrons do not experience any scattering except at the walls. Such ballistic cavities are convenient models for a class of nanostructures made by shaping a 2DEG with a set of top gates. The contacts between the cavity and the "bulk" 2DEG are defined by pairs of extra gates. Changing the gate voltages, one tunes the contacts from full isolation through the tunnel regime to quantum point contacts with many open channels. These devices can thus cross over between various regimes – ballistic transmission, resonant tunneling, and the interaction-dominated Coulomb blockade (see Chapters 3 and 5). Here we concentrate on the ballistic cavity regime when both contacts are open wide. As discussed in Section 2.5.2, the ballistic cavity can be regarded as a node if it is much bigger in size than the openings of the contacts. Also, the contacts have to be placed in such a way that there is no possibility of a direct ballistic transmission through both of them. Under these conditions, the electrons scatter many times when traversing the cavity and their motion is typically chaotic, so we have

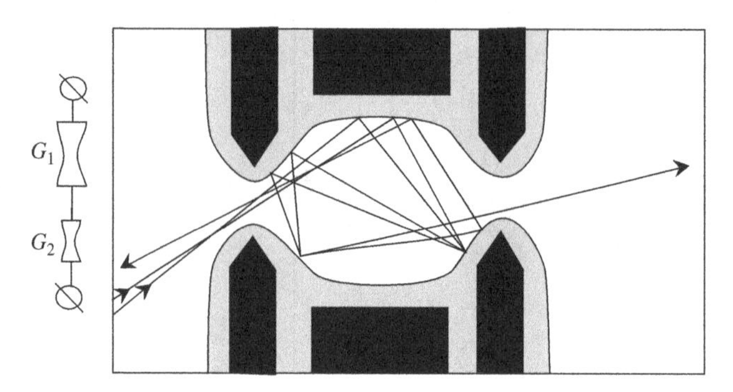

Fig. 2.10. A realization of a ballistic cavity in a 2DEG (top view). White: 2DEG; gray: depleted regions; black: top gates. The rectangular gates shape the cavity while the pairs of gates with sharp ends tune the openings of two ballistic contacts. The electron trajectories are scattered at the cavity boundaries only.

here a ballistic chaotic cavity. (We will give more details about chaos in Section 4.3.) Let us note that the size of the cavity exceeds the Fermi wavelength. The same is true for the size of the contact openings (this is consistent with the condition $G_{1,2} \gg G_Q$). Under these conditions, the electron motion is classical.

Now the paradox appears. If the motion is classical, *including* scattering at the boundaries, the future motion of an electron is completely certain given the initial conditions defining an electron trajectory, i.e. the starting point and the direction of motion. This is the same as saying a trajectory never branches. Let us look at the electrons coming from the left reservoir. Depending on the initial conditions, a given trajectory of such an electron either reflects back, ending up in the same reservoir, or traverses the cavity to end up in the right-hand reservoir (Fig. 2.10). Let us note that this corresponds to a very plain transmission distribution: the transmission eigenvalues are either 0 (trajectories ending up in the same reservoir) or 1 (trajectories crossing to another reservoir). This is *not* the transmission distribution given by Eq. (2.111). Paradoxically, we have two incompatible answers for the transmission distribution under seemingly identical assumptions of semiclassical transport. Both answers eventually give the same conductance given by Ohm's law and thus cannot be distinguished thereby. However, they predict different Fano factors.

Control question 2.13. What is the Fano factor corresponding to the plain transmission distribution?

Although the physics related to the paradox is notoriously difficult to quantify, it has been sufficiently understood at a qualitative level [49]. It only *seems* that both answers correspond to identical assumptions. To understand this, let us use an analogy between electron trajectories and optical beams. We know that the resolution of an optical device, say a microscope, is limited by diffraction. An optical beam does not propagate in a microscope at a well defined angle; rather, there is an angle uncertainty called the *diffraction limit* $(\delta\phi)_{\text{diff}} \simeq \lambda/L$, λ being the wavelength of the light, $L \gg \lambda$ being a typical spatial scale of an optical system, say a lens diameter.

This implies that the direction of a trajectory in or near the cavity is uncertain, with uncertainty $(\delta\phi)_{\text{diff}} \simeq (k_F L)^{-1}$, L being the cavity diameter. To see how this leads to a branching, let us replace a single trajectory with two close trajectories separated by a small angle $(\delta\phi)_{\text{diff}}$ (see Fig. 2.10). Naively, the close trajectories are expected to remain close to each other and therefore to end up in the same reservoir.

Let us note, however, that each reflection at the concave boundaries of the cavity approximately doubles the angle between close trajectories, so it grows exponentially with the number of reflections. The trajectories that are initially separated by a small angle $(\delta\phi)_{\text{diff}}$ will after a short time be separated by a large angle $\simeq \pi$, i.e. after $N_E \simeq \log(k_F L) \gg 1$ reflections at the boundaries. If the angle is large, the initially close trajectories can easily end up in different reservoirs. Therefore, after $\simeq N_E$ reflections a trajectory can branch: a quantum or diffraction effect which is easy to miss.

This implies that there may be two distinct regimes of semiclassical transport via a ballistic cavity. If a typical number of reflections N_r in a cavity is smaller than N_E, diffraction effects can be safely ignored. The transport is said to be purely classical, and the transmission distribution is plain. In the opposite limit, $N_r \gg N_E$, the transport is semiclassical, the trajectories branch, and the transmission distribution is given by Eq. (2.111). We stress that the latter limit is the most practical one. Since N_E only logarithmically depends on $k_F L$, it is a large number only in theoretical considerations and two to three reflections lead to the branching. In addition, extra scattering in the cavity not confined to the cavity boundary induces extra branching of the trajectories [50].

A time scale at which an electron experiences N_E reflections is called the Ehrenfest time, τ_E. In the early days of quantum mechanics, Paul Ehrenfest suggested that the diffraction of particle waves sets a time scale characterizing the departure of quantum dynamics for observables from classical dynamics.

2.6.3 Diffusive connectors

The traditional concept of diffusive transport is underlaid by the microscopic picture considered in Section 2.2: electrons are scattered at a large number of impurities that are distributed in the nanostructure more or less uniformly. This microscopic model has been the starting point at which to derive both the drift-diffusion and the Usadel equations. Solving the Usadel equation eventually yields the diffusive transmission distribution given by Eq. (1.43). However, the reader will have noted that we have introduced the distribution $\rho_D(T)$ very early in this book. This is not because its relation to the diffusive nature of transport is obvious. Quite the opposite: the distribution $\rho_D(T)$ was first put forward by Dorokhov [51] in a rather specific context, and its significance remained unappreciated for years.

Now we are ready to understand this significance and the true universality of $\rho_D(T)$: the generalized Ohm's law makes it obvious. It is crucial to note that, by virtue of Eq. (2.102), the diffusive transmission distribution gives rise to a *linear* I–ϕ characteristic. Since any I–ϕ characteristic must be periodic in ϕ, the linear relation holds in the interval $(-\pi, \pi)$.

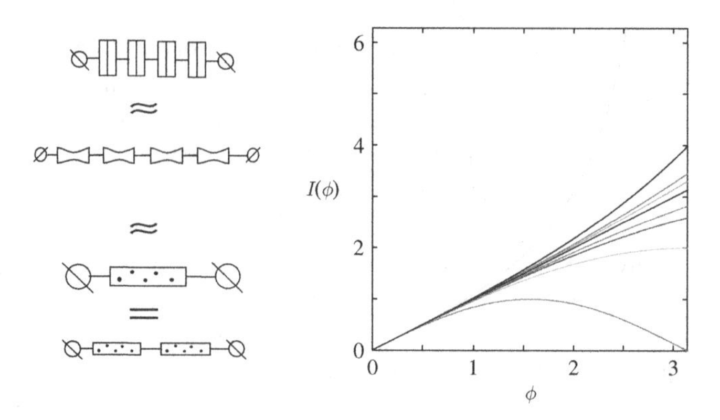

Fig. 2.11. Universality of diffusive transmission distribution $\rho_D(T)$. Any nanostructure consisting of diffusive elements retains $\rho_D(T)$. Any nanostructure consisting of several elements in series that have comparable resistance has a transmission distribution close to $\rho_D(T)$. In the plot, $I-\phi$ characteristics of the chains of ballistic contacts ($N = 1, 2, 3, 4$ from upper curve down) and tunnel junctions ($N = 1, 2, 3, 4$ from bottom curve up) approximate the linear I-ϕ characteristics of a diffusive conductor.

Let us now consider a large number N of similar connectors in series. For simplicity, we may assume that the I–ϕ characteristic is the same for every connector in the circuit and is given by $\mathcal{I}_0(\phi)$. Now we characterize $N-1$ nodes by the phases $\theta_1, \theta_2, \ldots, \theta_{N-1}$, which are found from the current conservation:

$$\mathcal{I}_0(\phi - \theta_1) = \mathcal{I}_0(\theta_1 - \theta_2) = \cdots = \mathcal{I}_0(\theta_{N-2} - \theta_{N-1}) = \mathcal{I}_0(\theta_{N-1}).$$

Since the connectors are the same, the phase drops over each connector are the same as well. Each drop thus equals ϕ/N. The overall $I - \phi$ characteristic is thus simply given by

$$\mathcal{I}(\phi) = \mathcal{I}(\phi/N).$$

The crucial observation is that the characteristic is linear in the limit of large N:

$$\mathcal{I}(\phi) = \mathcal{I}_0(\phi/N) \approx G\phi, \quad G \equiv N^{-1} \left. \partial\mathcal{I}/\partial\phi \right|_{\phi=0} = G_0/N, \tag{2.114}$$

irrespective of the type and details of the connectors. We thus prove that a series combination of a large number of connectors always gives rise to the diffusive transmission distribution $\rho_D(T)$. Thus, this is, in fact, a natural *definition* of diffusive transport. Any nanostructure that consists of many connectors of comparable conductance in series is effectively diffusive. This definition is universal since it is not underlaid by any specific microscopic model of scattering. An example, shown in Fig. 2.11, shows that both a long array of tunnel junctions or quantum point contacts are good approximations of a diffusive connector.

Control question 2.14. Let us take two different conductors, A and B (conductances G_A and G_B). What is the distribution of transmission eigenvalues of a long chain $A - B - A - B - A - B - \cdots$ comprising the serial connection of these conductors?

The transmission distribution for diffusive connectors has yet another remarkable property. Consider a combination of two diffusive connectors with conductances G_1 and G_2. Since both elements are linear, the resulting current–phase relation is also linear:

$$\mathcal{I}(\phi) = \frac{G_1 G_2}{G_1 + G_2}\phi \, .$$

Thus, the combination of diffusive connectors is a diffusive connector, with the conductance given by Ohm's law. This reasoning is valid for any number of connectors. Any nanostructure composed of diffusive conductors is just a diffusive conductor with transmission distribution ρ_D.

Exercise 2.16. Let us model a non-ideal quantum point contact with a circuit that consists of a ballistic connector (conductance G_B) in series with a diffusive connector of much smaller resistance (conductance $G_D \gg G_B$). (i) Find a correction to the I–ϕ characteristic $\propto G_B^2/G_D$. (ii) Demonstrate that the correction becomes relatively large at $\phi \to \pi$. (iii) Solve the circuit-theory equations at $\phi \approx \pi$ beyond the perturbation expansion in G_B/G_D. (iv) Find the distribution of the transmission eigenvalues and compare it with Eq. (2.112).

Exercise 2.17. Let us model a slightly non-ideal ballistic cavity as two ballistic connectors of conductance G_B in series with two diffusive connectors of conductance $G_D \gg G_B$. Make use of the results of Exercise 2.16 to find the correction to the distribution of transmission eigenvalues.

2.6.4 Tunnel junction and diffusive connector

All examples of transmission distributions considered so far can be separated into two classes. The first class is represented by two unequal tunnel junctions: there are no transmission eigenvalues near $T = 1$, and $\mathcal{I}(\phi = \pi) = 0$. For the second class, the ballistic cavity and the diffusive connector being two examples, the I–ϕ characteristic takes a finite value at $\phi = \pi$, and transmission distribution in accordance with Eq. (2.104) exhibits an inverse square-root singularity at $T = 1$. Nanostructures of the first class resemble a single tunnel junction: an electron traverses the most resistive part of the structure in one hop. In nanostructures of the second class, an electron first jumps into the structure and bounces a while inside it before making another jump and leaving it. Indeed, as we have seen, such bouncing provides transmission resonances, and therefore channels with high transmission appear.

Let us give an example of a structure that demonstrates a *transition* between two classes. It consists of a diffusive conductor in series with a tunnel junction (there can be another diffusive conductor in series on the other side of the tunnel junction – in circuit theory it is the total resistivity of two diffusive components that matters). Real systems of this type can be, for example, a sandwich comprising two metals separated by a tunnel layer.

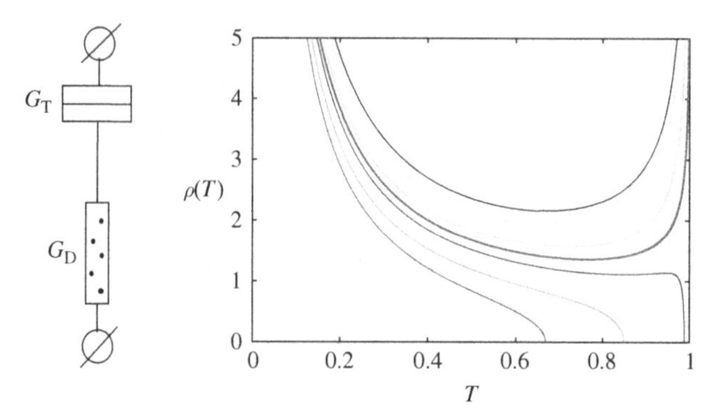

Fig. 2.12. Transmission distributions of the $T+D$ system at different values of G_T/G_D (1/5, 1/3, 2/3, 1, 3/2, 3, 5) increasing from the bottom to the upper curve. The distribution at $G_T < G_D$ does not contain open channels with $T \approx 1$, while the one at $G_T > G_D$ has an inverse square-root singularity at $T = 1$. The separating distribution at $G_T = G_D$ is denoted by a thicker line.

We have to match the currents through the diffusive conductor and the tunnel junction, i.e.

$$I = G_D\theta = G_T \sin(\phi - \theta),$$

where θ is, as usual, the phase of the node. This is enough to express the current in terms of ϕ:

$$I(\phi) = G_D X(\phi); \quad X(\phi) + \frac{G_T}{G_D}\sin(X(\phi) - \phi) = 0. \tag{2.115}$$

To extract the distribution, we make use of Eq. (2.104). The result comes in an implicit form, which suffices to plot and analyze it:

$$\begin{aligned} &\rho(T) = f(T)\rho_D(T); \; T = \frac{1}{\cosh^2(\mu/2)}; \\ &\mu = \operatorname{arccosh} Y - \frac{G_T}{G_D}\sqrt{Y^2 - 1}\cos(\pi f); \; Y = \frac{G_D \pi f}{G_T \sin(\pi f)}. \end{aligned} \tag{2.116}$$

The distribution is always suppressed in comparison with the one for diffusive connectors, $\rho(T) < \rho_D(T)$. The more resistive the barrier, the larger is the suppression. If $G_T > G_D$, the distribution diverges at $T \to 1$ as $(1-T)^{-1/2}$, indicating a nanostructure of the second class. This fraction turns to zero at $G_T = G_D$, where the transition occurs. At the transition point, the transmission distribution diverges still, $\rho(T) \propto (1-T)^{-1/4}$ for $T \to 1$. At the other side of the transition, for $G_T < G_D$, when the resistance of the system is dominated by the tunnel barrier, the maximum transmission available does not exceed a certain threshold value T_c. The nanostructure belongs therefore to the first class. At the transition point, $T_c = 1$, and T_c decreases upon increasing the tunnel resistance (see Fig. 2.12). At $G_T \ll G_D$ the transmission eigenvalues are concentrated near zero, $T_c = eG_T/G_D \ll 1$.

Exercise 2.18. Derive an implicit equation for Andreev conductance of the nanostructure under consideration using Eq. (2.97). Find the Andreev conductance in the limit $G_D \ll G_T$.

These features reveal an important new physics which is absent if tunneling occurs between two clean metals. In the latter case, transmissions through the barrier are always restricted by some maximal value depending on the structure of the barrier. Intuitively, one could expect that the same value restricts the maximal transmission through the compound system of the tunnel barrier and diffusive conductor. It does not happen to be the case. The average transmission is suppressed by the addition of extra scatterers, but some channels become even more transparent, having transmissions of the order of unity. This can be understood as a result of constructive quantum interference between different trajectories traversing the barrier. If the metals are clean, a typical trajectory reaches the barrier only once, either reflecting or getting through. However, if the metals are disordered, the typical trajectory gets back to the barrier to make another attempt to cross it. This enhances the possibilities for interference.

We finish this example with the following observation. Usually, in disordered systems, transmission eigenvalues close to unity are associated with delocalized (spread over the whole system) electron states. If the nanostructure in hand *were* uniform, the opening of the gap at $G_D = G_T$ would create the localization transition since the delocalized states disappear. It is clear that something drastic happens at this point, but one should be cautious about drawing direct conclusions. The mere introduction of the transmission distribution is an attempt to describe an almost classical system ($G \gg G_Q$) in quantum mechanical terms. This can make the results difficult to interpret. Indeed, neither the resistance that obeys Ohm's law, nor the Fano factor given by

$$F = \frac{1}{3}\left(1 + 2\left(\frac{G_D}{G_T + G_D}\right)^3\right)$$

in the system, exhibits any critical behavior around the transition point, at least within the framework of the semiclassical approach used.

2.7 Spin transport

This section is devoted to nanostructures that combine ferromagnets and normal (non-ferromagnetic) metals. We have already shown in Section 1.9 how these nanostructures can be described by the scattering theory. The structure of this kind that is the simplest to make is a layer of a normal metal between two ferromagnetic films. More normal and ferromagnet layers can be stacked together to form a multi-layer structure. Such structures are important due to their use in the applications of the giant magnetoresistance (GMR) effect – a strong dependence of the resistance on the mutual orientation of magnetizations in the magnetic layers. The effect arises from spin-dependent scattering at the

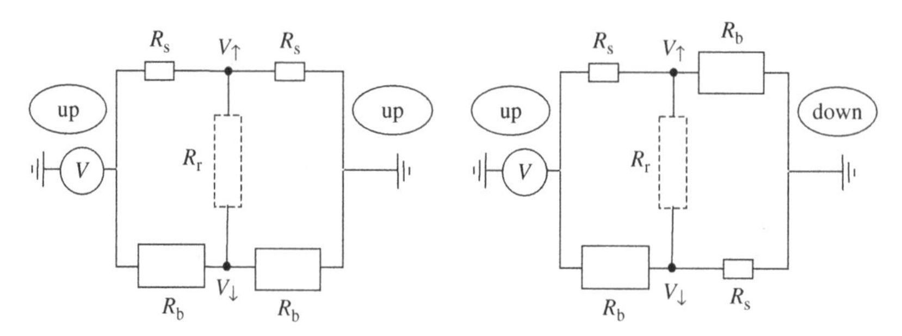

Fig. 2.13. **Classical spin balance explaining the GMR effect. Transport of electrons of two spin directions can be represented by two independent parallel circuits. The spin relaxation is described with a resistor R_r (dashed lines) connecting the circuits. Labels "up" and "down" denote the magnetization directions of the corresponding ferromagnetic reservoirs.**

interfaces between the layers. Scattering theory would provide the resistance in terms of a (complicated) scattering matrix.

In a typical GMR experiment, the nanostructure conductance greatly exceeds the conductance quantum simply because the structure is relatively large supporting a large number of transport channels. The scattering matrix would be too complicated. One needs a theory that gives an adequate description of the phenomenon, expressing all possible microscopic details in terms of a handful of parameters. If magnetizations of all magnets in the structure are collinear, that is, either parallel or antiparallel, such a theory is yet another application of classical balance reasoning. In this case, the electrons can be regarded as particles of two sorts – with spins "up" and "down." The electrons of two different types experience different scattering in and near ferromagnets, this results in different conductances for the two types. In the first approximation, the scattering does not change the numbers of particles of each sort. So one can work with two balance equations instead of one, or, equivalently, with *two* independent parallel ohmic circuits for two electron sorts (see Fig. 2.13). The effects of spin relaxation may be incorporated in the form of a "leakage" current that tries to decrease the difference of concentrations of particles of different sorts. This current can be represented by a connection between the nodes of the independent circuits (see R_r in Fig. 2.13). The origin of GMR can be then understood from the example sketched in. There, two ferromagnets are reservoirs connected to a normal metal node. In a reservoir, electrons of both spin directions are biased at the same chemical potential (= voltage). The interface resistance between a ferromagnet and the normal metal depends on electron spin, and we assume that it is smaller (R_s) for electrons polarized along the magnetization direction. If the magnetization direction changes to opposite, this swaps the resistances for a given spin direction, $R_s \Leftrightarrow R_b$. This gives the difference of resistances for parallel and antiparallel configurations of the magnets, the GMR[4] effect [52] (for a review, see Ref. [53]).

[4] "Giant" in GMR is relative to other contributions to the magnetoresistance, which are negligible at the relatively small magnetic fields used to swap magnetization directions. The actual scale of GMR commonly does not exceed 10%. This suffices for practical applications.

Exercise 2.19. Compute the resistance of the circuit in Fig. 2.13 for parallel and antiparallel orientation of magnetizations. Discuss the limits $R_r \to 0$ and $R_r \to \infty$.

It would seen to be a minor difference if the magnetizations of different magnets in the system were not perfectly aligned. However, this is the difference between quantum and classical mechanics. Suppose some of the magnets are polarized in the x direction while others are polarized in the z direction. An electron state polarized in the x direction is a *quantum* superposition of the states quantized in the z direction,

$$|x_\uparrow\rangle = \frac{1}{\sqrt{2}}\left(|z_\uparrow\rangle + |z_\downarrow\rangle\right),$$

and vice versa. This prohibits a classical balance reasoning. In the following, we demonstrate how the spin transport in this situation can be described by means of a very simple circuit theory based on conservation laws of charge and spin current, a 2×2 matrix current. It makes use of 2×2 matrix voltages that come from a 2×2 density matrix of spin-polarized electrons. The matrix structure of this voltage is related both to electrical voltage and to spin accumulation discussed in Section 1.9, $\check{V} = V + (\hbar/2e)(W_x\check{\sigma}_x + W_y\check{\sigma}_y + W_z\check{\sigma}_z)$.

The quantum circuit theory of spin transport falls into the general template outlined previously in this chapter, but simultaneously presents a simplification and a complication. The simplification comes from the fact that, upon substitution of the matrix voltages of the form given by Eq. (2.84), the circuit-theory, equations become *linear*. There is also no appreciable energy dependence, so that the equations can be trivially integrated over the energy within the energy strip relevant for the transport.

The complication is that the spin-dependent scattering is not accounted for in Eq. (2.90). Thus, the general expression for the Landauer connector has to be extended to incorporate interesting effects such as GMR. In addition, there are two kinds of nodes and reservoirs in theory: normal ones and ferromagnetic ones. In a ferromagnetic node, only two current components are conserved: the particle currents of electrons with spins "up" and "down" with respect to its magnetization direction. This is because the "up" and "down" electrons experience rather different scattering in the ferromagnet. This destroys the coherence at a time scale of the order of τ_P. In the normal metal node, the scattering is predominantly spin-independent, and all four components of the matrix current are conserved.

2.7.1 Spin currents and spin-dependent connectors

We have already discussed spin-dependent connectors in Section 1.9. An important case corresponds to a scattering matrix that is block-diagonal in spin space upon a proper choice of the spin quantization axis. This is the case for a connector between a normal and a ferromagnetic node (reservoir). The electric current, as well as the three components of the spin currents, are related to the voltage difference and spin accumulation by the linear equations given in Eqs. (1.188) and (1.189). Four contact-specific parameters enter these equations: two real spin conductances ($G^\uparrow$ and $G^\downarrow$), which are just the generalization of the concept of Landauer conductance to spin-dependent scattering, and the real and imaginary parts of the mixing conductance $G^{\uparrow\downarrow}$. These parameters are given by the spin-dependent scattering matrix.

Equations (1.188) and (1.189) are written in the fixed basis where the magnetization is directed along the z axis. Since in this section we are mostly interested in the non-collinear magnetizations, we can no longer use this convenient basis and we need to write the equations in an arbitrary basis. This is accomplished by the use of projector matrices $\check{u}^{\uparrow} = (\hat{1} + \hat{\boldsymbol{\sigma}} \cdot \boldsymbol{m})/2$ and $\check{u}^{\downarrow} = (\hat{1} - \hat{\boldsymbol{\sigma}} \cdot \boldsymbol{m})/2$, where $\hat{\boldsymbol{\sigma}}$ is the vector of Pauli matrices and $\boldsymbol{m}$ is a unity vector in the direction of the magnetization. Any matrix A that is block-diagonal in the fixed basis is given by

$$\check{A} = \check{u}^{\uparrow} A^{\uparrow\uparrow} + \check{u}^{\downarrow} A^{\downarrow\downarrow}$$

in an arbitrary basis.

Control question 2.15. Can you demonstrate that $\check{u}^{\alpha}$ is a projector, that is $(\check{u}^{\alpha})^2 = \check{u}^{\alpha}$?

For example, this holds for the scattering matrix given by $\check{s}_{pq} = \sum_{\alpha} \check{u}^{\alpha} s^{\alpha}_{pq}$, where $\alpha = \uparrow, \downarrow$.

Making use of $\check{u}^{\alpha}$, we come to a compact relation between matrix current and voltage in an arbitrary basis [54]:

$$\begin{aligned} \check{I} = {} & G^{\uparrow} \check{u}^{\uparrow} \left(\check{V}_{\mathrm{F}} - \check{V}^{\mathrm{N}} \right) \hat{u}^{\uparrow} + G^{\downarrow} \check{u}^{\downarrow} \left(\check{V}_{\mathrm{F}} - \check{V}^{\mathrm{N}} \right) \hat{u}^{\downarrow} \\ & - G^{\uparrow\downarrow} \check{u}^{\uparrow} \check{V}^{\mathrm{N}} \check{u}^{\downarrow} - (G^{\uparrow\downarrow})^{*} \check{u}^{\downarrow} \check{V}^{\mathrm{N}} \check{u}^{\uparrow}. \end{aligned} \tag{2.117}$$

The actual voltage in the node or in the reservoir is given by a halved trace of this matrix. In the normal node, the three remaining independent matrix elements give three independent components of spin accumulation. In any (normal or ferromagnetic) reservoir, the matrix distribution function $\check{f}^{\mathrm{F}}$ is proportional to the unit matrix $\check{1}$ (and is thus invariant with respect to the choice of axes).

Exercise 2.20. Consider a-so-called spin valve – a normal node between two ferromagnetic reservoirs with directions of magnetization $\boldsymbol{m}_{\mathrm{L}}$ and $\boldsymbol{m}_{\mathrm{R}}$. Assume, for simplicity, identical values of $G_{\uparrow}$, $G_{\downarrow}$, $G_{\uparrow\downarrow}$ for both connectors. (i) Make use of Eq. (2.117) and the matrix current conservation in the node to find the matrix voltage $\check{V}^{\mathrm{N}}$. (ii) Find the matrix current through the node. (iii) Find the conductance of the device as a function of the angle between the magnetization directions.

Spin-orbit interaction and possible magnetic impurities result in the loss of spin coherence, even in a normal metal. Quite generally, one characterizes such processes using the spin-flip time τ_{sf}. Since spin-orbit interaction is usually weak, and the number of magnetic impurities is less than the number of usual defects, $\tau_{\mathrm{sf}} \gg \tau_P$, Spin coherence is lost at a time scale exceeding the isotropization time. Spin-flip scattering causes relaxation of spin accumulation $\check{V}^{\mathrm{N}}$ in a normal metal:

$$\frac{\partial \check{V}^{\rm N}}{\partial t} = -\frac{1}{\tau_{\rm sf}}\left(\check{V}^{\rm N} - \frac{{\rm Tr}\,\check{V}^{\rm N}}{2}\check{1}\right).$$

As usual, we incorporate the effect into the circuit theory by assigning a spin leakage current to each normal node as follows:

$$\check{I}_{\rm lc} = -G_{\rm r}\left(\check{V}^{\rm N} - \frac{{\rm Tr}\,\check{V}^{\rm N}}{2}\check{1}\right);\; G_{\rm r} = G_{\rm Q}\frac{\hbar \nu \mathcal{V}}{\tau_{\rm sf}},$$

where $\mathcal{V}$ is the volume of the node. Since there is no electric leakage from the node, ${\rm Tr}\,\check{I}_{\rm lc} = 0$. This works only if the leakage from the node is not large so that the size of the node does not exceed $L_{\rm sf} = \sqrt{D\tau_{\rm sf}}$. If the size of the nanostructure does exceed this scale, it is separated into many nodes.

Exercise 2.21. Add spin-flip processes into the spin-valve setup considered in Exercise 2.20. Obtain the angular dependence of the conductance in the limit $G_{\rm r} \gg G^{\uparrow}, G^{\downarrow}, G^{\uparrow\downarrow}$.

We can set electron spins into precession by applying an external magnetic field $\boldsymbol{B}$. Since the precession frequency depends on neither the momentum nor the energy of the electrons, the spin accumulation in a given node would precess as a whole:

$$\frac{\partial \check{V}}{\partial t} = \frac{{\rm i}\mu_{\rm B}}{\hbar}[(\boldsymbol{B}\cdot\check{\boldsymbol{\sigma}}), \check{V}].$$

This suggests that the spin precession can also be incorporated into leakage current:

$$\check{I}_{\rm lc} = {\rm i}G_{\rm Q}\mu_{\rm B}\nu\mathcal{V}\left[(\boldsymbol{B}\cdot\check{\boldsymbol{\sigma}}), \check{V}^{\rm N}\right].$$

Quantum Kirchhoff equations in normal nodes assume the usual form:

$$\sum_{\rm connectors} \check{I}_c + \check{I}_{\rm lc} = 0.$$

In a ferromagnetic node (reservoir) with magnetization direction $\boldsymbol{m}$, one has to leave only two components in these equations of the four, which correspond to "up" and "down" currents. The lack of equations is compensated for by a lesser number of unknown matrix voltages in the ferromagnetic node; we have already mentioned that the spin accumulation in the node must be collinear to the magnetization axis.

With these additions, the solution of the Kirchhoff equations defines the matrix voltage – the actual voltage as well as the spin accumulation – in all nodes. Then we can find currents through each connector.

2.7.2 Spin transistor

We will now illustrate the theory by computing the electric currents in the three-terminal device shown in Fig. 2.14. A normal metal node (N) is connected to three ferromagnetic reservoirs (F1, F2, and F3) by arbitrary contacts parameterized by our spin conductances.

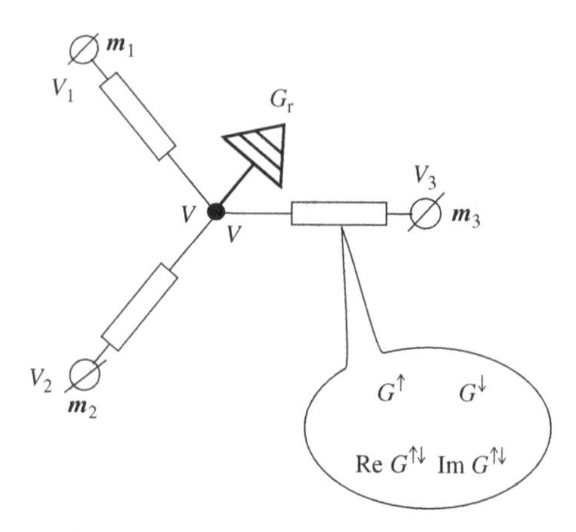

Fig. 2.14. A typical spin circuit. A normal node is connected to three ferromagnetic reservoirs characterized by their magnetizations. Each spin-dependent connector is characterized by four "conductances." Spin accumulation **V** in the node depends on all magnetizations and influences the currents through each connector. Spin-flip scattering in the node is described by "leakage" current.

The reservoirs are biased at voltages $V_{\mathrm{F}}^{1,2,3}$. Since the reservoirs are in equilibrium, the spin accumulation is absent there. The fourth connector, with conductance G_{r}, describes spin relaxation in the node.

To start with, let us separate the expression (Eq. (2.117)) for the current of a certain connector into electric and spin parts, substituting $2\check{I} = I + \boldsymbol{I} \cdot \hat{\boldsymbol{\sigma}}$, $\check{V} = V + \boldsymbol{V} \cdot \hat{\boldsymbol{\sigma}}$. We also introduce the compact notation $G = G^{\uparrow} + G^{\downarrow}$, $P = (G^{\uparrow} - G^{\downarrow})/G$. This yields

$$I = G(V_{\mathrm{F}} - V) - PG(\boldsymbol{m} \cdot \boldsymbol{V}), \tag{2.118}$$

$$\boldsymbol{I} = (V_{\mathrm{F}} - V)PG\boldsymbol{m} - \hat{M}\boldsymbol{V}, \tag{2.119}$$

where $\hat{M}$ is a 3×3 matrix given by

$$M_{ab} = \mathrm{Re}(G^{\uparrow\downarrow})\delta_{ab} + \left(\frac{G}{2} - \mathrm{Re}\, G^{\uparrow\downarrow}\right) m_a m_b + \mathrm{Im}\, G^{\uparrow\downarrow} e^{abc} m_c,$$

where Latin indices label Cartesian coordinates and e^{abc} is the asymmetric tensor of the third rank. We see from Eq. (2.118) that the current via a connector is sensitive to spin accumulation. The first term on the right-hand side of Eq. (2.119) shows that the same coefficient PG shows how much spin current is produced by a voltage drop across the connector. There is also a contribution to the spin current from the spin accumulation determined by the matrix $\hat{M}$, which is generally anisotropic in spin space.

The actual spin accumulation $\boldsymbol{V}$ in the node is determined from the balance of spin currents. Summing up Eqs. (2.119) for all three connectors α, one relates the spin accumulation to the voltage V in the normal node:

$$\boldsymbol{V} = (\hat{M}^{\mathrm{tot}})^{-1}\boldsymbol{I}_{\mathrm{tot}}; \quad M_{ab}^{\mathrm{tot}} = \sum_{\alpha} M_{ab}^{(\alpha)};$$

$$\boldsymbol{I}_{\mathrm{tot}} = \sum_{\alpha} G_{\alpha} P_{\alpha} \boldsymbol{m}_{\alpha} (V_{\mathrm{F}}^{\alpha} - V).$$

In the next step, we compute electric currents in each conductor in the presence of spin accumulation with the help of Eq. (2.118):

$$I_\alpha = \sum_\beta G_{\alpha\beta}(V_F^\beta - V);$$

$$G_{\alpha\beta} = \delta_{\alpha\beta} G_\alpha - P_\alpha P_\beta G_\alpha G_\beta \left(\boldsymbol{m}_\alpha \left[(\hat{M}^{\text{tot}})^{-1} \right]_{\alpha\beta} \boldsymbol{m}_\beta \right).$$

Note that the summation now also includes the fake terminal that we introduced to take spin relaxation into account.

From a purely electric point of view, we see a strange "non-local" effect – a current in a certain connector is not only proportional to the voltage drop across this connector, but also is contributed to by voltage drops across all other connectors of the circuit. The "non-local" cross-conductances $G_{\alpha\beta}, \alpha \neq \beta$, are entirely due to the spin-accumulation effect.

Finally one finds the voltage V in the normal node from the electric Kirchhoff rule $\sum_\alpha I_\alpha = 0$:

$$V = \sum_{\alpha\beta} G_{\alpha\beta} V_F^\beta \Big/ \sum_{\alpha\beta} G_{\alpha\beta}.$$

To obtain compact expressions, let us assume that spin-dependent scattering is relatively weak, so that, for each connector P, $\text{Im}\, G^{\uparrow\downarrow}/G \ll 1$. In this limit, one has $\text{Re}\, G^{\uparrow\downarrow} = G/2$ for each connector, and the matrix $\hat{M}$ can be regarded as isotropic:

$$\hat{M}^{\text{tot}} = (G_{\text{r}} + G_1 + G_2 + G_3)\hat{1}/2 \equiv G_\Sigma \hat{1}/2\,,$$

where the (cross)-conductances are proportional to the scalar products of magnetization vectors of proper reservoirs, i.e.

$$G_{\alpha\beta} = \delta_{\alpha\beta} G_\alpha - \frac{2 P_\alpha P_\beta G_\alpha G_\beta (\boldsymbol{m}_\alpha \cdot \boldsymbol{m}_\beta)}{G_\Sigma}\,,$$

and the currents in each connector manifest dependence on mutual orientation of magnetizations. A larger spin-flip rate increases G_{r} and G_Σ, and thus suppresses the effect.

Exercise 2.22. Find the (cross)-conductances in the above limit in the presence of a magnetic field $\boldsymbol{B}$ in the normal node.

2.8 Circuit theory of superconductivity

Superconductivity induces coherence between electrons and holes that persists at typical length scales that exceed by far the Fermi wavelength and, frequently, the mean free path. This is why superconducting systems are a natural field for the application of semiclassical methods. As a matter of fact, these methods were developed for bulk superconductors by Larkin, Ovchinnikov, Eilenberger, and Usadel much earlier than for normal metals. Subsequently, these methods were extended to constrictions, interfaces, and other heterogeneous structures and hybrid systems made artificially. Thereby the theoretical development along these lines has been essentially accomplished. We will refer to this piece of knowledge as the "full theory."

There used to be a sharp contrast between the complexity of the full theory and the relative simplicity of the final results. It has been a kind of disappointment to address, for instance, the problem of linear conductivity of a double tunnel barrier separating a normal metal and a superconductor, to spend weeks and weeks calculating and to obtain that, under very general assumptions, $G = G_N/\sqrt{2}$, G_N being the conductivity in the normal state. Such frustration prompted attempts to provide a reformulation of the "full theory" in simpler, or at least user-friendly, terms. One of the attempts was the implementation of the scattering approach in the form described in Section 1.8. Although useful for nanostructures with a small number of transport channels, the scattering approach has the same disadvantages as any fully quantum theory – it becomes less and less operational with the increase of the dimension of the scattering matrix. For many exemplary setups, these problems can be overcome with the use of random matrix theory (RMT, see Section 4.3), but in this case the relation of RMT to the full theory remains elusive. In this book, we will not consider the RMT application to hybrid structures [55]. Rather, we concentrate on a different semiclassical approach – circuit theory for hybrid systems. It is essentially a discretized version of the "full theory." Even the most sophisticated experimental layouts can be presented as a collection of a few circuit theory elements – reservoirs, nodes, and connectors. The extension of usual circuit theory presented in this section includes normal and superconducting circuit elements on an equal footing, accounts for the decoherence of electrons and holes, and thereby allows one to consider arbitrary connectors – superconducting junctions of any kind.

For the sake of simplicity, we only talk about *stationary* non-equilibrium superconductivity. For normal nanostructures, the transport is stationary if the voltages applied to all reservoirs are stationary. In contrast, if there are two superconducting terminals biased at different stationary voltages, the corresponding superconducting phases are linear in time and induce non-stationary Josephson currents, with the frequencies corresponding to the voltage difference. Thus, the stationary condition for superconducting and/or hybrid nanostructures implies that *all* superconducting terminals are at the same voltage. It is natural to set this voltage to zero.

2.8.1 Specifics

Let us introduce the specific rules and matrix structure of the theory. Under stationary conditions, the theory is for 4×4 Keldysh–Nambu matrix voltages – Green's functions – at each energy. There are two kinds of reservoirs, superconducting and normal. The matrix voltages are fixed in the reservoirs by Eqs. (2.87) and (2.88), respectively. There are also two kinds of nodes: superconducting and normal. They are distinguished by "leakage" currents, which describe decoherence of electrons and holes and are given by Eq. (2.80). For a normal node, the leakage is proportional to the energy and describes only decoherence. For a superconducting node, the leakage involves the superconducting order parameter Δ, providing coherent coupling of electrons and holes.

Generally, one needs to evaluate the matrix currents separately at each energy. The Kirchhoff rules – matrix current conservation – enable us to solve for matrix voltages at the nodes. Electric currents are obtained by integration of the $\hat{\tau}_3\hat{\eta}_3$ component of the matrix currents over the energy. It is interesting to note that the electric currents in the course of this procedure are *not* automatically conserved in the superconducting nodes: this leakage is the conversion of electron or hole quasiparticles into Cooper pairs that form a superconducting condensate. Indeed, it is known in superconductivity theory that the current is conserved only if the superconducting order parameter satisfies the self-consistency relation ([56]; see also the end of Section 2.3.5). This relation is not incorporated into the circuit theory automatically; for example, the order parameter in the node can be affected by non-equilibrium distribution of the carriers. Thus, to ensure current conservation, one should, in principle, incorporate the self-consistence relation into circuit theory. In practice, this non-conservation of current is rarely encountered, and we will not discuss the self-consistency in this section.

In this form, the circuit theory rules can already be implemented numerically. To make analytical progress, it is important to note that the Kirchhoff equations of the theory can be solved by separate blocks. This comes from the fact that in standard "non-extended" Keldysh theory only three of the four Keldysh blocks are independent. Traditionally, this relation is emphasized by applying the unitary transform $\check{U} = (\check{1} + i\tau_2)/\sqrt{2}$ to all matrices in Keldysh space. After this transform, the Keldysh Green's functions (our matrix voltages) acquire the following block form:

$$\check{G} = \begin{pmatrix} \hat{R} & \hat{K} \\ 0 & \hat{A} \end{pmatrix}. \tag{2.120}$$

The lower left block is always zero, and $\check{R}$, $\check{A}$, and $\check{K}$ are 2×2 matrices in Nambu space, corresponding to retarded, Keldysh, and advanced Green's functions, respectively.

Control question 2.16. What is the structure of the product of two Green's functions $\check{G}_1$ and $\check{G}_2$ with the block structure given in Eq. (2.120)?

Equations for $\check{R}$ and $\check{A}$ separate: they do not contain other elements in the Keldysh space. In particular, neither equation depends on the actual filling of the electron–hole states – this information is only contained in block $\check{K}$; $\check{R}$ and $\check{A}$ just determine spectral properties of these states. Solving these equations is the first step in solving the network. These equations for 2×2 matrices are, as we show in the following, relatively easy to solve. In fact, it is enough to solve for $\check{R}$ for all energies, and then find $\check{A}$ from the relation $\check{A}(\epsilon) = -\eta_3 \check{R}(-\epsilon)\eta_3$ (see Refs. [57] and [58]).

The next step is to solve for $\check{K}$. Equations for $\check{K}$ contain $\check{R}$ and $\check{A}$, obtained in the previous step. Importantly, the equations are *linear* in $\check{K}$. This is also a consequence of the structure of the Green's function: the Keldysh component of a product of any number of Green's functions only contains terms in which a product of several retarded blocks is followed by *one* Keldysh block, and then by a product of a number of advanced blocks (see Control question 2.16). In addition, the normalization condition $\check{G}^2 = \check{1}$ implies

$\check{R}\check{K} + \check{K}\check{A} = 0$. Therefore, matrix $\check{K}$ has only two independent components. Traditionally, block $\check{K}$ is presented in terms of two scalar "distribution functions" $f^{\mathrm{L,T}}$, as follows:

$$\check{K} = \check{R}\check{f} - \check{f}\check{A};\ \check{f} = f^{\mathrm{L}}\check{1} + f^{\mathrm{T}}\hat{\eta}_3. \tag{2.121}$$

Since the normalization also implies $\check{R}^2 = \check{A}^2 = \check{1}$, the constraint $\check{R}\check{K} + \check{K}\check{A} = 0$ is automatically satisfied in this way.

Similarly, the matrix current in block $\check{K}$ also has only two independent components, $I_0 = \mathrm{Tr}(\hat{\tau}_3\check{I})$, $I_3 = \mathrm{Tr}(\hat{\tau}_3\hat{\eta}_3\check{I})$. Circuit-theory equations for $\check{K}$ are therefore the Kirchhoff rules for these currents. For each connector, there is a convenient linear relation between $I_{0,3}$ and the distribution functions $f^{\mathrm{T,L}}$ at the ends of the connector:

$$\begin{aligned} I_3 &= g_{\mathrm{T}}(f_1^{\mathrm{T}} - f_2^{\mathrm{T}}) + g_{\mathrm{TL}} f_1^{\mathrm{L}} - g_{\mathrm{LT}} f_2^{\mathrm{L}}, \\ I_0 &= g_{\mathrm{L}}(f_1^{\mathrm{L}} - f_2^{\mathrm{L}}) + g_{\mathrm{TL}} f_2^{\mathrm{T}} - g_{\mathrm{LT}} f_1^{\mathrm{T}}. \end{aligned} \tag{2.122}$$

Four energy-dependent coefficients, "conductances" g_{T}, g_{L}, g_{TL}, and g_{LT}, are expressed in terms of $\check{R}$ and $\check{A}$ at both ends of the connector. The balance of currents $I_{3,0}$ at each energy provides us with the distribution functions $f^{\mathrm{T,L}}$ at each node, and eventually the currents. Solutions at positive energies are related to those at negative energies since I_0, f^{L}, g_{TL}, and g_{LT} are odd in energy while I_3, f^{T}, g_{T}, and g_{L} are even in energy. The quantity I_3 is immediately related to electric current, $I_{\mathrm{el}} = \int \mathrm{d}\epsilon\ I_3(\epsilon)/4e$.

An important case arises when this scheme can be simplified even further. If the "mixing" conductances g_{TL} and g_{LT} can be disregarded, the equations for I_0 and I_3 separate. In particular, the current I_3 only depends on f^{T}:

$$I_3(\epsilon) = g_{\mathrm{T}}(\epsilon)\left[f_1^{\mathrm{T}}(\epsilon) - f_2^{\mathrm{T}}(\epsilon)\right]. \tag{2.123}$$

In this case the coefficient $g_{\mathrm{T}}(\epsilon)$ can be viewed as an energy-dependent effective conductance of the junction. Indeed, the Kirchhoff rules that determine f^{T} at any given energy have the same structure as those in electric circuit theory, apart from the fact that the conductances depend on energy. Such effective conductance differs from the Landauer conductance of the connector, being renormalized by superconductivity.

This condition of "no mixing" is always satisfied at zero energy since g_{TL} and g_{LT} are odd in energy. In addition, g_{TL} and g_{LT} are associated with the drops of superconducting phase across the connectors. This is why this condition is always fulfilled in nanostructures with a single superconducting reservoir. It can also be satisfied if the normal and superconducting currents are geometrically separated.

2.8.2 Proximity effect and density of states

A normal metal is a material in which the superconducting order parameter Δ equals zero. If a superconductor is brought into contact with a normal metal, the density of states in the normal metal in the proximity of the superconductor will be modified – this is the *proximity effect*. Indeed, electrons in the normal metal do not stay in the same place; they travel, and sooner or later reach the superconductor and interact with its order parameter. This interaction is expected to reduce the density of states in the normal metal, since in a superconductor there are no quasiparticles below the gap. In particular, if a finite-size piece of normal metal is connected to a bulk superconductor *only*, the density of states is zero below a certain energy $\simeq E_{\text{Th}} \simeq \hbar/\tau$, τ being a typical time to reach the superconductor. Alternatively, if this piece is connected to a normal reservoir as well, there is a finite density of states, even at zero energy. This density is, however, reduced in comparison with the density of states in the absence of the proximity effect. Let us describe this effect quantitatively by circuit-theory means.

The density of states at a given point of the structure can be related to the retarded Green's function $\check{R}(\epsilon)$, $\nu(\epsilon)/\nu = \text{Re Tr}\{\hat{\eta}_3 \check{R}\}$, ν being the density of states in the normal state. Therefore, if we are after the density of states, we can work with 2×2 matrices $\check{R}$ rather than with the full Keldysh structure. Applying this relation to the superconductor itself (see Eq. (2.87)), we reproduce the BCS density of states,

$$\frac{\nu(\epsilon)}{\nu} = \Theta(\epsilon^2 - |\Delta|^2)\frac{|\epsilon|}{\sqrt{\epsilon^2 - |\Delta|^2}}, \tag{2.124}$$

which is zero for energies below $|\Delta|$, exhibits an inverse square-root singularity at $|\epsilon| = |\Delta|$, and approaches the normal-metal value ν at $|\epsilon| \gg |\Delta|$.

To implement the simplest model of the proximity effect, let us consider a network consisting of a single normal node (matrix voltage $\check{R}$) connected to a superconducting reservoir ($\check{R}_S$). The decoherence of electrons and holes is described by the leakage. The Kirchhoff equation in the node is given by $\check{I}_c + \check{I}_{\text{lc}} = 0$. We express the current through the connector using Eq. (2.93), while the leakage current is taken from the Usadel equation, $I_{\text{lc}} = \text{i}\nu\mathcal{V}_{\text{node}}[\check{E}, \check{R}]$, $\mathcal{V}_{\text{node}}$ being the volume of the node. Since, in the normal node, $\Delta = 0$, we have $\check{E} = \epsilon\check{\eta}_3$, and the current conservation becomes

$$\text{i}\mathcal{I}(-\text{i}\ln \check{R}_S\check{R}) + \text{i}E[\check{\eta}_3, \check{R}] = 0.$$

As we know, different types of connectors (tunnel, ballistic, and diffusive) yield different functions $\mathcal{I}$.

To simplify, we concentrate on low energies $\epsilon \ll |\Delta|$. In this case, we can approximate the Green's function in the superconducting reservoir by $\check{R}_S = \check{\eta}_1$ (Eq. (2.87)). Let us parameterize the matrix voltage in the normal node by $\check{R} = \sin\theta\ \hat{\eta}_1 + \cos\theta\ \hat{\eta}_3$. Both currents in this case are proportional to $\hat{\eta}_2$, and we have a scalar equation for an unknown function θ;

$$\frac{\mathcal{I}(\pi/2 - \theta)}{G\sin\theta} = \frac{\epsilon}{E_{\text{Th}}};\ E_{\text{Th}} \equiv \frac{G}{2\nu\mathcal{V}_{\text{node}}}, \tag{2.125}$$

where we write $\sin\theta - \text{i}\hat{\eta}_2\cos\theta$ as $\exp(-\text{i}\hat{\eta}_2(\pi/2 - \theta))$, take into account that $\mathcal{I}$ is always an odd function of its argument, and that for any odd function $f(x)$ one has $f(x\hat{\eta}_2) = \hat{\eta}_2 f(x)$.

In Eq. (2.125), we made use of E_{Th}, the Thouless energy of the node. We assume that $E_{\text{Th}} \ll |\Delta|$. The density of states in the node is given by $\nu(\epsilon)/\nu = \text{Re}\ \cos\theta$. The above equation has purely imaginary solutions for θ up to the gap energy $E_g \simeq E_{\text{Th}}$. Above the gap, the solution is complex, yielding a non-zero density of states. Let us illustrate this

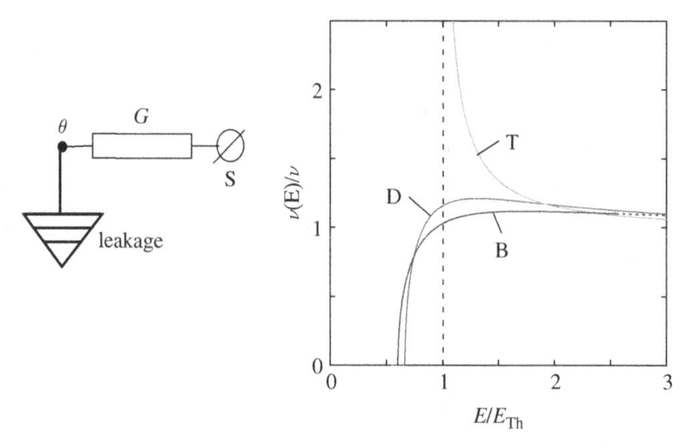

Fig. 2.15. **The density of states in a normal node connected to a superconducting reservoir (S) develops a gap. The plots are for ballistic (B), diffusive (D), and tunnel (T) connectors of the same conductance *G*.**

point using the example of a tunnel junction, $\mathcal{I} = G\sin\theta$,

$$\left(\frac{\nu(\epsilon)}{\nu}\right)_{\mathrm{T}} = \Theta(\epsilon^2 - E_{\mathrm{Th}}^2)\frac{|\epsilon|}{\sqrt{\epsilon^2 - E_{\mathrm{Th}}^2}}.$$

The tunnel connector mimics the BCS density of states with the gap reduced to E_{Th} (sometimes referred to as a *mini-gap*) and a divergent singularity at the gap energy. For other types of connectors, the mini-gap exists, but the singularity at the edge is of square-root type, as we see from Fig. 2.15.

Exercise 2.23. Find the density of states for the case where the connector is a quantum point contact with conductance G_{B}.

In the presence of a normal reservoir, the density of states remains finite, even at zero energy. Let us exemplify this with a long chain of nodes – a diffusive wire. The wire of length L in the x direction connects normal ($\check{R}_{\mathrm{N}} = \hat{\eta}_1,\ x = 0$) and superconducting ($\check{R}_{\mathrm{S}} = \hat{\eta}_3,\ x = L$) reservoirs. As mentioned, applying circuit theory to the wire amounts to solving the Usadel equation there. Conveniently, at zero energy there are no leakage currents. The matrix currents are conserved, and we can use the general solution of the Usadel equation, Eq. (2.44), obtained in Section 2.3.4. Adapting it for the problem in hand, we obtain the retarded Green's function $\check{R}$ at point x in the wire,

$$\check{R} = \exp\{u(x)\check{M}\}\check{R}_{\mathrm{S}}, \quad \check{M} = \ln\left(\check{R}_{\mathrm{S}}\check{R}_{\mathrm{N}}\right) = \ln(\mathrm{i}\check{\eta}_2) = \mathrm{i}\pi\check{\eta}_2/2,$$

and u solves the Laplace equation with the boundary conditions $u(0) = 0$, $u(x) = 1$. For a simple one-dimensional wire, $u = x/L$, and we immediately find the density of states at zero energy ϵ in each point x of the wire:

$$\frac{\nu(0)}{\nu} = \sin\left(\frac{x\pi}{2L}\right).$$

This changes from zero at the superconducting end to the normal value at the normal end. In the middle of the wire the density of states is close to the normal one.

Control question 2.17. A diffusive nanostructure made in the shape of a chicken has been measured in a two-terminal setup. While the left-hand electrode was grounded, a voltage V was applied to the right-hand electrode. Measurement with a non-obtrusive voltage probe has yielded 0.7 V at the tip of the beak. The voltage source has been switched off, and the temperature has been lowered so that the right-hand electrode has become a superconductor. What is the density of states at zero energy and at the top of the beak?

2.8.3 Circuit theory of Andreev conductance

As we have just seen, the limit of very low energy $\epsilon \ll E_{\mathrm{Th}}, |\Delta|$, allows for extensive simplifications, with which, one can build up a yet more compact circuit theory called the circuit theory of Andreev conductance [59]. When evaluating $\check{R}$, one can safely disregard leakage currents. Besides, $\check{R}$ is Hermitian, $\check{R} = \check{R}^{\dagger}$. This allows for an instructive geometric interpretation: $\check{R} = (\boldsymbol{s} \cdot \hat{\boldsymbol{\eta}})$, $\boldsymbol{s}$ being a real three-dimensional "spectral" vector. Due to the normalization condition, the vector $\boldsymbol{s}$ takes values on the northern hemisphere of a unit sphere, $\boldsymbol{s}^2 = 1$. In these terms, normal terminals correspond to the northern pole, superconducting terminals are mapped onto the equator, and their longitude is set by their superconducting phase. The spectral vector in the nodes lies somewhere between the equator and the northern pole, and its latitude determines the reduction of the density of states, $\nu(0)/\nu = \sin\theta$. The spectral vector in the nodes is determined from the balance of vector "spectral" currents in all connectors. The direction of the current is orthogonal to spectral vectors at both ends of the connector, and its absolute value is given by $\mathcal{I}(\theta_{12})$, θ_{12} being the geometric angle between the spectral vectors. Recall that $\mathcal{I}$ is expressed in terms of the transmission eigenvalues of a conductor (see Section 2.6).

The second simplification at vanishing energy is that $g_{\mathrm{TL}} = g_{\mathrm{LT}} = 0$ in Eq. (2.122), and electric properties of the circuit are determined by the coefficients g_{T} of the connectors only. The energy-dependence of g_{T} can be disregarded provided the voltages applied and the temperature are much smaller in energy units than both the Thouless energy E_{Th} and the superconducting gap $|\Delta|$. Under these conditions, we can do what we have done already many times in the book: we can integrate the equations over energy. We introduce thereby a voltage in each node by $e\tilde{V} = \int d\epsilon \; f_{\mathrm{T}}(\epsilon)$. We have so many matrix voltages in this chapter that we have to say now explicitly: this is really a voltage one would measure with a non-obtrusive voltage probe connected to this node.[5]

The electric currents in each connector are determined by the drop of these voltages $\tilde{V}$ across the connector, and the conductances g_{T} that depend on "spectral" vectors at the ends, and are related to the transmission eigenvalues of the connector by the *derivative* of $\mathcal{I}$:

[5] It is a fascinating fact that this voltage may differ from the *electrostatic* voltage measured with an electrometer close to the node [60].

$$g_T = \dot{\mathcal{I}}(\theta_{12}) \,. \tag{2.126}$$

The current conservation sets the voltages in each node and finally allows one to calculate electric currents in each connector.

Let us relate this circuit theory to the facts learned in the preceding sections. First, we note that if there are no superconducting terminals, all spectral vectors are at the north pole, and all coefficients g_T are equal to Landauer conductances. Thereby the results of the common circuit theory are reproduced. Secondly, if we deal with a two-terminal nanostructure with one superconducting and one normal terminal, there is no need for the special circuit theory of Andreev conductance at all: one can use the circuit theory of transmission distribution to access the I–ϕ characteristics of the whole nanostructure and then implement the relation for the Andreev conductance, Eq. (2.97). Indeed, it is in strict correspondence with Eq. (2.126): in this case, the angle θ_{12} is the angle between the vectors pointing to the north pole (normal reservoir) and to the equator (superconducting reservoir), that is, $\pi/2$. Consider, for example, two identical tunnel junctions with conductances G in series. The I–ϕ characteristic of a single tunnel junction is $I = G\sin\phi$. Since the setup is symmetric, the phase in the node between the junctions is $\phi/2$, and the current through either junction becomes $I = G\sin(\phi/2)$. From Eq. (2.126) we find the Andreev conductance, $G_A = \dot{I}(\pi/2) = (G/2)\cos(\pi/4)$. Taking into account that the conductance of the structure in the normal state is $G_N = G/2$, we obtain $G_A = G_N/\sqrt{2}$, as mentioned in the beginning of this section.

Control question 2.18. Derive the Andreev conductance for a diffusive connector in contact with a superconductor.

The circuit theory of Andreev conductance is indispensable for multi-terminal nanostructures. Let us give an example of a system which consists of a normal reservoir and two superconducting reservoirs, S1, with superconducting phase φ_1, and S2, with φ_2. All the reservoirs are connected by the tunnel junctions to the same node. For simplicity, we consider the case when the conductances in the normal state separating the node from both superconducting reservoirs are the same, G_S, and the tunnel junction leading to the normal node has a conductance G_N. We introduce the spectral vectors: for the normal reservoir $\boldsymbol{s}_N = \hat{\eta}_3$; for the two superconducting reservoirs $\boldsymbol{s}_{S1,2} = \cos\varphi_{1,2}\hat{\eta}_1 + \sin\varphi_{1,2}\hat{\eta}_2$; and in the normal node we have $\boldsymbol{s}_n$. Representing the currents in the connectors by $\mathcal{I}_a = g_{Ta}\boldsymbol{s}_n \times \boldsymbol{s}_a$, for each connector, a = N, S1, S2, making use of current conservation in the node, $\mathcal{I}_N + \mathcal{I}_{S1} + \mathcal{I}_{S2} = 0$, we find the spectral vector of the node. Indeed, since $s_n^2 = 1$, it has to be

$$\boldsymbol{s}_n = \frac{G_N\boldsymbol{s}_N + G_S(\boldsymbol{s}_{S1} + \boldsymbol{s}_{S2})}{\sqrt{G_N^2 + 2G_S^2(1+\cos\varphi)}} \,,$$

with $\varphi = \varphi_1 - \varphi_2$. The next step is to compute the Andreev renormalization of the conductances G_a. Equation (2.126) yields $\tilde{G}_a = G_a(\boldsymbol{s}_a \cdot \boldsymbol{s}_n)$. This yields

$$\tilde{G}_N = \frac{G_N^2}{\sqrt{G_N^2 + 2G_S^2(1+\cos\varphi)}}; \quad \tilde{G}_S = \frac{G_S^2(1+\cos\varphi)}{\sqrt{G_N^2 + 2G_S^2(1+\cos\varphi)}}.$$

Now we can find the Andreev conductance of the circuit. Since both superconductors are at zero voltage, both tunnel junctions that connect them to the node are connected in parallel, and the third junction is in series with both. Thus,

$$G_A = \frac{2\tilde{G}_N \tilde{G}_S}{\tilde{G}_N + 2\tilde{G}_S} = \frac{2G_N^2 G_S^2 (1+\cos\varphi)}{[G_N^2 + 2G_S^2(1+\cos\varphi)]^{3/2}}.$$

We see that the Andreev conductance of the setup essentially depends on the phase difference between the two superconducting reservoirs. At $\varphi = 0$, the conductance reaches maximum and vanishes for $\varphi = \pi$.

Exercise 2.24. Show that the conductance does not vanish at $\varphi = \pi$ if the conductances of the junctions to the superconductors are not the same. What is the minimum conductance?

Exercise 2.25. Replace the tunnel junction with G_N by a point contact of the same conductance and evaluate the phase-dependent conductance of the whole structure.

2.8.4 Double junction

Let us now go beyond the low-energy limit and give an example of a current calculation valid at any energy/voltage. To illustrate, let us turn again to the well used nanostructure, the double tunnel junction. We consider two tunnel junctions of the same conductance G_T with a normal metal node in between that separate a superconducting electrode and a normal metal biased at voltage V. The node is thus connected to two reservoirs and experiences "leakage." First, we find the retarded Green's function $\check{R}$ in the node. For a tunnel junction, the matrix current is proportional to the commutator of matrix voltages at the sides of the junction. The matrix current conservation thus reads as follows:

$$\left[\check{M}, \check{R}\right] \equiv \left[G_T \check{R}_S + G_T \check{R}_N + i\hat{\eta}_3 \epsilon / E_{Th}, \check{R}\right] = 0,$$

with $\check{R}_{S,N}$ being the retarded Green's functions describing the superconducting and normal reservoirs. Since $\check{R}$ is a 2×2 matrix, normalized to $\check{R}^2 = 1$, the solution is given by

$$\check{R} = \frac{\check{M}}{\mathrm{Tr}\sqrt{\check{M}^2/2}}.$$

To simplify the expressions, we write them explicitly only for $\epsilon \ll \Delta$:

$$\check{R} = \frac{1}{\sqrt{(1+i\epsilon/E_{Th})^2+1}} \begin{pmatrix} 1+i\epsilon/E_{Th} & 1 \\ 1 & -1-i\epsilon/E_{Th} \end{pmatrix}. \tag{2.127}$$

Since there is only one superconducting reservoir in the system, the "no mixing" condition applies. The currents at each energy are thus determined by the renormalized conductances g_T. For a tunnel junction, the renormalization is given by Eq. (2.134). Even though our tunnel junctions initially have identical conductance, the renormalization makes them unequal:

$$G_S(\epsilon) = G_T \mathrm{Re}\left(1/\sqrt{(1+\mathrm{i}\epsilon/E_{Th})^2+1}\right);$$
$$G_N(\epsilon) = G_T \mathrm{Re}\left((1+\mathrm{i}\epsilon/E_{Th})/\sqrt{(1+\mathrm{i}\epsilon/E_{Th})^2+1}\right). \tag{2.128}$$

The differential conductance of the system is then given by an Ohm's law expression:

$$\frac{\mathrm{d}I}{\mathrm{d}V} = \frac{G_N(eV)G_S(eV)}{G_N(eV)+G_S(eV)},$$

provided $k_B T \ll E_{Th}$. The plots of G_S, G_N, and $\mathrm{d}I/\mathrm{d}V$ are given in Fig. 2.16.

2.8.5 Details of conductance renormalization

In this technical subsection, we work with 4×4 matrices to derive the four energy-dependent conductances in Eqs. (2.122), to give concrete (and generally rather long) formulas in terms of $\check{R}$ and $\check{A}$ at the ends of the connector, and relate particular limits of these formulas to the cases already studied in the book. The starting point is the general expression for the effective action of a Landauer connector, Eq. (2.91). For 2×2 matrices, the expression is rather trivial, since for any 2×2 traceless matrices $\check{G}_{1,2}$ the anticommutator is proportional to the unity matrix:

$$\check{G}_1\check{G}_2 + \check{G}_1\check{G}_2 = \check{1}\ \mathrm{Tr}(\check{G}_1\check{G}_2 + \check{G}_1\check{G}_2)/2.$$

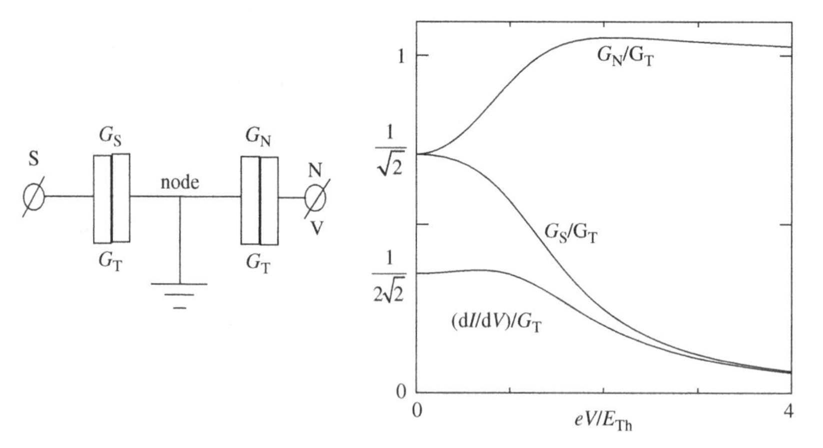

Fig. 2.16. Double junction between a normal metal (N) and a superconductor (S). The differential conductances of two identical junctions are the same at $V = 0$, $G_S = G_N = G/\sqrt{2}$. At higher energy, the normal node gets more "normal," and the proximity effect is suppressed. Consequently, G_S drops, while G_N approaches its normal value. The differential conductance of the composite system $\mathrm{d}I/\mathrm{d}V$ reaches maximum at $eV \approx 0.8E_{Th}$.

The first step of the derivation is to derive a similar relation for 4×4 matrices. To do so, we note that, in this case, the commutator has two distinct eigenvalues, which can be expressed in terms of the traces $\mathrm{Tr}(\check{G}_1\check{G}_2)$ and $\mathrm{Tr}(\check{G}_1\check{G}_2\check{G}_1\check{G}_2)$. We use these eigenvalues to arrive at the following:

$$\mathcal{S} = \sum_p \sum_{\pm} \ln\left(1 + \frac{T_p}{4}(A \pm B)\right); \quad A = \frac{\mathrm{Tr}[\check{G}_1\check{G}_2]}{2} - 2;$$
$$B^2 = \frac{\mathrm{Tr}[(\check{G}_1\check{G}_2)^2]}{2} + 2 - \frac{(\mathrm{Tr}[\check{G}_1\check{G}_2])^2}{4}.$$

We are interested in particle currents per energy interval, $I_0 = \mathrm{Tr}(\hat{\tau}_3\check{I})$, $I_3 = \mathrm{Tr}(\hat{\tau}_3\hat{\eta}_3\check{I})$. To this end, we calculate the current, making use of the variational relation in Eq. (2.74) and assuming that the variation is proportional to $\hat{\tau}_3$ (for I_0) or to $\hat{\tau}_3\hat{\eta}_3$ (for I_3). We obtain

$$I_\alpha = \sum_{p,\pm} \frac{T_p}{4} \frac{W^\alpha/2 \pm B^{-1}(U^\alpha - 2W^\alpha(A+2))/4}{1 + T_p(A \pm B)/4},$$

where W^α and U^α are variations of the two traces that enter Eq. (2.127),

$$W^\alpha = \mathrm{Tr}\left[\hat{\tau}_3\hat{\eta}_\alpha[\check{G}_1, \check{G}_2]\right]; U^\alpha = \mathrm{Tr}\left[\hat{\tau}_3\hat{\eta}_\alpha((\check{G}_1\check{G}_2)^2 - (\check{G}_2\check{G}_1)^2)\right].$$

Calculation of the traces yields

$$A \pm B = 2\Theta(\pm\epsilon) - 2; \; \Theta(\epsilon) \equiv \mathrm{Tr}\left[\check{R}_1\check{R}_2\right]/2; \; \Theta^*(\epsilon) = \Theta(-\epsilon),$$

so that the traces do not contain the actual filling of electron states given by matrices $\check{K}_{1,2}$ and are determined by retarded Green's functions $\check{R}$ at the two ends of the connector. To present the information about transmission eigenvalues in compact form, we introduce the function

$$\mathcal{Z}(\Theta) \equiv G_Q \sum_p \frac{T_p}{1 + (T_p/2)(\Theta - 1)},$$

which is related to the function $\mathcal{I}$ used in this chapter by $\sin\phi\mathcal{Z}(\cos\phi) = \mathcal{I}(\phi)$. In this notation, the currents are given by

$$I^\alpha = \frac{\mathrm{Re}\,\mathcal{Z}}{4}W^\alpha + \frac{\mathrm{Im}\,\mathcal{Z}}{16\,\mathrm{Im}\,\Theta}(U^\alpha - 4W^\alpha\,\mathrm{Re}\,\Theta). \tag{2.129}$$

The variations of the traces W^α, U^α depend on the functions $\check{K}_{1,2}$,

$$\check{K}_i = \check{R}_i \begin{bmatrix} f_i^{\mathrm{T}} + f_i^{\mathrm{L}} & 0 \\ 0 & f_i^{\mathrm{T}} - f_i^{\mathrm{L}} \end{bmatrix} - \begin{bmatrix} f_i^{\mathrm{T}} - f_i^{\mathrm{L}} & 0 \\ 0 & f_i^{\mathrm{T}} - f_i^{\mathrm{L}} \end{bmatrix} \check{A}_i, \tag{2.130}$$

and are linear in these matrices. So it is convenient to present those in the following form:

$$W^\alpha = \sum_{\beta=\mathrm{L,T}} W_1^{\alpha\beta} f_1^\beta + W_2^{\alpha\beta} f_2^\beta, \tag{2.131}$$

and similarly for U^α. Specific expressions are given by

$$\left\{ \begin{array}{c} W_1^{\alpha\beta} \\ -W_2^{\alpha\beta} \end{array} \right\} = \mathrm{Tr}\left[\eta^\alpha(\check{R}_1\eta^\beta\check{A}_2 + \check{R}_2\eta^\beta\check{A}_1)\right] - \left\{ \begin{array}{c} \mathrm{Tr}\left[\eta^\alpha\eta^\beta(\check{A}_1\check{A}_2 + \check{R}_2\check{R}_1)\right] \\ \mathrm{Tr}\left[\eta^\alpha\eta^\beta(\check{A}_2\check{A}_1 + \check{R}_1\check{R}_2)\right] \end{array} \right\};$$

$$\left\{ \begin{array}{c} U_1^{\alpha\beta} \\ -U_2^{\alpha\beta} \end{array} \right\} = \mathrm{Tr}\left[\eta^\alpha\check{R}_1(\check{R}_2\check{R}_1\eta^\beta + \check{R}_2\eta^\beta\check{A}_1 + \eta^\beta\check{A}_2\check{A}_1)\check{A}_2\right] + \mathrm{Tr}\left[\eta^\alpha\check{R}_2(\check{R}_1\check{R}_2\eta^\beta + \check{R}_1\eta^\beta\check{A}_2 + \eta^\beta\check{A}_1\check{A}_2)\check{A}_1\right] - \left\{ \begin{array}{c} \mathrm{Tr}\left[\eta^\alpha\eta^\beta((\check{A}_1\check{A}_2)^2 + (\check{R}_2\check{R}_1)^2)\right] \\ \mathrm{Tr}\left[\eta^\alpha\eta^\beta((\check{A}_2\check{A}_1)^2 + (\check{R}_1\check{R}_2)^2)\right] \end{array} \right\}.$$

In any case, $W_1^{\alpha\alpha} = W_2^{\alpha\alpha}$ and $W_1^{\alpha\beta}(\epsilon) = W_2^{\beta\alpha}(-\epsilon)$; the same relations also hold for $U_{1,2}^{\alpha\beta}$. Substituting this into Eq. (2.129), we obtain the "conductances" g_{T}, g_{L}, g_{LT}, and g_{TL}.

Let us illustrate these relations with simple examples. First, we obtain the Andreev conductance of a single connector. In this case, it is enough to set energy ϵ to zero, to consider reservoir 1 to be a normal metal ($R_1 = -A_2 = \eta_3$), and to consider reservoir 2 to be superconducting. Since for Andreev reflection the conductances g_{LT} and g_{TL} can be neglected ("no mixing" condition), it is enough to calculate only the conductance g_{T} which contains the matrices $W_1^{33} = W_2^{33}$ and $U_1^{33} = U_2^{33}$. Explicitly, we have

$$W^{33} = 0;\ U^{33} = 8;\ \Theta = 0;\ \frac{\mathrm{Im}\,\mathcal{Z}}{\mathrm{Im}\,\Theta} \to \mathcal{Z}'(0) = G_{\mathrm{Q}} \sum_p \frac{2T_p^2}{(2+T_p)^2}, \tag{2.132}$$

and we reproduce Eq. (1.166) derived from the scattering approach.

A complementary example is the equilibrium supercurrent equilibrium at zero temperature. In this case, we take $f_{1,2}^{\mathrm{T}} = 0$, $f_{1,2}^{\mathrm{L}} = \mathrm{sgn}(\epsilon)$. In this case, we only need the "mixing" conductances g_{TL}, g_{LT}, which are expressed in terms of the matrices W^{03} and U^{03}. We set two reservoirs to (see Eq. (2.87))

$$\check{R}_{1,2} = \frac{1}{\sqrt{(\epsilon + \mathrm{i}\delta)/\Delta - 1}} \begin{bmatrix} \epsilon/\Delta & \mathrm{e}^{\mp\mathrm{i}\varphi/2} \\ -\mathrm{e}^{\pm\mathrm{i}\varphi/2} & \epsilon/\Delta \end{bmatrix}, \tag{2.133}$$

corresponding to the difference of the superconducting phases φ. In this case, we have

$$\Theta = \frac{(\epsilon/\Delta)^2 - \cos\varphi}{(\epsilon/\Delta)^2 - 1},$$

and the denominators of Z approach zero at energy E_p given by

$$1 + \frac{T_p}{2}(\Theta - 1) = 0 \Rightarrow E_p = \pm\sqrt{1 - T_p \sin^2(\varphi/2)}\,.$$

We thus reproduce Eq. (1.170) for the energies of Andreev bound states.

Extending this for a tunnel junction, we derive a simple relation valid under "no mixing" conditions:

$$G(\epsilon) = \frac{G_T}{8} \operatorname{Tr}\left[(\check{R}_1 + \check{R}_1^\dagger)(\check{R}_2 + \check{R}_2^\dagger)\right]. \tag{2.134}$$

2.9 Full counting statistics

As the last application, we formulate in this section the circuit theory of the full counting statistics (FCS) of non-interacting electrons for a multi-terminal circuit. Let us first explain why we immediately go to the multi-terminal case. The point is that, by virtue of the Levitov formula, the FCS of a two-terminal circuit is readily expressed in terms of the transmission distribution $\rho(T)$ (Section 1.4). There is no need for a special circuit theory for the FCS: one just can use the circuit theory for transmission distribution given in Section 2.6. We have learned in Chapter 1 that the multi-terminal FCS can also be expressed in terms of the scattering matrix. However, this is a multi-terminal scattering matrix that cannot be characterized by a transmission distribution. As an example, we recall that noise in multi-terminal circuits is determined by the following combinations of scattering amplitudes (see Eq. (1.86)):

$$A_{\alpha\beta,\gamma\delta}(E) = \operatorname{Tr}\left\{\left[\delta_{\alpha\gamma}\delta_{\alpha\delta} - \hat{s}^\dagger_{\alpha\gamma}(E)\hat{s}_{\alpha\delta}(E)\right]\left[\delta_{\beta\delta}\delta_{\beta\gamma} - \hat{s}^\dagger_{\beta\delta}(E)\hat{s}_{\beta\gamma}(E)\right]\right\},$$

with the Greek indices labeling the terminals. Higher-order cumulants are defined by similar expressions involving higher powers of the scattering matrix. The number of possible combinations quickly grows with increasing order of the cumulant.

In fact, the circuit theory of multi-terminal FCS sorts out which combinations of scattering amplitudes are relevant for quantum transport and provides averages of these combinations over the phase shifts. We illustrate this at the end of this section by explicit calculation of $A_{\alpha\beta,\gamma\delta}$.

The approach is based on the extended Keldysh Green's function method as outlined in Section 2.5. At each energy, we operate with 2×2 Green's functions – matrix voltages. In the terminals, they are fixed to

$$\check{G}_\alpha(E) = \exp(\mathrm{i}\chi_\alpha\hat{\tau}_3/2)\check{G}^{(0)}_\alpha(E)\exp(-\mathrm{i}\chi_\alpha\hat{\tau}_3/2), \tag{2.135}$$

where $\check{G}^{(0)}_\alpha(E)$ corresponds to the equilibrium Keldysh Green's function taken sufficiently far in the terminal α and expressed in terms of the corresponding filling factor $f_\alpha(E)$:

$$\check{G}^{(0)}_\alpha = \begin{bmatrix} 1 - 2f_\alpha(E) & -2f_\alpha(E) \\ -2(1 - f_\alpha(E)) & 2f_\alpha(E) - 1 \end{bmatrix}. \tag{2.136}$$

The calculation proceeds according to the general rules: we find the matrix voltages at the nodes using the conservation of matrix currents. To find the generating function of the FCS, one substitutes the matrix voltages found into the total action, Eq. (2.72), and integrates over energy to obtain

$$\ln \Lambda(\{\chi_\alpha\}) = -\Delta t \int \frac{\mathrm{d}E}{2\pi\hbar} S(E; \{\chi_\alpha\}) \equiv -\Delta t\, S_{\mathrm{FCS}}(\{\chi_\alpha\}). \tag{2.137}$$

The probability of N_α electrons transferring to each reservoir is given by the inverse Fourier transform of the generating function:

$$P(\{N_\alpha\}) = \int_0^{2\pi} \prod_\alpha \mathrm{d}\chi_\alpha \; \Lambda(\{\chi_\alpha\})\, \mathrm{e}^{-\mathrm{i}\sum_\alpha \chi_\alpha N_\alpha} \,.$$

To find the probability, it is enough to evaluate the integral over the counting fields $\{\chi_\alpha\}$ in the saddle-point approximation, that is, to find its extremum over all χ_α. This is always valid for the limit of zero-frequency FCS, where the measurement time Δt is assumed to be much longer than the typical time between successive electron transfers. We express the probability in terms of currents measured during the time Δt, $I_\alpha^{\mathrm{el}} \equiv eN_\alpha/\Delta t$. In the saddle-point approximation, this amounts to minimizing the function

$$\ln P(I_\alpha^{\mathrm{el}}) = -\Delta t \left\{ S_{\mathrm{FCS}}(\{\chi_\alpha\}) + \mathrm{i} \sum_\alpha I_\alpha^{\mathrm{el}} \chi_\alpha \right\} \tag{2.138}$$

over the counting fields. The extremum is usually reached at *imaginary* counting fields.

2.9.1 Three-terminal setup

From now on, let us concentrate on the simplest example: a multi-terminal circuit with a single node. Then there is just a single matrix voltage to solve for: the one for the node, $\check{G}_{\mathrm{N}}$.

An easy analytical solution is available for a specific case when all contacts are tunnel junctions. In this case, the currents from the reservoirs to the node are proportional to the commutators of the Green's function $\check{G}_{\mathrm{N}}$ and the matrix voltages of the corresponding reservoir, and the Kirchhoff rule is given by

$$\left[\check{M}, \check{G}_{\mathrm{N}}\right] = 0; \; \check{M} \equiv \sum_\alpha G_\alpha^{(\mathrm{T})} \check{G}_\alpha,$$

where $G_\alpha^{(T)}$ are tunnel conductances of the connectors. For 2×2 matrices, the solution is proportional to $\check{M}$ and needs to be normalized as follows:

$$\check{G}_{\mathrm{N}} = \check{M}/\sqrt{\mathrm{Tr}\{\check{M}^2\}/2}\,.$$

Each tunnel junction contributes to the action with a term $\propto \mathrm{Tr}(\check{G}_\alpha \check{G}_{\mathrm{N}})$ (see Eq. (2.81)). Summing up, we readily find the effective action up to an additive constant:

$$\begin{aligned} S_{\mathrm{FCS}} &= \frac{1}{\pi\hbar G_{\mathrm{Q}}} \int \mathrm{d}E \left(\sqrt{\mathrm{Tr}\{\check{M}^2\}/2} - 1\right) \\ &= \frac{G_\Sigma}{2e^2} \int \mathrm{d}E \left[\left(1 + \sum_{\alpha\beta} 2 g_\alpha g_\beta \left[f_\alpha(1-f_\beta)(\mathrm{e}^{\mathrm{i}(\chi_\alpha - \chi_\beta)} - 1) \right.\right.\right. \\ &\quad \left.\left.\left. + \; f_\beta(1-f_\alpha)(\mathrm{e}^{-\mathrm{i}(\chi_\alpha-\chi_\beta)} - 1)\right]\right)^{1/2} - 1\right], \end{aligned} \tag{2.139}$$

where $G_\Sigma = \sum_\alpha G_\alpha^{(T)}$ is the total conductance and $g_\alpha = G_\alpha^{(T)}/G_\Sigma$ is a connector conductance relative to the total one.

In the limit of low temperatures (shot-noise limit), $k_B T \ll e|V_\alpha - V_\beta|$, the states in each reservoir are either occupied ($f = 1$) or empty ($f = 0$). Let us now restrict our considerations to a three-terminal setup ($\alpha = 1, 2, 3$). The contribution to the FCS is given only in the two following situations: either one of the reservoirs injects electrons ($f = 1$) and two remaining reservoirs drain them ($f = 0$), or one of the reservoirs injects "holes" ($f = 0$) and two remaining ones drain them ($f = 1$). This is why, in order to evaluate the FCS at any set of voltages $V_{1,2,3}$, it is enough to know the FCS in the situation where a given reservoir (say number 3) injects carriers being at either positive or negative bias V_3 and reservoirs 1 and 2 are grounded, as shown in the schematic in Fig. 2.17. For this situation, the FCS becomes

$$S_{\text{FCS}} = \frac{G_\Sigma V_3}{2e} \Big(1 + 2g_1g_3(e^{i(\chi_1-\chi_3)} - 1) + 2g_2g_3(e^{i(\chi_2-\chi_3)} - 1)\Big)^{1/2} - 1. \tag{2.140}$$

The terms proportional to g_1g_3 and g_2g_3 describe electron transfers from 3 to 1 or 3 to 2, respectively. If there were no square root sign in the above expression, these processes would be Poissonian and uncorrelated. The square root sign thus signifies the correlations between the transfers of two kinds. The physical origin of this correlation is intuitively clear – fluctuations of current in the third connector produce deviations of the filling factor in the node. Positive deviation results in more current to both 2 and 3, while negative deviation reduces both currents – this explains the correlations. Equation (2.140) quantifies them. In the limit $G_{1,2}^{(T)} \gg G_3^{(T)}$ or $G_{1,2}^{(T)} \ll G_3^{(T)}$ the filling factor deviations in the node are strongly shunted by the largest conductance, so we expect the correlations to vanish. Equation (2.140) can be expanded in terms of the small relative conductances to give

$$S_{\text{FCS}} = \frac{G_\Sigma V_3}{2e} \left(g_1g_3e^{i(\chi_1-\chi_3)} + g_2g_3e^{i(\chi_2-\chi_3)}\right), \tag{2.141}$$

and corresponds indeed to two independent Poissonian processes.

Control question 2.19. What is the distribution $P(I_1, I_2, I_3)$ corresponding to Eq. (2.141)?

Exercise 2.26. Derive an analog of Eq. (2.140) for a setup of N terminals connected to a single node.

For arbitrary connectors, one solves two independent non-linear equations to find $\check{G}_N$ at any set of G_α: a task best handed over to a computer. We present here a set of numerical results for $\ln P(I_1, I_2)$ obtained for the three-terminal structure in the shot-noise limit. We consider three particular types of connectors: tunnel (T), ballistic (B), and diffusive (D), and assume equal resistances of all three connectors of the circuit.

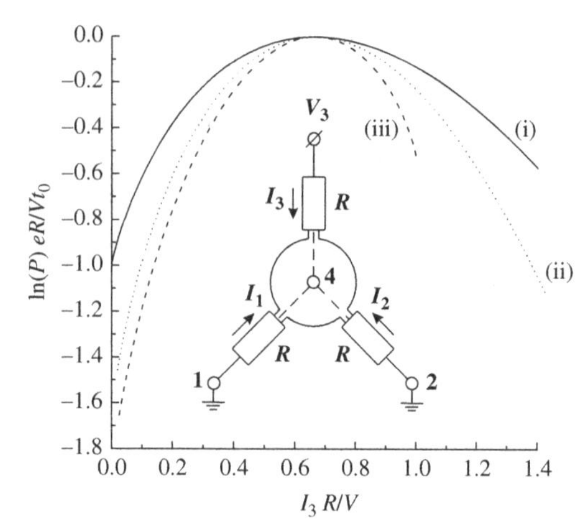

Fig. 2.17. Logarithm of the current probabilities in the three-terminal single-node nanostructure as a function of I_3, under the condition $I_1 = I_2$. The schematic presents the system configuration. The resistances $R = G^{(\mathrm{T})-1}$ of all connectors are assumed to be equal. (i) Tunnel connectors; (ii) diffusive connectors; (iii) ballistic connectors.

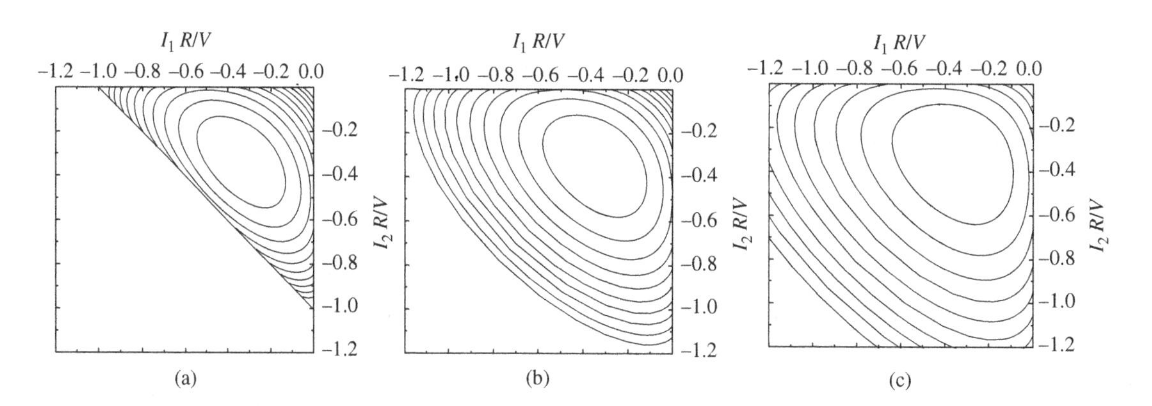

Fig. 2.18. The contour maps of the current distribution $\log[P(I_1, I_2)]$ in the three-terminal single-node nanostructure for different configurations of connectors: (a) ballistic; (b) diffusive; (c) tunnel.

As we mentioned above, it is sufficient to consider only the case $V_1 = V_2 = 0$, $V_3 = V$. The results of the calculations are presented in Figs. 2.17 and 2.18. The maximum of probability always occurs at $I_1 = I_2 = -V/3R$, $I_3 = 2V/3R$, which is exactly the set of currents obeying common electric Kirchhoff rules. As seen from the figures, the current distribution $P(I_1, I_2)$ for a ballistic system is bounded. The reason for this is that the current through any ballistic connector cannot exceed the maximum value $G_\mathrm{B} V$ that is set by the number of open channels. On the other hand, for tunnel and diffusive junctions every connector has, in principle, an infinite number of partially open transmission channels, and thus the current fluctuations can be arbitrary. One also sees that the relative probabilities of large current fluctuations increase in the sequence ballistic → diffusive → tunnel. Qualitatively, this is explained by increasing the Fano factor in this sequence: $F_\mathrm{B} = 0$, $F_\mathrm{D} = 1/3$, $F_\mathrm{T} = 1$, and the noise production is increased from ballistic to tunnel.

Exercise 2.27. Using Eq. (2.140) and taking identical tunnel connectors, calculate the matrix of third cumulants of the currents $\langle I_\alpha I_\beta I_\gamma\rangle - \langle I_\alpha\rangle\langle I_\beta I_\gamma\rangle - \langle I_\beta\rangle\langle I_\alpha I_\gamma\rangle - \langle I_\gamma\rangle\langle I_\alpha I_\beta\rangle + 3\langle I_\alpha\rangle\langle I_\beta\rangle\langle I_\gamma\rangle$.

2.9.2 Noise in multi-terminal circuits

Let us attempt to understand this point in more detail and concentrate on the current noises. It makes sense to do this in a more general setup where a single node is connected to many reservoirs by arbitrary connectors. This is also a good opportunity to illustrate the use of the variational principle in quantum circuit theory. Let us present the matrix voltage at the node in the form similar to that of the reservoirs:

$$\check{G}_{\rm N} = \exp(i\chi_{\rm N}\tau_3/2)\begin{bmatrix} 1-2f_{\rm N}(E) & -2f_{\rm N}(E) \\ -2(1-f_{\rm N}(E)) & 2f_{\rm N}(E)-1 \end{bmatrix}\exp(-i\chi_{\rm N}\tau_3/2),$$

where we introduce the filling factor $f_{\rm N}$ and counting field $\chi_{\rm N}$ in the node. By virtue of the variational principle, the total action at each energy is obtained by global minimization of the sum of connector actions with respect to these two parameters:

$$\mathcal{S}(\{\chi_\alpha\},\{f_\alpha\}) = \underset{f_{\rm N},\chi_{\rm N}}{\rm extr}\left[\sum_\alpha \mathcal{S}_\alpha(\chi_\alpha-\chi_{\rm N}, f_\alpha, f_{\rm N})\right],$$

where the sum is taken over all connectors (situated between the central node and reservoir α), and the connector action $\mathcal{S}_\alpha$ is given by Eq. (2.91) at the corresponding energy in accordance with the Levitov formula. Since we are interested in noises only, and these are given by second-order expansion in the vicinity of $\chi_\alpha = 0$, we only need to know the action near $\chi_\alpha \to 0$ with accuracy up to quadratic terms in $\{\chi_\alpha\}$. Therefore we expand each connector action, keeping the linear and quadratic terms only;

$$\mathcal{S}_\alpha \approx \frac{G}{G_{\rm Q}}\left\{ i(\chi_\alpha-\chi_{\rm N})(f_\alpha-f_{\rm N}) - \frac{1}{2}(\chi_\alpha-\chi_{\rm N})^2\left[f_\alpha(1-f_\alpha)+f_{\rm N}(1-f_{\rm N})+F_\alpha(f_\alpha-f_{\rm N})^2\right]\right\}.$$

We see that, with this accuracy, all information concerning the type of the connector is incorporated into its Fano factor $F_\alpha = \sum_n T_n(1-T_n)/\sum_n T_n$. The total action is obtained by minimization with respect to $f_{\rm N}, \chi_{\rm N}$ and is given by

$$\mathcal{S}(\{\chi_\alpha\},\{f_\alpha\}) = \underset{f_{\rm N},\chi_{\rm N}}{\rm extr}\sum_\alpha \frac{G_\alpha}{G_{\rm Q}}\left\{ i(\chi_\alpha-\chi_{\rm N})(f_\alpha-f_{\rm N}) - \frac{1}{2}(\chi_\alpha-\chi_{\rm N})^2\left[f_\alpha(1-f_\alpha)+f_{\rm N}(1-f_{\rm N})+F_\alpha(f_\alpha-f_{\rm N})^2\right]\right\}.$$

Varying this expression, we see that the extremum with this accuracy is achieved at $f_{\rm N} = \bar{f}$, $\chi_{\rm N} = \bar{\chi}$, where $\bar{f} = \sum_\alpha g_\alpha f_\alpha$, $\bar{\chi} = \sum_\alpha g_\alpha \chi_\alpha$. Evidently, they are just averages of the corresponding reservoir values with weights given by the relative conductances of the connectors. The quadratic part of the action taken at this point is given by

$$\mathcal{S} = \sum_{\alpha} \frac{(\chi_\alpha - \bar{\chi})^2}{2G_\mathrm{Q}} S^{(\alpha)}.$$

This defines the noises per energy interval in the reservoirs:

$$S^{\alpha\beta} = \frac{\partial^2 \mathcal{S}}{\partial\chi_\alpha \partial\chi_\beta} = \delta_{\alpha\beta} S^{(\alpha)} - \frac{G_\alpha}{G_\Sigma} S^{(\alpha)} - \frac{G_\beta}{G_\Sigma} S^{(\beta)} + \frac{G_\alpha G_\beta}{G_\Sigma^2} \sum_\gamma S^{(\gamma)}, \tag{2.142}$$

where

$$S^{(\alpha)} = G_\alpha \left(f_\alpha(1 - f_\alpha) + \bar{f}(1 - \bar{f}) + F_\alpha(f_\alpha - \bar{f})^2 \right)$$

are current noises per energy interval produced in each connector. Note that they have the same form as the two-terminal noise, Eq. (1.60), with the only difference being that one of the filling factors f_N is no longer the Fermi function. Since the conductances G_α do not depend on energies, Eq. (2.142) is also valid upon integration over energies. The interpretation of Eq. (2.142) is straightforward: each connector independently produces current fluctuations δI^α with intensity given by $S^{(\alpha)}$. The current fluctuation sent into the node returns to all the reservoirs (including the sender) being divided according to Ohm's law: $\delta I_\mathrm{N}^\alpha = -(G_\alpha/G_\Sigma)\sum_\alpha \delta I^\alpha$. The actual current fluctuation in each connector is the sum of the two: $\delta I_\alpha + \delta I_\mathrm{N}^\alpha$.

Control question 2.20. What does Eq. (2.142) look like in the two limits of (i) zero temperature and (ii) no voltages applied to the terminals?

Let us now recall Eq. (1.86), which gives the noise in each terminal in terms of the scattering matrix of the whole nanostructure. Comparing the two expressions, we discover that we have determined the coefficients $A_{\alpha\beta,\gamma\delta}$ that are characteristics of the scattering matrix of the whole nanostructure:

$$G_\mathrm{Q} A_{\alpha\beta,\gamma\delta} = \delta_{\alpha\beta}\Gamma^\alpha_{\gamma\delta} - g^\alpha \Gamma^\beta_{\gamma\delta} - g^\beta \Gamma^\alpha_{\gamma\delta} + g^\alpha g^\beta \sum_\varepsilon \Gamma^\varepsilon_{\gamma\delta},$$

where

$$g^\alpha = G_\alpha/G_\Sigma;\ \Gamma^\alpha_{\beta\delta} = G_\alpha \left((1 - F_\alpha)(\delta_{\alpha\gamma}\delta_{\alpha\delta} + g_\gamma g_\delta) + F_\alpha(\delta_{\alpha\gamma} g_\delta + \delta_{\alpha\delta} g_\gamma) \right).$$

Thus, FCS calculations enable the characterization of the scattering matrix of a complex nanostructure; we have found rather abstract "quantum" coefficients $A_{\alpha\beta,\gamma\delta}$ in terms of conductances and Fano factors of the connectors that form the nanostructure.

3 Coulomb blockade

A paradox of solid state physics is that electrons in conductors are almost exclusively regarded as non-interacting particles, even though they do interact. This comes both from physical reasons and from the human need for convenience. The physical reason is that the interacting electrons form a ground state, and charged elementary excitations above the state – quasielectrons – do not interact provided their energies are sufficiently close to the Fermi surface. This makes a model of non-interacting electrons completely adequate for quasielectrons, at least in the low-energy limit. This allows a scattering approach to quantum transport that assumes the absence of interaction. The convenience model is that the physics of non-interacting particles is much easier to understand and apply. Besides, sticking to a convenient picture usually goes unpunished. In fact, in solid state physics there are only a few rather exotic examples where interaction effects really reign and the non-interacting approach produces obviously erroneous results. These cases are notoriously difficult to comprehend and to quantify; some effects revealed almost a century ago (for instance, Mott insulator transition) are still on the front-line of modern research.

In contrast to solid state physics, there is a very common regime in quantum transport where interaction effects are dominant: the Coulomb blockade regime, and here the scattering approach fails. However, in contrast to the situation in solid state systems, Coulomb blockade systems are usually even simpler than those of Chapters 1 and 2 where interaction does not play an important role. Due to this, one can quickly grasp the fascinating features of Coulomb blockade physics, and begin to design and apply Coulomb blockade circuits and devices. This physics is based on *charge quantization*, giving rise to *charging energy*. It occurs in any nanostructure where a place to store an electron – an island – is fenced off with tunnel barriers, provided the tunnel conductances are sufficiently small, $G \ll G_Q$. This chapter is devoted to Coulomb blockade physics.

The charging energy E_C is a classical concept. The nanostructure is described with an equivalent capacitance circuit (Section 3.1). Elementary processes in Coulomb blockade systems are *single-electron transfers*, described by the master equation, which expresses the classical probability balance (Section 3.2). The simplest Coulomb blockade circuit with only one island is called a *single-electron transistor* (SET). It exhibits a number of signifying features of the Coulomb blockade – Coulomb diamonds, Coulomb oscillations, and the Coulomb staircase (Section 3.3). Quantum mechanics gives rise to cooperative tunneling of two electrons – *co-tunneling* – which becomes the dominant transport mechanism in the situation when single-electron transport is blocked (Section 3.4).

The combination of the Josephson effect and Coulomb blockade creates a quantum system, which can be in a quantum superposition of charge states, creating the base for solid state qubits. To describe this combination, we turn to superconducting islands connected to normal and superconducting electrodes. Each tunnel junction in this setup becomes a Josephson junction. The Josephson effect sets a new energy scale, the Josephson energy E_J. Most importantly, a superconducting state characterized by a phase, which is a conjugated variable to the charge: the state with a certain charge has a phase which is undetermined, and vice versa. This uncertainty is of a quantum nature. This *macroscopic quantum mechanics* is considered in Section 3.5. We start from the generic example, the Cooper pair box (CPB), quantize the phase and charge, and reveal the Schrödinger equation for the CPB. We introduce two complementary regimes where either Coulomb or Josephson energy is dominant and discuss macroscopic quantum tunneling. In Section 3.6 we come to a more complex system, the *Josephson junction array*, which is a collection of superconducting islands connected by Josephson junctions. If one superconducting island can be likened to a home-made artificial atom, then an array is a home-made solid, which has distinct quantum phases and phase transitions. We shortly review the most interesting aspects of Josephson junction array physics – vortices, charge–vortex duality, and the Berezinsky–Kosterlitz–Thouless transition.

In Section 3.7, we turn to superconducting islands *beyond the Josephson limit*. The Josephson limit implies that only Cooper pairs may be transferred through Josephson junctions, and the number of electrons at each island is even. In this final section of the chapter, we go to energies of the order of the superconducting gap. Then an odd number of electrons is also permitted at the islands, this number being fixed by the parity effect, and tunneling of quasiparticles can also be important. In this situation, some of the transfer processes are coherent and are characterized by quantum-mechanical amplitudes, whereas others are incoherent and are described by tunneling rates. We introduce the density matrix description appropriate for this situation, and solve the equation of motion for the density matrix, which replaces the Schrödinger equation.

Principles of single-electron tunneling have been reviewed in Ref. [61].

3.1 Charge quantization and charging energy

Let us start with an isolated metallic island hanging somewhere in space. The number of elementary particles in the island – positively or negatively charged – must be integer. Thus, its charge Q must be an integer amount of elementary charges, $Q = Ne$, N being the number of excess electrons in the island. Since the island is metallic, the actual charge is concentrated at the surface of the island; there is no charge in the bulk, although the extra electrons are spread over the volume of the island. The charge Q produces an electric field $\boldsymbol{E}$ around the island, this field accumulating some *electrostatic* energy. As we know from the electrostatics, this energy can be expressed via the capacitance C of the island,

$$E = \frac{Q^2}{2C} = \frac{e^2}{2C}N^2 \equiv E_C N^2. \tag{3.1}$$

Suppose we add an extra electron to the island, transferring it from a metal with the same Fermi energy. Since we are charging the island, we have to pay the energy E_C. This energy cost rises if we add more electrons: we must provide $E_C N^2 - E_C(N-1)^2 = E_C(2N-1)$

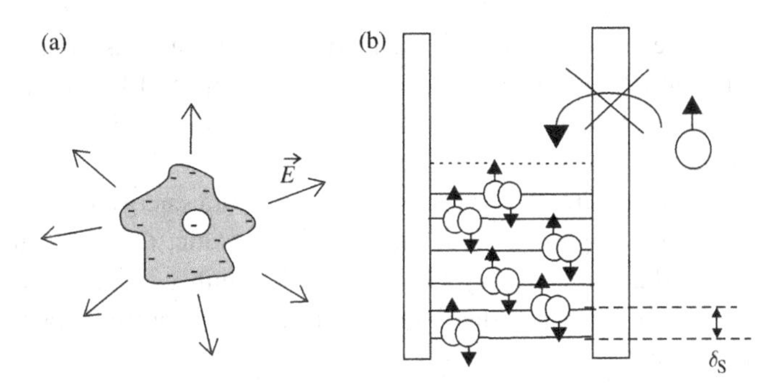

Fig. 3.1. (a) Excess charge in an isolated metallic island produces an electric field outside, thus accumulating charging energy. (b) The energy cost to put an electron into the island is not just a typical level spacing δ_S: it includes charging energy $E_C \gg \delta_S$.

extra energy to add the Nth electron. The same goes for extraction, since we also charge the island, although with positive charge. The fact that the addition energy depends on the number of particles added is in contradiction with the paradigm of non-interacting electrons: the charging energy produces an electron–electron interaction. The bigger the island, the bigger the capacitance. Because of this, the charging energy $\propto 1/C$ vanishes in the traditional thermodynamic limit when the energies are counted per particle. However, it presents a finite energy scale for any finite island.

We see that the charging energy E_C is in fact a classical concept since it is expressed in terms of classical capacitance of the island with no regard for quantum mechanics forming the electron states in there. The charging energy is also an electrostatic part of the *addition* energy: the energy required to add one extra electron to the neutral island. To see how good the classical description is, let us compare this part with quantum effects. Due to quantum mechanics, the electron levels in the island are discrete; there is a typical energy distance between the electron levels, the *mean level spacing* δ_S. This also contributes to the addition energy, since the added electron would go to the first unoccupied level separated by δ_S from an occupied one (see Fig. 3.1).

To get a feeling of the energy scales involved, let us take a cubic island of size L. It consists of $N_{at} \simeq (L/a)^3$ atoms, a being the interatomic distance. Typically, there is one valence electron per atom, and these electrons fill up the energy band E_F. The mean level spacing is therefore given by $\delta_S \simeq E_F/N_{at}$. From Coulomb's law, we estimate that the charging energy is of the order of e^2/L: the charge e is spread over a typical distance L. The ratio of the two energy scales is estimated as follows:

$$\frac{\delta_S}{E_C} \simeq \frac{E_F L}{e^2 N_{at}} = \frac{E_F a}{e^2} \frac{L}{a N_{at}} \simeq \frac{1}{N_{at}^{2/3}}.$$

Here we make use of the fact that $e^2/a \simeq E_F$. The conclusion is that δ_S/E_C is small provided the island consists of many atoms. To illustrate this with numbers, let us take $L = 100\,\text{nm}$ corresponding to $N_{at} = 10^9$, and $e^2/a \simeq E_F \simeq 10\,\text{eV}$. We obtain $E_C \simeq 1\,\text{mV}$, $\delta_S \simeq 10^{-8}\,\text{eV}$, a large difference. So, for practical purposes, $\delta_S \approx 0$, and one may safely disregard all effects related to the *discreteness* of the electron spectrum. This discreteness only becomes important for nanostructures consisting of several natural atoms, like a molecule, or those comprising an artificial atom, such as a quantum dot in a semiconductor heterostructure. For such nanostructures, $\delta_S/E_C \simeq 0.1$–0.5. The discrete spectrum will be discussed in detail in the context of quantum dots in Chapters 4 and 5. Here, we disregard the discreteness.

We have learned that it costs us some extra energy to add an electron to an isolated island, and to good accuracy it is an electrostatic, *Coulomb* energy. To transfer electrons to the island, this energy must be supplied either by an external voltage source or by a thermal fluctuation. Otherwise, the electron transport is blocked, and *Coulomb blockade* takes place.

3.1.1 Single-electron box

Not much would happen if we simply had a single isolated island: it would keep a given number of extra charges forever. Let us add some functionality to our setup and make an elementary Coulomb blockade circuit: a so-called single-electron box. The box consists of a single island and two bulk metallic electrodes: a source and a gate. We place the island very close to the source electrode so that the electrons can jump between the electrode and the island, changing its charge. We assume that the source electrode is grounded. The gate electrode is placed further away so that the electrons cannot be transferred between the island and the gate.

It turns out that we can tune the number of extra electrons in the island, changing the potential V_g of the gate electrode with respect to the source (so that $V_s = 0$). Physically, V_g shifts the electrostatic potential of the island with respect to the source; this helps electrons to enter the island. Alternatively, we can say that the gate potential induces charge in the island. Let us quantify this.

In a stationary state, the number of electrons in the island corresponds to the minimum electrostatic energy. Let us find this energy. We have now two capacitors: C, which is in between the island and the source, and C_g, which is in between the island and the gate. Let us denote the charges at the plates of these capacitors, respectively, as $\pm q_{1,2}$. The electrostatic energy consists of the energy accumulated in these capacitors and the work done by the voltage source to transfer the charge q_2 to the gate electrode, $-q_2 V_g$. This energy is given by

$$E_{el} = \frac{1}{2}\left(\frac{q_1^2}{C} + \frac{q_2^2}{C_g}\right) - q_2 V_g. \tag{3.2}$$

It can be expressed either in terms of $q_{1,2}$ or in terms of the corresponding voltage drops $V_{1,2}$ across the capacitors since they are related by $CV_1 = q_1, C_g V_2 = q_2$:

$$E_{el} = \frac{1}{2}\left(CV_1^2 + C_g V_2^2\right) - C_g V_2 V_g. \tag{3.3}$$

To determine these charges/voltage drops we use two conditions: (i) the sum of the voltage drops on the two capacitances is V_g, (ii) the charge in the island is quantized and equals eN. So, we have

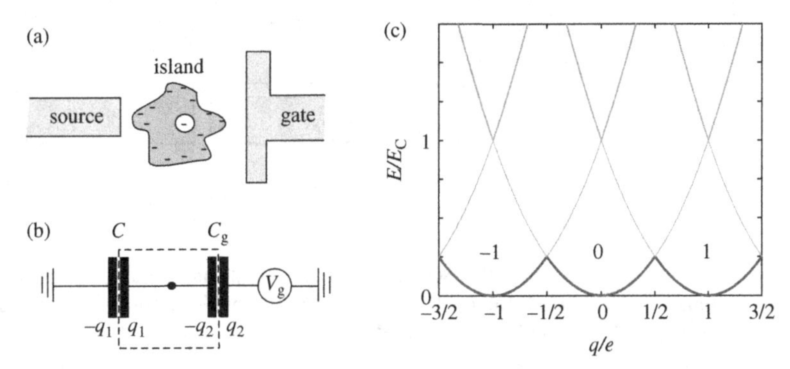

Fig. 3.2. Single-electron box. (a) Setup. (b) Equivalent capacitance circuit (the dashed box includes capacitance plates that belong to the island). (c) The charging energy of the single-electron box versus $q = -C_g V_g$. Each parabola corresponds to a charge state with N excess charges. The lowest segments of the parabolas give the minimum energy and the actual charge state.

$$V_1 + V_2 = q_1/C + q_2/C_g = V_g \text{ (i)};$$
$$-q_2 + q_1 = eN \text{ (ii)}.$$

We will now introduce a convenient notation: a charge induced in the island by the gate, $q \equiv -C_g V_g$. This induced charge is not quantized; it is a continuous quantity proportional to the gate voltage. What is quantized is the charge in the island. Using this notation, $q_1 = (eN - q)/(1 + C_g/C)$, $q_2 = -(eN - q)/(1 + C/C_g) - q$, and the energy is given by

$$E_{el} = E_C \left(N - \frac{q}{e}\right)^2 - \frac{q^2}{2C_g}; \; E_C \equiv \frac{e^2}{2(C + C_g)}. \tag{3.4}$$

We will disregard the final term in E_{el} since it does not depend on N and therefore does not influence the energy balance between the different charge states.

The next question is what is the number of extra electrons N that correspond to the minimum energy. To understand this, we plot the energy versus q. Each number N gives a parabola that reaches zero minimum at $q_N = eN$. For all possible values of N we obtain a series of shifted parabolas (see Fig. 3.2). The state with minimum energy corresponds to the lowest segments of the parabolas. The minimum energy thus occurs at different discrete N_q, $N_q = [q/e + 1/2]$, where the square brackets denote the floor of a continuous number. We see that N_q is a step-like function of q (or V_g), and one can tune the number of excess electrons in the island by varying V_g. We note the *e-periodicity* of the charging energy: the minimum energy is periodic in q with a period e. If the value of induced charge is q and the corresponding charge state is N, at the induced charge $q + |e|M$ (M being an integer), the charge state is $N + M$. This state will have the same properties with respect to Coulomb blockade as the original state.

The electrostatic energy defined above manifests charge quantization. If there were no quantization, N would be continuous, and the minimum energy would always be $E_{el} = 0$ at $eN = q$. This could actually happen if there is a metal shortcut between the island and the source electrode. In this case, the charge in the island does not have to take integer values since there is no natural separation between the electron states in the island and in the electrode, and an electron has to be localized neither in the island nor in the electrode.

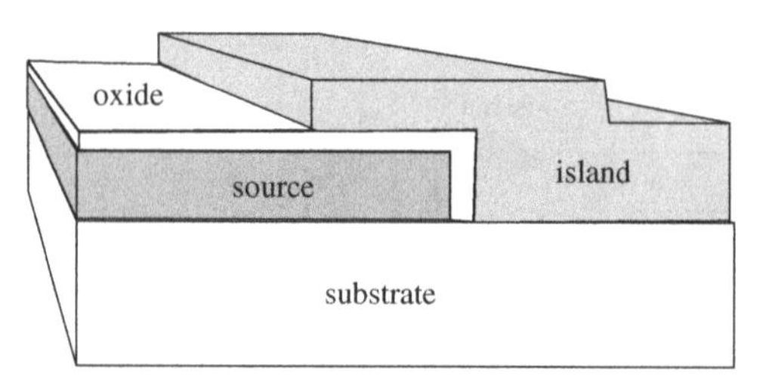

Fig. 3.3. **Overlap junction fabrication scheme. First, a metallic film (source electrode) evaporated on the substrate is oxidized. The oxide layer so formed provides a tunnel contact for the subsequently evaporated second electrode (island).**

3.1.2 Islands and barriers

This brings us to the following question: how good should the isolation be between the islands and electrodes in order to keep the charge quantization? A quick answer would be: the thicker, the better. However, we are interested in electron transport. A good isolation would imply that the electron transfers occur very rarely; this current would be too small to measure. On the other hand, a bad isolation would imply a shortcut, destroying the charge quantization.

In practice, the islands in metal-based Coulomb blockade systems are always placed very close to the electrodes. They are separated from them by tunnel barriers, the best choice of which is a thin film of an oxide. For aluminum, the native oxide film is just 1.5 nm thick. Complicated structures can be made by evaporation of metal film, oxidation, and yet another evaporation under a different angle: an overlap junction is an example of this (see Fig. 3.3). Such junctions have a high capacitance; it can be estimated as $\mathcal{A}/d$, $\mathcal{A}$ and d being the contact area and the thickness of the barrier, respectively. This is bigger than the geometric capacitance $\propto L$. This is not good, since the charging energy scales down. However, this is usually the price we must pay for a well controlled contact.

A tunnel junction is characterized by its conductance (or resistance), so the question is about the resistance of the junction. As an estimation, we use the Heisenberg uncertainty relation $\Delta E \Delta t \simeq \hbar$. Let us put an extra electron in the island. We have paid the energy E_{C}. What is a typical time that it remains there? We can obtain this time this in a very classical way: it should be of the order of the classical discharge time of the capacitor through the (tunnel) conductance, the so-called RC-time $\tau_{RC} = 1/R_{\mathrm{T}}C = G_{\mathrm{T}}/C$. This corresponds to an energy uncertainty $\hbar/\tau_{RC}$. For a state with an extra electron to be well defined, this uncertainty should not exceed the charging energy E_{C}, i.e. $E_{\mathrm{C}}\tau_{RC} \gg \hbar$. Remarkably, the capacitance does not enter this condition:

$$E_{\mathrm{C}}\tau_{RC} \gg \hbar \Rightarrow G \ll \frac{e^2}{\hbar} \simeq G_{\mathrm{Q}}. \tag{3.5}$$

Therefore, the junction must be sufficiently resistive, and the conductance should not exceed the conductance quantum G_{Q}. Recall that we encountered a similar condition when studying the scattering approach to quantum transport. In that case, it separated the "classical" conductors, with many transport channels participating in conductance, from the "quantum" ones where only a few channels contribute to the transport. Remarkably, the same typical value of conductance separates nanostructures with pronounced Coulomb blockade effects from those where these effects are negligible. Universality of scattering suggests that one does not need high tunnel barriers to provide a good isolation: any conductor with $G \ll G_Q$ would suffice. Charge quantization and Coulomb blockade at $G \sim G_Q$ will be discussed in Section 6.5.

From the above discussion, one could receive the impression that if we know the gate voltage, we also know how many excess integer charges there are in the island. In fact, this is not so in most practical situations. We only know the relative change of discrete charge; for example, we get one charge more by shifting the voltage by e/C_g. The point is that the gate electrode is not the only source of the induced charge. In a realistic nanostructure, there are always some stray charges located in the substrate around the island, and/or there may be charged immobile impurities in the oxide barriers of the tunnel junctions. If the nanostructure is composed of metals/semiconductors with different Fermi energies, this difference is compensated by charging. All together, these random sources produce a background charge q_i which adds to the charge induced by the gate electrode in the island i. Usually the absolute amount of the excess charges does not matter, since the charging energy is periodic in $N, q/e$. Therefore one can regard q_i to be uniformly distributed in the interval $(-|e|/2, |e|/2)$. In this interval, the background charge can be always compensated by the proper shift of the gate voltage.

3.1.3 Many-island capacitive circuits

The essential features of Coulomb blockade can be illustrated with a single-island system. However, one can make more complicated, interesting, and useful Coulomb blockade circuits by bringing more islands together. A new element is the interaction of electrons that are in different islands: excess discrete charge in island i changes the energy of the charge state in island j.

Charging effects in many-island systems can also be understood using equivalent capacitance circuits. Let us preview the answer. The state on any Coulomb blockade system is characterized by the number of excess charges in each island. The electrostatic energy of the system is given by

$$E_{\mathrm{el}} = \sum_{i,j} E^{(\mathrm{C})}_{ij}(N_i - q_i)(N_j - q_j), \tag{3.6}$$

where i, j number the islands, N_i is the quantized charge in island i, and q_i is the charge induced in the same island by any electrodes. The charging energy becomes a matrix, proportional to the inverse capacitance matrix of the islands. The diagonal elements of the charging energy matrix denote the energy cost of adding an excess electron to a given island. Non-diagonal matrixes denote a repulsive interaction between quantized charges in different islands.

To derive this from classical electrostatics, let us consider the most general capacitance circuit. We have some gate electrodes labeled k and biased at voltages $V^{(\mathrm{g})}_k$. We count the source electrode among the gate electrodes. There is a capacitive connection between each island i and gate electrode k with capacitance $C^{(g)}_{ik}$. In addition, the islands are connected to each other by capacitors $C^{(c)}_{ik}$. The voltages of the islands are given by V_i. The full energy consists of energies of the capacitors and the work done by voltage sources:

$$E_{\text{el}} = \frac{1}{2}\sum_{i>j} C^{(c)}_{ij}(V_i - V_j)^2 + \frac{1}{2}\sum_{i,k} C^{(g)}_{ik}(V_i - V^{(g)}_k)^2 - \sum_{ik} q_{ik} V^{(g)}_k.$$

Here, q_{ik} is the charge in the capacitance $C^{(c)}_{ik}$, so that

$$q_{ik} = C^{(g)}_{ik}(V^{(g)}_k - V_i).$$

We use this to bring the electrostatic energy to the following form:

$$E_{\text{el}} = \frac{1}{2}\sum_{i>j} C^{(c)}_{ij}(V_i - V_j)^2 + \frac{1}{2}\sum_{i,k} C^{(g)}_{ik} V_i^2 - \sum_{i,k} C^{(g)}_{ik}(V^{(g)}_k)^2. \tag{3.7}$$

As for the single-electron box, we will disregard the final group of terms since they do not depend on the charge state of the system. To proceed further, let us note that the full charge in island i is the sum over charges accumulated in all capacitors connected to this island:

$$N_i = \sum_j q_{ij} + \sum_k q_{ik} = \sum_j C^{(c)}_{ij}(V_i - V_j) + \sum_k C^{(g)}_{ik}(V_i - V^{(g)}_k).$$

This is most conveniently rewritten as

$$N_i - q_i = \sum_j C_{ij} V_j. \tag{3.8}$$

Here $q_i \equiv -\sum_k C^{(g)}_{ik} V^{(g)}_k$ is the *offset charge* induced on island i by all the gate electrodes, and C_{ij} is the capacitance matrix. Its diagonal elements are sums of all the capacitances connected to an island, and the non-diagonal ones are cross-capacitances with the *minus* sign:

$$C_{ij} = \begin{cases} \sum_j C^{(c)}_{ij} + \sum_k C^{(g)}_{ik} & i = j \\ -C^{(c)}_{ij} & i \neq j. \end{cases} \tag{3.9}$$

We invert this relation by expressing voltages in terms of island charges, $V_i = \sum_j (C^{-1})_{ij}(N_j - q_j)$, and substituting this into Eq. (3.7). We recover the charging energy given in Eq. (3.6), with

$$E^{(C)}_{ij} = \frac{e^2}{2}(C^{-1})_{ij}.$$

Let us illustrate this with a simple circuit consisting of two islands (see Fig. (3.4)). Each island is controlled by its own gate, so the charges induced at each island are $q_{1,2} = C^{(g)}_{1,2} V^{(g)}_{1,2}$. There is a capacitance C_{12} between the islands. The diagonal elements of the capacitance matrix are contributed to by all capacitors connected to an island, $C_{1,2} = C^{(g)}_{1,2} + C_{12} + C^{(s)}_{1,2}$ Obviously, $C_{1,2} > C_{12}$. Inverting the capacitance matrix, we obtain the charging energy matrix:

$$\hat{E}^{(C)} = \frac{e^2}{2(C_1 C_2 - C_{12}^2)} \begin{bmatrix} C_2 & C_{12} \\ C_{12} & C_1 \end{bmatrix}. \tag{3.10}$$

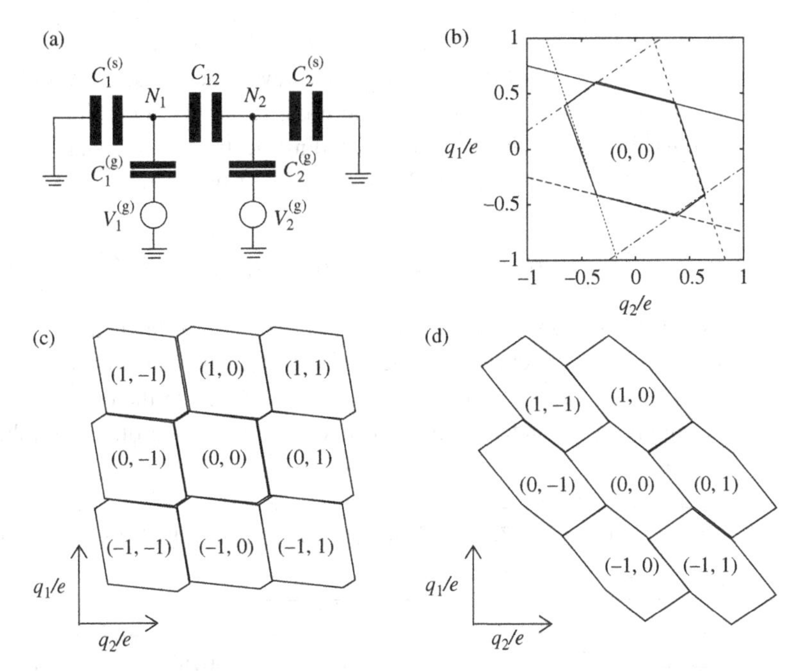

Fig. 3.4. **(a) Equivalent capacitance circuit of a two-island system. (b) The energies of two different charge states are equal along a line in the q_1–q_2 plane. Six lines define a region where the state (0, 0) has minimum energy. (c), (d) Periodic tilings of the q_1–q_2 plane.**

Let us make use of convenient dimensionless notations $C_{1,2} = C_{12}(1 + \alpha_{1,2})$.

In these notations, the charging energy is given by

$$E_{\rm el}(N_1, N_2) = E_2\left(\alpha_2\left(N_1 - \frac{q_1}{e}\right)^2 + \alpha_1\left(N_2 - \frac{q_2}{e}\right)^2 + \left(N_1 + N_2 + \frac{q_1}{e} + \frac{q_2}{e}\right)^2\right); \; E_2 \equiv \frac{e^2 C_{12}}{2(C_1 C_2 - C_{12}^2)}.$$

Our present goal is to find the charge state (N_1, N_2) of the system that corresponds to the minimum energy at given q_1, q_2, the ground state. Let us note the e-periodicity in both charges: if (N_1, N_2) is the ground state at q_1, q_2, the ground state at $q_1 + M_1 e, q_2 + M_2 e$ is $(N_1 + M_1, N_2 + M_2)$ and resembles the original state with respect to its charging properties. Therefore the regions where a given charge state occurs comprise a *periodic tiling* of the q_1–q_2 plane. Due to periodicity, it is enough to look at the tile corresponding to $(0, 0)$. To find the shape of this tile in the q_1–q_2 plane, we compare the energies of the $(0, 0)$ state with those made with the addition/extraction of one electron to/from each island:

$$E(0,0) < E(\pm 1, 0) \quad \rightarrow \quad \left| q_1 + \frac{q_2}{1+\alpha_2} \right| < \frac{e}{2};$$

$$E(0,0) < E(0, \pm 1) \quad \rightarrow \quad \left| q_2 + \frac{q_1}{1+\alpha_1} \right| < \frac{e}{2}.$$

Each condition from the four given above gives a line in the q_1–q_2 plane, so four lines carve a diamond in the q_1–q_2 plane. However, the periodically shifted diamonds overlap and thus do not provide the periodic tiling required! The point is that the charge state of two islands can be changed without adding electrons to the island, just by electron transfer between the islands. We check the energies of two states $(1, -1)$ obtained in this way as follows:

$$E(0,0) < E(1,-1), E(0,0) < E(-1,1) \rightarrow \frac{|\alpha_2 q_1 - \alpha_1 q_2|}{\alpha_1 + \alpha_2} < \frac{e}{2}.$$

This gives two extra lines that crop the diamond, resulting in a *hexagonal* tile (see Fig. 3.4). For $\alpha_{1,2} \gg 1$, the charges in the different islands almost do not interact, and the tile is almost a square $|q_1|, |q_2| < |e|/2$ (see Fig. 3.4(c)). In the opposite case of large C_{12}, the hexagons are extended in the $(-1, 1)$ direction and the tiling resembles a brick wall (see Fig. 3.4(d)).

Control question 3.1. What does the structure look like for (i) $C_1, C_2 \gg C_{12}$; (ii) $C_1 \gg C_2 \gg C_{12}$; (iii) $C_1 \gg C_{12} \gg C_2$?

Exercise 3.1. Consider a system of three identical islands in series; take into account the capacitances to the gate C_g and capacitances of the junctions C, but disregard the mutual capacitance of the two side islands. (i) Write down the charging energy in terms of three offset charges q_i. (ii) Starting from the state when all the islands are empty, increase q_1. At which value of charge q_1 does the first electron enter the system, and in which island will it reside?

3.1.4 Coulomb blockade in arrays

Technology allows us to fabricate large arrays of formally identical nanostructures. These nanostructures may be electrically connected to each other (as elements of microcircuits) or interact with each other by other means (as we will see below). The arrays can be regarded as *nanostructured materials*. Common materials are made of (periodically placed) atomic or molecular units, and their functionality is determined by solid state particles: electrons, holes, phonons, excitons, etc., localized on these units and/or propagating between them. The arrays are made of nanostructures – artificial atoms – and their properties are determined by the (excited) states of the nanostructures – artificial "particles." A potential advantage of arrays is that the nanostructures can be designed to achieve the desired properties, and the design is easier to implement than that of atoms.

The physics of the arrays can be quite complicated since many "particles" may be involved and their collective behavior may be important. Like common materials, the arrays can exhibit phase transitions and are generally described by models of (quantum) statistical physics. Since the focus of this book is on an individual nanostructure, statistical aspects of array physics are not discussed here at the quantitative level. We present, however, some qualitative discussion of this interesting physics [62].

Let us concentrate on the arrays of metallic islands connected by tunnel junctions. If the conductance of a single junction is much smaller than G_Q, the arrays are in Coulomb blockade regime. A state of the array is a charge state: it is defined by the number of excess charges in each island of the array, $\{N_i\}$. The energetics of charge states is governed by the same charging energy matrix as in Eq. (3.6). Let us consider uniform periodic arrays with capacitive connections between neighboring islands only. This implies that the important parameters – the charges induced in the islands – are the same throughout the array, $q_i = q$. The islands can either form a line (1d arrays) or cover a region in a plane (2d arrays). Three-dimensional (3d) arrays are difficult to fabricate in a controlled fashion; a "natural" realization of such an array is a thick film of granulated metal where the metal granules (islands) are separated by oxide. All islands usually share the same gate, so that the induced charge is the same for all islands of the array. In this case, the arrays are e-periodic with respect to this charge: changing it by e adds one excess electron to all islands of the array.

An important parameter is the ratio of junction capacitance C to the capacitance to the gate C_g. To reveal its physical significance, let us look at the charging energy matrix in a long uniform 1d array. We number the islands in order, so that island i is connected to islands $i \pm 1$. The matrix elements $E_{ij}^{(C)}$ in a uniform array depend on the distance $|i-j|$ between islands i, j only:

$$E_{ij}^{(C)} = E_C\, \mathrm{e}^{-|i-j|/\kappa};\ E_C = \frac{e^2}{2C_g\sqrt{4C/C_g+1}}; \tag{3.11}$$

$$\kappa^{-1} = \ln\left(\frac{\sqrt{4C/C_g+1}+1}{\sqrt{4C/C_g+1}-1}\right). \tag{3.12}$$

As we mentioned, the non-diagonal elements of the charging energy matrix denote the interaction between the charges situated at different islands. The parameter κ denotes the effective radius of this interaction in units of island spacing. For $C \ll C_g$, $\kappa \approx 1/\ln(C_g/2C) \ll 1$, and the charges hardly interact, even if they are at neighboring islands. In the opposite limit, $C \gg C_g$, $\kappa \approx \sqrt{C/C_g} \gg 1$. An extra charge placed on such an array is called a *charge soliton*, to stress the fact that its potential spreads over many islands.

Let us start with $q = 0$. (Because of the e-periodicity we can start with any multiple of e.) The charge state corresponding to the minimum energy of the array is very simple: $N_i = 0$. Let us increase the induced charge. It costs an energy E_C to add a charge into the array, so the first "particle" will enter the array only when it becomes energetically favorable. This occurs at some critical value q_c of the induced charge, defined by energy balance between the state with no (E_0) and one (E_1) extra charge. If the discrete charge appears in island j, the energy balance is given by

$$E_1 = E_0 \rightarrow E_{jj}^{(C)} = 2(q_c/e)\sum_i E_{ji}^{(C)}$$

$$\rightarrow E_C = 2(q_c/e)E_C\sum_i \mathrm{e}^{-|i|/\kappa} \rightarrow q_c = \frac{e}{2}\frac{1}{\sqrt{4C/C_g+1}}$$

Once q_c is crossed, discrete charges enter the array. Their concentration quickly grows until the average distance between the charges becomes of the order of the interaction radius κ.

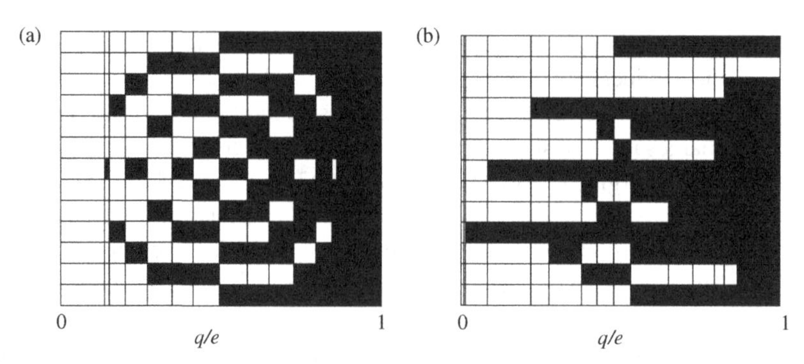

Fig. 3.5. The charges in a 13-island uniform array ($C/C_g = 3$). The rows in the diagrams correspond to the islands of the array. (a) Filling the island with extra charges upon increase of q. (Black pixels denote extra charge.) (b) The same in the presence of random background charges.

Then they start to repel each other, and this slows down further concentration growth. If the coordinates of these charges were continuous, the minimum-energy arrangement would be the periodic lattice of the charges, with the period given by their concentration, the so-called Wigner lattice. The coordinates are, however, discrete. There is an issue of commensurability of the periods of charge lattice and the original lattice of the islands: whether the period of the charge lattice is a multiple of the original period. With increasing induced charge, the system passes a series of commensurable configurations. Those with shorter period are more stable and correspond to wider intervals of the charge induced; $q = |e|/2$ is a symmetry point. At this point, exactly half of the islands have an extra charge, and the other half is empty. The charges thus form an ideal checkerboard pattern (obviously commensurable with the lattice of islands). The minimum-energy charge configuration at $|e|/2 < q < |e|$ is *mirrored* with respect to that at $q_m = |e| - q$: Empty (filled) islands of one configuration correspond to filled (empty) islands of the mirrored one. Instead of electron-like particles, one can think in terms of hole-like particles situated where $N = 0$. Upon further increase of induced charge, the concentration of holes decreases, with the last hole disappearing at $q = |e| - q_c$.

The qualitative part of this story is almost the same for both limits $C \gg C_g$ and $C \ll C_g$. The difference is the interval of induced charge where the story happens. If $C \ll C_g$ (weakly interacting charges), q_c is already very close to $|e|/2$, $|e|/2 - q_c \approx 2|e|C/C_q$. The charge state changes from $N = 0$ for all islands to $N = 1$ for all islands in a narrow interval $\simeq 4eC/C_g$ around $|e|/2$. In the opposite case, $C \gg C_g$ (charge solitons), $q_c \approx \sqrt{C_g/C}/4 \ll 1$, and transition from an empty to a completely filled state takes the whole interval of induced charge. In fact, the concentration of discrete charges in this case approximately follows the charge induced, with an accuracy of the order of q_c. Qualitatively, the above scenarios also apply to 2d arrays.

The above quantitative reasoning is illustrated with numerical results for a uniform 1d array of 13 islands with $C/C_g = 3$ (see Fig. 3.5). This gives $\kappa = 1.75$, so that the interaction radius is about 2 islands. The first charge enters the array at $q \approx 0.14$, which is very close to q_c of an infinite array. We note that the islands at the edge of the array miss a capacitive connection, so the charging energy is higher at the edges. This results in a potential profile for extra charges with a minimum in the center of the array. This is why the first charge appears in the center. The interval of q where one charge is energetically favorable is tiny; two charges arrive almost immediately. Their positions are determined by the balance of their mutual repulsion and the effect of potential profile, which pushes them to the center. Upon further increase of q, we observe a tendency for formation of commensurable charge lattices with a period of four and three islands. These lattices are, however, distorted. Finally, at $q \approx 0.4$, we reach the checkerboard pattern, and the picture is mirrored at $q > 0.5$.

The spoilers of array physics as described above are random background charges. In a Coulomb blockade system with several islands, one can, at least in principle, tune out the effect of the charges with gate electrodes. However, this implies that one should have at least one gate electrode per island, and such extensive wiring is not feasible for a big array. It is therefore reasonable to assume that, beside the charge induced by the gate, each island is shifted by a random uniformly distributed charge q_i. The effect of all background charges is to create a random potential relief for discrete charges. A typical amplitude of this random potential is the charging energy E_C. The effect of background charges is illustrated with a numerical example in Fig. 3.5. The random potential relief has minima near the fourth, seventh, and tenth islands and maxima near the twelfth, second, and eighth islands. The filling of the islands upon increasing the charge induced is mostly determined by this potential relief, and this masks any interaction effects. It is an important technological challenge to fabricate arrays with no background charges, since this problem hinders most of applications of complex Coulomb blockade systems.

We end by noting several analogies between the array physics and that of doped semiconductors. At $q = 0$ there is an energy cost E_C associated with adding either positive or negative charge, similar to the energy gap of an undoped semiconductor. The induced charge plays the role of a dopant forcing either positive or negative discrete charges into the array. Finally, background charges present the disorder that is also intrinsic for most doped semiconductors.

3.2 Single-electron transfers

The appeal of Coulomb blockade systems is that the electrons, under most general circumstances, are transferred one-by-one. This follows from the fact that the charge states are well defined, and most of the time the system is in a well defined charge state. This state, however, may change as a result of electron tunneling either to or from the leads from or to one of the islands or between the islands. Most probable changes are those involving only one tunneling electron – single-electron transfers.

The purpose of this section is to present the notion of single-electron transfers and to explain in general terms how one describes electron transport under these circumstances. The specific manifestations, signatures, and opportunities of single-electron transport will be discussed at length in Section 3.3. However, to avoid too much abstraction, we begin

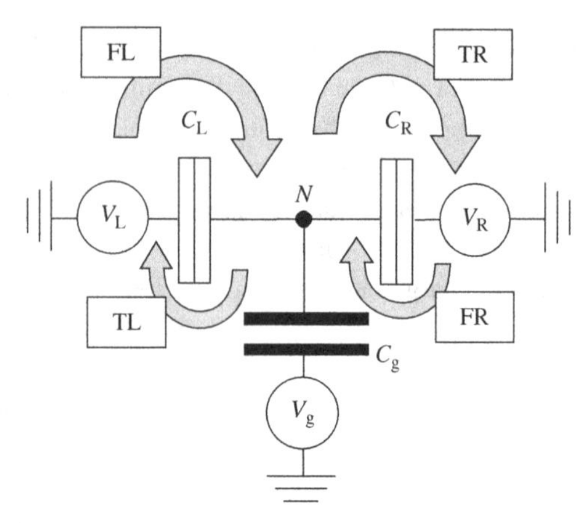

Fig. 3.6. Single-electron transistor (SET). The electron transport from left to right may be influenced by the gate voltage V_g. Arrows denote four possible single-electron transfers in the SET (FL = from the left; FR = from the right; TL = to the left; TR = to the right).

this section with the description of a particular device – the single-electron transistor. This device will provide the necessary illustrations of more general concepts discussed in this section, for example energetics of the transfers, the master equation describing transport in Coulomb blockade systems, and the energy dependence of tunneling rates.

3.2.1 Single-electron transistor

Let us introduce the generic design of a Coulomb blockade system: the so-called single-electron transistor, or SET.[1] Its design is very similar to that of a single-electron box. An important distinction is an extra transport electrode contacting the island (Fig. 3.6): like any transistor, a SET has three electrodes. Two electrodes, R and L, the source and drain, are the transport ones: one expects electric current from R to L if the voltage difference is applied. The electrons transferred are forced to move through the island. The third electrode is the gate: it does not have a direct electric contact either with the island or with the other two electrodes. To evaluate the electrostatic energy for a given charge configuration, we draw an equivalent capacitance circuit, with the capacitances C_R, C_L, C_g being present. The difference from the single-electron box seems minor: the electrostatic energy is given by the same expression:

$$E_{el} = E_C(N - q/e)^2; \ E_C = \frac{e^2}{2C_\Sigma} . \tag{3.13}$$

The difference is that the charging energy is inversely proportional to the total capacitance of the island $C_\Sigma = C_R + C_L + C_g$ contributed to by all capacitors. The same occurs with induced charge: it is contributed to by all three electrodes, $q = C_R V_R + C_L V_L + C_g V_g$.

[1] An alternative name frequently used in literature is single-electron *tunneling* transistor, or SETT.

Let us now inspect all possible processes of single-electron transfer in a SET. Suppose that initially there are N extra charges in the island: it is thus in charge state N. The number of charges can change in four ways. An extra electron can jump into the island either *from* the left or from the right electrode, and then N is replaced with $N + 1$. Alternatively, an electron can also leave the island either *to* the left or to the right, with the change $N \to N - 1$.

Each process at a given N is characterized by an energy difference between the final and initial states. It is important to recognize that this energy difference is not just the difference of electrostatic energy given by Eq. (3.13). Indeed, in the course of a transfer we add (extract) an electron to (from) the corresponding electrode i. The energy cost associated with this is given by eV_i ($-eV_i$ for extraction). We did not take that into account in the previous section since we considered all source electrodes to be at the same potential $V = 0$. In this case, we could figure out the equilibrium charge state to be one with minimum energy. It is very different for a SET. If $V_R \neq V_L$, the system cannot be in equilibrium. Nor can we speak of an energy of a given charge state! To understand this, let us start with a given charge state N, add an electron from the left, and extract it to the right. The sum of the energy differences of the two processes is $e(V_L - V_R)$, although the charge state after this operation is the same as before.

Taking this into account, we list the energy differences associated with all four processes:

$$\begin{aligned} \text{from the left: } \Delta E_{FL}(N) &= E_{el}(N+1) - E_{el}(N) - eV_L; \\ \text{to the left: } \Delta E_{TL}(N) &= E_{el}(N-1) - E_{el}(N) + eV_L; \\ \text{from the right: } \Delta E_{FR}(N) &= E_{el}(N+1) - E_{el}(N) - eV_R; \\ \text{to the right: } \Delta E_{TR}(N) &= E_{el}(N-1) - E_{el}(N) + eV_R. \end{aligned} \tag{3.14}$$

When we add (extract) a charge to (from) the island, the electrostatic energy change is given by

$$E_{el}(N \pm 1) - E_{el}(N) = 2(\pm N + 1/2 \mp q/e)E_C.$$

It is interesting to note that we can understand a lot about single-electron transport even if we do not know (or do not care) how the electrons are actually transferred. The only thing we have to know are the energy differences listed. Let us consider zero temperature so that the tunneling electron can gain no energy from heat. In this case, *an electron transfer can only occur if the corresponding energy difference is negative*, $\Delta E < 0$; indeed, the energy can only be dissipated, not gained. This simple inequality governs the transport in Coulomb blockade systems and defines the regions of the parameters – gate and transport voltages – where the electron transport proceeds in a certain way, the regions of different transport regimes.

The simplest, and most important, transport regime is the Coulomb blockade itself. In this case, in the given charge state N all four single-electron processes are forbidden. This requires all four energy differences to be positive:

$$\Delta E_{TL,FL,TR,FR}(N) > 0. \tag{3.15}$$

Since the charge state does not change, and no electron transfers take place, no current is expected in the Coulomb blockade regime.

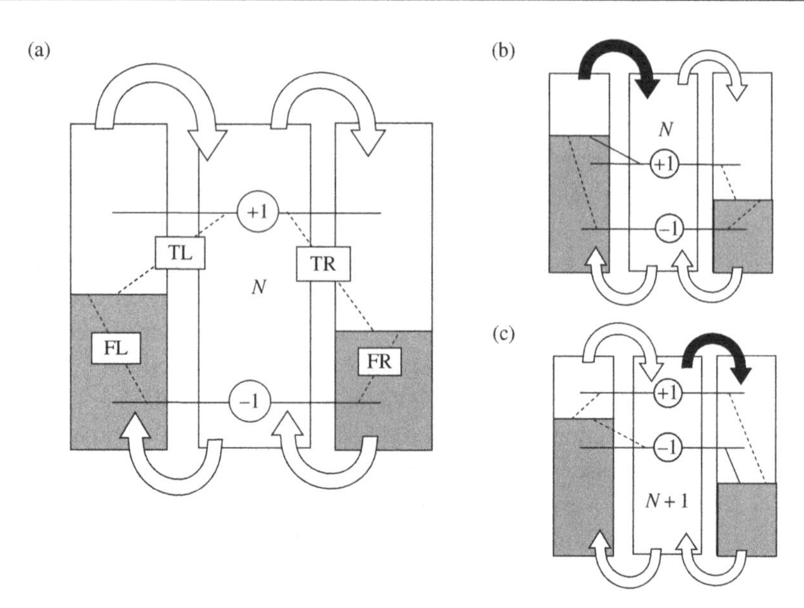

Fig. 3.7. **Energy differences of single-electron transfers in a SET. (a) The charge state *N* is blocked. (b), (c) One-by-one transfer. The SET switches between (b) *N* and (c) *N* + 1. Filled arrows indicate the transitions allowed in each state.**

Another example transport regime is when the electrons go one-by-one from the left to the right (see Fig. 3.7). The parameter region where such transport regime takes place is defined by the condition that only two processes are allowed: a transfer from the left at N and a transfer to the left at $N + 1$. Note this number $N + 1$: in order to go to the right, the extra electron must already be in the island! These two processes thus have negative energy differences given by

$$\Delta E_{\mathrm{FL}}(N) < 0; \; \Delta E_{\mathrm{TR}}(N+1) < 0,$$

whereas all other possible processes at N and $N + 1$ must be blocked and the corresponding energy differences must be positive. In particular,

$$\Delta E_{\mathrm{FL}}(N+1) > 0.$$

The latter condition ensures that the electrons really go one-by-one: the second electron cannot enter the island while the extra electron is there.

3.2.2 Coulomb diamonds in a SET

Let us now concentrate on parameter regions where Coulomb blockade of single-electron transport takes place. To reduce the number of parameters, let us assume symmetric capacitors $C_R = C_L$ and antisymmetric bias voltage $V_L = -V_R = V/2$ (in such a setup, the induced charge q conveniently does not depend on bias voltage). If $N = 0$, the Eqs. (3.15) become

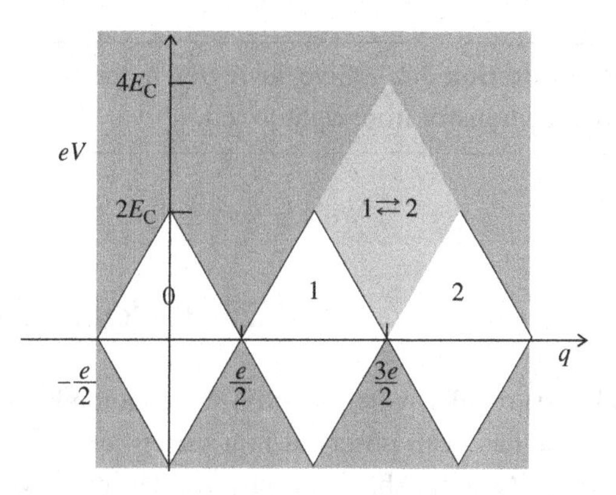

Fig. 3.8. Coulomb diamonds in a SET. The transport is blocked inside white diamonds. The light gray region denotes where the one-by-one transport cycle takes place.

$$\Delta E_{\rm FL}(N=0) = 2E_{\rm C}(1/2 + q/e) - eV/2 > 0;$$
$$\Delta E_{\rm TL}(N=0) = 2E_{\rm C}(-1/2 + q/e) + eV/2 > 0;$$
$$\Delta E_{\rm FR}(N=0) = 2E_{\rm C}(1/2 + q/e) + eV/2 > 0;$$
$$\Delta E_{\rm TR}(N=0) = 2E_{\rm C}(-1/2 + q/e) - eV/2 > 0.$$

In the V–q plane, each condition is represented by a slanted straight line. Four such lines bound a rhombus, a diamond, that is, a region in the V–q plane where all four conditions are fulfilled. For $N = \pm 1, \pm 2, ...$ we have the same diamond shifted by $\pm e, \pm 2e, ...$ along the q axis (see Fig. 3.8). At vanishing temperature there is no single-electron current inside the diamonds, and there is a noticeable current outside. This makes the diamond pattern easy to observe when sweeping the gate and bias voltages while measuring current in the SET. In reality, there is always some current inside the diamonds. Even if the temperature is negligible, a finite current may arise from the electron transfers involving *more than one* electron. These processes will be discussed in Section 3.4. Here we only need to know that such processes are much less probable than single-electron transfers provided $G \ll G_{\rm Q}$. Thus, there is enough current contrast between the inside and outside of the diamonds.

The diamond pattern is exactly periodic: all diamonds have the same shape, width, and height. The diamonds touch each other at $V = 0$. The touch points correspond to the values of the induced charge at which two charge states are degenerate having the same energy (compare with the crossing of parabolas in Fig. 3.2). Due to this degeneracy, in the vicinity of these points the Coulomb blockade is already lifted by a small bias voltage $V \ll E_{\rm C}$.

Exercise 3.2. Determine the shape of Coulomb blockade diamonds in the V–$V_{\rm g}$ plane for unequal capacitances $C_{\rm L} \neq C_{\rm R}$ assuming $V_{\rm L} = 0$, $V_{\rm R} = V$.

Where does the one-by-one transport regime occur? If we count the conditions to be fulfilled, we get eight, which might imply a complicated shape for this region. This is not the case: the corresponding region is also a diamond. It has the same shape as a Coulomb diamond and is shifted in such a way that it touches the Coulomb diamonds with N and $N + 1$.

Control question 3.2. Give the region in the V–q plane corresponding to the electron one-by-one transfer from right to left.

3.2.3 Coulomb shards

The (almost) regular pattern of SET-like Coulomb diamonds with equally spaced degeneracy points has been observed in a variety of nanostructures and is considered to be a fingerprint of Coulomb blockade physics. The point we discuss is that this pattern is characteristic to a SET only, and is by no means typical for more complicated Coulomb blockade systems.

More typical situations are presented in Fig. 3.9 for two- and four-island arrays. There are three immediately noticeable features in these diagrams. First, the pattern is no longer regular. The current threshold as a function of the charge induced by the gates zigzags wildly in a seemingly chaotic fashion. There are no nicely shaped diamonds; in fact, the regions look more like shards. This situation is sometimes called a *stochastic* Coulomb blockade. Secondly, there are obvious signatures of order in this chaos: the boundaries of the diamonds are straight lines, and there are many series of lines with the same slope. Finally, although the current threshold changes over a wide interval of voltages, it never hits the $V = 0$ axis. In distinction from the SET, the Coulomb blockade is never lifted at an arbitrary small voltage.

Let us discuss these features. The first feature, "chaoticity," turns out to be a matter of perception. The $(1 + 1)$-dimensional diagram in fact presents a cross-section of a structure which is regular and periodic in $(1 + n)$-dimensional space, n being the number of islands in the system. This follows from a fact mentioned in the preceding text: the electrostatic energy stays the same if $N_i \to M_i$ and $q_i \to q_i - eM_i$ for any integer M_i. The same applies to the energy differences for single-electron transfers, and therefore to the transport properties of the system, which have to be periodic in n-dimensional space of q_i. To give an example, for a two-island system each charge state is blocked within a three-dimensional body in (V, q_1, q_2) space. Due to periodicity, all bodies are of the same shape and size forming a regular structure. Figure 3.4 presents the cross-section of the structure in the $V = 0$ plane and is evidently regular and periodic. To construct a $(1 + 1)$-dimensional plot, we choose a line in (q_1, q_2) space: we change the gate voltage(s) in such a way that $q_2 = \text{const.} + \alpha q_1$ along this line. This usually corresponds to a realistic experimental situation where the voltage of a single gate is swept, but this gate voltage induces the charges in many islands. If α is a rational number, the resulting plot is periodic, with a period given by the denominator of α. For irrational α, the period tends to infinity, and the resulting plot is completely irregular. Even rational numbers with moderately large denominators give rise to visibly chaotic patterns. To stress this, Fig. 3.9(a) has been constructed for the rational value $\alpha = 3/5$. The whole picture is periodic in q_1 with a period $5e$. Since only one period has been shown, this regularity cannot be seen.

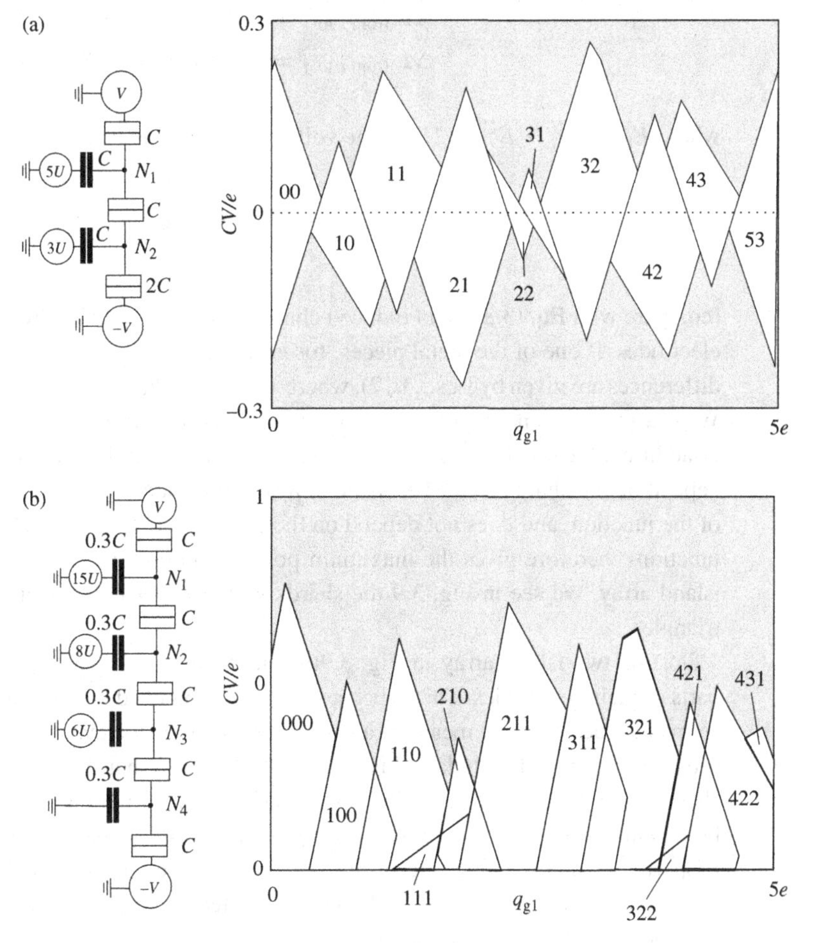

Fig. 3.9. **Coulomb shards in more complicated Coulomb nanostructures. The figures present the parameter regions (white) where a certain charge state is blocked for (a) two- and (b) four-island arrays biased as shown. There is some single-electron current in the gray-shaded regions. For a four-junction array, $N_4 = 0$ for all blocked states in the parameter region shown.**

This hidden order also explains the second feature: there is some visible order in the slopes of the boundary lines. To see the explicit origin of this, we recognize that the boundaries of each shard are, in principle, set by $2m$ conditions, m being the total number of junctions. Each junction connects two metal pieces i and j, i, j labeling both islands and electrodes. For each junction, two single-electron transfers must be blocked: from i to j and from j to i. Thus, each junction gives rise to two conditions as follows:

$$\Delta E_{\text{from } i \text{ to } j} > 0; \;\; \Delta E_{\text{from } j \text{ to } i} > 0. \tag{3.16}$$

Let us find the explicit form of these conditions for an arbitrary capacitance circuit. We first consider the case when both i and j are islands. The energy differences in this case are the differences of electrostatic energies given by Eq. (3.6):

$$\Delta E_{\text{from } i \text{ to } j} = E_{\text{th}} + e(V_j - V_i);$$
$$\Delta E_{\text{from } i \text{ to } j} = E_{\text{th}} + e(V_i - V_j); \tag{3.17}$$

where $E_{\text{th}} = E^{\text{C}}_{ii} + E^{\text{C}}_{jj} - 2E^{\text{C}}_{ij}$. The voltages of the islands are given by

$$eV_k = \sum_p E^{\text{C}}_{kp}(N_p - q_p/e) \tag{3.18}$$

(compare with Eq. (3.8)), and induced charges are contributed to by the gate and transport electrodes. If one of the metal pieces, for example j, is the transport electrode, the energy differences are given by Eqs. (3.17), where V_j is the voltage of the electrode and $E_{\text{th}} = E^{\text{C}}_{ii}$. We learn that the energy differences are linear functions of the gate and transport voltages. In addition, the coefficients in these linear relations depend on a junction only, and do not depend on the charge state. The slope of a shard boundary line is therefore a characteristic of the junction, and does not depend on the charge state bounded. The doubled number of junctions therefore gives the maximum possible number of shard sides: six for the two-island array. We see in Fig. 3.9 the shards with a fewer number of sides; some are even triangles.

For the two-island array in Fig. 3.9(a), the Coulomb shards do not overlap and share parts of their boundaries. This is because the states blocked in adjacent shards can be transformed to each other by means of a single-electron transfer, eventually via the junction that provides the shared boundary. Figure 3.9(b) illustrates that this is not true in general: the shards can overlap. In the overlap region, at least two different charge states are blocked, both being stable with respect to single-electron transfers. Such bistability is a remarkable property of Coulomb blockade systems, and can be used in principle for memory cells (see the discussion in Section 3.3). Thermally activated and multi-electron tunneling processes provide random switching between the blocked states.

To understand the third feature, let us concentrate on vanishing bias voltages. In this case, we can talk about energies of the charge states. We see that by varying induced charges we can always reach a degeneracy point where the energies of two charge states are the same. This corresponds to two shards sharing a part of the boundary. For a SET, this is sufficient to organize a transport cycle, since the degeneracy point happens to be at the boundaries corresponding to both the left and right junctions. Generally, this is not the case, and only one junction is involved, instead of, for example, the three needed to transfer an electron via a two-island array. For two degenerate states, the transport cycle at a vanishingly small voltage may only work if both transport electrodes are connected to the same island. In addition, the degenerate states should differ only by a single extra electron precisely in this island. The only other way to achieve a finite current at vanishing voltage is to have multiple degeneracy: three or more states at the same energy. A concrete example is the two-island system. We return to Fig. 3.4 and recognize that three-fold degeneracy should occur in the vertices of the hexagons where three of them come together. Indeed, at this point the current threshold goes to zero. A possible transport cycle is the successive electron transfer through all three junctions: $(0,0) \to (1,0) \to (0,1) \to (0,0)$.

Control question 3.3. For an array of three identical islands, can one achieve four charge states at the same energy by varying the voltage V and three background charges on the islands independently?

3.2.4 The master equation

The electron transport in the Coulomb blockade regime allows for very accurate quantitative evaluation based on a master equation. We note that whereas many parameters may characterize the electrons in a Coulomb nanostructure, only a few are relevant, these being the numbers of extra charges in the islands $\{N_i\}$. We can thus assume that the system at a given moment in time is in a certain charge state N. We stress that this charge state is not a *quantum* state, since the quantum coherence between the different states is absent, and the system can never be in a superposition of two charge states (see Section 5.5.1). The charge state can be regarded as a classical one, and this allows a simple classical approach.

The system goes from one state to another by means of electron transfers. These processes are random; if the condition $\Delta E < 0$ for a certain transfer is fulfilled, it does not occur instantly. A certain transfer is characterized by a *rate* $\Gamma(\{N_i\})$, the probability per unit time for this transfer to happen given the initial charge state $\{N_i\}$, so that it takes some random time, of the order of $1/\Gamma$, for this process to occur. This is why the dynamics of Coulomb blockade systems is random, and the approach to the dynamics has to be probabilistic. The system shall be described with probabilities $p_{\{N_i\}}(t)$ to be in the states $\{N_i\}$. The master equation is a straightforward balance equation for this probability distribution.

Let us write this equation for a SET first and then zoom in on various terms. A SET state is characterized by a single integer number of extra electrons N in the island, and the master equation for the probability distribution $p(N)$ reads as follows:

$$\frac{\mathrm{d}}{\mathrm{d}t}p_N(t) = -\left(\Gamma_{\mathrm{F}}(N) + \Gamma_{\mathrm{T}}(N)\right)p_N(t) + \Gamma_{\mathrm{F}}(N-1)p_{N-1}(t) + \Gamma_{\mathrm{T}}(N+1)p_{N+1}(t). \tag{3.19}$$

Here, $\Gamma_{\mathrm{F,T}}$ are the total rates of going from/to the island:

$$\Gamma_{\mathrm{F}} = \Gamma_{\mathrm{FL}} + \Gamma_{\mathrm{FR}};\ \Gamma_{\mathrm{T}} = \Gamma_{\mathrm{TL}} + \Gamma_{\mathrm{TR}}.$$

There is a time derivative on the left-hand side, and it presents the probability change per unit time. This change is contributed to by two groups of terms: outgoing and incoming. Outgoing terms are due to processes that transfer the system from state N to any other states. For our SET, these states can be either $N-1$ or $N+1$, the corresponding rates adding accordingly. The contribution to the derivative is negative and proportional to the probability of being in the initial state p_N; the system must be in this state for the process to occur.

Incoming terms are due to processes that transfer the system from any initial state to the state N. Their contribution to the derivative is positive and proportional to the probabilities of being in the initial states and the corresponding transition rates. Since we are dealing with single-electron transfers, the possible initial states are $N+1$ and $N-1$, with the corresponding rates given by $\Gamma_F(N+1)$ and $\Gamma_T(N-1)$.

If the parameters of the system (gate and bias voltages, capacitances) do not vary in time, the rates defined by these parameters do not vary either. Under these circumstances there is a stationary solution of the master equation, $p_N^{(0)}$. This can be readily obtained from the master equation with the left-hand side-set to zero ($\mathrm{d}p/\mathrm{d}t = 0$) and the condition that the probability of being in all possible states must total unity, $\sum_N p_N = 1$.

It is a separate problem to get a stationary current from $p_N^{(0)}$. Again we use the probability balance. For instance, the current through the left junction is contributed by "from the left" processes with positive sign and by "to the left" processes with negative sign, so that

$$I_L = e\sum_N [\Gamma_{FL}(N) - \Gamma_{TL}(N)]p_N. \tag{3.20}$$

Similarly, for the current through the right junction, we obtain

$$I_R = e\sum_N [\Gamma_{TR}(N) - \Gamma_{FR}(N)]p_N.$$

By virtue of charge conservation, $I_L = I_R$. This is automatically fulfilled if p_N is the stationary solution of the master equation.

Exercise 3.3. Prove that, in the stationary case, $I_L = I_R$.

How does this apply to a general Coulomb nanostructure? The charge states are many, so to simplify the notation we label them with Greek letters α, β. In principle, there can be transition rates from any α to any β, $\Gamma_{\alpha\to\beta}$. Making balance of the probabilities, we obtain the general form of the master equation as follows:

$$\frac{\mathrm{d}p_\alpha}{\mathrm{d}t} = -\sum_\beta \Gamma_{\alpha\to\beta}p_\alpha + \sum_\beta \Gamma_{\beta\to\alpha}p_\beta \equiv \sum_\beta \tilde{\Gamma}_{\alpha\beta}p_\beta. \tag{3.21}$$

As in the SET master equation, Eq. (3.19), the first group on the right-hand side is formed by outgoing processes, while the second group are incoming ones. The equation can be viewed as a linear algebra problem with a matrix $\tilde{\Gamma}$ and is solved with linear algebra methods. The stationary solution $p^{(0)}$ of the master equation is, in these terms, the eigenvector of the matrix $\tilde{\Gamma}$ that corresponds to the zero eigenvalue, i.e.

$$0 = \sum_\beta \tilde{\Gamma}_{\alpha\beta}p_\beta^{(0)}. \tag{3.22}$$

We note that each rate $\Gamma_{\alpha\to\beta}$ enters the matrix twice: it contributes to the diagonal term $\tilde{\Gamma}_{\alpha\alpha}$ with the minus sign and to the off-diagonal term $\tilde{\Gamma}_{\beta\alpha}$ with the positive sign. This guarantees that $\tilde{\Gamma}$ has a zero eigenvalue.

This form is, of course, too general as we are dealing with single-electron transfers. The transition between α and β can only happen if these states can be transformed one to another by such transfers. Each transfer is associated with a junction. Each junction c that connects islands i and j gives rise to two rates: an electron can be transferred in either the forward (from i to j) or the backward direction. These rates, $\Gamma_f^{(c)}(\alpha)$ and $\Gamma_b^{(c)}(\alpha)$, depend on the initial state α. The final state β is completely determined by the direction of the transfer: $N_i^{(\beta)} = N_i^{(\alpha)} \mp 1$; $N_j^{(\beta)} = N_j^{(\alpha)} \pm 1$; $N_k^{(\beta)} = N_k^{(\alpha)}$, if $k \neq i, j$. The same is true for a junction that connects island i and transport electrode j, the only difference being that the electron transferred to/from the electrode does not change the charge state, $N_i^{(\beta)} = N_i^{(\alpha)} \mp 1$. Here the upper (lower) sign is for the forward (backward) rate. If we know $p_\alpha^{(0)}$, we can compute the current via each junction c using a relation similar to Eq. (3.20):

$$I^{(c)} = \sum_\alpha \left(\Gamma_\mathrm{f}^{(c)}(\alpha) - \Gamma_\mathrm{b}^{(c)}(\alpha) \right) p_\alpha^{(0)}. \tag{3.23}$$

3.2.5 FCS of charge transfers and master equation

We stress that the above equations only apply to the averaged current, not to the instant one. It is clear that the charge is transferred in random single-electron shots. The typical estimation of the current noise is the Schottky value, $S(0) = 2eI$, corresponding to uncorrelated, Poisson statistics of electron transfers, as discussed in Section 1.4. However, the coefficient is not quite the same. The electron transfers in the Coulomb blockade regime correlate. Take, for instance, the one-by-one regime we found in a SET. The transfer through the left junction can only occur after the transfer through the right one, and vice versa. This makes it relevant to investigate the statistics of electron transfers in the Coulomb blockade regime as we did in Section 1.4 for electron transfers in the framework of the scattering approach.

The full counting statistics can be obtained with a minor modification of the master equation approach. Suppose we wish to study the statistics of the current through junction c. As discussed in Section 1.4, the statistics are defined by the generating function of the charges transferred, $\Lambda(\chi)$. To obtain this function, we modify the tunneling rates. As mentioned, each transfer rate via this junction enters $\tilde{\Gamma}$ twice, contributing to diagonal and non-diagonal elements of the matrix $\tilde{\Gamma}$. In non-diagonal elements, we modify these rates as follows:

$$\Gamma_\mathrm{f}^{(c)}(\alpha) \to e^{i\chi} \Gamma_\mathrm{f}^{(c)}(\alpha); \; \Gamma_\mathrm{b}^{(c)}(\alpha) \to e^{-i\chi} \Gamma_\mathrm{b}^{(c)}(\alpha). \tag{3.24}$$

We do not modify these rates in the diagonal elements. Thus the modified matrix $\tilde{\Gamma}(\chi)$ has the following set of eigenvalues:

$$\lambda^{(k)} p_\alpha^{(k)} = \tilde{\Gamma}(\chi)_{\alpha\beta} p_\beta^{(k)}. \tag{3.25}$$

We focus on the eigenvalue $\lambda^{(0)}(\chi)$ with the real part closest to zero. Since at $\chi \to 0$ the modification of $\tilde{\Gamma}$ vanishes, by virtue of Eq. (3.22) $\lambda^{(0)}(\chi) \to 0$ at $\chi \to 0$.

It can be proven that

$$\Lambda(\chi, \tau) = \exp\left(-\lambda^{(0)}(\chi)\tau \right); \tag{3.26}$$

this enables one to compute statistics with the master equation approach. Taking the derivative of Λ at $\chi \to 0$ reproduces Eq. (3.23) for average current. This method can be readily generalized for many junctions. In this case, counting fields are introduced for each junction, and the rates of transfer via a certain junction are modified with the corresponding counting field. A certain elegance of the method is that one does not have to compute the eigenvectors of $\tilde{\Gamma}$ – one eigenvalue suffices.

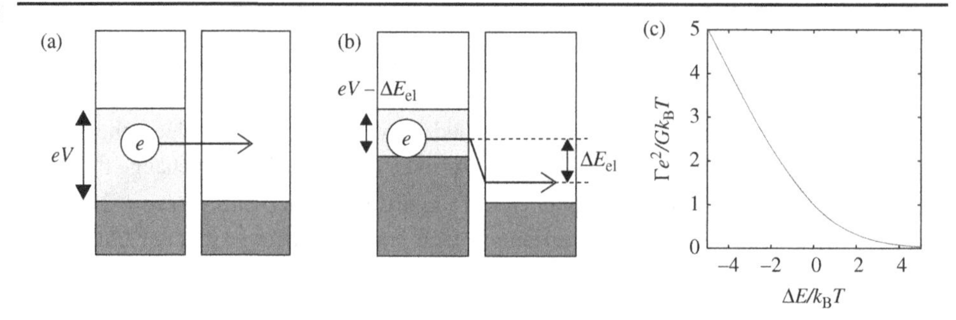

Fig. 3.10. Tunneling rates. (a) Without Coulomb blockade, an electron preserves its energy when tunneling. (b) With Coulomb blockade, it has to pay charging energy ΔE_{el}, this reduces the energy strip available for tunneling by a corresponding amount. (c) Energy dependence of the rate at finite temperature.

3.2.6 Tunneling rates

To use the master equation for the quantitative description of single-electron transport, we must know the rates $\Gamma(\{N_i\})$ as functions of the gate and bias voltages. The good news is that the rates depend on macroscopic parameters only – resistances and charging energies – that is, capacitances in the circuit.

To see this, we start from a tunnel junction between two reservoirs and forget about Coulomb blockade for a moment. If a voltage difference much bigger than $k_B T$ is applied to the reservoirs, the electrons in the energy strip of width eV can go from the left to the right (Fig. 3.10). The rate determines the number of electron transfers per second, and is related to the current by $\Gamma = I/e$. Since the electron transfers at different energies are independent, both the rate and the current are proportional to the width of the energy strip available for transfers, that is, to the voltage applied. By definition of conductance, $I = GV$. Therefore, $\Gamma = eVG/e^2$, and this is the forward rate. A rate is never negative, so for the forward rate we have $\Gamma_f = eV\Theta(eV)G/e^2$, whereas the rate of transfers in the backward direction is given by $\Gamma_b = -eV\Theta(-eV)G/e^2$.

Without a Coulomb blockade, the energy of the tunneling electron is the same on the left and on the right. What happens if a Coulomb blockade is present? Suppose the right-hand electrode is the island of a SET. In this case, the tunneling electron must perform some extra work ΔE_{el} to charge the capacitors of the SET. Therefore if it starts with energy E on the left, its energy on the right is $E - \Delta E_{el}$. For this tunneling to occur, the state at this energy has to be empty. For positive (negative) ΔE_{el} this reduces (increases) the width of energy strip available for tunneling by a corresponding amount. Therefore, $\Gamma = (eV - \Delta E_{el})\Theta(eV - \Delta E_{el})G/e^2$. We can merge eV and ΔE_{el} if we recall the definition of energy difference ΔE discussed above. We thus come to a straightforward formula, valid at zero temperature:

$$\Gamma = \frac{G}{e^2}(-\Delta E)\Theta(-\Delta E).$$

We already know that $\Gamma = 0$ if $\Delta E > 0$. Now we learn that the rate increases linearly with decreasing negative ΔE.

There is some implicit assumption hidden in the above straightforward reasoning. Namely, the charges in all the capacitors of the SET are supposed to adjust instantly to the new value of the extra charges in the island. They must at least do it faster than the time it takes an electron to accomplish the tunneling from left to right. The recharge time, a typical RC-time of this process, is indeed zero if the resistances of the island material and the lead are precisely zero. In reality these are small but finite, so the RC-time is also finite. If the electron tunnels faster than this time, the charges do not adjust. The electron would have to pay *more* energy than ΔE. It turns out that the proper estimation of tunneling time is given by the uncertainty relation, $\hbar/\Delta E$. In this case, the recharge time is bigger than the tunneling time provided the resistance of the lead/island is much smaller than $1/G_{\rm Q}$. We discuss this in detail in Section 6.5.

To calculate the rate at finite temperature, we note that the tunneling probability comes with the blocking factor $f(E)(1 - f(E')$, E, E' being energies before and after tunneling (see Eq. (1.54)). This reflects the probability of the state being filled at E and empty at E'. The rate at finite temperature is contributed to by all energies:

$$\Gamma = \frac{G}{e^2}\int \mathrm{d}E\ f(E)(1 - f(E - \Delta E)) = \frac{G}{e^2}\frac{\Delta E}{\exp\left(\Delta E/k_{\rm B}T\right) - 1} \tag{3.27}$$

The rate (see Fig. 3.10) is enhanced in comparison with its zero-temperature value. Most importantly, there is a finite rate at positive energy differences: the energy required is gained from thermal fluctuations. Quite generally, the rates satisfy the so-called detailed balance condition, given by

$$\Gamma(\Delta E)/\Gamma(-\Delta E) = \exp(\Delta E/k_{\rm B}T)\,. \tag{3.28}$$

This guarantees that, in the absence of bias voltages, the probability distribution assumes the Boltzmann form required by the laws of thermodynamics:

$$p(\{N_i\}) \propto \exp\left\{-\frac{E_{\rm el}(\{N_i\})}{k_{\rm B}T}\right\}. \tag{3.29}$$

The master equation with these rates provides the quantitative description of single-electron transport.

3.2.7 Tunneling Hamiltonian

We managed to obtain the tunneling rates without any calculation, just by pure reasoning. A "scientific" disguise of this reasoning is known as the *tunneling Hamiltonian method*. This method is convenient when analyzing more complicated tunneling processes.

Let us label all the electron states in the left (right) reservoir by l (r). In the first approximation, no tunneling occurs, so that these states are completely independent. The idea of the tunneling Hamiltonian method is to regard tunneling as a perturbation described by the term

$$\hat{H}_{\rm int} = \sum_{l,r} T_{lr}\hat{a}_l\hat{a}_r^\dagger + \text{h.c.} \tag{3.30}$$

Each term annihilates an electron on the left (right) side and creates an electron on the right (left) side, thereby transferring it from the left to the right (from the right to the left).[2]

The general rules of quantum mechanics imply that the rate between the initial state i and the final state f should be calculated from the Fermi Golden Rule:

$$\Gamma_{if} = \frac{2\pi}{\hbar} \left| \left\langle i \left| \hat{H}_{\text{int}} \right| f \right\rangle \right|^2 \delta(E_f - E_i). \tag{3.31}$$

The total rate from the left to the right is a sum over all possible one-electron states l, r, and is given by

$$\Gamma = \frac{2\pi}{\hbar} \sum_{l,r} |T_{lr}|^2 f(E_l)(1 - f(E_r))\delta(E_r - E_l + \Delta E).$$

It would be a good practical result if we knew T_{lr}. One could develop a microscopic model to derive them for a concrete nanostructure, but such a model, if even possible, would be necessarily restricted to a specific setup. Luckily, one does not have to know the concrete values of T_{lr}: there is a trick we can use to get around this. Let us introduce extra integration variables $E_{\text{L,R}}$ such that we can make the following replacement:

$$\sum_{l,r} \rightarrow \int \mathrm{d}E_{\text{R}}\, \mathrm{d}E_{\text{L}} \sum_{l,r} \delta(E_{\text{R}} - E_r)\delta(E_{\text{L}} - E_l).$$

Comparing this with the single-electron rate derived in Eq. (3.27) we find that

$$\sum_{l,r} |T_{rl}|^2 \delta(E_r - E_{\text{R}})\delta(E_l - E_{\text{L}}) = \frac{G}{2\pi^2 G_{\text{Q}}}. \tag{3.32}$$

This relates this particular combination of T_{rl} to the junction conductance and enables the practical use of the tunneling Hamiltonian method. If the result of a calculation can be expressed in terms of this combination, one just puts the junction conductance in place. The shorter version of the same trick is to express the combination in terms of the rate, rather than the conductance:

$$\sum_{l,r} |T_{lr}|^2 f(E_l)(1 - f(E_r))\delta(E_r - E_l + \Delta E) = \frac{\hbar}{2\pi}\Gamma(\Delta E), \tag{3.33}$$

the rate being given by Eq. (3.27). We will use yet another trick, which is useful if the level spacing in the islands is not negligible, $\delta_{\text{S}} \neq 0$. Let us evaluate the tunneling rate from a *given* state l to all possible states r. The Golden Rule readily yields

$$\Gamma_{\text{given}} = \frac{2\pi}{\hbar} \sum_{r} |T_{rl}|^2 \delta(E_l - E_r). \tag{3.34}$$

We will see in Chapter 5 that this rate, in principle, strongly fluctuates from state to state. However, if there are *many* states involved in electron transfer, the fluctuations are averaged away. To relate this averaged rate to junction conductance, we note that if the energy difference ΔE is available for transport, $\Delta E \delta_{\text{S}}$ discrete states are involved. Therefore

$$\left\langle \Gamma_{\text{given}} \right\rangle = (G/e^2)\delta_{\text{S}} \tag{3.35}$$

[2] The numbers T_{lr} of the Tunneling Hamiltonian method are *not* transmission amplitudes introduced in Chapter 1, although they may seem to perform the same function. Even their dimensions do not match.

and

$$\sum_r \left\langle |T_{lr}|^2 \right\rangle \delta(E - E_r) = \frac{G}{2\pi^2 G_Q} \delta_S. \tag{3.36}$$

3.3 Single-electron transport and manipulation

In this section, we aim for concrete manifestations of the Coulomb blockade phenomena in transport. We analyze in detail the I–V curves of the SET transistor and address noise in the device. The orderly and quantized nature of single-electron transport opens up the possibility of controlling single-electron transfers and manipulating charge states. We look into classic examples of this manipulation: memory cell, turnstile, and pump.

3.3.1 Experiment

Let us describe a pioneering experiment in single-electron transport: the first SET ever fabricated [63]. The device was made by two-step deposition of aluminum layers ($\approx$14 nm thick) on a silicon substrate. The island is fabricated in the first step. The subsequent oxidation covers the island with a native oxide barrier. During the second step, long strips of approximately 100 nm width – the electrodes – are deposited, contacting the island where they overlap it. Since the island is protected by oxide, the resulting junctions are of a tunnel nature. The technological challenge was to make these tunnel junctions of sufficiently small area. Indeed, to observe Coulomb blockade, the charging energy ($\simeq$ 0.4 meV) should exceed at least several times the thermal energy ($k_B T \simeq 0.1$ meV). For aluminum oxide junctions, $Ae/C \approx 100\,\text{mV}\,\text{nm}^2$, A being the junction area, so that the total area of the junctions connected to the island should not exceed 200 nm^2. This requires alignment with nanometer precision. This problem has been solved using a special technique put forward by the authors: they made a deposition mask using e-beam lithography and evaporated aluminum twice through the same mask at two different angles, thereby achieving precise alignment of the electrodes and the island. This method produces dummy copies of the electrodes in the first step and the island in the second step. These copies are in almost perfect electric contact with the originals since the tunnel junctions separating them are of very large area. Thus, one can just forget that there are copies. This technique is still commonly used to fabricate metallic Coulomb blockade systems.

The experimental results are presented in Fig. 3.11. At voltages above 0.5 mV, the I–V curves are almost linear, with differential conductance about $0.5G_Q$. At voltages below 0.5 mV, the current is suppressed, which is a combined effect of Coulomb blockade and superconducting energy gap (aluminum is superconducting at this temperature). The suppression is big but not absolute, since the temperature is still of the order of E_C and the conductance is of the order of G_Q. However, the most important feature observed is not the

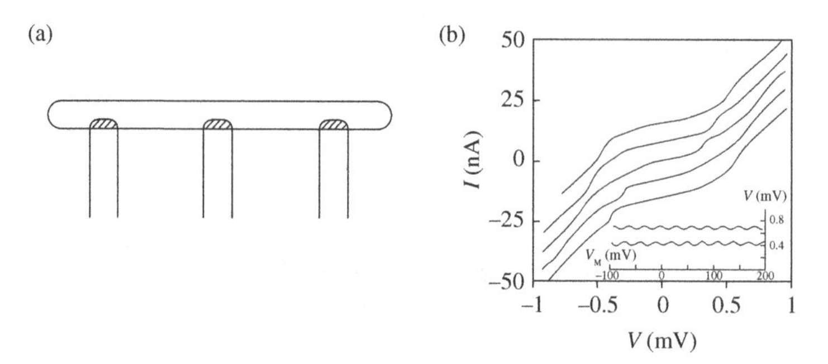

Fig. 3.11. Single-electron transistor (adapted from Ref. [63]). (a) Top view of three transport electrodes overlapping a Coulomb island. Length of the island is about 1 μm; the shaded areas of overlap are tunnel junctions. One of the electrodes is not used. (b) Periodic dependence on gate voltage. The I–V curves of the device are shown at equally spaced gate voltages covering 5/6 of the cycle. Curves are offset by increments of 7.5 nA. Inset: voltage response on the gate voltage at two fixed currents through the SET (10.5 and 26 nA).

blockade itself – rather, it is a pronounced periodic dependence of I–V curves on gate voltage V_M applied to the reverse side of the substrate. The period corresponds to the change in induced charge q by e. This experiment has provided the first direct evidence of charge quantization in artificially made nanostructures. Even the first SET fabricated was an astonishingly good *electrometer*, being sensitive to a fraction of elementary charge induced to its island.

3.3.2 SET in two states

To begin our detailed study of transport properties of SETs, we consider first an analytically tractable regime where the SET is, with overwhelming probability, in one of the two states that differ by one extra electron. To be specific, we concentrate on the states $N = 0$ and $N = 1$. From the analysis of Section 3.2 we see that this approximation is relevant in a parameter region composed of two Coulomb diamonds "0" and "1" and two adjacent one-by-one diamonds (see Fig. 3.8). In the master equation, Eq. (3.19), we thus may set $p_N = 0$ for $N \neq 0, 1$ to obtain

$$\frac{\mathrm{d}p_0}{\mathrm{d}t} = -\Gamma_\mathrm{T}(0)p_0 + \Gamma_\mathrm{F}(1)p_1;$$
$$\frac{\mathrm{d}p_1}{\mathrm{d}t} = -\Gamma_\mathrm{F}(1)p_1 + \Gamma_\mathrm{T}(0)p_0, \tag{3.37}$$

so that for the stationary probability and current we readily obtain

$$p_0^{(0)} = \frac{\Gamma_\mathrm{T}(0)}{\Gamma_\mathrm{T}(0) + \Gamma_\mathrm{F}(1)}; \; p_1^{(0)} = \frac{\Gamma_\mathrm{F}(1)}{\Gamma_\mathrm{T}(0) + \Gamma_\mathrm{F}(1)}; \tag{3.38}$$

$$I/e = \Gamma_\mathrm{TL}(0)p_0^{(0)} - \Gamma_\mathrm{FL}(1)p_1^{(0)}$$
$$= \frac{\Gamma_\mathrm{TL}(0)\Gamma_\mathrm{FR}(1) - \Gamma_\mathrm{FL}(1)\Gamma_\mathrm{TR}(0)}{\Gamma_\mathrm{T}(0) + \Gamma_\mathrm{F}(1)}. \tag{3.39}$$

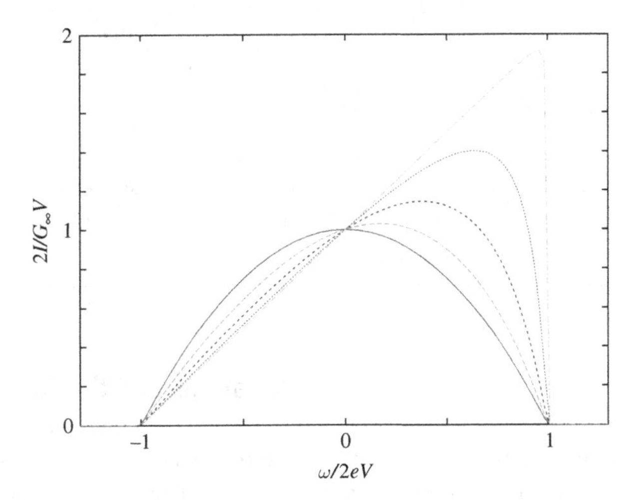

Fig. 3.12. **Current in a SET within a one-by-one diamond at $eV \gg k_BT$. The symmetric curve is for equal conductances of the junctions, $G_R = G_L$. Increasingly asymmetric curves correspond to $G_L/G_R = 2, 5, 20, 1000$.**

To quantify the rates, we recall the energy differences given by Eq. (3.14):

$$\Delta E_{TL}(0) = -\Delta E_{FL}(1) = w - eV/2; \tag{3.40}$$

$$\Delta E_{TR}(0) = -\Delta E_{FR}(1) = w + eV/2; \tag{3.41}$$

$$w \equiv 2E_C(1/2 - q/e) - e(V_R + V_L)/2. \tag{3.42}$$

Here we conveniently introduce an energy difference w that allows us to treat a non-symmetric SET with compact notation. For a symmetric asymmetrically biased SET, w does not depend on bias voltage, and, in any case, $dw/d(eV_g) = C_g/C_\Sigma$. The four-diamond region where we work is bounded by conditions $|w| + |eV/2| < E_C$. Let us assume $k_BT \to 0, eV > 0$, so that we are in the upper one-by-one diamond where $|w| > eV/2$. The only non-vanishing rates are $\Gamma_{TL}(0) = G_L(eV/2 - w)/e^2$ and $\Gamma_{FR}(1) = G_R(eV/2 + w)/e^2$, and the current is given by

$$I = \frac{G_\infty V}{2} \frac{1 - (2w/eV)^2}{1 - (2w/eV)(G_L - G_R)/(G_L + G_R)}, \tag{3.43}$$

where $G_\infty \equiv G_R G_L/(G_L + G_R)$ is the high-voltage differential conductance of the SET given by Ohm's law. The current thus vanishes at the boundaries separating the one-by-one diamond from adjacent Coulomb diamonds and depends linearly on V and w near the boundaries (see Fig. 3.12).

An interesting regime occurs near the degeneracy point $w = 0, V = 0$. Even if $k_BT \ll E_C$, the thermal activation of the rates becomes important at sufficiently small $w, eV \simeq k_BT$. All four rates are in play, and this disrupts the strict sequence of electron transfers typical in the one-by-one regime. At vanishing bias voltage the current disappears while the conductance remains finite. The conductance peaks sharply at $w = 0$, the width of the peak being determined by k_BT. The shape of the peak is given by

$$\frac{G}{G_\infty} = \frac{w/k_B T}{2\sinh(w/k_B T)}, \tag{3.44}$$

with a maximum conductance of $G_\infty/2$.

Exercise 3.4. Prove Eq. (3.44) for the peak shape. For this: (i) find the rates at finite temperature from Eq. (3.27); (ii) substitute those into Eq. (3.39) for the current; (iii) expand the result in small bias voltage.

3.3.3 Coulomb oscillations and the Coulomb staircase

Because of the e-periodicity, the zero-bias conductance peaks described above arise any time the charge states N and $N+1$ are degenerate. A periodic pattern of equidistant peaks is thus observed, with a period corresponding to $\Delta q = e$, so that $\Delta V_g = e/C_g$. With increasing temperature, the peaks first become wider and then merge at $k_B T \approx 0.15 E_C$ (Fig. 3.13). However, the periodic modulation – the Coulomb oscillations – remains visible up to $k_B T \simeq 0.7 E_C$. Coulomb oscillations serve as an experimental signature of the Coulomb blockade effects in various SET-like nanostructures. We note another feature visible in Fig. 3.13. At higher temperatures, the conductance, although not sensitive to the gate voltages, remains sensitive to temperature, saturating to G_∞ up to $k_B T \approx 4E_C$. Thus, the SET in this temperature range can be used as a very precise *thermometer*.

Control question 3.4. A periodic modulation of conductance by gate voltage in the range of 20–35 Ω^{-1} is observed during an experiment, and is interpreted as Coulomb oscillations. Would you support this interpretation?

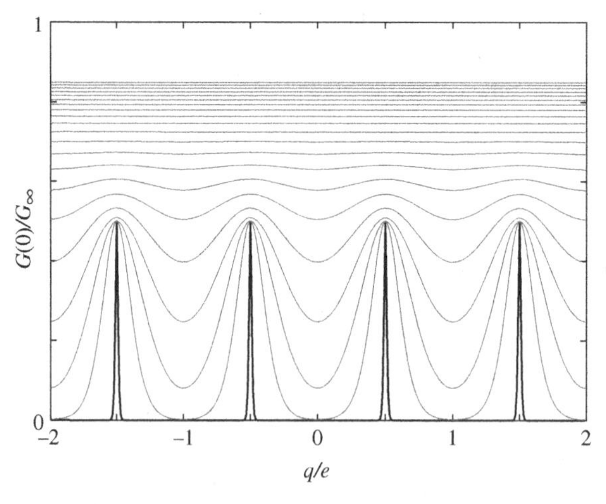

Fig. 3.13. Zero-voltage conductance of a symmetric SET versus the charge induced at different temperatures; $k_B T/E_C = 0.01$ for the thick curve and ranges from 0.1 to 2.0 with step 0.1 for subsequent curves.

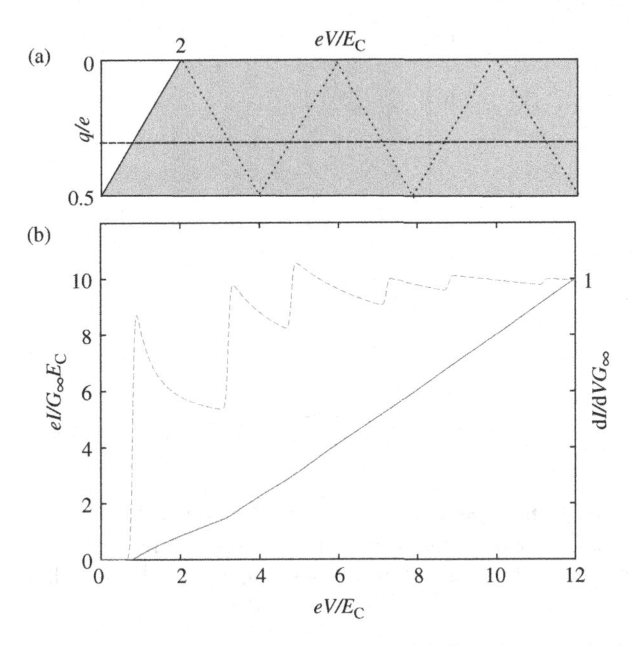

Fig. 3.14. **Coulomb staircase for $G_L = G_R$ at vanishing temperature. (a) There is a cusp in the current at special voltages shown by dotted lines (see Eq. (3.45)). The white region is the Coulomb blockade diamond. (b) The current and differential conductance versus V at $q = 0.3e$ (dashed line in (a)). The cusps in the current are almost invisible in the I–V curve (solid), although they are still noticeable in the differential conductance (dashed curve); $k_BT = 0.01E_C$.**

Another signature of Coulomb blockade phenomena is observed when one increases the bias voltage at fixed gate voltage. The number of charge states available for the SET increases with increasing voltage. At $k_BT \ll eV$ this results in a singularity of I–V characteristics at special values of bias voltage – this pattern is called a *Coulomb staircase*. At these values, a new single-electron transfer becomes possible, setting the system to a newly available charge state. The condition for this is that one of the energy differences involved equals zero:

$$\Delta E_{\mathrm{TL,TR,FL,FR}}(N) = 0.$$

For a symmetric SET, these conditions give rise to two series of special voltages:

$$eV_{\mathrm{special}} = 4E_C\left(M + \frac{1}{2} \pm \frac{q}{e}\right), \tag{3.45}$$

where M is integer and $\pm$ corresponds to a new transfer through the right (left) junction. To understand this qualitatively, let us choose $q = 0.3$ (Fig. 3.14(a)). At zero voltage, the blockade takes place and the only charge state available is $N = 0$. At $eV > 0.8E_C$ the electron transfer through the left junction becomes possible and electrons go one-by-one from the left to the right. The available states are $N = (0, 1)$. At higher voltage $eV = 3.2E_C$, it becomes possible to tunnel via the right junction while in state "0" going to

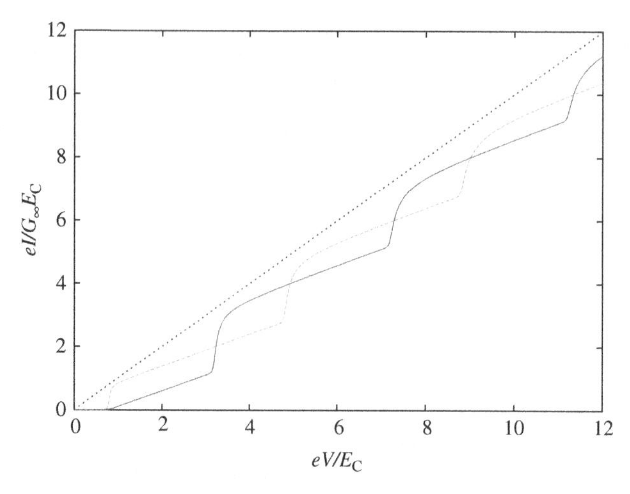

Fig. 3.15. Coulomb staircases with $G_L = 100G_R$ and $G_L = 0.01G_R$ ($q = 0.3$) at vanishing temperature for a symmetric asymmetrically biased SET. The cusp in the current is followed by the step-like increase. Only one series of special voltages is visible for each curve. Dotted line: Ohm's law.

state "-1," so that $N = (-1, 0, 1)$. The next two special voltages, $eV = 4.8E_C$ and $eV = 7.2E_C$, extend this set of states to $N = (-1, 0, 1, 2)$ and $N = (-2, -1, 0, 1, 2)$, respectively.

In principle, there is a cusp in the current – a jump in the differential conductance – at any special voltage for any SET. However, if the conductances of two tunnel junctions are comparable, most cusps remain as a theoretical possibility only. The corresponding conductance jumps are numerically small, and a small but finite temperature quickly washes them away (Fig. 3.14).

The Coulomb staircase reveals its full beauty if the junction conductances differ by more than an order of magnitude. This is illustrated by Fig. 3.15: the cusps are pronounced and develop into characteristic smooth steps separated by $4E_C/e$. Only one cusp series is visible, corresponding to new transfers via the most conducting junction.

To understand why this happens, we note that in the first approximation we can disregard the tunneling through the least-conducting junction (for concreteness, the right-hand one). In this case, the charge state is set by transfers through the most-conducting junction. Therefore,

$$N_c = \left[\frac{1}{2} + \frac{q}{e} + \frac{eV}{4E_C} \right] .$$

As for the current, it is contributed to by transfers through the least-conducting junction. Since the SET is almost always in state N_c, the current is just given by the rate $\Gamma_{TR}(N_c)$,

$$I = G_R \left(V/2 + 2(E_C/e)(N_c - 1/2 - q/e) \right) .$$

The number of electrons N_c increases by one at special voltages $eV_c = 4E_C(1/2 + N_c - q/e)$; therefore the current jumps by $2G_R E_C/e$ at these values of voltage.

Exercise 3.5. Find the I–V characteristics of a SET with $G_R \gg G_L$ in the vicinity of a special voltage V_c. (i) Show that the SET in this region can only be in two charge states, N_c and N_c+1. (ii) Determine two possible rates. (iii) Finally, make use of Eq. (3.39).

At greater voltages, above E_C/e, the current tends to Ohm's law, $I = G_\infty V$, since the large voltage drops involved mask any manifestation of single-charging phenomena. It is interesting to note that the best fit is provided by an offset Ohm's law $I = G_\infty(V - 2E_C/e \,\mathrm{sign} V)$. This is used for estimating the charging energy.

3.3.4 FCS and noise in SETs

To characterize the FCS in a SET, we use the method outlined in Section 3.2 and summarized by Eqs. (3.25) and (3.26). For simplicity, we consider only the situation described at the beginning of this section: the SET can only be in two possible charge states. The FCS in this case are readily given by an eigenvalue of a 2×2 matrix (see Eqs. (3.25) and (3.37)):

$$\lambda \begin{pmatrix} p_0 \\ p_1 \end{pmatrix} = \begin{bmatrix} -\Gamma_T(0) & \tilde{\Gamma}_F(1) \\ \tilde{\Gamma}_T(0) & -\Gamma_F(1) \end{bmatrix} \begin{pmatrix} p_0 \\ p_1 \end{pmatrix}, \tag{3.46}$$

where $\tilde{\Gamma}$ are the rates modified according to (3.24),

$$\tilde{\Gamma}_F(1) = \Gamma_{FL}(1)e^{i\chi} + \Gamma_{FR}(1),$$
$$\tilde{\Gamma}_T(0) = \Gamma_{TL}(0)e^{-i\chi} + \Gamma_{TR}(0).$$

We pick out an eigenvalue that vanishes at $\chi \to 0$, given by

$$\lambda(\chi) = \frac{\Gamma_T(0) + \Gamma_F(1)}{2}\left(-1 + \sqrt{1 + A^+(e^{i\chi} - 1) + A^-(e^{-i\chi} - 1)}\right); \tag{3.47}$$
$$A^- \equiv \frac{4\Gamma_{FL}(1)\Gamma_{TR}(0)}{(\Gamma_T(0) + \Gamma_F(1))^2};\ A^+ \equiv \frac{4\Gamma_{FR}(1)\Gamma_{TL}(0)}{(\Gamma_T(0) + \Gamma_F(1))^2}.$$

Expanding $\lambda(\chi)$ in series at $\chi \to 0$, we obtain the cumulants of the FCS: the current, noise, and the third cumulant. We give these only in the shot noise limit $eV \gg k_B T$, where only two rates, $\Gamma_{TL}(0)$ and $\Gamma_{FR}(1)$, are in play so that all electrons go from the left to the right. We introduce a convenient parameter that measures the relative magnitude of these rates, $b \equiv 1/(\sqrt{\Gamma_{FR}(1)/\Gamma_{TL}(0)} + \sqrt{\Gamma_{TL}(0)/\Gamma_{FR}(1)})$. If one of the rates is much bigger than the other one, $b \to 0$. If the rates are equal, b achieves its maximum, $b = 1/2$. Using this notation,

$$I = e\left(\frac{\Gamma_{TL}(0)\Gamma_{FR}(1)}{\Gamma_{TL}(0) + \Gamma_{FR}(1)}\right) = eb(\Gamma_{TL}(0) + \Gamma_{FR}(1)); \tag{3.48}$$

$$S = 2eI\left(1 - 2b^2\right); \tag{3.49}$$

$$\langle\langle Q^3 \rangle\rangle = e^2 \Delta t I\left(\frac{1}{6} - b^2 + 2b^4\right). \tag{3.50}$$

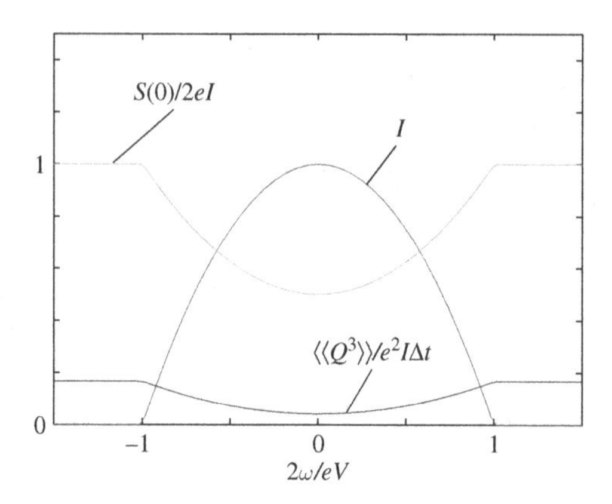

Fig. 3.16. **The first three cumulants of the charge transferred for conditions corresponding to the symmetric curve in Fig. 3.12. The noise and the third cumulant are normalized to the current. They approach their Poissonian values at the boundaries and beyond the one-by-one diamond.**

These relations are illustrated in Fig. 3.16 for a SET that corresponds to the symmetric curve in Fig. 3.12. Near the boundaries of the diamond one of the rates vanishes. At the boundaries, both the noise and the third cumulant approach their Poissonian values, indicating uncorrelated electron transfers. They are suppressed by factors $1/2$ and $1/4$, respectively, in the middle of the diamond, where the rates are equal.

This helps us to understand which correlations are actually reflected in the FCS of a SET. Two successive transfers through the left and right junctions are obviously correlated since one enables the other. This correlation, however, is not detected in the FCS: the statistics are only sensitive to completed transfers when an electron jumps through both junctions. What is detected is that two successive completed transfers cannot occur if they are as close in time as if they were uncorrelated. Indeed, it takes some time to complete a transfer so there must be some "dead time" separating the successive transfers. If both rates are comparable, the "dead time" is of the order of the average time between the transfers, and the rates exhibit strong correlations in time. If the rates are of very different magnitude, the average time is determined by the slowest rate, while the "dead time" is determined by the fastest one. In this case, the "dead time" is only a small fraction of the average time between two transfers, and those almost do not correlate.

The "dead time" FCS given by Eq. (3.47) seems to be different from those given by the Levitov formula, where correlations arise from the regularity of the stream of incident electrons. It may be tempting to explain the difference in terms of the opposition between the coherent transport of non-interacting electrons and the incoherent correlated tunneling in the Coulomb blockade regime. However, we will encounter similar statistics at least twice in this book: in the context of one-channel resonant tunneling and the multi-channel double junction, and the transport is coherent in both cases.

A peculiar fact is that if the two rates are the same, so that the correlations are at maximum, the FCS expression reduces to

$$\lambda(\chi) = \Gamma(e^{i\chi/2} - 1). \tag{3.51}$$

Comparing this with Eq. (1.52), we understand that this denotes the statistics of uncorrelated transfers of electron *halves*. One should not seek for physical implications here: obviously, a SET does not chop electrons into two pieces. Albeit, the above expression has a certain mnemonic value.

Exercise 3.6. Close to the degeneracy point, one always works in the Nyquist limit $k_B T \gg eV$. What does Eq. (3.47) yield for I, S, and $\langle\langle Q^3 \rangle\rangle$ in this case?

3.3.5 Memory cell

The energetics of the Coulomb blockade make it possible to enable and disable certain single-electron transfers at will just by applying voltages to the gate and/or transport electrodes. This allows the construction of electronic devices based on physical principles that are very different from those used in up-to-date commercial electronics. We review below the simplest versions of such devices.

We start with a memory cell. It consists of two islands connected in series to a single transport electrode (Fig. 3.17(a)). One can regard this circuit as either a single-electron box with a tunnel junction replaced by a SET or a two-junction array cut from one of the electrodes.

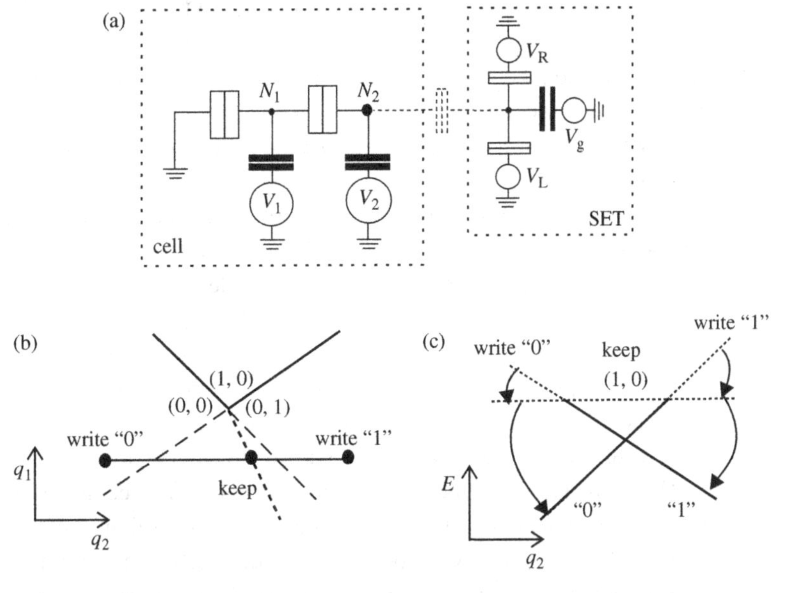

Fig. 3.17. Single-electron memory cell. (a) The information is encoded by the number of electrons in the second island ($N_2 = 0, 1$) and can be read with a SET capacitively connected (dashed lines) to this island. (b) Stability diagram and manipulation of the cell. (c) Energies of the three states versus q_2. The dashed line indicates that a state is unstable with respect to a single-electron transfer.

The information is encoded by the number of excess electrons in island 1: it can be either in charge state "0" or "1." To read the information, one can use a separate SET with a weak capacitive coupling to island 1, as shown in Fig. 3.17(a). The current through the SET will be sensitive to N_1. One operates the cell by changing q_2. For sufficiently large (small) q_2, the only stable state of the cell is "1" ("0"). There is an interval of q_2 where both states are stable. The information is kept if q_2 is within this interval. To write "1," we shift q_2 to the left, wait for the possible single-electron transfers to occur, and shift it back to the interval. To write "0," we do the same, shifting q_2 to the right.

To understand the energetics that enables such operations, we refer to the stability diagram of the two-island system (Fig. 3.4). We plot this again in Fig. 3.17(b), focusing on a small area where the states $(0,0)$, $(0,1)$, and $(1,0)$ come together. Since in the memory cell setup island 2 is not connected to any transport electrode, the line separating the domains $(0,0)$ and $(0,1)$ no longer corresponds to any single-electron transfer. It simply indicates where the energies of "0" and "1" are the same. Two remaining lines and their continuations (dashed) indicate where the energy of either "0" or "1" match the energy of $(0,1)$, and correspond to single-electron transfers through junction 2. The bistability interval of q_2 is bounded by dashed lines. Indeed, if we start with state "1" in the interval and shift q_2 to the left, crossing the left dashed line, we enable single-electron tunneling to the state $(1,0)$. From this state, the electron tunnels out through junction 1 so that we are in state "0." The energies of the three states versus q_2 and the single-electron transitions are shown in Fig. 3.17(c).

3.3.6 Turnstile

Suppose we connect the working island of the cell with double junctions to two electrodes, L and R. Next we assume that we are able to manipulate the conductances of the junctions by switching them on and off. Consider the following cycle. We start with cell in "0" state. We switch the left double junction on, the right double junction off, and write "1." An electron is taken from the left electrode. We reconnect the cell so that the right junction is on and the left one is off. We write "0." The extra electron goes to the right-hand electrode, and we are back at the starting point of the cycle. We have achieved a controllable transfer of precisely one electron between the left and right electrodes!

Such a transfer can be organized in a more elegant way. It turns out that we do not have to manipulate the conductances: we can let the Coulomb blockade do this for us. The corresponding device is called a turnstile [64]. The simplest turnstile is indeed a biased three-island array: a memory cell connected to two electrodes (see Fig. 3.18).

The turnstile is operated by changing the charge induced in the central island. The region of operation is where the Coulomb shards – stability regions – of states $(0,1,0)$ and $(0,0,0)$ overlap. The $(0,1,0)$ shard is bounded by a line that corresponds to the transition to $(0,0,1)$ through junction 3. The system does not stay long in $(0,0,1)$ since it is not stable in the operational region. The electron tunnels to the right, bringing the system to a stable state, $(0,0,0)$. Similarly, the $(0,0,0)$ shard is bounded by a line of transitions to $(1,0,0)$. This state is not stable, turning to $(0,1,0)$. The working cycle starts deep in the $(0,0,0)$ shard.

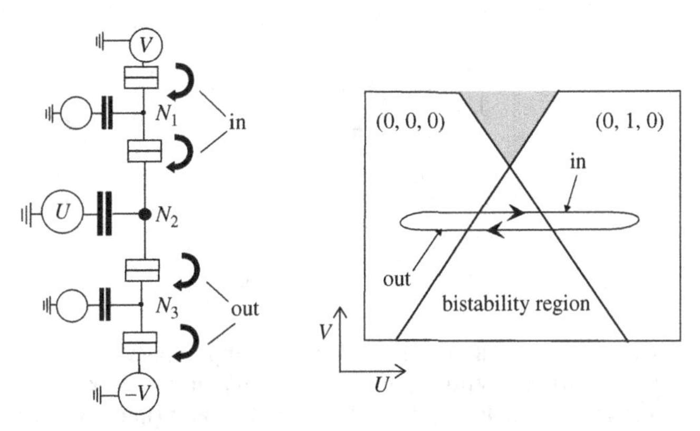

Fig. 3.18. **A single-electron turnstile is operated by the cycling gate voltage, *U*, of the central island. The cycle crosses the region where both states (0, 0, 0) and (0, 1, 0) are stable. The tunneling "in" and "out" occurs in two steps right after crossing the boundaries of bistability region. The gray area depicts the region where uncontrolled tunneling through the turnstile would take place.**

When we cross the boundary of this shard, an electron is transferred in two jumps from the left electrode to the central island. We turn back. When we cross the boundary of the $(0, 1, 0)$ shard, the electron trapped in the central island is released to the right electrode. In this region of operation, we cannot reverse the direction of electron flow by changing the cycle; in fact, the direction is set by the bias applied to the turnstile. This protects the turnstile against possible manipulation errors.

To demonstrate the operation, one applies to the central gate a time-dependent signal with frequency ω. If the turnstile works ideally, the current is precisely $I = e\omega/2\pi$, depending neither on the modulation amplitude nor the bias voltage. The accuracy of more complicated turnstiles allows one to use them in metrology as very accurate current and capacitance standards.

3.3.7 Single-electron pump

There is an alternative way to transport electrons one-by-one in a controllable fashion [65]. It greatly resembles the adiabatic pumping considered in the context of the scattering approach; see Section 1.7.4.

Let us once again turn to the stability diagram of a two-island array connected to two electrodes and concentrate on the vicinity of the common point where three different charge states come together (see Fig. 3.19). No bias voltage is applied. We now change both charges $q_{1,2}$ in a cyclic fashion. We do this slowly, giving all the allowed single-electron transfers time to occur. If the cycle does not enclose the common point, nothing special happens. Let us consider cycle A, which encloses the point. We start in the $(0, 0)$ region. After crossing the first line, an electron jumps into island 1 from the left. Crossing the second line promotes it into island 2, and, finally, crossing the third line brings the electron to the right-hand electrode. Thus, we have transferred an electron from the left to the right.

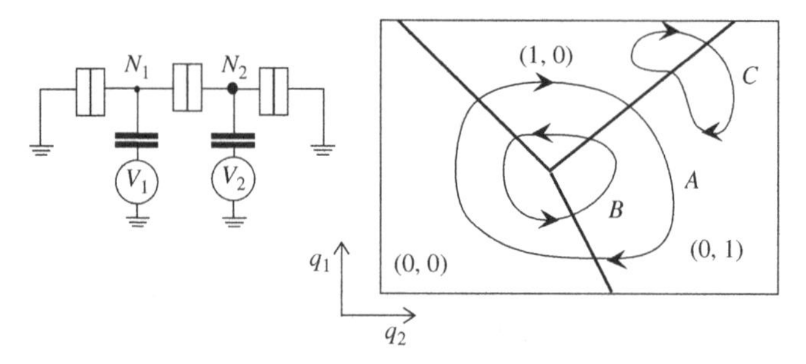

Fig. 3.19. **A single-electron pump, like all adiabatic pumps, operates when two parameters ($q_{1,2}$) are cycled. A cycle that encloses the common point in the clockwise (counterclockwise) direction transfers a single electron from the left to the right (from the right to the left).**

In contrast to the turnstile, the reverse cycle transfers an electron in the opposite direction. We see that the single-electron pump is advantageous in comparison with the adiabatic pumps considered in Section 1.7: the charge transferred is always quantized, not dependent on the actual shape of the cycle.

The same factors limit the applicability of all the devices considered – the memory cell, the turnstile, and the pump. They should not be operated too fast. The limitation is that all allowed single-electron processes should take place before the energy conditions have changed. The upper limit of the operation frequency is set by a typical single-electron rate $\Gamma_{se} \simeq GE_C/e^2 \simeq G/C$. The devices also must not be operated too slowly. The lower limit of the operational frequency is set by the small rates of unwanted processes that disrupt the normal operation. For example, an electron trapped in the memory cell could just tunnel away, and the information would be lost. Such unwanted processes originate from three sources. The first source is thermal activation, whereby the electrons can tunnel in the wrong direction with a rate estimated at $\Gamma_{se}\exp(-E_C/k_BT)$. Such rates are exponentially suppressed with decreasing temperature. The second source involves tunneling processes involving more than one electron. They are enabled even at vanishing temperature. They will be considered in detail in Section 3.4. Here we note a standard way of protecting against such processes: one replaces the double junctions of the cell and turnstile by many junctions in series, that is many-island arrays. This, however, increases the importance of the least fundamental and most limiting source: the random drift of offset charges. A given stable configuration of offset charges can, in principle, be tuned away by gates affecting all islands of the system. The change in offset charge, however, may disrupt all energy differences so that the device ceases to work normally unless it is retuned.

3.4 Co-tunneling

An old joke likens classical physics to a totalitarian regime, since the behavior of every classical particle is strictly determined by the regulations – the physical laws. What is not commanded is forbidden. Quantum physics in this joke represents a democratic society where everybody does what they want unless it is explicitly forbidden by some fundamental law. It turns out that everything not forbidden is compulsory. The tunnel processes involving many electrons – *co-tunneling* processes – under Coulomb blockade conditions exemplify this joke rather explicitly.

The concept of a Coulomb blockade is classical in nature: there are single-electron transitions between well defined charge states. At sufficiently low bias, these transitions are forbidden by virtue of energy conservation: one has to pay energy to charge the island. Does the concept precisely fit the reality, and is the transport really forbidden if quantum effects are taken into account? The answer is "no," since the energy restriction is not fundamental. In fact, two or more electrons may cooperate in the process of tunneling (co-tunnel) and thereby cheat the energy conservation [66].

As an example, we describe a SET in the Coulomb blockade regime and consider a process where the tunneling of an electron from the left is immediately followed by the tunneling of another electron to the right. In the resulting state, the island charge does not change. However, one electron charge has been transferred from the left to the right.

Instead of immediately presenting a quantitative theory of co-tunneling, let us start with a "cartoon," which will give us order-of-value estimations of co-tunneling rates. Electrons want to tunnel from the left to the right (see Fig. 3.20(a)). They would do this with a typical rate Γ_{FL}. They cannot tunnel by virtue of energy conservation: they have to pay the charging energy E_{C}, which they do not possess. The cartoonish feature we use is that the energy conservation is not imposed instantly: the electron may stay in the island during the Heisenberg uncertainty time $t_{\text{H}} \simeq \hbar/E_{\text{C}}$, even if the energy conservation forbids a permanent stay. This gives a small, but finite, chance for co-tunneling; during this short stay,

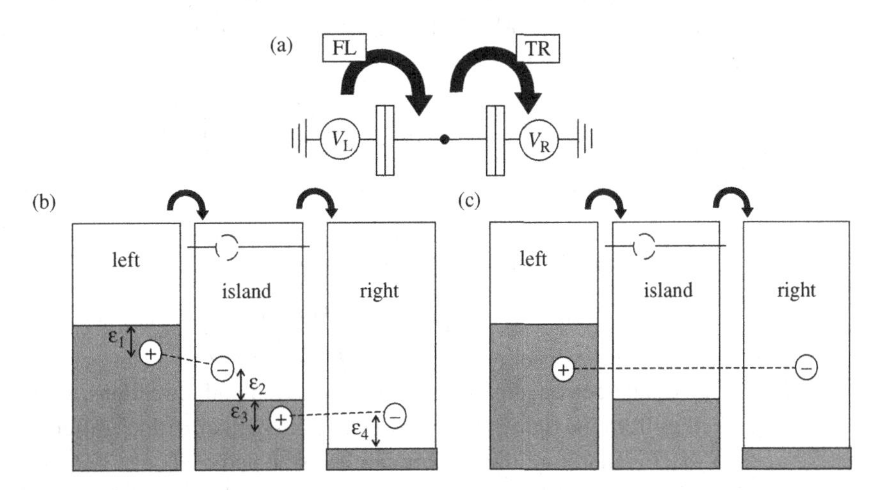

Fig. 3.20. Co-tunneling in a SET. (a) Co-tunneling: simultaneous transfer of two electrons in two junctions is not forbidden by energy conservation. (b) Energy diagram for inelastic co-tunneling: two electron–hole pairs remain in the final state. (c) Energy diagram for elastic co-tunneling: the transferred electron keeps its energy.

another electron from the island may jump to the right-hand junction. This is energetically favorable since, during the Heisenberg time, the island is in the charge state "1." This state is termed *virtual*, since the energy conservation forbids its realization. The rate in the right-hand direction in this virtual state is thus given by $\Gamma_{\rm TR}$. The chance to realize co-tunneling during the short stay is therefore given by $\Gamma_{\rm TR} t_{\rm H} \ll 1$.

Multiplying these chances, we obtain the following estimation:

$$\Gamma_{\rm cot} \simeq \Gamma_{\rm FL} \Gamma_{\rm TR} \frac{\hbar}{E_{\rm C}}.$$

Since a typical estimation of the single-electron rate is $\Gamma_{\rm se} \simeq G/C \simeq G E_{\rm C}/e^2$, we rewrite this as follows:

$$\Gamma_{\rm cot} \simeq \Gamma_{\rm se} \frac{G}{G_{\rm Q}}. \tag{3.52}$$

We see that the co-tunneling is essentially less probable than single-electron tunneling provided $G \ll G_{\rm Q}$. The same condition ensures the presence of a Coulomb blockade. At $G \simeq G_{\rm Q}$ the co-tunneling and similar multi-particle processes may become as important as single-electron ones; this results in the corruption and disappearance of the Coulomb blockade.

We can improve on this if we take into account the details of the final state and the energy conservation in this state. After two electron transfers, four excitations have been created in the system: a hole in the left electrode, an electron and a hole in the island, and an electron in the right electrode. We count the energies of these excitations from the corresponding Fermi levels (Fig. 3.20(b)), so that all $\epsilon_i > 0$. If we impose no energy restrictions, all ϵ_i are of the order of $E_{\rm C}$. Let us assume that the SET is biased by a voltage much smaller than the Coulomb blockade threshold, $eV \ll E_{\rm C}$. Energy conservation dictates that $\sum \epsilon_i = eV$. Therefore each $\epsilon_i < eV$. Since the number of electron states in an energy strip is proportional to the width of the strip, the number of states available is reduced by a factor $\simeq (eV/E_{\rm C})^3$. The same factor reduces the rate in comparison with the estimation given in Eq. (3.52):

$$\Gamma_{\rm cot} \simeq \frac{G_{\rm L}}{G_{\rm Q}} \frac{G_{\rm R}}{G_{\rm Q}} \left(\frac{eV}{E_{\rm C}}\right)^2 eV. \tag{3.53}$$

This consideration is relevant at vanishing temperature $k_{\rm B}T \ll eV$. In the opposite case, $k_{\rm B}T \gg eV$, the excitations can appear in the energy strip of the width of $k_{\rm B}T$; this gives the rate of thermally activated co-tunneling,

$$\Gamma_{\rm cot} \simeq \frac{G_{\rm L}}{G_{\rm Q}} \frac{G_{\rm R}}{G_{\rm Q}} \left(\frac{k_{\rm B}T}{E_{\rm C}}\right)^2 k_{\rm B}T. \tag{3.54}$$

The thermally activated processes can proceed in both directions, from the left to the right as well as from the right to the left. The rates in both directions differ only by a small factor $\simeq eV/k_{\rm B}T$. Therefore the current can be estimated as $I_{\rm cot} \simeq V(G/G_{\rm Q})^2(k_{\rm B}T/E_{\rm C})^2$. We conclude that two-electron co-tunneling is an *inelastic* process; this leads to non-linear and/or temperature-dependent I–V curves.

We can apply the above estimations to the simple memory cell considered in Section 3.3. If there is an energy difference ΔE between two stable states of the cell, co-tunneling through two junctions will bring the system to the state with the lowest energy. Since the energy available for co-tunneling is ΔE, the rate of this process is estimated by Eq. (3.53). This "bad" process thus determines the time of memory loss. One can substantially reduce this time by tuning ΔE to zero. In this case, only thermal energy $k_B T$ is available for co-tunneling, and the time of memory loss is determined by the rate given in Eq. (3.54). At larger time scales, the cell switches randomly between "0" and "1" with this rate.

We have already mentioned the use of multi-junction arrays to protect single-electron devices against undesired multi-electron processes. In the case of co-tunneling through an array of N junctions of equal conductance G, N electrons should co-operate, jumping simultaneously. Such processes create $2N$ excitations in the final state. Multiplying the rate suppression factors for each junction and for each excitation brings us to the following estimate (with the energy available for co-tunneling $\Delta E \simeq \max(eV, k_B T)$):

$$\Gamma_{\text{cot}} \simeq \left(\frac{G}{G_Q}\right)^N E_C/\hbar, \qquad \Delta E \simeq E_C; \tag{3.55}$$

$$\Gamma_{\text{cot}} \simeq \left(\frac{G}{G_Q}\right)^N \left(\frac{\Delta E}{E_C}\right)^{2N-2} \frac{\Delta E}{\hbar}, \qquad \Delta E \ll E_C. \tag{3.56}$$

Exercise 3.7. One of the "bad" processes that limits the accuracy of a four-junction single-electron turnstile is the co-tunneling through all four junctions in the bias direction. Give an estimate of the rate of this process assuming $E_C = 1\,\text{meV}$, $1/G = 1\,\text{M}\,\Omega$.

Control question 3.5. Why would co-tunneling through all junctions not disrupt the current in the single-electron pump? Give an example of a co-tunneling process that does affect the pump accuracy.

Until now we have assumed that at least two electrons participate in the co-tunneling process. Eventually, there is a chance that the *same* electron that enters the island from the left tunnels out to the right, thereby completing the co-tunneling process. How big is this chance? The electrons would tunnel to the right from an energy window of the order of E_C with the rate $\Gamma_{\text{se}} \simeq G E_C/e^2$. The electron states in the island are discrete, with average spacing δ_S. Thus, the rate of tunneling from any *given* state is $\Gamma_{\text{given}} \simeq (G/G_Q)\delta_S/\hbar$. It seems unlikely that the same electron would complete the co-tunneling: the chance of it happening is just δ_S/E_C. However, the process involving the same electron is *elastic*: no excitations are there in the island in the final state, and the electron transferred to the right has the same energy as the incoming electron. This removes the extra factor $(eV/E_C)^2$, which suppresses the inelastic co-tunneling at small bias:

$$\Gamma_{\text{el-cot}} \simeq G_{\text{el}} V/e; \; G_{\text{el}} \simeq \frac{G_L G_R}{G_Q} \frac{\delta_S}{E_C}. \tag{3.57}$$

Comparing this with Eq. (3.53), we find that the elastic co-tunneling dominates at sufficiently low energies $\Delta E \leq \sqrt{\delta_S E_C}$. If we apply the same reasoning to a multi-junction array, we obtain

$$G_{\mathrm{el}} \simeq \left(\frac{G}{G_{\mathrm{Q}}}\right)^{N} \left(\frac{\delta_{\mathrm{S}}}{E_{\mathrm{C}}}\right)^{N-1}. \tag{3.58}$$

Comparison with Eq. (3.56) yields that the cross-over between inelastic and elastic co-tunneling takes place at the same energy scale $\sqrt{\delta_{\mathrm{S}} E_{\mathrm{C}}}$ – the geometric mean between Coulomb energy and level spacing.

This result is conceptually important for all quantum transport. The first two chapters of this book have dealt mostly with elastic, coherent electron transport. In this chapter, we study Coulomb blockade systems which at first sight look entirely different. Now we can link these two: at sufficiently low energies, below $\sqrt{\delta_{\mathrm{S}} E_{\mathrm{C}}}$, the transport in Coulomb blockade nanostructures is essentially elastic and coherent, and can be described with the scattering approach of Chapter 1.

It would be unfortunate to complete the present qualitative discussion leaving the impression that the only possible role of co-tunneling is to spoil Coulomb blockade physics. Co-tunneling can certainly be both interesting and useful. To prove this, we discuss briefly the following device idea [67]. Let us place two one-dimensional Coulomb arrays close to each other, creating a substantial capacitive coupling between neighboring islands of opposite arrays (see Fig. 3.21(b)). There is no tunneling between the arrays, and each of them is connected to its own transport electrodes. With proper gate voltages, one induces charges of opposite polarities in these opposite arrays. Thereby one creates extra electrons in one array and extra holes in another one. The capacitive coupling between the arrays

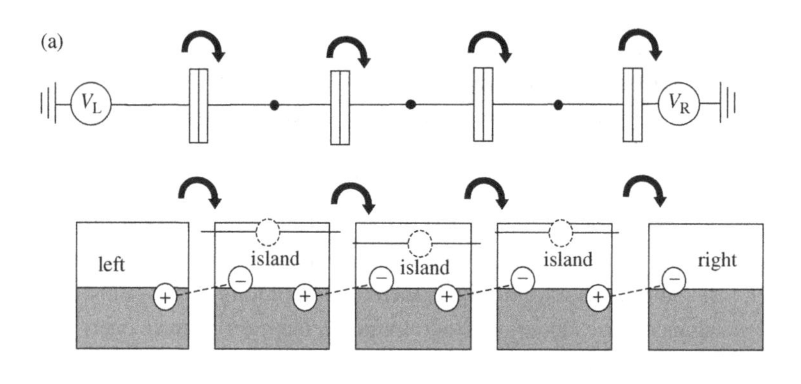

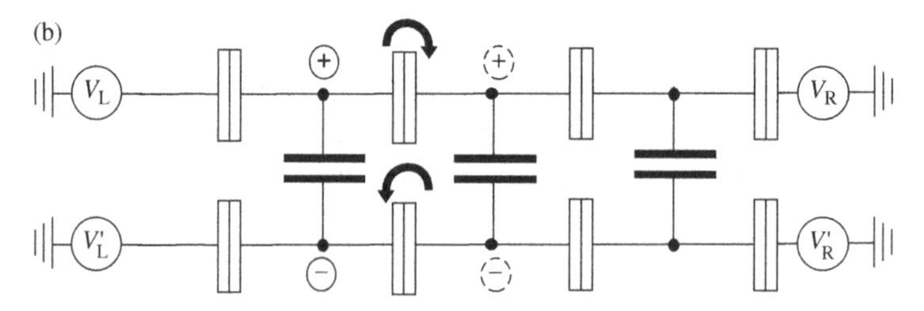

Fig. 3.21. (a) Complex co-tunneling process: charge transfer in four-junction array. Four electron–hole pairs are left in the final state. (b) Co-tunneling enables transport of artificial excitons – electron–hole pairs in capacitively coupled arrays.

enables Coulomb attraction between these charges of opposite sign. They come together on neighboring islands, forming bound states of electron and hole: they are called *excitons* in semiconductor physics. To set our artificial excitons into motion, we apply small bias voltages to the arrays. Single-electron processes cannot take place; as they would break the exciton apart, and this costs binding energy. The exciton moves in this double array as a whole by means of a co-tunneling process comprising two simultaneous electron transfers in opposite directions and opposite arrays. Over a wide range of bias voltages, the currents in both electric circuits are therefore the *same* in amplitude and opposite in direction. Therefore the device is called a *current copier*.

3.4.1 Quantum effects in single-electron box

Before we proceed with the quantitative theory of co-tunneling, let us check, by means of a rigorous quantum calculation, the most important feature of the cartoon in use. In the cartoon, we allow an electron to stay in the Coulomb island for time t_H even if energy conservation forbids this. For a single-electron box, this implies that there is a chance of finding the island in a charge state that is not the minimum-energy state of the system in the absence of tunneling. This chance is proportional to the intensity of tunneling and is of the order of $\Gamma_{se} t_H \simeq G/G_Q$.

In quantum language, this implies that the true ground state of the single-electron box is no longer an eigenstate of charge. Rather, the tunneling results in a small but non-vanishing admixture of other charge states. This admixture is readily evaluated using the general rule of quantum perturbation theory: an admixture of any state n to the unperturbed state g is given by

$$\psi_n = \sum_{n \neq g} \frac{H^{(\mathrm{int})}_{ng}}{E_n - E_g}, \tag{3.59}$$

where $E_{n,g}$ and $H^{(\mathrm{int})}_{ng}$ are the unperturbed energies and the matrix element of the perturbation.

In our case, $\hat{H}^{(\mathrm{int})}$ is the tunneling Hamiltonian given by Eq. (3.30). Let us assume that in the ground state $N = 0$, and label the electron states in the electrode (island) by r (i). The action of the terms of the tunneling Hamiltonian on the ground state produces two groups of states. If in the ground state r is empty and i is filled, the electron is transferred from i to r and the charge state of the island becomes $N = -1$. If i is empty and r is filled, the electron is transferred to the island and $N = 1$. Now we can evaluate the probability of finding the island in the charge state $N = 1$ by summing up over all possible r, i:

$$p_1 = \sum_{r,i} |\psi_{r,i}|^2 = \sum_{r,i} \frac{|T_{r,i}|^2 f(E_r)(1 - f(E_i))}{(E^{(+)} + E_i - E_r)^2}, \tag{3.60}$$

where $E^{(+)} \equiv E_{\mathrm{el}}(1) - E_{\mathrm{el}}(0) > 0$ account for the charging energy. Since we are working with pure states, we set $k_B T = 0$. A similar expression with $E^{(+)} \to E^{(-)} \equiv E_{\mathrm{el}}(-1) - E_{\mathrm{el}}(0)$ is valid for p_{-1}. Now we use the trick given in Eq. (3.33) to express the tunneling amplitudes in comprehensive terms of the single-electron rate ($\xi \equiv E_i - E_r$):

$$p_{\pm 1} = \frac{\hbar}{2\pi}\int_{\infty}^{\infty} \frac{\Gamma(-\xi)\mathrm{d}\xi}{(E^{(\pm)}+\xi)^2} = \frac{G}{2\pi^2 G_Q}\int_0^{\infty} \frac{\xi\,\mathrm{d}\xi}{(E^{(\pm)}+\xi)^2}. \tag{3.61}$$

A very rough order-of-value estimation gives $p_{\pm 1} \simeq G/G_Q$, in accordance with the cartoon. Close inspection of the expression, however, reveals two alarming features worth discussion.

The first, and seemingly acute, problem is that the integral in Eq. (3.61) diverges at large electron–hole pair energies ξ, so that the small probability of being in the states $N = \pm 1$ is formally infinite. Such *ultra-violet* divergency is commonplace in many-body quantum physics – very similar problems arise, for example, in quantum electrodynamics. The problem is philosophical rather than physical and arises from our human restrictions: we consider the "objective" quantum states of a single-electron box *with* tunnel coupling in the basis of "subjective" states *without* tunnel coupling. The quick remedy is to look at the "more physical value." If we concentrate on the average charge in the ground state, the integrand does not diverge at large energies. The integration yields

$$\delta\langle N\rangle = p_1 - p_{-1} = \frac{G}{2\pi^2 G_Q}\ln\left(\frac{1/2 + q/e}{1/2 - q/e}\right) \tag{3.62}$$

(see Fig. 3.22) and represents a smoothing quantum correction to a classical step-like $N - q$ relation.

Control question 3.6. In Fig. 3.22, which solid curve corresponds to which value of G?

The second feature is that the integral diverges at low energies provided $E^{\pm} \to 0$; this corresponds to $q \to \pm e/2$. This is also seen in the expression for the average charge: the correction formally diverges at $q \to \pm e/2$. This indicates that the perturbation theory developed is no good at tiny energy scales $E^{\pm} \simeq E_C \exp\{-G/(2\pi^2 G_Q)\}$. We will come back to this problem in Chapter 6.

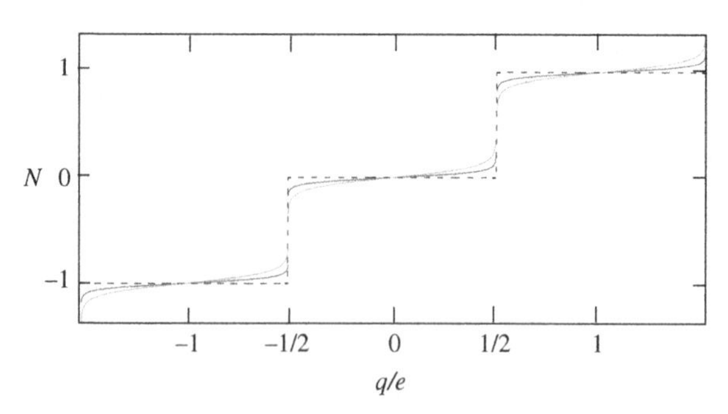

Fig. 3.22. Quantum fluctuations – virtual transitions to other charge states – smoothen the step-like features in the $N - q$ relation. The dashed line shows the $G = 0$ limits, and the two solid curves correspond to $G = 0.5G_Q$ and $0.25G_Q$. See Control question 3.6.

3.4.2 Co-tunneling rate

Let us quantify the co-tunneling rate in a SET setup. The general rules of quantum mechanics imply that the transition rate between the initial state i and the final state f is given by the Fermi Golden Rule, Eq. (3.31), the heart of which is the square of the transition matrix element M_{if} between the states. In a common situation, the perturbation $\hat{H}^{(\text{int})}$ has matrix elements between i and f and, quite simply, $M_{if} = H^{(\text{int})}_{if}$. This is not our situation. The perturbation is the tunneling Hamiltonian that can only provide a single-electron transfer through any junction. In order to get to the final state, we need two such transfers, and we have to visit intermediate virtual states where the charge of the island differs by $\pm e$ from its stable value.

The same general rules dictate that, for a complex process involving the intermediate virtual states, the matrix element M_{if} is more complicated: it is composed of matrix elements of perturbation involving all possible virtual states v and energy difference(s) between these states and the initial state:

$$M_{if} = \sum_v \frac{H^{(\text{int})}_{iv} H^{(\text{int})}_{vf}}{E_i - E_v}. \tag{3.63}$$

Let us specify this general relation to our SET. As we have seen at the beginning of this section, the final state of an elementary co-tunneling process, where an electron is transferred from state l in the left electrode to state r in the right electrode, also involves two electron states $i_{1,2}$ in the island. The electron tunneling from the left ends up in state i_1, while the electron tunneling to the right originates from another state i_2.

Each final state can be reached via two possible virtual states depending on the order of the two single-electron transfers. If the tunneling through the left junction comes first, the charge of the island changes from N to $N+1$. The energy of the resulting virtual state v_+ with respect to the initial one is given by

$$E_{v_+} - E_i = E_{\text{el}}(N+1) - E_{\text{el}}(N) + E_{i_1} - E_l - eV_{\text{L}} \equiv E^{(+)} + E_{i_1} - E_l$$

(see Eqs. (3.14)). Alternatively, the transfer through the right junction happens first, changing the charge of the island to $N-1$, so the energy of alternative state v_- is given by

$$E_{v_-} - E_i = E_{\text{el}}(N-1) - E_{\text{el}}(N) + E_r - E_{i_2} - E_l + eV_R \equiv \Delta E^{(-)} + E_r - E_{i_2}.$$

The matrix element is the sum of these two possibilities, given by

$$M_{if|i_1 i_2} = T_{li_1} T_{i_2 r} \left(\frac{1}{E^{(+)} + E_{i_1} - E_l} + \frac{1}{E^{(-)} + E_r - E_{i_2}} \right).$$

Since the rate is given by the square of the matrix element, the contributions from these two possible orderings of elementary transfers do not sum up in the rate. Rather, they interfere, very much like the amplitudes of the complex scattering processes discussed in Section 1.6. The total tunneling rate is given by the sum over all possible final states:

$$\Gamma_{\rm cot} = \sum_{l,r,s_1,s_2} |M_{if|i_1i_2}|^2 \delta(E_r - E_l + E_{s_1} - E_{s_2} - eV)$$
$$\times f(E_l)(1 - f(E_r))f(E_{s_2})(1 - f(E_{s_1})).$$

We implement the same trick (Eq. (3.33)) to replace the sum over the electron states by integration over two energies $\xi_{\rm L} = E_{s_1} - E_l$, $\xi_{\rm R} = E_r - E_{s_2}$, $\xi_{\rm L(R)}$ being the energy of electron–hole pair created by electron transfer in the left (right) junction. We obtain finally

$$\Gamma_{\rm cot} = \frac{\hbar}{2\pi} \int {\rm d}\xi_L \, {\rm d}\xi_R \, \Gamma_{\rm L}(-\xi_{\rm L})\Gamma_{\rm R}(-\xi_{\rm R})$$
$$\times \left(\frac{1}{E^{(+)} + \xi_{\rm L}} + \frac{1}{E^{(-)} + \xi_{\rm R}} \right)^2 \delta(\xi_{\rm R} + \xi_{\rm L} - eV). \quad (3.64)$$

This gives the co-tunneling from the left to the right; the co-tunneling in the opposite direction is obtained by the replacement $eV \to -eV$. Some features of the cartoon used at the beginning of this section are still recognizable in this exact expression if we associate the squares of the energy denominators with those arising in Eq. (3.60) for the chance of being in the wrong charge state.

Let us elaborate on two simple limits. For $eV \gg k_{\rm B}T$, the co-tunneling proceeds only in the energetically favorable direction. The integration domain is restricted by $\xi_{\rm L,R} > 0$ and the expression is valid in the whole Coulomb diamond restricted by $E^{(\pm)} > 0$. The single-electron rates in this situation are given by $\Gamma_{\rm L,R}(-\xi) = G_{\rm L,R}/e^2\xi$, and the integration can be performed analytically to give

$$I_{\rm cot}(V) = e\Gamma_{\rm cot} = \frac{\hbar G_{\rm L} G_{\rm R} V}{2\pi e^2} \left[\left(1 + \frac{2}{eV} \frac{E^{(+)}E^{(-)}}{E^{(+)} + E^{(-)} + eV} \right) \right.$$
$$\left. \times \ln(1 + eV/E^{(+)})(1 + eV/E^{(-)}) - 2 \right]. \quad (3.65)$$

This is illustrated in Fig. 3.23. Another significant limit is that of relatively small voltage $eV \simeq k_{\rm B}T \ll E^{(\pm)}$. To integrate, we disregard the dependence of energy denominators on $\xi_{\rm L,R}$, so the rate reduces to

$$\Gamma^{\rm LR}_{\rm cot} = \frac{\hbar}{2\pi} \left(\frac{1}{E^{(+)}} + \frac{1}{E^{(-)}} \right)^2 \int {\rm d}\xi_{\rm L} \, \Gamma_{\rm L}(-\xi_{\rm L})\Gamma_{\rm R}(\xi_{\rm L} - eV)$$
$$= \frac{\hbar G_{\rm L} G_{\rm R}}{12\pi e^4} \left(\frac{1}{E^{(+)}} + \frac{1}{E^{(-)}} \right)^2 \frac{(eV)((eV)^2 + (2\pi k_{\rm B}T)^2)}{1 - \exp(-eV/k_{\rm B}T)},$$

and $\Gamma^{\rm RL}_{\rm cot}(V) = \Gamma^{\rm LR}_{\rm cot}(-V)$. The co-tunneling current is the difference of these two rates:

$$I(V) = e(\Gamma^{\rm LR} - \Gamma^{\rm RL}) = \frac{\hbar G_{\rm L} G_{\rm R}}{12\pi e^2} \left(\frac{1}{E^{(+)}} + \frac{1}{E^{(-)}} \right)^2 V((eV)^2 + (2\pi k_{\rm B}T)^2), \quad (3.66)$$

in full accordance with the qualitative estimations given by Eqs. (3.53) and (3.54) at the beginning of the section. The zero-voltage conductance diverges at the crossing points of the diamonds (Fig. 3.23); we will deal with this divergency later on.

The rates of more complicated co-tunneling processes are quantified in the same way. Like a single-electron transfer, a general co-tunneling process is characterized by the electrostatic energy difference ΔE between the final and initial charge states. For the

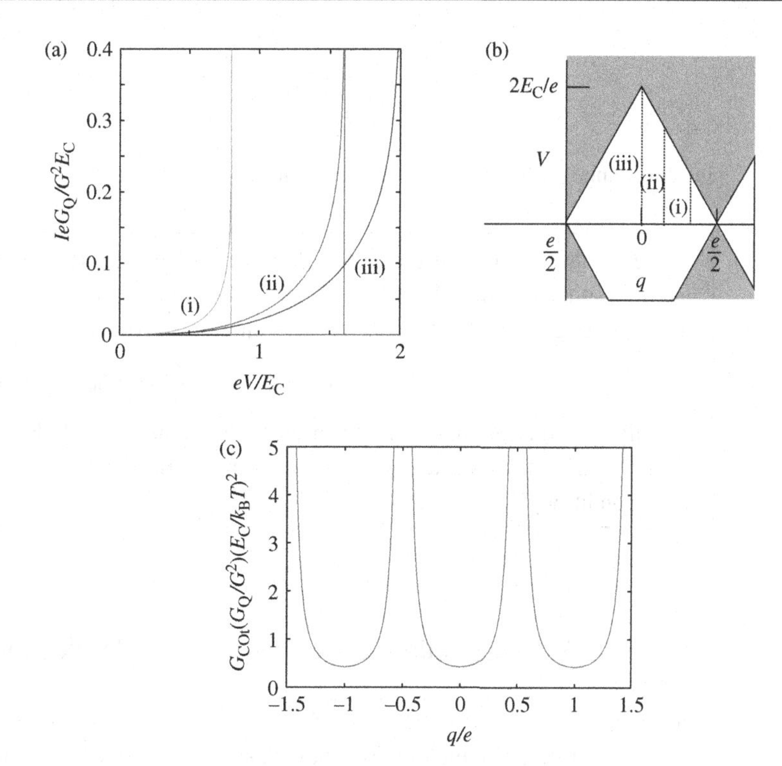

Fig. 3.23. (a) Co-tunneling current in symmetric SET versus voltage for several values of q. (b) Dotted lines show the q values used. (c) Zero-voltage co-tunneling conductance versus q.

simplest case of a single electron transferred between the leads it is simply $-eV$. There is an electron–hole pair produced in each junction, and the integration proceeds over all electron–hole pair energies ξ_j. Subject to the energy conservation condition $\sum_j^M \xi_j + \Delta E$,

$$\Gamma_{\text{cot}} = \frac{\hbar}{2\pi} \int \left(\prod_{j=1}^{M} \frac{\hbar \Gamma_j(-\xi_j) d\xi_j}{2\pi} \right) S(\{\xi_j\}) \delta \left(\sum_j^M \xi_j + \Delta E \right). \tag{3.67}$$

Here the factor S accounts for the contribution of energy denominators. For $M > 2$ it becomes very cumbersome, requiring scrupulous book-keeping. The point is that there are $M!$ possible sequences of M single-electron transfers, and S is the sum contributed by all sequences. Given a sequence $\{j_1, \ldots, j_M\}$, the system is subsequently transferred to $M-1$ virtual states $\{s_1, \ldots, s_k, \ldots s_{M-1}\}$, s_k being the charge state at the kth step. Each step brings a factor of inverse energy of this state, the latter comprising electrostatic energy and energies of the electron–hole pairs created up to this step, $E(s_k) + \sum_{i=1}^k \xi_{j_i}$. Therefore

$$S = \sum_{\{j_1,\ldots,j_M\}} \prod_{k=1}^{M-1} \frac{1}{E(s_k) + \sum_{i=1}^k \xi_{j_i}}.$$

The integration over ξ_j can be performed analytically in the limit $k_B T \to 0$ and small energy available for co-tunneling $\Delta E \ll E_s$. The rate is given by

$$\Gamma_{\text{cot}} = \frac{2\pi}{\hbar}\left(\prod_{j=1}^{M}\frac{G_j}{2\pi^2 G_{\text{Q}}}\right) S(\xi_j = 0)\frac{(-\Delta E)^{2M-1}}{(2M-1)!}, \tag{3.68}$$

in accordance with the estimation given in Eq. (3.56).

Exercise 3.8. Let us consider a long double array (Fig. 3.21(b)) with one extra electron and hole situated in the islands j, k, where $j(k)$ numbers the islands in the upper (lower) array. Let us assume that the electrostatic energy is given by

$$E(j,k) = \text{const.} - E_0 \exp(|j-k|/\kappa) - eV_{\text{up}}j - eV_{\text{down}}k.$$

Initially the electron and hole are in adjacent islands, $j = k$. Find the condition that forbids single-electron transfers at zero temperature. Find the co-tunneling rate under these conditions.

3.4.3 Co-tunneling co-existing with single-electron transfers

Upon inspecting Eqs. (3.65) and (3.66) for co-tunneling current, one notes an annoying detail: they diverge at the Coulomb blockade threshold where either $\Delta E^{(+)}$ or $\Delta E^{(-)}$ approaches zero. Above this threshold, single-electron tunneling is enabled. It is clear that the co-tunneling rate cannot exceed a typical single-electron rate. This implies that the divergences are not physical, rather they signal that the elementary perturbation theory ceases to work close to a Coulomb blockade threshold where co-tunneling and single-electron tunneling may co-exist. It happens frequently that such a breakdown of perturbation series signals novel, complicated, and sometimes incomprehensive physics. Fortunately, this is not the case at the Coulomb blockade threshold: one can sort out the situation quite simply by obtaining more insight into the quantum nature of both processes.

Let us sketch another cartoon to help in our qualitative understanding. Till now, we have assumed that if a tunneling event happens it is accomplished instantly. Let us assign a Heisenberg uncertainty duration to a tunneling event: we assume that if the energy difference ΔE is associated with the event, it takes time $t_{\text{H}} \simeq \hbar/\Delta E$ for the event to be accomplished. To check if this makes sense, let us apply the cartoon to generic single-electron tunneling, for example, in a SET. The single-electron transfers occur randomly in time, the typical time separation of two events being of the order of the inverse rate $1/\Gamma_{\text{se}} \simeq (G_{\text{Q}}/G)\hbar/\Delta E$. We observe that, for $G \ll G_{\text{Q}}$, the time separation greatly exceeds the duration of the events, so they do not overlap: an event is accomplished long before the next event starts. This enables the classical treatment of single-electron transfers in the Coulomb blockade regime. If G approaches G_{Q}, the events begin to overlap and there is no longer any strict separation between them. This indicates, as we have already noted, the break-down of the Coulomb blockade or at least its classical description.

Let us now specifically apply the cartoon at the Coulomb blockade threshold near a diamond boundary assuming $eV \gg k_{\text{B}}T(G/G_{\text{Q}})$. If we come to the threshold from the

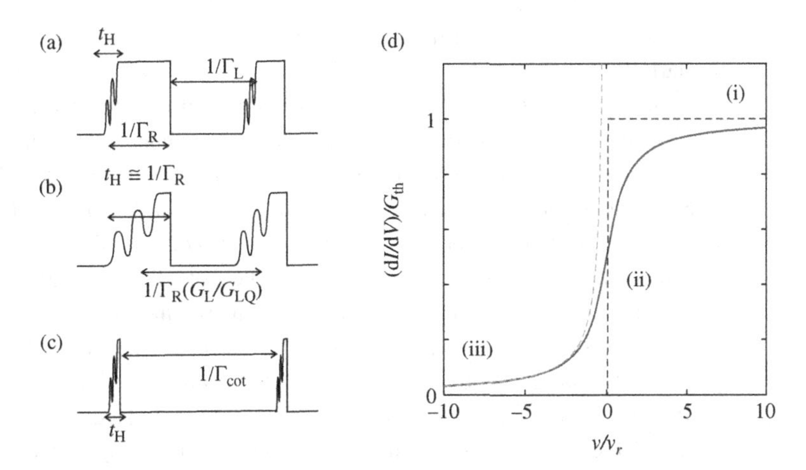

Fig. 3.24. **Co-tunneling coexisting with single-electron tunneling at the diamond boundary. Cartoonish time-line of tunneling events: charge in the island versus time. The Heisenberg duration of events is given by a wavy profile. The events are grouped in pairs. Within each group, the events are well separated on the single-electron side (a), overlap at the transition (b), and cannot be separated on the co-tunneling side (c). (d) Differential conductance in the vicinity of the boundary. The dashed lines present co-tunneling and single-electron tunneling asymptotes.**

single-electron side, we encounter the familiar one-by-one tunneling. To simplify the notation, we concentrate on electrons going from the left to the right with the transition via the left junction being enabled at the threshold. The charge states involved are "0" and "1." All other possible cases are obtained by changing left to right and shifting the charge states. The energy difference ΔE_L related to the tunneling via the left junction vanishes at the threshold with the corresponding rate Γ_L, while $\Delta E_R, \Gamma_R$ stay constant in the close vicinity of the threshold.

The tunneling events are grouped into pairs (see Fig. 3.24). Within the group, the events are separated by a short time interval $1/\Gamma_R$ while the groups are separated by $1/\Gamma_L$. Very close to the threshold, the events begin to overlap within each group – this happens at $|\Delta E_L|\Gamma_R \simeq \hbar$ – while the groups are still perfectly separated. Let us note now that on the other side of the threshold the events of the group become the constituents of a co-tunneling process and cannot be separated in time. The Heisenberg duration of this co-tunneling process is still of the order of $\hbar/|\Delta E_L|$, and successive co-tunnelings are separated by time intervals $\simeq 1/\Gamma_{cot} \gg t_H$. To give an overall picture, the charge near the threshold is transferred in well separated groups. Far above the threshold, each group can be regarded as two successive single-electron transfers. Far below, the group is a single co-tunneling process. At the threshold, single-electron tunneling and co-tunneling processes cannot be regarded separately: the group is neither co-tunneling nor single-electron tunneling.

An alternative scenario of co-existence is realized near the degeneracy point of two diamonds at $eV \ll k_BT$. If the energy difference w between two charge states is of the order of k_BT, the transport is mostly due to thermally activated single-electron processes. Each process disposes energy $\simeq k_BT$; this defines its Heisenberg duration, and occurs with the rate $\Gamma \simeq (G/G_Q)(k_BT)/\hbar$. This implies that the single-electron processes remain well separated in time, even at the degeneracy point $w = 0$, quite in contrast with the former scenario.

Is there a place for co-tunneling events in this regime? Yes, there is. As a matter of fact, the time interval remains random. There is always a chance of experencing the rare coincidence of two events when this time interval becomes comparable with $t_{\rm H}$. The fraction of rare coincidences in the event flow can thus be estimated as $t_{\rm H}\Gamma \simeq (G/G_{\rm Q})$. In this case, two events cannot be separated in time so one expects that the rate of each event is modified in comparison with that of well separated events: the correlation at short time interval $\simeq t_{\rm H}$ should take place. The co-tunneling in this regime accounts for this rate change for rare coincidences. So we expect it to give the corrections of the order of $G/G_{\rm Q}$. In principle, the rate change can be positive as well as negative, the latter corresponding to mutual inhibition of coinciding events. We will see that this is indeed the case near the degeneracy point. Further from the point where $w \gg k_{\rm B}T$, separate single-electron processes become rare. The rare coincidences, however, do survive: in this case, the electrons co-operate in the course of a co-tunneling event. To summarize, the single-electron tunneling and co-tunneling contributions to the transport can be separated; this is in contrast to the former scenario.

Now we are ready for a quantitative discussion, and we start with the first scenario. First we note that the rate $\Gamma_{\rm R}$ is the fastest of the two and the Heisenberg duration of the corresponding tunneling event is negligible. If the island is in the state "1," the rate $\Gamma_{\rm R}$ brings it to the state "0" with an electron transferred to the right electrode. This is why the current in this situation just measures the probability to be in the state "1":

$$\frac{I}{e} = p_1 \Gamma_{\rm R}.$$

On the co-tunneling side of the transition, this probability is due to virtual tunneling transitions involving the left junction and the excited state "1," and is given by (see Eq. (3.61))

$$p_1 = \frac{G_{\rm L}}{2\pi^2 G_{\rm Q}} \int_0^\infty \frac{\xi \, {\rm d}\xi}{(\Delta E_{\rm L} + \xi)^2}. \tag{3.69}$$

This expression diverges at $\Delta E \to 0$. We remove this problem by taking into account that the excited state "1" is not stable: it decays with the rate $\Gamma_{\rm R}$. We account for this in the simplest fashion possible: we multiply the wave function of the state by a time-dependent factor that describes the reduction of probability to be in this state, $\psi \propto \exp(-{\rm i}Et/\hbar) \to \exp({\rm i}Et/\hbar - t\Gamma_{\rm R}/2)$. The factor $1/2$ is here because it takes the square of a wave function to create a probability. This is equivalent to an imaginary shift of the energy difference $\Delta E_{\rm L}$, which we incorporate into the denominator of Eq. (3.69) as follows:

$$\frac{1}{(\xi - \Delta E_{\rm L})^2} \to \frac{1}{|\xi - \Delta E_{\rm L} + {\rm i}\hbar\Gamma_{\rm R}/2|^2}. \tag{3.70}$$

After this modification, the probability and the current are reduced to the expressions that contain no divergencies at the threshold, and the expressions make sense even at $\Delta E_{\rm L} < 0$, where single-electron tunneling is enabled.

We check the validity of this expression at the single-electron side of the transition, where $|\Delta E_L|\Gamma_R \ll \hbar$. In this case, we decompose the denominator into the leading delta-function term and the rest:

$$\frac{1}{|\xi - \Delta E + i\hbar\Gamma/2|^2} \approx \frac{2\pi}{\hbar\Gamma}\delta(\xi - \Delta E) + \mathrm{Re}\left(\frac{1}{(\xi - \Delta E + i0)^2}\right) + O(\hbar\Gamma/\Delta E). \quad (3.71)$$

The leading term reproduces the result of the master equation approach $p_1 = \Gamma_L/\Gamma_R$, $I = e\Gamma_L$, and therefore presents the contribution of single-electron tunneling. The whole expression in the threshold region reduces to (see Fig. 3.24)

$$\frac{\partial I}{\partial V} = \frac{\partial}{\partial V}\int d\xi \frac{eG_L\xi\Gamma_R(2\pi^2 G_Q)^{-1}}{|\xi - \Delta E_L + i\Gamma_R/2|^2} = G_{th}\left(\frac{1}{2} + \frac{1}{\pi}\arctan\left(\frac{v}{v_r}\right)\right). \quad (3.72)$$

In Eq. (3.72), we change to universal notations: G_{th} is the differential conductance at the single-electron side of the transition that takes place at $V = V_{th}$, $v \equiv V - V_{th}$. The cusp of the $I-V$ curve is rounded off at voltage scale $v_r \equiv \hbar\Gamma_R/(2\partial\Delta_E/\partial V) \simeq (G/G_Q)E_C/e$. We have therefore quantified the transport near the threshold that are due to events which are neither co-tunnelings nor single-electron tunnelings.

While the separation given in Eq. (3.71) is artificial near the diamond boundary, it allows us to quantify the co-tunneling contribution in situations where it can be actually separated from the single-electron one. Let us concentrate on the vicinity of the degeneracy point between the "0" and "1" diamonds, assuming the energy difference w between the states at zero voltage. On the "0" side of the transition, $|w| \gg k_B T$, the current is mostly due to the co-tunneling. The rate is given by Eq. (3.64), where we include only the dominant contribution of the virtual state "1" and note that $\Delta E_{L,R} = \pm w - eV/2$:

$$\Gamma_{cot}^{(L\to R)}(0) = \frac{\hbar}{2\pi}\int d\xi \frac{\Gamma_L(-\xi + eV/2)\Gamma_R(\xi - eV/2)}{(w - \xi)^2}.$$

There is a divergence at $\xi = w$ provided both single-electron rates are enabled at this energy, that is, $\Gamma_L(\Delta E_L)\Gamma_R(\Delta E_R) \neq 0$. The divergence is clearly a contribution of single-electron processes to be subtracted. We employ the decomposition in Eq. (3.71), skip the leading term, and keep the next-to-leading one. The resulting rate is plotted in Fig. 3.25 versus $w/k_B T$. While its asymptotics at $w \gg k_B T$ agree with the co-tunneling result in Eq. (3.66), the rate is strikingly negative at $w \simeq k_B T$, achieving its minimum $-(G_R G_L/2\pi^3 G_Q^2)(k_B T/\hbar)$ at $w = 0$. A negative rate does not seem to make any sense. However, this is a rate for two separate processes going together under conditions where both processes may go separately. Thus the defined rate may be perfectly negative, presenting a small negative correlation correction to the positive rates of the two constituent processes. Indeed, the co-tunneling rate evaluated turns negative only if the rates of constituent single-electron processes are sufficiently large. In this case, the co-tunneling rate presents a small (G/G_Q) correction to these rates.

Generally, there are four distinct co-tunneling rates in this situation: two for co-tunneling in state "0" in two opposite directions, $\Gamma_{cot}^{(L\to R)}(0)$ and $\Gamma_{cot}^{R\to L}(0)$ and two for co-tunneling

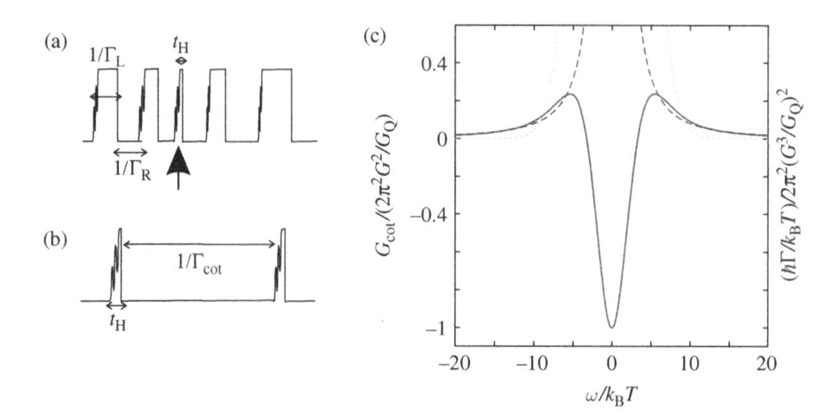

Fig. 3.25. Co-tunneling coexisting with single-electron tunneling near the degeneracy point. (a), (b) Cartoonish time-line of tunneling events. (a) At the Coulomb peak, ($w \simeq k_BT$) the transport is due to well separated single-electron tunnelings. The co-tunneling accounts for rare coincidences of tunneling events (shown by the arrow). (b) Far from the peak ($|w| \gg k_BT$), the transport is due to these rare coincidences – co-tunneling events. (c) Zero-voltage co-tunneling rate and correction to the conductance (solid curve). Dotted curve: single-electron contribution to the conductance given by Eq. (3.44) plotted for $G = 10G_Q$. Dashed curve: co-tunneling asymptotics corresponding to Fig. 3.23.

in state "1." The contributions of all four, weighted with the probabilities $p_{1,0}$ of being in a given state, make up the current:

$$\frac{I_{cot}}{e} = p_0\left\{\Gamma_{cot}^{(L\to R)}(0) - \Gamma_{cot}^{(R\to L)}(0)\right\} + p_1\left\{\Gamma_{cot}^{(L\to R)}(1) - \Gamma_{cot}^{(R\to L)}(1)\right\}.$$

Since the co-tunneling process does not change the charge state, the probabilities $p_{1,0}$ are determined by the balance of the single-electron processes. It turns out that, in the vicinity of the degeneracy point, $\Gamma_{cot}(0) = \Gamma_{cot}(1)$, so that $I_{cot} = e(\Gamma_{cot}^{(L\to R)}(0) - \Gamma_{cot}^{(R\to L)}(0))$. In the limit of vanishing voltage, $\partial\Gamma_{cot}/\partial(eV) = \Gamma_{cot}(V=0)/2k_BT$, so that the correction to zero-voltage conductance is $G_{cot} = \pi G_Q \hbar\Gamma_{cot}(V=0)/k_BT$. This enables us to plot the correction and $\Gamma_{cot}(V=0)$ in the same figure (see Fig. 3.25). At the peak, the correction is negative and equals $-G_RG_L/2\pi^2G_Q$. We have thus demonstrated that, in the vicinity of the degeneracy point, the co-tunneling contribution is separated from the single-electron tunneling processes and presents a correction of the order of G/G_Q. Formally speaking, this is not the only correction of this order. Quantum fluctuations change (renormalize) the energy differences between charging states by a factor $\simeq E_C(G/G_Q)$ with respect to energy differences without tunnel coupling. However, as we have already learned, the states without coupling are "subjective," presenting no valid reference, and renormalization corrections are hardly observed in experiment unless the tunneling conductance of the device can be changed during the course of the experiment.

The cross-over between the two scenarios discussed takes place when the temperature matches the fastest decay rate, $k_BT \simeq \hbar\Gamma_R$, that is, $eV \simeq (k_BT)(G/G_Q)$.

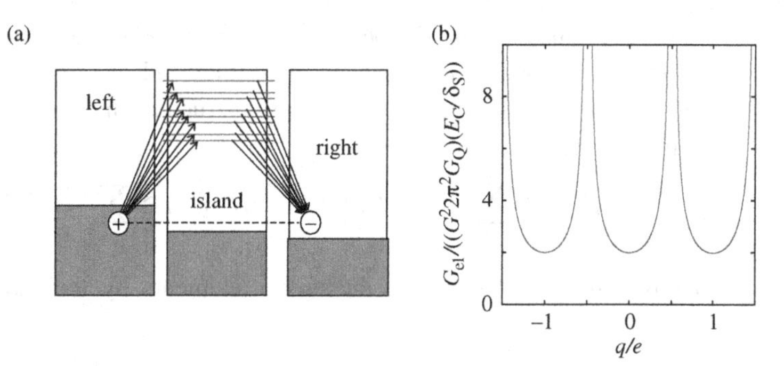

Fig. 3.26. **(a) The elastic co-tunneling amplitude is made up from the contributions of $\sim E_C/\delta_S$ discrete levels in the island. These contributions come with random phase shifts, resulting in a strong destructive interference. (b) Zero-voltage elastic conductance versus q.**

3.4.4 Elastic co-tunneling

So far, we have considered inelastic co-tunneling. In a SET, an elementary co-tunneling process transfers an electron from state l in the left lead to state r in the right, leaving an excitation in the island, with an electron in state s_1 and a hole in state s_2. An alternative is to set $s_1 = s_2$ so that no excitation is left in the island. This *elastic* co-tunneling is usually weak in metallic systems, where $E_C \gg \delta_S$, except when only small energy differences are available.

To quantify, we write the matrix element of the transition between the states l and r using Eq. (3.63):

$$\tilde{T}_{lr} = \sum_s T_{ls} F(E_s, E_l, E_r) T_{sr};$$
$$F(E_s, E_l, E_r) \equiv \left(\frac{1 - f(E_s)}{E^{(+)} + E_s - E_l} + \frac{f(E_s)}{E^{(-)} + E_r - E_s} \right). \tag{3.73}$$

Since the process is elastic, we can immediately relate these matrix elements to the rate and the corresponding conductance (see Eq. (3.32)):

$$G_{\rm el} = 2\pi^2 G_{\rm Q} \sum_{lr} |\tilde{T}_{lr}|^2 \delta(E_r - E)\delta(E_l - E).$$

Each electron state s in the box contributes to each $\tilde{T}_{lr}$, giving rise to two virtual states: one with $N = 1$ if the state is filled; one with $N = -1$ otherwise. It is constructive at this stage to recall the favorite model of Chapter 2: the double tunnel junction. As discussed, we encounter there a set of transmission resonances that become localized levels in the limit of low transparency of the junctions. These resonances are also found here in the resonant structure of the denominators $F(E_s)$. However, we see from Fig. 3.26 that the positions of the resonance are shifted from the Fermi level by the Coulomb energy $E^{(\pm)}$. This is why the transmission does not proceed through the resonance closest to E; rather, the transmission amplitude is contributed by the tails of many resonances, and the transmission probability is a result of interference of all pairs of resonances.

From the material of Chapter 2, we readily understand that this interference must be destructive: phase shifts cancel each other and average to zero. Let us assume for the moment that the interference is so destructive that the transmission probability is contributed by squares of amplitudes:

$$|\tilde{T}_{lr}|^2 = \sum_{s',s} T_{ls}T^*_{ls'}T_{sr}T^*_{s'r}F(E_s)F(E_{s'}) \approx \sum_s |T_{ls}|^2|T_{sr}|^2F^2(E_s, E).$$

In this case, we can replace the matrix elements by their averaged values, given by (Eq. 3.36):

$$\sum_r \delta(E - E_r)|T_{sr}|^2 = \delta_S \frac{G_R}{2\pi^2 G_Q};$$

similarly for the left junction. We also replace the summation over s by integration over E_s. We thus obtain ($k_B T, E \ll E_C$) the following

$$G_{el} = \frac{G_R G_L \delta_S}{2\pi^2 G_Q} \int dE_s \, F^2(E_s) = \frac{G_L G_R}{2\pi^2 G_Q}\left(\frac{\delta_S}{E^{(+)}} + \frac{\delta_S}{E^{(-)}}\right) \simeq \frac{G_R G_L}{G_Q}\frac{\delta_S}{E_C},$$

in agreement with previous estimations. The small factor δ_S/E_C in this expression comes from our assumption of destructive interference: if it were constructive, we would end up with $G_{el} \simeq G_R G_L/G_Q$. We revisit this assumption in Chapter 5, where it will become clear how it is related to the structure of wave functions of localized states.

The evaluation of multi-junction elastic co-tunnelling requires us to take steps similar to those we took to evaluate the inelastic co-tunneling in these structures. This includes the book-keeping of all intermediate virtual states and a subsequent summation over all possible sequences of elementary tunneling processes.

3.5 Macroscopic quantum mechanics

As we have learned, Coulomb blockade distinguishes states that differ by charge accumulated in an almost isolated metal island. At sufficiently low temperature and voltage, the island is in a certain charge state, while finite temperature and voltage causes transitions between different charge states. These states, however, are classical rather than quantum. The island is always supposed to be in a certain charge state, and never in a quantum superposition of the two.

The situation changes if both the island and the electrodes become superconducting. Superconductivity is essentially a coherent phenomenon. Gauge symmetry breaking associated with superconducting transitions gives rise to a new degree of freedom, the phase φ. A state of a superconductor, or of a piece of superconductor, is characterized by this phase. Manifestations of the coherence are superconducting currents in bulk superconductors caused by phase gradients and the Josephson current between two superconductors with different phases; the latter has been explained in Section 1.8.

What can we achieve by combining the Josephson effect and Coulomb blockade? Let us consider a small superconducting island connected by a Josephson junction with a bulk superconductor that is kept at zero phase. If an island is in a state with a certain phase, a Josephson current flows through the junction. This implies that the charge of the island cannot be constant, that is, certain. Vice versa, if the island is in the state with a certain charge, this implies that the phase of the island cannot be certain: a certain phase would produce a constant Josephson current to the island. It turns out that this dual uncertainty is of quantum nature, and is described by the Heisenberg uncertainty relation. As we will prove in this section, charge and phase are conjugated quantum variables, very much like the coordinate and the momentum of a quantum particle. Conjugated variables obey the Heisenberg uncertainty relation: if one variable is well defined, the other one is uncertain. Thus, if we combine the Josephson effect and Coulomb blockade, we make a quantum system which can be in a coherent superposition of the charge states.

Usually, discrete quantum states are associated with small particles: an example are the states of electrons in atoms. A quantum state arising from the combination of Coulomb blockade and the Josephson effect is associated with an object of micrometer scale: islands and junctions, which contain billions and billions of electrons and atoms. This is why it is frequently termed a *macroscopic* quantum state. Indeed, the micrometer scale is readily accessible for modern engineering: one could design circuits and manufacture devices at the micrometer scale as a hobby at home. The parameters and properties of such a quantum state, in sharp contrast with atomic states, can be changed both by nanostructure design and by turning handles of the device. This controllability brings about a potential for practical applications. Nanostructures of this type do not just realize a peculiar phenomenon of macroscopic quantum mechanics, they can also be used to realize artificial and controllable quantum states. Indeed, as we review in Chapter 5, the practical realization of qubits has been achieved by combining the Josephson effect and Coulomb blockade in several different ways.

3.5.1 Cooper-pair box and similar systems

We consider here the simplest system available to realize these macroscopic quantum states: a Cooper-pair box (CPB). The system is almost the same as the single-electron box discussed in Section 3.1: there is an island to store charges, a bulk electrode to provide these charges, and a gate electrode to shift the electrostatic potential of the island with respect to the bulk electrode and to tune the number of charges thereby. The crucial difference is that now both the island and source electrodes are superconducting, and the tunnel junction that connects them is in fact a Josephson junction (Fig. 3.27).

There are two energies that characterize the system. The first energy is the Josephson energy of the junction. As we know from Section 1.8, it is a periodic function of the superconducting phase difference across the junction, $-E_{\mathrm{J}} \cos \varphi$. It is convenient to set the phase of the lead to zero, so that the phase difference is just a phase that characterizes the superconducting state of the island. We also know from Section 1.8 that the relation between the Josephson energy and the transmission eigenvalues of the contact for the tunnel junction reduces to $E_{\mathrm{J}} = (G_{\mathrm{T}}/G_{\mathrm{Q}})\Delta/4$, where Δ is the energy gap in the superconductors. We recall that good isolation of the island, enabling Coulomb blockade effects, requires $G \gg G_{\mathrm{Q}}$, so that $E_{\mathrm{J}} \ll \Delta$.

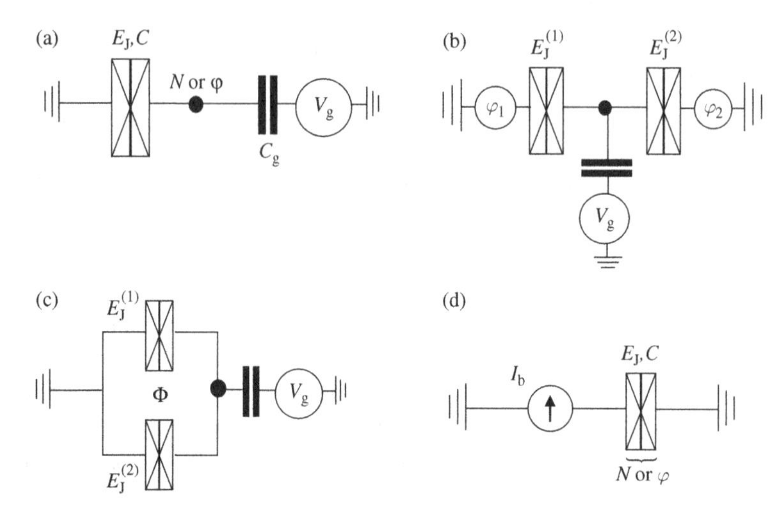

Fig. 3.27. **Cooper-pair box (CPB) and similar systems. A double-crossed box denotes a Josephson junction characterized by the Josephson energy E_J and capacitance C. (a) The CPB. (b) A superconducting SET is equivalent to a CPB with modified E_J. (c) This provides us with an opportunity to tune E_J with magnetic flux Φ. (d) A single current-biased Josephson junction is in many respects similar to a CPB. A crucial difference is that the charge in the capacitor does not have to be discrete.**

The second energy is the same as for the non-superconducting setup: the charging energy associated with discrete charge Ne in the island, given by

$$E_C(N - q/e)^2;\ E_C = \frac{e^2}{2(C + C_g)};\quad q \equiv C_g V_g.$$

Why is the system called a Cooper *pair* box? Each superconductor is a coherent reservoir of Cooper pairs rather then electrons, provided the energies involved are smaller than the superconducting energy gap. In this section, we will assume that there is always an *even* number of extra elementary charges in the island, and that charges are always transferred in Cooper pairs. More complicated situations are addressed in Section 3.7.

The CPB turns out to be close to two other physically different systems. What happens if we make yet another Josephson connection to another superconducting electrode? In doing so, we enable electron transfer between the electrodes through the island: we make a SSET, a superconducting version of a SET. A normal SET exhibits rather different physics from a single-electron box. For a SSET, an extra connection only means a modification in the Josephson energy of the CPB. Indeed, consider two superconducting leads with the phases $\varphi_{1,2}$ (Fig. 3.27(b)). The Josephson energy of two junctions is given by

$$-E_J^{(1)}\cos(\varphi - \varphi_1) - E_J^{(2)}\cos(\varphi - \varphi_2) = -\mathrm{Re}\left((\tilde{E}_J^{(1)} + \tilde{E}_J^{(2)})e^{i\varphi}\right),$$

where we have introduced the complex-valued Josephson amplitudes $\tilde{E}_J^{(1,2)} \equiv E_J^{(1,2)}\exp(-i\varphi_{1,2})$. We see that the amplitudes of different superconducting leads add up in the expression for energy, and the resulting energy is the same as for a single junction with the effective Josephson energy given by

$$E_\mathrm{J}^{(\mathrm{eff})}(\varphi_1 - \varphi_2) = \left|\tilde{E}_\mathrm{J}^{(1)} + \tilde{E}_\mathrm{J}^{(2)}\right|$$
$$= \sqrt{(E_\mathrm{J}^{(1)})^2 + (E_\mathrm{J}^{(2)})^2 + 2\cos(\varphi_1 - \varphi_2) E_\mathrm{J}^{(2)} E_\mathrm{J}^{(1)}}\,, \tag{3.74}$$

and a proper shift of the phase φ. This equivalence offers the practically important possibility of tuning the actual value of E_J by changing magnetic flux if necessary. To understand this, let us take a single superconducting lead connected to the island with two junctions (Fig. 3.27(c)). If there is a magnetic flux in the loop created by these junctions, it induces a phase difference $2\pi\Phi/\Phi_0$ between the points of the lead adjacent to the junctions. So these two junctions may be replaced by one with flux-dependent Josephson energy:

$$E^{(\mathrm{eff})}(\Phi) = \sqrt{(E_\mathrm{J}^{(1)})^2 + (E_\mathrm{J}^{(2)})^2 + 2\cos(2\pi\Phi/\Phi_0) E_\mathrm{J}^{(2)} E_\mathrm{J}^{(1)}}. \tag{3.75}$$

A less evident similarity is between the Cooper-pair box and a single Josephson junction under *current-bias* conditions (Fig. 3.27(d)). While the phase difference between two voltage-biased superconducting leads is set by these leads, the phase across the current-biased junction may change in time, following its own dynamics. The dynamics is governed by the same two energies: the Josephson energy and the charging energy of a capacitor associated with the junction. The difference between the current-biased junction and the Cooper pair box is that the charge accumulated on the capacitor does not have to be discrete as it is in the isolated island. In the following, we will see just how interesting the quantum manifestation of this difference is.

3.5.2 Second quantization

The two energy scales discussed above have been considered under the assumption that both the phase and the charge in the island behave as classical, well defined variables. But we have already learned that this cannot be true; they cannot be simultaneously well defined. So we have to understand the precise relationship between the phase and the charge. This is achieved by a *second quantization* procedure, which can be applied to any system, the classical dynamics of which is known. Examples of such systems include an electromagnetic field in a vacuum, governed by the Maxwell equations, and mechanical oscillations in atomic lattices, governed by elasticity. Both systems, and many others, can be successfully quantized: one builds up a quantum description of the system from the classical one. The most general quantization prescription is as follows.

(i) Replace classical variables by operators.
(ii) Postulate commutation relations between these operators.
(iii) Derive Heisenberg equations of motion from these commutation relations.
(iv) Check if these equations of motion correspond to classical ones.

The second point, (ii), is a creative one and requires some guesswork. Fortunately, for most systems the simplest guess works. For systems obeying linear equations of motion, the procedure is especially automated since, in this case, the problem is reduced to the quantization of a (large) number of harmonic oscillators.

To accomplish the quantization of our system, we should thus compare classical and quantum equations of motion. We will do this now explicitly for a simplified form of Josephson energy, $-E_J \cos\varphi \to \text{const.} + E_J\varphi^2/2$, which is obviously valid for $|\varphi| \ll \pi$. It implies that, with this accuracy, the Josephson junction can be replaced by an inductance. Indeed, in this approximation the current through the junction is given by$I = -I_c\varphi$ and $\dot{I} = I_c 2e/\hbar V$. By the definition of inductance, $L \equiv V/\dot{I} = \hbar/2eI_c$. In addition, we bypass the gate with the charge induced by it, so the charging energy is simply $e^2N^2/2C$. The reason for the simplification is that the resulting equations are nicely linear. The classical equations of motion comprise the charge conservation law,

$$\frac{\mathrm{d}Q}{\mathrm{d}t} = -I(\varphi) = -I_c\varphi, \tag{3.76}$$

and the Josephson relation,

$$\frac{\mathrm{d}\varphi}{\mathrm{d}t} = \frac{2eV}{\hbar} = \frac{2eQ}{C\hbar}, \tag{3.77}$$

where $Q = CV$ is the charge in the island.

Let us change the notation, making use of

$$Q = eN; \ E_C = \frac{e^2}{2C}; \ I_c = \frac{2e}{\hbar}E_J.$$

On doing so, we can bring these classical equations into a nice symmetric "quantum" form:

$$\frac{\mathrm{d}N}{\mathrm{d}t} = \frac{2E_J}{\hbar}; \ \frac{\mathrm{d}\varphi}{\mathrm{d}t} = \frac{4E_C}{\hbar}N. \tag{3.78}$$

Let us postulate the commutation relations. We assume

$$\left[\hat{N}, \hat{\varphi}\right] = \alpha,$$

where α is a constant that needs to be determined. As we have already mentioned, this resembles the commutation relations between the coordinate and momentum in one-particle quantum mechanics, $\left[\hat{p}, \hat{x}\right] = -\mathrm{i}\hbar$. The simplified Hamiltonian is the sum of Josephson and charging energy:

$$\hat{H} = E_J\frac{\hat{\varphi}^2}{2} + E_C\frac{\hat{N}^2}{2}.$$

Basic quantum mechanics states that the quantum equation of motion for an arbitrary operator $\hat{A}$ is given by

$$\frac{\mathrm{d}\hat{A}}{\mathrm{d}t} = \frac{\mathrm{i}}{\hbar}\left[\hat{H}, \hat{A}\right].$$

Applying this to our systems, we have to deal with the commutators $\left[\hat{\varphi}^2, \hat{N}\right]$ and $\left[\hat{N}^2, \hat{\varphi}\right]$, which are calculated as follows:

$$\begin{aligned}\hat{\varphi}^2\hat{N} - \hat{N}\hat{\varphi}^2 &= (\hat{\varphi}^2\hat{N} - \varphi\hat{N}\hat{\varphi}) - (\hat{N}\varphi^2 - \varphi\hat{N}\hat{\varphi}) \\ &= \hat{\varphi}\left[\hat{\varphi}, \hat{N}\right] - \left[\hat{N}, \hat{\varphi}\right]\hat{\varphi} = -2\alpha\hat{\varphi}.\end{aligned}$$

In this way, we obtain

$$\frac{dN}{dt} = \frac{i}{\hbar}[\hat{H}, \hat{N}] = -\frac{i\alpha E_J}{\hbar}\hat{\varphi};$$
$$\frac{d\varphi}{dt} = \frac{i}{\hbar}[\hat{H}, \hat{\varphi}] = \frac{i2\alpha E_C}{\hbar}\hat{N}.$$

Comparison with Eqs. (3.78) yields $\alpha = -2i$.

The commutation relation is very instructive, since we can use the whole apparatus of one-particle quantum mechanics and borrow physical analogies from this. Thus motivated, we introduce the wave function of the Cooper-pair box. We write it in the phase representation, $\Psi(\varphi)$, which is analogous to the coordinate representation in one-particle quantum mechanics. The operator N in this representation becomes $-2i\partial/\partial\varphi$ (recall the relation $\hat{p} \rightarrow -i\hbar\partial/\partial x$ for the momentum operator in one-particle quantum mechanics). We conclude that the wave function of a state with a certain charge N is a "plane wave" in phase representation:

$$|N\rangle \rightarrow e^{iN\varphi/2}. \tag{3.79}$$

What about the phase operator? First let us recognize that we do not need it directly. The actual Josephson energy is not quadratic in phase; rather, it is made of two exponents of the phase operator, $2\cos\varphi = \sum_{\pm} e^{\pm i\varphi}$. We see from Eq. (3.79) that these exponents either increase or reduce N by 2:

$$e^{\pm i\hat{\varphi}}|N\rangle = |N \pm 2\rangle. \tag{3.80}$$

The alternative representation of this operator is therefore given by

$$e^{\pm i\hat{\varphi}} = \sum_N |N \pm 2\rangle\langle N|. \tag{3.81}$$

We come to an interesting and physically appealing conclusion: the Josephson effect is associated with the coherent transfer of a Cooper pair (charge $2e$) via the junction in either ($N \rightarrow N \pm 2$) direction.

Given the Hamiltonian $\hat{H} = -E_J\cos\hat{\varphi} + E_C(\hat{N} - q)^2$, we come to the Schrödinger equation for the wave function Ψ. We can do this in either phase,

$$E\Psi(\varphi) = \hat{H}\Psi = \left[E_C\left(-2i\frac{\partial}{\partial\varphi} - \frac{q}{e}\right)^2 - E_J\cos\varphi\right]\Psi(\varphi), \tag{3.82}$$

or charge,

$$E\Psi(N) = \hat{H}\Psi = E_C\left(N - \frac{q}{e}\right)^2\Psi(N) - \frac{E_J}{2}\left(\Psi(N-2) + \Psi(N+2)\right), \tag{3.83}$$

representation, the relation between these representations being given by

$$\Psi(N) = \int d\varphi\, \Psi(\varphi)\exp(-iN\varphi/2).$$

The solutions of the Schrödinger equation determine the actual quantum states and energy levels of our superconducting system. We go on with the useful analogy between φ and a coordinate of a particle. The Schrödinger equation, Eq. (3.82), is that for a particle in cosine potential given by Josephson energy. The charging energy is associated with the kinetic energy of the particle with mass $\hbar^2/2m = 4E_C$. A smaller charging energy therefore corresponds to a heavier, "less quantum" particle. We will use this analogy a great deal and we will refer to the system as a "particle" without warning.

Before dealing with the equation, we must discuss one issue. The Josephson energy and the whole Hamiltonian of the system do not change if the phase φ is shifted by 2π. This brings up two possibilities. The first is that the quantum states that differ by phase 2π are just identical, $|\varphi\rangle = |\varphi + 2\pi\rangle$. This implies that the wave function in the phase representation must be *periodic* in phase. If we make use of our analogy between the phase and particle coordinate, this corresponds to a particle confined on a circle. The coordinate of the particle can be parameterized with polar angle θ, but θ shifted by 2π corresponds to the same point on a circle.

The alternative is that these states are different and distinguishable, so that $|\varphi\rangle \neq |\varphi + 2\pi\rangle$. The particle in this case is free to move in the whole one-dimensional space, but feels the periodic potential. It is discussed in solid state physics that the solutions of such an equation are Bloch functions that obey the Bloch boundary condition $\Psi(\varphi + 2\pi) = \exp(\mathrm{i}\pi\tilde{q})\Psi(\varphi)$. The parameter $\tilde{q}$ plays a similar role to the quasimomentum in solid state physics: it labels different groups of quantum states. This is why it is called *quasicharge*.

It turns out that both possibilities are physical. They describe two distinct physical situations and two different Josephson junction setups. For the Cooper-pair box, the choice is a periodic boundary condition. This makes the states discrete and provides the charge quantization. Indeed, if we disregard the Josephson energy for a while, we get an isolated island. The solutions of the Schrödinger equations are as follows:

$$\Psi_N = \exp(\mathrm{i}N\varphi/2),$$

and they satisfy periodic boundary conditions only if N is an even integer.

Bloch boundary conditions describe the quantum states of a single Josephson junction under current bias. The charge in this case is just the charge accumulated on the plates of the capacitor. The charge does not have to be integer: it is continuous. Indeed, in the limit of vanishing E_J, there are solutions to Schrödinger equation

$$\Psi_N = \exp(\mathrm{i}(N + \tilde{q})\varphi/2)$$

for any charge $(\tilde{q} + N/2)e$.

3.5.3 Charging energy dominating

There are two energy scales characterizing the Cooper-pair box, E_C and E_J, and the character of the quantum states depends on ratio of the two. We will look now at two opposite limits or regimes, $E_J \ll E_C$ and $E_J \gg E_C$. It is useful to note that the actual cross-over between the regimes occurs at $E_J/E_C \simeq 8$. This is due to a play of numerical factors: in fact, one has to compare diagonal and non-diagonal elements of the Hamiltonian given in Eq. (3.83), i.e. $4E_C$ and $E_J/2$.

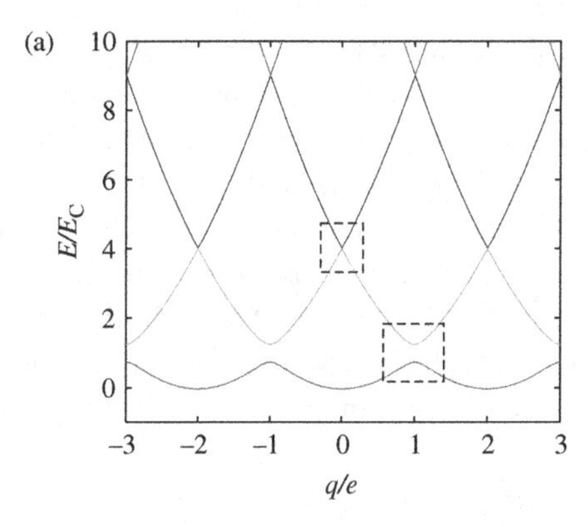

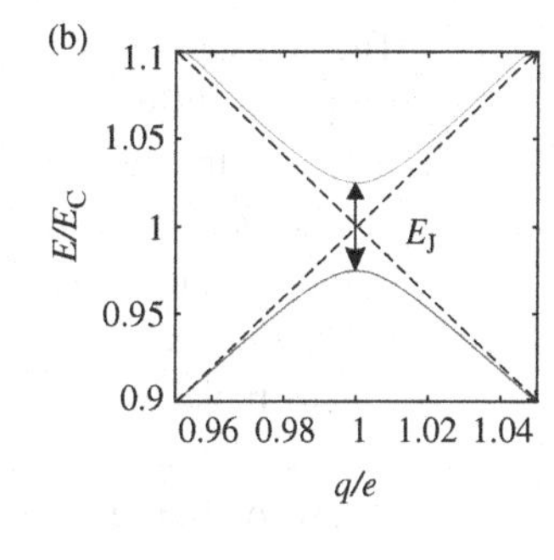

Fig. 3.28. (a) Energy levels in a Cooper-pair box at $E_J/E_C = 0.5$ versus q. The quantum states are almost pure charge states except at the parabola crossings. (b) The vicinity of the lowest crossing at $q = 1$ ($E_J/E_C = 0.05$). The eigenstates are superpositions of the two states that cross. A similar picture of repelling levels is valid for any crossing (for example at $q = 0$).

We start with the first regime when the charging energy dominates and the quantum states in the Cooper-pair box are almost pure charge states. Their energies are the same as for a non-superconducting island: if we plot them versus q (see Fig. 3.28(a)), we recognize the familiar sequence of shifted parabolas. The only detail is that the number of extra electrons is even, so the energies of the states are periodic in q with period $2e$ rather than e.

An interesting phenomenon takes place at critical values of q corresponding to the crossings of the parabolas. Small but finite E_J lifts the degeneracy in the crossing points and opens up the gaps between two charge states that are trying to cross. We have here the situation of *avoided level crossing*, which is commonplace in quantum physics. To look at this in more detail, we concentrate on the vicinity of the crossing between charge states $|0\rangle$ and $|2\rangle$ at the induced charge $q = e$ (see Fig. 3.28(b)). The Hamiltonian reduces to a 2×2 matrix in the basis of these two states, i.e.

$$\hat{H} = E_C + \begin{bmatrix} \epsilon & E_J/2 \\ E_J/2 & -\epsilon \end{bmatrix}, \tag{3.84}$$

where the Josephson energy provides the non-diagonal matrix elements between $|0\rangle$ and $|2\rangle$, and $2\epsilon \approx 4E_C(1 - q/e)$ is the splitting of the charge states in the absence of Josephson coupling. This reduction is valid provided $\epsilon \ll E_C$. Two eigenvalues and eigenvectors are given by

$$\begin{aligned} E_\pm &= E_C \pm \sqrt{\epsilon^2 + (E_J/2)^2}; \\ |+\rangle, |-\rangle &= \frac{1}{\sqrt{2(1 \pm \sin\theta)}}(|0\rangle \cos\theta + |2\rangle(\sin\theta \pm 1)); \\ \theta &= \arctan(2\epsilon/E_J). \end{aligned} \tag{3.85}$$

The minimum splitting (the gap) is E_J. At this point, the eigenstates are superpositions of two charge states with equal weights, $|\pm\rangle = (|0\rangle \mp |2\rangle)/\sqrt{2}$. Therefore, in the close vicinity of the crossings the quantum states are coherent superpositions of the states with different charge.

The crossings at higher energies are also avoided. However, the parabolas that cross at these points differ by more than two charges, so they cannot be directly coupled by $E_{\rm J}$. That is why the couplings appear in higher orders in $E_{\rm J}$; the estimation of the gap between parabolas that differ by $2m$ charges is $E_{\rm J}(E_{\rm J}/E_{\rm C})^{m-1} \ll E_{\rm J}$.

Let us evaluate the superconducting current through the SSET in this regime. As explained above, the energy of a SSET is obtained from the energy of a Cooper-pair box (CPB) by replacing $E_{\rm J} \to E_{\rm J}^{(\rm eff)}(\varphi)$ (Eq. (3.74)). We use the relation $I = -(2e/\hbar)(\partial E/\partial\varphi)$ to express the current through the phase-dependent part of the energy. The latter is given by the second-order perturbation correction to the energy of the ground state:

$$I(\varphi) = \frac{2e}{\hbar}\frac{\partial E^{(2)}}{\partial\varphi};$$

$$E^{(2)} = -\left(\frac{E_{\rm J}^{(\rm eff)}}{2}\right)^2\left(\frac{1}{E_+}+\frac{1}{E_-}\right); \tag{3.86}$$

$$I(\varphi) = \frac{e}{\hbar}E_{\rm J}^{(1)}E_{\rm J}^{(2)}\sin\varphi\left(\frac{1}{E_+}+\frac{1}{E_-}\right) \simeq I_{\rm c}(E_{\rm J}/E_{\rm C}), \tag{3.87}$$

where $E^{\pm}$ are the energy differences between the ground charge state and the closest excited states that differ by ± 2 elementary charges. Equation (3.87) bears a close resemblance to the co-tunneling matrix element given by Eq. (3.63): it comprises two Josephson amplitudes of two junctions and energy denominators of virtual states. Indeed, charge transfer in this situation can be viewed as a co-operative tunneling of two Cooper pairs via two junctions. This suppresses the superconducting current by a factor $\simeq E_{\rm J}/E_{\rm C}$.

Exercise 3.9. (i) Prove Eq. (3.86) and explain why it does not hold in the close vicinity of the crossing point. (ii) Find the current–phase relation in the vicinity of the crossing point using Eq. (3.85). (iii) Find the critical current from the current–phase relation obtained in (ii).

3.5.4 Josephson energy dominating

In the opposite limit, $E_{\rm J} \gg E_{\rm C}$, we expect the states to have a well defined phase. For the ground state and several excited states this is indeed the case. To get the picture, let us use the analogy with one-particle quantum mechanics and associate φ with the coordinate of the particle. The particle feels the potential profile $-E_{\rm J}\cos\varphi$. Since $E_{\rm C}$ is small, the particle is "heavy" and therefore almost classical. Therefore, in the ground state it is localized near the minimum of the potential at $\varphi = 0$, the energy of this state being $-E_{\rm J}$. The fluctuation of the phase in the ground state is small, $\sim(E_{\rm C}/E_{\rm J})^{1/4} \ll \pi$.

Low-energy excited states correspond to oscillations of the particle near this minimum. At energies much smaller than $E_{\rm J}$ the potential profile, as discussed, can be approximated by a parabola $E_{\rm J}\varphi^2/2$. The excited states are those of a harmonic oscillator: equidistant levels separated by $\hbar\omega_{\rm p} = \sqrt{8E_{\rm J}E_{\rm C}} \ll E_{\rm J}$, where $\omega_{\rm p}$ is the so-called "plasmon" frequency, the frequency of the oscillator. The higher the energy, the greater the amplitude of the

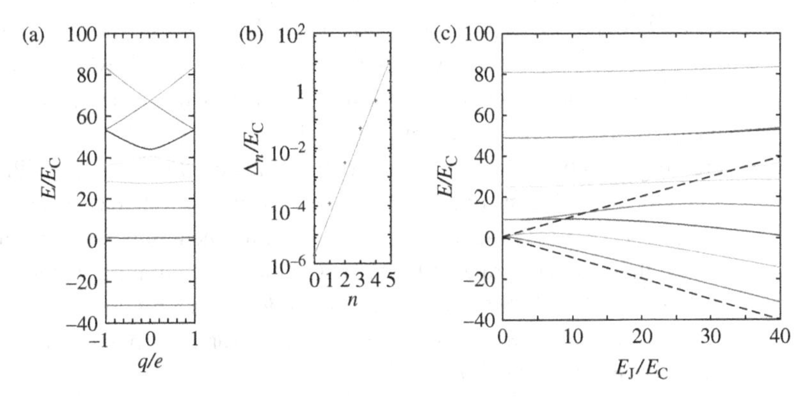

Fig. 3.29. (a) Energy levels in a CPB at $E_J/E_C = 20$ versus q. The states in the energy interval $(-E_J, E_J)$ are localized in phase. (b) The tunneling amplitudes determining the charge dependence of the phase-localized states. The solid line depicts the semiclassical result. (c) Evolution of the states with increasing E_J/E_C at $q = e$.

oscillations, so that the uncertainty of the phase gradually increases. The energies of all levels between $-E_J$ and E_J show an exponentially weak dependence on q. They are classically localized since the classical particle cannot reach the maximum of the potential E_J at $\varphi = \pm\pi$.

> **Exercise 3.10.** Produce a semiclassical estimate for the number of localized states. (i) Consider a particle in the periodic potential $U = -E_J \cos(x/a)$ and write down the Bohr–Sommerfeld quantization rule (see Appendix A) $\oint p \, dq = 2\pi\hbar(n + 1/2)$. (ii) Calculate the number of the state corresponding to the energy E_J (which gives the total number of states). (iii) Write down the equation for the CPB and, mapping it to the equation of motion of the classical particle, calculate the number of localized states. Hint: $N_c = (8E_J/\pi^2 E_C)^{1/2}$.

The q-dependence of the energy manifests if the state is localized or delocalized in phase space. The point is that q enters the Hamiltonian in very much the same way as magnetic flux enters the Hamiltonian of a particle on a circle, its effect being the interference of trajectories enclosing different flux (Section 1.6). All closed trajectories on the circle can be classified according to their winding number n_w: the number of full cycles the trajectory makes around the circle. The propagation amplitudes corresponding to these trajectories acquire the following phase factor:

$$F = \exp(i n_w \pi q/e). \tag{3.88}$$

For a localized state, classically allowed trajectories never cross the potential maximum E_J at $\varphi = \pm\pi$, so $n_w = 0$ for all trajectories. Indeed, the q-dependence is invisible for classically localized states in Fig. 3.29. In order to make a full circle in either direction ($n_w \pm 1$), a particle in a localized state has to *tunnel* through the potential maximum. The finite amplitude of this tunneling gives rise to an exponentially small q-dependence of the nth localized state: $E_n(q) = E_n + \Delta_n \cos(\pi q/e)$. In the limit $E_J \gg E_C$ the q-dependent part of the energy is given for the lowest localized state by $\Delta_0 = 16(E_J^3 E_C/2\pi^2)^{1/4} \exp(-\pi N_c)$. The exponential factor in this formula originates from the semiclassical expression for the transmission probability through the barrier, $T \propto \exp(-(2/\hbar)\int |p| dq)$, where p and q are, respectively, the classical momentum and coordinate for the CPB found by comparison with the equations of motion for the classical particle (see Exercise 3.10). Here, $N_c = (8E_J/\pi^2 E_C)^{1/2}$ is the semiclassical estimate for the number of localized states. For higher states, the following rule-of-thumb estimation is commonly applied: $\Delta_n = \Delta_0 \exp(\pi n)$. As shown in Fig. 3.29(b), it gives a good order-of-magnitude estimation. We note the rapid increase of the amplitude with increasing n: for each next excited state, the amplitude increases by $e^\pi \approx 23$ times.

A particle with energy higher than E_J crosses the potential maximum in the course of classical motion; this enables the dependence on q. Finally, at energies much bigger than $2E_J$ the energy of the particle is almost entirely kinetic (that is, charging) and it hardly sees the potential. The eigenfunctions are plane waves corresponding to states of certain charge, and the energy levels resemble the crossing parabolas.

The evolution of quantum states with increasing E_J/E_C ratio is illustrated in Fig. 3.29(c). The charge q is set to e, so the charging states are degenerate at $E_J = 0$. The degeneracy is visibly lifted only if E_J is of the order of the charging energy of a given state. At smaller E_J, the state is localized in charge. At larger E_J, the energy of the state goes down. The state becomes increasingly localized in phase and is finally converted into a plasmon excitation.

3.5.5 Macroscopic quantum tunneling

Let us turn from a CPB to a single current-biased Josephson junction. We have seen that the difference in the quantum description of these two systems is the phase range: while the phase of the superconducting island is confined to the interval from $-\pi$ to π, the phase across the single junction can take any value. Besides, if the bias current $I_b \neq 0$, it brings the extra term into the Hamiltonian:

$$H_b = \frac{\hbar}{2e}\hat{\varphi} I_b. \tag{3.89}$$

Indeed, only this term guarantees that the classical equation, Eq. (3.76), is correctly modified: $dQ/dt = -I_c \sin\varphi + I_b$. Thus, at non-zero bias current the particle feels a tilted cosine potential, or a *tilted washboard* potential (see Fig. 3.30). This potential has a series of local minima separated by 2π provided $|I_b| < I_c$.

The current-bias Josephson junction is an exemplary theoretical model of macroscopic quantum variable and conveys a clear and comprehensive physics. What is difficult to comprehend and explain is why the part of this physics – let us call it *coherent tunneling* – is almost impossible to realize, while the other – *incoherent* tunneling – has not only been realized, but still exemplifies our achievements in macroscopic quantum mechanics. This suggests the following plan: first we explain coherent tunneling, then we concentrate on the difficulties of its realization. This will bring us to incoherent tunneling.

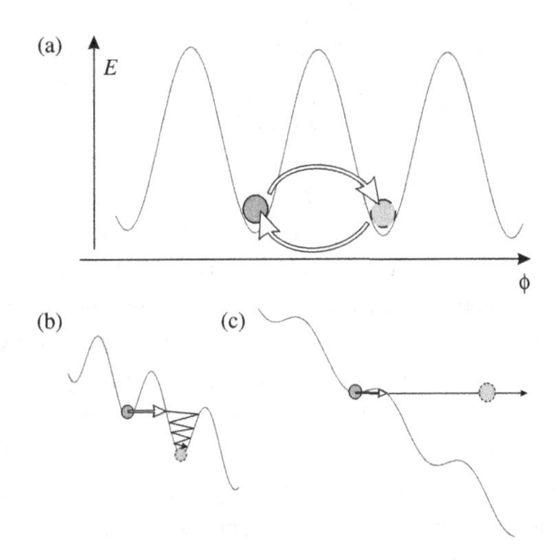

Fig. 3.30. **Macroscopic quantum tunneling in a single Josephson junction. White arrows: quantum tunneling; black arrows: classical motion. (a) Coherent tunneling forms Bloch states that are extended in phase space. (b) Incoherent tunneling in washboard potential at $I_b < 0.22I_c$. The particle is retrapped in the neighboring potential minimum. (c) Incoherent tunneling at $I_b > 0.22I_c$. After tunneling, the particle moves steadily to the right. The junction is switched to the finite-voltage state.**

Let us start with zero bias current. The Schrödinger equation is that of a particle in a periodic (cosine) potential: the eigenfunctions are thoroughly studied in solid state physics and are called Bloch states. They are labeled with the discrete band index n and continuous quasicharge $\tilde{q}$, $-e < \tilde{q} < e$, and are conveniently presented in the form

$$\Psi_{n,\tilde{q}}(\varphi) = u_{n,\tilde{q}}(\varphi)\mathrm{e}^{\mathrm{i}\pi\varphi\tilde{q}/e}, \tag{3.90}$$

where u_n is periodic in φ. These wave functions are extended in phase space and pretty much localized in charge space, irrespective of the ratio E_J/E_C. To understand this, we assume $E_J \ll E_C$, expecting the quantum state to be well localized in phase in one of the minima of the washboard potential. However, all minima are equivalent. The particle can tunnel to any neighboring minimum with no energy cost, and do it again and again (see Fig. 3.30(a)). Therefore, the particle is delocalized in phase space as a result of coherent tunneling between the minima.

We do not have to formulate a new calculation to determine the energy spectrum and the eigenfunctions; we can re-use the results for the CPB levels and states. The point is that u_n satisfies the same equation as the periodic wave function of CPB with $q = \tilde{q}$, n being the number of the CPB level. We stress that this does not imply the equivalence of the states of the CPB and the single junction. While the energy spectrum of the CPB is discrete, the spectrum of a single junction is continuous, and $\tilde{q}$ is not an external parameter set by the gate voltage: it is a dynamical variable characterizing the system.

The properties of Bloch states determine a simple and distinct physics of the current-biased quantum Josephson junction. The best way to present it is to use the *duality* between voltage-biased and current-biased situations. The voltage-biased Josephson junction is characterized by the phase-dependent energy $E(\varphi)$. At $V = 0$, the junction can be in a state with finite current $I = (2e/\hbar)(\mathrm{d}E/\mathrm{d}\varphi)$ not exceeding the critical current $I_c = (2e/\hbar)\max(\mathrm{d}E/\mathrm{d}\varphi)$. If voltage V is applied, the phase monotonically increases with time due to the Josephson relation, $\dot{\varphi} = 2eV/\hbar$. The current, being a periodic function of the phase, oscillates with the Josephson frequency $\omega_J = 2eV/\hbar$. The dual setup is the current-biased junction. It is characterized by a $\tilde{q}$-dependent energy $E_0(\tilde{q})$. ("0" refers here to the lowest-in-energy band.) At $I = 0$, the junction can be in a state with finite *voltage* $V = \mathrm{d}E/\mathrm{d}\tilde{q}$ not exceeding the critical voltage $V_c = \max(\mathrm{d}E_0/\mathrm{d}\tilde{q})$. If a (sufficiently small) current is applied, the quasicharge monotonically increases with time, $\dot{\tilde{q}} = I$. The voltage over the junction, being a periodic function of $\tilde{q}$ with period $2e$, oscillates with the Bloch frequency $\omega_B = I/2e$. This reveals the fact that the charge is transferred by discrete Cooper pairs, the transfers being strongly correlated in time.

Despite significant theoretical and experimental efforts, pure Bloch physics has never been realized in actual devices. The crucial difficulty is in achieving sufficiently good current bias. As discussed in Section 1.7, this can be easily achieved for low frequencies, while at higher frequencies the capacitance coupling between the leads effectively shunts the nanostructure. The resulting environmental impedance, Z_{env}, is typically much less than G_Q^{-1}. This impedance causes dissipation, which is absent in the model under consideration. Dissipative quantum mechanics will be considered in Chapter 6, and we will see that the coherence of many tunneling events in a single junction requires $Z_{\mathrm{env}} > G_Q^{-1}$ over a wide frequency region. This is similar to the condition of good isolation of a Coulomb island, $G_T \ll G_Q$, but the point is that *no* distinct island should be formed at either side of the junction: the charge in the island would become discrete rather than continuous as implied by Bloch physics, and we are back to either a SSET or a CPB configuration. A complicated, but technologically possible, way of approaching the isolation required is to place uniform high-resistive leads very close to the junction. In this way [68] the signatures of Bloch oscillations have been observed. The dissipation remaining in the setup suppresses the actual coherence of the tunneling events.

The current-bias condition is easier to fulfil for a single tunneling event not implying the coherence of successive events. Assuming $E_J \gg E_C$, we estimate the duration of a tunneling event as the inverse of the plasmon frequency, ω_p^{-1}. The impedance of the junction in this frequency range is estimated as $Z_j \simeq 1/\omega_p C \simeq G_Q^{-1}$. The current bias requires $Z_{\mathrm{env}} \gg Z_j$. This can be satisfied by $Z_{\mathrm{env}} \ll G_Q^{-1}$. It is interesting to note that the ratio Z_j/Z_{env} is simply the quality factor of the resonance at plasmon frequency: $\mathcal{Q} = \omega_p Z_{\mathrm{env}} C$. The current-bias condition therefore corresponds to negligible damping of the plasmon oscillations. The junction in this regime is called "underdamped."

We return to the washboard potential and concentrate on sufficiently big tilt $I \lesssim I_c$ (see Fig. 3.30(b)). Quantum effects aside, the particle is localized in one of the minima of the washboard potential. This is the zero-voltage state of the junction: on average, the particle does not move, and thus the time-averaged voltage is zero. Since the potential is tilted, it is energetically favorable to increase the phase. At sufficiently low temperature, this involves quantum tunneling through the potential barrier. After tunneling, the particle exerts classical motion, with the energy corresponding to the initial energy in the minimum. If this energy does not exceed the energy of the next potential minimum (this corresponds to $I < 0.22I_c$), the particle oscillates in the neighboring well. Eventually, it dissipates energy to the environment and stops at the bottom of the well. Then the next tunneling event takes place. Since the energy is dissipated between the tunneling events, they are evidently incoherent. If $I > 0.22I_c$, the classical motion after the tunneling event is unrestricted; the particle has sufficient energy to transverse all the potential barriers and to travel down the slope (see Fig. 3.30(c)). If the dissipation due to the environment is strictly zero, the particle always accelerates. The finite dissipation stabilizes its velocity at some finite average value. By virtue of the Josephson relation, this means the average voltage across the junction. Therefore, a single tunneling event at $I > 0.22I_c$ switches the junction between two stable regimes: zero-voltage and finite-voltage. This facilitates the observation of single-tunneling events and the accurate determination of the rate of these events. The usual measurement setup is as follows: the current is ramped up from a low value until the switching to the finite-voltage state takes place. After the finite voltage is detected, the current is brought back to a low value. The measurement is repeated many times until the switching current distribution is characterized. The tunneling rate is then extracted from this distribution. Macroscopic quantum tunneling was experimentally observed in 1981 [69] under conditions where quantum tunneling was affected by dissipation. Later experiments were performed [70] with an improved quality factor $Q \simeq 30$.

The rate is exponentially suppressed and can be estimated from the semiclassical formula as $\Gamma \propto \exp(-2\pi N_c)$, N_c being number of the levels in the inverted potential. In turn, N_c can be roughly estimated from the barrier height u_0 and the frequency of the plasmon oscillations around the minimum, since the latter determines the level separation, $N_c \simeq u_0/\hbar\omega_p$. This is why the exponential part can be presented as $\Gamma \propto \exp(-\alpha u_0/\hbar\omega_p)$, where α is a dimensionless coefficient of the order of 2π. The measurements at finite temperature distinguish between quantum tunneling and classical thermal activation. The latter is proportional to the probability of being at the potential maximum, $\simeq \exp(-u_0/k_B T)$. Comparing this with the quantum estimate, we see that quantum tunneling dominates at $k_B T < \hbar\omega_p/\alpha$.

The tunneling rate substantially increases close to the critical current, where $\tilde{I} \equiv (I_c - |I_b|)/I_c \ll 1$ since the potential barrier becomes lower and disappears at $\tilde{I} = 0$. Under these circumstances, the Josephson potential can be approximated by a cubic parabola ($\chi = \varphi - \pi/2, |\chi| \ll \pi$), given by

$$U(\chi) = \text{const.} + E_J\left(-\tilde{I}\chi + \chi^3/6\right). \tag{3.91}$$

Exercise 3.11. Evaluate the rate of macroscopic quantum tunneling in the cubic parabola approximation. (a) Compute ω_p and u_0 for the cubic parabola potential given in Eq. (3.91). Result: $u_0 = (2^{5/2}/3)\tilde{I}^{3/2}E_J$; $\omega_p = 2^{7/4}\sqrt{E_J E_C}\left(\tilde{I}\right)^{1/4}$. (b) Use the semiclassical formula to estimate the rate with exponential accuracy. Present the estimation in the form $\Gamma \propto \exp(-\alpha u_0/\hbar\omega_p)$ and give the expression for α. Result: $\alpha = 36/5$. (c) Estimate at which value of $\tilde{I}$ the above estimate becomes invalid. Explain why it is not possible to measure the tunneling rate in this range of $\tilde{I}$. Result: $\tilde{I} \lesssim (E_C/E_J)^{2/5}$.

(d) The asymptotically exact expression for the rate is given by

$$\Gamma = \frac{\omega_{\rm p}}{2\pi}\left(\frac{120\pi\alpha u_0}{\hbar\omega_{\rm p}}\right)^{1/2} \exp(-\alpha u_0/\hbar\omega_{\rm p}). \tag{3.92}$$

Given $E_{\rm J} = 300\,{\rm meV}$ and $E_{\rm C} = 0.1\,{\rm meV}$, find the bias current value at which the rate becomes $\approx 1\,{\rm s}$. What are the values of u_0 and $\omega_{\rm p}$? Result: $\tilde{I} = 0.219$, $u_0 = 58\,{\rm meV}$, $\omega_{\rm p} = 12\,{\rm meV}$.

3.6 Josephson arrays

It is remarkable that one can group many superconducting islands into an array, connecting them by Josephson junctions. Importantly, such a system is scalable: one can understand how the whole system works if one knows how its parts, the capacitors and Josephson junctions, work. Since the system is quantum, the full description is provided by a Hamiltonian. The degrees of freedom are either charges or phases of each island of the array. The full Hamiltonian consists of Josephson and charging parts. The Josephson part is just a sum over all junctions, each junction contributing a term proportional to the cosine of the difference of the phases of the islands that this junction connects. Similarly, the charging energy is the sum over energies accumulated in each capacitor. A minor complication is that the charge at the plates of each capacitor is not a variable characterizing the system, it is a linear function of these variables, integer and induced charges. Therefore the Coulomb energy is most conveniently expressed in terms of the inverse capacitance matrix C^{-1} that relates voltages and charges in the islands, $V_i = \sum_j (C^{-1})_{ij} Q_j$. The full Hamiltonian thus reads as follows:

$$\hat{H} = \sum_{i,j=\text{islands}} (e^2/2)(\hat{N}_i - Q_i)\hat{C}^{-1}_{ij}(\hat{N}_j - Q_j) + \sum_{k=\text{junctions}} E_{\rm J}^{(k)} \cos(\hat{\varphi}_1^{(k)} - \hat{\varphi}_2^{(k)}). \tag{3.93}$$

As in a single island, $\hat{N}_i$ and $\hat{\varphi}_i$ are the operators satisfying $[\hat{N}_i, \hat{\varphi}_j] = -2{\rm i}\delta_{ij}$.

Both our interest in the arrays and our difficulty in understanding the physics arise from the same source: the large number of quantum degrees of freedom involved. While a single superconducting island corresponds to one particle, a superconducting array corresponds to many quantum interacting particles. While a single superconducting island can be regarded as a home-made atom, an array presents a large collection of atoms, an artificial solid state system. For a system made of simple elements, the quantum physics of arrays is extraordinary rich and has analogies, and even direct similarities, with other complicated systems, for example quantum (anti)ferromagnets. The scalability of the arrays makes them suitable for quantum computer designs.

For further consideration of Josephson arrays, we focus on three topics: quantum phase transitions, vortices, and Kosterlitz–Thouless phenomena. A comprehensive review of the physics of Josephson arrays can be found in Ref. [71].

3.6.1 Quantum phase transition

Although the quantitative physics of the arrays is very much involved, the main qualitative feature is understood using a simple guide – the Heisenberg uncertainty relation. Let us consider an infinitely big uniform array that is characterized by the typical scales of Josephson and charging energies, E_J and E_C. If $E_C \gg E_J$, the charge in each island is well defined. It costs a charging energy E_C to add an extra hole (electron) into the array. The array is therefore similar to an insulator or semiconductor with a gap of order E_C. One can force charges to the array by applying a sufficiently large gate voltage; this is like doping a semiconductor. An important effect of the Josephson energy is that it makes these charges *movable*. Indeed, as discussed, a Josephson junction connecting two superconductors provides an amplitude for a Cooper pair hopping between them. The charges are forced to hop between the islands, forming extended quantum states, similar to those of charge carriers in semiconductors.

In the opposite case, $E_J \gg E_C$, the phase of each island is well defined. Currents induced by the phase differences emerge between the islands. The array supports the superconducting current and is like a bulk (inhomogeneous) superconductor. What happens if the ratio E_C/E_J changes from small to large values? A superconductor cannot be continuously changed into an insulator, so somewhere, at $E_C \sim E_J$, a sharp transition should occur separating these two distinct behaviors. This is called a *quantum phase transition*. Recent speculation identifies the complicated physics of high-T_c superconductors with the quantum phase transition in Josephson arrays.

Let us consider a more quantitative, though approximate, model of this transition [72]. We consider a uniform array where the dominant capacitance is between an island and the ground. As we have seen in Section 3.1, this allows us to disregard the interaction between the charges in different islands. The Hamiltonian in the phase representation is given by

$$\hat{H} = \sum_n \left(E_C \left(-2i\frac{\partial}{\partial \varphi_n} - \frac{q}{e} \right)^2 - \sum_{i \in Nb(n)} \frac{E_J}{2} \left(e^{i(\varphi_n - \varphi_i)} + e^{i(-\varphi_n + \varphi_i)} \right) \right),$$

where n numbers the islands, and $Nb(n)$ is a manifold of all neighbors of the island n connected by identical Josephson junctions E_J. This Hamiltonian cannot be solved exactly. Let us make an approximation in the spirit of the "mean-field" approach or the "Weiss field" method; these are routinely involved by analysis of the complicated models of (quantum) statistical mechanics. We concentrate on a given node and replace the terms that depend on the phases of neighboring islands by their *mean* values, $\exp(\pm i\varphi_i) \to \langle \exp(\pm i\varphi) \rangle$. This brings us to the approximate Hamiltonian of this particular node:

$$\hat{H} = 4E_C \left(\frac{\partial}{\partial \varphi} - \frac{q}{2e} \right)^2 - \frac{1}{2} \left(e^{i\varphi}\Delta^* + e^{-i\varphi}\Delta \right); \tag{3.94}$$

$$\Delta \equiv N_{nb} E_J \langle e^{i\varphi} \rangle, \tag{3.95}$$

where N_{nb} is the number of neighbors. We note that the CPB is described by almost the same Hamiltonian (see Eq. (3.82)), with Δ playing the role of effective Josephson energy. We are interested in the ground state of the array, and correspondingly in the ground state of

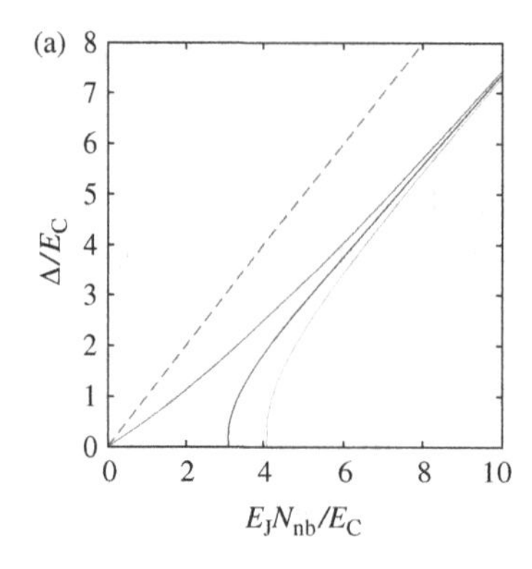

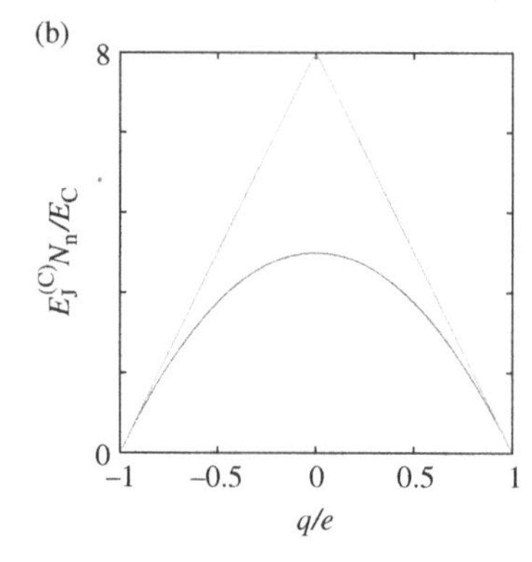

Fig. 3.31. **Quantum phase transition in Josephson array. (a) Effective Josephson energy Δ versus Josephson energy at three values of the charge induced ($q/e = 0, 0.5, 1$ from lower to upper curve). For $q \neq e$, finite Δ persists only above a critical Josephson energy. (b) Critical Josephson energy versus q for the two approximations described in the text.**

the Hamiltonian, Eq. (3.94). The phase factor averaged over this state is computed through the Δ-dependence of the ground-state energy:

$$\langle e^{i\varphi} \rangle = -\left\langle \frac{\partial \hat{H}}{2\partial \Delta^*} \right\rangle = -\frac{2\partial E_g}{\partial \Delta^*}.$$

Recalling Eq. (3.95), we observe that Δ is determined from the *self-consistency* equation as follows:

$$\Delta = -2N_{nb}E_J \frac{\partial E_g(|\Delta|)}{\partial \Delta^*}. \tag{3.96}$$

If E_J does not exceed a certain critical value $E_J^{(c)}$, the only possible solution is $\Delta = 0$. There are no hops between the islands, and the array is insulating. At $E_J > E_J^{(c)}$, the solution with the minimum energy corresponds to $|\Delta| \neq 0$: the array is superconducting. The transition takes place at $E_J = E_J^{(c)}$ and $|\Delta|$ approaches zero at this point: this is a *second-order* transition (see Fig. 3.31). To find the transition point, we need the dependence $E_g(\Delta)$ at small Δ; this is readily given by Eq. (3.86). We find that

$$E_J^{(c)} N_{nb} = 2\frac{E^+E^-}{E^+ + E^-}. \tag{3.97}$$

We recall that the method in use is approximate. To estimate the accuracy of this approximation, let us consider an alternative one. We inject a single Cooper pair into an insulating array. In the absence of Josephson energy, the resulting quantum states $|i\rangle$ are labeled by the number of the island i where the extra charge is situated. The Josephson energy mixes up these states, the Hamiltonian in this basis being

$$H = \text{const.} - \frac{E_J}{2} \sum_n \sum_{i \in Nb(n)} (|n\rangle\langle i| + |i\rangle\langle n|). \tag{3.98}$$

Such Hamiltonians are common in solid state physics. The eigenstates are Bloch states characterized by the quasimomentum $\boldsymbol{q}$, $|q\rangle = \sum_i \exp(i\boldsymbol{q}\boldsymbol{r}_i)$, $\boldsymbol{r}_i$ being the position of island i. They extend over all islands as promised above. Let us note that the lowest energy state corresponding to $\boldsymbol{q} = 0$ decreases with increasing E_J, $E_{\min} = -E_\mathrm{J} N_{\mathrm{nb}}/2$. Therefore, at sufficiently big E_J it becomes energetically favorable to add (extract) charges, $E_{\min} + E^+ < 0$ ($E_{\min} + E^- < 0$), and the movable charges flood the array, resulting in the superconducting state. The critical value of the Josephson coupling is given by

$$E_\mathrm{J}^{(\mathrm{c})} N_{\mathrm{nb}} = 2 \min(E_+, E_-). \tag{3.99}$$

The two approximations given in Eqs. (3.97) and (3.99) are plotted in Fig. 3.31(b) versus the induced charge. It is believed that the true transition line lies between these approximations.

Control question 3.7. Why is the Eq. (3.99) approximate?

Exercise 3.12. Given the results of Section 3.5.3, determine the asymptotics of Δ given by the self-consistency equation, Eq. (3.96), in the limit $E_\mathrm{C} \gg E_\mathrm{J}$.

Interaction between electrons in different islands complicates the phase diagram of the array. We have seen already in Section 3.1 that such interaction stabilizes the charge lattices with a period exceeding the period of the array, so that the charge distribution is no longer uniform. Such configurations are called "solid" phases since the emergence of these phases changes the symmetry of the charge configuration very much in the same way as a solid phase emerges from a liquid – the translational symmetry is broken. It appears that "solid" phases are not incompatible with superconductivity: the resulting "supersolid" phases combine superconducting currents with inhomogeneous periodic charge distribution over the islands. Needless to say, the random background charges mask all effects that depend on induced charge and therefore may significantly complicate the transition. Nevertheless, by virtue of the Heisenberg uncertainty principle, the transition should persist and was indeed observed in a real charge-disordered array [71].

3.6.2 Vortices

One cannot complete a discussion of the arrays without mentioning *vortices*. There is an extra tool we can use to tune the properties of Josephson arrays – the magnetic field. In the presence of a magnetic field, the energy of each Josephson junction c that connects islands i and j is modified to accommodate the phase shift due to the magnetic field:

$$E_J^{(c)} \cos(\varphi_i - \varphi_j) \rightarrow E_\mathrm{J} \cos(\gamma^{(c)});$$
$$\gamma^{(c)} \equiv \varphi_i - \varphi_j - \Phi_{ij}; \;\; \Phi_{ij} = \frac{2e}{c} \int_{\boldsymbol{r}_i}^{\boldsymbol{r}_j} \mathrm{d}\boldsymbol{r} \cdot \boldsymbol{A}(\boldsymbol{r}),$$

where $\boldsymbol{A}$ is the vector potential of the magnetic field, and $\boldsymbol{r}_i$ and $\boldsymbol{r}_j$ are the coordinates of the islands. The quantity $\gamma^{(c)}$ is known as the gauge-invariant phase. We see that in the presence of a magnetic field it is a characteristic of a junction rather than of the phases of two islands. The vortices in the arrays resemble Abrikosov vortices in bulk superconductors (see Appendix B). As in that case, the magnetic field favors the formation of vortices. However, there are differences due to the discrete nature of the arrays.

In a superconducting film, a vortex has a coordinate: a point where the superconducting order parameter becomes zero to accommodate the phase whirl. In the arrays, the superconducting order parameter in the islands never vanishes. Vortices in the arrays do not have a point-like core. Rather, they dwell in plaquettes: the smallest loops of the Josephson junctions. Let us consider the simplest plaquette formed by three junctions and a path encircling the plaquette $1 \to 2 \to 3 \to 1$ (see Fig. 3.32(a)). The sum of the magnetic phase shifts taken along this path is directly related to the total magnetic flux Φ encircled, i.e.

$$\Phi_{12} + \Phi_{23} + \Phi_{31} = 2\pi f; \qquad f \equiv \Phi/\Phi_0.$$

If we take the sum of three gauge-invariant phases, the island phases drop out, as follows:

$$\gamma_{12} + \gamma_{23} + \gamma_{31} = \Phi_{12} + \Phi_{23} + \Phi_{31} = 2\pi f.$$

Let us now note that the Josephson energies stay the same. This suggests that the states that differ by this shift are probably equivalent. To remove this ambiguity, let us consider *reduced* phases – those shifted by an integer number of 2π to the interval $(-\pi, \pi)$. Mathematically, they are given by $\text{mod}(\gamma) = \gamma - 2\pi[\gamma/(2\pi) - 1/2]$. The crucial observation is that the sum of reduced phases along the path is not the same as the sum of the phases; it may deviate by an integer multiple of 2π,

$$\gamma_{12} + \gamma_{23} + \gamma_{31} \neq \text{mod}(\gamma_{12}) + \text{mod}(\gamma_{23}) + \text{mod}(\gamma_{31}) = 2\pi f + 2\pi M, \tag{3.100}$$

where the integer M gives the number of vortices in the plaquette. Note that M can be negative, indicating antivortices. A critically inclined reader has to exclaim at this point: "But what is going on? They add a π, subtract a π, and sell it as a physical object." To see the vortices at work, let us consider the following example.

We take a ring of N identical Josephson junctions penetrated by flux. We disregard the charging energy for the time being. Let us find stationary states of the ring. The current must be the same in all junctions. Since the current is directly related to the gauge-invariant phase in each junction, $I = (eE_J/\hbar)\sin(\gamma)$, all reduced phases are the same. Equation (3.100) becomes

$$N\,\text{mod}(\gamma) = 2\pi f + 2\pi M; \qquad \text{mod}(\gamma) = 2\pi\frac{f - M}{N}.$$

We see that, at a given flux, the ring can be in *many* stationary states. These states differ in the number of vortices in the ring. Since $|\text{mod}(\gamma)| < \pi$, for a plaquette of N junctions M cannot deviate from f by more than $[N/2]$. The energy of a stationary configuration is given by $E = -N\cos(\text{mod}(\gamma)) = -N\cos(2\pi(f - M)/N)$. We plot these energies versus flux and note that, while each curve is periodic in flux with period $N\Phi_0$, the whole set of curves is periodic with the period Φ_0 as expected. The stationary configurations can be stable as well as unstable, corresponding to solid or dashed lines in Fig. 3.32(b). This brings

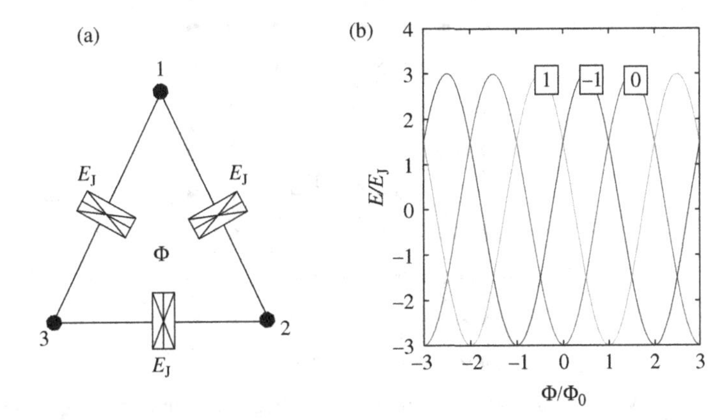

Fig. 3.32. (a) Three-junction Josephson ring: the simplest plaquette housing a vortex. (b) The energies of the states with different numbers of vortices (labels 1,-1,0 on the curves). Note the Φ_0 periodicity of the energy and the $3\Phi_0$ period of each curve.

us to the physical significance of the vortices: they label different stable configurations of a Josephson array in the limit of well defined phase, $E_J \gg E_C$.

There is an interesting similarity, or, to put it better, duality, between the discrete charges in the islands of an array and the discrete vortices in the plaquettes of an array. Let us recognize the close resemblance between Fig. 3.32(b) and Figs. 3.2 and 3.28. In those figures, each parabolic curve represented the energy of a certain charge state, while the whole set of curves displayed e ($2e$) periodicity in the induced charge. The discrete charge, corresponding to minimum energy, uniformly increases with q. In Fig. 3.32(b) we have the same situation with vortices instead of charges and flux instead of induced charge. The similarity is not exact: while at any given q the number of possible charge states in a CPB or single-electron box is, in principle, infinite, the number of stable vortex configurations in a ring of N junctions is restricted by $[N/2 + 1]$. In addition, the dependence of energy on f is not precisely parabolic.

The increase in charging energy makes the vortices quantum particles. We see from Fig. 3.32(b) that at half-integer values of flux the ground state is always degenerate: two states (say $|0\rangle$ and $|1\rangle$) that differ by one vortex have precisely the same energy. Making the analogy with a CPB (Fig. 3.28), we expect that quantum effects lift the degeneracy. Two resulting quantum states are superpositions $(|0\rangle \pm |1\rangle)/\sqrt{2}$ and are separated by small energy E_S; if $E_J \gg E_C$, E_s is exponentially small. Indeed, the mixing of two vortex states can only result from the macroscopic tunneling of phase between two distinct vortex configurations. This process for a single junction and for a CPB has been analyzed in section 3.5. The analytical estimate for E_S is more difficult to obtain since, in distinction from the single junction, the tunneling involves many degrees of freedom: all phases in the ring. However, estimations for the energy barrier ($\simeq E_J$) and attempt frequency ($\hbar\omega \simeq \sqrt{E_J/E_C}$ remain the same. Therefore, one expects $E_S \propto \exp(-\alpha\sqrt{E_J/E_C})$ with $\alpha \simeq 1$.

This consideration shows that for a finite array a vortex does not make a good particle. In the course of tunneling, a vortex just disappears from the plaquette: unlike the number

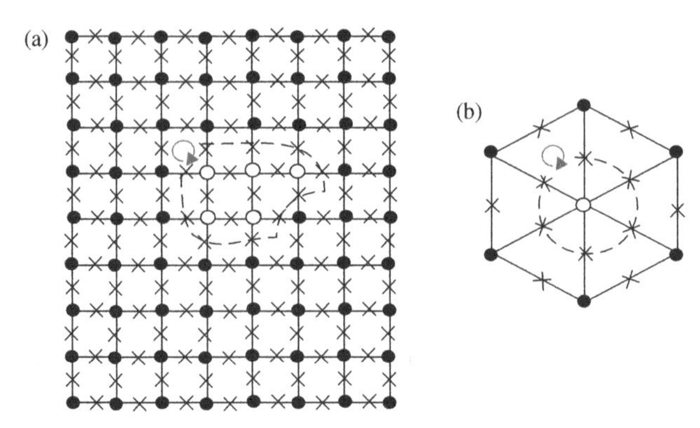

Fig. 3.33. (a) A vortex in an infinite array tunnels between neighboring plaquettes. Dashed line: a closed vortex trajectory encircling the islands (marked white). (b) A simple array to demonstrate the Aharonov–Casher effect.

of charges, the number of vortices in an isolated array is not conserved. The situation is different for a vortex in the middle of the array. To see why, let us note that the number of vortices in a given plaquette can only change if the reduced phase at one of the junctions reaches $\pm\pi$ (the Josephson energy of this junction passes through the maximum at this point). If the junction is on the array boundary, the vortex just disappears in vacuum. However, if the junction is in the middle of the array, and therefore is shared by two plaquettes, the π shift on the junction simultaneously shifts the sum of the reduced phases in both plaquettes. Therefore, the vortex moves to the neighboring plaquette and the vortex number is conserved. For small but finite $E_C \ll E_J$, this motion may proceed by quantum tunneling. Thus, we expect that, in a uniform array, a single vortex hops between neighboring plaquettes resulting in a delocalized state. The hopping amplitude E_s remains exponentially small. For a uniform square-lattice array, where the capacitance to the ground is a dominant one, one finds $E_s \propto \exp(-(\pi^3/2^{5/2})\sqrt{E_J/E_C})$. So we again encounter a duality: we have moving charges at $E_C \gg E_J$, on the insulating side of the quantum phase transition, and moving vortices at $E_J \gg E_C$, on the superconducting side of the transition.

One of the finest things in the array physics is quantum interference in the course of the vortex motion. We know that the phase of electrons moving in the solid and of charges moving in the array are affected by the magnetic field. The phase shift accumulated along a closed-loop trajectory is proportional to the flux penetrating the loop. Let us consider a vortex moving along a closed-loop trajectory. It encircles a certain number of islands (marked white in Fig. 3.33(a)). Although the phases of each island in the array are the same at the beginning and end of the motion, one can show that the phases of the encircled islands describe a full circle, passing all values in the interval $(0, 2\pi)$. The phases of the other islands are not shifted: they deviate (slightly) from their original values and then return. We know from our consideration of a CPB that this creates a phase shift proportional to q for each island (see Eq. (3.88)). This means that the total phase acquired by the vortex equals $2\pi \sum_i (q_i/2e)$, the summation running over the encircled islands i. This is the total charge "penetrating" the loop. This is yet another example of duality between vortices and charges. Aharonov and Casher [73] predicted a similar phase shift for magnetic dipoles moving in an electric field: this gave the name to the effect.

The Aharonov–Casher effect in the array is not only good at illustrating the duality between vortices and charges, but also illustrates how treacherous this approximate concept could be. To see this, let us add a vortex to a simple array (Fig. 3.33(b)) and induce a charge at the central island of the array. The vortex will hop around the island forming a delocalized state. The energy of the state is affected by the Aharonov–Casher phase shift. This results in a contribution to the energy of the array that is periodic in q with a period of $2e$. Can we regard this periodicity as a manifestation of the Aharonov–Casher effect? Evidentally not: we recall that the same periodic contribution persists in the energy of a CPB where no vortices are present (although the phase of the island can describe full circles). Let us now change the situation and induce the *same* charge to *all* islands of the array rather than to the central one. It looks like the periodic contribution to the energy should persist since it is not affected by the charge induced to the outer islands. We understand, however, that this cannot be true. The isolated array has a certain charge, and the q-dependence of its energy is purely parabolic in accordance with the reasoning of Section 3.1. It turns out that we miss a contribution from the process in which a vortex comes from the vacuum, moves round one of the outer islands, and disappears into the vacuum again. If all induced charges are the same, this contribution precisely cancels the contribution of the circulating vortex.

Exercise 3.13. Calculate the energies and eigenfunctions of a single vortex in the array in Fig. 3.33(b) assuming that the modulus of the tunneling amplitude between neighboring plaquettes equals E_s and that the array is symmetric.

3.6.3 Berezinsky–Kosterlitz–Thouless transition

In the 1970s, Berezinsky [74] and independently Kosterlitz and Thouless [75] found a phase transition of a new type (BKT transition). Their discovery amounted to a chapter of statistical mechanics, which we cannot cover in this book. Their ideas have had a significant impact on quantum transport, and will be discussed in several places in this book.

At first glance, the BKT transition appears to have little chance of being seen in practice since it occurs in rather specific situations: in systems of particles that move in two dimensions and are subject to an extremely long-range pairwise interaction that obeys the logarithmic law. It turns out that two-dimensional Josephson arrays present a unique system where the BKT transition can be directly observed [76, 77].

To proceed, let us calculate the energy of a single vortex in a two-dimensional square array with lattice period a in the absence of magnetic field. The presence of the vortex induces the phase differences on all junctions of the array. These phase differences are small for the junctions that are far from the plaquette where the vortex is situated. It turns out that the main contribution to the energy arises from these junctions. Their Josephson energy can be approximated by a quadratic expansion (as we did in Section 3.5). Also one can use the continuous coordinate $\boldsymbol{r}$ instead of discrete positions of the islands and replace the summation over the junctions by integration over $\boldsymbol{r}$. Thereby the energy is expressed in terms of phase gradients $\nabla\varphi$:

$$\mathcal{E} = \frac{E_{\mathrm{J}}}{2} \int \mathrm{d}\boldsymbol{r} (\nabla\varphi)^2 .$$

Let us enclose the vortex with a circle of radius r. The phase change along this path equals 2π, and the phase gradient is therefore given by $|\nabla\varphi| = 2\pi/2\pi r = 1/r$. The energy is expressed as an integral over r:

$$E = \pi E_{\mathrm{J}} \int \frac{\mathrm{d}r}{r} .$$

This integral diverges both at small and large r. While integration at small distances is cut off at $r \simeq a$, the upper limit is only restricted by the linear size of the array L. The energy of a single vortex thus diverges with increasing array size. To circumvent this problem, let us consider a vortex–antivortex pair separated by distance $R \ll L$. The phase differences induced by the vortex and the antivortex cancel each other at distances of the order of R. Therefore the energy of the pair is finite at $L \to \infty$ and can be estimated as follows:

$$E_{\mathrm{int}} = 2\pi E_{\mathrm{J}} \ln(R/a).$$

Next, we regard this pair as a thermal fluctuation and estimate the Boltzmann statistical weight of this fluctuation, summing over all possible positions of the vortex at $\boldsymbol{r}_1$ and the antivortex at $\boldsymbol{r}_2$:

$$W_1 \approx a^4 \int \mathrm{d}\boldsymbol{r}_1\, \mathrm{d}\boldsymbol{r}_2\, \exp -[E(|\boldsymbol{r}_1 - \boldsymbol{r}_2|)]/k_{\mathrm{B}}T \approx (L/a)^2 \int \frac{\mathrm{d}R\, R}{a^2} |R/a|^{-2\pi E_{\mathrm{J}}/k_{\mathrm{B}}T} .$$

At low temperatures, the integral converges at large distances. The statistical weight of the pairs is small and the thermal activation of pairs can be disregarded. The situation changes at sufficiently high temperature, greater than $T_2 = \pi E_{\mathrm{J}}$. The statistical weight diverges. This indicates that it is more probable that we have a pair in the array than we have none. Moreover, if a pair is present, the energy cost is compensated for by a diverging entropy factor.

Eventually, the BKT transition takes place at a lower temperature. Suppose that vortex–antivortex pairs are already present in the array. If we add another pair, the pairs already present screen the interaction between the vortex and the antivortex and reduce the energy cost we have to pay. The divergence takes place at lower temperatures. To quantify this, we estimate the statistical weight of N pairs separated by a typical distance R:

$$W_N \approx a^{4N} \int \mathrm{d}\boldsymbol{r}_1 \cdots \mathrm{d}\boldsymbol{r}_{2N} \exp\left(- \sum_{i<j\le 2N} M_j M_i (E(|\boldsymbol{r}_i - \boldsymbol{r}_j|)/k_{\mathrm{B}}T) \right)$$
$$\simeq (L/a)^2 \int \frac{\mathrm{d}R\, R^{4N-3}}{a^{4N-2}} |R/a|^{-2N\pi E_{\mathrm{J}}/k_{\mathrm{B}}T} .$$

The indices i and j label vortices and antivortices distinguished by their "charge" $M_i = \pm 1$. This statistical weight diverges at temperature $k_{\mathrm{B}}T_N = \pi E_{\mathrm{J}} 2N/(4N-2)$. As more pairs get involved, the temperature lowers and the transition eventually occurs at

$T_{BKT} = T_\infty = T_2/2 = \pi E_J/2$. Below this temperature no appreciable concentration of vortices is present in the array, and the array is superconducting. Above T_{BKT}, the array is filled with the gas of vortices and antivortices. The vortices move if the current is applied to the array. This results in dissipation and electric resistance. Thus, we conclude that the array is in a resistive state.

It is interesting to note that the BKT transition may take place for charges as well [77]. This requires negligible capacitance to the ground and $E_J \ll E_C$. If the junction capacitance C only is present, the interaction energy between the extra charges $2e$ and $-2e$, separated by R, is given by

$$E = \frac{2e^2}{\pi C} \ln(R/a).$$

Repeating all the above reasoning, we predict the BKT transition to occur at $k_B T = e^2/2\pi C$. Below this temperature, the array is in an insulating state. Above T_{BKT}, the gas of charges is present and the array conducts. Such a charge BKT transition also takes place in a normal uniform array. Since in this case the pair components are single charges, $\pm e$, the transition temperature is four times smaller.

The interaction between charges does not stay logarithmic at distances exceeding the screening length $a\sqrt{C/C_0}$. The same is true for the vortices: Josephson currents around the vortex generate a magnetic field not taken into account in our consideration; this magnetic field quenches the phase differences far from the vortex. This takes place at the screening length estimated as $\Phi_0^2/(E_J\mu_0)$. It is believed that the BKT transition persists provided the screening length is much bigger than a.

Exercise 3.14. The above estimate of transition temperature holds for a square array. Give T_{BKT} for triangular and hexagonal arrays with the same Josephson junction energy.

3.7 Superconducting islands beyond the Josephson limit

In Sections 3.5 and 3.6, we have studied Josephson physics in combination with the Coulomb blockade. Explicitly or implicitly it was assumed that the energy scales involved are much smaller than the superconducting energy gap Δ, at least as far as excitations are concerned. It allowed us to forget about quasiparticles – single-electron or hole excitations in the superconductors – and to deal only with coherent states formed by Cooper-pair transfer. It is simple to induce quasiparticles in experiments: one just applies a bias voltage to the nanostructure that is of the order of Δ. In this section, we concentrate on the interesting effects and processes involving single-quasiparticle excitations. The most fundamental, and first, to be considered is the *parity effect*. A superconducting island is sensitive to the parity of the number of electrons in it. Next, we list transport processes that involve single-electron transfer in superconductors. There are many, and this results in many transport regimes possible in SETs made of superconductors. In Section 3.7.3, we specifically describe the so-called JQP cycle: a transport regime where coherent Josephson tunneling

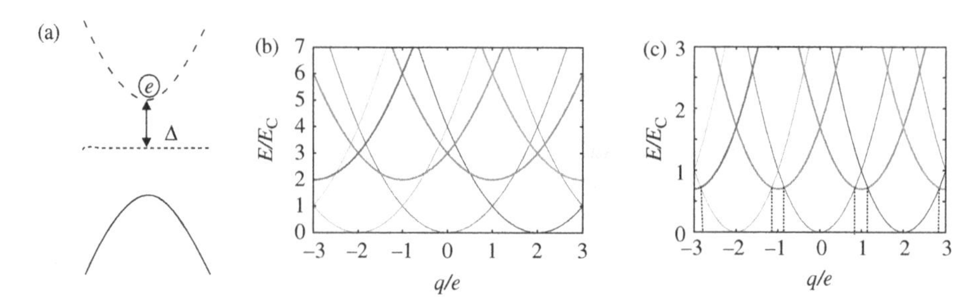

Fig. 3.34. **Parity effect. (a) An odd electron in a superconducting island. (b) The charge states for $\Delta > E_C$: there is always an even number of electrons in the ground state. (c) The charge states for $\Delta < E_C$: there is an odd number of electrons in the ground state in the intervals bounded by the dotted lines. The states with an odd number of electrons are depicted by thicker lines.**

co-exists with quasiparticle tunneling. This regime is important, both from the fundamental aspect of such a co-existence, and in its frequent usage for quantum measurements.

3.7.1 Parity effect

In Sections 3.5 and 3.6 we implicitly assumed that a superconducting island can only contain an even number of extra charges – Cooper pairs. Actually, we have not yet explained why.

The explanation comes from the microscopic theory of superconductivity (see Appendix B). According to this theory, in the ground state of a bulk superconducting island all electrons are combined in Cooper pairs: there is an *even* number of electrons in this state. The theory also states that quasi-particle excitations – either single electrons or single holes – are separated by energy Δ from the ground state. Let us carve a superconducting island from this superconductor. If there is an even number of electrons, the ground state of the island is similar to that of the bulk. If we wish to have an odd number, we add an extra electron (or hole) to the superconducting island. It stacks in the lowest state available for a quasiparticle; this state is separated by energy Δ from the Fermi level. Therefore, it costs the extra energy Δ to add one extra electron (or hole). Surprisingly, if one adds yet another electron, the energy of the system may become lower. Two electrons do not have to be quasiparticles: they can form a Cooper pair and return to the Fermi level. We conclude that the ground-state energy of a superconducting island has an addition Δ at any time that the number of electrons is odd. This is called the *parity effect*. Let us combine the parity effect with the familiar picture of crossing parabolas: the charge states versus q (see Fig. 3.34(b)). We understand that all parabolas with an even number of electrons must be shifted upwards by energy Δ. If $\Delta > E_C$, the ground state always corresponds to an even number of electrons. Let us recall that, for a well isolated island, $E_J \simeq (G_T/G_Q)\Delta \ll \Delta$. This implies that, in the interesting quantum regime where $E_C \simeq E_J$, one has $E_C \ll \Delta$, the parity effect is strong, and we can disregard the states with an even number of electrons.

If $\Delta < E_C$, the situation is slightly more complicated: the ground state has an even number of electrons in the vicinity of even integer values of q/e (see Fig. 3.34(c)). The Coulomb diamonds with an odd number of electrons are a factor of $(1 - \Delta/E_C)$ narrower than those with an even number. The overall pattern of diamonds is $2e$-periodic in q in contrast with the e-periodic pattern of a normal-metal island.

Control question 3.8. Which values of Δ/E_C have been taken for the plots in Figs. 3.34(b) and (c)?

Despite the simplicity of the parity effect and its seemingly robust manifestations, $2e$-periodicity of Coulomb blockade in superconducting systems was not observed until 1992. This caused significant confusion and even doubts about the validity of the traditional theory of superconductivity. The parity effect was first experimentally confirmed in Ref. [78]. The reference explains that the parity effect is only seen below a typical temperature T^*. To estimate this temperature, we note that the parity effect is due to a single quasiparticle trapped in the island. If there are several thermally excited quasiparticles in the island, in the state with either even or odd numbers of electrons, the difference between even and odd is gone. A quick estimation would be $k_B T^* \simeq \Delta$. However, the typical number of quasiparticles at this temperature is already big: they fill the energy levels in the energy strip of the order of Δ, that is, there are $\simeq \Delta/\delta_S$ levels and quasiparticles. To obtain a better estimate, let us calculate a statistical weight of two excited quasiparticles. It is given by the summation of Boltzmann factors over all possible quasiparticle states ϵ_p,

$$W_2 = \sum_{\epsilon_p^{(1)},\, \epsilon_p^{(2)}} \exp\left(\frac{\epsilon_p^{(1)} + \epsilon_p^{(2)}}{k_B T}\right) = \left(\delta_S^{-1} \int_\Delta d\epsilon\, \nu_{BCS}(\epsilon)\, e^{-\epsilon/k_B T}\right)^2$$

$$\approx \left(\delta_S^{-1} e^{-\Delta/k_B T} \int dE \sqrt{\frac{\Delta}{2E}} e^{-E^2/2\Delta k_B T}\right)^2 = e^{-2\Delta/k_B T}\, \Delta\sqrt{\Delta k_B T}/\delta_S^2 ,$$

where $\nu_{BCS} = |E|\Theta(|E| - \Delta)/\sqrt{E^2 - \Delta^2}$ is the BCS density of states (see Appendix B). In the preceding equation, we assume $k_B T \ll \Delta$ and $E \equiv \epsilon - \Delta \ll \Delta$. At $T = T^*$, $W_2 \simeq 1$, so we conclude that $k_B T^* = \Delta/2\ln(\Delta/\delta_S)$, $k_B T^* \approx 0.2\Delta$, for typical dimensions of metallic islands.

Under common experimental conditions, the obstacle to observating the parity effect is not the temperature of the nanostructure, but rather the undesired electromagnetic irradiation coming from the room-temperature environment ($k_B T \simeq 100\Delta$). A quantum of this irradiation produces several dozen quasiparticles if absorbed in the nanostructure. These quasiparticles are subsequently trapped in the islands and superconducting leads nearby and poison the parity effect. Indeed, it was demonstrated that, owing to the parity effect, superconducting SET transistors may be used as ultrasensitive detectors of microwave radiation for frequencies greater than or equal to $\Delta/\hbar \simeq 100$ GHz [79].

3.7.2 Transport processes involving quasiparticles

The convenient aluminum technology for Coulomb blockade systems results in superconducting devices, provided the superconductivity of aluminum is not suppressed by a magnetic field. Eventually, the first SET, described in Section 3.2, was all-superconducting – both leads and the island were in superconducting states. This did not matter much for I–V curves of the device in the regime of single-electron transfer. Since the superconducting gap Δ was several times smaller than E_C, the I–V curves look very similar to those of a normal SET. The situation changes if Δ is much bigger than E_C and/or at bias voltages smaller than Δ. A rich variety of specific single-electron and multi-electron processes is observed in this range. A complete description would make a chapter by itself, so the following discussion is very sketchy. We concentrate on transport in two symmetric SETs: an all-superconducting one, and one with a superconducting island and normal leads (NSN-SET). We give the estimations of the typical rates assuming $eV \simeq \Delta \simeq E_C$. In these estimations, two small parameters will play a role: G_T/G_Q and δ_S/Δ. For concreteness, we always assume $G_T/G_Q \gg \delta_S/\Delta$.

We start with single-charge transfers. A transfer from normal metal to a superconductor would create a quasiparticle; this costs at least Δ of extra energy. This fact is accounted for by the BCS density of states. The tunnel rate corresponding to the electrostatic energy difference ΔE is obtained by incorporating the BCS factor into Eq. (3.27) as follows:

$$\begin{aligned}\Gamma_{S-N}(\Delta E) &= \frac{G_T}{e^2}\int dE\; \nu_{BCS}(E) f(E)(1 - f(E - \Delta E)) \\ &= \frac{G_T}{e^2}\sqrt{(\Delta E)^2 - \Delta^2}\,\Theta(-\Delta E - \Delta) \text{ at } T = 0. \end{aligned} \tag{3.101}$$

If a single-charge transfer takes place between two superconductors, a quasiparticle is created at each side of the junction. The BCS factors appear on both sides. Assuming the same gap Δ for both superconductors yields

$$\Gamma_{S-S}(\Delta E) = \frac{G_T}{e^2}\int dE\; \nu_{BCS}(E)\nu_{BCS}(E - \Delta E) f(E)(1 - f(E - \Delta E)). \tag{3.102}$$

These two rates at vanishing temperature are plotted in Fig. 3.35. The most important feature is that the tunneling does not proceed below a certain energy threshold: Δ for Γ_{N-S} and 2Δ for Γ_{S-S}. The square-root singularity of the BCS factor causes the rates to rise quickly just above the threshold: while $\Gamma_{N-S} \propto \sqrt{\Delta - \Delta E}$ at $E \to \Delta + 0$, Γ_{S-S} suddenly rises by $(\pi/4)\,(G_T\Delta/e^2)$. This is why the rates are already very close to the normal rate already at $\Delta E > 2\Delta$ and can be estimated as $\Gamma_{se} \simeq \hbar\Delta(G_T/G_Q)$. Finite temperature smoothes the singularities. For both rates, the detailed balance relation in Eq. (3.28) holds, as it should.

Let us start with the NSN-SET [80]. At sufficiently high voltages and temperatures, the transport is determined by the single-charge rates Γ_{N-S}. The transport characteristics are readily obtained from the master equation with these rates. Upon decreasing temperature, Γ_{N-S} become exponentially small (proportional to $\exp(-\Delta/k_B T)$), and more complicated transport processes take over. At vanishing temperature, this happens below the threshold voltage $V_{th}^{(1)}$ that is shifted up by $2\Delta/e$ with respect to that of the normal SET. What are these processes? In the normal SET, inelastic co-tunneling would dominate since it can proceed at an arbitrary small bias voltage. In NSN-SET, inelastic co-tunneling must create two quasiparticles in the island so that it can proceed only above the threshold voltage $2\Delta/e$ (see the horizontal dashed lines in Figs. 3.36(a) and (b)). To estimate the rate, we assume $eV \simeq \Delta \simeq E_C$, so that $\Gamma_{cot} \simeq (G_T/G_Q)\Gamma_{se}$. The elastic co-tunneling still occurs

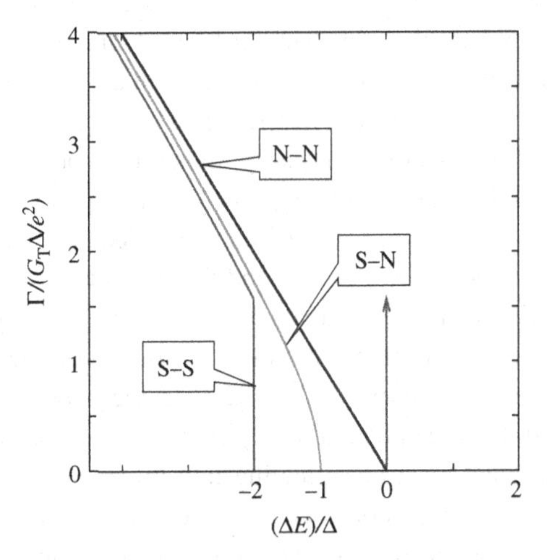

Fig. 3.35. Single-charge transfer rates versus energy difference ΔE: N–N = between normal metals; S–S = between superconductors; S–N = between a superconductor and a normal metal; arrow denotes the maximum supercurrent in the S–S tunnel junction ($I/2e$).

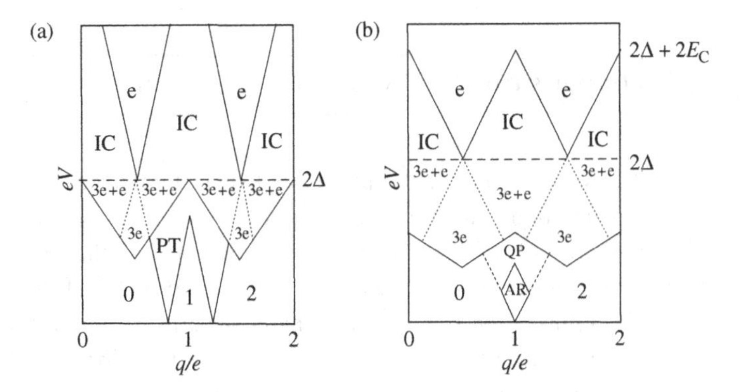

Fig. 3.36. Transport regimes in symmetric NSN-SET (e = single-electron transfers; IC = inelastic co-tunneling; PT = parity tunneling; 3e = three-electron tunneling; 3e+e = alternating single- and three-electron tunneling; AR = Andreev reflection; QP = quasiparticle poisoning; 0,1,2 = stable charge states). (a) At $\Delta = 2E_C/3 < E_C$. (b) At $\Delta = 3E_C/2 > E_C$.

at an arbitrary small voltage with a typical rate $(G_T/G_Q)(\delta_S/\Delta)\Gamma_{se}$. However, it does not necessarily dominate the transport at $V < 2\Delta/e$.

If $\Delta < E_C$, at zero bias voltage we encounter degeneracy points between the charge states with even and odd numbers of electrons, say 0 and 1. Near these points in the course of the transport cycle the SET switches between zero and one with the aid of a peculiar single-charge process that we call *parity tunneling*. We have seen that the state with an odd number of electrons has a single quasiparticle trapped at the lowest energy level available. During parity tunneling, this same quasiparticle leaves the island, so the SET goes from 1 to 2, where "2" denotes a state with two electrons. Since the quasiparticle tunnels from a given state, the rate of this process is rather small, $\Gamma_{given} \simeq (\delta_S/\Delta)\Gamma_{se}$ (see Eq. (3.36)), although it is still faster than elastic co-tunneling. In contrast to the parity tunneling, the switching between 0 and 1 proceeds with a quasiparticle created anywhere in the allowed energy strip. Therefore it happens at a much faster rate. The SET spends most of its time in an odd state waiting for parity tunneling to occur.

If $\Delta > E_C$, the zero-voltage degeneracy points separate states with even numbers of electrons, say 0 and 2. The switching between these states occurs by means of *two-electron tunneling*. Indeed, we understand that two electrons do not have to create a quasiparticle on entering a superconductor: they can form a Cooper pair at no extra energy cost. It is important to understand that this process is completely equivalent to Andreev reflection, considered in Sections 1.8 and 2.8. In the course of Andreev reflection, the electron is reflected from a superconductor, resulting in a hole. This hole can be regarded as the result of the second electron tunneling into the superconductor. In distinction from the elastic Andreev reflection considered previously, the difference between the electron and hole energies – or the sum of the energies of two incoming electrons – must be equal to the change of electrostatic energy. For order-of-value estimation of the two-electron tunneling rate, we can disregard the difference; this yields $(G_A/G_T)\Gamma_{se}$. The Andreev conductance G_A, as explained in Section 2.8, is usually determined by the effective resistance R_N of the normal lead: $G_A \simeq G_T^2 R_N$. Thus, we have $(G_T/G_Q) \ll (G_A/G_T) \ll 1$.

Control question 3.9. Explain the latter estimation.

If we forget about all the processes except two-electron ones, the energetics are the same as for single-electron tunneling in the normal SET, apart from the doubled charge. The two-electron transport cycles are formed beyond the Coulomb diamonds, which are four times higher and twice as wide as in the normal SET. They form a $2e$-periodic pattern. This makes the transport extremely sensitive to the parity. If a single quasiparticle enters the island, its effect on the two-electron transport is a shift of the induced charge by e. At sufficiently small voltages, this interrupts the two-electron tunneling cycle, and no transport takes place before a parity tunneling event removes the quasiparticle. Since two-electron rates are usually faster than the parity tunneling rate, the current is quenched, being determined by the parity rate. This is called *quasiparticle poisoning*. Such poisoning always takes place if one of the states, 0 or 2, is unstable with respect to single-charge tunneling since the latter process brings a quasiparticle to the island. As we can see from Fig. 3.36(b), this restricts the two-electron transport to a diamond with maximal height $2(\Delta - E_C)$. The current at the edge of this diamond sharply drops with *increasing* bias voltage.

Another low-voltage process is *three-electron* (3e) tunneling – co-tunneling of three electrons. To describe it, we start with state 0. First, two electrons enter the island, forming a virtual state with charge $2e$ and no quasiparticles. To complete the process, an electron leaves the island. The resulting final state has charge e and an excited quasiparticle. As for all co-tunneling processes, the rate estimate is reduced by a factor (G_T/G_Q) in comparison with the two-electron tunneling rate and is given by $(G_A/G_Q)\Gamma_{se}$. A similar process with the opposite order of tunneling events brings the SET back from 1 to 0, forming a transport cycle. To evaluate the threshold voltage, we note that the difference of the electrostatic energies of the initial and final states is same as for a single-charge tunneling. The energy given by the voltage source is, however, different. For a single-charge tunneling, one electron is extracted from a lead. This gives $eV/2$. For the 3e process, two electrons are extracted and an electron is added to the opposite lead, resulting in $3eV/2$. This is why the threshold voltage for 3e tunneling is three times smaller than that for single electrons, $V_{\mathrm{th}}^{(3)} = V_{\mathrm{th}}^{(1)}/3$. The cycle can also be formed by subsequent 3e and e tunneling processes (Fig. 3.36(b)). The resulting cycles do not depend on parity. There is a theoretical possibility for more complicated processes that switch between 0 and 1. The threshold voltage of a process involving N electrons is reduced by a factor N with respect to $V_{\mathrm{th}}^{(1)}$. However, the smaller rate of these processes presently forbids their experimental observation.

Let us turn to the all-superconducting SET (SSS-SET) [81]. At sufficiently high voltages, the single-charge tunneling once again dominates the transport. The threshold voltage for single-charge transport is shifted by $4\Delta/e$ with respect to the threshold voltage for the normal SET. The inelastic co-tunneling process produces at least four quasiparticles: two in the island and one in each lead. The threshold jump in the S–S rate results in the jump on co-tunneling current at this voltage. Careful theoretical consideration along the lines of Section 3.4 shows that the maximum conductance at the jump reaches the universal value $G_{\mathrm{Q}}/2$; this has been proven experimentally [82].

What determines the transport at lower voltages? A peculiarity of an SSS-SET is that at zero voltage a superconducting current can flow through it. It can be seen as a co-tunneling of a Cooper pair through both junctions that yields an effective Josephson coupling $E_{\mathrm{J}}^{\mathrm{SET}}$ between the leads, $E_{\mathrm{J}}^{\mathrm{SET}} \simeq (G_{\mathrm{T}}/G_{\mathrm{Q}})^2\Delta$ at $\Delta \simeq E_{\mathrm{C}}$. The effective rate corresponding to this current is rather large, $\simeq(G_{\mathrm{T}}/G_{\mathrm{Q}})\Gamma_{\mathrm{se}}$. However, a finite voltage applied to the SET causes the supercurrent to oscillate with the Josephson frequency $2eV/\hbar$. This results in zero dc current at finite voltage. It is instructive to regard this problem as an energy mismatch: a Cooper-pair transfer from one lead to another results in energy gain $2eV$; this energy cannot be easily disposed of into electronic degrees of freedom.

This energy can, in principle, be given to the electromagnetic environment of the SET. Such energy transfer indeed provides a finite dc current. However, it is not specific for the SET, it occurs in any Josephson junction, and depends on the properties of the environment rather than on those of the system under consideration. To this end, we postpone its consideration until Chapter 6. Another channel of the energy disposal is to create (at least) two quasiparticles. We note that this can be achieved at arbitrary low voltage, provided an elementary process involves a sufficient number of Cooper-pair transfers. Since the energy gain from N transferred Cooper pairs is $2eVN$, the corresponding threshold voltage is Δ/eN. Let us note the analogy between these processes and multiple Andreev reflection described in Section 1.8. Both processes transfer a number of Cooper pairs through the nanostructure, result in the creation of a quasiparticle pair, and have peculiarities at voltages Δ/eN.

In real Coulomb blockade systems, many-particle co-tunneling processes are hard to see due to their low rates. However, 3e tunneling is readily observed in most SSS-SETs and proceeds as follows. Given the initial state 0, a Cooper pair is transferred to the island, bringing it to the virtual state 2. During the second step, a quasiparticle tunneling takes

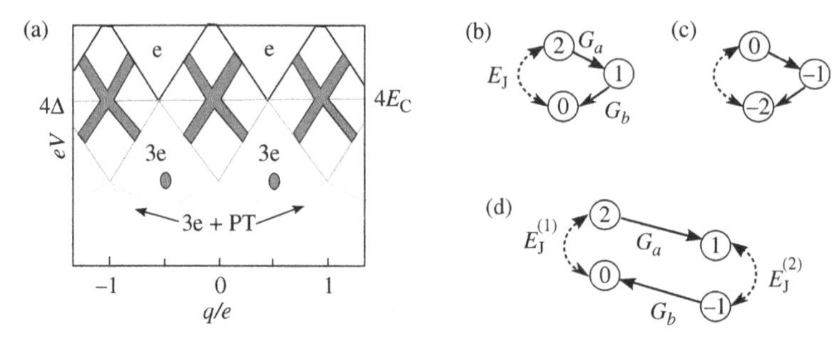

Fig. 3.37. **Transport regimes in a symmetric SSS-SET. (a) Overall picture at $\Delta \approx E_C$ (corresponding to Ref. [81]). (b)–(c) Josephson quasiparticle transport cycle. (d) Bright spot transport cycle.**

place in another junction. The resulting state has charge e and two quasiparticles: one in the island and one in the lead. Similar to NSN-SET, a reverse process may bring the SET from 1 back to 0, completing the transport cycle. The threshold voltage of this transport regime is one-third that of single-electron tunneling. Since three-electron transfers are involved, a typical rate is estimated as $(G_T/G_Q)^3\hbar\Delta \simeq (G_T/G_Q)^2\Gamma_{se}$. The three-electron transport cycle is not sensitive to parity and results in an e-periodic pattern. At lower voltages, the transport cycle is formed by 3e tunneling and parity tunneling. The latter process transfers the trapped quasiparticle to the lead. Since the number of quasiparticles in the initial and final states is the same, the threshold voltage of this process is the same as for single-electron tunneling in a normal SET (Fig. 3.37).

The I–V curve of a SSS-SET displays spectacular resonant features: high currents in the vicinity of certain lines or points in the V–V_g plane. At these lines or points, the energy difference associated with a Cooper-pair transfer through one or both junctions is close to zero. Therefore there is no energy mismatch for such a transfer, and the junction, in fact, supports supercurrent. The sharpness of resonant features enables the use of the SSS-SET in this regime as a sensitive detector, since small changes in gate/bias voltage result in large changes in current. Many applications of this kind make use of the *Josephson quasiparticle cycle* (JQP cycle), sometimes called *resonant pair tunneling*. It gives rise to the St Andrew cross features seen in Fig. 3.37(a).

To describe the cycle, let us start in state 0. The resonant Cooper-pair transfer brings the SET to state 2. This state is sufficiently high in energy for a quasiparticle to tunnel through another junction, bringing the SET to state 1. Yet this state has enough energy for another quasiparticle to tunnel through the same junction and bring the SET to 0 to complete the cycle (Fig. 3.37(b)). The resonance condition reads $E(2) - E(0) = eV$. An alternative cycle in the same diamond differs in the junctions where the transfers take place. First, the Cooper pair is transferred from the island, bringing the SET to state -2. Two successive quasiparticle additions through another junction bring the SET back to the beginning of the cycle (Fig. 3.37(b)). Two lines presenting the above condition cross at $4E_C/e$. JQP cycles can proceed only at voltages that are sufficiently high to enable quasiparticle processes. They are not sensitive to parity, forming the e-periodic pattern shown.

Another prominent resonant feature is the so-called "bright spot" (the high currents appear as brighter tones in gray-scale experimental plots). At the bright spot, the Cooper-pair transfer is resonant in both junctions but in different charge states. The transport cycle proceeds in four steps: Cooper-pair transfer, bringing the SET from 0 to 2; quasiparticle tunneling to 1; Cooper-pair transfer to -1; another quasiparticle tunneling to 0 (Fig. 3.37(d)). We will see in Section 3.7.3 that the maximum current for both resonant processes is estimated as $e\Gamma_{\text{se}}$. Since the features are sharp, the maximum differential conductance does not depend on the junction conductance, reaching values of the order of G_{Q}.

Control question 3.10. At which values of Q_{g} and V do the bright spots occur?

3.7.3 Coherence in the JQP cycle

To describe resonant processes taking place in a SSS-SET, we talked about Cooper-pair transfers as if they were classical processes switching the system between two well defined states. In fact, they are not. To see this, let us cut off the non-resonant junction. We get a Cooper-pair box in the limit described in Section 3.5.3. The resonant condition corresponds to the parabola crossing in Fig. 3.28. We recall that, in this case, the charging states are not well defined. Cooper-pair transfer is a coherent process and is characterized by an amplitude $E_{\text{J}}/2$ rather than by a rate. The transfer proceeds in both directions and results in new eigenstates that are the quantum superpositions of the two, $(|0\rangle \pm |2\rangle)/\sqrt{2}$.

The JQP cycle thus provides a generic example of quantum transport in a situation where coherent and incoherent electron transfers are equally important. We will encounter this many times in Chapter 5. Here we present the proper description of the transport. Before doing this, we quantify E_{J} under JQP cycle conditions.

The natural assumption would be that this E_{J} is the same as that discussed in Section 1.8 and given by the Ambegaokar–Baratoff formula $E_{\text{J}} = (G_{\text{T}}/4G_{\text{Q}})\Delta$. This is indeed true if $\Delta \gg E_{\text{C}}$, as assumed in Section 1.8. However, now we have $\Delta \simeq E_{\text{C}}$. To evaluate E_{J}, we make use of the tunneling Hamiltonian method. Cooper-pair transfer is a complex process, involving two-electron tunneling events. Suppose that in the course of the first tunneling event an electron is transferred from state l in the electrode to state i in the island. The system is brought thereby into a virtual state with the island charge e and two quasiparticles in states l and i. The amplitude is composed of the tunneling amplitude T_{il} and the coherence factors u_i, v_l that account for the excitation of quasiparticles (see Appendix B). In order to annihilate the quasiparticles, the second electron tunneling shall proceed between the same states and an electron with opposite spin is transferred. The corresponding amplitude is composed of T_{il} and u_l, v_i. According to the general formula given in Eq. (3.63), the total amplitude of the Cooper-pair transfer is given by

$$\frac{E_{\text{J}}}{2} = \sum_{i,l} T_{il} u_i v_l \frac{1}{E_i + E_l + E^{(+)}} T_{li} u_l v_i,$$

where $E_{i,l}$ are the quasiparticle energies and $E^{(+)}$ is the electrostatic energy paid for the addition of the electron. Since coherence factors depend on the energies only, $uv = \Delta/2E$, we can replace the summation over the states by integration over energies and use Eq. (3.63) to express the result in terms of the tunnel junction conductance. This yields

$$\frac{E_\text{J}}{2} = \Delta \frac{G_\text{T}}{2\pi^2 G_\text{Q}} \int \mathrm{d}E\,\mathrm{d}E' \,\frac{1}{E + E' + E^{(+)}}.$$

The charging energy suppresses E_J in comparison with the Ambegaokar–Baratoff formula.

Coming back to our general thread, we have to deal with the situation where some transitions are coherent and characterized by amplitudes whereas others are incoherent and characterized by rates. We can use neither Hamiltonian quantum mechanics nor the classical master equation. The proper tool to analyze the situation is the *density matrix* approach, which is a convenient quantum generalization of classical probabilities used in the master equation context. The matrix is defined in a basis where each element presents either a quantum state or a group of quantum states. Diagonal elements yield the probabilities to be in a certain (group of) states. Non-diagonal elements of this matrix describe the quantum coherence between the states involved.

For a completely coherent quantum system, the evolution of the density matrix is determined by a Hamiltonian:

$$\frac{\partial \hat{\rho}}{\partial t} = -\frac{i}{\hbar}[\hat{H}, \hat{\rho}]. \tag{3.103}$$

For our case, the Hamiltonian in the basis of $|0\rangle$ and $|2\rangle$ states is given by the 2×2 matrix in Eq. (3.84). The full density matrix certainly has elements $\rho_{00}, \rho_{22}, \rho_{02}, \rho_{20}$. There also exists a probability of finding the SET in the charging state 1, so that ρ_{11} must be present as well. We recognize that "1" represents a large group of quantum states; each state in the group is characterized by labels of all possible final states of two quasiparticles excited in the process of single-electron tunneling. Therefore there is no coherence between the states of the group "1" and "0" or "2," $\rho_{12} = \rho_{10} = 0$.

Let us concentrate on the evolution equations for the diagonal elements of the density matrix. We write down these equations (called Bloch equations) by summing up the terms originated from Eq. (3.103), with terms describing the probability balance in the master equation. Denoting the rates of first- and second-quasiparticle tunneling, respectively, by Γ_a, Γ_b, we obtain

$$\frac{\partial \rho_{00}}{\partial t} = -\mathrm{i}\frac{E_\text{J}}{2}(\rho_{20} - \rho_{02}) + \Gamma_b \rho_{11}; \tag{3.104}$$

$$\frac{\partial \rho_{22}}{\partial t} = \mathrm{i}\frac{E_\text{J}}{2}(\rho_{20} - \rho_{02}) + \Gamma_a \rho_{22}; \tag{3.105}$$

$$\frac{\partial \rho_{11}}{\partial t} = \Gamma_a \rho_{22} - \Gamma_b \rho_{11}. \tag{3.106}$$

(We drop $\hbar$ for brevity.)

The Hamiltonian brings about non-diagonal elements to the first two equations, so we need the equations for non-diagonal elements as well. We obtain them using the same procedure:

$$\frac{\partial \rho_{02}}{\partial t} = \mathrm{i}2\epsilon\rho_{02} - \mathrm{i}\frac{E_\mathrm{J}}{2}(\rho_{22} - \rho_{00}) - \frac{1}{2}\Gamma_a\rho_{02}. \tag{3.107}$$

Since the density matrix is Hermitian, $\rho_{20} = \rho_{02}^*$, and we do not need to write a separate equation for ρ_{20}: it is obtained by complex conjugation. The origin of the factor 1/2 in front of Γ_a is precisely the same as in Eq. (3.70). Heuristically, only one index ("2") is affected by quasiparticle decay, whereas two are affected in the equation for ρ_{22}.

To find the stationary solution of these equations, let us first exclude the non-diagonal elements with the help of Eq. (3.107). The resulting system is given by

$$0 = \frac{\partial \rho_{00}}{\partial t} = \Gamma_\mathrm{p}(\rho_{22} - \rho_{00}) + \Gamma_b\rho_{11}; \tag{3.108}$$

$$0 = \frac{\partial \rho_{22}}{\partial t} = -\Gamma_\mathrm{p}(\rho_{22} - \rho_{00}) + \Gamma_a\rho_{22}; \tag{3.109}$$

$$0 = \frac{\partial \rho_{11}}{\partial t} = \Gamma_a\rho_{22} - \Gamma_b\rho_{11}, \tag{3.110}$$

where

$$\Gamma_\mathrm{p} \equiv \frac{E_\mathrm{J}^2\Gamma_a}{16\epsilon^2 + \Gamma_a^2}. \tag{3.111}$$

We will refer to Γ_p as the *pseudorate*. Indeed, the above equations coincide in form with a master equation where incoherent switchings from "0" and "1" and back take place with equal rates Γ_p. This is of course not an adequate picture of quantum transport in this situation; one can see this, for example, from the fact that Γ_p depends on the intensity of another rate, Γ_a, and would vanish if $\Gamma_a = 0$. However, the use of pseudorates conveniently reduces the density matrix approach to equations for probability balance. Using the pseudorate, one readily solves for ρ_{11}, ρ_{22}, and ρ_{00}. The average current is found from the relation $I/e = \Gamma_b\rho_{11}$ and is given by

$$I/e = \frac{\Gamma_a\Gamma_b\Gamma_\mathrm{p}}{\Gamma_a\Gamma_b + \Gamma_\mathrm{p}(2\Gamma_b + \Gamma_a)} = \frac{E_\mathrm{J}^2\Gamma_a}{16\epsilon^2 + \Gamma_a^2 + E_\mathrm{J}^2(2 + \Gamma_a/\Gamma_b)}. \tag{3.112}$$

We see that the current reaches its maximum, $I_\mathrm{max} \simeq e\,\max(E_\mathrm{J}^2/\Gamma, \Gamma) \simeq e\Gamma_\mathrm{se}$, at the resonance $\epsilon = 0$ and falls off upon decreasing the detuning $|\epsilon|$. A current versus ϵ plot shows a Lorentzian peak, with the half-width of order $\max(E_\mathrm{J}, \Gamma) \simeq \Gamma_\mathrm{se}$. Since the detuning is shifted by the bias voltage, $\delta\epsilon = e\delta V$, a typical differential conductance near the peak is just G_Q.

To access the FCS in the density matrix approach, we adopt the method used for the master equation in Section 3.2.5. We monitor the rate of the Γ_b transition and modify the rate in Eq. (3.104): $\Gamma_b \to \Gamma_b \exp(\mathrm{i}\chi)$. We look at an eigenvalue λ of the evolution equations; this is determined by solving the following system:

$$\lambda \rho_{00} = -\mathrm{i}\frac{E_\mathrm{J}}{2}(\rho_{20} - \rho_{02}) + \Gamma_b \exp(\mathrm{i}\chi)\rho_{11}; \tag{3.113}$$

$$\lambda \rho_{22} = \mathrm{i}\frac{E_\mathrm{J}}{2}(\rho_{20} - \rho_{02}) + \Gamma_a \rho_{22}; \tag{3.114}$$

$$\lambda \rho_{11} = \Gamma_a \rho_{22} - \Gamma_b \rho_{11}; \tag{3.115}$$

$$\lambda \rho_{02} = \mathrm{i}2\epsilon \rho_{02} - \mathrm{i}\frac{E_\mathrm{J}}{2}(\rho_{22} - \rho_{00}) - \frac{1}{2}\Gamma_a \rho_{02}; \tag{3.116}$$

$$\lambda \rho_{20} = -\mathrm{i}2\epsilon \rho_{02} + \mathrm{i}\frac{E_\mathrm{J}}{2}(\rho_{22} - \rho_{00}) - \frac{1}{2}\Gamma_a \rho_{02}; \tag{3.117}$$

which is an analog of Eq. (3.25). The FCS is given by Eq. (3.26) and current cumulants are obtained by expanding $\lambda(\chi)$ near $\chi = 0$. The FCS obtained in such a way is *not* the FCS of the master equation with pseudorate given by Eq. (3.111), the difference can already be seen in the noise. However, there is no simple limit where the difference is qualitative. If one of the rates $\Gamma_\mathrm{p}, \Gamma_a$, or Γ_b is much smaller than the other two, the FCS is Poissonian with this smallest rate and is indistinguishable from a FCS of a master equation. We will see in Section 5.5 that applying time-dependent voltages – quantum manipulation – will make the quantum features explicit.

The density matrix approach can be easily expanded to more complicated systems. Let us illustrate this by giving the equations that describe the bright spot transport cycle (Fig. 3.37(d)). We have four states, $0, 1, -1$, and 2. There is coherence between 0 and 2 and between 1 and -1, giving rise to non-diagonal elements of the density matrix, $\rho_{02}, \rho_{20}, \rho_{1,-1}, \rho_{-1,1}$. The coherence within each pair of states is described by a 2×2 Hamiltonian matrix, with parameters $E_\mathrm{J}^{(1)}, \epsilon^{(1)}$ ($E_\mathrm{J}^{(2)}, \epsilon^{(2)}$) for $0, 2$ ($-1, 1$). Again, we sum up the contributions to evolution equations from coherent and incoherent transitions. For diagonal elements we obtain

$$\frac{\partial \rho_{00}}{\partial t} = -\mathrm{i}\frac{E_\mathrm{J}^{(1)}}{2}(\rho_{20} - \rho_{02}) + \Gamma_b \rho_{-1,-1}; \tag{3.118}$$

$$\frac{\partial \rho_{22}}{\partial t} = \mathrm{i}\frac{E_\mathrm{J}^{(1)}}{2}(\rho_{20} - \rho_{02}) - \Gamma_a \rho_{22}; \tag{3.119}$$

$$\frac{\partial \rho_{11}}{\partial t} = \mathrm{i}\frac{E_\mathrm{J}^{(2)}}{2} + \Gamma_a \rho_{22}; \tag{3.120}$$

$$\frac{\partial \rho_{-1,-1}}{\partial t} = -\mathrm{i}\frac{E_\mathrm{J}^{(2)}}{2}\Gamma_a \rho_{22} - \Gamma_b \rho_{-1,-1}. \tag{3.121}$$

For non-diagonal ones we have

$$\frac{\partial \rho_{02}}{\partial t} = \mathrm{i}2\epsilon^{(1)}\rho_{02} - \mathrm{i}\frac{E_\mathrm{J}^{(1)}}{2}(\rho_{22} - \rho_{00}) - \frac{1}{2}\Gamma_a \rho_{02}; \tag{3.122}$$

$$\frac{\partial \rho_{-11}}{\partial t} = \mathrm{i}2\epsilon^{(2)}\rho_{-11} - \mathrm{i}\frac{E_\mathrm{J}^{(1)}}{2}(\rho_{11} - \rho_{-1,-1}) - \frac{1}{2}\Gamma_b \rho_{-11}. \tag{3.123}$$

Exercise 3.15. Solve the above equations to find the stationary current in the center of the bright spot ($\epsilon^{(1)} = \epsilon^{(2)} = 0$). Hint: introduce pseudorates for both coherent transitions.

4 Randomness and interference

This chapter treats quantum interference effects in disordered conductors. We have already discussed interference phenomena for the few-channels case in Section 1.6, and in Chapter 2 we have seen that the scattering approach becomes increasingly complicated for many transport channels. This is why we need special methods to treat many-channel diffusive conductors. Some of their properties can be understood if we replace the Hamiltonian or a scattering matrix by a *random matrix*. In Section 4.1 we discuss random matrices as mathematical objects and review the properties of their eigenvalues and eigenvectors. We then use random matrix theory to describe the properties of energy levels (Section 4.2) and transmission eigenvalues (Section 4.3).

To describe interference effects in a very broad class of conductors, we develop in Section 4.4 the methods to handle *interference corrections for a circuit theory*, which allows us to take account of all nanostructure details. We evaluate universal conductance fluctuations and the weak localization correction to conductance.

We also find that, in some situations, electrons are localized – confined to small regions of space. Section 4.5 discusses under which conditions this *strong localization* occurs and briefly outlines transport properties associated with this regime.

4.1 Random matrices

In this section, we describe disordered and chaotic systems. *Disordered* systems contain some defects that scatter electrons; *chaotic* ones do not have any defects, but scattering at the boundaries induces a very different motion for particles with very close energies. In these systems, the positions of energy levels and transmission eigenvalues are random and vary from sample to sample. We have already encountered such a situation when discussing the transport properties of diffusive conductors in Chapter 2. Quantities of interest to quantum transport are not sensitive to the positions of the energy levels, and transmission eigenvalues and can be described statistically. It is natural, following the idea introduced by Wigner in the 1950s, to consider the Hamiltonian and the scattering matrix as *random matrices*. Energy levels as eigenvalues of the Hamiltonian, and transmission eigenvalues as eigenvalues of $\hat{t}^\dagger\hat{t}$, where $\hat{t}$ is the transmission block of the scattering matrix, are regarded as random entities with certain statistical properties. The mathematical technique dealing with random matrices is known as *random matrix theory* (RMT). In this section, we review the basic properties of random matrices as mathematical objects. Derivations and additional details can be found in Refs. [83] and [84]. The connection between RMT and quantum transport is studied in the following sections of this chapter.

4.1.1 Gaussian ensembles

Consider a set of square matrices $\hat{H}$ of size N whose elements are random numbers. Due to the central limit theorem, it is natural to assume that these elements are Gaussian-distributed. We consider three species of random matrices: (i) real and symmetric random matrices, so that $H_{mn} = H_{nm} = H^*_{mn}$; (ii) Hermitian random matrices without any further constraints, $H_{mn} = [H_{nm}]^*$; (iii) matrices of double size ($2N$). The latter, when expanded in Pauli matrices $\hat{\sigma}$,

$$\hat{H} = \hat{H}^{(0)}\hat{1} - \mathrm{i}\hat{\boldsymbol{H}} \cdot \hat{\boldsymbol{\sigma}}, \tag{4.1}$$

have the following symmetry: the $N \times N$ matrix $\hat{H}^{(0)}$ is real and symmetric, and the $N \times N$ components of the vector $\hat{\boldsymbol{H}}$ are real and antisymmetric, $H^{(\gamma)}_{mn} = -H^{(\gamma)}_{nm}$, $\gamma = x, y, z$. In Section 4.2, we will see that these types describe the Hamiltonian of a system with (i) time-reversal symmetry; (ii) broken time-reversal symmetry, for example because of an applied magnetic field; and (iii) presence of spin-orbit scattering. For each of the species, we assume that the probability density (in the space of all matrices with the required symmetry) of finding a matrix $\hat{H}$ is Gaussian, given by

$$P(\hat{H}) \propto \exp\left[-\frac{\beta}{2\delta_{\mathrm{S}}^2}\ \mathrm{Tr}\ \hat{H}^2\right], \tag{4.2}$$

where δ_{S} is the only parameter of the problem; we will see that it is equal to the average spacing between the eigenvalues of the random matrix. The variable β equals 1, 2, and 4 for the symmetries (i), (ii), and (iii), respectively. Note that β essentially counts a number of independent real numbers needed to describe an element of a random matrix – real, complex numbers, and 2×2 matrices for (i), (ii), and (iii), respectively. Random matrices are Gaussian-distributed around zero. In particular, for case (i) the probability can be written as

$$P(\hat{H}) \propto \exp\left[-\frac{1}{2\delta_{\mathrm{S}}^2}\sum_{mn} H_{mn}^2\right],$$

meaning that all matrix elements are independent random variables, Gaussian-distributed around zero with the same variance δ_{S}^2.

A set of random matrices described by Eq. (4.2) is known as a Gaussian ensemble, specifically a Gaussian orthogonal ensemble (GOE) for symmetry (i), a Gaussian unitary ensemble (GUE) for symmetry (ii), and a Gaussian symplectic ensemble (GSE) for symmetry (iii). Eigenvalues and eigenvectors of these random matrices are random quantities with statistical properties; RMT was designed to study these properties.

4.1.2 Eigenvalues of random matrices

A naive expectation would be that, since random matrices are Gaussian-distributed, their eigenvalues and eigenvectors are also Gaussian-distributed and independent. It turns out that this is not the case. For example, eigenvalues of random matrices repel each other. To understand the origin of this *repulsion*, we consider first an example of a random 2×2 matrix taken from a Gaussian ensemble. For instance, in a GUE a matrix must be real and symmetric. A general 2×2 real symmetric matrix has the following form:

$$\hat{H} = \begin{pmatrix} H_0 + H_z & H_x \\ H_x & H_0 - H_z \end{pmatrix}, \tag{4.3}$$

where H_0, H_x, and H_z are random real variables. From Eq. (4.2) we see that all three are *independent* Gaussian variables distributed around zero with the dispersion of $\delta_S^2/2$, $P(\hat{H}) \propto \exp(-\delta_S^{-2}(H_0^2 + H_x^2 + H_z^2))$.

The matrix in Eq. (4.3) can be easily diagonalized, producing the eigenvalues $E_{1,2} = H_0 \pm \sqrt{H_x^2 + H_z^2}$, and the spacing $\delta E = E_2 - E_1 = 2\sqrt{H_x^2 + H_z^2}$ does not depend on H_0. Since H_x and H_z are two independent Gaussian variables, we can treat them as two components of a random two-dimensional vector, with the eigenvalue spacing being twice the length of this vector. Introducing the polar coordinates – length $\delta E/2$ and the polar angle θ – and writing $\mathrm{d}H_x\, \mathrm{d}H_z = \delta E\, \mathrm{d}\delta E\, \mathrm{d}\theta/4$, we find the distribution function of the spacing δE, up to a normalization constant:

$$P_{GOE}(\delta E) \propto \delta E \exp\left(-(\delta E)^2/4\delta_S^2\right).$$

Note that the distribution function vanishes if the spacing is zero: eigenvalues cannot come close, or, in other words, they *repel each other*. This repulsion comes technically from the fact that the Jacobian of the transition from Cartesian to polar coordinates in two dimensions is proportional to the polar radius. In this sense, the repulsion is a purely geometric effect, originating from the symmetry constraints: the fact that we only have two relevant parameters is due to the symmetry.

Let us now look at a 2×2 matrix from a GUE, which depends on four real parameters:

$$\hat{H} = \begin{pmatrix} H_0 + H_z & H_x + \mathrm{i}H_y \\ H_x - \mathrm{i}H_y & H_0 - H_z \end{pmatrix}; \tag{4.4}$$

the eigenvalue spacing is given by $\delta E = 2\sqrt{H_x^2 + H_y^2 + H_z^2}$. Representing Gaussian variables H_x, H_y, and H_z as three components of a random vector and introducing spherical coordinates, we obtain the distribution function of the spacing:

$$P_{\mathrm{GUE}}(\delta E) \propto (\delta E)^2 \exp\left(-(\delta E)^2/4\delta_S^2\right).$$

Note that the repulsion is stronger than in a GOE: the distribution function is proportional to $(\delta E)^2$ rather than to δE.

For larger matrices, the calculations become progressively cumbersome. However, for $N \gg 1$ one can still obtain an expression for the joint probability of eigenvalues:

$$P_\beta(E_1 \cdots E_N) \propto \exp\left(-\frac{\beta}{2\delta_S^2}\sum_{n=1}^{N} E_n^2\right)\left|\prod_{n>m}\frac{E_n - E_m}{\delta_S}\right|^\beta$$
$$\propto \exp\left(-\beta\left[\frac{1}{2\delta_S^2}\sum_n E_n^2 + \sum_{m<n}\ln|E_n - E_m|\right]\right). \quad (4.5)$$

This formal expression has a transparent physical analogy. Let us identify E_i/δ_S as positions of classical charges. Then the probability P_β plays the role of the partition function, with effective temperature β^{-1}, and the exponent that of the free energy. The first term in the exponent is a sum over all positions and thus represents the confining potential $U \propto x^2$: the charges prefer to be close to the origin. The second term is a sum over all *pairs* of charges, and thus it represents interaction between the charges. This interaction is repulsive (positive potential) and logarithmically depends on the distance between the charges (like Coulomb interaction in two dimensions). Thus, the behavior of the eigenvalues of random matrices is the same as the behavior of N charges with logarithmic repulsion confined in a parabolic potential. The repulsion of the charges means that the eigenvalues of random matrices repel each other, as we have already seen in the example of a 2×2 matrix. The higher the temperature, the easier it is for the "charges" to overcome the level repulsion and come close to each other; this is why the level repulsion is weakest in a GUE ($\beta = 1$) and strongest in a GSE ($\beta = 4$).

Let us now investigate the quantitative consequences of Eq. (4.5). We look at the density of eigenvalues $\{E_n\}$, defined as

$$\nu(E) = \sum_n \delta(E_n - E),$$

and consider first the average density. Note that in Section 1.2 we investigated the density of two-dimensional electrons confined along one dimension. What we discovered is that the density vanishes at the edge of the electronic system as the square root of the distance to the edge. Similarly, eigenvalues of random matrices form a band, and the average density vanishes as a square root at the band edge. For all three ensembles, in the limit of large matrices $N \to \infty$, the following concise expression is valid:

$$\langle \nu(E) \rangle = \frac{\sqrt{N}}{\pi\delta_S}\sqrt{1 - \left(\frac{E}{2\sqrt{N}\delta_S}\right)^2}. \quad (4.6)$$

The density takes the shape of a semicircle centered around $E = 0$. There are no eigenvalues for $|E| > 2\sqrt{N}\delta_S$: the eigenvalues lie within a wide band. Close to the center of the band, the density of eigenvalues is constant and proportional to $\sqrt{N}$. This is easy to explain. Indeed, the eigenvalues are found as solutions of the equation $\det(\hat{H} - E) = 0$, which is a polynomial equation. The first two terms in the polynomial are E^N and $E^{N-1}\,\mathrm{Tr}\,\hat{H}$. The trace of $\hat{H}$ is a sum of N random terms, Gaussian-distributed around zero with variance δ_S. By virtue of the central limit theorem, $\mathrm{Tr}\,\hat{H}$ is also Gaussian with zero average and the variance of $N\delta_S^2$. Thus, the maximal eigenvalue is of the order of $\delta_S\sqrt{N}$. Since one has N eigenvalues, it follows that the density of states at the band center is of the order of $\sqrt{N}/\delta_S$.

It is more convenient to look at the range of eigenvalues close to the center of the band, and this is indeed what we need for the applications considered in this chapter. Below we only consider properties of the eigenvalues at the center of the band, $E \ll \sqrt{N}\delta_S$, and assume that the average density of eigenvalues is constant.

Next, we look at the *distribution function of spacings* δE between adjacent eigenvalues. This distribution can be calculated in a closed form, but the expression is too cumbersome. Instead, there is an convenient approximation known as the *Wigner surmise*, or the *Wigner–Dyson distribution*,

$$\rho_\beta(\delta E) = a_\beta \left(\frac{\delta E}{\delta_S}\right)^\beta \exp\left(-b_\beta \left(\frac{\delta E}{\delta_S}\right)^2\right), \tag{4.7}$$

where the coefficients a and b are given by

$$a_1 = \frac{\pi}{2}; \quad a_2 = \frac{32}{\pi^2}; \quad a_4 = \frac{32\,768}{729};$$
$$b_1 = \frac{\pi}{4}; \quad b_2 = \frac{4}{\pi}; \quad b_4 = \frac{64}{9\pi}.$$

We see that the probability of finding low spacings $\delta E \ll \delta_S$ is suppressed; the suppression increases from a GOE to a GUE and then to a GSE, due to the factor $(\delta E)^\beta$ in Eq. (4.7). This is the same eigenvalue repulsion that we have already encountered. Using the distribution, one can calculate all its moments. For example, the *average spacing between adjacent eigenvalues* in all three ensembles equals δ_S, as we anticipated. The root mean square of the spacing $\sqrt{\langle(\delta E)^2\rangle - \langle\delta E\rangle^2}$ is different in all three ensembles. Due to the weakest repulsion, it is largest in the GOE, where it equals $(4/\pi)^2\delta_S$.

Exercise 4.1. Calculate the root mean square of the spacing in a GUE and in a GSE.

Next, we calculate the (irreducible) *pair correlation function*, given by

$$R_2(\delta E) \equiv \frac{\langle \nu(E)\, \nu(E+\delta E)\rangle}{\langle \nu\rangle^2} - 1. \tag{4.8}$$

Since we are working in the middle of the band, this does not depend on the global position E, only on the separation δE between the levels.

Equation (4.8) is the correlation function. We set the energy of one eigenvalue at E and look at the probability that another eigenvalue is found at $E + \delta E$. It is important that these eigenvalues do not have to be adjacent, in contrast to Eq. (4.7); several or many eigenvalues can lie between these two. For large separations, $\delta E \gg \delta_S$, it is most probably the case; the two levels do not know anything about each other, the densities of eigenvalues at E and $E + \delta E$ are uncorrelated, and R_2 vanishes. For low separations, $\delta E \ll \delta_S$, there is a big chance that two eigenvalues are adjacent and they repel each other. Thus, for $\delta E \to 0$, the correlation function must approach -1.

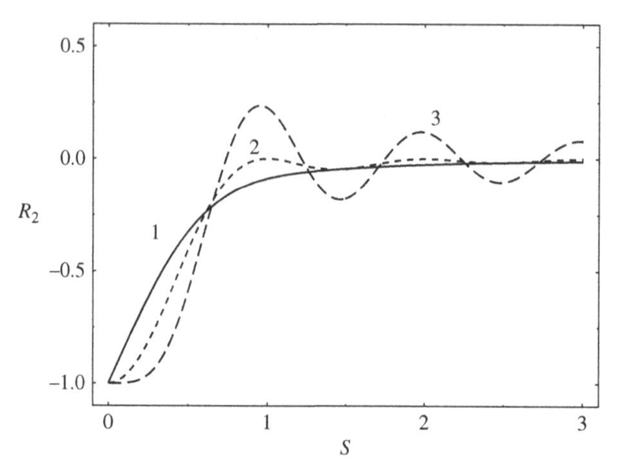

Fig. 4.1. Two-point correlation functions for three different ensembles: GOE (1), GUE (2), and GSE (3).

The RMT result for the correlation function is given by

$$R_2(\varepsilon) = \begin{cases} \delta(\varepsilon) - s^2(\varepsilon) - \dfrac{\mathrm{d}s}{\mathrm{d}\varepsilon}\displaystyle\int_\varepsilon^\infty s(\varepsilon')\mathrm{d}\varepsilon' & \text{(GOE)} \\ \delta(\varepsilon) - s^2(\varepsilon) & \text{(GUE)} \\ \delta(\varepsilon) - s^2(2\varepsilon) + \dfrac{\mathrm{d}s(2\varepsilon)}{\mathrm{d}\varepsilon}\displaystyle\int_0^\varepsilon s(2\varepsilon')\mathrm{d}\varepsilon' & \text{(GSE),} \end{cases} \tag{4.9}$$

where $\varepsilon \equiv |\delta E|/\delta_S$ and $s(x) \equiv \sin \pi x/(\pi x)$. These functions are shown in Fig. 4.1. The first term, with the delta-function, is the same in all three ensembles. It is just the correlation of an eigenvalue with itself and is required to ensure the conservation of numbers of eigenvalues: $\int \mathrm{d}\varepsilon\ R_2(\varepsilon) = 0$. The second term is negative and is thus a manifestation of eigenvalue repulsion. The strongest repulsion is for eigenvalues taken at the same value of E. The repulsion depends on $|\delta E|$ in an oscillatory manner. For example, in a GUE eigenvalues are not correlated if the separation between them is an integer multiple of δ_S. For separations much greater than the eigenvalue spacing, $\varepsilon \gg 1$, the smooth part of the pair correlation assumes a simple form:

$$R_2^{\text{smooth}}(\varepsilon) = -\frac{1}{\pi^2 \beta(\varepsilon + \mathrm{i}0)^2}. \tag{4.10}$$

A useful quantity is the *eigenvalue number variance* $\Sigma_2(\delta E)$: the variance of the number of eigenvalues between E and $E + \delta E$. It is meaningful provided the interval contains many eigenvalues, $\delta E \gg \delta_S$. The level number variance is related to the correlation function R_2 as follows:

$$\Sigma_2(\delta E) = \int_E^{E+\delta E} \mathrm{d}E_1\, \mathrm{d}E_2\ R_2(E_2 - E_1) = 2\int_0^{\delta E} (\delta E - E')R_2(E')\mathrm{d}E'.$$

The level number variance does not contain new information as compared with the correlation function, it just represents this information differently – the smooth part is better emphasized than the oscillations. The double integration of Eq. (4.10) gives, for the smooth part,

$$\Sigma_2^{\text{smooth}}(\delta E) = \frac{2}{\beta\pi^2} \ln \frac{\delta E}{\delta_S}. \tag{4.11}$$

Note that if the eigenvalues were independent (uncorrelated), the eigenvalue number variance would grow proportionally to the eigenvalue number itself, i.e. proportionally to δE, in accordance with the Gauss theorem. On the other hand, if the eigenvalues were equidistant, the variance obviously is less than one. RMT gives a result intermediate between these two extremes. This property is known as *spectral rigidity*: the spectrum consisting of equidistant levels is maximally rigid, whereas the RMT eigenvalue spectrum is more rigid than the one of independent eigenvalues.

Control question 4.1. On top of the smooth behavior given in Eq. (4.11), the level number variance also contains a contribution which oscillates with δE. Estimate the period and the magnitude of the oscillations.

4.1.3 Eigenvectors of random matrices

Properties of eigenvectors of random matrices are as simple as they could be: in the leading order in $1/N$, eigenvectors corresponding to different eigenvalues are independent. Thus, we only need to study properties of a single eigenvector $(\psi_1 \cdots \psi_N)$. We assume that it is normalized, $\sum |\psi_i|^2 = 1$. The distribution function for the amplitudes ψ_i is Gaussian, and for the *intensities* ψ_i^2 it is written as follows ($y = N|\psi_i|^2$):

$$\rho(y) = \begin{cases} \dfrac{1}{\sqrt{2\pi y}} \mathrm{e}^{-y/2} & \text{(GOE)} \\ \mathrm{e}^{-y} & \text{(GUE)}, \end{cases} \tag{4.12}$$

(the *Porter–Thomas distribution*). The distribution is normalized such as $\int \mathrm{d}y\, \rho = 1$.

Control question 4.2. What is the average intensity $\langle |\psi_i|^2 \rangle$? Explain the result.

Another remarkable fact is that, also in the leading order in $1/N$, the properties of eigenvalues and eigenvectors of random matrices are not correlated: they can be considered as independent entities.

4.1.4 Parametric statistics

In practice, we can have a control parameter that is varied in the experiment. It could, for example, be the external magnetic field or shape of the external potential. Let us denote this parameter as X. Each individual random matrix depends on X; this dependence is

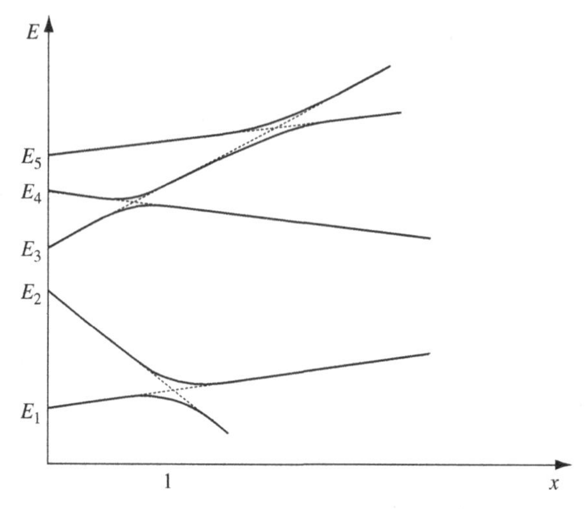

Fig. 4.2. **Parametric evolution of eigenvalues. Dashed lines show the motion of eigenvalues without the repulsion taken into account; solid lines include the repulsion.**

such that, for any value of X, the statistical properties of the Gaussian ensemble of random matrices stay the same. However, we can now ask new types of questions: What is the correlation between the properties of a random matrix at different values of X?

The first question concerns the evolution of the eigenvalues. By varying X, we shift all the eigenvalues of the random matrices simultaneously. Imagine we have taken a random matrix, and for $X = 0$ it has a set of eigenvalues $E_1(0), \dots, E_N(0)$. For a finite value of X, the eigenvalues have evolved to $E_1(X), \dots, E_N(X)$. Since the eigenvalues are random, they also evolve randomly. At small X, this evolution is characterized by the *eigenvalue velocity* – the derivative $\mathrm{d}E_n/\mathrm{d}X$. On average, the velocity vanishes, since, for some matrices, the nth eigenvalue moves up, and for others it moves down. Instead, one characterizes the eigenvalue motion by the mean square of eigenvalue velocity:

$$C(E) = \frac{1}{\langle \nu(E) \rangle} \left\langle \sum_n \left(\left. \frac{\mathrm{d}E_n}{\mathrm{d}X} \right|_{X=0} \right)^2 \delta(E - E_n) \right\rangle .$$

Close to the band center this variance does not depend on E. The parameter C characterizes the response of the random matrix ensemble to an external perturbation. It is convenient to rescale the variable X so that it takes the same dimension as eigenvalues, $x = X\sqrt{C}$. Expressed in terms of x, the properties of parametric statistics are universal [85] and do not really depend on the origin of the variable X. Note that x measures the *shift* of the eigenvalue due to the external perturbation.

If we take two adjacent eigenvalues, they both evolve with the parameter x; the spacing δE between them is also a function of x. For $x \sim 1$, the levels typically come close to each other. Further evolution (when the eigenvalues first become degenerate, and eventually pass each other) is prohibited by the level repulsion, so that they turn around and start moving in the opposite direction (see Fig. 4.2). This is the situation that we know from quantum mechanics as *avoided crossing*. For $x \gg 1$, the eigenvalues have already undergone many crossings, and thus the eigenvalues at the same value of E do not know anything about each other: they are uncorrelated.

An appropriate way to look at the parametric statistics is to use the analogy with a system of electric charges representing the eigenvalues, with a parabolic confining potential, logarithmic repulsion, and the effective temperature β^{-1}. To describe the level motion under external perturbation, one considers the *dynamics* of the charges. We introduce the fictitious time τ and look at the evolution of the positions of all charges. Since they are at finite temperature, they now perform random Brownian motion. The joint distribution function of the positions of all the charges (the distribution function of all eigenvalues), $P(\{E_n\}, \tau)$, obeys the Fokker–Planck equation:

$$\frac{\partial P}{\partial \tau} = \frac{\beta C}{2} \sum_{n=1}^{N} \frac{\partial}{\partial E_n} \left(\frac{1}{\beta} \frac{\partial P}{\partial E_n} + P \frac{\partial}{\partial E_n} \left(\frac{1}{2\delta_S^2} \sum_n E_n^2 - \sum_{m<n} \ln |E_n - E_m| \right) \right). \quad (4.13)$$

On the right-hand side, the first term (second derivative) represents "diffusion," whereas the other two are responsible for the "drift" in the effective potential – external parabolic confinement and logarithmic interaction potential created by all eigenvalues. In addition to the "effective temperature" β^{-1}, we also have the "viscosity" $2/C\beta$. Equation (4.13) tells us that the eigenvalues can perform Brownian motion in time τ, but, due to the logarithmic repulsion, they cannot come close to each other; then we have an avoided crossing. The time τ cannot be expressed in terms of the variable x; the rule is that one has to replace time differences $\tau - \tau'$ by $(x - x')^2$.

One proceeds by reducing Eq. (4.13) to an equation for the average density of eigenvalues, which has the form of the diffusion equation, and then derive various correlation functions. Thus, the distribution of the level velocities is Gaussian with zero average and variance equal to C:

$$\rho(v) = \frac{1}{\sqrt{2\pi C}} \exp\left(-\frac{v^2}{2C} \right), \quad v \equiv \frac{\mathrm{d}E_n}{\mathrm{d}X}.$$

Another question concerns the correlation function of eigenvalues at different values of the perturbation x, which is a generalization of the function R_2 in the case of parametric statistics. We define it in the following way:

$$R_2(\delta E, \delta x) = \frac{\langle \nu(E,x)\, \nu(E + \delta E, x + \delta x) \rangle}{\langle \nu(E) \rangle^2} - 1,$$

where we have taken into account that the average density of states does not depend on x. The previously studied function $R_2(E)$ is a particular case for $\delta x = 0$. The RMT result for this correlation function is rather cumbersome, and we only show the GUE expression, $\varepsilon = |\delta E|/\delta_S$,

$$R_2 = \frac{1}{2} \mathrm{Re} \int_{-1}^{1} \mathrm{d}z \int_{1}^{\infty} \mathrm{d}z' \exp\left\{ -\frac{\pi^2 (\delta x)^2}{2} (z'^2 - z^2) - \mathrm{i}\pi\varepsilon (z - z') \right\}.$$

For $\delta E = 0$, $\delta x \gg 1$, we obtain $R_2(0, \delta x) = -2/(\beta\pi^2(\delta x)^2)$ (this result is valid for all three Gaussian ensembles). The negative sign of the correlation is again a manifestation of the eigenvalue repulsion.

4.1.5 Circular ensembles

Gaussian ensembles are useful in discussing the positions of energy levels. If we want to discuss transport properties, we need to deal with scattering matrices rather than Hamiltonians. Scattering matrices are unitary, providing the motivation to study properties of random unitary matrices. Let us consider an ensemble of random matrices $\hat{s}$ that are unitary, $\hat{s}^\dagger\hat{s}$, and possess the following symmetry: they are either (i) symmetric, (ii) not symmetric, or (iii) each element of them is a 2×2 matrix with the symmetry properties given by Eq. (4.1). We assume that these random matrices are distributed uniformly: the probability of finding a matrix is always the same, provided the matrix is unitary and obeys the constraints for (i) and (iii). One says that random matrices are uniformly distributed in the *unitary group*. Such a collection of random matrices is known as a circular orthogonal ensemble (COE) for (i), a circular unitary ensemble (CUE) for (ii), and a circular symplectic ensemble (CSE) for (iii). These ensembles are similar to the corresponding Gaussian ensembles GOE, GUE, and GSE; in particular, they are characterized by the same "inverse temperature" $\beta = 1, 2, 4$.

To understand the properties of circular random matrix ensembles, let us start with an example of a 2×2 random unitary matrix from the orthogonal ensemble. A general symmetric unitary matrix can be parameterized as follows (see Eq. (1.39)):

$$\hat{s} = \begin{pmatrix} \sqrt{R}e^{i\theta} & \sqrt{T}e^{i\eta} \\ \sqrt{T}e^{i\eta} & -\sqrt{R}e^{i(2\eta-\theta)} \end{pmatrix},$$

with $T + R = 1$. It is characterized by the two independent phases, θ and η, and we assume that these phases are independent random numbers, *uniformly* distributed between 0 and 2π. The coefficient R is, in principle, also a random quantity, and its distribution must be found from the requirement that the matrix is distributed uniformly, but to illustrate our point it is enough to keep it fixed.

Let us calculate the eigenvalues of this matrix. Since the matrix is unitary, both eigenvalues lie on the unit circle in the complex plane:

$$\lambda_{1,2} = \pm e^{i(\eta\pm\phi)} \equiv e^{i\alpha_{1,2}}, \quad \sin\phi = \sqrt{R}\sin(\theta-\eta). \tag{4.14}$$

The difference between the phases $\alpha = \alpha_1 - \alpha_2 = \pi - 2\phi$ only depends on the difference of η and θ, not on their sum. Since the distribution function of $\theta - \eta$ is constant, $P(\theta - \eta) = 1/2\pi$, the distribution function of the phase difference is given by

$$P(\alpha) = \frac{1}{2\pi}\left|\frac{\partial(\theta-\eta)}{\partial\alpha}\right| = \frac{1}{4\pi}\frac{\sin(\alpha/2)}{\sqrt{R-\cos^2(\alpha/2)}}.$$

Thus, small phase differences are suppressed: the phases of the eigenvalues of the scattering matrix *repel each other*. Similarly for the eigenvalue repulsion in Gaussian ensembles, this repulsion is a geometric effect, related to the Jacobian of the transformation from phases of the scattering matrix to the phases of its eigenvalues.

The same repulsion occurs for matrices of a greater size. A unitary $2N \times 2N$ matrix has eigenvalues $\exp(i\phi_1), \ldots, \exp(i\phi_{2N})$. The joint distribution function of these eigenvalues has the form, up to a normalizing constant factor, given by

$$P(\phi_1, \ldots, \phi_N) \propto \prod_{m<n} \left| e^{i\phi_m} - e^{i\phi_n} \right|^{\beta} \propto \exp\left(\beta \sum_{m<n} \ln|\sin(\phi_m - \phi_n)| \right), \tag{4.15}$$

which, similarly to Gaussian ensembles, has an interpretation in terms of classical charges. Imagine a system of charges on a ring of unit radius. Their positions are characterized by the polar angles ϕ_m. Equation (4.15) states that these charges repel logarithmically, since $2|\sin(\phi_m - \phi_n)|$ is the distance between the two points on the ring. The effective "temperature" is β^{-1}; this results in a higher probability of small eigenvalue spacings in COE in comparison with other ensembles.

For the purpose of quantum transport, one needs to investigate (rather than the eigenvalues of the scattering matrix) the transmission eigenvalues T_n – eigenvalues of the matrix $\hat{t}^\dagger \hat{t}$, where t is a block of size $N_1 \times N_2$ in the representation of the random unitary matrix $\hat{s}$:

$$\hat{s} = \begin{pmatrix} \hat{r} & \hat{t} \\ \hat{t}' & \hat{r}' \end{pmatrix}.$$

Their joint distribution can be derived from Eq. (4.15) as follows:

$$P(\{T_n\}) \propto \prod_{n<m} |T_n - T_m|^{\beta} \prod_{n} T_n^{-1+\beta/2+(\beta/2)|N_1-N_2|}. \tag{4.16}$$

Transmission eigenvalues in the Gaussian ensembles also repel each other, and the repulsion is strongest in a COE and weakest in a CSE.

We discuss the consequences of Eq. (4.16) in Section 4.3.

4.2 Energy-level statistics

One of the main differences between classical and quantum mechanics is in the quantization of energy levels. In this section, we describe *quantum dots*, (home-made) systems where electrons are confined in *visibly* discrete levels. The word "visibly" is important here. By virtue of quantum mechanics, any confined motion gives rise to discrete levels. If an electron is confined within the Empire State Building, its states are characterized by discrete levels. However, these levels are separated by a vanishingly small energy, and although we theoretically know they are present and discrete, we are not able to verify this experimentally. Roughly speaking, with modern measuring techniques, one can resolve levels separated by approximately 0.1 meV. This value is not a fundamental physical constant, it just characterizes the apparatus currently available in a lab. Thus, the Empire State Building is not a quantum dot. Metallic islands described in Chapter 3 are not quantum dots either – the level separation there is way too small.

On the other hand, a hydrogen atom ideally suits the definition: it has discrete levels that are observed by optical means as spectral lines. With discrete atomic or molecular levels giving rise to thin spectral lines, one makes lasers. This was the original motivation to study quantum dots, and the research started in the 1970s. The discrete levels of atoms and molecules do not always lie at convenient frequencies, nor can they readily be tuned to those frequencies. The idea was to confine electrons in artificially made potential wells in order to make artificial atoms. By that time, it was already possible to engineer an electrostatic potential in superconducting heterostructures. The level distances in the wells can be engineered and sometimes even tuned. Using lithography, one can cover large areas with zillions of tiny quantum dots and use such surfaces for tunable lasers. These dots are *optical*: they are not connected to any electrodes, and they are measured only with light absorption and/or emission.

This idea never worked, and, despite the remarkable progress in the fabrication of quantum dots, it is still not clear whether it will ever work. Hydrogen atoms are all identical: the spectral line is thin for a single atom, and it remains thin if the light is emitted by a big atomic cloud. As for quantum dots, the good features immediately become drawbacks. If a level position can be engineered, it is sensitive to fabrication errors. If a level position can be tuned by, for example, gate voltage, it can be also "tuned" by any charged impurity in the vicinity of the dot. Since fabrication errors are inevitable, and impurity concentrations are different in different dots, the artificial atoms are never identical, and discrete levels in each quantum dot are different. Thus, there are no thin spectral lines for big ensembles of dots: the light from different dots is emitted at different frequencies. This is known as *inhomogeneous broadening*, and renders the discrete levels hardly visible.

If one can assess a single dot by optical means, one still sees a thin spectral line. This, however, became possible only a few years ago. It appears easier to observe discrete levels in *transport dots*, which are connected to the leads so that the electrons can be transferred from a lead to the dot and to the other lead. One understands that the coupling to the leads cannot be too good. If the coupling is too strong, discrete levels are mixed with continuous electron states in the leads and are no longer discrete. To keep the levels discrete, one should provide sufficient isolation of the dot states from the leads. We have already seen in Chapter 3 that the criterion for good isolation is a sufficiently low conductance of the contacts, $G \ll G_Q$. In this section, we assume that this condition is fulfilled.

Also, in Chapter 3, we saw that, for $G \ll G_Q$, charging effects may be important. In this section, we do not consider them – we only study the properties for which charging does not play any role. We return to the discussion of tunneling and Coulomb blockade in quantum dots in Section 5.4.

To prevent any misunderstanding, we note that in the literature on quantum transport one finds so-called *open quantum dots* with higher conductance, $G \geq G_Q$. We have considered these systems in Section 2.6, and will return to them in Section 4.3. They do not possess any discrete levels, and they do not conform to our definition.

Let us now start with a simple example, which we can fully understand. Consider a two-dimensional rectangular quantum dot of the size $L_x \times L_y$. The external potential inside the dot equals zero, and outside rises to infinity, so that the wave function vanishes at $x = 0, L_x$, $y = 0, L_y$. Solving the Schrödinger equation, we obtain the electron states labeled by two numbers $n_x, n_y = 1, 2, \ldots$, with wave functions given by

$$\psi(x, y) = \frac{1}{\sqrt{L_x L_y}} \sin \frac{\pi n_x x}{L_x} \sin \frac{\pi n_y y}{L_y} \tag{4.17}$$

and energies given by

$$E_{n_x,n_y} = \frac{\pi^2\hbar^2}{2m}\left[\left(\frac{n_x}{L_x}\right)^2 + \left(\frac{n_y}{L_y}\right)^2\right]. \tag{4.18}$$

First, these energy levels are perfectly regular. For example, if we fix $n_x = 1$, we get a series of levels with various n_y, arranged parabolically. Other values of n_x give similar series, offset by $(\pi\hbar n_x)^2/mL^2$. Secondly, we may want to plot all the levels without thinking about the states they describe. Then the system becomes irregular. For example, take $L_x = L_y = L$. There is a ground state – the state with energy equal to $\pi^2\hbar^2/mL^2$. The next state has energy $5(\pi\hbar)^2/2mL^2$ and is doubly degenerate. The next one is again non-degenerate and has energy $4(\pi\hbar)^2/mL^2$. The next one is doubly degenerate, and so on. However, for $L_x \neq L_y$ there is a mess – some spacings are large, some are small. For such a collection of energy levels, we can introduce the notion of *level statistics*.

Indeed, let us look at large scales. Consider the number of levels N in the energy interval between two energy values E_1 and E_2, which we choose large enough to contain many levels. In this case, it is not really important that the levels are discrete – continuous values of n_x and n_y would produce the same result and are much easier in practical calculations. Introducing the "wave vectors" of the standing wave $k_{x,y} = \pi n_{x,y}/L_{x,y}$, we obtain

$$N = 2_s\frac{L_xL_y}{\pi^2}\int_{\sqrt{2mE_1}/\hbar}^{\sqrt{2mE_2}/\hbar} k\,dk \int_0^{\pi/2} d\theta = \frac{2_s\mathcal{A}m}{2\pi\hbar^2}(E_2 - E_1),$$

where $\mathcal{A} = L_xL_y$ is the area of the dot, and we have represented the "vector" $\boldsymbol{k}$ in polar coordinates. The number of levels contained in the energy interval is proportional to the width of this interval and does not depend on the energies E_1 and E_2 otherwise. In other words, we can define the density of states per energy interval $\nu = N/(E_1 - E_2) = 2_s\mathcal{A}m/2\pi\hbar^2$ and the *mean level spacing* $\delta_S \equiv \nu^{-1}$.

In the same way, we can probe other properties of the system of levels. For example, one can calculate spacings between adjacent levels E_i and E_j. Plotting a histogram – the frequency of the appearance of the spacing $|E_i - E_j|$ – we discover that this dependence is exponential: larger spacings have a smaller probability, proportional to $\exp(-|E_i - E_j|/\delta_S)$. In other words, level spacings of a two-dimensional rectangular quantum dot obey a *Poisson distribution*.

4.2.1 Chaotic and disordered quantum dots

Things are never quite as easy as they seem. Let us consider a dot of arbitrary shape (not necessarily rectangular), often called a *billiard* in the literature. If the size of the dot is much greater than the wavelength, we can look at the properties of *classical* electron trajectories. Consider two classical trajectories originating from the same point at a small angle θ between them. These trajectories follow straight lines (ballistic propagation) until they hit the wall of a dot and are reflected specularly. If the wall is not flat, the angle between the trajectories changes. After many reflections from the wall, the angle can considerably

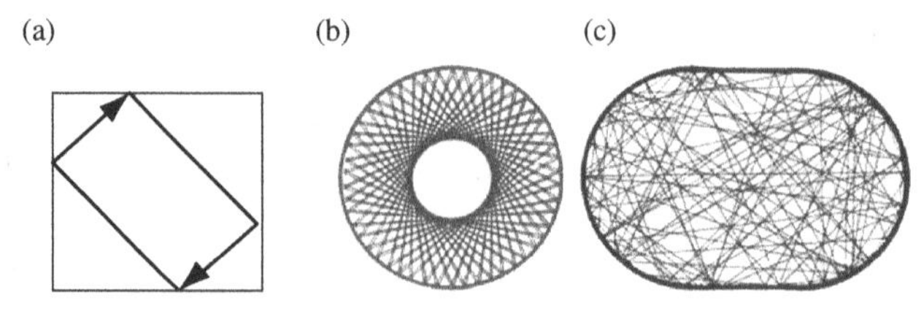

Fig. 4.3. (a) Rectangular and (b) circular quantum dots are integrable, and the classical trajectories are periodic with a finite period: short in (a) and longer in (b). The stadium billiard (c) is a textbook example of a chaotic billiard, and a classical trajectory fills all the space inside the billiard.

decrease, stabilize, or grow. The first two cases are realized in integrable dots – where separation of variables produces a number (more than one) of integrals of motion in the sense of classical mechanics. For instance, in a circular dot the energy and the angular momentum of electrons are always conserved. Rectangular dots are also integrable.

A different class is *chaotic* quantum dots. There, the classical trajectories are very irregular (Fig. 4.3). Each trajectory, if we let it run for a long time, will visit any point of the billiard and every point in the phase space allowed by the energy conservation. Chaos means that two close (in the phase space) trajectories after many reflections from the walls of the dot move apart from each other. Usually, if one takes a bundle of trajectories of width a_0, all moving in the same direction, the size of the bundle a increases exponentially with time, $a = a_0 \exp(\gamma t)$, $t \gg \tau_{\text{esc}}$, where $\tau_{\text{esc}} = v/L$ is the time it takes for an electron to escape the dot. The quantity γ is called the *Lyapunov exponent*. More precisely, there is a (usually infinite) set of Lyapunov exponents in a given chaotic system. They are related to the eigenvalues of the Liouville operator, defined in the phase space, $\hat{\mathcal{L}} = v\boldsymbol{n}\partial/\partial\boldsymbol{r}$, where n is the unit vector in the direction of motion and v is the absolute value of the velocity. The operator is supplemented by the boundary conditions that the reflection at the walls is specular. In chaos theory, this operator needs to be regularized by introducing some initial infinitesimal noise to make the chaotic dynamics irreversible. The regularized operator is known as the *Perron–Frobenius operator*.

Another typical situation occurs if there are impurities in the dot (*disordered quantum dots*). One situation is when they are concentrated at the surface – the reflection from the wall is not specular, but diffusive: an incoming electron can be reflected with certain probability in all directions. This is the case similar to classical chaos. The dynamics is again governed by the Liouville operator, and the Thouless time equals L/v, as in chaotic quantum dots. One can say that close trajectories lose track of each other after the first collision with the surface – "very strong" chaos.

Even more common is disorder spread over the dot. We already considered this situation in Section 2.3, and know that the classical dynamics of electrons is governed by the Usadel equation, which in the case of a normal metal just describes classical diffusion – the particle is scattered randomly at the impurities. The dynamics of an electron is governed by the diffusion operator, $-D\nabla^2$, where D is the diffusion coefficient. The boundary conditions constitute that there is no current through the impenetrable walls of the dot, that is $\nabla \cdot P = 0$ at the walls (see Section 4.4). All eigenvalues of the diffusion operator are real, in accordance with the fact that diffusion is irreversible. The escape time is $\tau_{esc} = L^2/D$ – this is how long it takes for an electron to diffuse across the dot.

Note that not any disorder leads to the diffusion: it must be strong enough. The strength of the disorder is characterized by the mean free path l; diffusion occurs for $l \ll L$. Another way to formulate this condition is to say that electrons experience many scattering events before actually being transmitted through the system – the average probability of transmission from the left to the right is very small. In the opposite case, electrons perform ballistic motion inside the dot, and impurities may lead (or may not lead) to classically chaotic motion.

We cannot predict positions of energy levels in chaotic or diffusive dots, since they depend on the fine details of the dot shape, or on the configuration of impurities. If any detail is changed, the position of a level can be changed at the scale of the mean level spacing δ_S. However, as in statistical mechanics, we can predict a probability of having a certain configuration of the levels. We can observe, in principle, this probability by considering a statistical ensemble of dots: a number of formally identical diffusive dots, or by taking a chaotic dot and changing its shape by nearby gates.

For quantum dots with different *classical* dynamics, one also has different properties of *quantum* levels. Surprisingly, the dots can be separated into two main groups according to the statistical properties of the energy levels. The spectrum of an integrable system consists of many series of levels. Within a series, the levels are separated by large, slowly varied intervals. Since there are many series, these intervals are much bigger than δ_S. Levels separated by a distance of the order of δ_S come from different series. Since the motion is separable, levels from different series are not aware of the presence of each other, and do not correlate their positions. They can occasionally become degenerate or cross under the influence of external perturbation such as a magnetic field. Typically, levels in an integrable system do not correlate at all, obeying Poisson statistics; sometimes long-distance correlations are observed. We will not consider integrable dots in this chapter further. However, the other hand, the statistics of both chaotic and disordered dots is very similar and is not Poissonian – levels interact with each other – as we describe next.

4.2.2 Level statistics in quantum dots

It is natural to assume that, from a statistical point of view, the Hamiltonian describing an electron in such quantum dots is a random matrix. In disordered dots, the randomness comes from disorder. In chaotic quantum dots, the randomness is not intrinsic. However, the properties of electron trajectories taken at very close energies are very different. Any real measurement necessarily involves electrons in a (narrow) window of energies, and all physical quantities should be averaged over the energy inside this window. This averaging – coarse-graining – plays the same role as ensemble averaging in disordered dots.

Each individual quantum dot is characterized by one member of an ensemble of random matrices. Energy levels are obtained as eigenvalues of these matrices. Of interest are the levels lying close to the Fermi surface. Assuming that the number of filled levels is very large, the Fermi surface is close to the center of the band, and the mean level spacing

δ_S is the only parameter describing statistical properties of energy levels. Then, using the statistical properties of these eigenvalues (Section 4.1), one calculates the physical quantities.

Which ensemble of random matrices one chooses depends on the time-reversal properties of the electron system. If there is no magnetic field, no spin-orbit or magnetic scattering, the Hamiltonian is real and symmetric. In this case, the random matrix must be chosen from a GOE. If a magnetic field is present, time-reversal symmetry is broken, and the Hamiltonian is Hermitian – one uses a GUE. To destroy the time-reversal symmetry, the magnetic field must have a value which produces the flux quantum Φ_0, piercing a typical trajectory. Finally, for the case of spin-orbit scattering, as explained in Section 1.9, the wave functions are spinors, and this symmetry corresponds to a GSE.

Random matrix theory immediately provides us with a result on the level statistics. We will discuss this result first and turn to the limits of applicability later. In RMT, the levels repel each other, and the distribution of spacings $|E_i - E_j|$ between adjacent levels obeys Wigner–Dyson statistics; it is proportional to $|E_i - E_j|^\beta$, with $\beta = 1, 2, 4$ in a GOE, GUE, and GSE, respectively. The two-point level correlation function,

$$\begin{aligned} R_2(E) &\equiv \delta_S^2 \langle \nu(E')\nu(E+E') \rangle - 1 \\ &= \delta_S^2 \left\langle \sum_{ij} \delta(E' - E_i)\delta(E + E' - E_j) \right\rangle - 1, \\ \nu(E) &= \sum_j \delta(E - E_j), \end{aligned} \tag{4.19}$$

with $\nu(E)$ being the density of states, and $\langle \nu(E) \rangle = \delta_S^{-1}$, has the easiest form in a GUE (Eq. (4.9)):

$$R_2(E) = \delta\left(\frac{E}{\delta_S}\right) - \left(\frac{\sin \pi E/\delta_S}{\pi E/\delta_S}\right)^2. \tag{4.20}$$

The first term describes a correlation of a level with itself and only exists at zero energy separation, and the second one oscillates with period δ_S. These oscillations are a remnant of the discrete structure of energy levels: for the levels with fixed positions one would obtain a set of delta-functions; for Poisson distribution the oscillations are washed out completely, and RMT describes an intermediate situation. The negative sign of this term indicates level repulsion. Note that if one averages out the oscillations, replacing sine squared with $1/2$, the correlation function takes an even simpler form, valid for $E \gg \delta_S$, $R_2(E) = -\delta_S^2/2\pi^2 E^2$; see Eq. (4.10).

Whereas in a GOE and a GSE all levels are doubly spin-degenerate, in a GUE the magnetic field produces Zeeman splitting δ_Z of energy levels. One can easily take this splitting into account, shifting corresponding distributions for different spin projections by δ_Z as follows:

$$R_2(E) \to \frac{1}{2} R_2(E) + \frac{1}{4} R_2(E + \delta_Z) + \frac{1}{4} R_2(E - \delta_Z). \tag{4.21}$$

Other results on level statistics in the framework of RMT are found in Section 4.1.

All results provided by RMT are *universal*: they only depend on the mean level spacing δ_S, not on the particular shape of the quantum dot, or on the degree of disorder. One can interpret this as a "zero-dimensional" result: the electron has enough time to explore the whole area of the dot. The characteristic time taken to explore the dot is the escape time τ_{esc}, and thus the RMT results such as Eq. (4.19) are only valid provided the level separation is not too large, $E \ll E_{Th}$, where $E_{Th} = \hbar/\tau_{esc}$ is known as *Thouless energy*. One always has $E_{Th} \gg \delta_S$. For example, in a disordered quantum dot of a square shape (size L), one has $\delta_S \sim \hbar^2/mL^2$, $E_{Th} \sim \hbar D/L^2$, and thus $E_{Th}/\delta_S \sim k_F l$. This quantity must always be much greater than unity, otherwise the disorder would be too strong and electron states would become localized (Section 4.5). If we attach electric contacts to the opposite sides of the quantum dot, we can formally calculate the conductance of the dot. (Note that it has nothing to do with real conductance, since the latter quantity is determined by the tunnel junctions separating it from the outside world.) Indeed, the average transmission probability is of the order of l/L (see Section 4.3), the number of transport channels is of the order of $k_F L$, and thus, by the virtue of Landauer formula, the conductance is given by $G \sim G_Q k_F l \sim G_Q E_{Th}/\delta_S$. Thus, our condition means that the (formally defined) conductance of the quantum dot is much greater than the conductance quantum. We return to this condition in Section 4.5.

At the level separation above the Thouless energy, level statistics is not universal. For the level correlation function, for energies $E \gg \delta_S$ the following formula is valid [86]:

$$R_2(E) = \frac{\delta_S^2}{8\pi^2} \operatorname{Re} \sum_{\mu} \left(\frac{1}{(\gamma_\mu - iE)^2} + \frac{3}{(\gamma_\mu - iE + 1/\tau_{SO})^2} + \frac{1}{(\gamma_\mu - iE + 1/\tau_H)^2} + \frac{3}{(\gamma_\mu - iE + 1/\tau_H + 1/\tau_{SO})^2} \right). \tag{4.22}$$

Here γ_μ are the eigenvalues of the operator responsible for the electron dynamics – the Perron–Frobenius operator in chaotic quantum dots or the diffusion operator in diffusive ones, and we have disregarded the Zeeman splitting. The cut-off time τ_H originates from the eigenvalues γ'_μ of the same operator modified by the magnetic field; for example, for the diffusion operator,

$$D\left(\nabla + 2ie\mathbf{A}(\mathbf{r})/c\right)^2 \psi_\mu = \gamma'_\mu \psi_\mu, \tag{4.23}$$

and $\mathbf{A}$ is the vector potential. In the absence of magnetic field, obviously $\gamma_\mu = \gamma'_\mu$. Working the equation out, we find that γ'_μ is replaced by $\gamma_\mu + 1/\tau_H$. We will see in Section 4.4 that the eigenvalues γ_μ and γ'_μ correspond to diffusons and cooperons, respectively. The form for τ_H depends on the geometry of the system; it is given by the cyclotron frequency $\tau_H^{-1} = 4DeH/\hbar c$ in diffusive three-dimensional and two-dimensional (provided the magnetic field is perpendicular to the plane) dots. In the one-dimensional and two-dimensional (longitudinal field) diffusive case this time is of the order of $\tau_H^{-1} \sim D(eHa)^2/(c\hbar)^2$, where a is the transverse dimension of the sample. The time τ_{SO} characterizes the spin-orbit scattering. Equation (4.22) is known as *Altshuler–Shklovskii formula*.

Note that in Eq. (4.22) the first term repeats in several variations. First, there are terms with coefficient 1 (first and third) and terms with the coefficient 3. They correspond to singlet and triplet pairing of electron spins, respectively; the coefficient 3 corresponds to three different triplet states. In zero magnetic field ($\tau_H \to \infty$), if the spin-orbit scattering is also weak, $\tau_{\rm SO} \to \infty$, singlet and triplet contributions are the same. Spin-orbit scattering suppresses the triplet contributions, and thus for strong spin-orbit scattering in zero magnetic field the correlation function R_2 is suppressed by a factor of 4. Another observation is that, if the magnetic field is finite, the terms that have $\tau_{\rm H}^{-1}$ in the denominator (cooperon contributions, as opposed to diffuson contributions, see Section 4.4), are also suppressed. Without spin-orbit scattering, this leads to the suppression of the correlation function by a factor of 2; in combination with strong spin-orbit scattering, this singles out the first term, and all other contributions are suppressed – the total suppression factor is 8.

Equation (4.22) also contains the summations over the modes γ_μ. It is important that, in closed systems, a *zero mode* $\gamma_\mu = 0$ is present, which corresponds to a uniform solution of the diffusion (or Liouville) equation. Terms with $\gamma = 0$ have the order of magnitude of E^{-2}, $\tau_{\rm H}^2$, or $\tau_{\rm SO}^2$, depending on the relation between these quantities. All other eigenvalues are of the order of the Thouless energy. Thus, at low energies, $E \ll E_{\rm Th}$, the main contribution to the Altshuler–Shklovskii formula originates from the zero mode. One has $R_2(E) = -\delta_{\rm S}^2/\beta\pi^2 E^2$, with $\beta = 1$ for $E \gg \hbar/\tau_{\rm H}, \hbar/\tau_{\rm SO}$, $\beta = 2$ for $\hbar/\tau_{\rm SO} \ll E \ll \hbar/\tau_{\rm H}$, and $\beta = 4$ for $\hbar/\tau_{\rm H} \ll E \ll \hbar/\tau_{\rm SO}$. Note that what we get are *universal* expressions, which are just smooth parts of the RMT results, Eqs. (4.9), with $\beta = 1, 2, 4$ for a GOE, a GUE, and a GSE, respectively. Physically, indeed, if no magnetic field and no spin-orbit interactions are present, we are always in the regime $E \gg \hbar/\tau_{\rm H}, \hbar/\tau_{\rm SO}$ and have to use an orthogonal ensemble of random matrices. Oscillations of the correlation function are not present in Eq. (4.22), and these expressions are not valid in any case for $E \sim \delta_{\rm S}$ (they diverge for $E \to 0$). On top of these contributions, there are small *non-universal* contributions originating from the eigenvalues $\gamma_\mu \neq 0$ [87]. The non-universal part has the following form, valid for $E \ll E_{\rm Th}$ (for simplicity, we assume zero magnetic field):

$$\delta R_2(E) = \frac{\delta_{\rm S}^2}{2\pi^2\beta} \,\mathrm{Re} \sum_\mu{}' \frac{1}{\gamma_\mu^2} \frac{\mathrm{d}^2}{\mathrm{d}E^2} \left(E^2 R_2^{\rm WD}(E) \right),$$

where the prime means that the zero mode $\gamma_\mu = 0$ is excluded from the summation, and $R^{\rm WD}$ is the correlation function provided by RMT. For a GUE it is given by Eq. (4.20); then one has

$$\frac{\mathrm{d}^2}{\mathrm{d}E^2} \left(E^2 R_2^{\rm WD}(E) \right) = \sin^2 \frac{\pi E}{\delta_{\rm S}},$$

and the smooth part (obtained by replacing the sine squared with 1/2) of R_2 is the same as the contribution to the Altshuler–Shklovskii result, Eq. (4.22), of the eigenvalues with $\gamma_\mu \neq 0$. The amplitude of these oscillations is small compared with RMT (factor $E_{\rm Th}/\delta_{\rm S}$). In contrast to RMT, this amplitude is energy independent. For $E \sim E_{\rm Th}$ it becomes of the same order as the RMT contribution.

Let us now investigate the Altshuler–Shklovskii expression for $E \gg E_{\rm Th}$. In this regime, the result is not universal: spectral statistics is determined by the eigenvalues γ_μ, and is different for chaotic and diffusive cases, or for systems of different dimensions. Equation (4.22) can be explicitly evaluated for a rectangular diffusive quantum dot, when the eigenvalues are given by $\gamma_\mu = \pi^2 n_i^2/L_i^2$, $n_i = 0, 1, \ldots$. For $E \gg E_{\rm Th}$, many eigenvalues γ_μ contribute, and the summation can be replaced by the integration, after which one finds for different dimensionalities,

$$R_2(E) \propto - \begin{cases} (E_{\text{Th}}/\delta_S)^{-3/2} E^{-1/2} & (3\text{d}) \\ (E_{\text{Th}}/\delta_S)^{-1/2} E^{-3/2} & (1\text{d}). \end{cases} \tag{4.24}$$

The level correlation function is always negative – levels repel each other even at separations higher than the Thouless energy. This is not a universal property, and in chaotic systems one may have level attraction. Also, the correlation function is again *smooth* and does not oscillate on the scale of δ_S. As expected, the results differ in different dimensions. Note also that, in two dimensions, the integration in Eq. (4.22) gives precisely zero (the integral is purely imaginary, and the real part vanishes). Thus, in this approximation in two dimensions, levels are not correlated. The correlation is in this case provided by quantum corrections, and the result reads $R_2(E) \propto -(E_{\text{Th}}/\delta_S)^{-2} E^{-1}$, and the levels still repel each other [88].

Exercise 4.2. Evaluate the level correlation function in one dimension for $E \gg E_{\text{Th}}$. Check the functional dependence of Eq. (4.24) and calculate the proportionality coefficient.

For $E \gg E_{\text{Th}}$ there is also an oscillating contribution to the correlation function – a remnant of the RMT oscillations – but it is non-universal, and, in diffusive dots, is exponentially suppressed by a factor of $\exp(-(E/E_{\text{Th}})^{D/2})$, $D = 1, 2, 3$, being the dimensionality. For chaotic dots, these oscillations can be significant.

From Eq. (4.22) it is also clear what role the magnetic field plays. Indeed, there is a natural scale for the field, H_c, at which the cut-off time τ_H becomes of the order of the escape time $\hbar/E_{\text{Th}}$. This corresponds to one flux quantum piercing a typical trajectory. For $H \ll H_c$ the contributions of all the terms in Eq. (4.22) are approximately the same; for $H \gg H_c$ only the first two terms contribute, and the two other terms are suppressed by the magnetic field. This is how the cross-over between orthogonal and unitary symmetries occurs at high separation between the energy levels.

The last remark is that, in diffusive quantum dots, all the above results are valid provided the level separation E is much less than $\hbar/\tau$, where τ is the elastic scattering time (momentum relaxation time). Indeed, otherwise we probe very short time scales, shorter than the time between collisions. At these scales, the electron motion is not diffusive – it is ballistic flight with little or no chance of scattering. Such systems must be considered as ballistic – in particular, for chaotic dynamics, similar formulas occur, but now with the diffusion operator replaced by the Perron–Frobenius operator.

Now we know the statistical properties of energy levels. In the following we show in a few examples how they help us to understand the statistical properties of physical quantities.

4.2.3 Persistent currents

Let us for a moment return to ideal systems (no disorder). Consider an ideal one-dimensional narrow ring of radius R. The motion of an electron in the transverse direction (across the ring) is quantized. We assume that only the lowest level of this quantization is

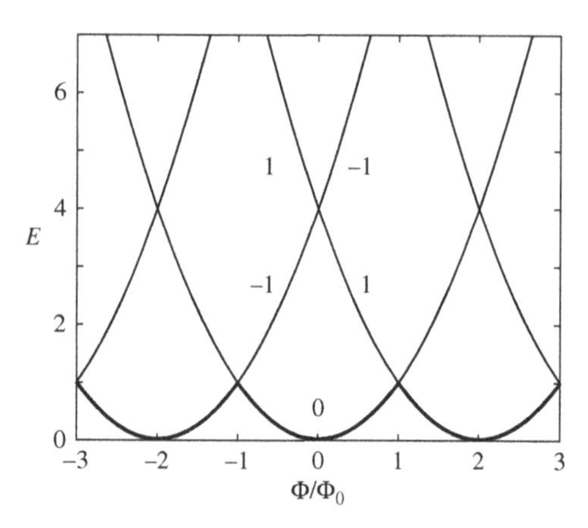

Fig. 4.4. **Energy levels in a ring (in units of $\hbar^2/2mR^2$) in dependence on applied flux. Five parabolas corresponding to $n = 0, \pm 1, \pm 2$ are shown. The ground state is indicated by the bold line.**

relevant and that all other levels have too high an energy to be considered. The Schrödinger equation for the wave function projected at this level of transverse motion is given by

$$-\frac{\hbar^2}{2mR^2}\left(\frac{\mathrm{d}}{\mathrm{d}\theta} - \mathrm{i}\frac{\Phi}{2\Phi_0}\right)^2 \psi(\theta) = E\psi(\theta), \tag{4.25}$$

where θ is the polar angle. We have assumed that the ring is placed in the magnetic field perpendicular to the plane of the ring; Φ is the magnetic flux through the ring and $\Phi_0 = \pi\hbar c/e$ is the flux quantum (see Section 1.6).

In zero magnetic field, the wave functions have to be periodic in θ with period 2π. Normalized solutions have the form $\psi_n(\theta) = (2\pi)^{-1/2}\exp(in\theta)$, $n = 0, \pm 1, \ldots$. The corresponding energies are $E_n = \hbar^2 n^2/(2mR^2)$.

In the magnetic field, the phase of the electron wave function is modified (Aharonov–Bohm effect, Section 1.6). In agreement with Eq. (4.25), we obtain

$$\psi_n(\theta) = \frac{1}{\sqrt{2\pi}}\mathrm{e}^{in(\theta - \Phi/2\Phi_0)}, \; n = 0, \pm 1, \ldots, \; E_n = \frac{\hbar^2}{2mR^2}\left(n - \frac{\Phi}{2\Phi_0}\right)^2. \tag{4.26}$$

The energy levels are plotted in Fig. 4.4.

Thus, the total energy E of the ring depends on the magnetic field. The field-dependent energy means that there is a current in the ground state, $I = -c\partial E/\partial\Phi$, known as the *persistent current*. It is a ground-state property and a thermodynamical quantity – in contrast to the transport current, considered elsewhere in this book. In our case, the explicit expressions for the persistent current are easily obtained. Indeed, the free energy of an electron system at zero temperature is just the sum of the energies of all filled levels – all levels below the Fermi energy E_{F}, i.e.

$$E = 2_s \sum_{n:E_n < E_{\mathrm{F}}} E_n, \tag{4.27}$$

and we find the current to be given by

$$I = -\frac{2_s c\hbar^2}{2mR\Phi_0}\sum_n\left(n - \frac{\Phi}{2\Phi_0}\right) = \frac{2_s e\hbar}{2\pi m R^2}$$
$$\times\begin{cases} (2N+1)\dfrac{\Phi}{2\Phi_0} & \left(N - 1 + \dfrac{\Phi}{2\Phi_0}\right)^2 < \dfrac{2mE_F R^2}{\hbar^2} < \left(N - \dfrac{\Phi}{2\Phi_0}\right)^2 \\ 2N\left(\dfrac{\Phi}{\Phi_0} - 1\right) & \left(N - \dfrac{\Phi}{2\Phi_0}\right)^2 < \dfrac{2mE_F R^2}{\hbar^2} < \left(N + \dfrac{\Phi}{2\Phi_0}\right)^2. \end{cases} \tag{4.28}$$

Here the upper and lower lines correspond to the case of odd $(2N - 1)$ or even $(2N)$ numbers of spin-degenerate filled levels, respectively, and for simplicity we have disregarded the Zeeman splitting of energy levels. Equation (4.28) is valid for $0 < \Phi < \Phi_0$; otherwise, the persistent current is odd in Φ and periodic with period $2\Phi_0$. It is positive for an odd number and negative for an even number of levels. If there is no flux, or precisely half a flux quantum per ring, the persistent current vanishes.

The scale of magnitude of the persistent current in Eq. (4.28) is ev_F/L. To understand this result, we need to realize that if there is a persistent current in the system, there is also magnetization. Indeed, the magnetization is defined as $M = -\partial E/\partial B = \mathcal{A}I/c$, where $\mathcal{A} = \pi R^2$ is the area enclosed by the flux. The magnetization is proportional to the mechanical moment exerted on electrons by the magnetic field, $\boldsymbol{M} = (e/mc)\boldsymbol{r} \times \boldsymbol{p}$, averaged over the area of the system. If electrons are at the Fermi surface, the magnetization is of the order of $eRk_F\hbar/mc$, which yields the value of the persistent current of the order of ev_F/R, in accordance with Eq. (4.28).

Without the magnetic flux, all energies except for $n = 0$ are doubly degenerate: the state with a certain value of n has the same energy as the state with $-n$. In a small magnetic field, the energies are slightly shifted, $E_n(\Phi) \approx E_n(0) - (\hbar^2 n/mR^2)(\Phi/2\Phi_0)$: the energies of levels with positive (negative) n decrease (increase). For the levels close to the Fermi surface, one estimates $n \sim k_F R$. Thus, at the value of the flux of the order of $\Phi \sim \Phi_0\delta_S/E_{Th}$, the levels come close to each other. In an ideal ring, described by Eq. (4.28), nothing happens: the levels cross and proceed further.

We can also add some external potential to our design. The energies of discrete levels still depend on the magnetic field, but in a more complicated manner. If the disorder is not too strong, the motion of the electrons is ballistic, and Eq. (4.28) holds as an order of magnitude estimate. However, the important difference now is that the levels in the process of evolution with magnetic flux cannot cross – they perform an avoided crossing, and a gap opens instead of every level crossing. Since $\delta_S/E_{Th} \ll 1$, a given level undergoes many avoided crossings if the flux varies by one quantum. This means that the simple classification of discrete levels associated with certain values of the momentum n at $\Phi = 0$ is only valid for very small fluxes, $\Phi \lesssim \Phi_0\delta_S/E_{Th}$, and for higher fluxes the states completely change their properties after many avoided crossings.

Note that the above considerations apply, strictly speaking, to one-channel single rings. The persistent current produced by such a ring is rather small, and one may wish to increase it by taking a ring with several transport channels (all contributing to the persistent current), or by measuring a response of an ensemble of the rings. It does not work this way. Indeed, the rings in the ensemble are never identical, and the persistent current given by Eq. (4.28) is very sensitive: it changes sign if just one electron is added. Thus, it is natural to assume that the persistent currents produced by different rings in the ensemble (or, for that matter, by different transport channels in the same rings) have random signs: the average persistent current is zero! However, the fluctuations are finite; if all M rings contribute independently, a typical current produced by such an ensemble would be given by $I \sim ev_F/(R\sqrt{M})$. It is suppressed by the factor of $1/\sqrt{M}$ as compared with the value for one single-channel ring.

The persistent current of a single ring in the ballistic regime was measured by Mailly, Chapelier, and Benoit [89], who fabricated a GaAlAs–GaAs ring approximately 3 μm in diameter, and measured the magnetization by a SQUID device situated on the same chip as the ring. The mean free path was estimated to be a factor 1.3 longer than the circumference of the ring. In the experiment, the ring was connected to two contacts, and a special gate was used to disconnect it. Thus, both conductance (Aharonov–Bohm effect) and magnetization (persistent current) were measured on the same ring. The conductance measurements indicated the presence of four conductance channels. The persistent current observed has a typical amplitude of 4 ± 2 nA, whereas the estimate $I \sim ev_F/R$ gave 5 nA. Given that the ring is not precisely ideal, the agreement can be regarded as good.

For strong disorder, the situation is similar to that of an ensemble of rings. The energies of the levels depend on the magnetic field in a random manner: some energies go up with the energy; others go down. The flux-dependent part of the total energy has a random sign, depending on the impurity configuration. This means that, on average, there is no persistent current in a disordered ring. One can characterize the magnitude of the persistent current only by its root mean square fluctuations, which have a finite value.

We should note an important detail, however. To calculate the persistent currents we need to use the canonical ensemble (CE) – the number of electrons is fixed for each number of the ensemble. Using a CE is appropriate for isolated rings. Usually, as soon as the total number of electrons is large, there is no difference between the CE and the grand-canonical ensemble (GCE, when the chemical potential is fixed for each ensemble member). The persistent-current problem is, as we see below, an example where the difference between the CE and the GCE is important: in the GCE the calculation is more simple, but would give zero for the average current. The CE calculation below gives a finite value and relates it to the level statistics [90, 91].

There are several ways we may describe the CE. For example, one can just follow thermodynamics: define the grand thermodynamic potential, express the chemical potential in terms of the number of particles, and proceed with the evaluation. We will take a different, conceptually simpler, route. Imagine that the Fermi level is pinned to one of the electron levels: $E_F = E_m + 0$. In this case, the number of electrons in each member of the ensemble is guaranteed to be integer. The energy E is still given by Eq. (4.27); however, now it depends on a random level position E_m. Averaging it over disorder, we obtain

$$
\begin{aligned}
E &= 2_s\delta_S \left\langle \sum_m \delta(E_F - E_m) E(E_m) \right\rangle \\
&= (2_s)^2 \delta_S \left\langle \sum_{mn} \int_0^{E_F} E_n \delta(E_F - E_m)\delta(E' - E_n) \mathrm{d}E' \right\rangle \\
&= (2_s)^2 \delta_S \int_0^{E_F} \langle \nu(E_F)\nu(E') \rangle E' \, \mathrm{d}E', \qquad (4.29)
\end{aligned}
$$

where we have used the definition of the density of states. The disorder average of the product of the densities of states is a sum of two parts: the product $\langle \nu(E') \rangle \langle \nu(E_F) \rangle$, which is independent of the flux and thus does not contribute to the persistent current, and the irreducible part $\langle \nu \rangle^2 R_2(E' - E_F)$. Thus, the persistent current only exists due to level correlations. Note that the integral is determined by the range of energies E–E_F of the order of the Thouless energy. In this case, we can employ the Altshuler–Shklovskii expression, Eq. (4.22), with γ_μ being the eigenvalues of the diffusion operator in the ring. The first two terms do not depend on the flux and thus do not contribute to the persistent current; for the two others, we assume that spin-orbit scattering is negligible. If the ring is thin enough, so that only the zero eigenvalue in the transverse direction is important, one finds, from Eq. (4.23), $\gamma'_\mu = (D/R^2)(n + \Phi/\Phi_0)^2$. Thus, the persistent current is given by

$$I = -\frac{(2_s)^2 c \delta_S E_F}{2\pi^2} \operatorname{Re} \int_0^{E_F} dE \frac{\partial}{\partial \Phi} \sum_{n=-\infty}^{\infty} \left(\frac{1}{-iE + (D/R^2)(n + \Phi/\Phi_0)^2} \right)^2 . \quad (4.30)$$

We see that the current is a periodic function of the flux with period Φ_0, and it is an odd function of flux. One can then expand as follows:

$$I = \sum_l I_l \sin 2\pi l \Phi/\Phi_0,$$

and find, for the amplitude,

$$I_l = -\frac{(2_s)^2 e \delta_S}{\pi^2 \hbar}. \quad (4.31)$$

Note that, for $\Phi \to 0$, all the harmonics in Eq. (4.30) give the same contribution to the current, of order of $(e\delta_S/\hbar)(\Phi/\Phi_0)$. The number of harmonics for which this equation holds is given by $\sqrt{E_{Th}/\delta_S}$, yielding the maximum value of the persistent current, $(e\sqrt{E_{Th}\delta_S}/\hbar)(\Phi/\Phi_0)$.

Measurements of persistent current (magnetization) were carried out both on a single isolated gold ring [92] and on an ensemble of 10^7 isolated copper rings [93]. Surprisingly, both experiments yield a magnetization approximately two orders of magnitude greater than the theoretical prediction.

Exercise 4.3. Estimate the magnitude of the mean square fluctuation of the persistent current. Use the grand canonical ensemble.

4.2.4 Magnetopolarizability

Can one actually measure correlations of energy levels? This question was raised in 1965 by Gor'kov and Eliashberg, and even though their quantitative conclusions were later disputed by many authors, the qualitative idea is correct. Consider a disordered isolated metallic grain in an external magnetic field. The field polarizes the grain – it induces the dipole moment. The dipole moment is proportional to the field, and the proportionality coefficient is called *polarizability*. To calculate the polarizability, one needs to take into account screening of electric charge by other electrons. This is a standard subject for an advanced solid state physics course, and a detailed description of screening goes outside the scope of this book, so we will only discuss it very briefly. It turns out that the polarizability is a classical quantity that depends only on the geometry of the system and knows nothing about discrete electron levels. However, there is a quantum correction to the polarizability, sensitive to the level statistics. Similar to the persistent current, it must be calculated in the canonical ensemble, and we show below that this quantum correction is sensitive to the three-point level correlation function. This correction is the only contribution to the polarizability that is sensitive to the external magnetic field, and we concentrate therefore on the *magnetopolarizability* – the difference of the polarizability with and without a magnetic field.

To give a concrete example, we consider a disordered metallic grain in the shape of a disk, radius R, in an external electric field $\boldsymbol{E}(t)$ applied in the plane of the disk. The field polarizes the particle, inducing the dipole moment. The density of charge $\rho(\boldsymbol{r})$ and the electrostatic potential $\Phi(\boldsymbol{r})$ obey the Poisson equation:

$$\Delta\Phi = -4\pi e\rho(\boldsymbol{r})\theta_\Omega\delta(z),$$

where θ_Ω is the function that equals unity inside the disk and zero outside, and z is the coordinate normal to the plane of the disk. In its turn, the density can be found in the Fourier representation as follows:

$$\rho(\boldsymbol{r},\omega) = -2e\int_\Omega \Pi(\boldsymbol{r},\boldsymbol{r}',\omega)\Phi(\boldsymbol{r}',\omega)\mathrm{d}\boldsymbol{r}',$$

where the integration is carried out over the area of the disk, and Π is the polarization operator, which, at distances longer than the wavelength, can be approximated as (see Eq. (6.29))

$$\Pi_0(\boldsymbol{r},\boldsymbol{r}',\omega) = \nu\left[\delta(\boldsymbol{r}-\boldsymbol{r}') - \frac{1}{V}\right]. \tag{4.32}$$

The boundary conditions for the electrostatic potential Φ are such that far from the particle it describes the external field, $\Phi = -\boldsymbol{E}\boldsymbol{r}$. The induced dipole moment,

$$\boldsymbol{d} = e\int_\Omega \boldsymbol{r}\rho(\boldsymbol{r})\mathrm{d}\boldsymbol{r},$$

for the disk geometry, since there is no preferential direction, must have the same direction as the field. In the Fourier representation, one writes $d(\omega) = \alpha(\omega)E(\omega)$. The response function $\alpha(\omega)$ is the polarizability. For the disk geometry, it does not depend on the frequency and equals $\alpha_0 = (4/3\pi)R^3$.

What does this nice exercise in classical electrodynamics have to do with the correlation of energy levels? The point is that Eq. (4.32) is only an approximation. The polarization operator can be expressed in terms of energy levels E_n and eigenfunctions of the energy states $\psi_n(\boldsymbol{r})$:

$$\Pi(\boldsymbol{r},\boldsymbol{r}',\omega) = \left\langle \sum_{m\neq n} \psi_m^*(\boldsymbol{r})\psi_n(\boldsymbol{r})\psi_n^*(\boldsymbol{r}')\psi_m(\boldsymbol{r}')\frac{f_\mathrm{F}(E_m) - f_\mathrm{F}(E_n)}{\hbar\omega - E_m + E_n}\right\rangle. \tag{4.33}$$

Apart from the short-range contribution given by Eq. (4.32), it also contains a long-range part originating from the correlation of the wave functions. This part is sensitive to the level statistics. It only produces a minor correction to the polarizability, small compared with the classical result $\alpha \sim R^3$, but is sensitive to the magnetic field applied – technically it is different in a GUE and a GOE. Thus, if one studies the magnetic-field dependence of the polarizability – the magnetopolarizability $\delta\alpha$ – one deals with the object sensitive to level and eigenfunction statistics.

One proceeds in the perturbation theory in the long-range part of the polarization operator. Since the grain is isolated, the canonical ensemble must be used, similar to the persistent current, and it is realized by pinning the Fermi level E_F to one of the single-particle states E_k. It is important that the statistics of the energy levels and the wave functions of the electrons are independent. We obtain

$$\delta\alpha = \frac{2e^2}{E^2\delta_S^2}\delta_B \int_{+0}^{\infty} dE \left(\frac{1}{E - \hbar\omega - i0} + \frac{1}{E + \hbar\omega + i0}\right) \times \left[R_2(E)\delta_S + \int_{+0}^{E-0} dE' R_3(E, E')\right] \left\langle \Phi_0^2 \right\rangle_E, \qquad (4.34)$$

where δ_B denotes the difference of the quantity calculated in a GUE and a GOE, that is, with and without a magnetic field. In the square brackets, the first term corresponds to the contributions with $n = k$, and therefore contains the two-point correlation function (correlations between the levels m and n), and the second term originates from the terms with $n \neq k$, and is expressed via the *three-point correlation function*, given by

$$\tilde{R}_3(E, E') = \delta_S^3 \left\langle \sum_{ijk} \delta(E'' - E_i)\delta(E'' + E - E_j)\delta(E'' + E' - E_k) \right\rangle,$$

which is independent of E'' and, in our case, describes the mutual correlations between the states m, n, and k. The three-point correlation function only appears here because we use the canonical ensemble. Note that, in distinction from R_2, $\tilde{R}_3$ is not a cumulant, i.e. the products of averages are not subtracted. Equation (4.34) is complex and contains both real (describing the phase shift) and imaginary (describing energy absorption) parts.

We denoted as $\langle \Phi_0^2 \rangle$ the average matrix element of the potential in the grain. This matrix element depends on the statistics of the wave functions, but not of the energy levels. The electric field is effectively screened by the electrons in the grain and decays rapidly away from the boundary. This is why the dielectric response of the grain only comes from a narrow boundary layer, and the magnetopolarizability is small compared with the classical value of the polarizability, of order R^3. Indeed, the calculation for the disk geometry gives a magnetopolarizability of the order of $R^2/(\kappa k_F l)$, where $\kappa = 4\pi e^2 \nu$ is the inverse screening radius and $\kappa R \gg 1$. What is interesting is the frequency dependence of $\delta\alpha$, expressed by the dimensionless function as follows [94]:

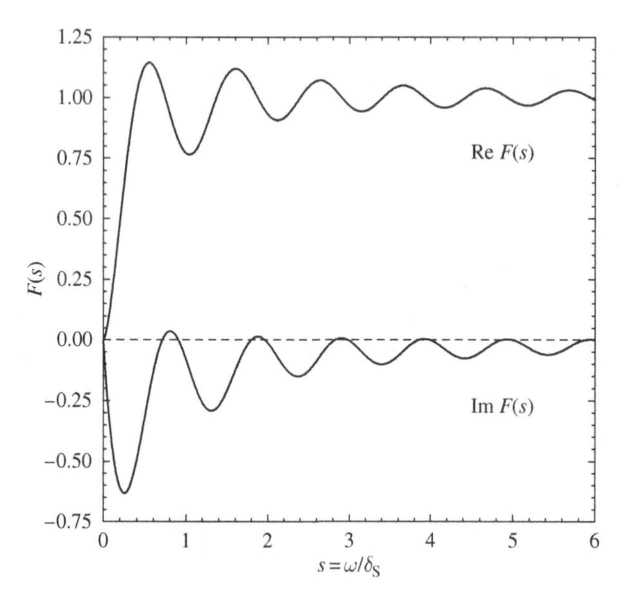

Fig. 4.5. Real and imaginary parts of the function $F(\omega)$ (Eq. (4.35)).

$$\delta\alpha \propto F(\omega) \equiv 1 + \int_{+0}^{\infty} \frac{\mathrm{d}E}{\delta_S} \left(\frac{1}{E - \hbar\omega} + \frac{1}{E + \bar{\omega}} \right) \times \left[\delta_S \boldsymbol{\delta}_B R_2(E) + \int_{+0}^{E-0} \mathrm{d}E' \, \boldsymbol{\delta}_B R_3(E, E') \right]. \tag{4.35}$$

Since the integrals in Eq. (4.35) for low frequencies are determined by the range of energies of the order of δ_S, one can substitute the expressions for the correlation functions produced by random matrix theory. Consequently, $F(\omega)$ is a universal function of the frequency, oscillating on the scale of δ_S. Real and imaginary parts of the polarizability are plotted in Fig. 4.5. They oscillate on the scale of δ_S, clearly indicating level correlations.

Control question 4.3. How low should the frequency ω be to justify the application of RMT in Eq. (4.35)?

4.3 Statistics of transmission eigenvalues

This section is about the transport properties of chaotic and disordered systems. We apply random matrix theory (RMT) to transmission eigenvalues, and find the statistical properties of conductance. For scattering matrices, the type of ensemble depends on the properties of the contacts connecting the system with the outside world. We first present the general treatment of the situation and then specialize to two cases – closed cavities (essentially, quantum dots, separated from the reservoirs by tunnel junctions), and (open) chaotic cavities – connected to the reservoirs by quantum point contacts. Finally, we show how RMT applies to a more complicated object – a disordered wire.

4.3.1 Chaotic cavities: Poisson kernel

Consider a chaotic cavity, separated from the reservoirs by arbitrary scatterers. We saw in Section 1.6 that the scattering matrix of such an object $\hat{s}$ is a combination of scattering matrices of the cavity itself, $\hat{s}_0$, and of the junctions connecting the cavity to the reservoirs. Indeed, the scattering matrices can be combined (see Eq. (1.131)) in the following way:

$$\hat{s} = \hat{r} + \hat{t}'\hat{s}_0 \left(1 - \hat{r}'\hat{s}_0\right)^{-1} \hat{t},$$

where, in our case, $\hat{r}$ and $\hat{t}$ characterize the junctions. The matrix $\hat{s}_0$ is random. In this subsection, we assume that we know the statistical properties of the matrix $\hat{s}_0$ and use them to characterize the properties of the matrix $\hat{s}$, determining the transport characteristics of the chaotic cavity.

Let us first characterize the matrix $\hat{s}_0$. The classical dynamics inside the cavity is chaotic. This means that an electron entering from one contact follows a very complicated classical trajectory, and eventually either exits back (reflection) or proceeds to one of the other leads (transmission). The key assumption for the calculation of transport properties is that the scattering matrix of a chaotic cavity is random and is a member of a circular ensemble of random matrices (see Section 4.1). If there is no magnetic field present in the system, we choose a COE ($\beta = 1$), and if magnetic field is present we choose a CUE ($\beta = 2$); a CSE ($\beta = 4$) can be realized if we assume spin-dependent scattering from the walls of the cavity.

Now we take the matrices $\hat{r}$ and $\hat{t}$ as not being random and depending only on the transmission eigenvalues of the barriers. The matrix $\hat{s}$ is random, but, in contrast to s_0, is not a member of a circular ensemble (unless $\hat{r} = 0$ – open cavities, see below). One can show [95] that $\hat{s}$ is not distributed uniformly, but rather has the distribution known as a *Poisson kernel*, given by

$$P(\hat{s}) \propto \left|\det\left(1 - \hat{r}^{\dagger}\hat{s}\right)\right|^{-(\beta N+2-\beta)}, \tag{4.36}$$

where N is the dimension of the matrix $\hat{s}_0$ (the sum of the numbers of transport channels in all the leads). If the leads are ideal, $\hat{r} = 0$, and the matrix $\hat{s}$ is distributed uniformly.

Equation (4.36) is a mathematical statement that one now has to use to extract the joint distribution of transmission eigenvalues of the matrix $\hat{s}$. To the best of our knowledge, this distribution is unknown – the problem is just too complicated. Even calculating averages of various quantities is sometimes difficult and requires a development of special techniques of integration of the unitary group [95]. It is not our goal to describe these techniques in this book, and we only give a couple of formulas to illustrate the point.

For these illustrations, we consider an almost closed chaotic cavity, connected with two reservoirs (left and right) by two tunnel junctions with low transparency, $T_L, T_R \ll 1$. Left and right leads support N_L and N_R transport channels, respectively; the total number of channels, $N = N_L + N_R$, we assume to be much bigger than unity.

Let us start from the distribution function of transmission probabilities, $\rho(T)$. The RMT result for this function is, for $N \gg 1$,

$$\rho(T) \propto \frac{1}{T^{3/2}\sqrt{1-T}},$$

which is exactly Eq. (2.108) we produced in Section 2.6 by means of the circuit theory. Let us try to understand its physical meaning.

For this purpose, we use again the example of a one-channel double barrier. The transmission probability is given by Eq. (2.108),

$$T = \frac{T_L T_R}{1 + R_L R_R - 2\sqrt{R_L R_R}\cos\chi},$$

where χ is the phase the electron acquires during the round trip between the barriers. We can consider the double junction as a combined system consisting of left and right barriers, and the "cavity" as the space between the barriers. The cavity is ideal and is characterized by one parameter – the phase χ. We assume that the scattering matrix of the cavity is a member of the circular ensemble. Then χ is a random variable uniformly distributed between 0 and 2π. Now we can easily find the distribution of the transmission probability T as follows:

$$\rho(T) \propto \left|\frac{\mathrm{d}T}{\mathrm{d}\chi}\right|^{-1} \propto \frac{1}{T\sqrt{4R_L R_R T^2 - (T + T R_L R_R - T_L T_R)^2}}.$$

In particular, for a symmetric opaque barrier, $T_L = T_R \ll 1$, we obtain Eq. (2.106).

Another example is the average conductance of the cavity. If there are two terminals, from Eq. (4.36) we obtain, for $N \gg 1$,

$$G = \frac{G_L G_R}{G_L + G_R} + G_Q\left(1 - \frac{2}{\beta}\right)\frac{G_R^2 Q_L + G_L^2 Q_R}{(G_L + G_R)^3}, \quad Q_{L,R} \equiv G_Q \sum_n T_{L,R}^2. \qquad (4.37)$$

The first term is just a series resistance addition – it reflects purely classical physics. For $\beta \neq 2$, i.e if the scattering matrix of the cavity is chosen from a COE or a CSE, there are quantum corrections to the classical result – *weak localization* (WL) corrections, originating from the quantum interference of different trajectories. We have already discussed WL corrections in a slightly different context in Section 1.6, and we will discuss them in more detail below. They are negative for a COE and positive for a CUE, and are in both cases much weaker than the classical contribution in the large number of channels limit. A magnetic field destroys the interference, and this is why the correction is absent in a CUE. Such a structure of the result is not specific for the conductance. Provided all transmission eigenvalues T_n are of the same order, the first term is proportional to the number of open transport channels N, and the second term is of the order of unity. The terms of order N^{-1} and higher are not written out in Eq. (4.37). Note also that the WL correction cannot be expressed in terms of the average conductance only – it contains some unusual combinations of transmission eigenvalues $Q_{L,R}$ that are required for the calculation of the shot noise, not of the average conductance. Conductance fluctuations require, in addition to $G_{L,R}$ and $Q_{L,R}$, another combination, $\sum T_{L,R}^3$. Thus, knowledge of the average classical conductance is generally not sufficient to obtain information on the quantum effects in transport.

Exercise 4.4. Calculate the quantities $Q_{L,R}$ and the weak localization correction to the conductance assuming that left and right junctions are diffusive conductors.

Let us now consider the case of an open chaotic cavity connected to the reservoirs by quantum point contacts with ideal transmission. For simplicity, we assume that there are only two contacts, one (left) supports N_L open transport channels and the other one (right) supports N_R open channels. The joint distribution of transmission eigenvalues follows from Eq. (4.36) as follows:

$$P(\{T_n\}) \propto \prod_{n<m} |T_n - T_m|^\beta \prod_n T_n^{-1+\beta/2+(\beta/2)|N_L - N_R|}$$
$$\propto \exp\left(\beta \sum_n \left(-\frac{1}{\beta} + \frac{1}{2} + \frac{|N_L - N_R|}{2}\right) \ln T_n + \beta \sum_{n<m} \ln |T_n - T_m|\right). \quad (4.38)$$

As on previous occasions, the joint distribution function has the form of a partition function of a system of classical particles at temperature β. These particles are placed in a confining potential (first term in the second equation), which, depending on the symmetry of the cavity, may favor either transparent ($T = 1$) or opaque ($T = 0$) channels. The particles also repel each other in a logarithmic fashion.

Equation (4.38) is a complete description of the transport properties of an open chaotic cavity. The results for average conductance, conductance fluctuations, counting statistics, and other relevant properties can be obtained by taking the averages with Eq. (4.38).

First, one can produce the distribution of transmission eigenvalues $\rho(T)$, given by

$$\rho(T) = \frac{N_L + N_R}{2\pi T}\sqrt{\frac{T - T_c}{1 - T}} + \frac{1 - 2/\beta}{4}\left(\delta(1 - T) - \delta(T - T_c)\right), \quad (4.39)$$
$$T_c = \left(\frac{N_L - N_R}{N_L + N_R}\right)^2,$$

which is valid for $N_L + N_R \gg 1$. The first term represents classical values and can be obtained by means of circuit theory (see Eq. (2.111)). However, Eq. (4.38) provides us with additional understanding of this result. Indeed, Eq. (4.39) is the average density of the transmission eigenvalues, and we already saw in Section 1.2 (and also in Section 4.2) that the density of confined two-dimensional particles interacting by logarithmic law (two-dimensional Coulomb particles) has a square-root singularity at the edge of the distribution. This is precisely what we see from Eq. (4.39). The delta-functions are quantum (weak localization) corrections to this classical result.

The simplest quantity one can calculate is the average conductance, given by

$$G = G_Q \frac{N_L N_R}{N_L + N_R - 1 + 2/\beta}. \quad (4.40)$$

Let us see whether we can comprehend this result. Consider the limit of the large number of channels, $N_L, N_R \gg 1$. We calculate now the transmission through the cavity by summing up the trajectories, in the same way as in Section 1.6. An incoming electron in a certain channel m in the left lead follows one of infinitely many sophisticated *classical* trajectories. Some of them eventually, after many reflections from the walls of the cavity, get to the lead on the right; some bring the electron back to the same channel in the left lead, and others return to the left lead but change the transport channel. A trajectory leading from channel m to channel n is characterized by an amplitude $A_i(mn) = a_i(mn)\exp(i\theta_i(mn))$. Here the index i labels the trajectories; we have separated the absolute value a_i and the phase θ_i, and the indices m and n both run from 1 to $N_\mathrm{L} + N_\mathrm{R}$ – thus, we describe both reflection and transmission. The phases acquired by the trajectories depend on the details of the classical motion. However, we assume that the trajectories are so sophisticated that these phases can be regarded as *random uncorrelated quantities*.

Let us now calculate the probability of transmission/reflection from m to n:

$$\left\langle |s_{mn}|^2 \right\rangle = \left\langle \left| \sum_i A_i \right|^2 \right\rangle = \left\langle \sum_i a_i^2 \right\rangle + \left\langle \sum_{i \neq j} a_i a_j \cos(\theta_i - \theta_j) \right\rangle ; \tag{4.41}$$

the summation is performed over all possible electron trajectories originating in transport channel m and leading to channel n. The second term depends on random phases, and the phases are random, and thus it averages to zero. As for the first term, it is a sum of contributions of many trajectories, and thus we assume that this sum does not depend on the initial and final channels: all $\langle |s_{mn}|^2 \rangle$ are the same. From the condition that the scattering matrix is unitary,

$$\sum_m \langle |s_{mn}|^2 \rangle = 1,$$

we find $\langle |s_{mn}|^2 \rangle = (N_\mathrm{L} + N_\mathrm{R})^{-1}$. Using the Landauer formula, we find the average conductance (Section 1.6) as follows:

$$G_\mathrm{cl} = G_\mathrm{Q} \sum_{m \in \mathrm{L}, n \in \mathrm{R}} \left\langle |s_{mn}|^2 \right\rangle = G_\mathrm{Q} \frac{N_\mathrm{L} N_\mathrm{R}}{N_\mathrm{L} + N_\mathrm{R}} = G_\mathrm{Q} \frac{G_\mathrm{L} G_\mathrm{R}}{G_\mathrm{L} + G_\mathrm{R}}. \tag{4.42}$$

The third equation highlights the fact that our result is merely a series resistance of two QPCs, with the conductances $G_{\mathrm{L,R}} \equiv G_\mathrm{Q} N_{\mathrm{L,R}}$, respectively. The classical result we obtained is proportional to the number of transport channels and is essentially Ohm's law.

However, Eq. (4.42) is not exactly the same as Eq. (4.40): they only coincide for the unitary ensemble ($\beta = 2$), but not for the orthogonal one ($\beta = 1$). Indeed, if we think more about it, we find a mistake in our consideration. For the unitary ensemble, the system possesses time-reversal symmetry. This means that, for any trajectory leading from m to n, there is also a time-reversed trajectory that leads from n to m, with the same amplitude A_i and phase θ_i. This fact does not affect the calculation of $|s_{mn}|^2$ with $m \neq n$. However, if we look at $|s_{mm}|^2$, for each trajectory $A_i(nn)$ contributing to Eq. (4.41) one has the time-reversed trajectory $A_i'(nn)$, which is different from A_i but has the same amplitude and phase. The contribution from the interference of these two trajectories is given by

$$\left| A_i + A_i' \right|^2 = 4a_i^2 \, ,$$

which is twice as high as the classical contribution $|A_i|^2 + |A'_i|^2$. The two trajectories interfere *constructively*, enhancing the probability of reflection. We obtain now, instead of Eq. (4.41),

$$\left\langle |s_{nn}|^2 \right\rangle = \left\langle \left| \sum_i A_i \right|^2 \right\rangle = 2 \left\langle \sum_i a_i^2 \right\rangle + \left\langle \sum_{i,j} a_i a_j \cos(\theta_i - \theta_j) \right\rangle , \qquad (4.43)$$

where the second term in the second equation only sums over two trajectories with different amplitudes and/or phases (it does not include a pair of a certain trajectory and its time-reversed analog). After averaging, the last term disappears, and one has $\langle |s_{nn}|^2 \rangle = 2\langle |s_{mn}|^2 \rangle$, $m \neq n$. Using the unitarity, we find $\langle |s_{mn}|^2 \rangle = (N_\mathrm{L} + N_\mathrm{R} + 1)^{-1}$, and for the conductance we have Eq. (4.40). We must keep in mind, of course, that our derivation is only valid in the limit of a large number of channels, and thus it is more correct to write the following:

$$G \approx G_\mathrm{cl} + G_\mathrm{Q} \frac{G_\mathrm{L} G_\mathrm{R}}{(G_\mathrm{L} + G_\mathrm{R})^2} \left(1 - \frac{2}{\beta} \right) .$$

The first term is the classical contribution, and the second term represents the weak localization correction, negative for a COE and positive for a CUE. In both cases the correction is of the order of G_Q (much weaker than $G_\mathrm{cl} \sim G_\mathrm{Q} N$ in the large number of channels limit).

Let us now look at a quantity that cannot be expressed as a sum of contributions of independent channels – the variance of the conductance. Indeed,

$$\begin{aligned} \mathrm{Var}\, G &\equiv \left\langle G^2 \right\rangle - \langle G \rangle^2 = G_\mathrm{Q}^2 \sum_{mn} \left(\langle T_m T_n \rangle - \langle T_n \rangle^2 \right) \\ &= \sum_{m,m' \in L, n,n' \in R} \left\{ \left\langle |s_{mn}|^2 \, |s_{m'n'}|^2 \right\rangle - \left\langle |s_{mn}|^2 \right\rangle \left\langle |s_{m'n'}|^2 \right\rangle \right\} , \end{aligned}$$

and we need information about the correlation of various elements of the scattering matrix. Let us first calculate the average fourth power of the scattering matrix element, $\langle |s_{mn}|^4 \rangle$. Using Eq. (4.41), we write

$$\left\langle |s_{mn}|^4 \right\rangle = \left\langle \left(\sum_i a_i^2 \right) + \sum_{i \neq j, k \neq l} a_i a_j a_k a_l \cos(\theta_i - \theta_j) \cos(\theta_k - \theta_l) \right\rangle , \qquad (4.44)$$

where we have discarded terms proportional to the cosine of the phase difference, since these terms always average to zero, provided we neglect weak localization corrections. The first term in Eq. (4.44) is just a product of averages, equal to $(N_\mathrm{L} + N_\mathrm{R})^{-2}$ in the limiting case of $N_\mathrm{L}, N_\mathrm{R} \gg 1$. In a CUE, in the second term the only contribution comes from the terms with $i = k$, $j = l$ and $i = l$, $j = k$. The cosine squared averages to $1/2$, and one finds the following:

$$\left\langle |s_{mn}|^4 \right\rangle_\mathrm{CUE} = 2 \left\langle |s_{mn}|^2 \right\rangle^2 = 2\,(N_\mathrm{L} + N_\mathrm{R})^{-2} .$$

In a COE, the time-reversal symmetry is preserved, and $s_{mn} = s_{nm}$. Then, in the second term of Eq. (4.44) there is another pair of contributions, when the trajectory k is the time-reversed trajectory of i, and l is the time-reversed trajectory of j, or i and l are time-reversed, and j and k are time-reversed. These contributions are exactly the same as in a CUE, and thus the fluctuations in a COE are twice as high as those in a CUE:

$$\left\langle |s_{mn}|^4 \right\rangle - \left\langle |s_{mn}|^2 \right\rangle^2 = \frac{2}{\beta} \frac{1}{(N_\mathrm{L} + N_\mathrm{R})^2} . \quad (4.45)$$

In the leading order in N_L and N_R all matrix elements, diagonal and off-diagonal, have the same fluctuation.

Naively, we could think that this gives the only contribution to the conductance fluctuations, since different matrix elements are determined by different trajectories and are thus uncorrelated. This is, indeed, partially correct – different matrix elements are not correlated in the leading order in $(N_\mathrm{L} + N_\mathrm{R})^{-1}$, i.e. their correlations are much weaker than the values of these matrix elements themselves. However, due to the large number of these contributions, they become important. What can RMT say about these correlations?

Consider the condition of the unitarity of the scattering matrix, written in the following way:

$$\sum_{kl} \left\langle |s_{mk}|^2 \, |s_{nl}|^2 \right\rangle = 1. \quad (4.46)$$

In RMT, in the leading order in $N_\mathrm{L}, N_\mathrm{R}$ one can only have three distinct correlation functions: $\langle |s_{mk}|^2 \rangle$ (all the same for any m and n), $\langle |s_{mk}|^2 |s_{ml}|^2 \rangle$ for $k \neq l$, and $\langle |s_{mk}|^2 |s_{nl}|^2 \rangle$ for $m \neq n, k \neq l$. The first one we have already calculated, and the other two are immediately expressed through the first one via the unitarity condition given by Eq. (4.46), where we need once to take $m = n$, and eventually $m \neq n$. As the result, we obtain

$$\left\langle |s_{mk}|^2 \, |s_{ml}|^2 \right\rangle - \left\langle |s_{mk}|^2 \right\rangle \left\langle |s_{ml}|^2 \right\rangle = -(N_\mathrm{L} + N_\mathrm{R})^{-3}, \ \ k \neq l;$$
$$\left\langle |s_{mk}|^2 \, |s_{nl}|^2 \right\rangle - \left\langle |s_{mk}|^2 \right\rangle \left\langle |s_{nl}|^2 \right\rangle = (N_\mathrm{L} + N_\mathrm{R})^{-4}, \ \ k \neq l, \ m \neq n.$$

Substituting this result into the expression for conductance fluctuations, one obtains

$$\mathrm{Var}\, G = \frac{G_\mathrm{Q}^2 (2/\beta) N_\mathrm{L}^2 N_\mathrm{R}^2}{(N_\mathrm{L} + N_\mathrm{R})^4}.$$

One could of course derive this expression directly from Eq. (4.38).

Control question 4.4. Explain why in a very asymmetric cavity, $N_\mathrm{L} \gg N_\mathrm{R}$, $N_\mathrm{L} \gg 1$, the variance of the conductance vanishes.

For symmetric barriers, one has $\mathrm{Var}\, G = G_\mathrm{Q}^2/(8\beta)$. This result is known as *universal conductance fluctuations* (UCF). We have already encountered UCF in Section 1.6, along with the weak localization; UCF have an order of magnitude of G_Q^2 – fluctuations of the conductance are of the same order as WL corrections. Thus, this is a quantum effect and is missed in any classical approach. The effect is twice as great in a COE than in a CUE. This is because the repulsion of transmission eigenvalues is stronger in a COE.

In the same way, we can, using the unitarity constraints, obtain other correlation functions, for example

$$\left\langle s_{mk}^* s_{nk} s_{nl}^* s_{ml} \right\rangle = -(N_\mathrm{L} + N_\mathrm{R})^{-3}, \ m \neq n, \ \ k \neq l,$$

and calculate the Fano factor for the shot noise, $F = N_\mathrm{L} N_\mathrm{R}/(N_\mathrm{L} + N_\mathrm{R})^2$. This result conforms to the notion of two QPCs in series: for $N_\mathrm{L} \gg N_\mathrm{R}$, the noise is controlled by the more resistive contact, and a noise of a QPC is zero; thus, the Fano factor becomes zero.

Control question 4.5. Is there a weak localization correction to shot noise of the order of $1/N$ relative to the classical value? Is there such a correction to the Fano factor?

Let us now look at the experiments. Conductance fluctuations in chaotic cavities were investigated in Ref. [96]. The authors studied the conductance of a "stadium" billiard connected to the reservoirs by two tunnel contacts. This "stadium" is one of the exemplified systems exhibiting classically chaotic dynamics. The main problem is how to create an *ensemble* to measure the UCF. Theoretically, we assumed that one has many samples identical in parameters, but different in phase shifts; this is not convenient experimentally. What experimentalists do in practice is to manipulate the phase shifts by a magnetic field. The same billiard for two close values of the field behaves as two different chaotic systems. Conductance as a function of the field shows aperiodic modulation, from which one extracts the statistics of the conductance fluctuations. For a stadium billiard, the measured magnitude of the fluctuations is exactly the same as is predicted theoretically.

4.3.2 DMPK equation: diffusive wires

Another system successfully described by RMT is a disordered wire. We have already derived the distribution function of transmission eigenvalues in Section 2.6 by means of circuit theory. Here, we will show how (i) one derives weak localization corrections and (ii) one calculates correlations of several transmission eigenvalues, for example to evaluate the conductance fluctuations.

Consider a wire of length L with random impurities. The scattering matrix of such a wire is random, but it obviously cannot be taken from a circular ensemble; for example, the average transmission probability through the wire depends on the length, and this dependence cannot originate from the circular ensemble of RMT.

To calculate the joint distribution of transmission eigenvalues, we apply the following trick [55]. Imagine we know this distribution for a wire of length L. Let us append a short disordered slice of length δL and see how the distribution is modified. In the end, we will be able to write an equation that determines the evolution of this distribution function with the length of the wire. From the equation, we will determine various statistical properties of transmission eigenvalues. Usually this is done by using the transfer matrices, with the advantage that the resulting transfer matrix is just a product. However, we did not consider statistical distributions for transfer matrices, and instead we have chosen to combine the scattering matrices of the wire and the slice.

Before addressing the general problem, let us first prepare for this by considering the simplest example of one transport channel $T_0 = 1 - R_0$, R_0 being random. We append to the wire a short slice of disordered material. This last piece is so short that it is almost ideally transmitting. Its reflection probability, $R_{\text{add}} = 1 - T_{\text{add}} \equiv \delta s \ll 1$, is obviously proportional to the length δL of the added slice. We assume that, due to the random scattering, the phase of the scattering matrix of the added piece is randomly uniformly distributed between 0 and 2π. Combining the scattering matrix of the wire and the added piece, we obtain for the transmission probability of the composed system (Eq. (1.106)) the following:

$$T = \frac{T_0 T_{\text{add}}}{1 + R_0 R_{\text{add}} - 2\sqrt{R_0 R_{\text{add}}}\cos\theta},$$

where θ is the phase acquired by an electron during the round trip. It is a linear function of the phase of the scattering matrix of the added slice, and thus can be taken to be a uniformly distributed random variable. Averaging over θ, we obtain

$$\langle T \rangle = \frac{T_0 T_{\text{add}}}{1 - R_0 R_{\text{add}}} \approx T_0 - T_0^2 \delta s,$$

where the second equation is the result of the expansion in δs up to first order.

One can also calculate higher-order cumulants of the transmission probability. The variance in the first order in δs is given by

$$\left\langle \delta T^2 \right\rangle \equiv \left\langle T^2 \right\rangle - \langle T \rangle^2 = 2T_0^2(1 - T_0)\delta s,$$

and all higher cumulants vanish (i.e. only contain second or higher orders in δs).

Thus, if we identify the parameter s, proportional to the length of the system, with a fictitious "time," we realize that the random variable T performs *Brownian motion* characterized by the average "velocity" $\langle T - T_0 \rangle / \delta s$ and the "diffusion coefficient" $\langle \delta T^2 \rangle / \delta s$. Such a motion, as we know from statistical mechanics, is described by the *Fokker–Planck equation* for the distribution function of the random variable $P(t)$:

$$\frac{\partial P}{\partial s} = \frac{1}{\delta s}\left\{ -\langle T - T_0 \rangle \frac{\partial P}{\partial T} + \left\langle \delta T^2 \right\rangle \frac{\partial^2 P}{\partial T^2} \right\}. \tag{4.47}$$

We thus obtain

$$\frac{\partial P}{\partial s} = \frac{\partial}{\partial T}\left[T^2 P + \frac{\partial}{\partial T} T^2 (1 - T) P \right]. \tag{4.48}$$

Such an approach be generalized to the case of many transport channels. We assume the joint distribution function of transmission eigenvalues $P(T_1 \cdots T_N)$ to be known for the wire of length L and we add a slice that is characterized by a random scattering matrix, uniformly distributed in the unitary group, with the only constraint that the average reflection coefficient is fixed to be $\delta s \ll 1$. Combining the scattering matrices, averaging over the unitary group, and extending the results for different symmetries with respect to the time-reversal, after some lengthy calculations one obtains the multi-dimensional Fokker–Planck equation:

$$\frac{\partial P}{\partial s} = \sum_{n=1}^{N} \frac{\partial}{\partial T_n} \left\{ A\{T_n\} P + \frac{1}{2} \frac{\partial}{\partial T_n} B\{T_n\} P \right\}, \tag{4.49}$$

with the coefficients given by

$$A = -T_n + \frac{2}{\gamma} T_n \left(1 - T_n + \frac{\beta}{2} \sum_{m \neq n} \frac{T_n + T_m - 2T_n T_m}{T_n - T_m}\right),$$

$$B = \frac{4}{\gamma} T_n^2 (1 - T_n), \quad \gamma \equiv \beta N + 2 - \beta.$$

Equation (4.49), known as the *Dorokhov–Mello–Pereira–Kumar (DMPK) equation*, looks more compact after the change of variables, $T_n = (1 + \lambda_n)^{-1}$, is performed (note that one also has to transform the distribution function to keep the normalization: integration over all variables gives the number of transport channels). We give it here for reference:

$$\frac{\partial P}{\partial s} = \frac{2}{\gamma} \sum_{n=1}^{N} \frac{\partial}{\partial \lambda_n} \lambda_n (1 + \lambda_n) J\{\lambda_n\} \frac{\partial}{\partial \lambda_n} \frac{P}{J\{\lambda_n\}}, \tag{4.50}$$

with

$$J = \prod_{n=1}^{N} \prod_{m=n+1}^{N} |\lambda_n - \lambda_m|^{\beta}.$$

The most practical form of the DMPK equation is that using the variables x_n, $T_n = \cosh^{-2} x_n$, and $\lambda_n = \sinh^2 x_n$. Note that now almost closed (almost open) channels correspond to $x_n \gg 1$ ($x_n \ll 1$), respectively. If, in addition, we perform the transformation of the function, $P = \Psi \exp(-\beta\Omega/2)$,

$$\Omega = -\frac{1}{2} \sum_{m \neq n} \ln \left|\sinh^2 x_n - \sinh^2 x_m\right| - \frac{1}{\beta} \sum_{n} \ln |\sinh 2x_n|,$$

the resulting equation,

$$-\frac{\partial \Psi}{\partial s} = H\Psi,$$

$$H = -\frac{1}{2\gamma} \sum_{n} \left(\frac{\partial^2}{\partial x_n^2} + \frac{1}{\sinh^2 2x_n}\right) + \frac{\beta(\beta - 2)}{4\gamma} \sum_{m \neq n} \frac{\sinh^2 2x_n + \sinh^2 2x_m}{(\cosh 2x_n - \cosh 2x_m)^2}, \tag{4.51}$$

has an easy physical interpretation. The equation describes the evolution of the system of classical particles in the imaginary time s. The first two terms on the right-hand side represent the kinetic and potential energy of the particles, and the last term denotes the interaction between them (note that the interaction is not translationally invariant, in the sense that it does not depend on the difference $|x_n - x_m|$). The interaction is attractive for $\beta = 1$ (orthogonal ensemble) and repulsive for $\beta = 4$ (symplectic ensemble). For $\beta = 2$ (unitary ensemble) there is no interaction between the particles. For this reason, the unitary ensemble is the simplest to handle: there is an exact solution of the DMPK equation, Eq. (4.49), for an arbitrary number of channels and the length of the wire s. It is too cumbersome to be written down here, but it is good to know that all approximate solutions can be checked against the exact one. For the two other symmetries, no exact solutions exist.

Control question 4.6. Can the DMPK equation as given in Eqs. (4.51) describe the situation when all transport channels are almost open?

The only remaining problem is that the meaning of the variable s is unclear: we know that it is proportional to the length of the wire L, but the proportionality coefficient is not specified. For the moment, we write $s = L/l$ and call l the mean free path. From the comparison with the solution of the DMPK equation in what follows we will relate the mean free path to the average conductance of the wire. In particular, for the metallic regime (large number of channels N, so that $\langle G \rangle \gg G_Q$) one finds $\langle G \rangle = G_Q Nl/L = G_Q N/s$; see the following text.

One can make further progress in the two cases. For one channel, the DMPK equation can be solved exactly. For a low number of transport channels, various quantities can be calculated without knowing the exact solution for the joint distribution function. It turns out that in this regime (we remind the reader that this is the regime opposite to the one we treated using circuit theory in Section 2.6) the electrons in the wire are *localized* – all transmission eigenvalues are small, and the transport is suppressed. The characteristic feature of this regime is that Ohm's law no longer holds; The resistance of the wire is an exponential function of its length, rather than a linear one. We will address the issue of localization in Section 4.5.

In the following, we concentrate on a different regime – a *metallic diffusive wire*, when a considerable fraction of almost open channels exists. This regime is realized for the large number of transport channels $N \gg 1$. In this case, one can build up an expansion of the results in a power of $G_Q/\langle G \rangle$, calculating first the classical effects of order C_J, and then the quantum corrections of order G_Q.

Imagine we want to calculate the average of some function $F\{T_1 \cdots T_N\}$. From the DMPK equation, it is easy to derive an equation for the "time" derivative of $\langle F \rangle$: indeed, we multiply the DMPK equation by F and integrate over all variables. On the left-hand side, we get $\partial \langle F \rangle / \partial s$, and on the right-hand side, after partial integrations, a cumbersome expression involving averages of first and second derivatives of the function F over the T_n values, multiplied by different combinations of transmission eigenvalues. This is not a closed equation. To close it, one also needs equations for the evolution of the combinations appearing on the right-hand side, and so on. The calculations require an infinite chain of equations, and become progressively messy. However, we can use the condition $N \gg 1$. In this case, the equations close up, and one can calculate the averages.

Let us illustrate this with the calculation of the average conductance $G = G_Q \sum_n T_n$. The equation is given by

$$\frac{\partial}{\partial s}\langle G \rangle = -\frac{\beta}{\gamma}\left\langle \frac{G^2}{G_Q} - G_Q\left(1 - \frac{2}{\beta}\right)\sum_n T_n^2 \right\rangle. \tag{4.52}$$

For $N \gg 1$, we can drop the second term in the brackets (it is proportional to N as compared with N^2 for the first term; these relations are checked after the calculation), and obtain $\partial \langle G \rangle / \partial s = -N^{-1} \langle G^2 \rangle / G_Q$. Similarly, one writes equations for the average higher powers of the conductance as follows:

$$\frac{\partial}{\partial s}\left\langle G^k \right\rangle = -\frac{k}{NG_Q}\left\langle G^{k+1}\right\rangle .$$

This chain of equations must be supplemented with the "initial" condition. We employ a formal trick: for $s = 0$, the wire has zero length and thus must be ideally transmitting. (Note that the DMPK equation is not valid in this regime, since there is no scattering, and formally one can only start from the wire of length of the order l, $s \sim 1$. It does not change our conclusions.) This means that, at $s = 0$, $T_n = 1$, and $\langle G^k \rangle = G_Q^k N^k$. The solution corresponding to this initial condition is given by

$$\left\langle G^k \right\rangle = G_0^k N^k (1+s)^{-k}.$$

Finally, we are only interested in long wires, $s \gg 1$, and thus $\langle G^k \rangle = (G_Q N/s)^k = (G_Q Nl/L)^k$.

The average conductance equals $\langle G \rangle = G_Q Nl/L$, and thus

$$\left\langle G^k \right\rangle = \langle G \rangle^k .$$

This is remarkable, and means that, in the leading order in $\langle G \rangle / G_Q$, the conductance does not fluctuate. To obtain fluctuations, or WL corrections, to the conductance, we have to keep the next terms of the expansion in $G_Q/\langle G \rangle$. One obtains, for the weak localization correction, $\delta G = (G_Q/3)(1 - 2/\beta)$, which is positive for an orthogonal ensemble, vanishes for a unitary ensemble, and is of the order of G_Q. Conductance fluctuations, $\text{Var}\, G = G_Q^2(2/15\beta)$, are also of the order of G_Q, and the value for an orthogonal ensemble is twice as high as for the unitary one.

Exercise 4.5. Keep the next-order terms in $G_Q/\langle G \rangle$ and check that the preceding expression for the conductance fluctuations is correct.

Let us now compare the average conductance with the WL correction. For $G_Q \ll \langle G \rangle$, the average conductance is much greater than the WL correction. This constitutes the metallic regime. On the other hand, for $\langle G \rangle \ll G_Q$ the correction is formally greater. This signals a problem: indeed, the expansion in $G_Q/\langle G \rangle$ formally does not work, and all higher corrections will be of the same order or higher. The average conductance is not proportional to L^{-1} – Ohm's law does not hold. We will see in Section 4.5 that this situation corresponds to *electron localization*. The metallic (diffusive) regime is thus only possible for $l \ll L \ll Nl$ (the condition $L = Nl$ corresponds to $\langle G \rangle = G_Q$). Electrons in longer wires are localized. Note that the diffusive regime only has a chance to exist for the large number of transport channels N – in full agreement with the assumptions we have made at the beginning of this subsection.

In the same manner, one can calculate more sophisticated averages in the diffusive regime. As an example, we give the expression for the distribution function of the transmission eigenvalues (it is more convenient to use the variable x, $T = (\cosh x)^{-2}$):

$$\rho(x) = \frac{\langle G \rangle}{G_Q} + \left(1 - \frac{2}{\beta}\right)\left(\frac{1}{4}\delta(x) - \frac{1}{4x^2 + \pi^2}\right). \tag{4.53}$$

The first term is equivalent to the familiar distribution function, Eq. (1.43), of the transmission eigenvalues, $P(T) = (G/2G_Q)(1/T\sqrt{1-T})$. The second term represents the WL correction, is of the order of G_Q and vanishes for the unitary ensemble, $\beta = 2$. Higher-order terms are not included.

4.4 Interference corrections

We have seen that RMT can provide a lot of information about the properties of the transmission eigenvalues. However, we have also seen that calculations become progressively cumbersome if a large number of elements of the scattering matrix are involved. In particular, RMT is not very well suited for the calculation of distribution functions.

In this section, we show how various quantum corrections to transport properties can be calculated with the help of circuit theory, which we introduced in Chapter 2 for classical values. The general argument proceeds as follows. Circuit theory treats quantum circuits as analogs of classical electric circuits. Currents in electric circuits are represented in circuit theory as matrix currents, obeying Ohm's law, and voltages are represented as matrix voltages. However, apart from the average currents, classical electric circuits exhibit current fluctuations – noise. We will show in this section that the analog of noise in quantum circuits takes the form of quantum corrections to transport – weak localization and conductance fluctuations.

Let us illustrate this with the following schematic consideration. An electric circuit at thermal equilibrium can be characterized using a set of fluctuating particles with coordinates V_i (potentials of the nodes). The partition function of the system is given by

$$Z = \prod_i \mathrm{d}V_i \, \exp(-S\{V_i\}), \quad S = E\{V_i\}/k_B T,$$

where E is the energy of the system. We need to minimize the energy with respect to all coordinates to find the equilibrium solutions V_i^{eq}. In the first approximation, the partition function (up to the normalization) is $Z = \exp(-E_{\mathrm{eq}}/k_B T)$, with $E_{\mathrm{eq}} = E\{V^{\mathrm{eq}}\}$. So far, we have disregarded the fluctuations. Expanding the energy to the second order in V_i^{eq}, $E - E_{\mathrm{eq}} = \sum_{ij} \partial^2 E/\partial V_i \partial V_j (V - V_i)(V - V_j)$, and on integrating over V_i we obtain

$$Z = \frac{1}{|\det \partial^2 E/\partial V_i \partial V_j|} \exp(-E_{\mathrm{eq}}/k_B T).$$

We see that there is a *fluctuation correction* to the free energy of the equilibrium state, $F \equiv k_B T \ln Z = \mathrm{E}_{\mathrm{eq}} + \ln|\det \partial^2 E/\partial V_i \partial V_j|$. This correction yields fluctuation contributions to physical observables.

In this section, we try to follow a similar program for quantum circuits out of equilibrium. In circuit theory, the effective action given by Eq. (2.91), depends on the matrix voltages $\check{G}$ in the nodes. "Fluctuation corrections" to the effective action, as we check after the calculation, yield a correction to the transport coefficients of the order of G_Q – weak localization and universal conductance fluctuations.

It is much easier said than done. First, the matrix "voltages" are constrained by the condition $\check{G}_i^2 = 1$, and thus some of the second derivatives turn to infinity. One has to perform the expansion carefully. Secondly, and more important, quantum corrections represent various limiting cases of a general problem – parametric correlations, for example $\langle G(x)G(x')\rangle$ – the correlation function of two conductances taken for two different values of the parameter x, which could be, for example, a magnetic field. These two values of the parameter represent two different "worlds" – say black and white – and the correlation between these worlds is provided by the averaging over the impurities. Thus, in order to describe parametric correlations, one needs to double the dimension of the matrix G and to introduce "black" and "white" elements. We show in the following that the quantum corrections are expressed through certain objects, known as diffusons and cooperons, and to define the objects in terms of the Green's function, we need to build up a systematic theory. Let us see how this is all done in practice.

4.4.1 Diagram technique: diffusons and cooperons

Here we do the preparatory work; we develop a technique of averaging over a random potential, the so-called *diagram technique*. We already introduced some components of it in Section 2.3, but here we explain them in a systematic way.

Our starting point is Eq. (2.26) for the Green's function. In this section, we only consider normal-state systems in the linear regime, and thus we do not need Nambu and spin decompositions. For Keldysh decomposition, we only need to find advanced and retarded Green's functions, for which we have the following equation:

$$(E \pm \mathrm{i}0 - \hat{H}_{\boldsymbol{r}})\check{G}_{\mathrm{R,A}}(\boldsymbol{r}, \boldsymbol{r}'; \epsilon) = \delta(\boldsymbol{r} - \boldsymbol{r}'), \tag{4.54}$$

where the upper and lower signs correspond to advanced and retarded functions, respectively, and $\hat{H}_{\boldsymbol{r}} = \hat{H}_0 + U(\boldsymbol{r})$, with U and $\hat{H}_0$ being the random potential and the Hamiltonian in the absence of random potential, respectively.

Now we are going to calculate the average Green's functions and average products of several Green's functions. This is done in the way outlined in Section 2.3. Denote the unperturbed Green's function (the solution of Eq. (4.54) for $U = 0$) as $G^{(0)}(\boldsymbol{r}, \boldsymbol{r}', E)$. In the simplest case, when H_0 only contains the kinetic energy, we obtain

$$G^{(0)}_{\mathrm{R,A}}(\boldsymbol{k}, \omega) = (\epsilon - \xi(\boldsymbol{k}) \pm \mathrm{i}0)^{-1},$$

where we have moved to the momentum representation (over $\boldsymbol{r} - \boldsymbol{r}'$), $\epsilon = E - \mu$, and $\xi(\boldsymbol{k})$ is the kinetic energy counted from the Fermi level, $\xi \approx v_{\mathrm{F}}|\boldsymbol{k} - \boldsymbol{k}_{\mathrm{F}}|$. Following tradition, in this section we use the system of units where $\hbar = 1$.

Now one develops a perturbation theory in U for Eq. (4.54), solving it formally, $G = (G_0^{-1} - U)^{-1}$, and then expanding the denominator in powers of U. To keep track of the expansion, we represent it graphically in the following way. Let us draw the unperturbed Green's function $G^{(0)}$ as an oriented straight line (straight line with an arrow). The arrow

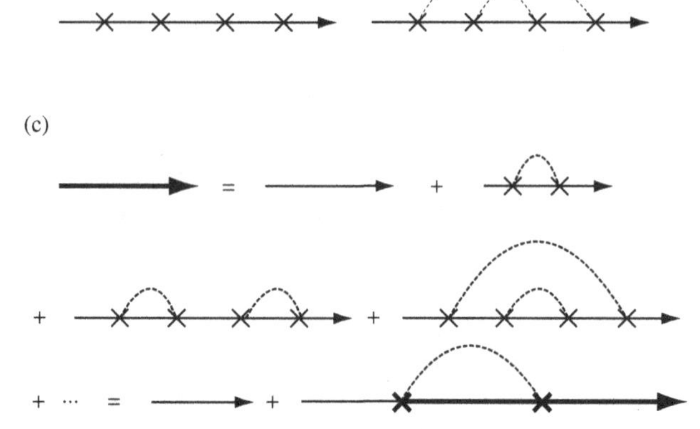

Fig. 4.6. (a) Green's functions with several crosses; (b) the simplest crossing diagram; (c) non-crossing approximation for the Green's function.

supports the momentum $\boldsymbol{k}$ and the energy ϵ. Alternatively, if one wishes to develop the perturbation theory in the space-time representation, the line connects the point $\boldsymbol{r}$, t with the point $\boldsymbol{r}'$, t'. The first-order correction to the Green's function is given by

$$G^{(1)}_{\mathrm{R,A}}(\boldsymbol{r},\boldsymbol{r}';\epsilon) = \int \mathrm{d}\boldsymbol{r}''\; G^{0}_{\mathrm{R,A}}(\boldsymbol{r},\boldsymbol{r}'';\epsilon)U(\boldsymbol{r}'')G^{0}_{\mathrm{R,A}}(\boldsymbol{r}'',\boldsymbol{r}';\epsilon).$$

In space-time representation, this is drawn as a straight line, a cross with an attached dashed line (indicating the random potential $U(\boldsymbol{r}'')$ – an electron is scattered at an impurity at the intermediate point $\boldsymbol{r}''$), and another straight line. This is the simplest *Feynman diagram* for the electron in a random potential. In the momentum representation, crosses also carry momentum (which is conserved: the sum of the momenta carried by two Green's functions and the cross equals zero). Note that a cross does not carry energy, as is obvious from the expression for $G^{(1)}$, since our disorder potential is time-independent.

Further orders of the perturbation theory create new diagrams. The full series of the perturbation theory for the Green's function is as follows: each term is a straight line intercepted by a number of crosses; all these terms come with equal weights (see Fig. 4.6(a)).

So far, we have not averaged anything over disorder. All our diagrams depend on the positions of the impurities involved in the scattering process. The averaging is performed using the rules $\langle U(\boldsymbol{r})\rangle = 0$, $\langle U(\boldsymbol{r})U(\boldsymbol{r}')\rangle = w\delta(\boldsymbol{r}-\boldsymbol{r}')$, and for simplicity we assume w to be position-independent. In the language of the diagram, this means that we have to connect *pairwise* all pieces of dashed lines attached to the crosses. After this averaging, all diagrams with an odd number of crosses vanish, and we get a huge number of diagrams with an even number of crosses connected by dashed lines. Each dashed line denotes the correlation between the potentials U, i.e. the scattering of an electron at the same impurity. A dashed line carries a momentum (but not an energy), but is momentum-independent and is equal to w in the momentum–energy representation.

It is impossible to sum up all the diagrams. Fortunately, one can check by a direct calculation that all the diagrams where the dashed lines cross (*crossing diagrams*) are small compared with the other (non-crossing) ones. The estimation of their relative contributions is given by $(kl)^{-1}$, where l is the mean free path. For $kl \gg 1$, one can disregard these and work in the non-crossing approximation. For the Green's function, in the non-crossing approximation we can write the following equation (see Fig. 4.6(c)):

$$G_{\mathrm{R,A}}(\boldsymbol{r},\boldsymbol{r}';\epsilon) = w \int \mathrm{d}\boldsymbol{r}''\, G^0_{\mathrm{R,A}}(\boldsymbol{r},\boldsymbol{r}'';\epsilon) G_{\mathrm{R,A}}(\boldsymbol{r}'',\boldsymbol{r}'';\epsilon) G_{\mathrm{R,A}}(\boldsymbol{r}'',\boldsymbol{r}';\epsilon),$$

which is most conveniently solved in the momentum–energy representation (we drop the angular brackets for the average Green's function) as follows:

$$G_{\mathrm{R,A}}(\boldsymbol{k},\epsilon) = (\epsilon - \xi(\boldsymbol{k}) \pm \mathrm{i}/2\tau)^{-1}\,. \tag{4.55}$$

This is what we saw in Section 2.3: for the average Green's function, the effect of impurities is to introduce the self-energy, equal in our case to $1/2\tau$, where the *(momentum) relaxation time* is related to the strength of the random potential w as $\tau^{-1} = 2\pi \nu w^2$. Note that, as it should be, retarded and advanced Green's functions only have poles in the upper lower and upper half-planes of the complex variable ϵ, respectively.

Exercise 4.6. Calculate the simplest crossing diagram (Fig. 4.6(b)) and verify that it is indeed small in comparison with the non-crossing one, being of second order in $(kl)^{-1}$. Where does this small factor come from?

In the coordinate representation, the averaged Green's function becomes short-ranged and decays proportionally to $\exp(-|\boldsymbol{r} - \boldsymbol{r}'|/l)$. This just expresses the fact that the electron can propagate without scattering to a distance not longer than l – eventually, it experiences elastic scattering, and its momentum is not conserved.

We proceed with complex averages. In this section, we will need the following object, known as the *diffuson*,

$$P^{BB'}_{\mathrm{diff}}(\boldsymbol{r},\boldsymbol{r}') = w\tau \left\langle G^{\mathrm{R}}(\boldsymbol{r},\boldsymbol{r}', E+\omega, B) G^{\mathrm{A}}(\boldsymbol{r}',\boldsymbol{r}, E, B')\right\rangle, \tag{4.56}$$

where B and B' are two values of a certain parameter characterizing the system, for example the magnetic field. These values play the roles of "black" and "white" indices as discussed. The diffuson has an optimum appearance in the momentum representation:

$$P^{BB'}_{\mathrm{diff}}(\boldsymbol{q}) = w\tau \int \frac{\mathrm{d}^d \boldsymbol{p}}{(2\pi)^d} \left\langle G^{\mathrm{R}}(\boldsymbol{p}+\boldsymbol{q}, E+\omega, B) G^{\mathrm{A}}(\boldsymbol{p}, E, B')\right\rangle. \tag{4.57}$$

Let us now perform the averaging over disorder. We need to draw crosses and connect them pairwise (there is also a contribution that is a product of two Green's functions G_0 with no crosses at all). We employ again the non-crossing approximation. There are only two types of dashed lines. Some lines connect two crosses at the same Green's function. As we have seen, these scattering processes just renormalize the average Green's function – each line becomes $\langle G \rangle$ rather than G_0. Now, there are also dashed lines connecting two different Green's functions (retarded and advanced ones). In the non-crossing approximation, all

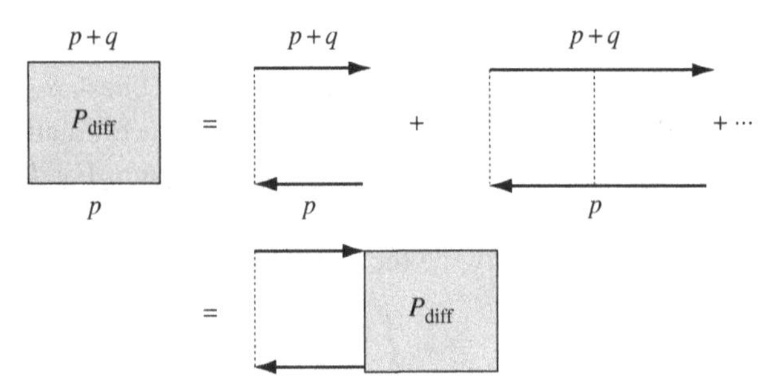

Fig. 4.7. Diffuson equation and averaging. The impurity crosses are not shown for simplicity.

these lines are parallel, like a ladder (see Fig. 4.7). This "ladder" series of diagrams is easy to sum up. Indeed, let us move to the momentum representation. The first diagram in the diffuson series equals $w\tau$ times $J(\boldsymbol{q}, \omega)$, where J is just the integrated product of retarded and advanced Green's functions. This product equals $2\pi\nu V\tau(1 - Dq^2\tau + \mathrm{i}\omega\tau - \tau/\tau_D)$, where the term $1/\tau_{\mathrm{D}}$ originates from the averaging of the square of the "long" momentum, $-\mathrm{i}\nabla - eB/c$, over the cross-section of the system perpendicular to the magnetic field. We have $1/\tau_{\mathrm{D}} = (e/\hbar)^2 D\langle A^2 - A'^2\rangle$, where the averaging is performed over the volume of the system. In particular, for a one-dimensional wire of square cross-section (side a) and the magnetic field directed along the wire, one has $\tau_{\mathrm{D}}^{-1} = e^2a^2(B - B')^2/(12c^2)$. The diffusion coefficient is $D = v_{\mathrm{F}}^2\tau/d$, d being the dimensionality. This expression is only good for $Dq^2\tau, \omega\tau, \tau/\tau_{\mathrm{D}} \ll 1$. The next term in the series equals $w^2\tau J^2$, and so on. Summing up the geometric series, we obtain $P = w\tau J(1 - wJ)^{-1} \approx (Dq^2 - \mathrm{i}\omega + 1/\tau_{\mathrm{D}})^{-1}$. Note that the series is converging very slowly and that all terms in the series contribute equally – in particular, omission of the first few terms would not change the result.

This expression is best written as the solution of the following equation in space-time representation:

$$\left(\frac{\partial}{\partial t} - D\nabla_r^2 + \frac{1}{\tau_{\mathrm{D}}}\right) P_{\mathrm{diff}}^{BB'} = \delta(\boldsymbol{r} - \boldsymbol{r}')\delta(t - t'). \tag{4.58}$$

(Note that even though we derived this for a translationally invariant system by moving into the momentum representation, Eq. (4.58) is more general.) For $B = B'$, $\tau_D^{-1} = 0$, and the diffuson obeys the diffusion equation: it is not affected by the magnetic field. The equation is only valid provided the time difference $t - t'$ is much longer than τ and the distance $|\boldsymbol{r} - \boldsymbol{r}'|$ is much longer than the mean free path.

One can also account for spin-orbit scattering [97]. In this case one has four diffusons: a singlet one, with time $\tau_{\mathrm{D}}^{-1} = e^2a^2(B - B')^2/(12c^2)$, and three identical triplet ones, with τ_{D}^{-1} replaced by $\tau_D^{-1} + \tau_{\mathrm{SO}}^{-1}$, see Eq. (4.22).

It is clear what the meaning of diffuson is from Eq. (4.58) – it is the probability of finding a particle at point $\boldsymbol{r}$ at time t provided it was at point $\boldsymbol{r}'$ at time t'. Our ladder diagrams show how it occurs at the microscopic level. Indeed, we saw that the average Green's function is short-ranged and decays exponentially, provided the distance between the end points is greater than the mean free path l. However, the sum of ladder diagrams is long-ranged: diffusive motion occurs due to multiple impurity scattering. The product of two averaged

Green's functions cannot produce such an object. Note that this conclusion only makes sense at long spatial scales, much longer than the mean free path ($ql \ll 1$). At shorter scales, the diffusion approximation does not apply.

One can also define a *cooperon* as follows:

$$P_{\text{coop}}^{BB'}(\boldsymbol{r},\boldsymbol{r}') = w\left\langle G^{\text{R}}(\boldsymbol{r},\boldsymbol{r}', E+\omega, B) G^{\text{A}}(\boldsymbol{r},\boldsymbol{r}', E, B')\right\rangle. \tag{4.59}$$

It obeys a similar equation, which we write again in the time representation (the cooperon depends on two times t, t' rather than on the frequency ω):

$$\left(\frac{\partial}{\partial t} - D\nabla_r^2 + \frac{1}{\tau_{\text{C}}}\right) P_{\text{coop}}^{BB'} = \delta(\boldsymbol{r}-\boldsymbol{r}')\delta(t-t'), \tag{4.60}$$

where the time τ_{C} is obtained from τ_{D} by the substitution $B - B' \to B + B'$. This means that for $B = B'$ it survives; for example, for the same wire geometry, we have $\tau_{\text{C}}^{-1} = \tau_{\text{H}}^{-1} \equiv e^2a^2B^2/3c^2$. For $B = B' = 0$, the cooperon obeys the diffusion equation and is identically equal to the diffuson. In contrast to the diffuson, the cooperon is sensitive to the time-reversal symmetry and is suppressed by the applied magnetic field. For the record, we give here the equation for the cooperon in both position- and time-dependent fields, described by the same vector potential $\boldsymbol{A}$:

$$\left(\frac{\partial}{\partial t} + D\left(\mathrm{i}\nabla_r + \frac{e}{c\hbar}\boldsymbol{A}(\boldsymbol{r},t) + \frac{e}{c\hbar}\boldsymbol{A}'(\boldsymbol{r},t')\right)^2\right) P_{\text{coop}}^{AA} = \delta(\boldsymbol{r}-\boldsymbol{r}')\delta(t-t'). \tag{4.61}$$

If spin-orbit scattering is present, the structure of the cooperon is similar to that of the diffuson.

It is also important that both the diffuson and the cooperon contain the average of the product of advanced and retarded Green's functions. It is easy to check that other averages – of the type $\langle G^{\text{R}}G^{\text{R}}\rangle$ or $\langle G^{\text{A}}G^{\text{A}}\rangle$ – are essentially equal to the products of averages, $\langle G^{\text{R}}G^{\text{R}}\rangle \approx \langle G^{\text{R}}\rangle\langle G^{\text{R}}\rangle$, and therefore decay exponentially at scales longer than the mean free path.

4.4.2 Quantum corrections to the effective action

Let us now return to the general matrix case and consider the Green's functions of an arbitrary additional matrix structure with N_{ch} indices, and later on apply this general treatment to various problems of quantum transport, as in Chapter 2. This action has cooperon and diffuson contributions to the action which are representing G_{Q} corrections. These contributions are given by wheel diagrams, made up of either cooperon or diffuson ladder sections in a straightforward way (see Fig. 4.8).

We now introduce the *operators* $\hat{K}$ presenting a section of a corresponding ladder (wJ in Section 4.4.1) as follows:

$$\hat{K}_{\text{diff}}^{ab,cd}(\boldsymbol{r},\boldsymbol{r}') = w(\boldsymbol{r})\langle G_{\text{b}}^{ac}(\boldsymbol{r},\boldsymbol{r}')\rangle\langle G_{\text{w}}^{db}(\boldsymbol{r},\boldsymbol{r}')\rangle, \tag{4.62}$$

$$\hat{K}_{\text{cooper}}^{ab,cd}(\boldsymbol{r},\boldsymbol{r}') = w(\boldsymbol{r})\langle G_{\text{b}}^{ac}(\boldsymbol{r},\boldsymbol{r}')\rangle\langle G_{\text{w}}^{bd}(\boldsymbol{r},\boldsymbol{r}')\rangle, \tag{4.63}$$

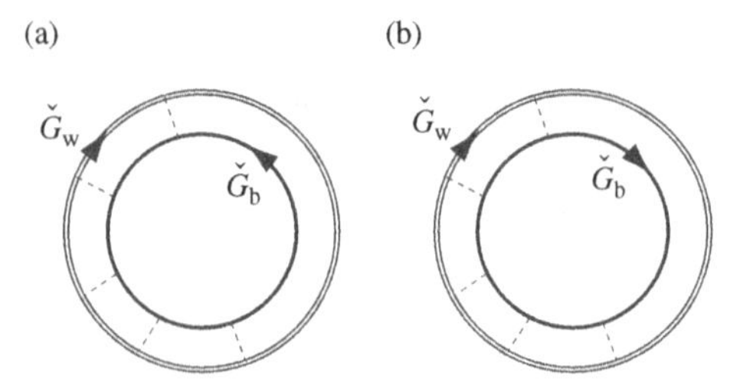

Fig. 4.8. Wheel diagrams that determine G_Q corrections to the action. A single (double) line represents the Green's functions from the black (white) block, while dashed lines represent the averaging over disorder. The diffuson (a) and cooperon (b) wheels differ by mutual orientation of the lines.

where Latin letters represent "check" indices (including the parameters and energies). Now the "white" Green's function is transposed for the cooperon. Diffuson and cooperon sections are operators in the space spanned by the coordinates and the two check indices.

Summing up all the diagrams, we find the formal operator expressions for the contributions to the action [98]. One contribution corresponds to fluctuations and is given by

$$\mathcal{S}_{G_Q} = \mathrm{Tr}\left[\ln(1 - \hat{K}_{\mathrm{coop}})\right] + \mathrm{Tr}\left[\ln(1 - \hat{K}_{\mathrm{diff}})\right]. \tag{4.64}$$

Another contribution, corresponding to the weak localization correction, takes into account the fact that the last ladder section is twisted. We do this by introducing the permutation operator $\hat{P}$, which exchanges the "check" indices,

$$\hat{P}\,\hat{K}^{ab,cd} = \hat{K}^{ba,cd}. \tag{4.65}$$

The corresponding contribution to the action becomes

$$\mathcal{S}_{\mathrm{WL}} = \frac{1}{2}\mathrm{Tr}\left[\hat{P}\ln(1 - \hat{K}_{\mathrm{coop}})\right]. \tag{4.66}$$

The factor 1/2 is included in Eq. (4.66) to take into account the fact that "black" and "white" Green's functions, in the case of weak localization, are no longer independent. We note that $\hat{K}$ for the cooperon is symmetric with respect to index exchange, so that $\hat{K}$ and $\hat{P}$ commute. The eigenfunctions and the eigenvalues of $\hat{K}$ are therefore either symmetric (K^+) or anti-symmetric (K^-) with respect to permutations. We can rewrite Eq. (4.66) as a sum over these eigenvalues:

$$\mathcal{S}_{\mathrm{WL}} = \frac{1}{2}\sum_n \left[\ln(1 - K_n^+) - \ln(1 - K_n^-)\right]. \tag{4.67}$$

In order to calculate the G_Q corrections, one has to evaluate the eigenvalues of the ladder section $\hat{K}$, both for the cooperon and the diffuson. We introduce a method to compute this matrix easily. The observation is that the ladders under consideration are not specific for G_Q corrections: the same ladders determine a response of *semiclassical* Green's functions upon variation of $\check{\Sigma}$.

To see this, let us go back to non-averaged Green's functions. We keep in mind that we have doubled the "check" space to include the white and black sectors. We add by hand a source term, the self-energy, which mixes up black and white Green's functions, $\delta\check{\Sigma}_{\mathrm{bw}}(\boldsymbol{r})$.

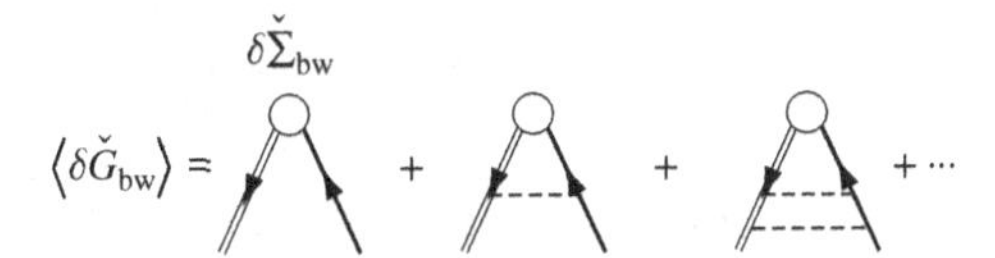

Fig. 4.9. **The response of Green's function $\delta\check{G}_{\rm bw}$ on self-energy $\check{\Sigma}_{\rm bw}$ in a semiclassical non-crossing approximation is determined by the sum of the ladder diagrams. This allows us to express the wheel diagrams in terms of the eigenvalues of the response kernels.**

This source term gives rise to a correction to the Green's function in the same black–white sector. In the first order, we have

$$\delta\check{G}_{\rm bw}(\boldsymbol{r},\boldsymbol{r}') = -\int {\rm d}\boldsymbol{r}_1\,{\rm d}\boldsymbol{r}_2\,\check{G}_{\rm b}(\boldsymbol{r},\boldsymbol{r}_1)\delta\check{\Sigma}_{\rm bw}(\boldsymbol{r}_1,\boldsymbol{r}_2)\check{G}_{\rm w}(\boldsymbol{r}_1,\boldsymbol{r}'), \tag{4.68}$$

which is best illustrated by the diagram in Fig. 4.9. The next step is to include the effect of the random potential $U(\boldsymbol{r})$. We average Eq. (4.68), discarding all diagrams with crossed dashed lines (non-crossing approximation) and obtaining a set of ladder diagrams (Fig. 4.9.) By summing up all the contributions, we obtain the correction – taken in coinciding points – to the Green's function:

$$\langle\delta\check{G}_{\rm bw}(\boldsymbol{r},\boldsymbol{r})\rangle = \frac{1}{w(\boldsymbol{r})}\frac{\hat{K}_{\rm diff}}{1-\hat{K}_{\rm diff}}\delta\check{\Sigma}_{\rm bw}\,(\boldsymbol{r}). \tag{4.69}$$

Equation (4.69) is very valuable: it demonstrates that the response of the Green's function to the source term $\delta\check{\Sigma}_{\rm bw}$ is determined by the same ladder operator $\hat{K}$, which we need to compute the $G_{\rm Q}$ corrections.

At the space scale of isotropization length (mean free path), one has $\hat{K}\sim 1$. Usually one is interested in the contribution arising from the larger space scale where a cooperon–diffuson approximation is valid. At this scale, the eigenvalues of $\hat{K}$ are either zero, or very close to unity. To see this, we cite the results for the homogeneous case with $\boldsymbol{r}$-independent $\check{\Sigma}$. A convenient basis in "check" space is one where $\check{\Sigma}$ is diagonal, the eigenvalues being Σ_n. The Green's function is diagonal in this basis as well, $G_n = s_a \equiv$ sign Im Σ_a. Due to homogeneity, the section operator is diagonal in the wave vector representation, with eigenvalues equal to $K^{nm}(\boldsymbol{q})$. Direct calculation yields $K^{nm}(\boldsymbol{q}) = 0$ provided $s_n = s_m$. For $s_n \neq s_m$,

$$1 - K^{ab}(\boldsymbol{q}) \approx \tau\left[{\rm i}s_b(\Sigma_a - \Sigma_b) + D\boldsymbol{q}^2\right] + \cdots \tag{4.70}$$

for $\Sigma\tau, ql \ll 1$. This equation forms the relation between our technique and the common technique for cooperons and diffusons in homogeneous media, as outlined above.

Now we note that zero eigenvalues contribute neither to Eq. (4.64) nor to Eq. (4.69). As for those close to unity, we may replace $\hat{K}$ by 1 in the numerator of Eq. (4.69). We also note that $\delta\check{G}$ can be presented as the derivative of the action Therefore, we can write the $G_{\rm Q}$ corrections due to the diffuson modes to the action in terms of a determinant comprising derivatives of the semiclassical action with respect to $\check{\Sigma}_{\rm bw}$, $\check{\Sigma}_{\rm wb}$:

$$\mathcal{S}_{G_Q,\text{diff}} = -\ln\det{}'\left(-w(\boldsymbol{r})\frac{\delta^2\mathcal{S}}{\delta\check{\Sigma}_{\text{wb}}\delta\check{\Sigma}_{\text{bw}}}\right). \tag{4.71}$$

The "prime" sign of the determinant signals that the zero eigenvalues are excluded: det′ is defined as the product of all non-zero eigenvalues. Indeed, as we have seen, some variations of self-energies do not change the Green's functions, giving rise to zero eigenvalues. We also note that the G_Q corrections are not affected by the specific form of $w(\boldsymbol{r})$: since the determinant of a product of two matrices is a product of their determinants, this matrix gives a constant contribution to the action that does not affect any physical quantities.

4.4.3 Finite-element approximation

Equation (4.71) is microscopic and holds in all cases. We now adapt it to the finite-element approach outlined in Chapter 2. We replace the actual $\boldsymbol{r}$-dependent matrices $\check{G}$ and $\check{\Sigma}$ by constants in each node. To obtain the action in these terms, one integrates over the volume of each node so that the correction to the action is given by

$$\delta\mathcal{S} = -\sum_{\alpha}\frac{i\pi}{\delta_\alpha}\,\text{Tr}\left[\check{G}_\alpha\delta\check{\Sigma}_\alpha\right], \tag{4.72}$$

where the summation is over the nodes, and $\delta_\alpha = (\nu\mathcal{V}_\alpha)^{-1}$ is the mean level spacing in the node. The discrete analog of the determinant relation, Eq. (4.71), is now given by

$$\begin{aligned}\mathcal{S}_{G_Q,\text{diff}} = &-\ln\det{}'\left(-\frac{w_\alpha}{\mathcal{V}_\alpha}\frac{\delta^2\mathcal{S}}{\delta\check{\Sigma}_{\text{wb}}\delta\check{\Sigma}_{\text{bw}}}\right)\\ &-\ln\det{}'\left(\frac{\pi}{2\tau_\alpha\delta_\alpha}\frac{\delta^2\mathcal{S}}{\delta\check{\eta}_{\text{wb}}\delta\check{\eta}_{\text{bw}}}\right) = \text{const.} - \ln\det{}'\left(\frac{\delta^2\mathcal{S}}{\delta\check{\eta}_{\text{wb}}\delta\check{\eta}_{\text{bw}}}\right),\end{aligned} \tag{4.73}$$

where we have introduced the dimensionless response matrix $\check{\eta}_\alpha \equiv i\pi\check{\Sigma}/\delta_\alpha$. The response matrix is determined from the solution of the Kirchhoff equations at the vanishing source term $\check{\eta}_\alpha^{\text{bw}}$. It has $N_{\text{nodes}}\cdot N_{\text{ch}}^2/2$ non-zero eigenvalues and the same number of zero eigenvalues. We observe that at $\Sigma_{\text{w,b}} = 0$ the eigenvalues of this matrix do not depend on the volume of the nodes $\mathcal{V}_\alpha$, rather they are determined by the transmission eigenvalues of the connectors only, and are of the order of G/G_Q. Since rescaling of all conductances gives only an irrelevant constant contribution to the action, the G_Q corrections depend only on the ratios of the conductances of the connectors: this brings about the universality of these corrections.

The circuit theory action is given in terms of $\check{G}_\alpha$. It is advantageous to present the answer for $\mathcal{S}_{G_Q}$ in terms of the expansion coefficients of the action around the saddle-point – the solution of the semiclassical circuit-theory equations, that is, to use $\delta^2\mathcal{S}/\delta\check{G}_{\text{wb}}\delta\check{G}_{\text{bw}}$ instead of $\delta^2\mathcal{S}/\delta\check{\eta}_{\text{wb}}\delta\check{\eta}_{\text{bw}}$. If the latter matrix were invertible, we would make use of the fact that $\delta^2\mathcal{S}/\delta\check{G}_{\text{wb}}\delta\check{G}_{\text{bw}} = (\delta^2\mathcal{S}/\delta\check{\eta}_{\text{wb}}\delta\check{\eta}_{\text{bw}})^{-1}$. In fact, due to the constraint $\check{G}^2 = 1$, there is a large number of zero eigenvalues in the response matrix, and the task in hand is not completely trivial.

We proceed as follows. We expand the action by replacing each $\check{G}$ in each node by

$$\check{G} = \check{G}_0 + \check{g} - \check{G}_0\check{g}^2/2 + \cdots \tag{4.74}$$

and collecting terms of the second order in $\check{g}$ and $\check{\eta}$. Equation (4.74) satisfies the constraint $\check{G}^2 = 1$ up to second-order terms, provided $\check{g}\check{G}_0 + \check{G}_0\check{g} = 0$. Let us work in the $N_{\rm ch}^2 N_{\rm nodes}$–dimensional space indexed with the "bar" index $\bar{a}$ composed of two "check" indices and one node index, $\bar{a} \equiv (a, b, \alpha)$. We present the result of the expansion as follows:

$$\delta\mathcal{S} = g_{\bar{a}}^{\rm wb} M_{\bar{a}\bar{b}} g_{\bar{b}}^{\rm bw} - \eta_{\bar{a}} g_{\bar{a}}^{\rm bw}. \tag{4.75}$$

The variation of Eq. (4.75) under the constraint $\check{g}\check{G}_0 + \check{G}_0\check{g} = 0$ gives the response matrix $\delta^2\mathcal{S}/\delta\eta_{\bar{a}}\delta\eta_{\bar{a}}$. Next we consider the matrix $\Pi_{\bar{a}\bar{b}}$ defined through the following relation:

$$\Pi_{\bar{a}\bar{b}} g_{\bar{b}}^{\rm bw} \rightarrow \frac{1}{2}\left(\check{g} - \check{G}_0\check{g}\check{G}_0\right) \rightarrow \frac{1}{2}\left(\check{g}^{\rm bw} - \check{G}_{\rm b}\check{g}^{\rm bw}\check{G}_{\rm w}\right); \tag{4.76}$$

the final equation makes the white–black block separation explicit. We note that $\Pi_{\bar{a}\bar{b}}$ is a *projector*: it separates "bar" space on two subspaces where $\check{g}$ either commutes or anticommutes with $\check{G}_0$, and projects an arbitrary $\check{g}$ onto the anticommuting subspace. Applying this projector to Eq. (4.75), we show that the projected matrix $\Pi_{\bar{a}\bar{b}} M_{\bar{b}\bar{c}} \Pi_{\bar{c}\bar{b}}$ is an inverse of the response matrix *within* the anticommuting subspace. Thus,

$$\mathcal{S}_{G_{\rm Q}} = \ln\det{}'\left(\Pi_{\bar{a}\bar{b}} M_{\bar{b}\bar{c}} \Pi_{\bar{c}\bar{b}}\right) = \ln\det\left(\Pi_{\bar{a}\bar{b}} M_{\bar{b}\bar{c}} \Pi_{\bar{c}\bar{b}} + \delta_{\bar{a}\bar{b}} - \Pi_{\bar{a}\bar{b}}\right). \tag{4.77}$$

In the final part of Eq. (4.77) we added the matrix $1 - \hat{\Pi}$. This procedure replaces all zero eigenvalues with 1, so that one can evaluate a usual determinant.

We recall that, as far as fluctuations are concerned, there are two contributions of this sort coming from diffuson and cooperon ladders, respectively. The weak localization correction involves the permutation operator, which sorts out eigenvalues involved according to Eq. (4.67). With this, Eqs. (4.77) and (4.73) give the $G_{\rm Q}$ corrections in an arbitrary circuit-theory setup in the most general form.

Up to now we have assumed that the Hamiltonian commutes with the "check" structure and is invariant with respect to time reversal. This implies strict coherence of waves with different "check" indices, which propagate in the disordered media described by this Hamiltonian. Even small "check"-dependent perturbations of the symmetric Hamiltonian give accumulating phase shifts to these waves and may significantly change their interference patterns at long distances. Due to its random nature, such phase shifts can be regarded as decoherence, although this should not be confused with a real decoherence coming from interaction-driven inelastic processes, treated in Chapter 6. In real experimental situations, two sources of such decoherence are usually of importance – spin-orbit scattering and the magnetic field, corresponding to unitary and symplectic ensembles in the RMT language (Section 4.3).

The most convenient way to incorporate magnetic field and spin-orbit scattering into our scheme is to present them as perturbative corrections to $\check{G}$-dependent action (Fig. 4.10).

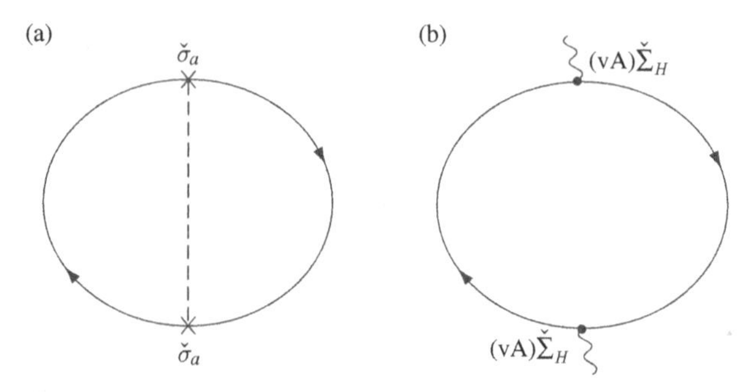

Fig. 4.10. Second-order diagrams in spin-orbit interaction (a) and magnetic field (b) provide decoherence terms in the action. These terms describe transitions between pure RMT ensembles.

The spin-orbit scattering enters the Hamiltonian in the form $\check{H}_{SO} = \hat{\sigma}_a H^a(\boldsymbol{r},\boldsymbol{r}')$, $H^a(\boldsymbol{r},\boldsymbol{r}') = -H^a(\boldsymbol{r}',\boldsymbol{r})$, $\hat{\sigma}_a$ representing spin Pauli matrices in "check" space. In second order in H^a, the averaging gives (Fig. 4.10)

$$\mathcal{S}_{SO} = \int d\boldsymbol{r} \, \frac{\pi \nu}{8\tau_{SO}(\boldsymbol{r})} \operatorname{Tr}\left[\check{G}(\boldsymbol{r})\check{\sigma}_a\check{G}(\boldsymbol{r})\check{\sigma}_a\right]. \tag{4.78}$$

At the level of microscopic approach, the spin-orbit scattering takes place anywhere in the nanostructure. In the finite-element approach, it is advantageous to ascribe spin-orbit scattering to nodes rather than to connectors. This is consistent with the main idea of our scheme: random phase shifts take place in the nodes. The spin-orbit contribution in each node α is obtained by integrating Eq. (4.78) over the node:

$$\mathcal{S}_{SO} = \frac{\eta_{SO}}{4} \operatorname{Tr}\left[\check{G}_\alpha\check{\sigma}_a\check{G}_\alpha\check{\sigma}_a\right], \tag{4.79}$$

where $\eta_{SO} \equiv \pi/2\tau_\alpha^{SO}\delta_\alpha$.

The magnetic field is incorporated into the Hamiltonian through the modification of the derivative,

$$H \to H - (e/c)(\boldsymbol{v}\boldsymbol{A})\Sigma_H$$

where $\boldsymbol{A}$ is the vector potential, $\boldsymbol{v}$ is the electron velocity, and $\check{\Sigma}_H$ ($\check{\Sigma}_H^2 = 1$) describes the interaction of different "check" waves with the magnetic field. In its simplest form, $\check{\Sigma}_H$ is the matrix in the white–black structure introduced such that $\check{\Sigma}_H^{\mathrm{b}} = 1$, $\check{\Sigma}_H^{\mathrm{w}} = -1$, *provided* we describe a cooperon. This is consistent with the requirement that one of the Hamiltonians must be transposed to describe a cooperon ladder. This is not the only plausible form of this matrix. For example, in non-equilibrium superconductivity $\check{\Sigma}_H$ involves an electron–hole Nambu structure.

The magnetic field decoherence contribution can also be assigned to a node and is given by

$$\mathcal{S}_H = \frac{\eta_H}{2} \operatorname{Tr}\left[\check{G}_\alpha\check{\Sigma}_H\check{G}_\alpha\check{\Sigma}_H\right], \tag{4.80}$$

where $\eta_{\mathrm{H}} = \pi/2\tau_{\mathrm{H}}\delta_\alpha$ and τ_{H} has been introduced in Section 4.2. It depends on the geometry of the node and its characteristics. By the order of magnitude, one has

$1/\tau_H\delta_S \simeq (\Phi/\Phi_0)^2(G_{\text{node}}/G_Q)$, where Φ is the magnetic flux through the node, $\Phi_0 \equiv \pi\hbar/e$ is the flux quantum, and G_{node} is a typical conductance of the node. The latter is limited by its Sharvin value in the ballistic regime, where the isotropization length is of the order of the node size.

To find the effect of the decoherence terms, Eqs. (4.79) and (4.80), on the eigenvalues forming the localization correction, we expand the action as it was done to obtain Eq. (4.75). The decoherence contribution to $\hat{M}$ is diagonal in the node index, and can be made diagonal in the "bar" index by a proper choice of the basis in "check" space. For instance, if no external spin polarization is present in the structure, the spin-orbit contribution is diagonal in the basis made of singlets and triplets in spin space. The simple realization of Σ_H mentioned is automatically diagonal. If, in addition, this diagonal contribution is the same in all nodes, both decoherence effects just shift the eigenvalues of $\hat{M}$ corresponding to the symmetric Hamiltonian. This provides us with an extremely convenient model of decoherence effects.

The action for fluctuations is modified as follows:

$$\mathcal{S}_{\text{diff}} = \sum_n \{\ln(M_n) + 3\ln(M_n + \eta_{\text{SO}})\}, \tag{4.81}$$

$$\mathcal{S}_{\text{coop}} = \sum_n \{\ln(M_n + \eta_H) + 3\ln(M_n + \eta_{\text{SO}} + \eta_H)\}, \tag{4.82}$$

where the summation runs over the non-degenerate eigenvalues of $\hat{M}$ and the factor 3 comes from three-fold degeneracy of the triplet (see Eq. (4.22)). To derive the modification for the weak localization contribution, we note that the singlet and the triplet are, respectively, antisymmetric and symmetric with respect to the permutation. Therefore, triplet extensions of symmetric and antisymmetric eigenvalues are, respectively, symmetric and antisymmetric. The weak localization correction is thus given by

$$\mathcal{S}^{\text{WL}} = \frac{1}{2}\sum_n \left\{\ln\frac{M_n^- + \eta_H}{M_n^+ + \eta_H} + \frac{3}{2}\ln\frac{M_n^+ + \eta_{\text{SO}} + \eta_H}{M_n^- + \eta_{\text{SO}} + \eta_H}\right\}, \tag{4.83}$$

$M^{+(-)}$ denoting the (anti)symmetric eigenvalues of $\hat{M}$.

Since the eigenvalues of $\hat{M}$ are of the order of G/G_{Q}, the decoherence effects become important at $\eta_{SO}, \eta_H \simeq G/G_{\text{Q}}$, that is when inverse decoherence times match the Thouless energy: $E_{\text{Th}} = (G/G_{\text{Q}})\delta_{\text{S}}$ of the node, $1/\tau_{\text{SO}}, 1/\tau_{\text{H}} \simeq E_{\text{Th}}$.

The magnetic field produces not only random, but also deterministic phase shifts, giving rise to the Aharonov–Bohm (AB) effect discussed in the following. Let us show how to incorporate the AB effect into our scheme. For simplicity, we now disregard all the orbital effects of the magnetic field and assume that only the topological effects – due to non-trivial geometry – are important. Consider a nanostructure in the shape of a closed ring threaded by a magnetic flux Φ. Neglecting orbital effects, one can get rid of the vector potential in the Schrödinger equation by a gauge transformation. Let us make an ideal cut in the nanostructure that breaks the loop (see Fig. 4.11). The topological effect of the flux can be incorporated into the boundary condition for the wave-function $\psi_{\text{L,R}}$ on two sides of the cut, $\psi_{\text{L}} = \exp(\text{i}\phi_{\text{AB}}\Sigma_H)\psi_{\text{R}}$. The phase of the wave function therefore

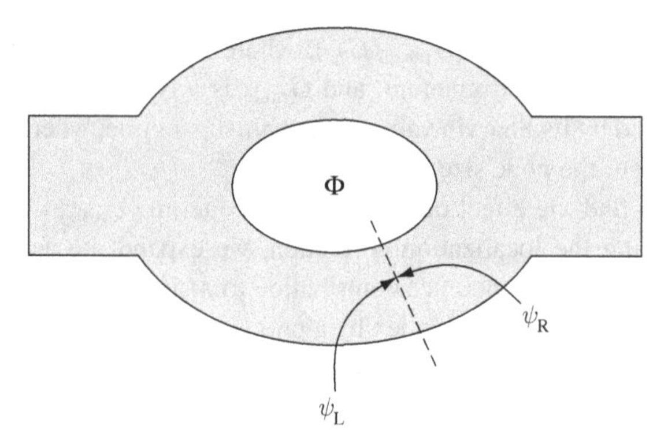

Fig. 4.11. **Aharonov–Bohm effect: the topological phase arising from the magnetic flux Φ is incorporated into the boundary conditions that relate the wave functions on two sides of an imaginary cut. The cut can be made in any place.**

presents a discontinuity at the cut that is equal to $\pm\phi_{\text{AB}}, \phi_{\text{AB}} = \pi\Phi/\Phi_0$. Since the transformation does not explicitly depend on the coordinate, it can be immediately extended to semiclassical Green's functions. These functions at two sides are related as follows:

$$\check{G}_{\text{L}} = \exp(\text{i}\phi_{\text{AB}}\check{\Sigma}_H)\check{G}_{\text{R}}\exp(-\text{i}\phi_{\text{AB}}\check{\Sigma}_H). \tag{4.84}$$

This solves the problem at the microscopic level. Once the nanostructure has been discretized to finite elements, we note that the cut always occurs *between* a connector and a node. The most convenient way to deal with the gauge transformation, Eq. (4.84), is to put it into the action of the corresponding connector. To do this, we observe that the Green's function at the right-hand end of the connector is not $\check{G}_{\text{R}}$ of the node anymore: since the cut is crossed, it is eventually $\check{G}_{\text{L}}$, given by Eq. (4.84). The connector action in the presence of flux is therefore given by

$$\mathcal{S}_c = \frac{1}{2}\sum_n \text{Tr}\ln\left[1 + \frac{T_n}{4}(\check{G}_{c1}\check{G}_{c2}(\phi_{\text{AB}}) + \check{G}_{c2}(\phi_{\text{AB}})\check{G}_{c1} - 2)\right], \tag{4.85}$$

with

$$\check{G}(\phi_{\text{AB}}) = \exp(\text{i}\phi_{\text{AB}}\check{\Sigma}_H)\check{G}\exp(-\text{i}\phi_{\text{AB}}\check{\Sigma}_H). \tag{4.86}$$

One checks that the variation of the action modified in such a way reproduces the Kirchhoff laws for matrix current. Due to global gauge invariance, it does not matter to which connector and to which end of the connector the A–B phase is ascribed. If there are more loops in the nanostructure, more connector actions have to be modified in this way.

Now, we have an action that contains all G_{Q} corrections. We still need to produce expressions for the transmission distributions $\rho(T)$ that eventually give weak localization correction to the conductance and conductance fluctuations. According to Section 2.6, the transmission distribution is given by

$$\rho(T) = -\rho_0(T)\,\text{Re}\left[\frac{\partial\mathcal{S}}{\partial\phi}\left(\pi + 2\text{i}\cosh^{-1}\frac{1}{\sqrt{T}} - 0\right)\right], \tag{4.87}$$

where the transmission distribution for the diffusive conductor $\rho_0(T)$ is given by Eq. (1.43). Classical circuit theory, developed in Section 2.6, provides the result in the limiting case $G \gg G_Q$. The weak localization contribution $\mathcal{S}_{\rm WL}$ provides the G_Q correction to the transmission distribution. The fluctuation contribution $\mathcal{S}_{G_Q} = \mathcal{S}_{\rm diff} + \mathcal{S}_{\rm coop}$ that depends on two parameters $\phi_{\rm w,b}$, gives *correlations* of transmission distributions:

$$\mathcal{S}_{G_Q}(\phi_{\rm b},\phi_{\rm w}) = \int dT\, dT' \langle\langle \rho(T)\rho(T')\rangle\rangle \times \ln\left[1 - T\sin^2\frac{\phi_{\rm b}}{2}\right]\ln\left[1 - T'\sin^2\frac{\phi_{\rm w}}{2}\right]. \tag{4.88}$$

A simple application of the above formulas are quantum corrections to the conductance. These are given by the derivatives of corresponding actions at $\phi_{\rm b,w} = 0$:

$$\frac{\delta G_{\rm WL}}{G_Q} = -2\left.\frac{\partial^2 \mathcal{S}_{\rm wl}(\phi)}{\partial\phi^2}\right|_{\phi_{\rm w}=-\phi_{\rm b}=\phi=0}; \tag{4.89}$$

$$\frac{\langle\langle G_{\rm b}G_{\rm w}\rangle\rangle}{G_Q^2} = 4\left.\frac{\partial^4 \mathcal{S}_{G_Q}(\phi_{\rm b},\phi_{\rm w})}{\partial\phi_{\rm b}^2\partial\phi_{\rm w}^2}\right|_{\phi_{\rm b},\phi_{\rm w}=0}. \tag{4.90}$$

Control question 4.7. Imagine we want to study the correlations of shot noise – a quantum correction, defined analogously to conductance fluctuations. What parts of the effective action should we use and how?

4.4.4 Application: junction chain

To compute the distribution of transmission eigenvalues in Section 2.6, we made use of 2×2 matrix voltages. Let us now apply our general expressions for the G_Q corrections to specific systems. We start with a chain of N identical tunnel junctions. As we remember from Section 2.6, in the limit of large N this chain is equivalent to a diffusive wire. In the semiclassical approximation, the "phase" ϕ drops by the same amount at each junction, and one eventually reproduces the transmission distribution $\rho_0(T)$.

To calculate the G_Q corrections, we augment this matrix by adding the "black" and "white" structure. The technical details of the calculation can be found in Ref. [98]. The parameter ϕ gets a "color" index "b" or "w". The semiclassical solution for the resulting 4×4 matrix is non-zero in bb and ww blocks:

$$\check{G}_k^0 = \begin{pmatrix} \check{G}(k\phi_{\rm b}/N) & 0 \\ 0 & \check{G}(k\phi_{\rm w}/N) \end{pmatrix}. \tag{4.91}$$

Now we need to derive the matrix $\hat{M}$, the eigenvalues of which determine the G_Q corrections. It is advantageous to use a parameterization of the deviations from the semiclassical solution, $\check{g}$, which automatically satisfy $\check{g}\check{G} + \check{G}\check{g} = 0$ in each node. To this end, we rewrite the semiclassical action in a special basis, where $\check{G}_k^0$ are diagonal in each node:

$$\check{G}_k^0 = \begin{pmatrix} 1 & 0 & 0 & 0 \\ 0 & -1 & 0 & 0 \\ 0 & 0 & 1 & 0 \\ 0 & 0 & 0 & -1 \end{pmatrix}. \tag{4.92}$$

Then the deviation of the form given by

$$\check{g}_k = \begin{pmatrix} 0 & 0 & 0 & g_{k,p}^{\mathrm{bw}} \\ 0 & 0 & g_{k,m}^{\mathrm{bw}} & 0 \\ 0 & g_{k,p}^{\mathrm{wb}} & 0 & 0 \\ g_{k,m}^{\mathrm{wb}} & 0 & 0 & 0 \end{pmatrix} \tag{4.93}$$

satisfies the above condition.

In this basis the semiclassical action is given by

$$\mathcal{S} = \frac{G_{\mathrm{T}}}{2G_{\mathrm{Q}}} \sum_{k=0}^{N-1} \mathrm{Tr}\left(\check{G}_k \check{L} \check{G}_{k+1} \check{L}^{-1}\right) - \frac{\mathrm{i}\pi}{\delta_{\mathrm{S}}} \sum_{k=1}^{N-1} \mathrm{Tr}\left(\check{\Sigma}_k \check{G}_k\right), \tag{4.94}$$

where the bb (ww) block of $\check{L}$ is given by

$$L^{\mathrm{bb(ww)}} = \begin{pmatrix} \cos(\phi^{\mathrm{b(w)}}/N) & \mathrm{i}\sin(\phi^{\mathrm{b(w)}}/N) \\ \mathrm{i}\sin(\phi^{\mathrm{b(w)}}/N) & \cos(\phi^{\mathrm{b(w)}}/N) \end{pmatrix}. \tag{4.95}$$

We expand the Green's matrices according to Eq. (4.74), write the quadratic form in terms of $\check{g}$, and diagonalize it to find the following set of eigenvalues, $l = 1, \ldots, N-1$:

$$\begin{aligned} \frac{4G_{\mathrm{Q}}}{G_{\mathrm{T}}} M_l^{\pm}(\phi_{\mathrm{b}}, \phi_{\mathrm{w}}) &= 2\cos\frac{\phi_{\mathrm{b}}}{2N}\cos\frac{\phi_{\mathrm{w}}}{2N}\cos\frac{\pi l}{N} - \cos\frac{\phi_{\mathrm{b}}}{N} \\ &- \cos\frac{\phi_{\mathrm{w}}}{N} \mp \sqrt{4\sin^2\frac{\phi_{\mathrm{b}}}{2N}\sin^2\frac{\phi_{\mathrm{w}}}{2N}\cos^2\frac{\pi l}{N} + \epsilon^2}, \end{aligned} \tag{4.96}$$

and $\epsilon \equiv 2\pi G_{\mathrm{Q}}(\Sigma_{\mathrm{b}} - \Sigma_{\mathrm{w}})/G_{\mathrm{T}}\mathrm{i}\delta_{\mathrm{S}}$ measures the difference of Green's function energy parameter in bb and ww blocks in units of a single-node Thouless energy. To obtain the eigenvalues that determine the weak localization contribution, we set $\phi_{\mathrm{w}} = -\phi_{\mathrm{b}} = \phi, \epsilon = 0$. This yields

$$M_{\mathrm{WL},l}^{+}(\phi) = \frac{G_{\mathrm{T}}}{2G_{\mathrm{Q}}}\cos\frac{\phi}{N}\left[\cos\frac{\pi l}{N} - 1\right], \tag{4.97}$$

$$M_{\mathrm{WL},l}^{-}(\phi) = \frac{G_{\mathrm{T}}}{2G_{\mathrm{Q}}}\left[\cos\frac{\pi l}{N} - \cos\frac{\phi}{N}\right]. \tag{4.98}$$

Let us discuss the weak localization correction first. If we disregard the decoherence factors, we can sum up over l to find a compact analytical expression:

$$\mathcal{S}_{\mathrm{WL}} = \frac{(N-2)}{2}\ln\left(\cos\frac{\phi}{N}\right) + \frac{1}{2}\ln\left(\frac{\sin 2\phi/N}{\sin\phi}\right). \tag{4.99}$$

In the limit $N \to \infty$ this becomes

$$\mathcal{S}_{\mathrm{WL}} = \frac{1}{2}\ln\left(\frac{\phi}{\sin\phi}\right). \tag{4.100}$$

It is interesting to note that the weak localization correction is absent for $N = 2$. This is a general property of a single-node tunnel-junction system.

Exercise 4.7. Does this property hold if a magnetic field or spin-orbit scattering is present?

A part of the weak localization correction in diffusive conductors is universal [99]: it depends neither on the shape nor on the dimensionality of the conductor. The universal part is concentrated near transmissions close to unity and is given by

$$\delta\rho_{\mathrm{WL}}(T) = -\frac{1}{4}\delta(T-1), \tag{4.101}$$

while the non-universal part is a smooth function of T. Equation (4.99) possesses this property at any N, since the universal part comes from the divergency in Eq. (4.99) at $\phi = \pi$, where the eigenvalue $M^{-}_{\mathrm{WL},1}$ goes to zero. Our approach proves that this correction is universal for a large class of the nanostructures, not limited to diffusive ones, for any nanostructure where transmission eigenvalues approach unity. This is guaranteed by the logarithmic form of the action. If $M^{-}_{\mathrm{WL},1} \propto (\pi - \phi)$ at $\phi \to \pi$, the correction is given by Eq. (4.101) irrespective of the proportionality coefficient.

Expanding Eq. (4.99) at $\phi \to 0$, we find the correction to the conductance of the tunnel junction chain:

$$\frac{\delta G_{\mathrm{WL}}}{G_{\mathrm{Q}}} = -\frac{1}{3}\frac{(N-1)(N-2)}{N^2}. \tag{4.102}$$

This is written for an orthogonal ensemble; for other pure ensembles the factor $(1 - 2/\beta)$ must be added. For a diffusive conductor, $N \to \infty$, Eq. (4.97) yields the weak localization correction $\delta G_{\mathrm{WL}} = -(1 - 2/\beta)(1/3)$, in accordance with the conclusions of random matrix theory (Section 4.3). The effect of spin and magnetic decoherence can be taken into account by shifting the eigenvalues, Eqs. (4.97) and (4.98), according to Eq. (4.83) since the decoherence factors $\eta_{H,\mathrm{SO}}$ in each node are the same. The correction to the transmission distribution is plotted in Fig. 4.12 for different strengths of spin-orbit coupling to illustrate the transition between orthogonal and symplectic ensembles.

Let us discuss the parametric correlations. Without decoherence factors and at the same energy ($\epsilon = 0$) one can still sum up over the modes to obtain an analytical expression:

$$\begin{aligned} \mathcal{S}_{\mathrm{diff}} = \mathcal{S}_{\mathrm{coop}} &= (N-1)\ln\left(\cos\frac{\phi_{\mathrm{b}}}{N} + \cos\frac{\phi_{\mathrm{w}}}{N}\right) \\ &+ \sum_{\pm}\ln\left(\sin\frac{\phi_{\mathrm{b}} \pm \phi_{\mathrm{w}}}{2} - \sin\frac{\phi_{\mathrm{b}} \pm \phi_{\mathrm{w}}}{2N}\right). \end{aligned} \tag{4.103}$$

The fluctuation of conductance obtained using Eq. (4.90) is given by

$$\frac{\langle\langle \delta G^2 \rangle\rangle}{G_{\mathrm{Q}}^2} = \frac{2}{15}\frac{N^4 + 15N - 16}{N^4}, \tag{4.104}$$

and converges to the expression $\langle\langle \delta G^2 \rangle\rangle = (2/15)G_{\mathrm{Q}}^2$ for $N \to \infty$, again in accordance with RMT. We note that this convergence is rather quick; the fluctuation at $N = 5$ differs

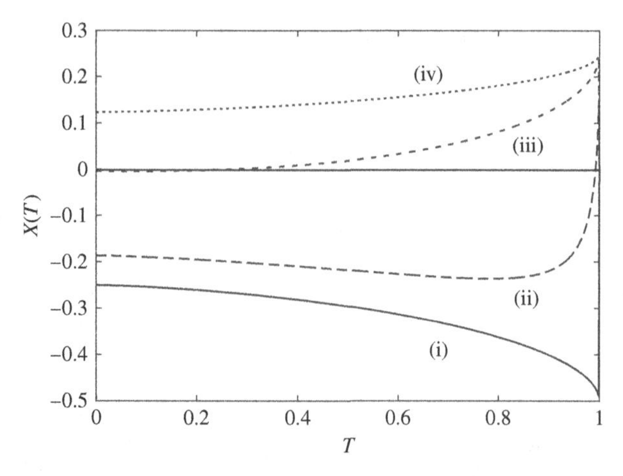

Fig. 4.12. Weak localization correction to the transmission distribution of a system of four identical junctions at different values of spin-orbit parameter η_{SO}. We plot the cumulate correction $X(T) \equiv \int_T^1 dT' \, T'\delta\rho(T')$; $X(1)$ represents the universal singular part of the correction (see Eq. (4.101)), while $X(0)$ gives the correction to the conductance. (i) The lowest curve corresponds strictly to zero η_{SO} and therefore represents a pure orthogonal ensemble. Its negative value at T=1 is partially compensated for by a positive non-universal contribution coming from $T \simeq 1$ so that the resulting correction to conductance $\delta G_{WL}/G_Q = X(0) \approx 0.2$. (ii), (iii) These curves correspond to relatively small values of η_{SO}, 0.05 and 0.4. While they are close to the orthogonal ensemble result at $T \simeq 1$, their behavior at $T \approx 1$ is quite different: the universal correction is that of a symplectic ensemble and is of positive sign. Curve (iv), corresponding to $\eta_{SO} = 10$, is close to the cumulate correction of pure symplectic ensemble, $X_{sym}(T) = -X_{ort}(T)/2$.

from asymptotic value by only 10%. We see thus that the diffusive wire, that in principle contains an infinite number of cooperon and diffuson modes, can be, with sufficient accuracy, described by the finite-element technique, even at a low number of elements.

Another point to discuss concerns the correlations of transmission eigenvalues T_n, which can be obtained by analytic calculation of Eq. (4.88). It is instructive to concentrate on relatively small eigenvalue separations, those much smaller than unity but still exceeding the average spacing (of order of G_Q/G) between the eigenvalues, $G_Q/G \ll |T - T'| \ll 1$. We observe that the correlation in this case is determined by the divergence of $\mathcal{S}$ at $\phi_b - \phi_w \to \pm 2\pi$. Indeed, M_1^- approaches zero in this limit. This again suggests the universality of these correlations. Indeed, this corresponds to the notion of RMT that, for diffusive conductors, the correlations in this parameter range are determined by universal Wigner–Dyson statistics and reduce to

$$\langle\langle \rho(T)\rho(T')\rangle\rangle = -\frac{2}{\pi^2\beta}\,\mathrm{Re}\left(\frac{1}{(T - T' + \mathrm{i}0)^2}\right). \tag{4.105}$$

Since the conductance fluctuations are contributed by correlations of T_n at scale of order 1 as well, they are not universal. We plot in Fig. 4.13 the correlator of conductance fluctuations as a function of energy difference at several N.

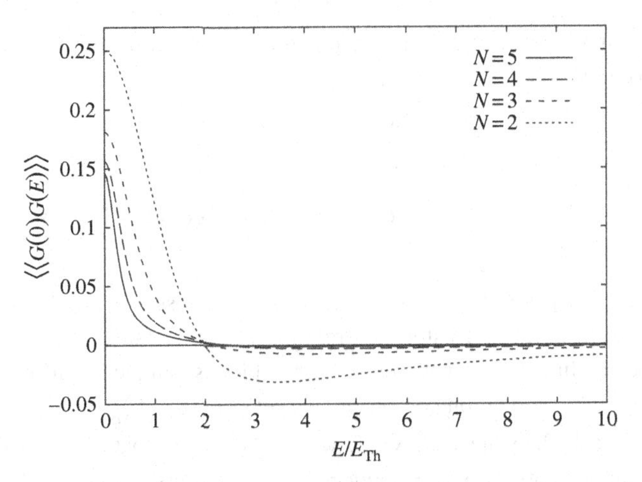

Fig. 4.13. Energy dependence of the correlator of conductance fluctuations $\langle\langle G(0)G(E)\rangle\rangle$ for chains with different numbers of junctions N. The energy difference is normalized to the Thouless energy of the whole chain, $E_{Th} \equiv \delta_S G_T/2\pi G_Q N^2$. Note the fast convergence of the correlator to that of diffusive wire for large N and negative correlations at large E.

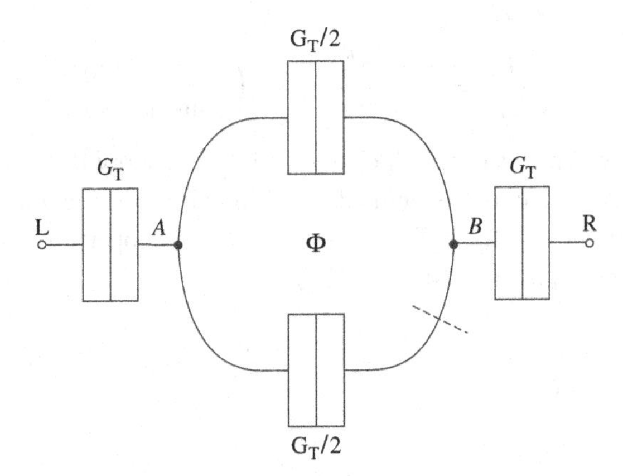

Fig. 4.14. A simple circuit to exemplify the AB effect consists of two nodes and four (tunnel) connectors. The AB phase modifies the Green's function on the tight side of the lowest connector with respect to $\check{G}_B$. The cut is given by the dashed line.

4.4.5 Example: Aharonov–Bohm ring

Now we exemplify evaluation of the AB effect within our scheme. We concentrate on the simple circuit presented in Fig. 4.14. It contains four tunnel junctions and two nodes labeled A and B. The conductances of the junctions are chosen as equal in order to re-use the results of the previous subsection for a chain of three tunnel junctions: the solution of semiclassical circuit theory equations is given by Eq. (4.91) for $N = 3$. The action is given by

$$S = \frac{G_T}{4G_Q} \,Tr\left\{\check{G}_L\check{G}_A + \frac{1}{2}\check{G}_A\check{G}_B + \frac{1}{2}\check{G}_A\check{G}_B(\phi_{AB}) + \check{G}_B\check{G}_R\right\} - i\frac{\pi}{\delta_S}\sum_{i=A,B}\mathrm{Tr}\,\check{\Sigma}_i\check{G}_i, \tag{4.106}$$

where $\check{G}_{L,R}$ are the Green's functions in the reservoirs and $\check{G}_B(\phi_{AB})$ is modified according to Eq. (4.84). To study correlation of conductance fluctuations, we consider different Green's functions for white and black blocks, subject to different fluxes $\phi^b_{AB} \neq \phi^w_{AB}$. For the weak localization correction, we set $\phi^b_{AB} = -\phi^w_{AB} = \phi_{AB}$.

To calculate the matrix $\hat{M}$, we again use the basis where $G_i^{(0)}$ are diagonal and the parameterization for $\check{g}$ introduced in the previous subsection. It is given by

$$M = \begin{pmatrix} M_d & M_{od} \\ M^*_{od} & M_d \end{pmatrix},$$

where 2×2 blocks $M_{d,od}$ are given by

$$\frac{G_T M_d}{4G_Q} = \begin{pmatrix} -\cos\phi_b - \cos\phi_w + \epsilon & 0 \\ 0 & -\cos\phi_b - \cos\phi_w - \epsilon \end{pmatrix},$$

$$\frac{G_T M_{od}}{4G_Q} = \frac{1 + e^{i(\phi^b_{AB} - \phi^w_{AB})}}{2}\begin{pmatrix} \cos\phi_b/2\cos\phi_w/2 & \sin\phi_b/2\sin\phi_w/2 \\ \sin\phi_b/2\sin\phi_w/2 & \cos\phi_b/2\cos\phi_w/2 \end{pmatrix}.$$

The parameter ϵ, which characterizes the energy difference between the black and white Green's function, is defined in the previous subsection. At $\epsilon = 0$, we obtain an explicit expression for the diffuson eigenvalues (the cooperon ones are obtained by the substitutions $\phi_w \to -\phi_w$ and $\phi^w_{AB} \to -\phi^w_{AB}$),

$$\frac{4G_Q}{G_T}M^+_{1,2} = -\cos\frac{\phi_b}{3} - \cos\frac{\phi_w}{3} \pm \cos\left(\frac{\phi^b_{AB} - \phi^w_{AB}}{2}\right)\cos\left(\frac{\phi_b - \phi_w}{6}\right),$$

$$\frac{4G_Q}{G_T}M^-_{1,2} = -\cos\frac{\phi_b}{3} - \cos\frac{\phi_w}{3} \pm \cos\left(\frac{\phi^b_{AB} - \phi^w_{AB}}{2}\right)\cos\left(\frac{\phi_b + \phi_w}{6}\right).$$

The weak localization correction to the action is given by

$$\mathcal{S}_{WL}(\phi) = \frac{1}{2}\ln\left(\frac{\cos^2(\phi/3)(4 - \cos^2(\phi_{AB}))}{4\cos^2(\phi/3) - \cos^2(\phi_{AB})}\right), \tag{4.107}$$

from which we calculate the correction to conductance as the function of the flux,

$$\frac{\delta G_{WL}}{G_Q} = -\frac{4\cos^2(\phi_{AB})}{9(7 - \cos(2\phi_{AB}))}. \tag{4.108}$$

We see that the weak localization correction cancels at half-integer flux $\phi_{AB} = \pi$. This is because the junctions forming the loop are taken to be identical. The flux dependence exhibits higher harmonics, indicating semiclassical orbits that encircle the flux more than once. We discuss this in more detail below for diffusive conductors.

For the correlation function responsible for conductance fluctuations we obtain

$$\frac{\langle \delta G^2 \rangle}{G_Q^2} = \sum_{\pm} \frac{259 - 4\cos(\phi_{AB}^{w} \pm \phi_{AB}^{b}) + \cos 2(\phi_{AB}^{w} \pm \phi_{AB}^{b})}{81(\cos(\phi_{AB}^{w} \pm \phi_{AB}^{b}) - 7)^2}, \tag{4.109}$$

where plus and minus signs indicate the cooperon and diffuson contributions, respectively. Higher harmonics are present as well.

4.4.6 Example: two connectors and one node

Probably the simplest system to be considered by circuit-theory methods consists of a single node and two connectors (Fig. 4.15). Since in this case there are only N_{ch} eigenvalues, one can straightforwardly elaborate on complicated arbitrary connectors. For this setup we are still able to find an analytical expression for cooperon and diffusion eigenvalues. This allows us to obtain an expression for the weak localization correction to the conductance which was vanishing in the case of two tunnel junctions. Each connector is, in principle, characterized by the distribution of transmission coefficients $\{T_n^R\}$, $\{T_n^L\}$, or, equivalently, by the functional form of the connector action, Eq. (2.91), $\mathcal{S}_L$ and $\mathcal{S}_R$. The action for the whole system is given by

$$\mathcal{S} = \mathcal{S}_L(\check{G}_L, \check{G}) + \mathcal{S}_R(\check{G}, \check{G}_R) - i\frac{\pi}{\delta_s}\mathrm{Tr}\,\check{\Sigma}\check{G}, \tag{4.110}$$

where $\check{G}$ is the Green's function of the node. We employ 2×2 matrices and set the Green's functions in the left and right reservoirs to $\check{G}(-\phi/2)$ and $\check{G}(\phi/2)$, respectively. The saddle-point value of $\check{G}$ is given by the phase χ. This phase, for the general choice of $\mathcal{S}_L$ and $\mathcal{S}_R$, does depend on ϕ, $\chi \equiv \chi(\phi)$. The total action in the saddle point is therefore $\mathcal{S}(\phi) = \mathcal{S}_L(\chi + \phi/2) + \mathcal{S}_R(\chi - \phi/2)$. We expand the Green's function according to Eq. (4.74). The second-order correction to the action in this case is given by

$$\begin{aligned}\mathcal{S}^{(2)} = &-i\frac{\pi}{\delta_s}\mathrm{Tr}\,\check{\Sigma}\,\check{g} - \frac{1}{2}\sum_n \mathrm{Tr}\Bigg\{\frac{T_n^L}{4 + T_n^L(\{\check{G}^0, \check{G}_L\} - 2)}\check{g}^2\check{G}^0\check{G}_L \\ &+ \left[\frac{T_n^L}{4 + T_n^L(\{\check{G}^0, \check{G}_L\} - 2)}\right]^2 (\check{g}\check{G}_L\check{g}\,\check{G}_L + \check{g}^2) + (L \leftrightarrow R)\Bigg\}.\end{aligned} \tag{4.111}$$

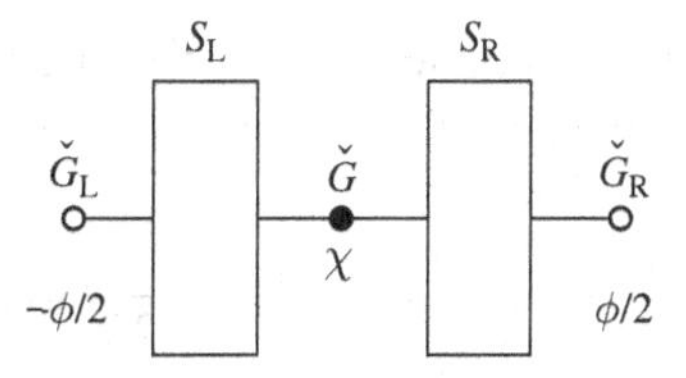

Fig. 4.15. **The simplest possible circuit comprises one node, two connectors (L and R), and two reservoirs where Green's functions $\check{G}_{L,R}$ are fixed.**

As in the previous subsections, we find two diffuson eigenvalues (the cooperon ones are obtained by the substitution $\phi^{\rm w} \to -\phi^{\rm w}$) as follows:

$$
\begin{aligned}
M^{\pm}(\phi_{\rm b}, \phi_{\rm w}) = & \sum_{i={\rm b,w}} I(\phi^i)\left[\cot\left(\chi(\phi^i) - \frac{\phi^i}{2}\right) - \cot\left(\chi(\phi^i) + \frac{\phi^i}{2}\right)\right] \\
& - \frac{1}{2}\left(1 - \cos\left[\chi(\phi_{\rm b}) + \frac{\phi_{\rm b}}{2} \mp \chi(\phi_{\rm w}) \mp \frac{\phi_{\rm w}}{2}\right]\right) \\
& \times \sum_{i={\rm b,w}} \frac{1}{\sin^2(\chi(\phi^i) + \phi^i/2)}\left[\frac{I'(\phi^i)}{\chi'(\phi^i) + 1/2} - I(\phi^i)\cot\left(\chi(\phi^i) + \frac{\phi^i}{2}\right)\right] \\
& + \frac{1}{2}\left(1 - \cos\left[\chi(\phi_{\rm b}) - \frac{\phi_{\rm b}}{2} \mp \chi(\phi_{\rm w}) \pm \frac{\phi_{\rm w}}{2}\right]\right) \\
& \times \sum_{i={\rm b,w}} \frac{1}{\sin^2(\chi(\phi^i) - \phi^i/2)}\left[\frac{I'(\phi^i)}{\chi'(\phi^i) - 1/2} - I(\phi^i)\cot\left(\chi(\phi^i) - \frac{\phi^i}{2}\right)\right].
\end{aligned}
\tag{4.112}
$$

Here we introduced $I(\phi) \equiv \partial \mathcal{S}/\partial \phi$ to characterize the derivative of the total semiclassical action. We see that M^- approaches zero in the limit $\phi_{\rm b} \to \pm\pi$, $\phi_{\rm w} \to \mp\pi$, provided $I(\phi)$ stays finite. As discussed, this divergency guarantees the universality of the correlations of the transmission eigenvalues. In the following, we specify to two different cases.

Symmetric setup

If we set $\mathcal{S}_{\rm R} = \mathcal{S}_{\rm L}$, $\chi(\phi_{\rm b(w)})$ is zero regardless of the concrete form $\mathcal{S}_{\rm L}$. The total action therefore reads $\mathcal{S}(\phi) = 2\mathcal{S}_{\rm L}(\phi/2)$. The eigenvalues in Eq. (4.113) take a simpler form. To compute the weak localization correction to the action we set $\phi^{\rm b} = -\phi^{\rm w} = \phi$ to find

$$
\mathcal{S}_{\rm WL} = \frac{1}{2}\ln\left(\frac{I'(\phi)}{I(\phi)}\tan\frac{\phi}{2}\right) + \text{const.} \tag{4.113}
$$

For tunnel junctions, $I(\phi) \propto \sin(\phi/2)$ and the correction disappears. Expanding Eq. (4.113) in a Taylor series near $\phi \to 0$ we find

$$
\frac{\delta G_{\rm WL}}{G_{\rm Q}} = -\frac{t_2}{4t_1}, \quad t_p = \sum_n T_n^p\,. \tag{4.114}
$$

A similar expansion of diffusion and cooperon eigenvalues yields

$$
\frac{\langle(\delta G)^2\rangle}{G_{\rm Q}^2} = \frac{3t_2^2 + 2t_1^2 - 2t_1(t_2 + t_3)}{8t_1^2}\,. \tag{4.115}
$$

It is instructive for understanding the circuit theory of $G_{\rm Q}$ corrections to elaborate on Eq. (4.113) for the specific case of diffusive connectors. Since in this case $I(\phi) \propto \phi$, we obtain

$$
\mathcal{S}_{\rm WL,node} = \frac{1}{2}\left[\ln(\tan(\phi/2)) - \ln\phi\right]. \tag{4.116}
$$

A two-connector, single-node situation can be easily realized in a quasi-one-dimensional wire with an inhomogeneous resistivity distribution along the wire. A low-resistivity region would make a node if bounded by two shorter resistive regions comprising the connectors. On the other hand, the existing literature shows that the weak localization correction in inhomogeneous wires does not depend on the resistivity distribution. Therefore, it has to be universally given by Eq. (4.100), $\mathcal{S}_{\mathrm{WL,1d}} = (1/2)\ln(\phi/\sin\phi) \neq \mathcal{S}_{\mathrm{WL,node}}$. How should this apparent discrepancy be understood?

This illustrates a very general point: G_{Q} corrections may be accumulated at various space scales ranging from mean free path to sample size. The experimental observation of the corrections relies on the ability to separate the contributions coming from different scales, for example by changing the magnetic field. With our approach, we evaluate the part coming from the interference at the scale of the node. The part coming from the interference at shorter scales associated with the connectors is assumed to be included into the transmission distribution of these connectors.

For our particular setup, this extra contribution comes from two identical connectors. Since only half of the phase ϕ drops at each connector, the contribution equals $2\mathcal{S}_{\mathrm{WL,1d}}(\phi/2)$. Summing up both contributions, we obtain

$$\mathcal{S}_{\mathrm{WL,node}} + 2\mathcal{S}_{\mathrm{WL,1d}}(\phi/2) = \mathcal{S}_{\mathrm{wl,tot}} = \mathcal{S}_{\mathrm{wl,1d}}(\phi).$$

That is, the weak localization correction in this case remains universal, provided the contribution of the node is augmented by the contributions of two connectors.

Control question 4.8. What do we obtain for the weak localization correction for the node between two quantum point contacts?

Non-ideal quantum point contact

The transmission distribution of an ideal multi-mode quantum point contact with conductance $G_B \gg G_{\mathrm{Q}}$ is highly degenerate since all $T_n = 1$ or 0. This degeneracy is lifted if the QPC is adjacent to a disordered region, even if the scattering in this region is weak. This can be modeled as a connector with conductance $G_{\mathrm{D}} \gg G_B$ in series.

Below, we show that the weak localization correction to the conductance has the order of $\delta G_{\mathrm{WL}} = -G_{\mathrm{Q}}(G_B/G_{\mathrm{D}})$, i.e. it is small compared to G_{Q}. The usual way to verify the applicability of the semiclassical approach to quantum transport is to compare the conductance of a nanostructure with the weak localization correction to it. For a generic nanostructure, this gives $G \gg G_Q$. However, for our particular example, one has $\delta G_{\mathrm{WL}} \ll G_B$ even for a few-channel QPC with $G_B \simeq G_Q$. Thus, one may ask the following question: is the semiclassical approach really valid at $G_B \simeq G_{\mathrm{Q}}$?

To produce an answer, we compute the weak localization correction to the transmission distribution. Since the system is not symmetric, we make use of Eq. (4.113). In the limit of $G_{\mathrm{D}} \gg G_{\mathrm{B}}$, the relevant values of ϕ are close to π. We stress it by shifting the phase, $\mu = \pi - \phi, |\mu| \ll 1$.

The circuit-theory analysis in the semiclassical limit yields

$$I(\mu) = G_{\rm D}\left(-\frac{\mu}{2}+\sqrt{\frac{\mu^2}{4}+R_{\rm c}}\right), \qquad \chi = \frac{\pi}{2}+\sqrt{\frac{\mu^2}{4}+R_{\rm c}},$$

where $R_{\rm c} \equiv 4G_B/G_{\rm D} \ll 1$. This gives to the following distribution of reflection coefficients (see Eq. (4.87)):

$$\rho(R) = \frac{G_D}{2\pi G_{\rm Q}}\theta(R_{\rm c}-R)\sqrt{\frac{R_{\rm c}}{R}-1}.$$

We use the above relations with Eq. (4.113) to find the cooperon eigenvalues:

$$M^+(\mu) = -G_{\rm D}\left(-\frac{1}{2}+\frac{\mu/4}{\sqrt{\mu^2/4+R_{\rm c}}}\right)\frac{\mu^2/4+R_{\rm c}}{R_{\rm c}/2}, \tag{4.117}$$

$$M^-(\mu) = G_{\rm D}\left(-\frac{\mu}{2}+\sqrt{\frac{\mu^2}{4}+R_{\rm c}}\right)\frac{\mu}{2R_{\rm c}}. \tag{4.118}$$

This yields the weak localization correction to the current,

$$I_{\rm WL}(\mu) = \frac{2R_{\rm c}}{\mu(\mu^2+4R_{\rm c})}.$$

The resulting correction to the transmission distribution consists of two delta-functional peaks of opposite sign, located at the edges of the semiclassical distribution:

$$\delta\rho_{\rm WL}(R) = \frac{1}{4}\left[\delta(R-R_{\rm c})-\delta(R)\right]. \tag{4.119}$$

In particular, the weak localization correction is $\delta G = -G_{\rm Q}G_B/G_{\rm D}$ and is, indeed, anomalously small.

To estimate the conditions of applicability, we smoothen the correction at the scale of R_c. This gives $|\delta\rho|/\rho \simeq G_{\rm Q}/G_B$, and the semiclassical approach does not work at $G_B \simeq G_{\rm Q}$.

4.4.7 Extended diffusive conductors

As a particular case of a junction chain (N arbitrary junctions for $N \gg 1$) we treated a one-dimensional diffusive conductor. The results obtained for a weak localization correction to the conductance and the conductance fluctuations are identical to those following from the DMPK equation. In this subsection, we concentrate on extended diffusive conductors. Starting from circuit theory, we shift to the more traditional cooperon–diffuson approach and review the results that depend on the effective dimensionality of a conductor taking magnetic and spin-orbit decoherence into account.

Let us derive the one-dimensional (junction chain) results in a slightly different way. For $N \gg 1$, the phase difference at each connector, $\phi^{\rm w,b}/N$ is small. We can thus expand the effective action in series of ϕ/N. In particular, it is obvious from Eq. (4.89) that we need to expand up to the second order for the weak localization, and from Eq. (4.90) that fourth-order terms are needed for conductance fluctuations. Let us start with the weak localization. Expanding Eqs. (4.97) and (4.98) to second order in ϕ and substituting the eigenvalues into the effective action, Eq. (4.83), and differentiating twice over ϕ, we obtain

$$\frac{\delta G_{\mathrm{WL}}}{G_{\mathrm{Q}}} = -\frac{4}{N^2}\sum_{n=1}^{N-1}\left(1-\frac{1}{\cos \pi n/N}\right) \approx -\frac{2}{N^2}\sum_{n=1}^{\infty}\left(\frac{\pi n}{N}\right)^{-2}, \tag{4.120}$$

where we have taken into account that only eigenvalues with $n \ll N$ contribute. Note that the sum is actually taken over the *eigenvalues* of the operator M. Performing the summation, $\sum n^{-2} = \pi^2/6$, we restore the result $\delta G_{\mathrm{WL}} = -G_{\mathrm{Q}}/3$.

Now we see how this result can be generalized to two and three dimensions: we just have to sum over all available eigenvalues. Indeed, generally the eigenvalues are labeled by three indices n_x, n_y, and n_z, and one writes (assuming we have N_i junctions in the i direction),

$$\frac{\delta G_{\mathrm{WL}}}{G_{\mathrm{Q}}} = -\frac{2}{N_x^2}\sum_{n_x,n_y,n_z}\frac{1}{(\pi n_x/N_x)^2 + (\pi n_y/N_y)^2 + (\pi n_z/N_z)^2}. \tag{4.121}$$

The summation is carried out for non-negative values of integers n_x, n_y, n_z, with $n_x = n_y = n_z = 0$ excluded, and the current flows in the x direction. Note that in Eq. (4.121) the distinction between one-, two-, and three-dimensional behavior is simple. Indeed, for $N_x \sim N_y \sim N_z$, all values of n_x, n_y, n_z contribute, and the system is two-dimensional. For $N_x, N_y \gg N_z$, only the value $n_z = 0$ gives a significant contribution to the weak localization correction, and the system is effectively two-dimensional. Finally, for $N_x \gg N_y, N_z$ only the values $N_y = n_z = 0$ are significant, and we return to the one-dimensional result given in Eq. (4.120).

Let us now take into account the decoherence factors η_{SO} and η_H. We shift the eigenvalues as prescribed by Eq. (4.83) to arrive at the following:

$$\begin{aligned}\frac{\delta G_{\mathrm{WL}}}{G_{\mathrm{Q}}} = -\frac{1}{N_x^2}\sum_{n_x,n_y,n_z}\Bigg(&\frac{3}{(\pi n_x/N_x)^2 + \left(\pi n_y/N_y\right)^2 + (\pi n_z/N_z)^2 + [2G_{\mathrm{Q}}(\eta_{\mathrm{SO}}+\eta_H)]/G_{\mathrm{T}}} \\ &- \frac{1}{(\pi n_x/N_x)^2 + \left(\pi n_y/N_y\right)^2 + (\pi n_z/N_z)^2 + 2G_{\mathrm{Q}}\eta_H/G_{\mathrm{T}}}\Bigg).\end{aligned} \tag{4.122}$$

To proceed, we write the weak localization correction as a sum over discrete wave vectors $q_i = \pi n_i/N_i$:

$$\frac{\delta G_{\mathrm{WL}}}{G_{\mathrm{Q}}} = -\frac{D}{L_x^2}\sum_{\mathbf{q}}\left(\frac{3}{Dq^2 + 1/\tau_{\mathrm{H}} + 1/\tau_{\mathrm{SO}}} - \frac{1}{Dq^2 + 1/\tau_{\mathrm{H}}}\right). \tag{4.123}$$

Equation (4.123) is general for diffusive conductors of any dimension, for arbitrary magnetic field and spin-orbit interaction. If the system is big enough, $\min(\tau_{\mathrm{H}}, \tau_{\mathrm{SO}}) \ll L_i^2/D$, we can replace the summation over discrete modes by integration over wave vectors. It could be one-dimensional integral (if only modes with $n_y = n_z = 0$ are important), or two- or three-dimensional. Writing the result for the conductivity (rather than conductance), we obtain

$$\delta\sigma_{\mathrm{WL}} = -G_{\mathrm{Q}}D \int \frac{d\boldsymbol{q}}{(2\pi)^3} \left(\frac{3}{Dq^2 + 1/\tau_{\mathrm{H}} + 1/\tau_{\mathrm{SO}}} - \frac{1}{Dq^2 + 1/\tau_{\mathrm{H}}} \right). \tag{4.124}$$

Control question 4.9. What is the condition under which the system can be considered one-, two- or three-dimensional, in terms of L_i, τ_{H}, and τ_{SO}?

Equation (4.124) can be evaluated explicitly. In particular, for one-dimensional diffusive conductors we obtain ($L = L_x$)

$$\delta G_{\mathrm{WL}} = G_{\mathrm{Q}} \frac{L_H}{2L} \left(1 - \frac{3}{\sqrt{1 + \tau_{\mathrm{H}}/\tau_{\mathrm{SO}}}} \right), \quad L_H = \sqrt{D\tau_{\mathrm{H}}}. \tag{4.125}$$

Comparing this with the result $\delta G_{\mathrm{WL}} = -G_{\mathrm{Q}}/3$ valid for $L \ll L_H, L_{\mathrm{SO}} = \sqrt{D\tau_{\mathrm{SO}}}$, we find that in one dimension the WL correction by order of magnitude is given by the product of G_{Q}/L, with the lowest of the lengths L, L_H, and L_{SO}.

We also write an order-of-magnitude estimate in two and three dimensions. Note that the integrals diverge at high q, and had to be cut off at the scale $q \sim 1/l$, since, at shorter length scales, the diffusion approximation does not work. Taking all spatial dimensions equal to L, we obtain

$$\frac{G_{\mathrm{WL}}}{G_{\mathrm{Q}}} \sim \begin{cases} L_{\min}/L & 1\mathrm{d} \\ \ln L_{\min}/l & 2\mathrm{d} \\ L_{\min}/l & 3\mathrm{d}, \end{cases} \tag{4.126}$$

where $L_{\min} = \min(L, L_H, L_{\mathrm{SO}})$.

We now provide a physical interpretation of the WL results, already obtained by three different methods. We have seen in Section 1.6 that the weak localization correction originates from the interference of two electron trajectories, similar to a double-slit experiment. Generally, if one has two trajectories with amplitudes A_1 and A_2, the quantum-mechanical probability is given as the squared absolute value of their sum,

$$P_{\mathrm{q}} = |A_1 + A_2|^2 = |A_1|^2 + |A_2|^2 + 2\mathrm{Re}\,(A_1^* A_2).$$

Here the first two terms (squared absolute values of the amplitudes A_1 and A_2) give the probability in the classical picture, and the final term is the interference – quantum-mechanical – contribution, giving rise to weak localization. In a diffusive system, one has to average this contribution over disorder. Naively, due to the dependence of the amplitudes of random phase shifts, one concludes that the interference contribution averages to zero. However, there are special pairs of the trajectories (see Fig. 4.16) – those spanning clockwise and counterclockwise the same loop – that have the same phase (see the consideration which led us to Eq. (4.43)). For these pairs, the interference term is phase-independent and does not average to zero. It is clear that the appearance of such a term leads to the decrease in conductance, since the particle spends more time in the loop. The correction is also small compared with the classical value of the conductance, since the probability of finding such an intersecting trajectory is small. One can write the result given by Eq. (4.124)

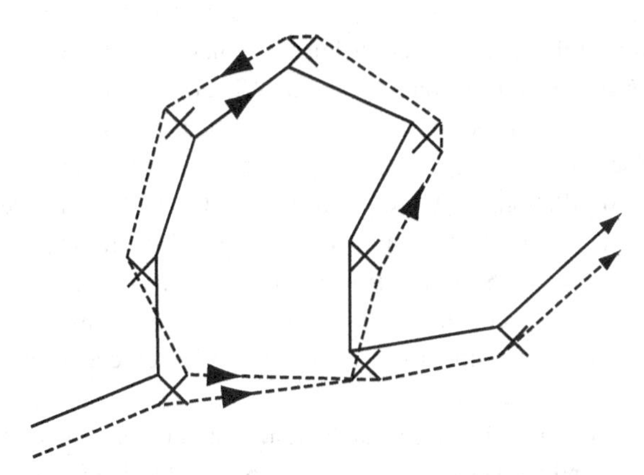

Fig. 4.16. Two classical trajectories contributing to the weak localization corrections. Crosses indicate the impurities; the trajectories with amplitude A_1 (A_2), shown by a solid (dashed) line, traverse the same loop clockwise (counterclockwise).

for the weak localization correction in terms of the cooperon P – the probability that the trajectory returns to the same point $\boldsymbol{r}$ during the time interval t:

$$\delta\sigma_{\text{WL}}(B) = -G_{\text{Q}}D\int_0^\infty \mathrm{d}t\; P^{BB}_{\text{coop}}(\boldsymbol{r},t'+t;\boldsymbol{r},t')\,(3\exp(-t/\tau_{\text{SO}})-1)\,. \tag{4.127}$$

Here, P is the solution of Eq. (4.60) – without spin-orbit decoherence, and the decoherence corrections are taken into account as the additional factor in the integrand.

If a magnetic field is present, the trajectories A_1 and A_2 acquire different phases. The total phase of the interference term equals the magnetic flux through the loop divided by the flux quantum (see Section 1.6). This phase depends on the area of the loop, and the interference term is suppressed after the impurity averaging. This is why the weak localization correction is suppressed by the external magnetic field.

Let us now return to the pioneering experiments described in Section 1.6. To understand the results of Sharvin and Sharvin [20], we consider an Aharonov–Bohm effect in a diffusive ring. In principle, one has AB oscillations containing all possible harmonics, $\sin \pi n\Phi/\Phi_0$, corresponding to the periods $2\Phi_0/n$ in the flux Φ, n being an integer. However, the ring supports many transport channels (otherwise it would not be diffusive), and in each channel the coefficients with which a given harmonic appears have random signs. Thus, in a diffusive ring all oscillations average out – the average current is magnetic-field-independent.

We saw above (see, for example, Eq. (4.108)) that a straightforward calculation does not confirm this conclusion – the flux dependence of the conductance comes from the WL correction. Indeed, the WL correction is due to the special pairs of trajectories that follow the same loop in opposite directions. We have already discussed this for an AB ring (Section 1.6). The difference with the clean ring is that all trajectories are diffusive. In particular, the phase shift between the trajectories in Figs. 1.36(a) and (b) is random and averages out. There is no conventional Aharonov–Bohm effect in diffusive rings. The simplest flux-sensitive contribution comes from the pairs of the trajectories shown in Figs. 1.36(d)–(g), corresponding to the WL correction. The magnetic flux dependence of the WL correction is due to the phase differences between these two trajectories, and this phase difference equals $2\pi\Phi/\Phi_0$. Thus, the conductance of a diffusive AB ring contains *only* oscillations with the period of Φ_0. Conversely, if one observes these oscillations and there is no sign of oscillations with period $2\Phi_0$, it means the WL correction is observed.

Sharvin and Sharvin measured the resistance of metallic cylinders (which, for our purposes can be considered as rings with many transport channels) in the magnetic field parallel to the cylinder axis. They clearly observe a Φ_0-periodicity, indicating the presence of a WL correction.

Let us now consider conductance fluctuations. The expressions in this case become heavy, and we keep the discussion to a qualitative level. For a quantitative discussion, the reader is referred to Ref. [100]. First, there are always two contributions to conductance fluctuations – one originating from the diffuson and one originating from the cooperon eigenvalues. If there is no magnetic field, $\tau_C^{-1} = \tau_D^{-1} = 0$, these two contributions are identical, and, at zero temperature, one obtains in all dimensions var $G = \alpha G_Q^2$, where α is a numerical coefficient of order 1, determined by the geometry of the system. We have seen that, for a one-dimensional system, $\alpha = 2/15$. This is the regime where conductance fluctuations are universal.

Next, let us switch on the magnetic field, $B = B'$. Diffusons are not sensitive to the time-reversal symmetry, and their contribution to the conductance fluctuations remains the same. On the other hand, cooperons are suppressed by a magnetic field, and in the strong-field limit, the cooperon contribution dies out: conductance fluctuations are reduced by a factor of 2. This fact, which we have already observed in the framework of RMT, now has an easy qualitative explanation. Indeed, as we discussed in Section 4.3, the interference of any two trajectories A_1 and A_2 (not necessarily the time-reversed ones) contributes to conductance fluctuations. For zero magnetic field not only the interference of A_1 and A_2 contributes, but also interference of A_1 and the time-reversed trajectory of A_2. Magnetic field suppresses this last contribution, and thus the conductance fluctuations decrease by a factor of 2. The typical scale of the field B_c is again given by $L \sim (D\tau_C)^{-1}$, similar to the weak localization correction.

In the same formalism, we can also discuss *parametric correlations*, taking $B \neq B'$. Now all diffusons decay with the time τ_D, and cooperons with the time τ_C. Both diffuson and cooperon contributions are suppressed in comparison with their universal values in a zero field. The asymptotic behavior of this decay for a one-dimensional system (we assume $B, B' > 0$) is $\langle G(B)G(B')\rangle - \langle G(B)\rangle\langle G(B')\rangle \sim G_Q^2(B_c/|B - B'|)^3$.

Let us now discuss the temperature dependence of the conductance fluctuations. For the weak localization correction, we did not find any temperature dependence (it only comes via the decoherence length, discussed in Chapter 6). The situation with the fluctuations is different: one finds a visible temperature dependence at the scale of the Thouless energy $E_{Th} = D/L^2$, var $G \sim G_Q^2(E_{Th}/k_BT)^{1/2}$ for the case $k_BT \gg E_{Th}$. This dependence occurs for the following reason. Diffusons and cooperons connect the Green's functions of two different electrons. These electrons may have different energies, up to a difference of k_BT. In Eq. (4.58), the time derivative would yield a factor of order k_BT, and

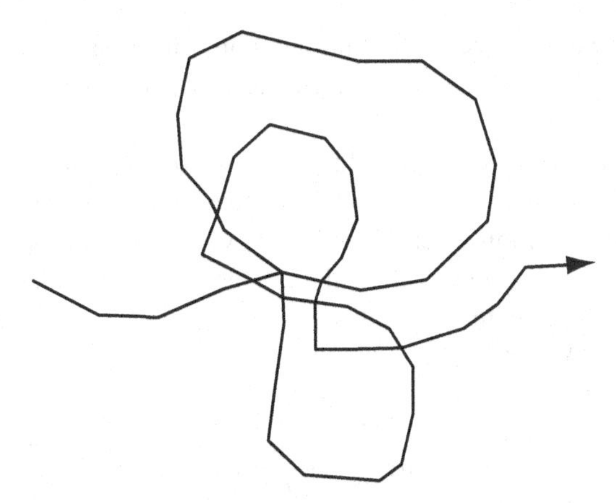

Fig. 4.17. An example of the classical trajectory returning to the same point several times.

the term with the gradient provides D/L^2. Thus, the behavior of a diffuson and a cooperon essentially depends on the ratio between the thermal and Thouless energies. Conductance fluctuations are sensitive to this behavior: obviously, conductances produced by electrons moving at different energies are less correlated than if the electrons had the same energy.

The experiment by Webb *et al.* [20], shown in Fig. 1.40, measured the conductance of small metallic rings with a relatively weak disorder – the mean free path was comparable with the length of a ring, and AB oscillations of all the periods were observed, not just Φ_0-oscillations. On top of the oscillations, we clearly see an aperiodic pattern in the conductance – fluctuations of the order of G_Q. These fluctuations are reproducible – if the same sample is measured again, the fluctuations are precisely the same – and thus represent the "fingerprints" of the diffusive system.

4.5 Strong localization

4.5.1 DMPK equation and localization in wires

We have already demonstrated in Section 4.3 (and confirmed in Section 4.4) that in a sufficiently long one-dimensional system the weak localization correction, calculated according to the perturbation theory, becomes formally greater than the classical value of the conductance. This occurs if the length of the wire exceeds lN, where l and N are the mean free path and the number of transport channels, respectively. Another way of describing the same condition is to say that the average conductance of the wire becomes of the order of G_Q. What happens for longer wires?

One way of reasoning is the following. The weak localization correction in the diffusive regime originates from the trajectories that return to the same point. We disregard the trajectories that return to the same point twice or several times (Fig. 4.17). Using the language of diagrams, these trajectories correspond to several diffusons and cooperons. For long wires, trajectories returning many times to the same point become non-negligible. This means that the electron likes to stay in the vicinity of the same point – it is said to be *localized*.

Let us now investigate this phenomenon by means of the DMPK equation, Eq. (4.50), valid for wires of any length. For long wires we make an Ansatz, verified a posteriori: all transmission eigenvalues T_n are very small. In terms of the variables x_n, $T_n = (1 + \sinh^2 x_n)^{-1}$, it means that all x_n values, as well as the distances between them, are very large. Approximating $\sinh x_n \approx \cosh x_n \approx \exp(x_n)/2$, we obtain the following equation ($\gamma = \beta N - \beta + 2$):

$$\frac{\partial P}{\partial s} = \frac{1}{2\gamma} \sum_{n=1}^{N} \frac{\partial}{\partial x_n} \left[\frac{\partial P}{\partial x_n} - 2(\beta n - \beta + 1) P \right]. \tag{4.128}$$

Note that this equation is much simpler than the full DMPK equation (Eq. (4.50)), since on the right-hand side different variables x_n are not mixed. The solution can be therefore written as a product,

$$P(s; x_1 \cdots x_n) = \prod_n P_n(x_n, s),$$

and each of P_n solves the following equation:

$$\frac{\partial P_n}{\partial s} = \frac{1}{2\gamma} \frac{\partial^2 P_n}{\partial x_n} - \frac{s}{\mu_n} \frac{\partial P_n}{\partial x_n}, \quad \mu_n \equiv \frac{\gamma}{1 + \beta n - \beta}.$$

The solution, obeying the initial condition $P_n(s = 0) = \delta(x_n - 1)$ (ideal transmission for an infinitely short wire) is given by

$$P = \prod_n \sqrt{\frac{\gamma}{2\pi s}} \exp\left(-\frac{\gamma}{2s} \left(x_n - \frac{s}{\mu_n} \right)^2 \right). \tag{4.129}$$

Each of the x_n values is a Gaussian variable distributed around $s/\mu_n \gg 1$ (for $n \gg 1$ this becomes sn/N), having a root mean square fluctuation of order of $(s/\beta N)^{1/2}$. Thus, the transmission eigenvalues are "crystallized": their averages are roughly equally spaced, and fluctuations are much less than the spacing between the neighboring values.

We now look at the conductance, determined by the lowest value of all the x_n values:

$$G/G_Q = \sum_n T_n = \sum_n \left(1 + \sinh^2 x_n\right)^{-1} \approx \exp(-2x_1).$$

The average conductance is readily calculated using the distribution function, Eq. (4.129),

$$\langle G \rangle = G_Q \sqrt{\frac{\gamma}{2\pi s}} \mathrm{e}^{-s/2\gamma}. \tag{4.130}$$

Thus, we see that the average conductance does not obey Ohm's law and does not scale inversely proportionally to the length L. (In other words, the parameter s is no longer proportional to the average conductance; it is, in leading order, proportional to the logarithm of the average conductance.) Instead, it decays exponentially at the length scale $\xi = \gamma l$ – the *localization length*. A long disordered wire is not a metal – it is an insulator. The characteristic feature of this regime is that the average conductance is much smaller than the conductance quantum G_Q.

Note that for a large number of transport channels the localization length becomes $\xi = lN\beta$. This means that the localization length in a magnetic field is twice as short as without it. In contrast, for one channel, the localization length is $\xi = 2l$ and does not depend on the symmetry of the system.

Exercise 4.8. Calculate the variance of the conductance in the localized regime.

We also see from Eq. (4.129) that the logarithm of conductance has a normal distribution. The conductance itself has a *log-normal* distribution. This is a situation very much different from what we encountered earlier. Indeed, in the diffusive regime, the fluctuations are always small compared with the typical (average) value of physical quantities. Distribution functions can be reasonably well approximated by (sharp) Gaussians. This is very convenient: for instance, the disorder average of any complicated function $\langle f(x)\rangle$, where x is an observable, can be replaced by $f(\langle x\rangle)$. This property is known as *self-averaging*. In contrast, conductance in the localized regime is not a self-averaged quantity. Indeed, the average logarithm of the conductance is given by

$$\langle \ln G/G_Q\rangle = -2\langle x_1\rangle \approx 4\ln\langle G\rangle/G_Q.$$

The average $\exp(\langle \ln G\rangle)$ decays faster than the average conductance $\langle G\rangle$. Consequently, each function of the conductance has to be averaged individually. We will give more examples later in this section.

4.5.2 Scaling theory of localization

Let us now look at the results for the weak localization correction in two and three dimensions (see Eq. (4.126)). In two dimensions, the correction grows logarithmically with the length. Thus, large two-dimensional systems are localized. One can estimate the localization length ξ as the scale at which the WL correction becomes of the same order as classical conductance. This yields $\xi \sim l\exp(\pi k_F l)$. This result is obtained under the assumption $k_F l \gg 1$, and for typical values of k_F and l the localization length is greater than the size of the Universe. Thus, large two-dimensional systems are theoretically localized, but practically they always remain metallic.

Control question 4.10. Estimate the localization length for an aluminum film taking $l = 100\,\text{nm}$.

In three dimensions, the WL correction decreases with the size of the system and is always small compared with the classical conductance – three-dimensional systems in the regime $k_F l \gg 1$ are always metallic, independent of their size.

What happens if the condition $k_F l \gg 1$ does not hold? In this case, the energy of the electrons at the Fermi surface is comparable with or less than the typical amplitude of random potential. Thus, an electron is typically trapped in a "potential well" formed by

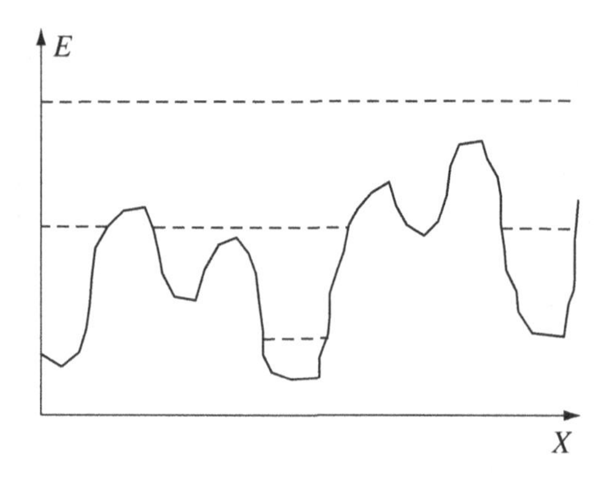

Fig. 4.18. Random potential landscape. For low energies (dashed lines) electrons are localized in the minima of the potential, whereas for high energies (solid line) the states are extended, and electrons see the random potential as a source of (weak) scattering.

one or several impurities. The states in neighboring potential wells typically have very different energies, and states with the same or close energies are located in the wells very far apart in space. Thus, these states do not hybridize and do not form extended states: the electrons states are localized. The phenomenon of electron localization for strong disorder is known as *Anderson localization* and occurs for any dimension of the system – at $d = 1$, $d = 2$, and $d = 3$.

It is important to realize that delocalized (extended) and localized states may correspond to the same disorder potential. The difference is in electron energy (Fig. 4.18). For high energies, exceeding the typical height of the random potential ($kl \gg 1$), the potential can be treated as a perturbation, and one is in the weak localization (metallic) regime. For the same potential, but low energies, electrons are localized. If one can change the electron density, one can go all the way from high densities (metallic) to low densities (localized).

It is interesting to understand how a metallic state evolves into an insulating state. In the following we first argue that any system enters the localization regime when its average conductance becomes of the order of the conductance quantum G_Q. This criterion already appeared in the comparison of the average conductance and the weak localization correction in the metallic regime, and here we present more general and transparent arguments. Then, we investigate what systems can become localized according to this criterion.

The following argument is due to Thouless [101]. Consider a d-dimensional piece of disordered metal. The electron levels in the metal are discrete, with the mean level spacing given by $\delta_S = (\nu L^d)^{-1}$, with L being the linear size of the piece of metal. Let us now put 2^d such pieces together, so that they form a piece of the size $2L$. Each eigenstate of an electron in a $(2L)^d$ system is a linear combination of the states in the L^d systems. There are two scenarios. If the resulting state is extended – all 2^d blocks contribute with appreciable probabilities – we have a metal. On the other hand, if the contribution of one of the blocks is of the order 1, and contributions of all other blocks are negligible, the electron is localized in this one block – we have an insulator. Obviously, the choice between metal and insulator depends on the properties of the L^d blocks. The question is: What are the properties determining the choice?

Let us look at the hybridization of states in different blocks more closely. From perturbation theory, we find that the correction to the wave function of an electron in one block due to the hybridization with the other block has an amplitude given by the ratio of the overlap integral between the blocks divided by the difference of energies of the states in the two blocks. The energy difference is of the order of the mean level spacing δ_S (it is the same in all blocks). For a localized state, the overlap t obviously would be exponentially small, since the wave functions are concentrated in different areas of the space. For an extended state, an electron can diffuse everywhere in the sample, and a good estimate for an overlap integral would be the inverse time for an electron to diffuse out of the sample, $t \sim E_{Th}$, with E_{Th} being the Thouless energy of one block.

Thus, the choice between metallic and insulating behavior is determined by just one dimensionless parameter t/δ_S. For $t \ll \delta_S$ we have an insulator, and, for $t \gg \delta_S$, a metal. Note that we can formally define the conductance of a piece L^d – imagine, for example, that ideal metallic contacts are attached to the opposite faces. The conductance is also determined by the same quantities – the overlap integral t and the mean level spacing δ_S – and thus it must be related to the parameter t/δ_S. In particular, it is easy to check that, in the metallic regime, the conductance is of the order of $G \sim G_Q E_{Th}/\delta_S = G_Q t/\delta_S$. Thus, for $G \gg G_Q$ one has a metal, and for $G \lesssim G_Q$ one has an insulator.

Let us now actually perform putting these pieces together, a procedure known as the *scaling theory of localization*. The smallest size of the cube we can describe in this way is the mean free path l – shorter systems do not exhibit electron scattering. Consider a square plaquette of size l; it is ideal, and thus its conductance equals $G_0 \sim G_Q N$, with N being the number of transport channels, that is $G_0 \sim G_Q k_F l$ or (the resistance) $R_0 = R_Q(k_F l)^{-1}$. For $k_F l \gg 1$ this piece is obviously metallic, since its conductance is large. Let us now put $M = L/l$ such pieces next to each other, to construct a one-dimensional conductor. The resistances add in series, and we obtain $R \sim R_Q L/(k_F l^2)$. The resistance grows with the length, and for $L \sim k_F l^2$ it becomes of the order of the resistance quantum – the system is localized, with localization length of order $\xi \sim k_F l^2$. This corresponds to the results we have obtained above by means of the DMPK equation. For $k_F l \ll 1$ the small pieces are insulating from the very beginning, and the composite system (wire) exhibits localization for any length.

We look now at two dimensions. Again, for $k_F l \ll 1$ any two-dimensional system would be insulating. Assume $k_F l \gg 1$ and let us count the plaquettes in circles of radius r, $l < r < L$. The plaquettes within each circle are connected in parallel, the resistance of a circle is $R_0 l/r$, and circles are connected in series. Performing the summation over the circles and replacing it by integration, we find $R \sim (k_F l)^{-1} \ln L/l$. This means that sufficiently large two-dimensional systems are always insulating. Writing down the localization length, we arrive at the same result, $\xi \sim l \exp(\pi k_F l)$, that we obtained previously from the comparison of the weak localization correction to classical conductance.

In three dimensions, the resistance only decreases provided we sum the resistances. This means that a three-dimensional system of any size is metallic provided $k_F l \ll 1$. Only if one varies the parameter $k_F l \gg 1$, for example, by varying the electron concentration, does one arrive at the insulating regime – the *Anderson phase transition*. An Anderson transition in three-dimensional systems has been observed experimentally, in doped semiconductors [102] and in the $Al_{0.3}Ga_{0.7}As$ compound [103]; however, it still remains an exotic phenomenon.

4.5.3 Localized regime: optimal fluctuation and hopping

In the rest of this section, we consider the localized regime. It goes well beyond the scope of this book, since it is not really transport (the resistance of an insulator is exponentially high) and not really quantum (the main features can be explained by purely classical physics; the only quantum concept important for localization is tunneling). The important ideas were elaborated a long time ago in the context of transport properties of disordered semiconductors, and currently form textbook material [104]. We will not therefore attempt a systematic discussion of transport at strong disorder, and only focus on the conceptually important ideas that made their way into the theory of quantum transport.

So far, the only tool we have had for a quantitative description of transport at strong disorder is the DMPK equation. However, its use is limited: it only describes one-dimensional situation, and only disordered wires. Let us look at a localized system from a different point of view: localized states. These states have random locations in space and random energy. The precise properties of the system may strongly depend on the type of disorder; it is not unreasonable to assume that the average density of states is constant both in space and energy. To get from the left side of the sample (attached to an electrode) to the right side (attached to another electrode), the electron must tunnel sequentially through these localized states (Fig. 4.19) – this process is known as *hopping conduction*.

The main issue in hopping conduction is to characterize the route the electron would follow. Indeed, we already mentioned that the states close together in energy are typically far away in space, and vice versa. In particular, this means that to be able to hop between neighboring (in space) localized states, an electron must change its energy. An alternative would be co-tunneling, but since *all* the states have different energy, the electron must co-tunnel from the left to the right through many states (see Section 3.4), and this is a quantum-mechanical process of a very high order of the perturbation theory. Disregarding the co-tunneling, we find that hopping conduction is an *inelastic process*. It is only possible at finite temperatures.

Next, we see that there is a conceptual difference between one dimensional hopping and hopping in two and three dimensions. Indeed, in one dimension the electron has no choice – it must sequentially tunnel through all available states. We have already seen from the DMPK calculations that the conductance is not self-averaging – the average conductance may differ strongly from the typical value of conductance in a given disorder configuration. We might expect that the average conductance is determined by the disorder configuration maximizing the conductance. This is not actually correct, since the probability of finding such a configuration is typically very small, and the configuration does not contribute to the average value. On the other hand, it is easy to find a disorder configuration yielding low conductance, but the contribution of these configurations to the average is small. One has to find the *optimal configuration* of disorder, maximizing the product of the conductance and the probability of finding such a configuration. Note that, for example,

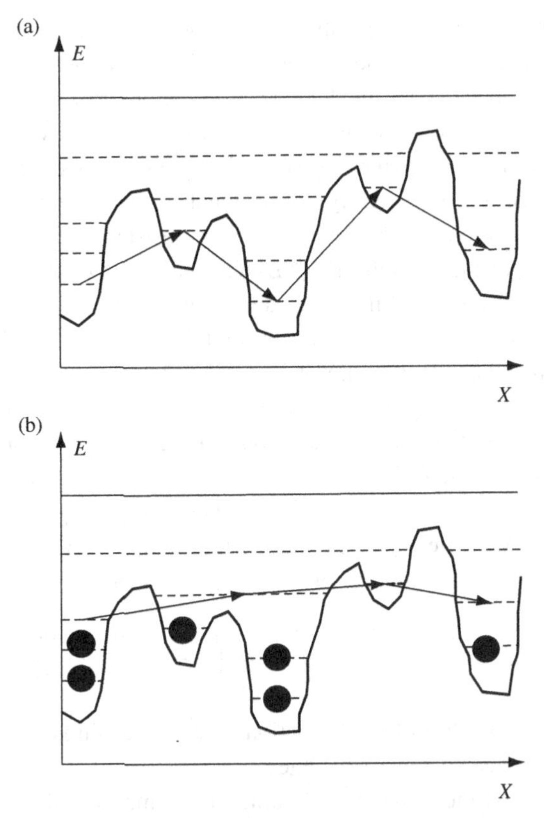

Fig. 4.19. Hopping in one dimension. (a) At low concentrations, electrons have to hop over the states with minimum energy in each potential well. (b) For higher concentrations, many of these states are already occupied, and the hopping electrons do not see some of the minima. Eventually, for even higher concentrations, all localized states are occupied, and electrons travel in extended states rather than hop.

the fluctuations of the conductance will also be determined by the same approach, but the optimal configurations for conductance and for conductance fluctuations are generally different. This is a direct consequence of non-averaging: every function of the conductance has to be calculated individually [105].

First, let us consider hopping in one dimension. As we have already discussed, in a given disorder configuration, the electron hops over available localized states. However, the probability of hopping is different for different disorder configurations. Let us guess which configuration of the localized energy levels is the most profitable for transport. We assume that the probability of a hop between two states located at points $\boldsymbol{r}_1$ and $\boldsymbol{r}_2$, having energies ϵ_1 and ϵ_2, measured from the Fermi surface, is given by

$$T_{12} = T_0 \exp\left(-\frac{|\boldsymbol{r}_1 - \boldsymbol{r}_1|}{a} - \frac{|\epsilon_1| + |\epsilon_2| + |\epsilon_1 - \epsilon_2|}{2k_B T}\right). \tag{4.131}$$

The second term in the exponent is just the Boltzmann factor, reflecting the fact that to create a particle above the Fermi surface, as well as the hole below the Fermi surface, costs energy. At zero temperature, the tunneling is only possible if both states are precisely at the Fermi energy. The first factor models the transmission of the potential barrier between the localized states; the length a is of the order of the localization length.

For our first guess, we take $N-1$ electron states forming a chain. The total resistance of the chain is given by the sum of resistances corresponding to all hops, and the resistance of each hop is inversely proportional to the probability of the hop T_{12}. For a given N, the maximum conductance (minimum resistance) of the chain is obviously realized for all sites lying at equal distances L/N, L being the total length of the system, and at the Fermi energy. The probability of finding such a chain is zero. This is why it does not contribute to the resistance. What we can do, however, is to consider lattice-like configurations: let the positions deviate slightly from the ideal lattice sites, and let the energies deviate from the Fermi energy. If the chain is along the z axis, we denote the possible deviation of the position in the longitudinal (along z) and transverse directions as δz and $\delta \boldsymbol{r}$, respectively, and the deviation in energy as $\delta\epsilon$. The total resistance of the chain is determined by the highest of all the resistances of the hops. Assuming the deviations to be small, $|\delta z|, |\delta \boldsymbol{r}| \ll L/N$, we write $|\boldsymbol{r}_1 - \boldsymbol{r}_1| \approx L/N + \delta z + (\delta \boldsymbol{r})^2 N/(2L)$, and the resistance of the chain up to a pre-exponential factor ($R_0 \propto T_0^{-1}$) is given by

$$R \approx R_0 \exp\left(\frac{L}{Na} + \frac{\delta z}{a} + \frac{(\delta \boldsymbol{r})^2 N}{2La} + \frac{\delta\epsilon}{k_B T}\right).$$

The probability of the formation of such a chain is proportional to $w_N = (\nu \delta z (\delta \boldsymbol{r})^2 \delta\epsilon)^N$, where ν is the density of states.

To find the optimal configuration, we multiply the resistance of a given configuration (characterized by the values of δz, $\delta \boldsymbol{r}$, $\delta\epsilon$, and N) by the probability of such a configuration, and calculate the resulting integral by the saddle point method. Technically, it means that we have to write

$$w_N = \exp(N \ln(\nu \delta z (\delta \boldsymbol{r})^2 \delta\epsilon)),$$

take the product $w_N R$, and find what values of the parameters optimize the exponent. For a given N, we obtain $\delta z = Na$, $\delta\epsilon = N k_B T$, and $|\delta \boldsymbol{r}| = \sqrt{2La}$. Optimizing the resulting expression with respect to N, we find $N = \sqrt{L/(a\lambda)}$, $\lambda \approx -\ln(4\nu k_B T L^2 a)$. Thus, the conductance decays exponentially with the size of the system L, $G \propto \exp(-\sqrt{L\lambda/a})$. Since the parameter λ logarithmically depends on the temperature, we also predict a stretched exponential temperature dependence of the conductance, $\ln G \propto \sqrt{\ln T}$.

Let us see under what conditions our calculation is valid. We have assumed that the displacement δz is small in comparison with the period of the chain $L/N = \sqrt{La\lambda}$. The optimal value of δz we found is $Na = \sqrt{2aL/\lambda}$. Thus, the condition is $\lambda \gg 1$, or $k_B T \ll (\nu L^2 a)^{-1}$. Other assumptions yield the same criteria. This condition is violated for high enough temperatures or in long enough samples.

Control question 4.11. At very low temperatures, we formally have $N \gg 1$. Find the temperature at which this occurs and characterize the behavior of the conductance at lower temperatures.

What happens at higher temperatures? Since the assumption of small displacement around the sites of the lattice no longer holds, electron hops are characterized by the exponential spread of resistances. In this situation, the resistance is determined by the largest of the resistances of all the hops. In Eq. (4.131), the term r/a in the exponential obviously favors short hops. However, as we already discussed, states close together in distance are far away in energy, and the second term, $\delta\epsilon/k_B T$, favors long hops. Assuming that the density of states is constant in space and energy, we can estimate, in three dimensions, $\delta\epsilon \sim (\nu r^3)^{-1}$. Writing $G(r) \propto \exp(-r/a - (\nu r^3 k_B T)^{-1})$, we find that it is maximal for the distance $r \sim a(T_0/T)^{1/4}$, where up to a numerical factor we have $T_0 \sim (\nu k_B a^3)^{-1}$. Note that the probability of finding such a pair of states is constant and thus does not affect the optimal hopping distance. For the average conductance, we obtain the dependence $G \propto \exp(-(T_0/T)^{1/4})$, known as *Mott's law*, or the *variable-range hopping conductance*.

Control question 4.12. Produce estimates for the variable-range hopping conductance in one and two dimensions.

Note that, in this picture, the radius of the hop decreases with temperature, and at a temperature of order T_0 this becomes of order a. For $T \sim T_0$, the hops are no longer tunnel processes; indeed, the transmission probability associated with each hop is not small. Thus, one has a metallic behavior at higher temperatures. Another way to arrive at metallic behavior is to increase the electron density, as discussed earlier in this section. Indeed, electrons in a strong disorder potential occupy minima of the potential. If the density is increased, electrons eventually become closer (they "fill" the potential minima up to a higher level), and then the minima eventually merge (Fig. 4.19). Once all the minima have merged, one has a metal rather than an insulator. In one dimension this occurs when the highest barrier between the minima has been overcome.

The picture is more complicated in two and three dimensions. Indeed, an electron has to choose from many paths connecting the leads, each path being a series of hops. Since the hops are incoherent, the resistance of each path is given by the sum of the resistances, and is thus determined by the greatest resistance along this path. The paths are connected in parallel, but their resistances differ exponentially, and thus only the one with the lowest resistance matters: with the exponential precision, the resistance of the insulator is determined by the highest resistance of all the hops in the lowest resistance path.

Let us now see what happens if we increase the electron density. Similarly to one dimension, the localization length increases, and electrons come "closer" to each other – the resistance of each hop goes down. Eventually, some of the former localized states merge, and hops disappear. Finally, at some concentration *all* states along a certain path merge – the last barrier has been overcome. A commonly used analogy is with water percolation. Imagine that we have a random landscape and fill it with water. At a low level of water (corresponding to low electron concentration), the water only fills isolated trenches to form lakes. Water cannot flow from the left to the right in this landscape. In terms of electrons, to make it from left to right, electrons have to hop between the "lakes" – the resistance does not obey Ohm's law, and we have an insulator. If the water level increases, the lakes become bigger, new lakes appear, and, at some level, a waterway between the left and the right sides is established. This is known as the *percolation threshold*, and the waterway is called the "percolating path." At the threshold, there is strictly one way of getting across the system.

Above the percolation threshold, electrons move along the paths connecting the two sides of the system without tunneling. This is usual diffusive propagation, and one has a metal. Thus, the percolation transition separates a metal and an insulator (the Mott transition). The conductance below the transition is exponentially small; above the transition it obeys Ohm's law. Such problems are studied by mathematical physicists, and analytical results on the percolation theory are limited. For instance, the conductance of a large system near the percolation threshold increases as a power law, $(n - n_c)^\rho$, where the exponent ρ depends on the dimensionality of the system, n is the electron concentration, and n_c denotes the percolation threshold. In two dimensions, the threshold n_c corresponds to the filling factor $1/2$ – "lakes" and "soil" occupy equal areas. For details, see Ref. [104].

4.5.4 Interaction and the Coulomb gap

What is the role of interactions in the localized regime? Interactions must be important for insulators, since there are no free charges to screen interactions, and they remain long-ranged. First, close to the percolation threshold, an insulator is a collection of "islands", which may contain many electrons and behave as small pieces of metal. Due to interactions, each island is characterized by the charging energy, and at low voltages the transport is modified similarly to arrays of Coulomb islands (see Section 3.1).

For low energies, Coulomb interactions between electrons have a strong effect on the electron properties. Let us return to the picture of localized levels E_i, located at the points $\boldsymbol{r}_i$, uniformly distributed in energy and space. In the ground state, all levels below the Fermi level are filled and all other levels are empty. Without interaction, the energy of an electron sitting in level i is E_i. With interaction, we define the *quasiparticle energy*,

$$\xi_i = E_i + \sum_{j \neq i} V_{ij}, \tag{4.132}$$

where the sum is taken over all occupied levels j (except for $i = j$, if i is occupied), and $V_{ij} = e^2/|\boldsymbol{r}_i - \boldsymbol{r}_j|$ is the Coulomb interaction. The energy cost of transfering a quasiparticle from an occupied state i to an empty state k is $\Delta_{ik} = \xi_k - \xi_i - V_{ik}$. The final term here appears because in Eq. (4.132), for an empty state k the quasiparticle energy includes the interaction with all occupied states, whereas for an occupied state i it includes interaction with all occupied states but i.

In the ground state, all *excitation energies* Δ_{ik} must be positive. This means that if states i and k are both close to the Fermi level, so that the difference $\xi_k - \xi_i$ is small, these states must be located very far from one another in space, so that the Coulomb energy does not exceed this difference. In particular, there are *no* states precisely at the Fermi level. This suppression of the density of states at the Fermi level by long-range Coulomb interactions is known as the *Coulomb gap*.

The condition $\Delta_{ik} > 0$ can be used to quantify the energy dependence of the density of states close to the Fermi level. For example, one can start with a state with a uniform distribution of energies ξ_i, which is not the ground state of the system. In this state, some of the conditions $\Delta_{ik} > 0$ are not fulfilled. The ground state is obtained from this initial state by transpositions of electrons within pairs of states, one of which is occupied and one empty, so that some of the excitation energies Δ_{ik} become positive. At the end of the process, one finds the ground state, and the evolution of the density of states in the process of transpositions is described by an equation similar to the Boltzmann equation [106] (with the distance between the states in the transposed pairs playing the role of time). It turns out that the density of states vanishes logarithmically for $D = 1$, and as $\nu \propto |E|^{D-1}$ for $D = 2, 3$. If the interaction is short-ranged (screened) rather than Coulomb, instead of a Coulomb gap one has a slight suppression of the density of states at the Fermi level [107].

This energy dependence affects Mott's law for hopping conductivity: if the Coulomb gap is effective, the same argument that led us to Mott's law now, with the energy-dependent density of states, yields $G \propto \exp(-(T_0/T)^{1/2})$.

The energy-dependent density of states can be measured in the tunneling experiment. Note that the thermodynamic properties of the system are not sensitive to interactions. This is why in interacting systems one can introduce two densities of states: the tunneling one, sensitive to interactions and determining the tunneling properties, and the thermodynamic one, insensitive to interactions and determining the thermodynamic properties.

5 Qubits and quantum dots

It is difficult nowadays to graduate from a department of natural sciences and not hear anything about quantum computing, most likely about the fascinating prospects of it. Quantum computing by its origin is a rather abstract discipline, a branch of math or information science. It has emerged from a persistent search for more efficient ways to process information when quantum mechanics made it to the scope of the search. Being an abstract discipline, quantum computing approximates a physical quantum system with a number of axioms and explores the consequences of these axioms, precisely as conventional math has been. These activities begun in the 1970s, and for a long time it was not obvious why quantum calculational schemes have to be any better than common computer algorithms. This extensive work was rewarded with a breakthrough in 1994, when Peter Shor discovered a remarkable quantum algorithm for the factorization of large numbers into prime ones. This sounds quite abstract, but many public key cryptosystems will become obsolete if Shor's algorithm is ever implemented in a practical quantum computer. Modern communication security is based on the fact that the factorization is a tough problem for a classical computer: it takes too long to crack a code. The proposed algorithm speeds up the factorization enormously. The discovery by Shor has brought quantum computing to the scientific and even public attention.

This progress has motivated a massive research effort towards the manipulation of individual quantum systems, the practical realization of quantum computing schemes being one of the most attractive goals. This chapter describes this goal, the quantum transport systems suitable for it, and the achievements made so far.

Section 5.1 describes a quantum computer: an information processor made of *qubits*, two-level quantum manipulable systems. "Qubits" is a shorthand notation for "quantum bits." We will explain the difference between the presentation of information by classical bits and qubits and formulate principles and algorithms of a quantum computer. Section 5.2 is devoted to fascinating and counter-intuitive aspects of quantum mechanics important for information. We recommend the book by Nielsen and Chuang [108] for a more advanced outline of these subjects. Next, we discuss the manipulation of physical quantum systems – single qubits and pairs of qubits – in Section 5.3. The second part of the chapter is devoted to quantum transport systems that realize the qubits and quantum gates. *Quantum dots*, frequently called artificial atoms, are generic objects that realize discrete quantum degrees of freedom, and deep discussion (Section 5.4) of their states and transport properties provides the necessary introduction to the practical realizations of qubits. We separate these onto charge (Section 5.5), phase and flux (Section 5.6), and spin (Section 5.7) qubits.

5.1 Quantum computers

In this section, we describe a quantum computer as an abstract scheme. The minimal element of the computer is a qubit – a quantum system characterized by a two-component wave function and outlined in detail below. The quantum computer encompasses N qubits and thus is characterized by a 2^N-component wave function. It is essential that the wave function is processed by the computer according to its program: a quantum algorithm. An important idealization is that any quantum algorithm is given by a $2^N \times 2^N$ unitary matrix $\hat{U}$ that relates the final and initial wave functions: $|f\rangle = \hat{U}|i\rangle$. This corresponds to a physical assumption that no other degrees of freedom are involved in the operation; that the quantum computer is an isolated system during the computation. After the computation, the resulting wave function is read out and the processed information is transferred to other degrees of freedom such as classical computers or our brains. To achieve efficient programming, an algorithm is decomposed into its elementary operations. These involve either a single qubit (qubit rotations) or pairs of qubits (two-qubit gates). From these elementary operations, one constructs the advanced and advantageous algorithms described at the end of the section.

Sometimes students ask: How long will it take to make a working quantum computer? Some do it rather ironically, implying that this will never happen. Others imply that if this happens before his/her graduation, he/she would not be able to contribute to this piece of progress and his/her quantum-mechanical studies would not make any sense. The analogy with classical computers may be relevant for an estimation of the time scales involved. The basic principles of modern digital computer architecture were established by Charles Babbage around 1825. However, it took 120 years to build a practical computer and yet another 40 years to make it an everyday item. One might think that in 1825 the technology was not sufficiently developed and/or that computers were not in demand, but this would not be true. There was a strong demand for automated calculations: the Royal Navy, for example, would have been a major customer. And the technology was sufficiently advanced for the mechanical implementation intended. Charles Babbage had the blueprints of a computer, and had received several grants to build it.

5.1.1 Qubits

The basic difference between classical and quantum computers is in the presentation of information. An elementary unit of a classical computer is a *bit*: a system that can be in two states, $|0\rangle$ or $|1\rangle$. Any information can be presented in bits as a long series of "0" and "1" digits, and processed by elementary operations whereby a bit is set to a state that depends on the state of other bit(s). A quantum analog of a bit is a qubit, a quantum two-state system. The trick is that an arbitrary state $|s\rangle$ of a qubit is not necessarily "0" or "1." It is given by its wave function, which can be any linear superposition of two basic states $|0\rangle$ and $|1\rangle$:

$$|s\rangle = \alpha\,|0\rangle + \beta\,|1\rangle\,; \; |\alpha|^2 + |\beta|^2 = 1, \tag{5.1}$$

where α and β are complex numbers satisfying the normalization condition. Frequently for a single qubit only the relative phase of α, β is of importance; then the wave function can be parameterized with two real floating-point numbers. For example, we may choose

$$\alpha = \cos(\theta/2)\mathrm{e}^{\mathrm{i}\phi/2}; \ \beta = \sin(\theta/2)\mathrm{e}^{-\mathrm{i}\phi/2}, \tag{5.2}$$

with real θ, ϕ. How many classical bits does it take to present a real floating-point number? The answer is "an infinite number," if we want to achieve an infinite accuracy. For a reasonable accuracy of 10^{-19}, 64 bits would suffice. So one needs $128 = 2 \times 64$ classical bits to present a qubit: not that many, in fact. The next question is: How many bits does it take to present a state of N qubits? A quick answer is that $128 \times N$ would be wrong. The point is that the wave function of N qubits has 2^N independent complex components. The basis states $|j\rangle$ are the states where each qubit is either in the state $|0\rangle$ or $|1\rangle$. They thus correspond to all possible sequences of "0" and "1" of the length N, that is, to integer non-negative numbers $0 \leq j \leq 2^N - 1$ written in binary notation. This integer is called the *binary code* of state $|j\rangle$. For example, for the basis state of four qubits, $|1\rangle|0\rangle|1\rangle|1\rangle \equiv |1011\rangle$, the binary code is given by $2^3 \cdot 1 + 2^1 \cdot 1 + 2^0 = 11$. An arbitrary wave function of N qubits is a superposition of the basis states,

$$|f\rangle = \sum_{j=0}^{2^N-1} f_j |j\rangle,$$

with complex f_j. There is a single normalization condition $\sum_j |f_j|^2 = 1$ imposed, and the global phase of all f_j is irrelevant, so that a wave function is generally characterized by $2 \times 2^N - 2$ real numbers. We conclude that the correct answer is $M = 128 \times (2^N - 1)$.

This sounds as if it is too much. The exact representation of the quantum state of several qubits requires plenty of classical memory. The RAM of a typical computer can represent 26 qubits only. A detailed simulation of a quantum system with a moderate number of quantum degrees of freedom (say 20 atoms) is a tough problem for classical computers because of the enormous memory required and the enormous number of operations needed to process the information in this memory. This is the negative side, and Richard Feynman, in 1981, was the first to point out the positive aspects. Feynman noticed that 20 atoms can simulate themselves perfectly in a fraction of a nanosecond, and they encode thereby an enormous information. A popular example intended for mp3 consumers says that with 47 qubits one may encode 1743 years of living music (transferred at 40 kB/s).

Let us turn back to the representation of a single-qubit wave function. A popular way to visualize it is to use the *Bloch sphere*. The wave function is a point on a unit-radius sphere, θ and ϕ being polar and azimuthal angles in spherical coordinates. The convenience of the representation comes from the fact that the expectation values of the three Pauli matrices $\sigma_{x,y,z}$ are given by the x, y, z components of the corresponding three-dimensional real vector (Fig. 5.1):

$$\langle s|\sigma_x|s\rangle = \sin\theta\cos\phi; \ \langle s|\sigma_y|s\rangle = \sin\theta\sin\phi; \ \langle s|\sigma_z|s\rangle = \cos\theta.$$

Control question 5.1. Can you prove the above formula?

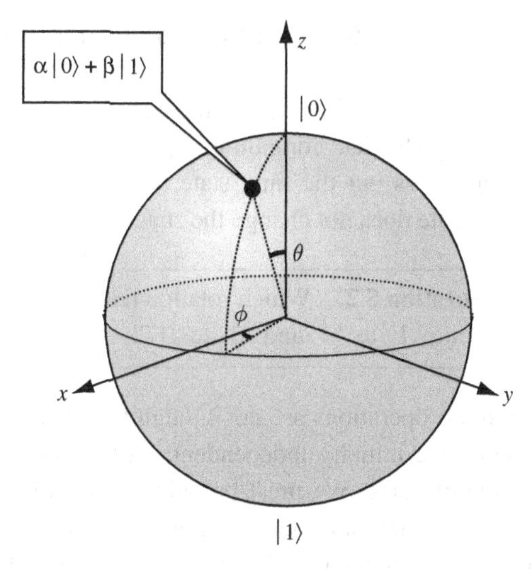

Fig. 5.1. Bloch sphere. A qubit wave function can be represented as a point on the unit sphere (see Eq. (5.2)).

The pure basis states $|0\rangle$ and $|1\rangle$ are at the north and south pole, respectively. A "natural" qubit, a quantum two-level system, is an electron with a spin degree of freedom. The electron spin is a three-dimensional (pseudo)vector, the expectation values of the components being proportional to the expectation values of the corresponding Pauli matrices $\mathbf{s} = (\hbar/2)\langle\boldsymbol{\sigma}\rangle$. The spin is transformed under rotations in real three-dimensional space precisely like a vector, so that these rotations are directly visualized as rotations on the Bloch sphere. Since any point on a sphere can be transformed to any other point by a proper rotation, and a wave function of a qubit is a point on the Bloch sphere, an arbitrary transformation of the qubit wave function is called a *rotation*. The rotation about the $\boldsymbol{n}$ axis by angle γ is represented by the following unitary 2×2 matrix:

$$\hat{R}_{\boldsymbol{n}}(\gamma) = \exp\left(\frac{\mathrm{i}\gamma}{2}\boldsymbol{n} \cdot \boldsymbol{\sigma}\right). \tag{5.3}$$

The factor $1/2$ is crucial here, and at this point the Bloch sphere visualization is slightly misleading. It looks as if a rotation by 360 degrees should transform a wave function to itself, as it does with the points on the sphere. This is not correct: the rotation by 360 degrees flips the sign of the wave function. One can check this, for example using the parameterization in Eqn. (5.2): the 360 degree rotation about z shifts ϕ by 2π. Only 4π rotation transforms the wave function to itself.

Exercise 5.1. Take the pure basis state $|0\rangle$. To which states does it transform upon three (separate) rotations (see Eq. (5.3)) by an arbitrary angle about the x, y, and z axes?

5.1.2 Principles of quantum computers

The rotation of a qubit provides a simple example of a quantum operation, or, as it is called in analogy with classical computing, a *gate*. Such a single-qubit gate takes the initial qubit state $|i\rangle$ and gives out the final state $|f\rangle = \hat{R}|i\rangle$, $\hat{R}$ being a unitary 2×2 matrix. The single-qubit gate does not change the states of the other qubits in the quantum computer.

Control question 5.2. Which rotation realizes a negation single-qubit gate, that is the gate changing "1" to "0" and "0" to "1"?

Single-qubit operations are not enough to unleash the power of quantum computation. If the qubits are initially independent, they remain so upon application of any number of single-qubit gates. We must organize an interaction to occur between the qubits. In distinction from naturally occurring quantum systems, the qubits in a quantum computer should interact in a controlled fashion; that is, only during an operation involving several qubits. For N qubits, such an operation is called a N-qubit gate.

Since the wave function of N qubits contains 2^N components, the N-qubit gate is represented by a $2^N \times 2^N$ unitary matrix $\hat{U}$. Sequential application of the gates $\hat{U}_1$ and $\hat{U}_2$ yields a gate represented by the matrix product $\hat{U}_2\hat{U}_1$, which is also unitary. The whole N-qubit quantum computer programmed in a certain way provides an N-qubit gate. This is why quantum computation is always reversible: there is always a unitary matrix that is the inverse of a given unitary matrix. This is not so in a classical computer: many classical gates are not reversible.

Most fortunately, one does not have to program a quantum computer using unitary matrices of high dimensions as elementary building blocks. It is proven in quantum computing that any complex many-qubit gate can be made combining single-qubit gates (different for different qubits involved) and an elementary two-qubit gate. Thus, in order to demonstrate the feasibility of a certain physical realization of a quantum computer, one concentrates on the realization of single-qubit and two-qubit gates. In mathematical terms, finding the correct combination of elementary gates proceeds in two steps. First, one decomposes the $2^N \times 2^N$ unitary matrix into two-level unitary matrices, i.e. matrices diagonal in all 2^N components except for two. Second, each of these matrices is represented as a combination of single-qubit and two-qubit gates. It is worth noting that this combination for an arbitrary quantum algorithm involves an exponentially large number of elementary gates, of the order of $N^2 4^N$. This slows the computation down and eventually cancels out the advantage of a large memory, $128 \cdot 2^N$ of the quantum computer, as compared with the classical one. We conclude that efficient quantum computation requires clever algorithms, involving less than an exponential number of elementary gates.

Usually a so-called CNOT (controlled-NOT) gate is chosen as an elementary two-qubit gate. A classical analog of the CNOT gate would do the following: it negates the value of a second (target) bit if the first (control) bit is in state "1" and does nothing if the first bit is in state "0." The resulting state of the target bit is the result of an XOR operation on two input bits. A quantum CNOT does the same to two qubits. There are four basis states for

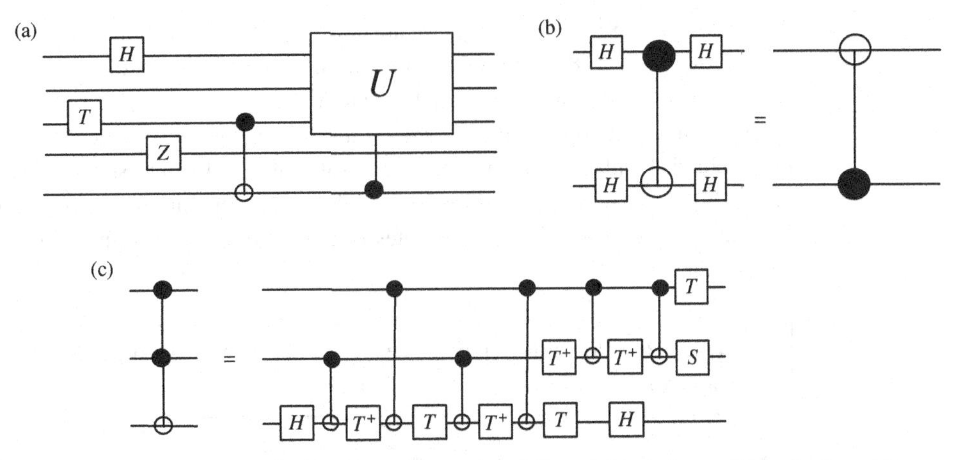

Fig. 5.2. Quantum circuits. (a) Graphical representation. From left to right we have: three single-qubit gates; a CNOT; a three-qubit gate controlled by the first qubit. (b) Composition of elements. The circuit on the left is equivalent to a CNOT. (c) Decomposion of Toffoli gate in terms of single-qubit gates and CNOTs.

two qubits given by a direct product of single-qubit basis states, $|0\rangle|0\rangle$, $|1\rangle|0\rangle$, $|0\rangle|1\rangle$, and $|1\rangle|1\rangle$. In terms of these basis states, the CNOT works as follows:

$$\begin{cases} |0\rangle\,|0\rangle \to |0\rangle\,|0\rangle \\ |0\rangle\,|1\rangle \to |0\rangle\,|1\rangle \\ |1\rangle\,|0\rangle \to |1\rangle\,|1\rangle \\ |1\rangle\,|1\rangle \to |1\rangle\,|0\rangle. \end{cases}$$

It is therefore represented by the following unitary matrix in the basis of the four states:

$$U_{\rm CNOT} = \begin{pmatrix} 1 & 0 & 0 & 0 \\ 0 & 1 & 0 & 0 \\ 0 & 0 & 0 & 1 \\ 0 & 0 & 1 & 0 \end{pmatrix}. \tag{5.4}$$

Let us describe a convenient graphical representation of quantum algorithms. In this representation, each qubit is shown as a horizontal line (see Fig. 5.2(a)). The horizontal axis represents time: symbols to the right denote operations taking place later. Single-qubit gates are naturally given by squares located on the corresponding line. It is customary to use the following single-qubit gates:

$$\text{Hadamard: } H = \frac{1}{\sqrt{2}}\begin{pmatrix} 1 & 1 \\ 1 & -1 \end{pmatrix} = \hat{R}_{\boldsymbol{h}}(\pi/2), \quad \boldsymbol{h} = \frac{\boldsymbol{x}+\boldsymbol{z}}{\sqrt{2}};$$

$$\text{phase: } S = \begin{pmatrix} 1 & 0 \\ 0 & \mathrm{i} \end{pmatrix} = \mathrm{e}^{-\mathrm{i}\pi/4}\hat{R}_z(\pi/2);$$

$$\pi/8\text{: } T = \begin{pmatrix} 1 & 0 \\ 0 & \mathrm{e}^{\mathrm{i}\pi/4} \end{pmatrix} = \mathrm{e}^{-\mathrm{i}\pi/8}\hat{R}_z(\pi/4);$$

$$Z = \hat{\sigma}_z = -\mathrm{i}\hat{R}_z(\pi).$$

The gates $Z = S^2$, $S = T^2$, and T are the z rotations multiplied by the phase factors $(-\pi/2, -\pi/4, -\pi/8)$. While these global phase factors are irrelevant for a single qubit, they become important when the qubits interact.

The CNOT gate operating on qubits a and b is given by a vertical line connecting the corresponding qubit lines. The target (control) qubit is distinguished by the empty (filled) circle (Fig. 5.2(a)). Some combinations of elementary operations can be readily simplified; for example, with four Hadamard gates one can flip the target and control qubits of the CNOT gate.

Exercise 5.2. Prove the equivalence shown in Fig. 5.2(b) explicitly, writing down the corresponding matrices.

More complicated and more functional gates can be constructed in this way. Figure 5.2(c) shows the decomposition of the *Toffoli* gate. The Toffoli gate has two control qubits that do not change during the operation. The target qubit is negated provided both control qubits are set to "1." The Toffoli gate is used to prove that any *classical* computation algorithm can be implemented with quantum software. At first sight, this does not seem possible. Classical computation necessarily involves irreversibility: one cannot figure out the input of a classical computer from the output and the program. As we will learn, quantum computation is always reversible. A simple trick to overcome this is to use extra qubits with fixed initial states called *ancilla* qubits. For example, the classical NAND gate outputs the negated product of the two input bits a and b. To realize it in a quantum circuit, we take a Toffoli gate with control qubits a and b and the target qubit set to $|1\rangle$. As a result, the target qubit is in the state corresponding to the NAND output with inputs a and b.

How do we access the results of quantum computation? Upon completion of a quantum algorithm, the resulting wave function must be converted into classical information: this is called a *read-out*. In fact, this is the measurement of the wave function as described in textbooks on quantum mechanics. The measurement described there is *projective*: after the measurement of all its qubits, the quantum computer appears to be in a certain state $|j\rangle$. This occurs with probability $P_j = |f_j|^2$. Therefore, the measurement is obtrusive, and the final quantum state is destroyed by it. A single measurement cannot provide accurate information about f_j. To obtain accurate information, one should re-run the quantum algorithm achieving the same state $|f\rangle$. We estimate that one needs $\simeq (1/s^2)P_j$ runs to measure the probability P_j with precision s.

Control question 5.3. Explain the above estimate.

If all probabilities P_j are of the same order, one needs at least $2^N/s^2$ runs: yet another factor that may slow down the efficient quantum computation. An advanced quantum algorithm must therefore output information that can be read quickly. Examples of such algorithms are given below.

We conclude by listing the provisions imposed by the abstract quantum computer scheme described in any practical realization of such a device.

(i) *Qubits* The quantum information must be stored in the computer: there must be physical quantum two-level systems of some sort.

(ii) *Operation of the qubits* The quantum information shall be processed. For physical qubits, this is achieved by quantum manipulation, as described in Section 5.3.

(iii) *Read-out* There must be means to convert the quantum state of the qubits to a classical signal.

(iv) *Coherence* An ideal quantum computer isolated from any environment would keep the quantum information stored forever. It would remain *coherent*. A real physical system cannot be completely isolated from the environment. Interaction with environmental degrees of freedom results in decoherence (Section 6.7). This can be seen as errors in the course of quantum computation: we expect a certain wave function after our manipulation, but it appears to be different. A quantum computer is much more sensitive to environmental effects than a classical one. Since the errors are accumulated in time, the coherence issue would make quantum computing useless. Fortunately, the errors of quantum computation can be corrected at "software level" by clever algorithms [108]. The example error correction algorithms ensure that a quantum computer is operational provided the probability of error of each elementary gate does not exceed 10^{-4}.

(v) *Scalability* Like a classical computer, a quantum computer must have a large number of elementary logical units in order to be useful. This presumes scalability: one is able easily to increase the performance of a quantum computer by adding new qubits and/or new blocks of qubits.

5.1.3 Advanced algorithms

An *advanced* quantum algorithm, by definition, requires much fewer operations than the best classical algorithm solving the same problem. At the time of writing, there are several such known algorithms. They can be separated into two groups. The algorithms of the first group use the *quantum Fourier transform*, while those of the second group are based on *quantum search*.

Let us start with the quantum Fourier transform (QFT). A "classical" discrete Fourier transform is a transformation of $M = 2^N$-component complex vectors x_j to the vectors y_j given by

$$y_k = \frac{1}{\sqrt{M}} \sum_{j=0}^{M-1} e^{2\pi ijk/M} x_j \,. \tag{5.5}$$

There is an efficient classical algorithm – the fast Fourier transform – that solves the sum. One uses the same algorithm for a continuous Fourier transform making a discrete approximation. The QFT is a unitary operator $\hat{F}$ that transforms any basis state $|j\rangle$ as follows:

$$\hat{F}|j\rangle = \frac{1}{\sqrt{M}} \sum_{j=0}^{M-1} e^{2\pi ijk/M} |k\rangle; \tag{5.6}$$

the numbers j, k being the binary codes of the states $|j\rangle$, $|k\rangle$. It is easy to see that the QFT transforms the superposition with coefficients x_j to the superposition with y_j given by Eq. (5.5); that is, it performs a classical Fourier transform.

The QFT can be implemented with $\simeq N^2$ single- and two-qubit quantum gates. This is an enormous increase in speed in comparison with the classical fast Fourier transform, which requires $\simeq N \cdot 2^N$ operations. To understand why the QFT can be this fast, one notes that there is a useful product representation for $\hat{F}$:

$$\hat{F}|j\rangle = \prod_{n=1}^{N} \frac{\left(|0\rangle + e^{i\phi_n}|1\rangle\right)}{\sqrt{2}}.$$

Here n numbers the qubits, and the product is the direct product of their wave functions. Each term in the product resembles the action of the Hadamard gate on a basis qubit state apart from the phase factor. The phase ϕ_n, however, is not determined by the state of the qubit n only. It is expressed in the form of the binary code, $\phi_n = 2\pi j/2^n$. Since 2π additions to the phase do not change the phase factor, the phase ϕ_1 for the first qubit depends on the state of the last qubit j_N only, the phase ϕ_2 depends on the states of the two last qubits, and so on. The quantum algorithm consists of single-qubit Hadamard gates followed by conditional two-qubit phase gates R_k that transfer information about the state of the control qubit to the phase shift of the target qubit:

$$\hat{R}_k = \begin{pmatrix} 1 & 0 & 0 & 0 \\ 0 & 1 & 0 & 0 \\ 0 & 0 & 1 & 0 \\ 0 & 0 & 0 & e^{2\pi i/2^k} \end{pmatrix}.$$

The gates are applied as shown in Fig. 5.3. Each qubit is affected once by its Hadamard gate. In addition, the first qubit acquires $N-1$ phase factors from R gates, the second one is affected by $N-2$ gates, and so on. In total, this gives $N(N+1)/2$ gates. However, we are not quite done. In comparison with Eq. (5.6), the order of the qubits is reversed. We need to apply two-qubit SWAP gates that swap the wave functions of two qubits. A SWAP gate can be implemented with three CNOT gates (Fig. 5.3(b)) and one needs $N/2$ SWAPs. In total, the QFT requires $N(N+4)/2$ operations.

This sounds like an exponential speed-up in comparison with the classical algorithm. Owing to widespread applications of the Fourier transform in science and statistical analysis, that would be really great news. Unfortunately, the read-out of the result presents a difficulty. Generally speaking, the resulting Fourier harmonics are of the same order. As discussed, it takes more than 2^N measurements to read the result. Therefore, the QFT by itself is not an advanced algorithm. However, it is used as an essential part of a set of advanced algorithms. Since these algorithms make heavy use of advanced math, in particular number theory, we just name them.

Since the major obstacle is the read-out, the output of an advanced algorithm would be better concentrated on a few basis states of the computer or its part to be read out. Ideally,

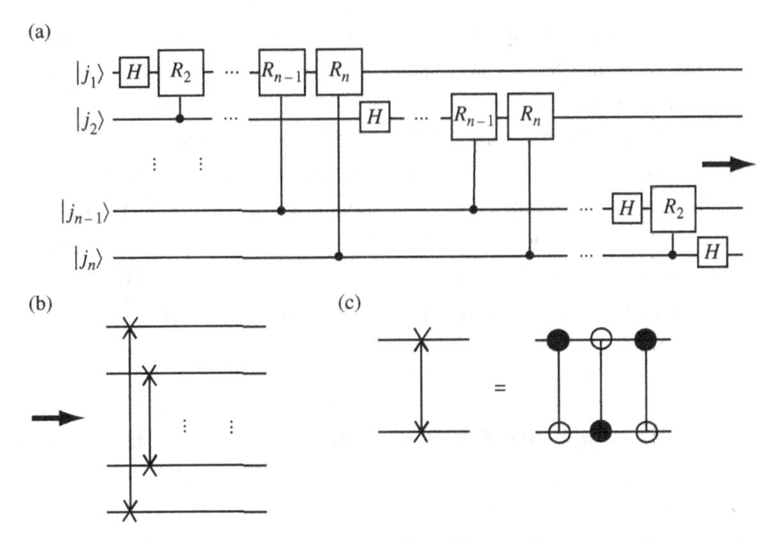

Fig. 5.3. Quantum Fourier transform. (a) Each qubit is affected by a Hadamard gate and the phase shifts conditioned on the state of "lower" qubits. (b) The pairwise swap of the qubits gives the QFT of the input. (c) A swap gate can be made from three CNOTs.

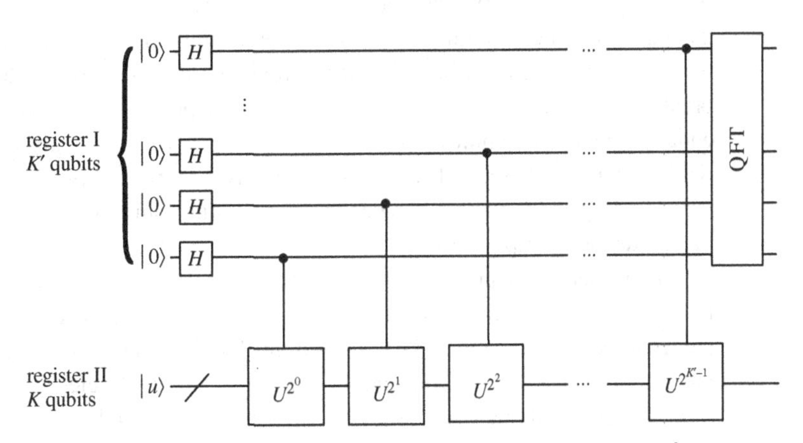

Fig. 5.4. Phase estimation. The second register is in an eigenstate $|u\rangle$ of the operator $\hat{U}$. The output of the first register gives a K'-bit approximation of the phase of the corresponding eigenvalue.

it would be just a single basis state: in this case, the result could be read in a single shot. The *phase estimation* algorithm provides such a result. The problem it solves looks rather abstract: given a unitary operator $\hat{U}$ of dimension 2^K and its eigenvector $|u\rangle$, find a good approximation of its eigenvalue. Since the eigenvalues of a unitary matrix are of the form $e^{i2\pi\phi}$, with real ϕ, this is the estimation of the phase ϕ. The computer consists of two sets, or registers, of qubits: $K' < K$ qubits of the first register give the K'-bit approximation of the phase, and the K qubits of the second register are transformed by the powers of $\hat{U}$ conditioned on the states of the qubits of the first register (Fig. 5.4). One uses only powers of the form $\hat{U}^{2^k}, k < K'$. The algorithm is advanced, provided each power can be computed for less than $2^{K'}$ operations and therefore does not work for arbitrary $\hat{U}$. The algorithm is sketched in Fig. 5.4 and uses QFT on the first register. The output of the first register is exactly the binary representation of ϕ, provided it takes less than K' binary digits to write ϕ exactly. Otherwise, it gives a good approximation of ϕ. This solves the read-out problem: it can be read quickly. Suitable operators $\hat{U}$ arise in the problem of *order-finding*. This is a standard problem of number theory: given two positive integers x, M, $x < M$, find the least integer r such that $x^r \bmod M = 1$, that is the remainder on dividing x^r by M is 1. For example, if $x = 3$ and $M = 16$, $r = 4$ since $3^4 = 5 \times 16 + 1$ and smaller powers of 3 give other remainders. The quantum order-finding involves $\hat{U}$ defined as $\hat{U}|y\rangle = |(xy) \bmod M\rangle$ for $y = 0, \ldots, M - 1$.

Control question 5.4. What are the non-zero matrix elements of $\hat{U}$ for $x = 3$ and $M = 2^4$? Can you show that the matrix is unitary? How many qubits are needed to represent the matrix?

The eigenvalues of this matrix are related to r, and the phase estimation algorithm can be used to find r efficiently: it requires $\simeq N^3$ gates, N being the number of bits required to represent M. This is exponentially faster than any classical known algorithm. Shor [109] has published the quantum order-finding algorithm and noted that the *factoring* integers into prime numbers can be reduced to order-finding. Since the factoring may be used to break the RSA cryptosystem – the most popular public key cryptosystem – and actually endanger modern communication security, this result has brought quantum computing to public attention. Kitaev [110] has demonstrated that all applications of QFT, including order-finding and factoring, can be viewed as partial solutions to a single general problem – the hidden subgroup problem.

Another set of advanced algorithms is based on the *quantum search* algorithm described by Grover [111]. Suppose we have M stones in a sack and that we know that $K \ll M$ stones are diamonds. Our task is to find all the diamonds. We cannot look into the sack, but we can get the stones out one-by-one. It is intuitively clear that we have to take out $\sim M$ stones. (Although we could be extremely lucky and find all the diamonds in the first K attempts). Indeed, the classical search algorithm requires $\sim M$ steps. The quantum search can be done in $\sqrt{M/K}$ steps. The increase in speed is not exponential, but it is still considerable.

In distinction from QFT methods, we can understand the quantum search speed-up in physical terms. The classical probability of finding a diamond in a single attempt is K/M. Let us consider a wave function defined on an M-dimensional basis, where K basis states are "diamonds." If we assume that the function is uniformly spread over all basis states, a typical *amplitude* corresponding to a diamond state is $\simeq 1/\sqrt{M}$. Since the quantum dynamics involve the amplitudes rather than the probabilities, the diamond states can be found faster.

To provide a quantitative illustration, we take the continuous version of the quantum search algorithm, not separating it into different discrete operations. Let us have N qubits with $M = 2^N$ basis states. We label the basis "diamond" states by $|x_k\rangle$, $k = 1, \ldots, K$. We will consider a special state: the equal-weight superposition of all basis states (including the "diamond" ones), $|c\rangle = (\sqrt{M})^{-1}\sum_j |j\rangle$. To perform the quantum search, we need the following Hamiltonian:

$$\hat{H} = \hbar\omega_0 \left(\sum_{k=1}^{K} |x_k\rangle\langle x_k| + |c\rangle\langle c| \right). \tag{5.7}$$

Initially, the wave function is chosen to be $|c\rangle$. Its time evolution is governed by the Hamiltonian. At any time moment, the wave function can be represented by two-time dependent amplitudes α, β,

$$|\psi(t)\rangle = \alpha(t) \sum_{k=1}^{K} |x_k\rangle + \beta(t)|c\rangle.$$

The resulting time-dependent Schrödinger equation for α, β,

$$\begin{aligned} \mathrm{i}\frac{\mathrm{d}\alpha}{\mathrm{d}t} &= \omega_0(\alpha + \beta/\sqrt{M}), \\ \mathrm{i}\frac{\mathrm{d}\beta}{\mathrm{d}t} &= \omega_0(\beta + K\alpha/\sqrt{M}), \end{aligned}$$

can be readily solved to yield $\alpha(t) = -\mathrm{i}\mathrm{e}^{-\mathrm{i}\omega_0 t}\sin(\omega_0 t\sqrt{K/M})/\sqrt{K}$; $\beta(t) = \mathrm{e}^{-\mathrm{i}\omega_0 t}\cos(\omega_0 t\sqrt{K/M})$.

Control question 5.5. Can you derive the above Schrödinger equation? Beware of the fact that $\langle c|x_k\rangle \neq 0$.

Let us note now that, at some value of $t = (\pi/2\omega_0)\sqrt{M/K}$, the coefficient β becomes zero. At this moment, we interrupt the computation and end up with an equal-weight superposition of "diamond" states". The measurement of the output thus produces a diamond state, each state sought arising with probability $1/K$. So we need $\simeq K$ measurements to find them all; this costs $\simeq\sqrt{KM}$ gates.

We still need to explain the discrete version of the algorithm that conforms to the principles of a quantum computer. There is, however, a seeming logical inconsistency. We need the Hamiltonian, Eq. (5.7), to perform the search of $|x_k\rangle$, and it explicitly contains the states sought. It seems that we cannot realize this Hamiltonian without knowing the search result! The point here is the difference in efforts between *finding* the result and *checking* if the result is correct. This difference is rather apparent for diamonds; a more mathematical example is the difference between the calculation of an indefinite integral and checking the correctness of the results. The discrete version of the algorithm involves a *quantum oracle*. Basically, it checks if a basic state is a diamond. A quantum oracle is a gate. It involves its own qubit register. The qubits of the oracle register receive the information about the state of the main register where the "diamond" states are sought. The algorithm of the oracle has to be designed separately; for example, one implements different algorithms to search for diamonds and gold nuggets. This is of no interest to us provided this algorithm is efficient. What is of interest is what the oracle does to the qubits of the main register. By convention, it flips the signs of all amplitudes of the diamond basis states while leaving other amplitudes intact. In operator terms, the oracle gate corresponds to $\hat{O} = \hat{1} - 2\sum_{k=1}^{K} |x_k\rangle\langle x_k|$. The application of the oracle gate is followed by an operation that corresponds to $2|c\rangle\langle c| - \hat{1}$. This is implemented as two Hadamard transforms of all qubits, separated by an operation that shifts the phase of all states except $|0\rangle$ by π. All together, it makes up a single Grover iteration. The quantum search algorithm consists of a finite number, $\approx\sqrt{M/K}$, of Grover iterations. Before the iterations are applied, the state of the main register is initialized to state $|c\rangle$ by the Hadamard transform of all qubits. In the limit $K \ll M$, the effect of the iterations corresponds to the dynamics governed by the Hamiltonian, Eq. (5.7). We note that for successful operation of the quantum search algorithm we need to know the total number of diamonds, K. This number can be efficiently found using another advanced algorithm – quantum counting – that combines the quantum search and the phase estimation.

5.2 Quantum goodies

In this section we present a set of concepts from the field of *quantum information*. We cannot say that at the time of writing these concepts form an indispensable part of the field of quantum transport. However, there is a substantial current interest towards the implementation and development of the quantum transport devices suitable of transmitting quantum information or at least of illustrating the concepts in question. There are several theoretical proposals along these lines, which will be discussed briefly at the end of the section. One cannot exclude the possibility that quantum information ideas will become the mainstream of quantum transport in near future. This is why we include this section.

There is a large, well developed field of information science, the analysis of communication, with the information transfer being an integral part of it. It is important for us to note that communication necessarily involves at least two *subjects*, say persons that exchange the information for some reason and are interested in faultless transmission and reception of the messages. This is in contrast to physical communication channels – copper wires or glass fibers – that transmit electricity or light without caring about the possible informational content. Communication science can be very abstract, so the subjects may be presented as mathematical objects. Albeit they remain the subjects. On the other hand, there is a well developed science called quantum mechanics that operates with a linear equation – the Schrödinger equation for the wave function $|\Phi\rangle$ – and expresses all physical quantities in terms of expectation values $\langle\Phi|\hat{A}|\Phi\rangle$ defined by the wave function and the corresponding Hermitian operator $\hat{A}$. We should note that such pure quantum mechanics has no subjects involved.

Loosely speaking, the field of quantum information comes about from a combination of quantum mechanics and information-hungry subjects. These subjects are customarily called *Alice* and *Bob*, and they exchange information using both quantum mechanical and classical means. The abstract representation of the quantum mechanical part includes an exchange of qubits and performing quantum gates. It has been proven that quantum mechanics provides novel opportunities for information exchange that are absent in classical world. This makes quantum ways potentially *better* than classical ones. This is why we call the concepts arising in this context *goodies*: they involve the notion of *better* that makes strictly no sense in the absence of subjects. For instance, one of the goodies is *quantum entanglement*, which is commonly considered as a *resource* for an efficient quantum information transfer. From the point of view of pure quantum mechanics, entanglement cannot even be associated with a physical quantity: there is no expectation value giving the entanglement.

When we deal with classical information, it is a very good assumption that once stored it remains uncorrupted and can be retrieved at any time. In addition, the classical information can be read and/or copied to another location without destroying it. An example is the information stored and processed in the Internet. These properties, which we often take for granted, eventually allow us to consider information separately from the physical devices used to store or transmit it. These properties do not hold for quantum information that is stored in a wave function, say that of a qubit. It is common to assume the projective read-out. If the initial wave function is $\alpha|0\rangle + \beta|1\rangle$, such a read-out is probabilistic, giving "0" with probability $|\alpha|^2$ and "1" with probability $|\beta|^2$. Also, the read-out distorts the wave function: it is either $|0\rangle$ or $|1\rangle$ after the measurement. Thus the superposition coefficients (i) are erased by the measurement and (ii) cannot be precisely determined by this measurement. These two very unpleasant obstacles could have been circumvented if we were able to *clone* the unknown quantum information without reading it. We could take a single qubit and produce two with identical wave functions. Repeating the process, we could produce as many copies of the quantum state as required to measure the superposition coefficients with the desired precision.

Control question 5.6. How many qubits are required to measure $|\alpha|^2$ with 1% accuracy?

Unfortunately, cloning is prohibited by quantum mechanical laws, and this is known as the *no-cloning theorem*. This is a simple consequence of the fact that the Schrödinger equation is linear. The resulting wave function after the first cloning would be given by

$$(\alpha|0\rangle + \beta|1\rangle)(\alpha|0\rangle + \beta|1\rangle) = \alpha^2|00\rangle + \alpha\beta(|01\rangle + |10\rangle) + \beta^2|11\rangle,$$

that is, quadratic in α, β. Therefore, no quantum algorithm can clone a qubit. It is important that the quantum state to be cloned is unknown. If we know α, β in advance, there is no problem to set to this state as many qubits as we want.

5.2.1 Quantum teleportation and key distribution

All this so far sounds hopeless. The good news is that the unknown quantum information can be transmitted from a subject to another subject without losses. The simplest scheme that allows for it bears the rather misleading name *quantum teleportation*. There are two subjects, Alice and Bob. Alice has a piece of unknown quantum information. It is stored in a qubit in the form of the superposition coefficients. She wants to transfer the information to Bob. The simplest thing to do would be just to send the qubit to him with DHL. This is impossible: there would be no story otherwise. However, Alice does have classical means to communicate with Bob (a cell phone). At first glance, it does not help much. If Alice measures the qubit, she would destroy it without learning α, β. She would have almost nothing to report to Bob, only the "0" or "1" outcome of the probabilistic read-out.

Alice and Bob can solve the problem in hand if they prepare themselves for such a situation. Namely, they have to come together and prepare in advance a pair of the qubits in the state

$$|B_{00}\rangle = \frac{1}{\sqrt{2}}\left(|00\rangle + |11\rangle\right).$$

Then they separate. Bob takes the first qubit while Alice gets the second one. Upon acquiring the third qubit in the unknown state, she inserts the second and the third qubit into her personal quantum computer. The quantum state of the three qubits is given by

$$|i\rangle = |B_{00}\rangle(\alpha|0\rangle + \beta|1\rangle).$$

The further discussion uses a special basis for any two qubits: the Bell basis. By definition, the four Bell states are as follows:

$$\begin{aligned} |B_{00}\rangle &= \frac{1}{\sqrt{2}}(|00\rangle + |11\rangle); \; |B_{01}\rangle = \frac{1}{\sqrt{2}}(|01\rangle + |10\rangle); \\ |B_{10}\rangle &= \frac{1}{\sqrt{2}}(|00\rangle - |11\rangle); \; |B_{11}\rangle = \frac{1}{\sqrt{2}}(|01\rangle - |10\rangle). \end{aligned} \tag{5.8}$$

State $|i\rangle$ already uses the Bell basis for the first and second qubit. However, these qubits are now separated. The idea is to write the same wave function using the Bell basis of Alice's qubits, the second and the third. Some algebra yields

$$\begin{aligned} |i\rangle &= \frac{1}{2}(\alpha|0\rangle + \beta|1\rangle)|B_{00}\rangle + \frac{1}{2}(\beta|0\rangle + \alpha|1\rangle)|B_{01}\rangle \\ &+ \frac{1}{2}(\alpha|0\rangle - \beta|1\rangle)|B_{10}\rangle + \frac{1}{2}(-\beta|0\rangle + \alpha|1\rangle)|B_{11}\rangle. \end{aligned}$$

It is now advantageous for Alice to measure her qubits in this Bell basis. To enable this measurement, she applies the quantum gate that transforms the Bell states into the basis states of her quantum computer, $|B_{ab}\rangle \to |ab\rangle$, $a, b = 0, 1$. This gate is a combination of a CNOT and a Hadamard gate (Fig. 5.5). The result of the measurement, (r_1, r_2), which can be "00," "01," "10," or "11," is now communicated to Bob over the cell phone. Now Bob has to apply a single-qubit gate X to his qubit. The crucial part is that the gate he must apply depends on Alice's report, $X = X(r_1, r_2)$. Let us look at the above equation. If the reported result is "00," Bob does not have to do anything; the state of his qubit is $(\alpha|0\rangle + \beta|1\rangle)$, precisely the unknown state to be transmitted to him. Otherwise he needs to perform the following gate: $-iR_x(\pi)$ if the report is "01," $-iR_z(\pi)$ if "10," and $-R_y(\pi)$ if it is "11." This accomplishes the quantum teleportation: the unknown state is copied into Bob's qubit. It is necessary to note that Alice has destroyed the initial unknown state by her measurement: otherwise it would be cloned.

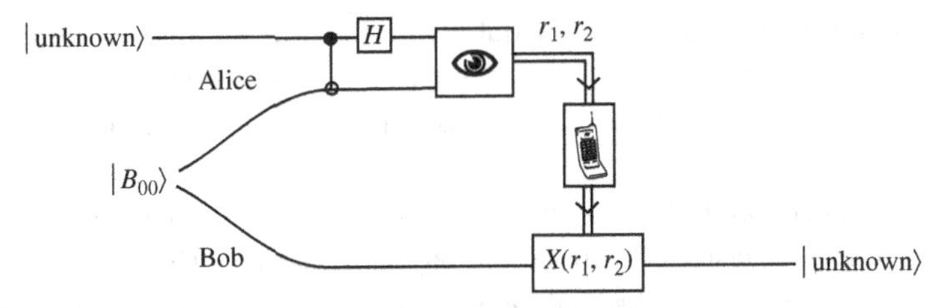

Fig. 5.5. **Quantum teleportation protocol. Alice can transmit an unknown state to Bob by using a classical communication line (shown as a double line) to report the results of her measurement (eye symbol). The condition is that Alice and Bob share a two-qubit entangled state $|B_{00}\rangle$.**

Quantum teleportation makes use of a special property of the Bell state $|B_{00}\rangle$: it is *entangled*, that is, it cannot be represented as a direct product of the quantum states of the two qubits. This is in distinction from the basis states of the qubits $|ab\rangle$; they are indeed the products, $|ab\rangle \equiv |a\rangle|b\rangle$. Not every entangled state will do the job: $|B_{00}\rangle$ belongs to a manifold of *maximally entangled* states. Any two-qubit state may be presented as the superposition of the basis states, $|\Psi\rangle = \sum_{ab} \Psi_{ab}|ab\rangle$. For such a state, one can look at the *reduced density matrix* of one of the qubits, say the first one. In this case, the possible states of the second qubit are of no interest to us, so the density matrix is obtained by summing up over them:

$$\rho_{ab} = \sum_c \Psi_{ac}\Psi^*_{bc} \Rightarrow \hat{\rho} = \hat{\Psi}\hat{\Psi}^\dagger. \tag{5.9}$$

In Eq. (5.9) we represent the wave function Ψ_{bc} by the matrix $\hat{\Psi}$. For a maximally entangled state, this density matrix is diagonal.

Exercise 5.3. Show that all the states of the Bell basis are maximally entangled. Show that a state $|\Psi\rangle$ is maximally entangled if the matrix is proportional to a 2×2 unitary matrix. Explicitly write a general state satisfying this condition. How many independent parameters (except the global phase factor) enter this expression?

The diagonal density matrix means that if Bob, instead of teleportation, measures his qubit, he gets "0" or "1" with equal probability, that is, completely randomly. The same holds for Alice. However, let us note that for the state $|B_{00}\rangle$ the outputs are *fully correlated*: Alice and Bob always measure the same result. For an arbitrary maximally entangled state, this is not the case; the probability of having the same outputs is given by $|\Psi_{00}|^2 + |\Psi_{11}|^2 < 1$. Let us show how this can be improved. Both Alice and Bob can apply a gate to their qubit. Generally, their gates are 2×2 unitary matrices, $\hat{u}^A$ for Alice and $\hat{u}^B$ for Bob. The state after applying the gates is given by

$$\Psi'_{ab} = \sum_{cd} u^B_{ac} u^A_{bd} \Psi_{cd} \Rightarrow \hat{\Psi}' = \hat{u}^B \hat{\Psi} \left(\hat{u}^A\right)^{\mathrm{T}}, \tag{5.10}$$

where we switch to matrix notations in the second equation. We note that in these notations $|B_{00}\rangle$ is proportional to the unity matrix $|B_{00}\rangle \to \hat{1}/\sqrt{2}$. It follows that either Alice or Bob can apply a gate to their qubit to transform the maximally entangled state to $|B_{00}\rangle$. A correcting gate is $(\hat{\Psi})^{-1}/\sqrt{2}$ for Bob (if Alice does nothing) and $(\hat{\Psi}^{\mathrm{T}})^{-1}/\sqrt{2}$ for Alice (if Bob does nothing). After this correction, the resulting state is $|B_{00}\rangle$ and their outputs coincide.

Quantum teleportation proves that the unknown state, a piece of quantum information, can be transferred without the physical transfer of its carrier. Though interesting, it does not seem to be practical. What is the use of transferring the information if it cannot be read unambiguously? Unexpectedly, the use of quantum information relies on just this drawback; it is almost impossible to intercept it.

Let us illustrate this point by discussing the *EPR quantum key distribution* protocol. We find Bob and Alice in a very unfortunate situation. All informational channels they may use are listened in to by a third subject, Eve. Bob and Alice can code their messages. To this end, they need to share a key: a long sequence of bits known to both, but not to Eve. This is, however, precisely the problem. If Bob devises the key he has to transmit it to Alice, and Eve would intercept it. The problem with the key is solved if they have prepared in advance a number of entangled pairs, say in the state $|B_{00}\rangle$. Let us first describe the protocol and then understand why they need every line of it.

- Alice and Bob communicate which pair they measure.
- Each decides randomly whether to apply the Hadamard gate to his/her qubit before the measurement. It is vital that they do not communicate the decision prior to the measurement.
- They perform the measurement of their qubits and keep the result secret.
- They communicate to figure out if the gate has been applied. If both have either applied or not applied the gate, they keep the result as a bit of the key: they are sure that the readings are the same. Otherwise, the result is discarded.
- They repeat the above procedure with the next pair until they get enough bits for the key.

If neither Alice nor Bob applies the gate, the equal outcome of the measurement is the property of $|B_{00}\rangle$ discussed above. If they both apply a gate, the resulting state is $\hat{u}^B(\hat{u}^A)^{\mathrm{T}}/\sqrt{2}$ (see Eq. (5.10)). It is equivalent to $|B_{00}\rangle$ if $\hat{u}^A = (\hat{u}^B)^*$. Since the Hadamard gate is a real matrix, this condition is fulfilled, and Alice and Bob's results are guaranteed to be identical. They could use any gates satisfying the above condition. Thus, they can establish a private key.[1]

The good news is that the key established by Bob and Alice is secure. Eve does not know the key provided she only listens in their communication channel. If she also tampers with the supply of their entangled pairs, her efforts will result in a *corrupt* key: some of the bits acquired by Bob and Alice will be different. Alice and Bob will notice that their versions of the key do not coincide. Suppose Bob transmits a message in plain English. If the keys differ, Alice would get gibberish. To check this for sure, they need to sacrifice a part of their key and communicate it to each other. They will pinpoint the difference and thereby detect Eve's evil activity.

[1] Alice and Bob cannot use a public key as modern Internet users do. As explained, a public key can be broken by a quantum computer, and Eve has one.

To see how this works, let us investigate the opportunities open to Eve. She can connect her quantum computer to the qubits of the entangled pair and transform $|B_{00}\rangle$ to any other state. However, she must avoid outputs that contain either $|01\rangle$ or $|10\rangle$: this yields the probability of Alice and Bob measuring different bits in case none of them applies the gate. If they never apply the gate, Eve's tactics are simple: she chooses the bit she wants the parties to receive. Then she transforms $|B_{00}\rangle$ to either $|00\rangle$ or $|11\rangle$ depending on her choice. This would result in an uncorrupted key, designed and perfectly known by Eve. To exclude this, Alice and Bob must use the gate from time to time. If only Bob applies the gate, the product state $|00\rangle$ received from the corrupted pair is transformed into $(|00\rangle + |10\rangle)/\sqrt{2}$. The key is corrupted with one-half probability. However, if Eve knows whether or not they use the gates, she can adjust the results. She simply corrects the state of the corresponding qubit with the inverse gate if necessary. This is keeping the decision about the gates secret is an important part of the protocol. The only way Eve can conceal her tampering is to transform $|B_{00}\rangle$ to the same $|B_{00}\rangle$ apart from a phase factor. But in this case she does not get any information concerning the key!

The "EPR" in the name of the protocol stands for Einstein, Podolsky, and Rosen, who jointly published a seminal article in 1935 [112]. The authors believed that a wave function does not provide a complete description of reality. In particular, they disliked the fact that the outcome of a measurement is not unambiguously defined by the initial state, and they explicitly gave a wave function that does not possess this property. In modern terms, their wave function was a maximally entangled state. In the decades that followed, scientists were ready to assume the existence of "hidden" variables that are not in the Schrödinger equation but nevertheless determine the measurement outcomes. The probabilistic nature of the outcomes just comes from our ignorance of the hidden variables. In the protocol described, a hidden variable is the choice of Eve. The variable determines the outgoing state – $|00\rangle$ or $|11\rangle$ – and thereby the measurement outcomes.

The EPR key distribution protocol would not work if hidden variables exist, at least not in a simple form. Bell [113] and Clauser, Horne, Shimony, and Holt (known as CHSH) [114] have quantified the statement, enabling its experimental verification. They suggested a criterion commonly referred to as the Bell inequality. Let us consider Alice and Bob again. Now they do not have any secrets. Rather, they investigate a source of qubit pairs, trying to pinpoint its state. They vary $\hat{u}^A$ and $\hat{u}^B$ – their gates to be applied to the qubits before the measurement. They repeat the measurement many times, and finally they determine the three independent probabilities of four possible outcomes. These probabilities depend on the gate settings: $P_{ab}(\hat{u}^A, \hat{u}^B)$. The criterion makes use of a single parameter, given by

$$E = P_{00} + P_{11} - P_{01} - P_{10},$$

which determines the correlation of outcomes: $-1 < E < 1$, the limiting values corresponding to absolute (anti) correlation. We have seen that if the state of the source is maximally entangled, Alice and Bob can always adjust the gates to achieve the absolute correlation, $E = 1$. However, recalling Eve, we understand that these settings also give coinciding outcomes for two product states: $((\hat{u}^B)^{-1}|0\rangle)((\hat{u}^A)^{-1}|0\rangle)$ and

$((\hat{u}^B)^{-1}|1\rangle)((\hat{u}^A)^{-1}|1\rangle)$. The source can give out these states randomly. In this case, Alice and Bob will measure the same probabilities as for the maximally entangled state.

The EPR protocol suggests that each observer should use at least two settings, $\hat{u}$ and $\hat{u'}$. Let us define the Bell parameter as follows:

$$\mathcal{B} = E(\hat{u}^A, \hat{u}^B) + E(\hat{u'}^A, \hat{u'}^B) + E(\hat{u'}^A, \hat{u}^B) - E(\hat{u}^A, \hat{u'}^B);$$

it is determined by four possible settings. It has been proven that, for any hidden variable scheme $\mathcal{B}^2 \leq 4$, while quantum mechanics can provide bigger values, $\mathcal{B}^2 \leq 8$.

If there are hidden variables λ, they define the result of the measurement $X(\lambda)$, $Y(\lambda)$ for Alice and Bob, respectively, $X, Y = 0, 1$. We assume that the hidden variables are local so that the measurement result for Bob does not depend on the settings of Alice, and vice versa. In quantum terms, this implies that, at a given value of λ, the source gives out a product state $((\hat{u}^B)^{-1}|Y(\lambda)\rangle)((\hat{u}^A)^{-1}|X(\lambda)\rangle)$. Let us note that

$$E(\hat{u}^A, \hat{u}^B) = \langle (2X-1)(2Y-1)\rangle \equiv \int \mathrm{d}\lambda\; P(\lambda)(2X(\lambda;\hat{u}^B) - 1)(2Y(\lambda;\hat{u}^A) - 1).$$

Here, $P(\lambda)$ is the probability distribution of the hidden variable. The Bell parameter is written as follows:

$$\mathcal{B} = 2\langle (X + X' - 1)(2Y' - 1) - (X - X')(2Y - 1)\rangle; \tag{5.11}$$

$$X \equiv X(\hat{u}^B),\; X' \equiv X(\hat{u'}^B),\; Y \equiv Y(\hat{u}^A),\; Y' \equiv Y(\hat{u'}^A). \tag{5.12}$$

Since the outcomes X, X', Y, and Y' only assume the values of 0 and 1, the expression between the angular brackets in Eq. (5.11) can only assume values ± 1. Since it is integrated with positive measure $P(\lambda)$, $-2 < \mathcal{B} < 2$, and we prove that, in the hidden-variable scheme, $\mathcal{B}^2 \leq 4$. In quantum terms, the scheme is related to product states. Indeed, for any product state of the two qubits, $|\phi^B\rangle|\phi^A\rangle$, $-2 < \mathcal{B} < 2$.

Let us access the quantum boundary of $\mathcal{B}$. Making use of the matrix notation for a two-qubit wave function, we write

$$E(\hat{u}^A, \hat{u}^B) = \mathrm{Tr}\left(\hat{\sigma}_z \hat{\Psi}' \hat{\sigma}_z \hat{\Psi}'^{\dagger}\right),$$

where $\hat{\Psi}'$ is given by Eq. (5.10). We express it in terms of $\hat{\Psi}$, $E = \mathrm{Tr}\left(\hat{\sigma}^B \hat{\Psi} \hat{\sigma}^A \hat{\Psi}^{\dagger}\right)$, where the settings of Alice and Bob are represented by the matrices $\hat{\sigma}^A = (\hat{u}^{A\dagger}\hat{\sigma}_z\hat{u}^A)^{\mathrm{T}}$ and $\hat{\sigma}^B = \hat{u}^{B\dagger}\hat{\sigma}_z\hat{u}^B$, respectively. These matrices are unitary rotations of $\hat{\sigma}_z$, and are best written in the form $\hat{\sigma}^{A,B} = \boldsymbol{n}^{A,B}\cdot\hat{\boldsymbol{\sigma}}$, $\boldsymbol{n}^{A,B}$ being three-dimensional vectors of unit length. We know that if $\hat{\Phi}$ is a maximally entangled state, an extra gate of either Alice or Bob can bring it to $|B_{00}\rangle$. Assuming this extra gate has been applied, we replace $\hat{\Psi}$ by $\hat{1}/\sqrt{2}$ to obtain $E = (1/2)\mathrm{Tr}\left(\hat{\sigma}^B\hat{\sigma}^A\right) = \boldsymbol{n}^A\cdot\boldsymbol{n}^B$. The Bell parameter becomes

$$\mathcal{B} = \boldsymbol{n}^A\left(\boldsymbol{n}^B - \boldsymbol{n}'^B\right) + \boldsymbol{n}'^A\left(\boldsymbol{n}^B + \boldsymbol{n}'^B\right),$$

where $\boldsymbol{n}^{A(B)}, \boldsymbol{n}'^{A(B)}$ denote two settings of Alice (Bob) corresponding to the gates $\hat{u}^{A(B)}, \hat{u'}^{A(B)}$. The rest is a geometry exercise. To find the upper boundary, we need to maximize the Bell parameter with respect to four vectors of unit length. The maximum with respect to $\boldsymbol{n}^A, \boldsymbol{n}'^A$ is reached at $(\boldsymbol{n}^A, \boldsymbol{n}'^A) = \left(\boldsymbol{n}^B \mp \boldsymbol{n}'^B\right)/|\boldsymbol{n}^B \mp \boldsymbol{n}'^B|$ so that $\mathcal{B} = |\boldsymbol{n}^B + \boldsymbol{n}'^B| + |\boldsymbol{n}^B - \boldsymbol{n}'^B|$.

Control question 5.7. Can you explain the geometric meaning of the above expressions?

The expression for $\mathcal{B}$ reaches a maximum for any $\boldsymbol{n}^B \perp \boldsymbol{n}'^B$, $\mathcal{B}_{\max} = 2\sqrt{2}$. The vectors $\boldsymbol{n}^A, \boldsymbol{n}'^A$ are also orthogonal, are in the same plane as $\boldsymbol{n}^B, \boldsymbol{n}'^B$, and are rotated by 45 degrees with respect to them.

5.2.2 Entanglement

The common denominator of the above examples is the notion of entanglement. It is time now to talk about this in the context of a general quantum system. Let us consider a full Hilbert space H of all quantum states of the system. Next, consider a *bipartition* of the space into subspaces A and B, so that $H = A \otimes B$. We have already seen the space of the two qubits: it is naturally[2] partitioned into Alice's and Bob's qubit space. A more general example is the Hilbert space of two particles with coordinates $x_{1,2}$. An element of the space is a wave function $\psi(x_1, x_2)$. The natural partition is into the Hilbert spaces of each particles, represented by single-particle wave functions $\phi_1(x_1)$, $\phi_2(x_2)$. Indeed, any two-particle wave function can be represented as a superposition of the product states $\phi_i(x_1)\phi_j(x_2)$, $\psi(x_1, x_2) = \sum_{ij} c_{ij}\phi_i(x_1)\phi_j(x_2)$, $\phi_i(x)$ being the basis states of the single-particle Hilbert space.

Formally, entanglement is defined by negation. A wave function is not entangled if it is a product of the wave functions $|\phi^A\rangle$, $|\phi^B\rangle$ from the partitions A and B, respectively, $|\psi\rangle = |\phi^A\rangle|\phi^B\rangle$. The entanglement is called a resource. Many resources are physical quantities, like energy. It would be nice to have an operator of entanglement $\hat{E}_{\mathrm{nt}}$ such that its expectation value $\langle\psi|\hat{E}_{\mathrm{nt}}|\psi\rangle$ gives us the degree of entanglement. It is essential for our understanding of entanglement that this is impossible in principle: the entanglement is not a physical quantity in this respect. One could note that the Bell parameter is an expectation value of an operator, though a rather complicated one. If $\mathcal{B}^2 > 4$, the wave function cannot be a product state and the Bell operator is said to *witness* entanglement. The problem is that the Bell operator essentially depends on the settings of Alice and Bob. Given the settings, the Bell operator gladly witnesses the entanglement of some wave functions whilst ignoring that of others. Generally, for each wave function there exists such a witness operator, whereas no operator witnesses all the entangled wave functions.

A true measure of entanglement cannot be quadratic in the wave function. Therefore it cannot be linear in the density matrix. For pure quantum states, the measure is the *von*

[2] Naturally, that is, for us subjects. From a purely quantum point of view bipartition is not natural. There is an important "democratic" symmetry of quantum description: it is valid in any basis of the Hilbert space, so all bases are proclaimed equal. The bipartition breaks this symmetry, making some bases better than others. Thereby it smuggles the subjects into quantum mechanics.

Neumann entropy of the density matrix reduced on a partition. For the general density matrix $\hat{\rho}$, the von Neumann entropy is given by

$$S = -\mathrm{Tr}\left(\hat{\rho}\log_2(\hat{\rho})\right) = -\sum_i p_i \log_2 p_i, \tag{5.13}$$

where p_i are the eigenvalues of the density matrix. By definition of the density matrix, these are the probabilities of finding the system in the corresponding eigenstate $|i\rangle$. They have a well defined classical correspondence: the probabilities of being in a certain classical state i. Indeed, the von Neumann relation is the quantum generalization of the classical Shannon entropy, widely used in information theory.

Let us consider a pure state. Its entropy in the whole Hilbert space is zero, since the density matrix of a pure state $|\Psi\rangle$ is given by $\hat{\rho} = |\Psi\rangle\langle\Psi|$ and has a single non-zero eigenvalue 1 corresponding to the eigenvector $|\Psi\rangle$. Recalling the matrix representation of $|\Psi\rangle$ valid for any bipartition, and Eq. (5.9), we find the entropy of the reduced density matrix in terms of the eigenvalues p_i^A of the matrix $\hat{\Psi}\hat{\Psi}^\dagger$:

$$S_A = -\sum_i p_i^A \log_2 p_i^A.$$

It is known from basic statistical mechanics that the entropy is maximized by an equal distribution, $p_i^A = 1/M$, M being the dimension of subspace A. The maximum entropy $S_{\max} = \log_2 M$, the reduced density matrix of such a state being a diagonal matrix. We have seen previously that this defines a maximally entangled state: such states thus maximize the von Neumann entropy. For a product state, $S_A = 0$. Thus we demonstrate that the entropy provides the measure for entanglement.

With this in mind, we are ready for the final example: *entanglement distillation*. Alice and Bob share a two-qubit state. They wish to transform it into another state, preferably with a higher degree of entanglement. They can play with their single-qubit gates as usual. The transformed wave function is given by Eq. (5.10). Since the gates in use are unitary matrices, the eigenvalues of $\hat{\Psi}'\hat{\Psi}'^\dagger$ are the same as of $\hat{\Psi}\hat{\Psi}^\dagger$, resulting in the same entropy. So it looks like Alice and Bob cannot change the degree of entanglement of the pair in their possession.

There is a way, however, to achieve this. Alice entangles her qubit with another one, performs a two-qubit gate, and measures the auxiliary qubit (see Fig. 5.6). She communicates the result of the measurement to Bob. He may want to apply a gate based on the Alice's report. Depending on the output of the measurement, the transformed pair state has either higher or lower entropy than the initial one. Let us examine this in detail.

Alice starts with the wave function $|\Psi\rangle|0\rangle$, the last ket being the initial state of the auxiliary qubit. After the action of a general gate, the state is given by

$$\hat{U}_0|\Psi\rangle|0\rangle + \hat{U}_1|\Psi\rangle|1\rangle,$$

and the 2×2 matrices $\hat{U}_{0,1}$ act in the space of Alice's qubit. It is important to note that $\hat{U}_{0,1}$ are no longer unitary: they are two blocks of a 4×4 unitary matrix describing the two-qubit gate. Two other blocks not used here describe the transformation from the $|1\rangle$ state of the auxiliary qubit. The only constraint imposed by the unitarity of the larger matrix is

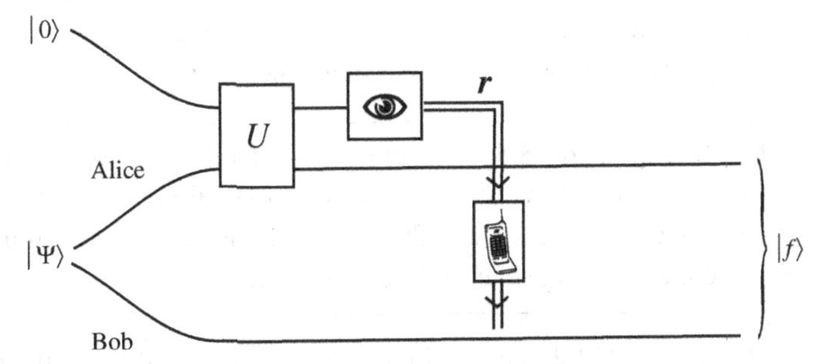

Fig. 5.6. A simple distillation algorithm. The goal is to make $|B_{00}\rangle$ from a known state $|\Psi\rangle$. To probe her qubit, Alice uses a designed gate U and an ancilla to be measured. If the measurement result is $r = 0$, the goal is achieved. If $|\Psi\rangle$ is not maximally entangled, the probability of success is given by $p_0 < 1$.

$$\hat{U}_0^\dagger \hat{U}_0 + \hat{U}_1^\dagger \hat{U}_1 = \hat{1}, \tag{5.14}$$

similar to the condition on the reflection and transmission matrices discussed in Section 1.3. We can say that after the measurement of the auxiliary qubit, the pair wave function becomes either

$$|\Psi_0'\rangle = \frac{\hat{U}_0|\Psi\rangle}{\sqrt{p_0}}; \; p_0 = \langle\Psi|\hat{U}_0^\dagger \hat{U}_0|\Psi\rangle,$$

if the measurement outcome is "0," or

$$|\Psi_1'\rangle = \frac{\hat{U}_1|\Psi\rangle}{\sqrt{p_1}}; \; p_1 = \langle\Psi|\hat{U}_1^\dagger \hat{U}_1|\Psi\rangle,$$

if the measurement outcome is "1," $p_{0,1}$ being the probabilities of the corresponding outcomes. The aim is to reach the maximally entangled state, $|B_{00}\rangle$ for simplicity, for one of the outcomes, say "0." If the state is known, there is no problem. In matrix notation,

$$\hat{1}/\sqrt{2} = \hat{\Psi}' = \hat{u}_B \hat{\Psi}(\hat{U}_0)^{\mathrm{T}}/\sqrt{p_0}.$$

Our aim is thus reached if Alice uses the gate with $\hat{U}_0 = (\hat{\Psi}^{-1})^{\mathrm{T}}\sqrt{p_0/2}$. Bob does not have to do anything.

At the moment, it looks like Alice can design the gate where the probability of favorable outcome reaches unity (or even more). This is because we did not look at the constraint given by Eq. (5.14), which implies that $\hat{U}_0$ is a kind of reflection matrix, that is, that the eigenvalues of $\hat{U}_0^\dagger \hat{U}_0$ – the reflection coefficients – cannot exceed unity. These eigenvalues for the matrix of interest are readily expressed through those of Alice's reduced density matrix $\hat{\rho}^A = (\hat{\Psi}\hat{\Psi}^\dagger)^{\mathrm{T}}$, $p^A_{1,2}$. Since $\hat{U}_0$ is proportional to the transposed inverse of $\hat{\Psi}$, the eigenvalues are given by $p_0/2p^A_{1,2}$. The condition they do not exceed unity provides the upper bound for the probability of the favorable outcome:

$$p_0(|\Psi\rangle) = 2\min(p^A_{1,2}).$$

The bound p_0 is a monotonic function of the entropy. If the initial state is a product state, that is, has zero entropy, p_0 is zero, and the creation of the Bell state is impossible. If the state is maximally entangled, so the entropy reaches maximum, $p_0 = 1$, and the Bell state can be created with certainty. In general, Alice takes a gamble in her quest for $|B_{00}\rangle$: she succeeds with probability p_0 and fails otherwise.

Control question 5.8. Assume that Alice has optimized the gate so that p_0 reaches the upper boundary. Express the entropy of the initial state in terms of p_0. Give the entropy of the final state in the cases of both success and failure. Does it make sense to reuse the resulting state in the case of failure?

If Alice and Bob have a single copy of the initial state, their strategy is too risky. The situation changes drastically if there are many copies of the same state. They repeat the procedure again and again, taking new copies, until they get $|B_{00}\rangle$. Moreover, in this case they can distill $|B_{00}\rangle$ from an *unknown* state or from a source that does not give a pure state but is characterized by a density matrix. To tune the gates, they have to spare several first attempts on the characterization of the state. There, Bob's measurements and communications become crucial (we note he played a rather passive role in the example considered). However, there may be a situation when all their attempts are futile. If the source gives the product states only, that is, does not contain entanglement, no entangled state can be produced by the methods described. The entanglement distillation plays an important role in quantum information. For instance, quantum error correction algorithms are based on a closely related concept.

All the above examples are basic in the rapidly expanding field of quantum information. The underlying theory is mathematically involved and aims to address the relevant problems of information technology using adequate language of devices and protocols. Many ideas – the violation of the Bell inequality, various protocols of quantum key distribution, quantum teleportation, entanglement distillation, and even quantum games – have been successfully realized experimentally (see Ref. [115] for a recent review). So far, the experiments are performed with photons generated in non-linear media, the quantum information being encoded in its polarization.

There is potential to realize these ideas in the field of quantum transport, and this is outlined in a significant number of theoretical proposals. However, the experiments have not been accomplished so far; it is interesting to discuss why. To start with, quantum transport systems readily provide vast "natural" resources of entanglement. The basis wave functions of identical particles – either fermions or bosons – are either antisymmetric or symmetric combinations of the product state, and are therefore entangled. The simplest example is the Pauli principle: two electrons sharing the same orbital state are entangled with respect to spin. Their wave function is the spin singlet,

$$|S\rangle = \frac{1}{\sqrt{2}}(|\uparrow\downarrow\rangle - |\downarrow\uparrow\rangle),$$

corresponding to the $|B_{11}\rangle$ state. Two electrons of a Cooper pair in a superconductor are also in a spin-singlet state. Moving electrons together and apart and letting them interact allows for control over their quantum state and therefore entanglement of the orbital and spin degrees of freedom. It is feasible to arrange a controlled sequence of quantum states that simulates any quantum information protocol and measures the results demonstrating the realization of the protocol. The reason this has not yet been done is rather subjective, to conform to the subjective nature of quantum information. There is not yet a natural realization of subjects – Alice and Bob – available in quantum transport. Without subjects, all quantum behavior can be understood, predicted, and measured without invoking quantum information concepts. In addition, a concrete experimental realization is frequently involved and requires much technological effort, while its results would not necessarily be superior to the advances in quantum optics. However, it appears that these subjective obstacles will be removed in the near future.

Most proposals naturally fall into three groups. Proposals of the first group use the interaction between electrons to move them in a controlled fashion: to extract them from a reservoir, to move them close together in order to entangle, and to separate them again. This imitates the situation of Alice and Bob. One can, for example, push the electrons into different reservoirs. The correlations of currents in these reservoirs may be related to the correlation of Alice's and Bob's outcomes and may be measured relatively easy. Most proposals from this group are reviewed in Refs. [116] and [117].

For the proposals of the second group, interactions are not necessarily essential. They use the analogy between the photons propagating in media or waveguides and electrons propagating and being scattered in nanostructures. The difficulty here is that there are many electrons doing this simultaneously. The multi-particle flow does not automatically and naturally separate into individual events involving either one or two, or even a few, particles. We encountered this before in Section 1.4. Confusingly, in many cases, the flow can be interpreted in terms of events, but the interpretation is ambiguous. To give a simple example, an electron tunneling between two reservoirs can be seen as a transfer of a *single* electron from the filled state l in the left reservoir to the empty state r in the right one. However, an equally legitimate interpretation is the creation of an electron and a hole at the tunnel junction at the moment of tunneling: the electron goes to state r, while the hole goes to l. So there are *two* particles simultaneously involved in the event. The proposals elaborated go as far as to demonstrate quantum teleportation and the violation of the Bell inequalities. They also prove that many electrons present in nanostructures potentially allow for more interesting and involved schemes than that of quantum optics. Most proposals are reviewed in Ref. [118].

The proposals of the third group combine photonics and electronics. Quantum dots, for instance, can be used as single-photon sources and even as sources of entangled photon pairs. Quantum dots can also detect photons. Thereby one can create quantum information in a nanostructure and transfer it using photons to another nanostructure.[3]

5.3 Quantum manipulation

In the preceding section, we discussed in detail the paradigm of quantum computing. Ideally, the paradigm implies our ability to perform an arbitrary unitary transformation $\hat{S}$ of a 2^N-component wave function of the N-qubit computer. It has been explained how this

[3] Such proposals are outlined in a private communication from L. P. Kouwenhoven.

general transformation can be carried out as a sequence of single-qubit and two-qubit unitary transformations. In this section, we consider how these transformations are realized in physical systems. This consideration is incomplete: we treat the computer as an isolated N-qubit system. No *actual* system is isolated; any computer is a part of the whole world. We postpone the discussion on the important effects of decoherence and dissipation created by outside agents until Chapter 6. The closed system is completely characterized by the 2^N-component wave function. The evolution of the wave function is determined by the time-dependent Schrödinger equation with a (time-dependent) Hamiltonian. Generally speaking, the Hamiltonian is determined by a set of external parameters – *handles*. We can change at will the values of the parameters (turn the handles) in time. This is known as quantum manipulation. The subject of this section is to show how we turn the handles to achieve the desired result. There are many possible physical realizations of qubits and handles, albeit the present discussion hardly depends on such realizations. The point is that the qubit as a physical system is ultimately simple, so that all possible qubits are described by the Hamiltonian of the same form. A sufficiently general Hamiltonian of an N-qubit computer consists of single-qubit Hamiltonians $\hat{H}_i$ and the pairwise interactions $\hat{H}_{ij}$ between different qubits:

$$\begin{aligned} \hat{H} &= \sum_{i=1}^{N} \hat{H}_i + \sum_{i<j}^{N} \hat{H}_{ij}, \\ \hat{H}_i &= h_a^{(i)} \hat{\sigma}_a^{(i)}, \quad \hat{H}_{ij} = U_{ab}^{(ij)} \hat{\sigma}_a^{(i)} \hat{\sigma}_b^{(j)} . \end{aligned} \tag{5.15}$$

Here the summation over repeated Cartesian indices $a, b = x, y, z$ is implied, and $\hat{\sigma}_a^{(i)}$ are Pauli matrices representing the three spin-projection operators of the qubit i. In principle, the Hamiltonian could also contain terms corresponding to the interaction of three or more qubits. For instance, a three-qubit term would be proportional to $\hat{\sigma}_a^{(i)} \hat{\sigma}_b^{(j)} \hat{\sigma}_c^{(k)}$. However, these terms are hardly useful for any manipulation, and their physical realization is less than obvious.

Single-qubit handles $h_a^{(i)}$ (to be referred to in this section as x-, y-, and z-handles) and qubit coupling handles $U_{ab}^{(ij)}$ correspond to very different physical quantities for different qubit realizations. We consider various realizations of solid state qubits in Sections 5.5, 5.6, and 5.7.

Since any quantum algorithm consists of elementary operations only involving single-qubit and two-qubit operations, for each elementary manipulation it is enough to consider either a single-qubit Hamiltonian, or a Hamiltonian involving two qubits. Thus, the big Hamiltonian, Eqs. (5.15), can be handled part by part for quantum information purposes – an enormous simplification that we take the advantage of.

5.3.1 Single-qubit manipulation

We concentrate now on a single qubit, given by

$$\hat{H}(t) = h_a(t) \hat{\sigma}_a. \tag{5.16}$$

To start with, let us assume that we can set all handles to zero, $h_a = 0$. Conveniently, the wave function of the qubit then stays constant. Now, we turn one of the handles (say z) to position h_z. The wave function does not change instantly; it evolves continuously in time. On the Bloch sphere, this evolution corresponds to a rotation around the z axis with angular velocity $-h_z/\hbar$. We keep h_z constant during the time interval τ and set the handle back to zero. Thereby we achieve a unitary transformation of the wave function (see Eq. (5.3)):

$$\hat{R}_z(\alpha) = \mathrm{e}^{\mathrm{i}\alpha\hat{\sigma}_z/2} = \cos\frac{\alpha}{2} + \mathrm{i}\sin\frac{\alpha}{2}\,\hat{\sigma}_z, \quad \alpha = -h_z\tau/\hbar. \tag{5.17}$$

To prove the above relations, one works in the basis where $\hat{\sigma}_z$ is diagonal and notes that, for any diagonal matrix $\mathrm{diag}(a_1, \ldots, a_N)$ the function of this matrix is given by $f(\mathrm{diag}(a_1, \ldots, a_N)) = \mathrm{diag}(f(a_1), \ldots, f(a_N))$. So,

$$\mathrm{e}^{\mathrm{i}\alpha\hat{\sigma}_z/2} = \exp\left[\mathrm{i}\alpha\begin{pmatrix} 1 & 0 \\ 0 & -1 \end{pmatrix}\right] = \begin{pmatrix} \mathrm{e}^{\mathrm{i}\alpha/2} & 0 \\ 0 & \mathrm{e}^{-\mathrm{i}\alpha/2} \end{pmatrix} = \cos\frac{\alpha}{2} + \mathrm{i}\sin\frac{\alpha}{2}\,\hat{\sigma}_z.$$

If we use other handles x, y and proper products of pulse height and duration, we implement the rotations $\hat{R}_x(\alpha)$, $\hat{R}_y(\alpha)$ around the corresponding axes. If we give a simultaneous pulse h_x, h_y, h_z of duration τ to all three handles, we can rotate the wave function about the axis $\mathbf{n} = (h_x, h_y, h_z)/h$ by the angle $-h\tau/\hbar$ ($h \equiv \sqrt{h_x^2 + h_y^2 + h_z^2}$). Since an arbitrary single-qubit operation is a rotation, such a *rotation by pulse* solves the manipulation problem in a straightforward way.

The real difficulty is our inability to switch off the Hamiltonian by setting the handles to zero. In most practical qubits, not all three handles are available for turning. The most common situation is that the evolution of the wave function cannot be stopped at will. In the absence of manipulation, the evolution is determined by a time-independent "background" Hamiltonian $\hat{H}_0$, and the turning of the handles adds a time-dependent part to it.

Resonant manipulation

Fortunately, the manipulation for this case does not appear to be much more complicated, and one may still use pulses. However, now they are pulses of an ac rather than a dc field. The frequency of the field matches the energy difference between two levels of the Hamiltonian $\hat{H}_0$.

To prove this, we present a celebrated approach to the time-dependent Schrödinger equation applicable close to the resonance – the *rotating wave approximation* (RWA). In this book, the RWA comes in two variants, which we call A and B.

We start with RWA-A. Let us work in the basis where the background Hamiltonian H_0 is diagonal. If the frequency matching the energy difference between its levels is Ω, $\hat{H}_0 = \mathrm{diag}(\hbar\Omega/2, -\hbar\Omega/2)$.

Control question 5.9. Why is it always possible to present $\hat{H}_0$ in this form?

The time-dependence of the handles is now oscillatory with frequency $\omega \approx \Omega$,

$$h_a(t) = \mathrm{Re}(\tilde{h}_a(t)\mathrm{e}^{-\mathrm{i}\omega t}),$$

$\tilde{h}_a(t)$ being the complex amplitudes that vary slowly at the time scale ω^{-1}. The Schrödinger equation is given by

$$i\hbar\frac{\partial}{\partial t}\begin{pmatrix}\psi_+(t)\\ \psi_-(t)\end{pmatrix} = \begin{pmatrix}\hbar\Omega/2 + h_z(t) & h_x(t) - ih_y(t)\\ h_x(t) + ih_y(t) & -\hbar\Omega/2 - h_z(t)\end{pmatrix}\begin{pmatrix}\psi_+(t)\\ \psi_-(t)\end{pmatrix}.$$

Without the manipulation, the general solution of the equation is given by $(\psi_+ e^{-i\Omega t/2}, \psi_- e^{i\Omega t/2})$. Let us search for the solution with the manipulation making $\psi_\pm$ time-dependent and replacing Ω by a close value ω, $(\psi_+(t)e^{-i\omega t/2}, \psi_-(t)e^{i\omega t/2})$. The resulting equation for this "new" wave function is as follows:

$$i\hbar\frac{\partial}{\partial t}\begin{pmatrix}\psi_+(t)\\ \psi_-(t)\end{pmatrix}$$
$$= \begin{pmatrix}\hbar(\Omega-\omega)/2 + h_z(t) & (h_x(t) - ih_y(t))e^{i\omega t}\\ (h_x(t) + ih_y(t))e^{-i\omega t} & -\hbar(\Omega-\omega)/2 - h_z(t)\end{pmatrix}\begin{pmatrix}\psi_+(t)\\ \psi_-(t)\end{pmatrix}.$$

This equation is still exact. The coming approximation relies on the separation of the frequency scales: we assume that the new wave function varies slowly, not changing significantly over time scale ω^{-1}. To this end, we average the time-dependent Hamiltonian over the oscillation period. For example,

$$\langle h_x(t)e^{i\omega t}\rangle = \frac{1}{2}\langle \tilde{h}_x(t) + \tilde{h}_x^*(t)e^{2i\omega t}\rangle = \tilde{h}_x(t),$$

where we have kept the slow term $\tilde{h}_x$, and the average of the rapidly oscillating second term proportional to $\exp(2i\omega t)$ averages to zero. The time-averaged Hamiltonian is given by (we omit the time dependence of the amplitudes for brevity)

$$\hat{H}_{\mathrm{RWA-A}} = \begin{pmatrix}\hbar\delta\omega/2 & \tilde{h}_x(t) - i\tilde{h}_y(t)\\ \tilde{h}_x^*(t) + i\tilde{h}_y^*(t) & -\hbar\delta\omega/2\end{pmatrix}, \tag{5.18}$$

where $\delta\omega \equiv \Omega - \omega$ is the frequency mismatch between the oscillation frequency of the manipulation handles ω and the working frequency of the qubit Ω.

Note first that the averaged Hamiltonian given by Eq. (5.18) is independent of the handle h_z. Furthermore, if the amplitudes $\tilde{h}_x$ and $\tilde{h}_y$ are time-independent, the Hamiltonian is stationary and has the following eigenvalues:

$$\pm\hbar\omega_{\mathrm{R}}/2 \equiv \sqrt{\hbar^2\omega^2/4 + |\tilde{h}_x|^2 + |\tilde{h}_y|^2}\,.$$

This defines a new frequency scale ω_{R} for slowly varying wave functions. By virtue of scale separation, we must require $\omega_{\mathrm{R}} \ll \Omega, \omega$. This sets the limits of applicability of RWA: small frequency mismatch $|\delta\omega| \ll \Omega$ and small oscillation amplitudes $|\tilde{h}_x|, |\tilde{h}_y| \ll \hbar\Omega$.

Let us set the frequency mismatch $\delta\omega$, as well as the amplitudes $\tilde{h}_x$ and $\tilde{h}_y$, to zero. We see then that the averaged Hamiltonian vanishes – we have managed to stop the evolution of the wave function, as discussed above. Let us assign pulses to these amplitudes. Such pulses are sharp at the slow time scale ω_{R}^{-1}, but smooth at the fast scale Ω^{-1}. These oscillation pulses work precisely as dc pulses described above for $\hat{H}_0 = 0$, and thus realize rotations of the wave functions. Now $\tilde{h}_x$ and $\tilde{h}_y$ correspond to the handles h_x and h_y. It might seem that, for the resonant manipulation, we miss out handle h_z. This is not strictly correct: the third handle is the frequency mismatch. A rotation around the z axis can be achieved by a slow variation of $\delta\omega$; see Eq. (5.18).[4]

We now fix certain initial conditions for the wave function, for example $(1, 0)$. We assume $\delta\omega = 0$. During the pulse with constant amplitude, the wave function evolves in time, $(\cos(\omega_R t/2), -\mathrm{i}\sin(\omega_R t/2))$. The probability of remaining in the initial state oscillates as a function of time with frequency ω_R:

$$P_+ = \cos^2(\omega_R t/2). \tag{5.19}$$

This phenomenon is known as *Rabi oscillations*, and ω_R as the *Rabi frequency*. For zero frequency mismatch, it is proportional to the amplitude of the resonant field.

The second variant of RWA, RWA-B, is most convenient to apply in the basis where a single oscillating handle $h_z(t) = \tilde{h}\cos\omega t$ is diagonal. The Hamiltonian $\hat{H}_0$ does not have to be diagonal in this basis and can be generally written as follows:

$$\hat{H}_0 = \frac{1}{2}\begin{pmatrix} \varepsilon & \mathcal{T} \\ \mathcal{T}^* & -\varepsilon \end{pmatrix}. \tag{5.20}$$

Let us search for the solution in the form $(\psi_+(t)\mathrm{e}^{-\mathrm{i}\phi(t)/2}, \psi_-(t)\mathrm{e}^{\mathrm{i}\phi(t)/2})$, where

$$\dot{\phi}(t) = \omega + 2h_z(t)/\hbar.$$

The idea behind this substitution is that, at the resonant frequency $\omega = \varepsilon/\hbar$ and $\mathcal{T} = 0$, the Schrödinger equation is satisfied with constant $\psi_\pm$. So we expect these wave functions to vary slowly near the resonance point. After substitution, we obtain the Hamiltonian, which we average over the period as above:

$$\hat{H}_{\text{RWA-B}} = \frac{1}{2}\begin{pmatrix} \hbar\delta\omega & \mathcal{T}\langle \mathrm{e}^{\mathrm{i}\phi(t)}\rangle \\ \mathcal{T}^*\langle \mathrm{e}^{-\mathrm{i}\phi(t)}\rangle & -\hbar\delta\omega \end{pmatrix} \langle \mathrm{e}^{\mathrm{i}\phi(t)}\rangle = -J_1(2\tilde{h}/\omega\hbar). \tag{5.21}$$

If the oscillation amplitude $\tilde{h}$ does not depend on time, the Hamiltonian given by Eq. (5.21) is stationary. As for $\hat{H}_{\text{RWA-A}}$, the diagonal elements are proportional to the frequency mismatch, $\delta\omega = \varepsilon/\hbar - \omega$. In distinction from $\hat{H}_{\text{RWA-A}}$, the non-diagonal elements depend on the oscillation amplitude in a rather complicated way – through the first-order Bessel function. We have already encountered Bessel functions in Section 1.7, where we saw that they represented photon-assisted tunneling. The same applies here: the Bessel function gives the amplitude of the transition between the levels + and − that is accompanied by the emission/absorption of one photon with energy $\hbar\omega$. The time-dependent wave functions driven by the Hamiltonian given in Eq. (5.21) oscillate with the Rabi frequency

$$\omega_R = \sqrt{(\delta\omega)^2 + |(\mathcal{T}/\hbar)J_1(2\tilde{h}/\omega\hbar)|^2}.$$

The approximation works provided $\omega_R \ll \omega$, that is $\max(\delta\omega, |\mathcal{T}|/\hbar) \ll \omega$. We note that RWA-A and RWA-B, although similar in spirit, are different approximations suitable for

[4] One might be surprised by the fact that one can rotate the wave function around the z axis by shifting the frequency ω, even if the amplitudes of all oscillations are strictly zero. We note that a rotation $\hat{R}_z$ corresponds to shifting the phase difference between the components ψ_+ and ψ_-. This phase difference is, however, purely conventional. For example, we could use other definitions of these two components, up to a time-dependent phase factor, $\psi_\pm(t) \rightarrow \psi_\pm(t)\exp(\pm\mathrm{i}\varphi(t))$.

different regimes. Whereas both require a small frequency mismatch $\delta\omega$, RWA-A is only good for the manipulation amplitudes that are small in comparison with $\hat{H}_0$ but do not have the same "direction" in spin space. RWA-B applies to the case where the manipulation part and H_0 have almost the same direction ($\mathcal{T} \ll \varepsilon$), but is not restricted to small values of the amplitudes.

Whatever the RWA, it proves that the resonant manipulation can be achieved using pulses, each rotating the wave function around a certain axis. The precise shape of the pulse is not important. More complicated manipulation protocols are realized using sequences of pulses, usually separated by time intervals greatly exceeding their duration so that they do not overlap or interfere.

Ramsey sequence

Let us initialize the qubit to the ground state $|-\rangle$. We first apply a short pulse that rotates the spin by $\pi/2$ around the x axis:

$$R_x(\pi/2) = \frac{1}{\sqrt{2}}\begin{pmatrix} 1 & \mathrm{i} \\ \mathrm{i} & 1 \end{pmatrix}.$$

After the pulse, the wave function becomes a superposition $(|-\rangle + \mathrm{i}|+\rangle)/\sqrt{2}$, and the corresponding spin is along the y axis. After a delay τ, another $\pi/2$ pulse is applied. Naively, one expects the two to add: the resulting state would be $\mathrm{i}|+\rangle$. This is indeed the case if the oscillation frequency exactly matches the working frequency Ω of the qubit. In practice, this match is never perfect: even if we match the two in a given moment of time, the frequency Ω may drift in time due to slow fluctuations of the parameters caused by "unfriendly" agents. Let us see how the frequency mismatch $\delta\omega$ affects the result of the Ramsey sequence. Between the pulses, the spin performs a free rotation along the x axis with frequency $\delta\omega$. The corresponding rotation matrix for the wave function is $\exp(-\mathrm{i}\delta\omega\tau\hat{\sigma}_z/2)$; the superposition created by the first pulse thus evolves in time, $(\mathrm{i}\exp(-\mathrm{i}\delta\omega\tau/2)|+\rangle, \exp(\mathrm{i}\delta\omega\tau/2)|-\rangle)/\sqrt{2}$. Application of the second pulse again rotates the wave function around the x axis, and the final state is $\mathrm{i}(\cos(\delta\omega\tau/2)|+\rangle + \sin(\delta\omega\tau/2)|-\rangle)$. The probability of finding the qubit in state $|+\rangle$ after the Ramsey sequence is thus given by $P_+ = \cos^2(\delta\omega\tau/2)$. This is an oscillating function of the pulse duration and the mismatch – the so-called "Ramsey fringes." The effect is used to fine tune quantum resonant systems, for example the atomic clocks used as a modern time standard.

Spin echo

There is an elegant way to cancel the effect of the frequency mismatch. Let us sophisticate the Ramsey sequence by applying an extra π-rotation pulse around the x axis in between the $\pi/2$ pulses. The time intervals between the pulses are τ_1 and τ_2. The principle is best understood in terms of spin. After the first $\pi/2$-pulse, the spin is in the y direction. Owing to the frequency mismatch, it rotates around the z axis. The rotation angle just before the π-pulse is $-\delta\omega\tau_1$. The pulse rotates the spin such that its x component does not change, while its y component flips sign. The rotation angle becomes $\pi + \delta\omega\tau_1$, so that before the last pulse it is $\pi - \delta\omega(\tau_2 - \tau_1)$. If the time intervals are precisely the same, the subsequent $\pi/2$ pulse brings the qubit back to $|-\rangle$, irrespective of the frequency mismatch: its effect is canceled by the manipulation. The term "spin echo" comes from the experiment with a large ensemble of (nuclear) spins. In an ensemble, there are inevitable fluctuations of Ω from spin to spin. They are manifested as an apparent decay of total spin at a time scale determined by a typical fluctuation (see Section 6.7 for details). The π-pulse applied after the time interval τ cancels the effect of the fluctuations so that the signal mysteriously reappears at time moment 2τ as an echo.

Control question 5.10. What is the probability P_- after the spin-echo sequence described if $\tau_1 \neq \tau_2$?

Diabatic and adiabatic manipulation

There are two useful alternatives to the resonant manipulation. We exemplify those with the Hamiltonian given by Eq. (5.20), assuming that we can vary ε in time in any way we please. Let us suddenly, that is *diabatically*, change ε from large ($\gg |\mathcal{T}|$) positive to large negative values. Initially, the wave function corresponds to the ground state $|-\rangle$ of the Hamiltonian given by Eq. (5.20) with large positive ε. The change does not immediately affect the wave function: is it a manipulation? In fact, it is: the Hamiltonian has changed. The wave function corresponds now to the excited state of the new Hamiltonian. The same result would be achieved by applying a π-pulse without the Hamiltonian change. Diabatic manipulation should be performed with caution. Usually, the two states representing a qubit do not constitute the full spectrum of the quantum system. There are more states with higher energy, which are usually disregarded. The diabatic manipulation may provide the energy required to reach these states: this temporarily destroys the qubit.

The opposite of diabatic manipulation is *adiabatic* manipulation. Here, we change ε from large positive to large negative values in a slow fashion. In this case, if the initial wave function is an eigenfunction of the initial Hamiltonian, in the moment t it remains very close to the eigenfunction of the instant Hamiltonian $\hat{H}(t)$. If we start in the ground state at positive ε, we end up in the ground state at negative ε – we turn the spin by π. Although it seems less intuitive, the same applies to the excited state (assuming no relaxation process has taken place). The point of caution concerning the adiabatic manipulation are the phase shifts acquired by the eigenstates in the process. These phase shifts are not generally the same; this corresponds to the rotation around the "direction" of the Hamiltonian.

Generally, one can vary ε neither infinitely fast (diabatically), nor infinitely slow (adiabatically). In practice, there is always a finite "speed" $\dot{\varepsilon}$ with which the variation is performed. In this case, if the qubit starts in the ground state of the initial Hamiltonian, there exist finite probabilities of it ending up in both the ground and the excited states of the final Hamiltonian. The probability of it switching from the ground to the excited state, as well as from the excited to the ground state, is given by the celebrated Landau–Zener formula:

$$P_{-\to+} = P_{+\to-} = \exp\left(-\frac{\pi|\mathcal{T}|^2}{2|\dot{\varepsilon}|\hbar}\right), \tag{5.22}$$

and the switching process is known as *Zener tunneling*.

We can now understand the condition under which the manipulation is diabatic (adiabatic). It all depends on the minimum energy difference $|\mathcal{T}|$ between the states. If $|\dot{\varepsilon}| \ll |\mathcal{T}|^2/\hbar$, the probability of Zener tunneling is exponentially small, and the manipulation is adiabatic. Otherwise, the manipulation is diabatic.

5.3.2 Two-qubit manipulation

Two-qubit manipulation is performed on a four-component wave function of two qubits. A convenient basis to work with comprises direct products of $|\pm\rangle$ states of two qubits, $|++\rangle, |+-\rangle, |-+\rangle, |--\rangle$. A manipulation involves the turning of a coupling handle U_{ab}; see Eqs. (5.15). The most convenient handle U_{zz} couples only the z components of the qubits, so that the coupling Hamiltonian reads $U_{zz}\hat{\sigma}_z^{(1)}\hat{\sigma}_z^{(2)}$, but this handle is not always available. Frequently for symmetry reasons the handles are grouped. As we will see in Section 5.7, the coupling of spin qubits is isotropic in spin space, $U\hat{\sigma}_a^{(1)}\hat{\sigma}_a^{(2)}$, so the turning of U affects three couplings. The inductive coupling of the charge qubits groups the two together, that is $V(\hat{\sigma}_x^{(1)}\hat{\sigma}_x^{(2)} + \hat{\sigma}_y^{(1)}\hat{\sigma}_y^{(2)})$.

Let us consider a meaningful, but simple, example involving the most convenient handle U_{zz}. Initially, the handle is set to zero so that the qubits are decoupled. A pulse of U_{zz} of duration τ gives a unitary transformation, as follows:

$$R_{zz}^{(12)}(\theta) = \exp(\mathrm{i}\theta\hat{\sigma}_z^{(1)}\hat{\sigma}_z^{(2)}) = \mathrm{diag}(\mathrm{e}^{\mathrm{i}\theta}, \mathrm{e}^{-\mathrm{i}\theta}, \mathrm{e}^{-\mathrm{i}\theta}, \mathrm{e}^{\mathrm{i}\theta});\ \theta = -U_{zz}\tau,$$

diagonal in the basis in use. This is thus a pure phase shift, opposite for parallel and antiparallel qubit spins. It cannot be represented as a sum of individual qubit phase shifts, and therefore can be utilized for more complex two-qubit gates.

Let us consider a CNOT gate (see Eq. (5.4)) since it is a building block for most quantum computation schemes. In terms of qubit Pauli matrices, it is given by

$$U_{\mathrm{CNOT}} = \frac{1+\hat{\sigma}_z^{(1)}}{2} + \frac{1+\hat{\sigma}_z^{(1)}}{2}\hat{\sigma}_x^{(2)}.$$

This is not diagonal in our basis. Let us try to make it diagonal, first with single-qubit rotations. It is the x-Pauli matrix of the second qubit which seems to break ranks, so we should rotate the second qubit. This matrix can be diagonalized by two rotations by opposite $\pi/2$ angles applied before and after the CNOT gate:

$$-\hat{\sigma}_z^{(2)} = R_y^{(2)}(-\pi/2)\hat{\sigma}_x^{(2)}R_y^{(2)}(\pi/2).$$

Note that a single rotation would not work, since it makes non-diagonal a part of the CNOT that is proportional to the unit matrix of the second qubit.

Therefore we can represent the CNOT by a diagonal matrix $\hat{U}_0$:

$$U_{\mathrm{CNOT}} = R_y^{(2)}(\pi/2)\hat{U}_0R_y^{(2)}(-\pi/2), \qquad \hat{U}_0 = \mathrm{diag}(1, 1, -1, 1). \tag{5.23}$$

To realize the CNOT gate we thus need to realize the diagonal matrix $\hat{U}_0$. We have three means of doing this: two single-qubit rotations around the z axis, $R_z^{(2)}(\phi_1)$ and $R_z^{(2)}(\phi_2)$, respectively, and a two-bit rotation $R_{zz}^{(12)}(\theta)$. All three commute, and thus their order is not important. The product of the three is the diagonal matrix given by

$$\text{diag}\left(e^{i(\phi_1/2+\phi_2/2+\theta)}, e^{i(\phi_1/2-\phi_2/2-\theta)}, e^{i(-\phi_1/2+\phi_2/2-\theta)}, e^{i(-\phi_1/2-\phi_2/2+\theta)}\right).$$

It has to be matched with $\hat{U}_0$ by a proper choice of the three phases. Given that the three phases must reproduce four elements of $\hat{U}_0$, this can only be done up to a (generally unimportant) phase factor. Choosing $\phi_1 = -\phi_2 = \pi/2$, $\theta = \pi/4$, we express the CNOT as follows:

$$U_{\text{CNOT}} = e^{-i\frac{\pi}{4}} R_y^{(2)}\left(\frac{\pi}{2}\right) R_z^{(1)}\left(\frac{\pi}{2}\right) R_{zz}^{(12)}\left(\frac{\pi}{4}\right) R_z^{(2)}\left(-\frac{\pi}{2}\right) R_y^{(2)}\left(-\frac{\pi}{2}\right).$$

That is, the CNOT is realized by a five-pulse manipulation sequence: two rotations of the second qubit are followed by a turn of the coupling handle, and subsequently the first and the second qubit are rotated. If the handles are grouped as above, a single turn of the coupling handle does not suffice: one needs at least two turns separated by a qubit rotation.

We have described a pulse manipulation. All single-qubit manipulation techniques; resonant, diabatic, and adiabatic, can be applied to turn the coupling handles. We will give another example in Section 5.6. In addition, the single- and two-qubit manipulations do not have to be separated in time. Although this complicates the analysis of the results, it may significantly increase the manipulation speed.

5.3.3 Manipulation for quantum state tomography

Given a quantum system with M basis states labeled by $|i\rangle$, the probability P of finding the system in any given state $|\psi\rangle$ is determined from its $M \times M$ density matrix $\hat{\rho}$ as follows:

$$P = \langle\psi|\hat{\rho}|\psi\rangle = \sum_{ij}\langle\psi|i\rangle\rho_{ij}\langle j|\psi\rangle.$$

We have encountered the density matrix in Chapters 2 and 3. If we know all the elements of a density matrix, we can predict all the probabilities. How do we reconstruct this density matrix from the measuring results? First of all, let us note that such a reconstruction implies our ability to initialize the quantum system and to manipulate it in a completely reproducible way, so that the same density matrix is generated as a result of quantum manipulation at any time. Reconstruction of the density matrix, as well as the measurement of a probability, cannot be performed in a single shot and requires data accumulation from many elementary measurements.

The full characterization of the density matrix is known as *quantum state tomography*. The term comes from the analogy with the computer tomography for medical applications where the rotation of a radiation source allows three-dimensional imaging of an inner body. Similarly, the quantum state tomography involves rotations – unitary transformations of the wave function or density matrix before measurement, that is the quantum manipulation.

Let us illustrate this for a single qubit. The 2×2 density matrix of the qubit can be decomposed into four terms corresponding to the Pauli matrices:

$$\hat{\rho} = \frac{1}{2} + \frac{1}{2} \sum_{i=x,y,z} r_i \hat{\sigma}_i, \quad r_i = \mathrm{Tr}\, \hat{\sigma}_i \hat{\rho},$$

where we have used the fact that Tr $\hat{\rho} = 1$, expressing that the sum of all probabilities equals one. Thus, to characterize the density matrix fully, one needs to measure three real parameters $r_{x,y,z}$, corresponding to expectation values of the three spin components.

The easiest way to carry out such a characterization would be to perform three sets of projective measurements of x, y, z components of the spin. The probability extracted from each set yields the corresponding r_i. However, this measurement setup is an unaffordable luxury. In most cases, one can only measure a single component, say z. To perform the tomography, one rotates the qubit spin immediately before the measurement. For example, to measure the x component, one uses the rotation $R_y(-\pi/2)$ so that the x axis comes to the former position of the z axis. Experimentally, one measures the probability of finding the system in state $|0\rangle$, varying the angle θ of the y rotation,

$$P_0 = \frac{1}{2} \left[1 + (\rho_{00} - \rho_{11}) \cos\theta - (\rho_{01} + \rho_{10}) \sin\theta\right].$$

The result depends on both spin components, $r_x = \rho_{01} + \rho_{10}$ and $r_z = \rho_{00} - \rho_{11}$, that are extracted from the measurements with the fit of θ-dependence with the above formula. Similarly, the y component $r_y = i(\rho_{01} - \rho_{10})$ is determined using an R_x rotation.

An N-qubit quantum computer is characterized by a $2^N \times 2^N$ density matrix. Its diagonal elements are determined from the simultaneous measurement of the z components of all N qubits. To determine the off-diagonal elements, one performs the rotations. For example, the element of the density matrix between two states differing by a flip of a single qubit can be determined by rotating this qubit as described above. If the two states differ by two flips, one needs to perform more measurement sets that differ in the rotation of both qubits, and so on.

Suppose now one performs quantum state tomography initializing the quantum system to various initial states. This allows one to carry out quantum process tomography: to crack the quantum protocol – the unitary transformation that gives the final state for an arbitrary linear superposition of possible initial states.

5.4 Quantum dots

We defined quantum dots in Section 4.2 as artificially made systems where electrons are confined in visibly discrete levels. We have discussed at large the properties of the levels both in the dots of regular shape ("integrable") and generic chaotic dots. The discreteness of levels in principle brings discrete quantum states, so the quantum dots are natural candidates for qubit applications. One can think of an electron that can be in two discrete levels (and the coherent superposition of the two) as a qubit. A quantum computer would be obtained by placing many electrons in the dot with a sufficient number of discrete levels, or, equivalently, patching many dots together. It would be nice if in such a setup we could, for simplicity, disregard the interaction between electrons, as is the custom in solid state physics and implemented in Chapters 1, 2, and 4. Unfortunately we are not able to make a quantum computer in this way. The states of many non-interacting particles are too simple, and this forbids the quantum computation; indeed, as discussed, a requirement for efficient quantum computation is that these states are complex and entangled.

This is why the knowledge of levels does not suffice if we want to implement the dots as elements of a quantum computer. Nor does it suffice to understand the physics of realistic quantum dot devices where Coulomb blockade is of importance. Indeed, as discussed, the connections to a quantum dot should have conductance $G \ll G_Q$ for the levels to be discrete. The same condition, as discussed in Chapter 3, enables strong charging effects. This sets the topic of the present section.

We start by filling the levels and making quantum states with a given number of electrons in the dot. We discuss the specifics that the interaction brings to the picture of these states. The charging energy – the most important interaction effect – only comes into play if the number of electrons is changed. This usually happens in the course of transport. We further discuss single-electron transport in quantum dots, mention relaxation, and review resonant tunneling and co-tunneling. Concrete quantum-dot-based qubit realizations are considered in Sections 5.5 and 5.7.

5.4.1 From levels to states

Levels are quantum states for a single electron. What are discrete quantum states in a dot that contains many electrons? We begin our discussion of the states by disregarding the electron–electron interaction. We will see later in this subsection what the interaction adds to the picture of the states.

For non-interacting electrons, a many-body state is a way of distributing electrons over the levels. Since electrons are fermions, each spin-split (with a certain spin direction) level can be either empty or filled with an electron: more electrons in the same level are forbidden by the Pauli exclusion principle. Let us label the levels by k and introduce the number of electrons in each level, n_k, $n_k = 0$ or 1. Then the state of the dot is given by a set, $\{n_k\}$. Such a set, a sequence of "0" and "1," is very much like a number in binary representation. We also made use of such sequences to label the states of qubit registers in Section 5.1.3. The energy of this state is given by $E = \sum_k E_k n_k$ in terms of the level energies E_k, provided the interaction is disregarded.

For quantum dot applications, we need to sort these states in a certain way. Let us consider a dot with a given number of electrons N ($N = 6$ in Fig. 5.7 (a)). In the lowest energy state – the ground state – they occupy the lowest levels available. The levels are assumed to be spin-degenerate in Fig. 5.7 so that each horizontal line gives two levels, each occupied by an electron with a certain spin direction; these directions must be opposite on each line.

Let us turn to the excited states of the dot with the same number of electrons. The simplest way to make an excited state is to take an electron from an occupied level k' and to put it in an empty level k. One can regard it differently: we have created an electron excitation in level k and a hole excitation in level k'. So, empty levels are places for electron

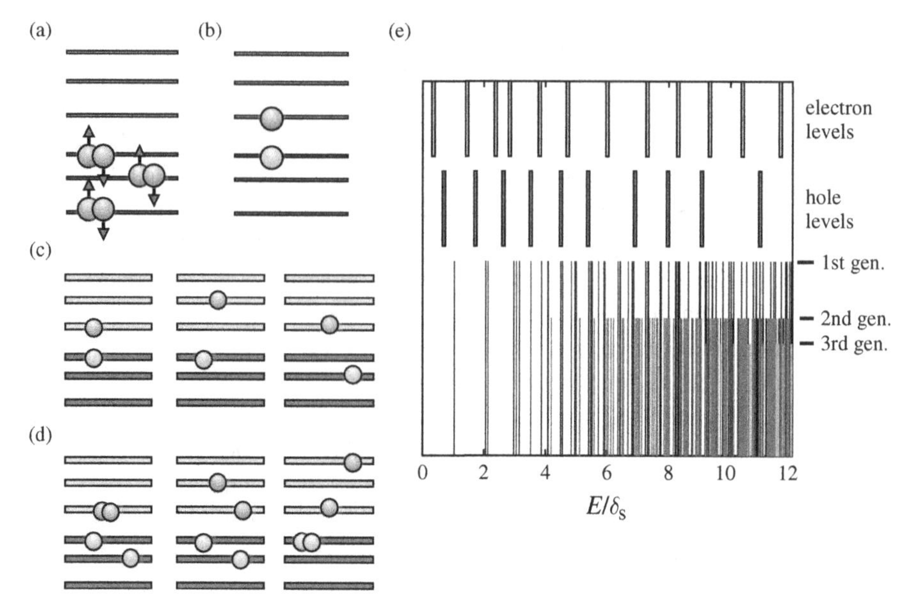

Fig. 5.7. **Levels and states in quantum dots. (a) The ground state of the dot with $N = 6$ electrons. (b) The first excited state. (c) Some states of the first generation. (d) Some states of the second generation. (e) The number of states exponentially increases with energy.**

excitations, while occupied levels are places for hole excitations (Fig. 5.7(b)). All states with one electron and one hole placed in levels k and k' are states of the *first* generation (Fig. 5.7(c)). The excitation energy – the energy of the state counted from the ground state – is given by

$$E_s = E_k - E_{k'}.$$

It is positive, with a notable exception. Suppose the levels are degenerate with respect to spin and there are an odd number of electrons in the dot. The ground state is a spin doublet: it is doubly degenerate. It costs no energy to flip the spin of the electron at the uppermost level. Such a flip is also equivalent to an electron-pair creation.

Control question 5.11. Can you redraw Fig. 5.7 for $N = 5$ and represent the spin-flip as an electron–hole pair creation?

A state of the second generation may be created from any state of the first generation by adding yet another electron–hole pair involving two more levels. Adding more pairs, one constructs the states of the third generation, and so on (Fig. 5.7(d)). The excitation energy of any such state is just the sum over energies of the corresponding levels:

$$E_s = \sum_k E_k - \sum_{k'} E_{k'},$$

where k (k') labels the levels of excited electrons (holes), so that $E_k - E_{k'} \geq 0$ for any pair k, k'.

The lowest excited state is separated from the ground state by an energy $\simeq\delta_S$. One wonders how many states are there in an energy strip several times bigger than δ_S, say $10\delta_S$. Naively, we would suggest several, ten to twenty maybe? That would be a typical human error: we tend to underestimate the speed of exponential multiplication. Eventually, the answer would be several thousand. In fact, the states multiply very quickly: there are many more states of the second generation than of the first one, and so it goes on. This is plotted in Fig. 5.7(e) for some random choice of the levels (the electron levels and the hole levels are shown). The states of the first generation (longest lines) start, as promised, at $E \approx \delta_S$, and are rather sparse in the beginning, although the spacing between them decreases with increasing energy. The states of the second generation (shorter lines) start at $E \approx 4\delta_S$, but by $E \approx 8\delta_S$ the spacing is so tiny that we cannot visually resolve the neighboring states. This is where the first states of the third generation appear. We summarize this with two formulas, for a typical number of levels and states below a certain energy $E \gg \delta_S$:

$$\begin{aligned} N_{\text{levels}} &\approx 2E/\delta_S; \\ N_{\text{states}} &\approx \exp\left(\pi\sqrt{2E/3\delta_S}\right). \end{aligned} \tag{5.24}$$

(Since we count both electron and hole levels, there is a factor of 2 in the first formula.)

Exercise 5.4. (The exercise requires knowledge of some elements of statistical mechanics.) Prove Eqs. (5.24). (i) Consider the partition function of the dot, given by

$$Z = \sum_s \mathrm{e}^{-E_s/k_B T}.$$

Implement Fermi statistics and give the expression for Z in terms of the electron and hole levels E_k, $E_{k'}$. (ii) Replace the summation over levels by integration over energies to arrive at $\ln Z = (\pi^2/12)k_B T/\delta_S$. (iii) The above expression for the partition function can be represented in the following form:

$$Z = \int \mathrm{d}E(\mathrm{d}N(E)/\mathrm{d}E)\exp(-E/k_B T).$$

Find $N(E)$ keeping the factors in the exponent only and taking the integral in the saddle-point approximation.

These considerations still concern non-interacting electrons. What does interaction do to these states? It turns out that it does surprisingly little as long as we do not change the number of electrons in the dot. There are still discrete states if interaction is taken into account, and their number and abilities to multiply are approximately the same as without interaction. Indeed, at an increasing number of levels and energy scale the excitation spectrum is continuous, and the dot is nothing but an isolated piece of a metal. We know that interaction effects in metals can be disregarded, and expect the same for discrete levels.

There are, however, several issues where the discreteness opens up the way for interactions to play a role (for a review, see Refs. [119] and [120]). Before discussing them, let us consider a formal Hamiltonian where the interactions are taken into account and spin degeneracy is assumed:

$$H = \sum_{i\sigma} E_i \hat{a}^{\dagger}_{i\sigma}\hat{a}_{i\sigma} + \sum_{ijlm}\sum_{\sigma,\sigma'}(\hat{a}^{\dagger}_{i\sigma}\hat{a}_{j\sigma})U_{ij,lm}(\hat{a}^{\dagger}_{l\sigma'}\hat{a}_{m\sigma'}). \tag{5.25}$$

Here, i, j, m and l label orbital levels, and $\hat{a}_{i\sigma}$ is the electron annihilation operator in the level i with spin σ. The first term gives the contribution of non-interacting particles, E_i being the energies of the levels. The second term gives the contribution of the Coulomb interaction, $U_{ij,lm}$ being the matrix elements of the interaction $U(|\boldsymbol{r}_1 - \boldsymbol{r}_2|)$:

$$U_{ij,lm} = \int \mathrm{d}\boldsymbol{r}_1\, \mathrm{d}\boldsymbol{r}_2\, \varphi_i(\boldsymbol{r}_1)\varphi^*_j(\boldsymbol{r}_1)U(|\boldsymbol{r}_1 - \boldsymbol{r}_2|)\varphi_l(\boldsymbol{r}_2)\varphi^*_m(\boldsymbol{r}_2), \tag{5.26}$$

where $\varphi_i(\boldsymbol{r})$ are the wave functions of the corresponding levels.

The interactions are responsible for the removal of extra spin degeneracy. Let us come back to the excited states of the first generation and look at the lowest excited state (Fig. 5.7(b)). We have "forgotten" that it is not a single state if the levels are spin-degenerate. With this, there are four states at the same energy corresponding to two possible spin orientations of the electron and the hole, $|\uparrow\rangle_{\mathrm{e}}|\uparrow\rangle_{\mathrm{h}}$, $|\uparrow\rangle_{\mathrm{e}}|\downarrow\rangle_{\mathrm{h}}$, $|\downarrow\rangle_{\mathrm{e}}|\uparrow\rangle_{\mathrm{h}}$, $|\downarrow\rangle_{\mathrm{e}}|\downarrow\rangle_{\mathrm{h}}$. The relevant part of the interaction is, in this case, the *exchange* interaction that arises from Coulomb repulsion between electrons and usually favors the states with higher total spin. So the four states are split into two groups – three components of a spin triplet $(S = 1)$, $|\uparrow\rangle_{\mathrm{e}}|\downarrow\rangle_{\mathrm{h}}, |\downarrow\rangle_{\mathrm{e}}|\uparrow\rangle_{\mathrm{h}}, (|\uparrow\rangle_{\mathrm{e}}|\uparrow\rangle_{\mathrm{h}} + |\downarrow\rangle_{\mathrm{e}}|\downarrow\rangle_{\mathrm{h}})/\sqrt{2}$ and a spin singlet $(S = 0)$, $(|\uparrow\rangle_{\mathrm{e}}|\uparrow\rangle_{\mathrm{h}} - |\downarrow\rangle_{\mathrm{e}}|\downarrow\rangle_{\mathrm{h}})/\sqrt{2}$.

Exercise 5.5. Find the excitation energy splitting of the state discussed treating the Coulomb interaction as a perturbation.

The same happens with all states: all extra degeneracy is removed and the states are sorted by the groups that differ by full spin S, the remaining degeneracy in the group being $2S + 1$. A weaker spin-orbit interaction may remove some degeneracy within spin multiplets. If no spatial symmetry is present in the dot (that is, the dot is sufficiently disordered), the remaining degeneracy is between the time-reversed states of opposite spin direction. This degeneracy can only be lifted by the magnetic field causing Zeeman splitting of the doublets.

Another important issue is that the electron–electron interaction removes the "barriers between generations." To recognize the barriers, let us assume that the interaction is irrelevant and that the dot is in the ground state and is excited by light. One expects the excitation to succeed if the light frequency matches an energy difference: but is this the difference between the levels or the states? If the interaction is neglected, only the first generation states are excited (in the first order in the light intensity): the absorption of a light quantum creates a single electron–hole pair. If the light frequency corresponds, say $10\delta_{\mathrm{S}}$, the absorption spectrum still consists of well separated lines. The interaction mixes up the generations. Indeed, the second term in Eq. (5.25) gives rise to matrix elements that are not diagonal in the number of electron–hole pairs, since some levels i, j, l, m are occupied.

Control question 5.12. What is the matrix element between the ground state with $N = 6$ and the lowest excited state of the first generation?

Owing to this, the true discrete states no longer belong to a certain generation – they are coherent superpositions of the states with different numbers of electron–hole pairs. If the matrix element is small in comparison with the level spacing, the states of different generations connected by the matrix elements form an interesting hierarchical network theoretically studied in Ref. [121].

Since the states of all generations are mixed, the absorption of a light quantum can commence any time the frequency matches the difference between the *states*. Recalling the estimations of $N_{\text{states}}(E)$, we recognize that the absorption spectrum becomes quasicontinuous; the spacing between the adjacent absorption lines is exponentially small. This is a quantitative effect brought about by interactions.

Interactions of a different kind do not involve the electrons in the dot only – the dot is open to an environment. The possibility to transfer energy to the environment gives rise to the finite lifetime, τ_{r}, of the excited states. The discrete states are no longer precisely discrete: they are broadened by $\delta E \simeq \hbar/\tau_{\text{r}}$. Comparing this with the exponentially small spacing between the states, we find the continuous spectrum of the dot excitations at $E \simeq \delta_{\text{S}} \ln(\delta_{\text{S}}\tau_{\text{r}}/\hbar)$. We stress that this estimation is very naive, and the border between the discrete and continuous spectrum, as well as the character of the transition between the two, is still under active scientific debate.

This concludes the discussion of interaction in quantum dots for a given number of electrons.

Charging effects

If we compare the energies of the states with different numbers of electrons, we encounter the most important interaction – the charging energy. It provides the dominant contribution to the energy difference between the states (either ground or excited) that differ in the number of electrons. In terms of Hamiltonians, this means that the dominant terms in Eq. (5.25) are reduced to charging energy, which is a function of the total number of electrons in the dot:

$$H = E_{\text{ch}}(\hat{N}) = E_{\text{ch}}\left(\sum_{i\sigma} \hat{a}^{\dagger}_{i\sigma}\hat{a}_{i\sigma}\right).$$

The discussion of the Coulomb interaction basically repeats one for a metallic Coulomb island (Chapter 3). The simplest setup consists of a quantum dot, a bulk electrode separated from the dot by a tunnel barrier, and the gate electrode. The electrostatic energy of the system depends on the integer charge, eN, of the dot:

$$E = E_{\text{C}}\left(N - \frac{C_{\text{g}}V_{\text{g}}}{e}\right)^2.$$

The equilibrium number of electrons corresponds to the minimum energy and can be tuned with the gate voltage:

$$N = \left[\frac{C_g V_g}{e} + \frac{1}{2} \right].$$

There are several notable differences, the first being the separation of energy scales. When dealing with a metallic island, the discreteness of the levels can be safely forgotten, $\delta_S \ll E_C$. For a quantum dot, the charging energy typically still exceeds the level spacing, but only by several times, $E_C = 2 - 10\delta_S$. For a metallic island, the periodicity of all Coulomb blockade features, for example the positions of Coulomb peaks, is close to ideal. For a dot, the Coulomb blockade peaks are randomly shifted by a noticeable fraction of the period. This is because the energy difference between the levels contributes to the addition energy of the dot. Indeed, since the internal interactions in the dot are weak, the added electron either comes to the next empty orbital level (for an even number of electrons in the dot) or to the same orbital level with opposite spin direction (for an odd number). In the first case, the addition energy is contributed by the spacing between the two levels; in the second case it comes from the exchange energy.

Secondly, the capacitances, and therefore E_C, in quantum dots depend on the number of electrons stored. To understand this, we recall that the capacitance is related to the geometric size of a dot/island. We may increase the size of the dot by putting more electrons into it.

Exercise 5.6. Consider a three-dimensional quantum dot created by a spherically symmetric parabolic potential $U = e^2 r^2 / r_0^3$. The gates are sufficiently far from the dot. Assume that the kinetic energy of the electrons can be disregarded and that the full energy of the dot is due to charging. Electrons can be thus regarded as a continuous charged liquid. (i) Find the charge density $q(r)$ in the dot at a given number of electrons N. (ii) Find the charging energy at a given number of electrons in the dot. What is the capacitance? What is the addition energy? (iii) Estimate the validity of the approximation. For this, calculate the energy levels of the electrons in the above parabolic potential and the full kinetic energy of the electrons.

The third and most obvious difference is that the dot can be emptied by a sufficiently large gate voltage, something we would never achieve for a metallic island due to the high number of electrons it contains.

5.4.2 Single-electron transport

There is a variety of transport regimes in quantum dots, the overall picture being more complex than that for Coulomb blockade devices described in Chapter 3. Let us present a brief overview of all regimes before concentrating on each one. At temperatures smaller than the charging energy, one sees (see, for example, Fig. 5.10(c)) a familiar picture of Coulomb diamonds in the V–V_g plane (or Coulomb shards, if we have several dots in series). Each diamond corresponds to a certain fixed number of electrons in the dot. The current in the diamonds is strongly suppressed, and the dominating transport mechanism is co-tunneling. Above a certain threshold voltage, single-electron transfers are available. We have already encountered this in Chapter 3.

A characteristic distinction between dots and other Coulomb blockade systems is the presence of a great number of extra lines in the V–V_g plane. The lines are mostly seen in the single-electron regions outside the diamonds and run (almost) parallel to the diamond edges. There are also lines inside the diamonds running parallel to the V_g axis at finite bias. Sometimes one even sees such lines at zero bias. In addition, there may be some lines running parallel to the V axis. Obviously, the lines manifest discrete states in the dot, both ground and excited. By measuring and analyzing their positions, we are performing *transport spectroscopy*: the characterization of the states in an (almost) isolated dot. The lines have a finite width in energy units, and this is either determined by temperature $k_B T$ or, at sufficiently low temperatures, by decay processes.

In short, the transport regimes are plain if they are *not* in the immediate vicinity of the lines with width *not* determined by temperature. A plain regime describes is either single-electron tunneling (to be understood via the master equation) or co-tunneling.

At the lines with width *not* determined by temperature, one encounters more "quantum" regimes of varying complexity. If the line is parallel to a diamond edge, it manifests the alignment of the energy difference between the states of the dot with N and $N+1$ electrons with the Fermi level. In this case, we have *resonant tunneling*, which we consider in Section 5.4.2. The resonant tunneling is simple if it involves a single spin-split level and presents a complicated and generally unsolved many-body problem if two spin-split levels are involved. The lines at zero bias manifest the *Kondo regime*, also related to many-body interactions involving electrons in the leads and discussed in Section 6.6. The lines parallel to the V_g axis usually indicate inelastic co-tunneling. The lines parallel to the V axis indicate the level crossings inside the dot and the importance of quantum superpositions of the discrete states crossed. We have a mixture of coherent and incoherent elementary transport processes to be understood using the density matrix and the Bloch equation (see Section 3.7.3). We address an example of such crossing, the *double* quantum dot, in Section 5.5.

This subsection concerns the single-electron transport through the dot. We concentrate on the most common situation, familiar from Chapter 3: the dot is connected to two electrodes, making a SET. In this regime, the electrons that go from one electrode to another have to pass the dot, changing its charge by $\pm e$. The single-electron transport in the dot is more complicated than in the metallic island of a common SET. This complication arises from the fact that the state of the dot cannot just be characterized by the number of electrons N inside. At a given N, the dot can be in one of many discrete states s. We will show how important it is for the transport. We label the states with a composite index $\alpha = (N, s)$.

As soon as we recognize this, we can proceed with the master equation as in Section 3.2.4. The master equation is a balance equation for the probabilities p_α for the dot to be in a certain state,

$$0 = \frac{\partial p_\alpha}{\partial t} = -p_\alpha \sum_\beta \Gamma(\alpha \to \beta) + \sum_\beta p_\beta \Gamma(\beta \to \alpha),$$

where $\Gamma(\alpha \to \beta)$ is the rate of the transition between states α and β.

The rates are of two sorts: tunneling and relaxation. The tunneling rates, as in metallic Coulomb blockade systems, correspond to a single-electron transfer through either the left or right junction in either direction:

$$\Gamma_{\mathrm{L,R}}(N, s \to N \pm 1, s').$$

The difference from metallic systems is that these rates in principle depend on the initial and final discrete states of the dot, s and s'. Only tunneling rates contribute to the current. We reason, as in Section 3.2.4, that

$$I_{\mathrm{L}} = \sum_{N,s,s'} p_{N,s} \left\{ \Gamma_{\mathrm{L}}(N, s \to N + 1, s') - \Gamma_{\mathrm{L}}(N, s \to N - 1, s') \right\},$$

and similarly for the right junction. Under stationary conditions, $I_{\mathrm{R}} = I_{\mathrm{L}}$ by virtue of current conservation.

In addition to the tunneling rates, there are transition processes that do not change the number of electrons in the dot, $\Gamma(\alpha \to \beta) = R(N, s \to N, s')$. These are usually relaxation processes: the dot driven to an excited state by a tunneling process, goes to the state of lower energy. At finite temperature, the transitions can also proceed in the opposite direction, with increasing energy. In both cases, these rates do not involve any tunneling and do not immediately contribute to the current. However, they change the probability distribution $p_{N,s}$ and thereby influence the transport.

Let us first discuss the tunneling rates. At vanishing temperature, the energy consideration applies: a transition occurs only if the energy difference between the final and initial states is negative. As in the case of metallic Coulomb blockade systems, this energy difference includes the difference of the charging energies and the contribution from the voltage of a corresponding lead (Section 3.2, Eqs. (3.14)). Unlike in the metallic island, this energy difference also includes the energy difference between the discrete states. The transition rate $\Gamma_{\mathrm{L,R}}(N, s \to N \pm 1, s')$ is thus allowed if the corresponding energy difference, given by

$$\Delta E = E(N \pm 1, s') - E(N, s) \mp eV_{\mathrm{L,R}},$$

is negative.

If the rate is allowed, it hardly depends on the energy difference, at least at vanishing temperature. This is in distinction from single-electron rates in a SET that are proportional to ΔE. The point is that this tunneling rate is to/from a given discrete state (Eq. (3.34)) and involves the tunneling matrix element between the localized level in the dot and the extended electron state in a lead. In principle, this matrix element is a characteristic of the localized level and may vary strongly from level to level. Its statistics are related to the statistics of wave functions studied in Section 4.1.3. We have seen in Chapter 3 that the typical magnitude of the rate is related to the conductance of the tunnel barrier separating the dot from the lead, $\hbar\Gamma_{\mathrm{L,R}} = (G^{(\mathrm{T})}_{\mathrm{L,R}}/G_{\mathrm{Q}})\delta_{\mathrm{S}}$.

At finite temperature, the ΔE-dependence is given by the Fermi distribution function f_{F}:

$$\Gamma(\Delta E) = f(\Delta E)\Gamma; \Rightarrow \Gamma(\Delta E) = \Theta(-\Delta E)\Gamma \text{ at } T = 0.$$

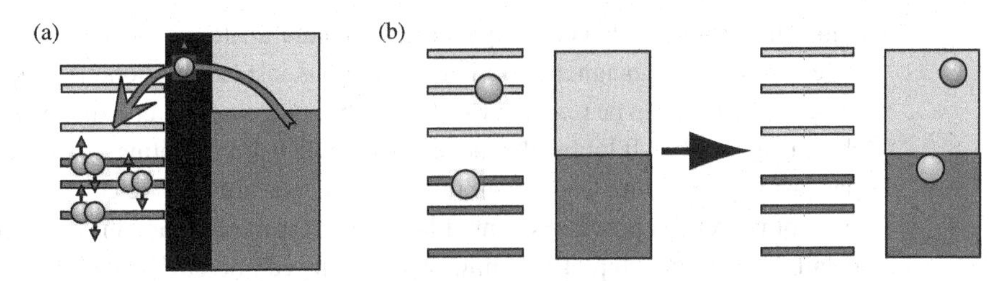

Fig. 5.8. (a) Tunneling to a discrete state of the dot. (b) Relaxation process in the dot: the electron–hole pair in the dot can transform itself into an electron–hole pair in a nearby lead. No tunneling is involved.

We put all this in a comprehensive diagram (Fig. 5.8(a)). The lead on the right is filled with electrons. On the left, we picture the discrete levels in the dot. If the dot before the tunneling contains N electrons, its energy E_N consists of the charging energy $E_C(N)$ and the sum of energies of all levels occupied by electrons, $\sum_k n_k E_k$. After the tunneling, the energy of the dot is increased by the *addition* electron energy ΔE, which has the electrostatic part, $E_C(N+1) - E_C(N)$, and the energy E_k of the level that the new electron occupies. The level positions drawn in Fig. 5.8(a) incorporate the change of charging energy corresponding to these addition energies.

For $N+1$ electrons, the dot is in the ground state provided the new electron comes to the lowest unoccupied level. This transition, $\Gamma_R(N, g \to N+1, g)$, is shown by the arrow in Fig. 5.8(a). (Note that the ground states are different for N and $N+1$ electrons, and for an odd number of electrons there are two ground states due to spin degeneracy.) Since the tunnel process can only occur if there is an electron in the lead with the corresponding energy ΔE, the rate is proportional to the filling factor $f_F(\Delta E)$ in the lead at the corresponding energy. This picture is actually a cartoon since it disregards the mixing of the states of different generations by electron–electron interaction. If the interaction is weak, the picture illustrates selection rules for tunneling rates: the most probable transitions are selected. At the cartoon level, the incoming electron either jumps in an empty level or annihilates a hole that can be present in an occupied level. Therefore, as a result of a tunneling event, the generation number either changes by ± 1 or remains the same. If the mixing of the generations is taken into account, the selection rules are violated and more complex processes can take place, enabling transitions between any (N, s) and $(N+1, s')$. The rates of such processes are smaller, including the small interaction parameter.

Let us turn to the relaxation rates $R(N, s \to N, s')$. They do not involve any tunneling to the leads. It might thus seem that, for these rates only, the dot itself is of importance; the transition seems to be an "internal affair" of the dot. However, for a completely isolated dot these transitions just would not happen! A completely isolated dot is a closed quantum system where energy conservation applies. If put in a discrete excited state, it remains there forever. Therefore, the relaxation in the dot requires an environment. The situation is the same as for a generic qubit, and the relaxation is detailed in Section 6.7. Here we just list possible sources of the relaxation: the natural environment of the dots. In many cases, the environment is the substrate where the dots are manufactured. The oscillations of the atoms of the substrate, phonons, can take the energy of the excited state: the dot emits a phonon with a transition to a lower-energy state. Like natural atoms, the dots can dispose of energy by emitting electromagnetic irradiation, photons instead of phonons. The photon does not have to go far: it can be readily absorbed in the adjacent lead, creating an electron–hole pair there (Fig. 5.8(b)). It looks like the electron–hole pair presenting an excited state of the dot has just slipped to the lead, albeit this process does not involve any tunneling. The typical rates of relaxation processes cannot be readily estimated since they depend greatly on the details of the dot setup. Two limiting situations can be envisaged: all relaxation rates are much slower and all relaxation rates are much faster than the tunneling rates Γ involved. It is important that there are selection rules for relaxation rates with respect to spin: the rates between the states of the same total spin are much faster than between the states of different spin. The latter processes require rather exotic mechanisms, such as spin-orbit interaction and hyperfine coupling with nuclear spins [122, 123].

We know enough about the rates to start with the transport. Let us go to the crossing of two diamonds. At zero bias voltage, the ground states with N and $N+1$ electrons have the same energy in the crossing point. At sufficiently low bias voltage the excited states do not appear in the transport cycle (see region I in Fig. 5.9(b)). The state of the dot changes between the ground states with N and $N+1$ electrons. Let us assume the spin degeneracy: the ground state with an odd $(N+1)$ number of electrons is double-degenerate. The probabilities of these degenerate states are the same, so the master equation can be written in terms of probabilities p_N (to be in the ground state with N) and p_{N+1} (to be in either of the ground states with $N+1$):

$$0 = \mathrm{d}p_N/\mathrm{d}t = -2\Gamma_\mathrm{L}\, p_N + \Gamma_\mathrm{R}\, p_{N+1},$$
$$0 = \mathrm{d}p_{N+1}/\mathrm{d}t = -\Gamma_\mathrm{R}\, p_{N+1} + 2\Gamma_\mathrm{L}\, p_N.$$

Here the electrons are transferred from the left to the right; at opposite bias voltage the transfer direction is opposite and $\Gamma_\mathrm{R} \leftrightarrow \Gamma_\mathrm{L}$. The factors of 2 reflect the double degeneracy: the electron may come from the left lead with two possible spin directions. It goes to the right lead from a state with certain spin, so that there is no factor of 2 in front of Γ_R.

This yields the following current:

$$I_\mathrm{I} = e\frac{2\Gamma_\mathrm{R}\Gamma_\mathrm{L}}{\Gamma_\mathrm{R} + 2\Gamma_\mathrm{L}} \qquad (5.27)$$

for electrons going from the left to the right. So current is constant in region I, where both rates are allowed, and zero in the diamonds. There is a sharp step in current at the lines corresponding to $\Delta E_\mathrm{L} = 0$ or $\Delta E_\mathrm{R} = 0$. This sharp step in the current gives a high differential conductance at the lines. Let us note that at opposite voltage the current is not precisely opposite: it is given by Eq. (5.27) with a minus sign and R, L *swapped*.

Control question 5.13. Describe what changes occur if N is odd and $N+1$ is even.

In reality, of course, no sharp feature is really sharp: it must be smoothed somehow. The natural smoother is temperature, in our case, the electron temperature in the leads. We expect the width of the step in energy units to be of the order of $k_\mathrm{B}T$,

$$\text{width} \sim k_B T;\; G_\mathrm{max} \sim I/(e \cdot \text{width}) \sim G_\mathrm{Q}\hbar\Gamma/k_\mathrm{B}T. \qquad (5.28)$$

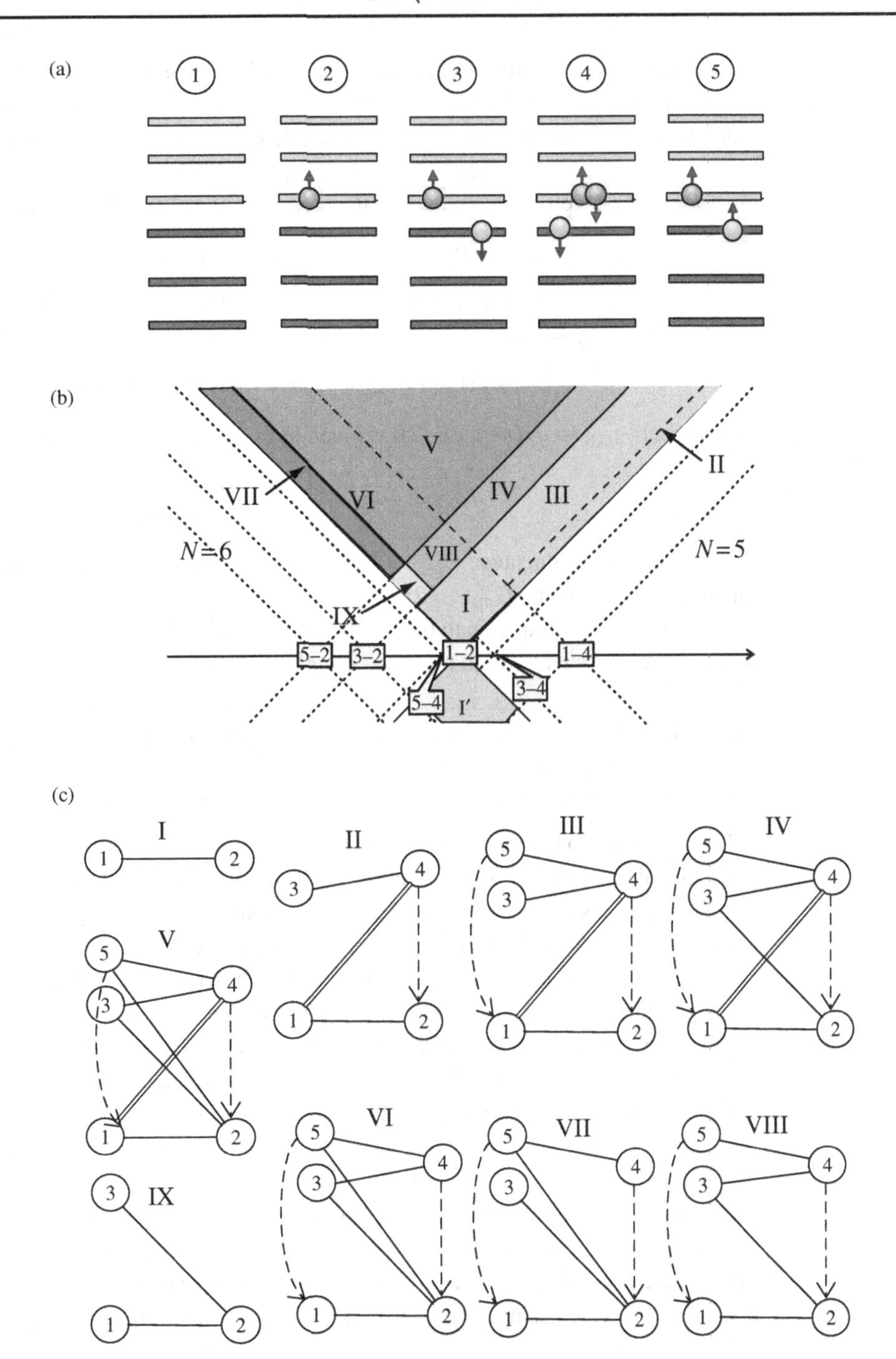

Fig. 5.9. Single-electron transport in a quantum dot. (a) Five states involved in transport for the example considered. (b) Transport regions in the V–V_g plane. White rectangules with labels denote energy crossings of the corresponding states. (c) Transport cycles for all regions shown. The line between an odd and an even state shows single-electron transfers in both directions; dashed lines show relaxation. States not participating in transport are not shown.

We see that the differential conductance grows with decreasing temperature. It would reach values of the order of G_Q at $k_B T \simeq \hbar\Gamma$. We will discuss this limit in Section 5.4.3. To quantify, one has to take into account all four rates allowed at finite temperature (see Eqs. (3.37) and (3.39)). Recalling convenient energy differences w, eV (w also incorporates the energies of the levels so that $w = 0$ at the crossing), we obtain the master equation:

$$
\begin{aligned}
0 &= -p_N 2(\Gamma_L f_F(w - eV/2) + \Gamma_R f_F(w + eV/2)) \\
&\quad + p_{N+1}(\Gamma_L f_F(-w + eV/2) + \Gamma_R f_F(-w - eV/2)); \\
0 &= p_{N+1}(\Gamma_L f_F(-w + eV/2) + \Gamma_R f_F(-w - eV/2)) \\
&\quad - p_N 2(\Gamma_L f_F(w - eV/2) + \Gamma_R f_F(w + eV/2)).
\end{aligned}
$$

The current ($f_{L,R} \equiv f_F(w \mp eV/2)$) is given by

$$
I_1/e = \frac{2\Gamma_L\Gamma_R(f_L - f_R)}{\Gamma_L(1 + f_L) + \Gamma_R(1 + f_R)}.
$$

Zero-voltage conductance ($V \to 0$) exhibits a slightly asymmetric Coulomb peak; the maximum conductance $G_{max} = 2\pi(3 - 2\sqrt{2})G_Q(\hbar\Gamma_R\Gamma_L)/k_B T(\Gamma_L + \Gamma_R)$ is achieved at $w \approx 0.347 k_B T$. A simple analytical expression can also be obtained for the conductance at the step at sufficiently high bias voltages $eV \gg k_B T$.

Exercise 5.7. Find analytical expressions for zero-voltage conductance and conductance at the step at $eV \gg k_B T$.

Let us now go to higher positive bias voltage where single-electron transfers may lead to the appearance of excited states. We assume the bias is sufficiently high to neglect thermal effects: a process is only possible if the corresponding ΔE is negative. Let us take into account the lowest excited states for both N (even) and $N + 1$. For N, this amounts to a triplet and a singlet state, as discussed in Section 5.4.1. For $N + 1$, the excited state is a doublet. Thus we have five states participating in transport, some being degenerate (see Fig. 5.9). We number them as follows: 1 – ground state for N; 2 – ground state for $N + 1$; 3 – triplet excited state; 4 – the excited doublet; 5 – singlet excited state. The excitation energies are sorted as follows: $E_3 - E_1 < E_4 - E_2 < E_5 - E_1$. The single-electron transfers take place only between the states with numbers of different parity. In addition, there are relaxation processes $4 \to 2$, $5 \to 1$. The relaxation processes $5 \to 3$ and $3 \to 1$ are very slow since they do not conserve spin. Things become complicated at high pace: this gives rise to nine distinct transport regimes at positive bias voltage (Figs. 5.9(b),(c)). The lines separating the regimes cross at $V = 0$ at the point where $E_m - E_n = \mu$. The energy distance between crossings is immediately related to the excitation energy.

Control question 5.14. What is the energy distance between the crossings 2–3 and 2–5? When does the transition between 1 and 4 become possible?

Now we can determine which tunneling processes can take place in each region bounded by the lines. Not every allowed process will contribute to the transport. No lines are visible

Table 5.1

States	$\vert\uparrow\rangle$	$\vert\downarrow\rangle$
$\vert\uparrow\rangle_e \vert\downarrow\rangle_h$	Γ'_L	0
$\vert\downarrow\rangle_e \vert\uparrow\rangle_h$	0	Γ'_L
$\frac{\vert\uparrow\rangle_e\vert\uparrow\rangle_h+\vert\downarrow\rangle_e\vert\downarrow\rangle_h}{\sqrt{2}}$	$\Gamma'_L/2$	$\Gamma'_L/2$

inside the diamonds. For example, let us look at the bias voltages above the 4–1 crossing. The dot would switch between 4 and 1 there, but why would it come to any of these states? It must be in state 2: no tunneling or relaxation would lead from 2 to either 1 or 4. This also explains why the lines coming from the 4–5 and 4–3 crossings are not visible in region I: no process leads from 1 or 2 to any of the excited states.

The transport cycles for all the regions are given in Fig. 5.9(c). All five states are involved in transport, except regions II and IX. Let us concentrate on region IX. The tunneling between 1 and 2 is characterized by the familiar rates $\Gamma_{L,R}$. The transitions between 2 and 3 involve another level, characterized by other rates Γ'_L, Γ'_R. The electrons are still transferred from the left to the right. The actual tunneling rates have to be evaluated, taking the spin structure of the triplet and doublet state into account. For tunneling from the triplet to the doublet we have three possible initial states: $(|\uparrow\rangle_e|\downarrow\rangle_h, |\downarrow\rangle_e|\uparrow\rangle_h, (|\uparrow\rangle_e|\uparrow\rangle_h + |\downarrow\rangle_e|\downarrow\rangle_h)/\sqrt{2})$, and two final states: $|\uparrow\rangle, |\downarrow\rangle$. The rates are summarized in Table 5.1.

Summing up over the final states, we understand that the rate is the same for all initial triplet states, Γ'_L, as it should be due to spin symmetry. The total rate of the reverse processes is evaluated in a similar way, yielding $3\Gamma'_R/2$.

Control question 5.15. Can you draw the corresponding table for the reverse processes?

The solution of the master equation with three states 1, 2, 3 yields the following current:

$$I_{IX}/e = \frac{\Gamma_L\Gamma'_L(2\Gamma_R + 3\Gamma'_R)}{2\Gamma_L\Gamma'_L + 3\Gamma_L\Gamma'_R + \Gamma_R\Gamma'_L},$$

which differs from the current in the adjacent region I. Usually, $I_{IX} > I_I$, as one expects from a current at higher voltage. To give an example: if all rates are the same, $\Gamma_{L,R} = \Gamma'_{L,R} = \Gamma$, $I_{IX} = (5/6)e\Gamma$, and $I_I = (2/3)e\Gamma$, a 25% increase. However, there is no general rule that guarantees this increase. For example, if Γ'_L is much smaller than all the other rates, I_{IX} is small, being proportional to Γ'_L. The dot is trapped in state 3. This implies that the conductance at the line is *negative*, corresponding to the current decrease with increasing voltage.

If we look at the transport in region II, we note that the rates between 1 and 4 are small provided the interdot interaction is weak. Indeed, in the energy-level diagram the addition of an electron to any empty level of 1 does not give 4: one needs an extra electron–hole pair. This is why the lines coming out of crossing 4–1 are dashed in Fig. 5.9(c). Is there a visible current step on the line? Somewhat unexpectedly, this is determined by the ratio of the slow rate $\Gamma_{1\to4}$ to the relaxation rate $R_{4\to2}$. If the relaxation is faster, the dot switches all the time between states 1 and 2; if it is occasionally brought to state 4, it quickly relaxes to 2. If the relaxation is slower, the dot spends comparable times carrying out one of the two transport cycles: either $1\leftrightarrow2$ (current $I_{12}/e = 2\Gamma_L\Gamma_R/(\Gamma_R + 2\Gamma_L)$) or $3\leftrightarrow4$ (current $I_{34}/e = 3\Gamma_L\Gamma_R/(3\Gamma_R + 2\Gamma_L)$). The slow rates $\Gamma_{1\to4}$, $\Gamma_{4\to1}$ switch between the cycles, and result in an average current given by

$$I_{\mathrm{II}} = \frac{\Gamma_{1\to4}I_{34} + \Gamma_{4\to1}I_{12}}{\Gamma_{1\to4} + \Gamma_{4\to1}} \neq I_{\mathrm{I}}.$$

The relative increase or decrease of the current does not have to be small in this case.

To summarize, the lines visible in the V–V_g plane provide information about the energies of the discrete states in the dot, the ground state as well as excited states. The relative value of the current in the bounded regions and visibility of lines depends on rather fine details of tunneling and relaxation, and thus supplies information about these details.

Exercise 5.8. Evaluate the current in region VI. Take into account the relaxation rates $R_{4\to2}$ and $R_{5\to1}$, disregarding the relaxation between singlet and triplet states. Consider two limiting cases of fast and slow relaxation.

Exercise 5.9. Assume fast relaxation to the ground state for N and $N + 1$, including singlet–triplet transitions. Determine the current in all the regions. Which lines in the diagram remain visible?

Experiment

We review here the first observation of discrete states in quantum dots by means of transport measurement [124]. The dot was defined by metal gates on top of a GaAlAs–GaAs structure as discussed in Sections 1.2.3 and 2.6.2. The dot diameter was estimated as 100 nm, and it typically housed 25 electrons. The curves (see Fig. 5.10(a)) present the current traces at several constant bias voltages. At the lowest voltage, two Coulomb peaks are seen separating three diamonds and corresponding to the tunneling between the ground states in each diamond. The trace at intermediate voltage cuts region I, where tunneling still takes place between the same states. The current is seen in the interval $w = eV$. It displays a significant smooth dependence on gate voltage in this interval owing to the voltage dependence of the tunneling amplitudes. The trace at higher voltage displays an extra feature near the middle of the increased interval. This indicates the appearance of a discrete excited state in the tunneling cycle, for both transition regions between the diamonds. The discrete state positions are best visualized by the traces of differential conductance at constant gate voltage. Figure 5.10(b) shows the set of such traces at different magnetic fields. The voltage positions of the slanted lines are given by either peaks or dips of the differential conductance. They depend on magnetic field, which significantly affects the energies of the levels. Let us recall that the voltage position of a line gives the energy difference between

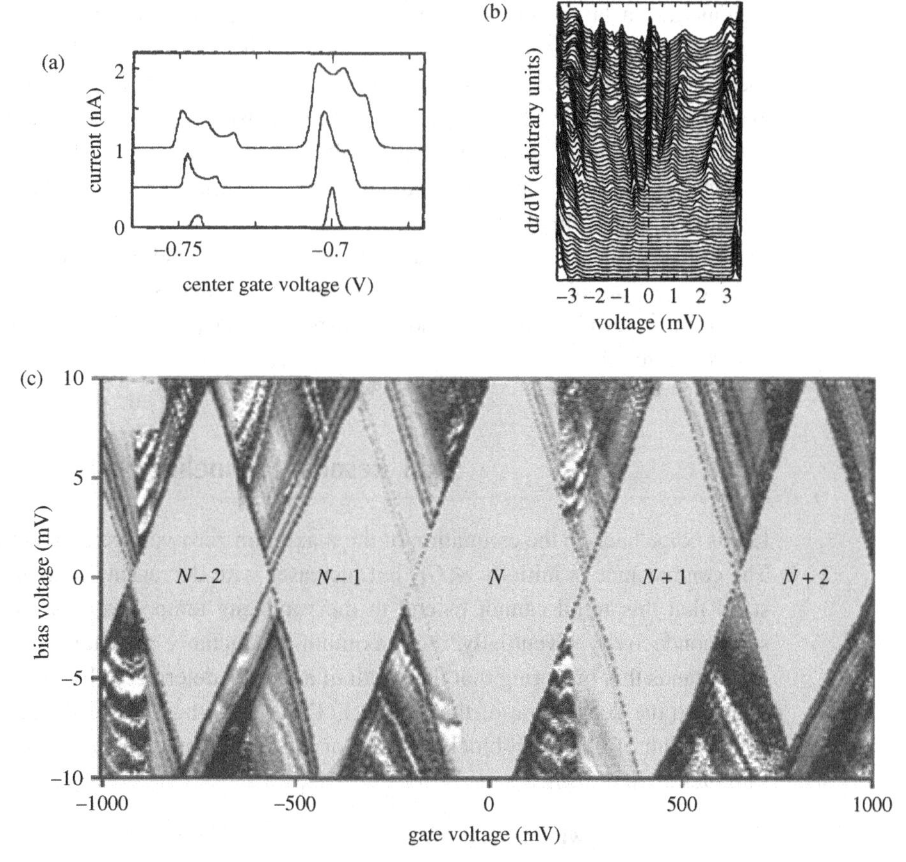

Fig. 5.10. **Discrete states in quantum dots: experiments. (a) Current versus gate voltage for three diamonds at increasing bias voltage. The highest trace shows a current step associated with a discrete state [123]. (b) Positions of conductance peaks and dips indicate the energies of the discrete states. The evolution of these energies in a magnetic field is shown [124]. (c) Overall picture of the discrete states across five diamonds [125].**

two discrete states corresponding to N and $N+1$. The magnetic-field dependence of this energy difference gives the change of the magnetization of the dot upon transition between the states.

If we move to higher voltages $\simeq$3–5 δ_S/e, more and more excited states of the dot come into play, creating new transport cycles and breaking old ones. The appearance of each new cycle is manifested as a sharp current step at a slanted line. With more states, the picture becomes increasingly complicated. We have learned that the number of states exponentially increases with energy. Since the width of a current step is finite, the states can no longer be resolved at voltages $eV \simeq \delta_S \ln(\delta_S/\text{width})$. Conclusions can be drawn from the experimental figure (Fig. 5.10(c)) presenting the transport spectroscopy in a quantum dot made in a carbon nanotube [125]. The gray scale reflects the differential conductance: lighter regions are domains of almost constant current, whereas dark lines indicate current steps with enhanced differential conductance. Large diamonds correspond to Coulomb-blockade regions, and a large number of lines at higher voltage manifest discrete states at each charge configuration. As well as demonstrating the feasibility and accuracy of transport spectroscopy, Fig. 5.9(c) also demonstrates how messy the transport becomes at energies exceeding the 3–5 level spacings due to the increasingly high number of states involved.

One could think that at yet higher voltages the exponentially large number of states renders the transport totally incomprehensible. In fact, we have the opposite situation: it becomes simpler. There are so many discrete states that they cannot be resolved, and all the formerly sharp features in transport become smooth. The effects of discreteness are gone, and the spectrum of electron states is effectively continuous. Thus we return to Chapter 3, which illustrates Coulomb-blockaded transport in an island with a continuous spectrum of electron states.

5.4.3 Resonant tunneling

Let us come back to the estimation of the maximum zero-voltage conductance, Eq. (5.28). The conductance is initially $\ll G_Q$ but increases with decreasing temperature. We understand that this trend cannot extend to the vanishing temperatures since it would imply superconductivity. Eventually, the maximum conductance saturates at values $\simeq G_Q$. One understands this by noting that the width of a level is determined by its energy uncertainty related to the decay time of the level $\simeq 1/\Gamma$. The width of the level sets the lowest limit of the width of Coulomb-blockade peak or the current step at finite bias. Thus, the width saturates at $k_B T \simeq \hbar\Gamma$:

$$\text{width} \sim \hbar\Gamma;\ G_{\max} \sim I/(e \cdot \text{width}) \sim e^2\Gamma/\hbar\Gamma \sim G_Q.$$

Let us investigate the situation in a low-voltage regime near a diamond crossing. In this case, there is a single level in the energy strip where the states in the left lead are filled and those in the right one are empty. Such a situation is called *resonant tunneling*.

We consider now a non-degenerate level with a certain spin direction ("up"), assuming that the spin degeneracy has been lifted by, for example, an external magnetic field. Taking a non-degenerate level provides us with a rare opportunity to disregard the Coulomb interaction in the dot and to think in terms of non-interacting electrons and their scattering matrix as detailed in Chapter 1. Let us pause to explain the reason for this. If we concentrate on a single level and treat the low-lying electrons as a background, the dot can only be in two states with two possible numbers of extra particles, $|0\rangle$ and $|1\rangle$. An interaction, however, is always between the particles and therefore can be safely skipped for no and one particles. Why does this not work for a degenerate level? In this case, the non-interacting electrons give rise to an extra state, with both spin directions filled, that lies at the same energy. This is, however, nonsense for a dot subject to Coulomb blockade: such a state would have extra energy $\simeq E_C$. The doubly occupied state thus has to be excluded from our consideration: unfortunately, this forbids the use of non-interacting electrons.

Making use of non-interacting electrons, we associate the resonant level in the quantum dot with a single transmission resonance between two tunnel junctions, as considered in Sections 1.6 and 2.1.1. If we concentrate on energies close to the peak transmission, we can disregard the contributions of other resonances. The energy-dependent transmission is given by the Breit–Wigner formula, Eq. (1.109):

$$T(E) = \frac{\Gamma_L \Gamma_R}{((E - E_0)/\hbar)^2 + ((\Gamma_R + \Gamma_L)/2)^2}. \tag{5.29}$$

Equation (5.29) does not only apply at low voltages, it works perfectly for any voltage and temperature at which we can disregard the presence of other levels. For high voltages, $eV \gg \Gamma_{L,R}$, the current is given by

$$I = \frac{G_Q}{2} \int_0^{eV} T(E) \mathrm{d}E \approx \frac{e}{2} \frac{\Gamma_L \Gamma_R}{\Gamma_L + \Gamma_R}, \tag{5.30}$$

provided $0 < E_0 < eV$. To calculate the integral, we note that the width of the transmission peak is much smaller than the integration range and thus we extend both limits to infinity. The result is consistent with Eq. (5.27) if we correct for the spin degeneracy. The Landauer formula also allows for an evaluation of the precise shape of the current step at the diamond edges.

Exercise 5.10. Calculate the maximum differential conductance at the diamond edge assuming that the bias voltage is much bigger than $\hbar\Gamma/e$.

Qualitatively, the situation is the same for a spin-degenerate level, and also if we take more excited states into account. The width of any sharp feature, such as a Coulomb peak, Coulomb diamond edges, or current steps at the lines separating different regions of single-electron transport, is of the order of the tunneling rate. Accordingly, the maximum conductance at any feature is of the order of G_Q. However, an accurate evaluation of the precise shape of these steps involves many-body interactions with electrons in the leads and is currently not accessible by analytical methods – the Landauer formula does not apply. Much progress with many-body problems has been achieved for the Anderson model, which involves one spin-degenerate level (see Section 6.6). The striking prediction of this model is that a narrow ($\ll \hbar\Gamma$) Kondo resonance is formed at the Fermi level, visible as a line at zero bias voltage at all extent of the Coulomb diamonds corresponding to an odd number of electrons.

5.4.4 Co-tunneling

Let us consider co-tunneling in quantum dots; we start with elastic co-tunneling. In this case, the electron is transferred at once via both tunnel junctions that separate the dot from the leads. There is no change in the dot state in the course of the process; the dot is void. Indeed, how would we distinguish a quantum dot from any other nanostructure between the junctions? We can only do this if the transport excites some states in the dot that are different from the ground state. Otherwise, we deal with a generic nanostructure with no interesting internal degrees of freedom. Such nanostructures are described in the framework of the scattering approach, which therefore suits our purposes for elastic co-tunneling.

We will consider elastic co-tunneling inside the Coulomb diamonds with an even number of electrons. The point is that, although the formulas for the co-tunneling rate may look similar for even and odd diamonds, the dot in the odd diamond is not void unless the spin degeneracy is lifted by a spin magnetic field. If spin degeneracy is present, the co-tunneling may switch the dot between two distinct ground states – two spin directions. This sets up a new situation where the Kondo phenomenon may develop (see Section 6.6.2). Equation (6.128) gives a concrete expression for co-tunneling in the odd diamond.

Elastic co-tunneling in even diamonds is, in fact, identical to this process in a SET (see Section 3.4.4). The difference is that in the dot the discrete states require more attention. Following Section 3.4.4 we obtain the tunneling matrix element for the transition between electron states l, r in the left and right lead, as follows:

$$\tilde{T}_{lr} = \sum_n T_{ln} \left(\frac{1}{E_n^{(+)} - E_l} + \frac{1}{E_n^{(-)} + E_r} \right) T_{nr}. \qquad (5.31)$$

Here, the summation runs over the discrete states in the dot. There is a contribution from the states with $N+1$ electrons, these states (typically doublets) have positive addition energies $E_n^{(+)}$. In the energy-level diagram, these energies are $E^{(+)} + E_n$, where E_n is the energy of an empty level and $E^{(+)}$ is the electrostatic energy. Another contribution comes from the states with $N-1$ electrons; the positive extraction energies are $E^{(-)} - E_n$, where E_n are the energies of occupied levels. In the energy-level diagram, this coincides with Eqs. (3.73), where Fermi factors discriminate between the empty and occupied levels. The amplitude contains the virtual states that in principle can lie very high in energy, so one may wonder if these states provide a decisive contribution to the amplitude. One could estimate this contribution by setting *equal* tunneling matrix elements T_{ln}, T_{nr} for all the levels. In this case, the sum over the levels can be replaced by integration over the energy and $\tilde{T}_{lr} \simeq T_r T_l \ln(E_\mathrm{F}/E_\mathrm{C})/\delta_\mathrm{S}$. The co-tunneling conductance is then estimated as $G \simeq G_\mathrm{Q}(\hbar^2 \Gamma_\mathrm{R}\Gamma_\mathrm{L}/\delta_\mathrm{S}^2)\ln^2(E_\mathrm{F}/E_\mathrm{C})$. This strange expression hardly depends on E_C, and is in fact a large overestimate of the actual co-tunneling effect. It was a mistake to regard $T_{ln}T_{nr}$ as level-independent.

To see what is going wrong, let us recall a simple rectangular two-dimensional dot. The levels are labeled by two integers $n_{x,y}$, and their wave functions are given by (see Eq. (4.17))

$$\psi(x, y) = \frac{1}{\sqrt{L_x L_y}} \sin\frac{\pi n_x x}{L_x} \sin\frac{\pi n_y y}{L_y},$$

where the edges of the dot correspond to $x = 0, L_x$ and $y = 0, L_y$. Let us assume that the left lead is connected to the $x = 0$ edge and that the right lead is connected to the $x = L_x$ edge. The tunneling amplitude T_{ln} is proportional to the overlap of the wave functions of the lead and of the dot. It is convenient to choose the wave functions in both leads to be real, then the amplitudes T_{ln} and T_{nr} are also real. The sign of the product $T_{ln}T_{nr}$ is then proportional to the product of the signs of ψ_{n_x,n_y} at two different edges, $\propto(-1)^{n_x+1}$. The sign therefore alternates with increasing n_x. This forbids the approximation of $T_{ln}T_{nr}$ with the same value for all levels. In a more realistic dot, the sign of this product does not precisely alternate owing to the random nature of the levels. However, for a given level, the positive and negative sign of $T_{ln}T_{nr}$ occurs with the same probability, so the contributions of different levels to $\tilde{T}_{lr}$ cancel each other, resulting in destructive interference. In the limit of continuous spectrum (see Section 3.4.4), we could replace the square of the amplitude with the sum of squares of the amplitudes of the individual levels. This yields $G_{\rm el} \simeq G_{\rm Q}\hbar^2\Gamma_{\rm R}\Gamma_{\rm L}/(\delta_{\rm S}E_{\rm C})$, assuming $\delta_{\rm S} \ll E_{\rm C}$. In quantum dots, where $\delta_{\rm S}$ is in principle comparable with $E_{\rm C}$, the interference contributions cannot be disregarded. The same pertains to the contribution of the lowest-energy excited states, especially in the vicinity of Coulomb peaks.

To show this, let us evaluate the elastic co-tunneling in the previously used setup. We are in the N diamond and only take into account the virtual states of lowest energy: the ground states with $N+1$ and $N-1$ electrons. We denote their energies as $E_{\pm}$, the smallest addition and extraction energies. This choice is the most convenient one: $E_{+}(E_{-})$ become zero at the edges of the Coulomb diamond.

Control question 5.16. Can you express these energies in terms of the electrostatic energies $E^{(\pm)}$ and the energies of electron levels? How do these energies depend on gate voltage?

We adapt Eq. (5.31) to the current case of two levels characterized by $\Gamma_{\rm R,L}$ and $\Gamma'_{\rm R,L}$ and make use of Eqs. (3.32) and (3.34) to express the tunneling matrix elements in terms of conductance and Γ values. This yields the elastic co-tunneling conductance at a given electron energy E, as follows:

$$G_{\rm el}(E) = G_{\rm Q}\left(\frac{\Gamma_{\rm R}\Gamma_{\rm L}}{(E_{+}-E)^2} + \frac{\Gamma'_{\rm R}\Gamma'_{\rm L}}{(E_{-}+E)^2} \pm A_{\rm if}\frac{\sqrt{\Gamma_{\rm R}\Gamma_{\rm L}\Gamma'_{\rm R}\Gamma'_{\rm L}}}{(E_{+}-E)(E+E_{-})}\right). \quad (5.32)$$

The coefficient $A_{\rm if}$ accounts for interference. It cannot be directly expressed in terms of the Γ values since they are contributed by all possible electron states with generally different coefficients. A notable exception is the case where the tunneling to the dot is dominated by a single transport channel in each lead. In this case, T_{ln}, T_{nr} do not depend on the states in the channel and the coefficient $A_{\rm if}$ achieves its maximum value of 2. The $\pm$ factor reflects the possibility of different signs of the product $T_{ln}T_{nr}$ for the two levels involved.

Exercise 5.11. Express the coefficient $A_{\rm if}$ in terms of T_{ln}, T_{nr} to prove the above statement. What is the value of $A_{\rm if}$ if a large number of channels contributes to the tunneling?

The co-tunneling conductance diverges at $E_{+} = E$ and $E_{-} = -E$; we understand these as the conditions for resonant tunneling described in Section 5.4.3. Indeed, an electron coming at energy E_{+} may become stuck in the dot bringing it to the ground state with $N+1$ electrons. Similarly, if an electron state with energy $-E_{-}$ is empty in a lead, the dot can go to the ground state with $N-1$ electrons, putting the electron in this empty state. We note that the co-tunneling conductance in the leading term (either $(E_{+}-E)^{-2}$ or $(E_{-}+E)^{-2}$) matches the asymptotics of the resonant tunneling formula, Eq. (5.29), $T(E) \approx \hbar^2\Gamma_{\rm R}\Gamma_{\rm L}/(E-E_0)^2$. Although we had resonant tunneling for a non-degenerate

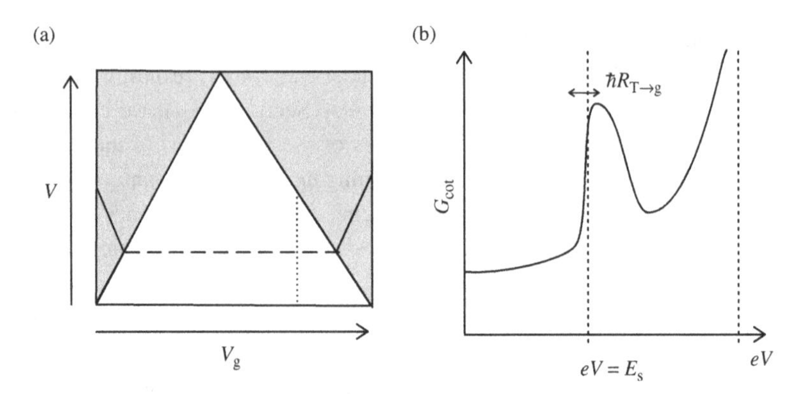

Fig. 5.11. **Inelastic cotunneling in quantum dots. (a) The even diamond. The inelastic co-tunneling threshold is given by the dashed line. It crosses the diamond edges at the same position as the lines of current cusps in single-electron transport. (b) Differential conductance in the diamond along the dotted line in (a). The jump at the threshold is frequently seen as a conductance peak. The width of the jump is determined by $\hbar R_{T\to g}$.**

level, the tails of the resonant peak simply add for two spin directions, resulting in a double conductance, conforming to Eq. (5.32).

Let us consider *inelastic* co-tunneling in quantum dots (see Fig. 5.11). This is very different from the inelastic co-tunneling in a SET (see Section 3.4.2) since it involves discrete states: in this case, excitations with the same N. For instance, an electron with energy E coming to the dot can give it energy E_s to bring it into a triplet excited state (state 3 from Section 5.4.2) and go to another lead with energy $E - E_s$. These processes are therefore enabled at sufficiently large bias $e|V| > E_s$. This allows us to witness another peculiarity in transport spectroscopy: a line parallel to the gate voltage axis seen inside the diamond. The energy consideration immediately means that this line crosses the diamond edge at the same point as the slanted line that corresponds to the excitation of the triplet state in the course of single-electron transport. Let us evaluate the rate of this switching process. The virtual states arising are identical to those already used: the ground states with $N \pm 1$ electrons. For both virtual states, the tunneling involves the matrix elements T_l, T'_r corresponding to two levels in play. If an electron comes with spin "up", there are two possible final states: (i) $|\uparrow\rangle_{\rm e}|\downarrow\rangle_{\rm h}$ and electron with spin "up" in the right lead and (ii) $(|\uparrow\rangle_{\rm e}|\uparrow\rangle_{\rm h} + |\downarrow\rangle_{\rm e}|\downarrow\rangle_{\rm h})/\sqrt{2}$ and electron with spin "down" in the right lead. The rates of these processes differ by a factor of 2. The total rate includes all possible spin configurations in the initial and final states and is given by (see Eq. (3.64))

$$\Gamma_{g\to T} = \frac{3\hbar}{2\pi}\int dE\, dE'\, f_L(E)(1 - f_R(E'))\delta(E - E' - E_s)\Gamma_L\Gamma'_R \times \left(\frac{1}{E_+ - E} + \frac{1}{E_- + E'}\right)^2 . \tag{5.33}$$

Slightly above the threshold voltage, $\Gamma_{g\to T} = (G_j/e^2)(eV - E_s)$, $G_j = 3G_Q\hbar^2\Gamma_L\Gamma'_R (E_+^{-1} + E_-^{-1})^2$ at $E_\pm \gg E_s$. This gives the conductance jump G_j at the threshold voltage; the magnitude of the jump can be comparable with G_{el}, the conductance below the threshold, or even exceed this conductance by a factor of 2–3. It is interesting to note that experimentally one frequently sees a rather broad conductance peak rather than a jump (see Fig. 5.11(b)). This is related to the interesting dynamics of the dot at voltages exceeding the threshold. The master equation describing the dynamics concerns the probabilities of being in the ground singlet and excited triplet states, $p_{\rm g}$ and $p_{\rm T}$:

$$0 = {\rm d}p_{\rm T}/{\rm d}t = -R_{\rm T\to g}p_{\rm T} + \Gamma_{\rm g\to T}p_{\rm g};$$
$$0 = {\rm d}p_{\rm g}/{\rm d}t = -\Gamma_{\rm g\to T}p_{\rm g} + R_{\rm T\to g}p_{\rm T},$$

where $R_{\rm T\to g}$ is the relaxation rate from the excited triplet to the ground state. This, in a typical situation, is not due to environment: this is yet another co-tunneling rate comparable with the rates discussed. In distinction from these rates, both elementary tunneling events can happen in the same junction, either left or right.

Exercise 5.12. (Extensive calculation!) Compute the rate $R_{\rm T\to g}$ in the model formulated.

Let us note that the co-tunneling current is different in two states available above the threshold, $I_{\rm g} \neq I_{\rm T}$. In addition, $\Gamma_{\rm g\to T}$ and a part of the relaxation rate, $\tilde{R}_{\rm T\to g}$, also contribute to the current that reads $I = p_{\rm T}(I_{\rm T} + \tilde{R}_{\rm T\to g}) + p_{\rm g}(I_{\rm g} + \Gamma_{\rm g\to T})$. There is a substantial increase of $p_{\rm T}$ immediately above the threshold that inhibits the current increase and results in lower conductance (see Fig. 5.11).

What determines the width of the conductance jump at vanishing temperature? The situation unexpectedly resembles that in a SET at the cross-over between co-tunneling and single-electron tunneling, although there is no single-electron transfer at the threshold voltage. To understand the analogy, let us look at Fig. 3.25. Above the threshold, we have switching events between g and T with rates $\Gamma_{\rm g\to T}$ and $R_{\rm T\to g}$ (see Fig. 3.25(a)). It is essential that $\Gamma_{\rm g\to T}$ goes to zero upon approaching the threshold. The switching events are thus grouped in pairs. They are separated by a time interval $\simeq 1/R_{\rm T\to g}$ that competes with the Heisenberg uncertainty time $t_{\rm H} \simeq \hbar/(eV - E_s)$ (see Fig. 3.25(b)). This gives the estimation of the energy width of the cross-over, width $\simeq \hbar R_{\rm T\to g}$. One can adapt Eq. (3.72) to quantify the cross-over at this scale. Slightly below the threshold, there is a current increase due to a complicated process made of two co-tunnelings switching between g and T (see Fig. 3.25(c)).

5.5 Charge qubits

In Sections 5.5–5.7, we give several examples of "working" solid state qubits, that is, those experimentally realized up to date (2009). We do not review the numerous, and possibly even more successful, realizations and proposals not related to quantum transport, such as nuclear magnetic resonance schemes, ultra-cold atom manipulation, or optically driven qubits. Solid state qubits always involve electron transfers, either coherent or incoherent. Since electrons are charged, it looks natural to employ a *charge* degree of freedom to represent the quantum information. These qubits are discussed in the present section. It turns out that so far the most successful realizations have involved superconductivity. Superconducting nanostructures can be exploited in a different regime where the superconducting phase, rather than charge, is the working degree of freedom. These *phase* and *flux* qubits will be considered in Section 5.6. The very notion of a qubit historically originates from our experience with 1/2 electron spin. Naturally enough, there are solid state qubits utilizing the spin degree of freedom – *spin* qubits (see Section 5.7).

5.5.1 Double quantum dots

We learned in Chapter 3 that a normal-metal Coulomb blockade can be used to create and manipulate states of fixed charge. However, they are not suitable when realizing a qubit: the transitions between those states involve infinitely many degrees of freedom of the electrons in the reservoirs and therefore do not preserve quantum coherence. Neither can such charge states form a quantum superposition. The simplest way out is not to involve the reservoirs in the electron transfer. Let us consider two quantum dots placed close to each other and, for the moment, not connected to the leads. The energy is dominated by a Coulomb interaction, and the charge states of both dots are controlled by two gates. We have considered a similar situation in Section 3.1. Adjusting the two gate voltages (see Fig. 3.4(a)), one can align the Coulomb energies of states $(1,0)$ and $(0,1)$. (In Fig. 3.4(a), (N_1, N_2) denotes the state with N_1 (N_2) excess electrons in the first (second) dot.) If the spectrum of the electron states in the dot were continuous, many of these states would have close energies, and the degeneracy would not lead to quantum coherence. The discreteness of the states in the dots makes a crucial difference. If the energy mismatch of two charge states $(1,0)$ and $(0,1)$ is sufficiently small, only the *ground* states of both dots are close in energy, while other states can be disregarded. This is the case provided this Coulomb energy mismatch does not exceed the mean level spacing δ_S in either dot. Thereby we achieve one-to-one correspondence between the charge configuration and the quantum-mechanical state of the system: we have the simplest-charge qubit. (For the sake of compactness, we denote the states as $(1,0) \to |0\rangle$ and $(0,1) \to |1\rangle$ in the following.) This setup is commonly known as a *double* quantum dot. This stresses the fact that two quantum dots brought together still comprise an isolated discrete system, that is, a quantum dot (see Fig. 5.12).

The qubit Hamiltonian involves the energy mismatch ε. In addition, we recognize that, for sufficiently close dots, electrons can tunnel through the barrier separating the dots. Such tunneling switches the double dot between the charge states, mixes them, and is represented by the non-diagonal matrix element $\mathcal{T}/2$ in the basis of these charge states. If the tunneling is not affected by interference effects induced by an external magnetic flux, the phase of $\mathcal{T}$ is irrelevant and $\mathcal{T}$ can be chosen as real. The Hamiltonian assumes the common qubit form:

$$\hat{H} = \frac{1}{2}\begin{pmatrix} \varepsilon & \mathcal{T} \\ \mathcal{T} & -\varepsilon \end{pmatrix} = \frac{1}{2}\left(\epsilon\hat{\sigma}_z + \mathcal{T}\sigma_x\right). \tag{5.34}$$

Two eigenstates are separated by energy $\hbar\Omega = \sqrt{\varepsilon^2 + \mathcal{T}^2}$.

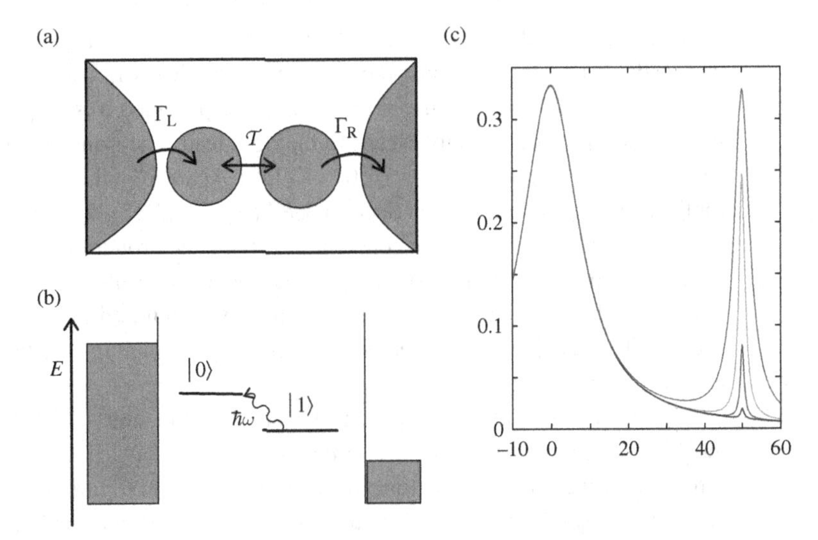

Fig. 5.12. Double-dot charge qubit. (a) Double dot between two leads. The arrows show the tunnel processes: coherent ($\mathcal{T}$) and incoherent ($\Gamma_{R,L}$). (b) Energy diagram of the effective qubit. Irradiation enables manipulation. (c) Current versus energy mismatch ε at different irradiation intensities: $\tilde{\varepsilon}/\varepsilon\Gamma_R\hbar = 1, 3, 9, 27$ from lower to upper curves. $\mathcal{T}/\hbar\Gamma_R = 10$, $\hbar\omega = 5\mathcal{T}$. The broad peak (see Eq. (5.36)) is the qubit "leakage." The narrow peak (see Eq. (5.38)) is the result of the resonant irradiation growing with intensity.

This qubit can be easily *prepared* in a certain state. If this state is given by

$$|\psi\rangle = -(|0\rangle + |1\rangle)\frac{\sin(\theta/2)}{\sqrt{2}} + (|1\rangle - |0\rangle)\frac{\cos(\theta/2)}{\sqrt{2}},$$

this is the ground state of the system provided $\epsilon/\mathcal{T} = \tan\theta$. We tune the gate voltages to satisfy this condition, and wait long enough for the qubit to relax: the state is prepared. The *manipulation* can be achieved by applying the resonant oscillating fields with frequency $\omega \approx \Omega$. It is simpler to vary the potential of the gate electrodes, that is, to vary ε. We note that the efficiency of this action becomes increasingly lower with $\varepsilon \to 0$ since, in this case, the direction of the manipulating "field" coincides with the quantization axis. To circumvent this, one may modulate the height of the potential barrier separating the dots with the optional extra gate electrode. This modulates $\mathcal{T}$. The *read-out* of the qubit is the measurement of charge. For example, one can measure the charge in a quantum dot by measuring the current in the nearby quantum point contact (see Section 5.7 for details).

There are some practical reasons that mean that this qubit realization is limited in its applications. The main reason is the large charge noise affecting the qubit arising from the slow motion of the background charges (see Chapter 3) in the structure. One can reduce the effect of the noise by manipulating the qubit at $\varepsilon \approx 0$. At this "magical" point, the qubit energy splitting $\hbar\Omega$ does not depend on the external charge/field, $\partial\Omega/\partial\varepsilon \approx 0$. Alas, this makes it difficult to manipulate the qubit using the external field. However, the double-dot qubits were realized even before qubits came into fashion. The scheme, however, was not as good as that first described as it involved the electron tunneling to/from the dots [126].

Let us describe this "worse" double-dot qubit (see Fig. 5.12). We set the dots between two leads, L and R. We adjust the voltages of the leads in such a way that the electron from the left lead can only tunnel to the first dot (provided it is empty), while the electron from the second dot can only escape to the right lead. The transport cycle is between three charge states: $(1,0) \rightarrow (0,1) \rightarrow (0,0) \rightarrow (1,0)$. Apart from the notation, this is the same as for the JQP cycle considered in Section 3.7.2 (see Fig. 3.37). The transitions between $(0,0)$ and the two other states are not coherent, involving electron transfer to/from the reservoirs, while those between $(0,1)$ and $(1,0)$ are clearly coherent. As we learned in Section 3.7.3, to describe the situation we need a density matrix with non-diagonal elements between $|0\rangle$ and $|1\rangle$. To specify the parameters, we denote the tunneling rate from the left and right leads as Γ_L and Γ_R, respectively.

Why does this make a qubit? Let us start with the state $(0,0)$. Upon tunneling from the left lead, the system is *"prepared"* in the state $|0\rangle$. It will not stay in this state forever, but we can use the time it stays there for the *manipulation*. With a resonant pulse, we may bring the system to state $|1\rangle$ or to a superposition of $|0\rangle$ and $|1\rangle$. The *read-out* is the electron tunneling to the right lead: the current through the right junction is proportional to the probability p_1 of being in state $|1\rangle$, $I = e\Gamma_R p_1$. The read-out is destructive: after the tunneling, the system is in the state $(0,0)$ and not a qubit at all. However, the tunneling through the left lead creates a (new) qubit and initializes it to $|0\rangle$ again.

Why is this a bad qubit? Ideally, the qubit is initialized to an eigenstate of the qubit Hamiltonian and a probability of being in an eigenstate is read. This is not the present case: the eigenstates of Eq. (5.34) differ from $|0\rangle$, $|1\rangle$. Therefore, the current in the double-dot cannot be always identified with the result of the manipulation. Since each eigenstate is a superposition of the two charge states $|0\rangle$ and $|1\rangle$, each one can decay by tunneling through the junction on the right. Therefore the current flows even in the absence of the resonant excitation of the qubit that switches the eigenstates.

To make this quantitative, we consider the average current through the double-dot subject to constant external irradiation. The equations for the density matrix are similar to Eqs. (3.108)–(3.110) and are given in the present notation by

$$\begin{aligned} \frac{\partial \rho_{00}}{\partial t} &= \Gamma_L(1-\rho_{00}-\rho_{11}) - \frac{i\mathcal{T}}{2\hbar}(\rho_{10}-\rho_{01}); \\ \frac{\partial \rho_{11}}{\partial t} &= -\Gamma_R\rho_{11} + \frac{i\mathcal{T}}{2\hbar}(\rho_{10}-\rho_{01}); \\ \frac{\partial \rho_{01}}{\partial t} &= -\frac{1}{2}\Gamma_R\rho_{01} + \frac{i\varepsilon(t)}{\hbar}\rho_{01} - \frac{i\mathcal{T}}{2\hbar}(\rho_{11}-\rho_{00}). \end{aligned} \tag{5.35}$$

Here, $\varepsilon(t) = \varepsilon + \tilde{\varepsilon}\cos\omega t$, where ε is the time-averaged energy mismatch, and $\tilde{\varepsilon}$ and ω are the amplitude and frequency of the irradiation, respectively.

Let us consider moderate amplitudes of the irradiation, $\tilde{\varepsilon} \ll \max(\varepsilon, \mathcal{T})$. In this case, the irradiation can be disregarded except for in the vicinity of the resonance $\omega \approx \Omega$. The current $I = e\Gamma_R\rho_{11}$ is determined by the stationary solution of Eqs. (5.35):

$$I/e = \frac{\mathcal{T}^2\Gamma_R}{\mathcal{T}^2(2+\Gamma_R/\Gamma_L)+\hbar^2\Gamma_R^2+4\varepsilon^2}. \tag{5.36}$$

The current as a function of ε has a Lorentzian-shaped peak with width of the order of $\max(\mathcal{T}, \hbar\Gamma_R, \mathcal{T}\sqrt{\Gamma_R/\Gamma_L})$. We further concentrate on the limit $\mathcal{T} \gg \hbar\Gamma_R$, where the lifetime of the qubit exceeds the period of the coherent oscillations $2\pi/\Omega$. The current vanishes at $|\varepsilon| \gg \mathcal{T}$ since in this case the eigenstates of the Hamiltonian given by Eq. (5.34) almost coincide with $|0\rangle$, $|1\rangle$. At $\varepsilon \approx 0$, the eigenstates are equal mixtures of $|0\rangle$ and $|1\rangle$.

Control question 5.17. What is the decay rate for each of the eigenstates?

The current reaches a maximum possible value given by $eI = \Gamma_R\Gamma_L/(2\Gamma_L+\Gamma_R) < \Gamma_R/2$.

Let us concentrate on the vicinity of the resonance. In Section 5.3 we described two methods to deal with the qubit Hamiltonian in the presence of resonant excitation. In both methods, the time-dependent Hamiltonian is replaced by a quasistationary one, $\hat{\tilde{H}}$. The difference between the methods is the time-dependent unitary transformation made: it is diagonal either in the basis of the time-averaged Hamiltonian or in the basis where the time-dependent modulation is diagonal. We opt for the second method. The effective Hamiltonian is given by

$$\hat{\tilde{H}} = \frac{1}{2}\begin{pmatrix} \hbar\delta\omega & \tilde{\mathcal{T}}(t) \\ \tilde{\mathcal{T}}^*(t) & -\hbar\delta\omega \end{pmatrix}, \tag{5.37}$$

where $\hbar\delta\omega = \varepsilon - \hbar\omega$ is the frequency mismatch and

$$\tilde{\mathcal{T}} = -\mathcal{T}J_1(\tilde{\varepsilon}/\hbar\omega) \approx -\mathcal{T}\tilde{\varepsilon}/2\varepsilon$$

is the effective tunneling amplitude. The validity of the method is restricted by $\max(\delta\omega, \tilde{\mathcal{T}}/\hbar) \ll \omega$. The density matrix equations are quasistationary and are readily obtained by replacing the Hamiltonian given in Eq. (5.34) by $\hat{\tilde{H}}$, that is, making the replacements $\varepsilon(t) \to \hbar\delta\omega$, $\mathcal{T} \to \tilde{\mathcal{T}}$. The current in the vicinity of the resonance is obtained from Eq. (5.36) by changing notation:

$$I/e = I_{\max}\frac{1}{1+(\delta\omega/w)^2}, \tag{5.38}$$

with the height $I_{\max} = \Gamma_R(2+\Gamma_R/\Gamma_L+4\hbar^2\Gamma_R^2\varepsilon^2/(\mathcal{T}^2\tilde{\varepsilon}^2))^{-1}$ and the width $w^2 = (\mathcal{T}^2\tilde{\varepsilon}^2)(2+\Gamma_R/\Gamma_L)/(16\hbar^2\varepsilon^2)+\Gamma_R^2/4$. We note the emergence of the second current peak at $\varepsilon = \hbar\omega$. This peak is narrow and well separated from the non-resonant peak (see Fig. 5.12(c)) provided $\omega \gg \mathcal{T}/\hbar, \Gamma_R$. For $\Gamma_R \ll \mathcal{T}/\hbar$, it becomes as big as the non-resonant peak at relatively small irradiation amplitudes $\tilde{\varepsilon}/\varepsilon \simeq \Gamma_R\hbar/\mathcal{T}$. We note that under these conditions the resulting qubit is not as bad as it looked at the beginning: it obviously responds to the manipulation. The result of the manipulation is clearly visible in the current, so the read-out works as well. The next step of the improvement would be to pulse the irradiation and possibly the bias voltage. This, however, was first realized for a superconducting qubit rather than for the normal-electron double-dot [127].

5.5.2 Superconducting charge qubits

The heart of a superconducting charge qubit is the Cooper-pair box described in detail in Section 3.5.1. It consists of a superconducting Coulomb island connected to a superconducting lead (or leads) by a Josephson junction (or junctions). This gives rise to a rich energy spectrum. The charge qubits exploit the limit of large charging energies $E_C \gg E_J$ (to avoid complications with quasiparticles, one also requires that the superconducting energy gap exceeds the charging energy, $\Delta_S > E_C$).

As discussed in Section 3.5.3, if the charging energy dominates, the charge of a Coulomb island is a well defined variable. The energies of the states are shifted parabolas if plotted versus the gate voltage. The notable exceptions are the values of the gate voltage where two parabolas cross. The Josephson energy gives rise to a single Cooper-pair transfer, changes the charge by 2, and therefore mixes up the crossing states. We concentrate here on a single crossing. The charge states crossing correspond to $n = 0$ and $n = 2$ extra charges. To comply with the standard qubit notation, we call these states $|0\rangle$ and $|1\rangle$, respectively. Let us first recognize that the Hamiltonian of the superconducting charge qubit can be presented in precisely the same form as Eq. (5.34) provided $\mathcal{T} \equiv E_J$. This is not especially surprising since the Hamiltonians of all qubits are the same. Surprisingly, there were more similarities between the double-dot setup described above and the pioneering experiment by Nakamura *et al.* [127], where the coherent control of a superconducting charge qubit was achieved. The read-out in this experiment basically followed the JPQ cycle (see Section 3.7.2). The extra probe junction at sufficiently high bias provided the two-stage quasiparticle decay from state $|1\rangle$ to state $|0\rangle$. The rates of the decays correspond to Γ_R and Γ_L of the double-dot setup. Experimental values were $(6\text{ ns})^{-1}$ and $(8\text{ ns})^{-1}$, respectively. The qubit is thus completely described by Eqs. (5.35).

Let us describe the working sequence used in Ref. [127]. In the beginning of the sequence, the qubit is tuned far from the degeneracy point. The read-out is on during the whole sequence. This means that the qubit sooner or later is found in the state close to $|0\rangle$: the final stage of the read-out induced decay. Then a dc voltage pulse is applied. The height of the pulse is such that the pulse brings the qubit to the degeneracy point $\varepsilon \approx 0$. The pulse is *not* adiabatic: it is so fast that the qubit remains in the state $|0\rangle$. However, at the crossing point the state $|0\rangle$ is not an eigenstate of the Hamiltonian. The actual eigenstates are $(|0\rangle \pm |1\rangle)/\sqrt{2}$, corresponding to energies $\pm\hbar\Omega/2$. The wave function evolves in time according to the following expression:

$$|\psi(t)\rangle = \frac{1}{2}\left((|0\rangle + |1\rangle)\mathrm{e}^{-\mathrm{i}\Omega t/2} + (|0\rangle - |1\rangle)\mathrm{e}^{\mathrm{i}\Omega t/2}\right);\ |\psi(0)\rangle = |0\rangle. \tag{5.39}$$

The probabilities of being in certain charge states, ρ_{00}, ρ_{11}, oscillate in time. The pulse lasts for the time interval τ. After this, the qubit returns to the preparation point, which is far from the degeneracy point. If the pulse is sufficiently short, $\Gamma_R\tau \ll 1$, the coherent dynamics of the qubit during the pulse is not affected by the read-out. The read-out takes place afterwards. If the qubit after the manipulation is in state $|0\rangle$, nothing happens: it remains in $|0\rangle$ till the next pulse. If the qubit is in state $|1\rangle$, two quasi-particle tunnel processes take place: the current they produce is detected. Thereby the current is proportional to the

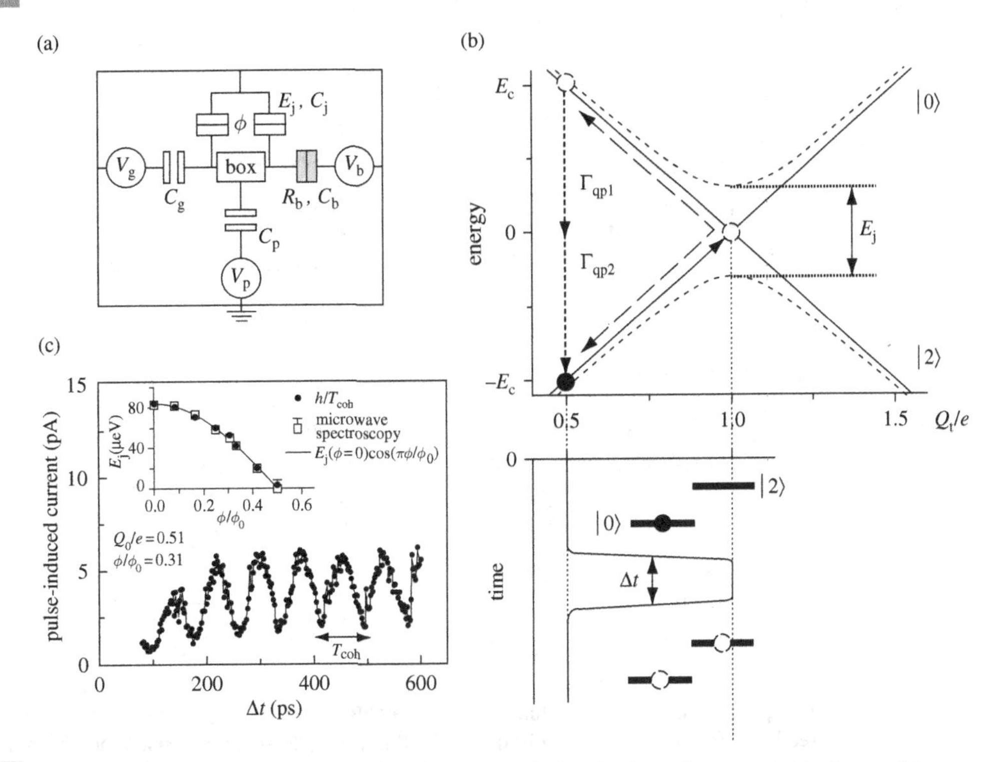

Fig. 5.13. Cooper-pair box as a charge qubit. (a) The setup includes the CPB, the gate electrodes, and the read-out junction (shaded). The flux loop is used to tune E_J. (b) Working sequence and energy diagram. (c) The pulse-induced current measures the probability p_1 and displays Rabi oscillations as a function of the pulse duration τ. Inset: E_J tuned. Reprinted by permission from Macmillan Publishers Ltd: Ref. [127], copyright (1999).

probability of being in state $|1\rangle$. After the tunneling, the qubit is back in state $|0\rangle$ and ready for the next pulse. Equation (5.39) yields $p_1 = \sin^2(\tau\Omega/2)$. Thus, the charge transferred per sequence period, $Q = 2ep_1$, exhibits oscillations as a function of pulse duration. Such oscillations have indeed been observed. The ratio of the signal observed to the theoretical maximum $Q = 2e$ – often called "visibility" – was about 25% (see Fig. 5.13). The low visibility was explained by the finite fall and rise times of the pulse and the rather short sequence period not exceeding by much the inverse tunneling rates (faulty preparation). The visibility was greatly increased in the subsequent experiments of the group, in which two coupled superconducting charge qubits have been measured and the entanglement of the two was accessed.

The next step in the charge qubit design – the so-called "quantronium" [128] – was to improve the coherence times by two orders of magnitude and to characterize different decoherence mechanisms (see Fig. 5.14). The heart of the quantronium is the same Cooper-pair box (CPB). However, the Josephson energy was increased to the order of the charging energy. As mentioned, in this situation neither charge nor phase is a good quantum number. The CPB eigenstates are coherent superpositions of several charge states. Therefore, it is not plausible to choose the basis states of a qubit as the states with fixed charge. Rather, the

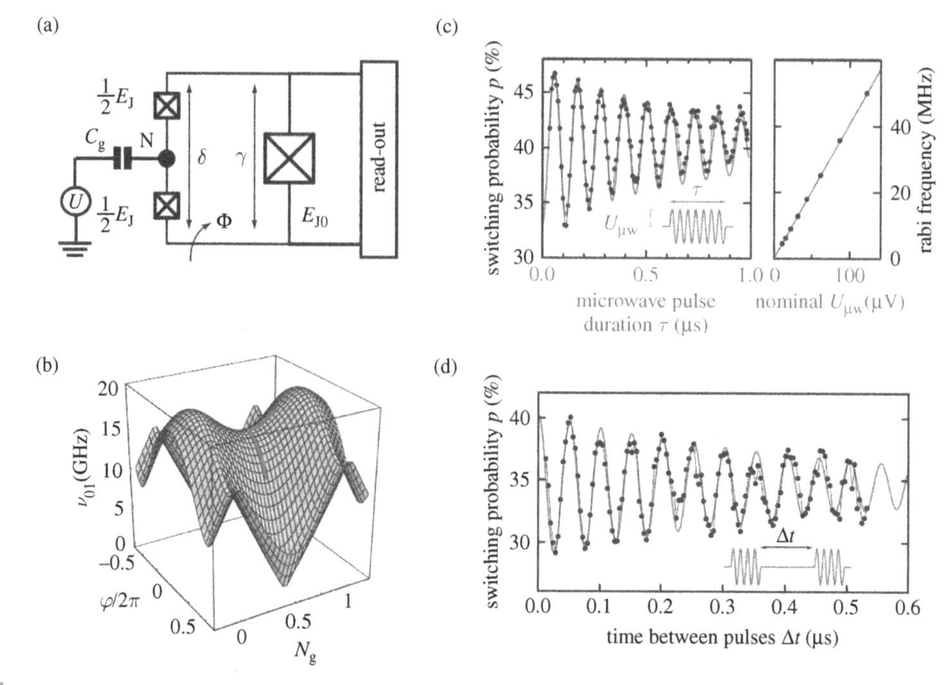

Fig. 5.14. The quantronium. (a) The setup is basically a CPB. The extra loop and junction are used for the read-out. (b) The operational frequency of the qubit ($(E_1 - E_0)/\hbar$) versus the phase difference and the charge induced $N_g = q/2e$. The saddle point is "magic": the operational frequency is least sensitive to both charge and flux noise. (c) Rabi and (d) Ramsey oscillation experiments. From Ref. [128]. Reprinted with permission from AAAS.

basis states of the qubit, $|0\rangle$ and $|1\rangle$, are represented by the ground state and first excited state of the CPB, respectively. The Hamiltonian is thus diagonal in this basis.

The increased Josephson energy increases supercurrents flowing through the CPB. This makes it possible to read out the qubit states in a less obtrusive and more controlled way. The read-out variable is this supercurrent. The CPB is embedded into a larger superconducting current-biased loop closed by a more conducting junction with Josephson energy $\simeq 20E_J$. To obtain the read-out, a current pulse with a value slightly below the critical current of the conducting junction is applied to the loop. The CPB current adds to this bias current in the conducting junction. Now let us note that the CPB current is different for different qubit states. It is given by the derivative of the energy of the state with respect to the superconducting phase, $I_{0,1} = (2e/\hbar)\partial E_{0,1}/\partial\varphi$ (see Section 1.8). The maximum difference of the currents, $I_1 - I_0$, was as big as 12 nA. Adjusting the height of the read-out pulse, one makes sure that the total current through the conducting junction exceeds its critical current if the qubit is in state $|1\rangle$ and does not exceed it for state $|0\rangle$. Adjusting the duration of the pulse, one tries to make sure that the conducting junction switches to the resistive state if the qubit is in state $|1\rangle$ and does not switch otherwise. The voltage pulse generated by the junction switching is detected.

The quantronium has *two* control parameters, discussed in Section 3.5.1: the induced charge q and the superconducting phase difference φ, and the energies and wave functions of the states depend on the two. The trick that allowed us to achieve long coherence times is a specific choice of the working point in this two-dimensional parameter space (see Fig. 5.14(b)). This "magic" point is chosen from the saddle-point condition:

$$\frac{\partial(E_0 - E_1)}{\partial\varphi} = \frac{\partial(E_0 - E_1)}{\partial q} = 0.$$

The idea behind this choice is that the main sources of noise are fluctuations of background charges and flux in the loop. Thus both can be seen as random deviations of the control parameters q and φ. At the magic point, the energy difference of the qubit states is not sensitive to these deviations, at least in linear order. The qubit was thus prepared and manipulated at the magic point. The read-out would not work at the magic point since $I_1 = I_0$. However, the current pulse used for the read-out automatically shifts the phase, taking the qubit away from the magic point.

The quantronium is best manipulated by pulses of irradiation at resonant frequency $\Omega \equiv (E_0 - E_1)/\hbar$, (which yields the value 1.04×10^{11} Hz). The irradiation can be applied to the gate electrode as well as to the coil controlling the phase. What matters is that the interaction of the qubit and irradiation contains operators σ_x, σ_y orthogonal to the quantization z axis of the qubit. This enables the resonant manipulation described in Section 5.3. The x axis can always be chosen such that the interaction is proportional to $\hat{\sigma}_x$.

Control question 5.18. Explain why at the magic point the interaction of the qubit and irradiation does not contain terms $\propto\hat{\sigma}_z$.

To probe coherent oscillations of the qubit, two different experiments were performed. In the Rabi oscillations experiment, the qubit, initially in state $|0\rangle$, was subject to a long irradiation pulse of duration τ. During the pulse, the qubit oscillates with Rabi frequency $\omega_R = \sqrt{U^2 + (\delta\omega)^2}$, where U is the pulse amplitude and $\delta\omega \equiv \Omega - \omega$ is the frequency mismatch. The probability distribution after the pulse is given by Eq. (5.39) and exhibits periodic oscillations in $\omega_R\tau$ (see Fig. 5.14(c)). Experimentally, these oscillations are weakly damped owing to decoherence. This allows the measurement of coherence time $T_2 \approx 1$ μs.

Another measurement has been taken by applying the Ramsey pulse sequence (Section 5.3). Since the $\pi/2$ pulses in the sequence are short in comparison with the interval τ between them, the measurement probes the free evolution of the qubit during this interval. The Ramsey oscillations with period $2\pi/(\delta\omega)$ are damped; this damping gives a somewhat shorter decoherence time $\tilde{T}_2 = 0.5$ μs (see Fig. 5.14(d)).

The explanation of this fact implies that the main source of decoherence is still low-frequency fluctuations of Ω induced by the charge and flux noise. The free evolution of the qubit is immediately subject to these fluctuations. As for Rabi oscillations, their frequency is less susceptible to the fluctuations of Ω since $\partial\omega_R/\partial\Omega = (\delta\omega)/\omega_R < 1$ so that the effect of the fluctuations is insignificant provided $|\delta\omega| \ll U$.

The important advantage of a superconducting charge qubit is that it has two control parameters – charge and flux – although at the same time it makes the qubit vulnerable to both noises. Corresponding to the two control parameters, qubits can be coupled in two different ways. Theoretical proposals [129] stressed the advantages of inductive coupling where the qubits share a common loop so their fluxes interact. This also seems easier to fabricate. However, an alternative – capacitive coupling – has been realized experimentally [130]. In that case, the interaction is between the charges of adjacent CPB islands.

5.6 Phase and flux qubits

So far, the most successful realizations of solid state qubits have been made with using "home-made" quantum mechanics based on Josephson- and Coulomb-blockade phenomena. There are two reasons for this. First, the systems involved are rather big and the manufacturing technology is mature. Secondly, the system is an integral part of an electric circuit; this simplifies the manipulation and read-out. As well as superconducting charge qubits, where the ratio of the Josephson and charging energies is either small or moderate, successful designs have been realized with $E_J \gg E_C$. Such designs come in two classes: phase and flux qubits, which we consider in this section.

5.6.1 Phase qubits

A phase qubit is conceptually very simple: the physical system where it is realized is just a (relatively) big Josephson junction with a capacitance. The Josephson energy dominates, $E_J/E_C > 10^4$. The junction is biased by the current I_b close to the critical value. The goal of the qubit design is to make the Josephson potential well so small that it houses only a few quantum levels. In this case, the potential is well approximated by a cubic parabola (see Eq. (3.91)).

The two lowest levels are used to make up a qubit with states $|0\rangle$ and $|1\rangle$ [131]. Formally speaking, they are not true quantum states. The cubic parabola Hamiltonian supports the continuum spectrum at the same energy so that all "discrete" states have finite lifetimes due to macroscopic quantum tunneling (see Section 3.5.5). The ingenious qubit design implemented is based on the fact that these lifetimes differ by two or three orders of magnitude for neighboring levels. The lifetime of the ground state $|0\rangle$ by far exceeds the measurement time.

The faster decay rate of the higher states has been used for *read-out*. There are two types of read-out proposed. In each case, the well during the qubit manipulation houses three levels (see Fig. 5.15(a)). The first read-out scheme uses the irradiation pulse, with frequency matching the energy difference of the upper level $|2\rangle$ and the excited qubit level $|1\rangle$. The pulse performs a SWAP operation between $|1\rangle$ and $|2\rangle$, so the probability ρ_{11} is transferred to state $|2\rangle$. The upper state $|2\rangle$ is close in energy to the top of the barrier, so it lives for a very short time, while lifetimes in $|0\rangle$ and $|1\rangle$ by far exceed the measurement time. Therefore, the probability of the junction switching is ρ_{11}. The second read-out scheme involves lowering the potential barrier by an adiabatic change of the bias current after manipulation. The reduced potential well houses only two levels, so that state $|1\rangle$

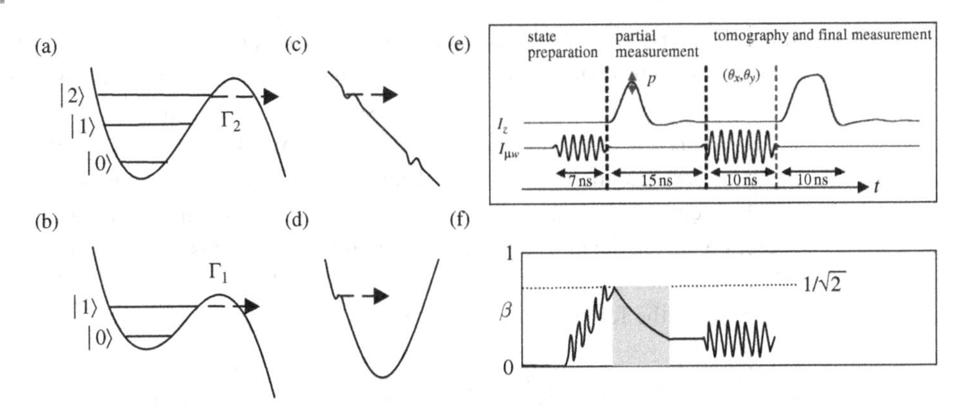

Fig. 5.15. **Phase qubits. (a) Three- and (b) two-level wells. (c), (d) Two setups differ in their histories after the switching. (e) The working sequence of the "partial collapse" experiment. (f) Corresponding time-line of the amplitude β. It experiences a "non-unitary" evolution (Eq. (5.41)) during the first read-out pulse (gray region). From Ref [132]; reprinted with permission from AAAS.**

decays fast (Fig. 5.15(b)). Therefore, in this case the switching probability is also ρ_{11}. We note that, in contrast to the schemes described in Section 5.5, the read-out variable is just the level number, that is, it commutes with the Hamiltonian.

The history of the system after switching differs for the two schemes used. In an older scheme, the Josephson potential was a classical tilted washboard (see Fig. 5.15(c)). After switching, the junction was retrapped in the adjacent potential minimum and relaxed to the ground state $|0\rangle$. This is how *preparation* was implemented. In a newer scheme, the qubit junction is embedded into a SQUID loop, so that the Josephson potential is an almost parabolic well with a minute metastable minimum at one side (see Fig. 5.15(d)). After switching, the junction relaxes to the bottom of the well. Subsequent adiabatic shift of the flux in the SQUID loop distorts the potential and finally brings the junction back to the minute minimum. After this is completed, the *manipulation* of the qubit is performed in a standard way, by applying irradiation pulses with frequency matching the energy difference between $|0\rangle$ and $|1\rangle$.

Technical advances have allowed high quality read-out and manipulation at the fast time scale of 10 ns. Importantly, one can have two read-out pulses within the working sequence of the qubit. This made possible the following interesting experiment [132]. The qubit was prepared in the state $|0\rangle$. The first irradiation pulse changed it into the superposition $\cos(\theta_0/2)|0\rangle + \sin(\theta_0/2)|1\rangle$. After this, the first read-out pulse of variable duration was applied. The second irradiation pulse served to perform tomography of the superposition after the first read-out pulse. The second read-out pulse was subsequently applied, and the switching probability was measured (see Fig. 5.15(e)). The subject of interest is the dependence of this probability on the rotation angles of the second pulse. This allows one to perform the tomographic scan and accurately determine the coefficients in the resulting superposition.

Let us try to predict the results of the experiment. There seem to be two outcomes of the working sequence. If the junction has switched during the first read-out pulse, our qubit is destroyed (none of the levels $|0\rangle$, $|1\rangle$ is populated, and it is not even clear if the levels still exist). The second irradiation pulse works on nothing, and this outcome produces no contrast in the tomography measurement. The second outcome is no switching during the first read-out pulse. It is very natural to assume that in this case the superposition remains intact, taking a possible decoherence effect apart. We therefore predict that the tomography measurement would give the original superposition coefficients $\cos(\theta_0/2), \sin(\theta_0/2)$.

This seemingly indisputable reasoning appears to be wrong. Let us describe the quantum evolution during the first read-out pulse in more detail. This evolution involves *three* states: $|0\rangle, |1\rangle$, and the state $|s\rangle$ into which $|1\rangle$ can decay. The density matrix of the systems evolves according to the following system of equations:

$$\begin{aligned} \frac{\partial \rho_{00}}{\partial t} &= 0; \quad \frac{\partial \rho_{11}}{\partial t} = -\Gamma \rho_{11}; \quad \frac{\partial \rho_{ss}}{\partial t} = \Gamma \rho_{11}; \\ \frac{\partial \rho_{01}}{\partial t} &= -\frac{\Gamma}{2} \rho_{01}; \end{aligned} \tag{5.40}$$

with Γ being the decay rate of state $|1\rangle$. We disregard the small decay rate of state $|0\rangle$. The first three equations are in fact classical and describe the probabilities of being in the corresponding states. The last equation for the non-diagonal matrix element is quantum and is a simplified version of the equations used earlier in the book, for example Eqs. (5.35). The equations have to be solved with initial conditions corresponding to the superposition made by the first irradiation pulse: $\rho_{00} = \cos^2(\theta_0/2)$; $\rho_{11} = \sin^2(\theta_0/2)$; $\rho_{01} = \rho_{10} = \sin(\theta_0)/2$; $\rho_{ss} = 0$. The solution for the density matrix after the pulse of duration τ reads: $\rho_{00} = \cos^2(\theta_0/2)$; $\rho_{11} = \sin^2(\theta_0/2)\exp(-\Gamma\tau)$; $\rho_{01} = \rho_{10} = \exp(-\Gamma\tau/2)\sin(\theta_0)/2$. We normalize the solution to $\rho_{11} + \rho_{00} = 1$. This is so we focus on the case when switching did not take place. We thus exclude state $|s\rangle$ and look at the qubit density matrix obtained. It appears that $\hat{\rho}^2 = \rho$, indicating the pure state to be characterized by a wave function. This is given by

$$\begin{aligned} |\psi\rangle = & \frac{\cos(\theta_0/2)}{\sqrt{\cos^2(\theta_0/2) + \sin^2(\theta_0/2)\exp(-\Gamma\tau)}} |0\rangle \\ & + \frac{\exp(-\Gamma\tau/2)\sin(\theta_0/2)}{\sqrt{\cos^2(\theta_0/2) + \sin^2(\theta_0/2)\exp(-\Gamma\tau)}} |1\rangle. \end{aligned} \tag{5.41}$$

We see that the coefficients in the superposition are modified by the first read-out pulse. This conclusion has been experimentally confirmed in Ref. [132]. The authors interpret this as a "non-unitary evolution" of the state in the course of "partial collapse measurement." This is very expressive; however, the evolution of the whole quantum system is certainly unitary, according to the basics of quantum mechanics. And Eqs. (5.40) do not contain any terms specifically responsible for the measurement and/or the collapse. Rather, they describe the natural evolution of the system. The result is easy to understand if one recalls that the components of a wave function are not just amplitudes but probability amplitudes. The change of the probabilities during the first read-out pulse must therefore necessarily be accompanied by the corresponding change in the probability amplitudes.

Exercise 5.13. Evaluate the probability p_1 after the tomography pulse. Assume that the pulse makes a rotation about the x axis by the angle ϕ. Use Eqs. (5.40) and (5.41).

The quantum gate for two phase qubits has been realized by coupling two Josephson junctions capacitively [133].

5.6.2 Flux qubits

Under the same condition that Josephson energy dominates the charging energy, there is also an alternative design, known as a *flux qubit*. The idea of this design originates from a device, an rf (radio-frequency) SQUID, developed back in the 1960s and successfully used for measurements of low magnetic fields – a Josephson junction embedded in a superconducting loop with inductance L. To understand the principle, let us write the energy of the rf SQUID in terms of the phase φ of the Josephson junction:

$$H = -E_{\mathrm{J}} \cos\varphi + \frac{(\varphi\Phi_0/2\pi - \Phi_{\mathrm{ext}})^2}{2L} + \frac{Q^2}{2C_{\mathrm{J}}}.$$

The first term is the Josephson energy of the junction; the second one represents the magnetic energy of the loop, with Φ_{ext} being the flux created by an external magnetic field in the loop area. The final term is the charging energy responsible for the quantum effects and is disregarded for the moment. Without the charging energy, the phase is a classical variable, and the energy of the ground state is found by minimization of the energy over the phase φ:

$$E(\Phi_{\mathrm{ext}}) = \min_{\varphi} H(\varphi, \Phi_{\mathrm{ext}}).$$

The differential inductance of the whole system,

$$\frac{1}{L_{\mathrm{sys}}} = \frac{\partial^2 E(\Phi_{\mathrm{ext}})}{\partial \Phi_{\mathrm{ext}}^2},$$

is measured from the SQUID response on an external rf signal; this is why it is called an *rf SQUID* and differs from the two-junction dc SQUID considered in Section 1.8.

Let us now turn to the idea of the qubit. We use the rf SQUID in a new fashion: we would like to tailor the potential for φ in such a way that it has two close symmetric minima (see Fig. 5.16(a)). This is achieved by matching the inductances of the ring and the junction. We note that the inductance of the junction becomes *negative* for $\cos\varphi < 0$. We choose parameters in such a way that the positive inductance of the ring almost compensates for the negative inductance of the junction near $\varphi \approx \pi$; this requires $(2\pi/\Phi_0)^2 E_{\mathrm{J}} L \approx 1$. Then at $\Phi_{\mathrm{ext}} \approx \Phi_0/2$ we reproduce the (approximately symmetric) double-well potential.

Let us switch the charging energy, on enabling the quantum-mechanical effects. Discrete levels appear in each well. For a symmetric situation, the energies of the states $|0\rangle$ and $|1\rangle$, localized in the right/left well, respectively, are precisely the same. The tunneling through the barrier mixes the states and thereby lifts their degeneracy. The true eigenstates of the Hamiltonian are symmetric and antisymmetric combinations $(|0\rangle \pm |1\rangle)/\sqrt{2}$, separated by

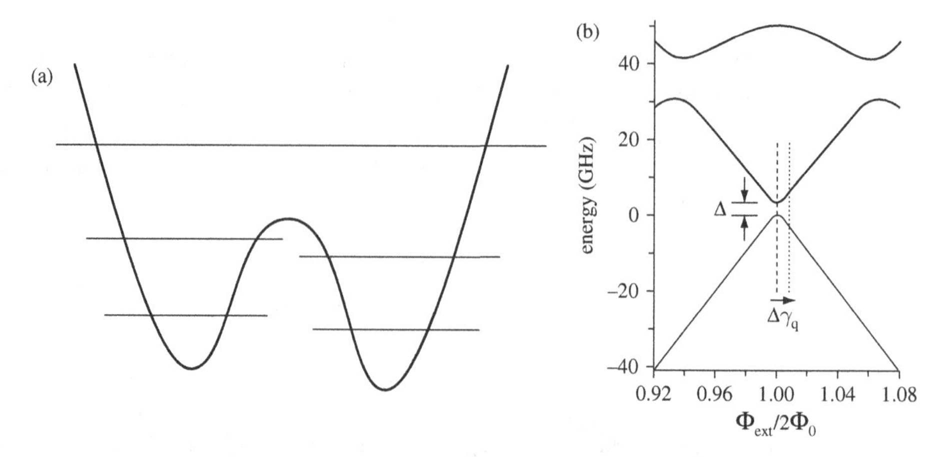

Fig. 5.16. **(a) An almost symmetric double-well potential of an rf SQUID with matching inductances near $\varphi = \pi$. The lowest states localized in different wells form a qubit. (b) Energy levels of a more complicated three-junction flux qubit versus flux. The two lowest states are similar to those in the double-well potential. Part (b) taken from Ref. [134]; reprinted with permission from AAAS.**

the energy difference $\mathcal{T}$. If the external flux deviates slightly from $\Phi_0/2$, the potential is slightly warped so that one of the wells becomes higher in energy. The energy difference between the bottoms of the wells is proportional to this deviation, $\varepsilon \propto (\Phi_{\text{ext}}/\Phi_0 - 1/2)$. This provides a convenient handle for quantum manipulation.

The two states, $|0\rangle$ and $|1\rangle$, localized in two wells, are used to represent the qubit. The Hamiltonian in the basis of these two states acquires the standard form as in Eq. (5.34):

$$\hat{H} = -\frac{1}{2}\left(\varepsilon\hat{\sigma}_z + \mathcal{T}\hat{\sigma}_x\right).$$

The qubit states $|0\rangle$ and $|1\rangle$ correspond to the opposite directions of the magnetization current in the qubit loop and can be distinguished thereby.

The design of the actual qubit [134] is slightly more sophisticated. To start with, it comprises three Josephson junctions combined in a single loop. Two extra junctions replace the inductance loop: it appears to be easier to fabricate the two extra junctions than to adjust the inductance of the loop. The system contains two Coulomb islands and therefore encompasses two quantum degrees of freedom. Diagonalizing the Hamiltonian, we obtain the energy levels, conveniently represented as functions of the external phase φ, over the three junctions, which now plays the role of the external flux. Although the actual system is more complicated than the rf SQUID, as far as the two lowest levels are concerned, it is very similar. The main features remain intact: a symmetric two-well potential is formed near $\varphi = \pi$, the two states in different wells correspond to opposite current circulations, and the potential is utilized to represent a qubit manipulated by modulations of the external flux with the frequency matching the energy difference.

The read-out of a flux qubit is basically the flux difference induced by the currents of two opposite circulations. First, measurements have been performed by a dc SQUID, whose loop was placed around the structure to pick up this flux difference. The value measured was the critical current of the dc SQUID, similar to the readout of the quantronium (see Section 5.5). Such a read-out scheme has some disadvantages due to strong influence of the dc SQUID on the qubit during the switching process. Another way to read-out is similar to the traditional use of the rf SQUID: the measurement of differential inductance of the qubit loop by means of an ac response. The very fast and discriminating read-out makes use of the bistability of a non-linear oscillator driven by the external source with frequency close to the resonant frequency of the oscillator. Such a driven oscillator, if the magnitude of the driving current lies within a certain range, may perform stable oscillations with two different amplitudes ("high" and "low"). The non-linear oscillator has been realized using another Josephson junction biased by ac current. Inductive coupling between this circuit and the qubit results in a small shift of the frequency of the resonator; this shift depends on the state of the qubit. For the setup of the experiment [135], the shift was about 1%. However, this was enough to ensure a discriminating and fast read-out. The qubit state $|0\rangle$ caused the oscillator to switch into the "high" state from the "low" state within 5 ns, while state $|1\rangle$ failed to induce the switching. It is interesting to note that it takes more time ($\approx$25 ns) to read out the amplitude of the oscillator, despite the fact that it is a rather classical object.

Such an improved read-out made it possible to demonstrate the concept of *quantum non-demolition* (QND) *measurement*. By definition, a QND measurement is arranged in such a way that the interaction of the quantum system measured with the measuring device does not cause any transitions between the eigenstates of the system measured. In more formal terms, the Hamiltonian of the system measured commutes with $\hat{H}_{\text{int}}$ which describes the interaction with the measuring device. For example, if the qubit Hamiltonian is proportional to $\hat{\sigma}_z$, the read-out will discriminate between the states $\sigma_z = \pm 1$. The most important property of QND measurement is its *repeatability*: if a QND measurement gives the outcome 0 (1), the next QND measurement will reproduce the same result with 100% probability. This, of course, disregards any possible relaxation of the qubit between two measurements, which can be caused by the agents not related to the measurement. QND is very much reminiscent of the text-book notion of the *projective* measurement. The projective measurement is supposed to cause the collapse of the wave function so that the latter is projected on the basis state corresponding to the measurement result. As far as realistic measurements are concerned, the QND and projective measurements differ. We will discuss the difference in Section 6.7.

The fast and relatively reliable read-out implemented in Ref. [135] made it possible to perform two measurements of the qubit state, separated by a time delay, as in the phase qubit experiment discussed previously. The improvement is that the result of *each* measurement is read out. In this way, three independent quantities can be extracted: the probability P_0 of the qubit of being in state $|0\rangle$ after the first measurement, and two conditional probabilities $P_{0|0}$ ($P_{1|1}$) of measuring the same result $|0\rangle$ ($|1\rangle$) in both measurements. All other probabilities are fixed by the normalization requirements, for example, $P_0 + P_1 = 1$.

In the first experiment described in Ref. [135], the two measurements were separated by a time delay of the order of the relaxation time of the qubit. The experimental curves are shown in Fig. 5.17(b). Plotted are the probabilities P_0, $P_{0|0}$, and $P_{1|1}$. The probability P_0 does not depend on the delay time due to the trivial reason that the first measurement has

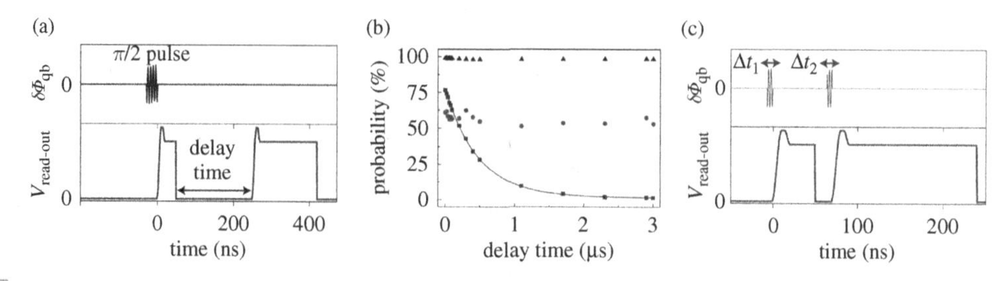

Fig. 5.17. QND measurement with flux qubits. (a) The working sequence with two read-out pulses provides two successive measurements of the qubit. (b) Results: probabilities P_0, $P_{0|0}$, $P_{1|1}$ (from upper to lower curve) versus the delay time. (c) The working sequence with an additional pulse proves the quantum coherence during the sequence. Taken from Ref. [135]; reprinted by permission from Macmillan Publishers Ltd, © (2007).

already occurred. The fact that $P_{0|0}$ also does not depend on the time delay indicates indeed that the measurement is repeatable, as one expects from a QND measurement. The conditional probability $P_{1|1}$ of measuring the excited states in both measurements displays an exponential fall-off with time constant, corresponding to the relaxation time as yet another proof of the repeatability.

The decisive proof comes, however, from the second experiment, where the delay time was short enough to allow for coherent evolution of the qubit between the measurements. Two manipulation pulses with durations t_1 and t_2 were applied before the first and the second measurement, respectively (see Fig. 5.17(c)). The pulses change the pure states $|0\rangle$ and $|1\rangle$ into superpositions, with the coefficients depending on the pulse durations. Owing to this, the outcome of the first measurement did depend on t_1. With experimental accuracy, the conditional probabilities did not depend on t_1, that is, on the wave function before the first measurement. This proved that the first measurement is indeed projective: once a definite outcome (say, $|0\rangle$) is obtained, the information about the superposition coefficients of the measured wave function is lost. The conditional probabilities did depend on t_2 in the expected fashion: the second measurement is sensitive to the superposition made by the second pulse rather than to the outcomes of the first measurement.

As a step towards the realization of quantum algorithms, several flux qubits, eventually whole arrays of them, can be inductively coupled. The resulting interaction is proportional to the currents $\propto \hat{\sigma}_z$ in each qubit, so that the interaction Hamiltonian of two qubits 1, 2 is given by $H_{\text{int}} \propto \sigma_z^{(1)} \sigma_z^{(2)}$. The strength of this coupling can be varied by varying the flux; this provides the means for two-qubit manipulation. The main decoherence source also comes through the $\hat{\sigma}_z$ channel: the qubits, as well as the interaction between them, are sensitive to the noise of the external flux.

To present the state of the art of the flux qubit array, we describe the design proposed in Ref. [136]. This design aims to reduce the sensitivity of the qubits to the external flux noise. To achieve this, the two coupled qubits are biased in the corresponding "magic" points, $\varepsilon_{1,2} = 0$. The Hamiltonian of the two is thus given by ($\mathcal{T}_2 > \mathcal{T}_1 > 0$)

$$\hat{H} = \frac{\mathcal{T}_1}{2} \hat{\sigma}_x^{(1)} + \frac{\mathcal{T}_2}{2} \hat{\sigma}_x^{(2)} + J_{12} \hat{\sigma}_z^{(1)} \hat{\sigma}_z^{(2)}. \tag{5.42}$$

Owing to vanishing currents for both qubit states in this point, the inductive coupling J_{12} has no direct effect and can be disregarded. So the qubits are (almost) decoupled by default. The variable coupling is implemented via a third qubit, which is not used to store the quantum information but to provide a non-linear inductance that can be modulated by the microwave irradiation. The idea is to match the frequency of such irradiation and the energy difference of one of the transitions $|10\rangle \leftrightarrow |01\rangle$ or $|00\rangle \rightarrow |11\rangle$. These energy differences are $T_2 \mp T_1$. The part of the coupling mediated by the third qubit is therefore *resonantly* enhanced. Let us keep in the Hamiltonian the only resonant terms corresponding to all four resonant frequencies:

$$
\begin{aligned}
\hat{H} = &\frac{T_1}{2}\hat{\sigma}_x^{(1)} + \frac{T_2}{2}\hat{\sigma}_x^{(2)} \\
&+ \tilde{\varepsilon}_1 \cos(T_1 t/\hbar + \phi_1)\hat{\sigma}_z^{(1)} + \tilde{\varepsilon}_2 \cos(T_2 t/\hbar + \phi_2)\hat{\sigma}_z^{(2)} \\
&+ \tilde{J}_{12}^{+} \cos((T_2 + T_1)t/\hbar + \phi_+)\hat{\sigma}_z^{(1)}\hat{\sigma}_z^{(2)} \\
&+ \tilde{J}_{12}^{-} \cos((T_2 - T_1)t/\hbar + \phi_-)\hat{\sigma}_z^{(1)}\hat{\sigma}_z^{(2)}.
\end{aligned}
$$

Let us rewrite this in the rotating frame and swap the quantization axes $x \to z, z \to y$, $y \to x$. The resulting Hamiltonian is given by

$$
\begin{aligned}
\hat{H}/\hbar = &\Omega_1 \left(\hat{\sigma}_y^{(1)} \cos\phi_1 + \hat{\sigma}_x^{(1)} \sin\phi_1\right) + \Omega_2 \left(\hat{\sigma}_y^{(2)} \cos\phi_2 + \hat{\sigma}_x^{(2)} \sin\phi_2\right) \\
&+ \Omega_+ \left((\hat{\sigma}_y^{(1)}\hat{\sigma}_y^{(2)} - \hat{\sigma}_x^{(1)}\hat{\sigma}_x^{(2)}) \cos\phi_+ - (\hat{\sigma}_x^{(1)}\hat{\sigma}_y^{(2)} + \hat{\sigma}_y^{(1)}\hat{\sigma}_x^{(2)}) \sin\phi_+\right) \\
&+ \Omega_- \left((\hat{\sigma}_x^{(1)}\hat{\sigma}_x^{(2)} + \hat{\sigma}_y^{(1)}\hat{\sigma}_y^{(2)}) \cos\phi_- - (\hat{\sigma}_x^{(1)}\hat{\sigma}_y^{(2)} - \hat{\sigma}_y^{(1)}\hat{\sigma}_x^{(2)}) \sin\phi_-\right),
\end{aligned}
$$

where the Rabi frequencies of the four transitions are proportional to the amplitudes of the corresponding modulations, $\Omega_{1,2} = \tilde{\varepsilon}_{1,2}/2\hbar$, $\Omega_\pm = \tilde{J}_{12}^{\pm}/4\hbar$.

We see that one achieves full control over the coupled qubits: the pulses with frequencies $T_1/\hbar$ and $T_2/\hbar$ perform separate unitary transformations for the first and second qubits, respectively, while the pulses with frequencies $(T_2 \pm T_1)/\hbar$ provide two-qubit gate operations.

In the experiment [136] one realizes the unitary gate

$$
U_+ = \exp\left(-\mathrm{i}\Omega_+\tau \left(\hat{\sigma}_x^{(1)}\hat{\sigma}_x^{(2)} - \hat{\sigma}_y^{(1)}\hat{\sigma}_y^{(2)}\right)\right)
$$

by applying a pulse of duration τ at the sum frequency $(T_1 + T_2)/\hbar$. In particular, for $\Omega_+\tau = \pi/4$, the gate performs the double-CNOT operation (up to two single-qubit rotations). Three applications of this operation, supplemented by proper single-qubit rotations, are sufficient to implement any two-qubit gate. For $\Omega_+\tau = \pi/2$, the gate is diagonal, $U_+(\pi/2) = \mathrm{diag}(-1, 1, 1, -1)$. This means that, if the input state of the gate is an eigenstate, the corresponding oscillations of the probabilities have a period of $\tau_P = \pi/2\Omega_+$. However, $U_+(\pi/2)$ is not an identity gate involving the phase shift of π. The true identity gate is $U_+^2(\pi/2) \equiv U_+(\pi)$, so that the true period of the oscillations is $2\tau_P$. This can be probed if, instead of taking the eigenstate, one takes an input state that is a superposition of two eigenstates.

The read-out in the experiment was implemented with a switching SQUID sensitive to the states of both qubits. The read-out was not designed to discriminate between the

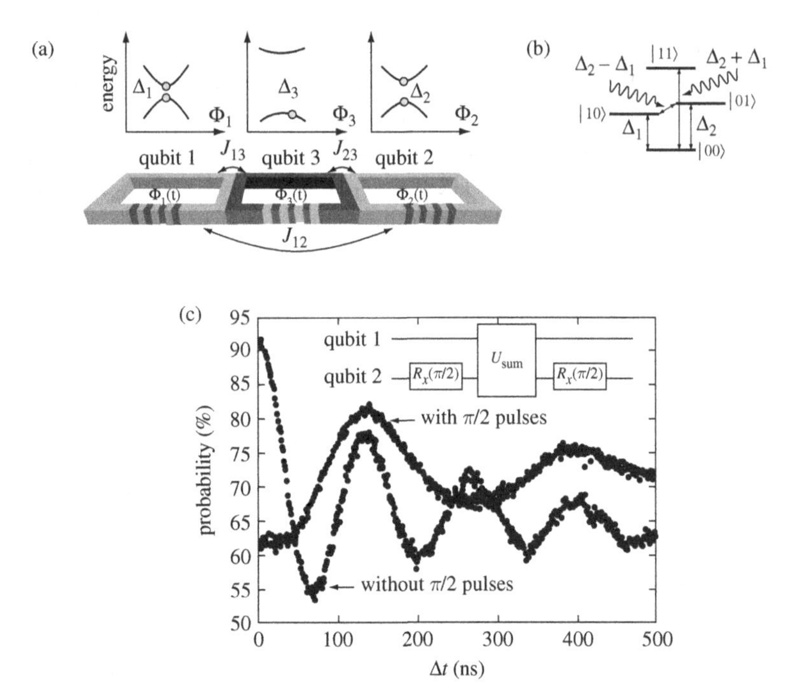

Fig. 5.18. (a) Two-qubit gate implementation involves three flux qubits. The central qubit only serves to provide a controllable coupling between the left and right qubits. The energy diagrams show the bias points of the qubits. (b) The levels of the two qubits and possible channels of the resonant manipulation. (c) Experimental demonstration of the $U_+(\Omega_+\tau)$ two-qubit gate. The curves corresponding to the two working sequences differ in period. Taken from Ref. [136]; reprinted with permission from AAAS.

states. Instead, the switching probability measured by the SQUID was proportional to the expectation value of the weighted sum of the qubit currents, $\langle a_1\sigma_z^{(1)} + a_2\sigma_z^{(2)}\rangle$, in the final state of the qubits. The coefficients $a_{1,2}$ characterize the qubit coupling to the SQUID loop. The typical measurement results are plotted in Fig. 5.18 (c). Two different working sequences have been explored. In both cases, the qubits are initialized to the ground state with $\sigma_z^{(1)} = \sigma_z^{(2)} = -1$. During the first sequence, the $U_+(\Omega_+\tau)$ gate has been applied with no qubit rotations. In this case, the input state for the gate is an eigenstate, and the periodicity is τ_P. This corresponds to the curve labeled "without $\pi/2$ pulses." During the second working sequence, the second qubit was subject to two $\pi/2$ rotations with pulses at frequency $T_2/\hbar$, before and after the gate operation. The corresponding curve apparently has a double period.

Exercise 5.14. Compute the theoretical outcome of the measurement for both sequences, i.e. $\langle a_1\sigma_z^{(1)} + a_2\sigma_z^{(2)}\rangle$ as a function of $\Omega_+\tau$. Compare this qualitatively with the measurement results. Can you draw the conclusion about the ratio a_2/a_1 in the experiment?

5.7 Spin qubits

Spin qubits are very distinct from superconducting qubits. They exploit a very straightforward qubit realization: electron spin. In distinction from the macroscopic quantum states described in preceding sections, the spin is a generic property of an elementary particle. We concentrate on spin qubits that are based on controllable quantum dots (for a review, see Ref. [137]). We start this section by outlining the influential proposal of Loss and Di Vincenzo back in 1998 [138]. Most of the proposal may look naive nowadays, but it played a catalytic effect in the field of solid state spin manipulation.

The electrons are confined in quantum dots formed in a semiconductor heterostructure. Each dot contains strictly one electron, due to the high charging energy. The spin of each dot represents a qubit. The possibility of manipulation is provided by an external magnetic field interacting with the spin. The single qubit Hamiltonian is given by

$$\hat{H} = -\mu_e \hat{\boldsymbol{\sigma}} \cdot \boldsymbol{B}, \tag{5.43}$$

where μ_e is the effective magnetic moment of an electron (which in heterostructures can be very different from the value in vacuum, $\mu_B/2$). We thus see that three components of the magnetic field, $B_{x,y,z}$, serve as three independent handles. Note that to manipulate several qubits separately, one needs separate sources of magnetic fields acting on each qubit. For single qubits, one can immediately implement the variety of manipulation techniques elaborated in the field of spin resonance. A static magnetic field along the z axis also serves for the preparation of the qubit: if one applies this field and waits long enough, the qubit will, with certainty, be found in the spin-up state.

The coupling of two or more qubits is more difficult. Interacting qubits, i.e. quantum dots, are supposed to be very close to each other, separated only by a tunnel junction which can be tuned by a nearby gate. If the tunnel barrier is set to "high," the spins of the qubits do not interact; setting the barrier to "low" allows the overlap of the localized electron states in the dots. This results in an exchange interaction between the spins:

$$H_{\text{int}} = J(t)\hat{\boldsymbol{\sigma}}^{(1)} \cdot \hat{\boldsymbol{\sigma}}^{(2)}.$$

The specific feature of this interaction is the isotropy in the spin space, i.e. the matrices $\hat{\sigma}_{x,y,z}$ come with the same weights. One can make pulses of the exchange interaction, to be used as an elementary unitary transformation to enable more complicated quantum gates. Thus, all pulses of the coupling with $\int J(t)\mathrm{d}t = \theta$ define the quantum gate $U_s(\theta)$. The SWAP operation is obtained as $U_{\text{sw}} = U_s(\pi/4)$. By itself, the SWAP operation is not enough to set up a functioning quantum computer. One requires a CNOT gate, which can be expressed in terms of our SWAP gate as follows:

$$U_{\text{CNOT}} = R_z^{(1)}(\pi/2)R_z^{(2)}(-\pi/2)U_{\text{sw}}^{1/2}R_z^{(1)}(\pi)U_{\text{sw}}^{1/2}, \tag{5.44}$$

where we have defined $U_{\text{sw}}^{1/2} \equiv U_s(\pi/8)$.

Exercise 5.15. Prove Eq. (5.44).

An obvious disadvantage of the proposed two-qubit gate is that the qubits have to be very close to each other, that is separated by a distance of the order of the electron wavelength. One might conclude that it is impossible to build a scalable quantum computer based on the principle of two-qubit operation; indeed, it is difficult to imagine that all elements of such computer would be in such a close proximity. However, it has been shown that a one-dimensional array of quantum dots would work as a scalable quantum computer. The only disadvantage would be that the operation time is longer. A gate operation involving two qubits, which are not in immediate proximity, can always be implemented as a sequence of two-qubit operations involving only neighboring qubits. If these two qubits are separated by a long distance, the operation takes a long time indeed.

The read-out of the spin is a challenging task, and currently one cannot measure the electron spin remotely without destroying it. In the original proposal, the read-out was based on the spin-valve effect (see Section 1.9): a dot is connected by a controllable junction to a ferromagnetic reservoir. When the junction is open, during the read-out, the transmission probability is higher provided the spin of the dot is oriented parallel to the magnetization of the ferromagnetic reservoir. The qubit is destroyed thereby, and its spin is converted into electric current.

It is clear that quantum dots in principle allow for more flexibility than the above scheme. For instance, the scheme assumes that one qubit is represented by a single quantum dot with precisely one electron. In fact, one can put any odd number of electrons in each dot: the resulting ground state will be degenerate in spin anyway. Moreover, the ground states of the dots may correspond to higher spin states; this allows for multi-state qubits of higher degeneracy.

Any quantum computer must be optimized with respect to decoherence. The most attractive feature of spin qubits is their long coherence time. This is due to the fact that the coupling of electron spin to any noisy field and/or any orbital degree of freedom is small: this coupling is in principle a relativistic correction negligible for non-relativistic electrons in solid state systems. Spin is thus a well isolated degree of freedom. The point is that this good isolation of spin from any environment simultaneously hinders its detection and therefore the qubit read-out.

This is why initial experimental efforts were directed at achieving a reliable read-out. The measurements with the read-out realized have confirmed that the spin relaxation times are long and have quantified them. We review here the research described in Ref. [139]. The idea implemented requires the conversion of spin into charge. The charge in a dot is measured by a non-ideal quantum point contact (see Section 1.2) placed near the dot. Both the dot and the QPC are defined by means of the metallic gates depleting the two-dimensional gas (see Fig. 5.19). The transmission coefficient T_0 of the contact depends slightly on the charge in the dot n, $T_0(n)$, since the electric field of the charge affects the potential relief in the QPC. The difference between the transmission coefficients for the adjacent charge configurations n and $n+1$ is usually small, $\delta T \equiv (T_0(n+1) - T_0(n)) \simeq 0.01$. However, it produces an extra current $\delta I = G_Q \delta T\ V$ proportional to the voltage over the QPC. At sufficiently large voltages (1 mV), the extra current is big enough ($\delta I \simeq 1$ nA) to be measured quickly (0.2 ms). One has to be cautious about the feedback of such a

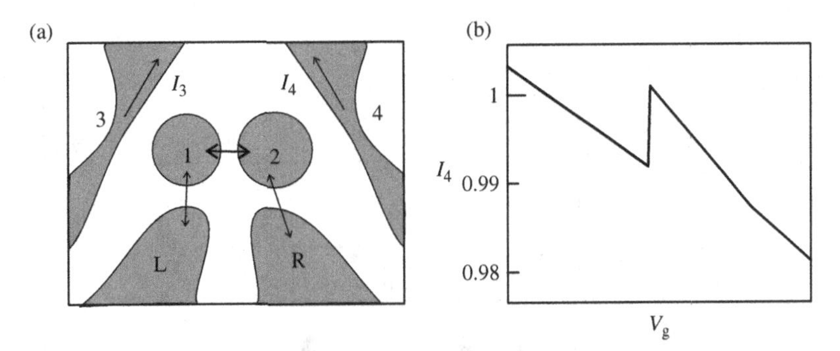

Fig. 5.19. (a) Double-dot layout used in experiments on spin-charge conversion. Gray areas are filled with electrons of two-dimensional gas. Gate electrodes are not shown. QPCs 3 and 4 formed near dots 1 and 2 measure the charge of the dots. (b) A typical trace of the QPC current versus a gate voltage. The jump gives the change of the charge configuration in the dot.

detector: the charges traversing the QPC can bring an energy of the order of eV to the object measured. This effect, however, could be disregarded in the experiments described below.

Control question 5.19. Estimate theoretically the minimal duration of the measurement required to resolve the extra current ΔI given the background current I. Hint: Schottky formula (Section 1.4).

Let us have a single electron in the dot and apply a magnetic field that splits levels corresponding to the direction of spin, $E_\downarrow > E_\uparrow$. Let us bring the dot into contact with the lead at a chemical potential that lies in between the split levels. If an electron is in the upper energy state $|\downarrow\rangle$, it tunnels to the lead and the charge of the dot is changed to zero. Otherwise, it remains in the dot. This is how we intend to convert spin into charge. The potential shift between the dot and the lead can be readily tuned by a gate electrode. During the working sequence (Fig. 5.20) the dot was first emptied. Then both levels are shifted below the chemical potential of the lead. An electron with unknown spin enters the dot. This is followed by a waiting time t_w, during which the electron can relax to the ground state $|\uparrow\rangle$ with rate Γ_s. The read-out pulse brings the $|\downarrow\rangle$ state above the chemical potential, while the $|\uparrow\rangle$ state remains below to enable the spin detection as explained. If there is a spin-up electron in the dot, the dot charge remains equal to e during the read-out. If there is a spin-down electron, it will tunnel out during a random time of the order of Γ^{-1}, Γ being the tunneling rate from the level. The charge of the dot is zero. It takes time for the electron from the lead to come and fill the $|\uparrow\rangle$ level, bringing the charge back to e. The pulse of the charge is detected by the QPC and the presence or absence of such a pulse allows one to determine the electron spin. At the end of the sequence, the dot is again emptied. Measuring the dependence of the fraction of the spin-down electrons on the waiting time, one extracts the spin relaxation rate Γ_s. The main relaxation mechanism was predicted to be the spin-orbit interaction (Section 1.9). The energy difference $2\mu_e B$ between the excited and ground states must be given to an environment, in this case radiated as a phonon. This produces

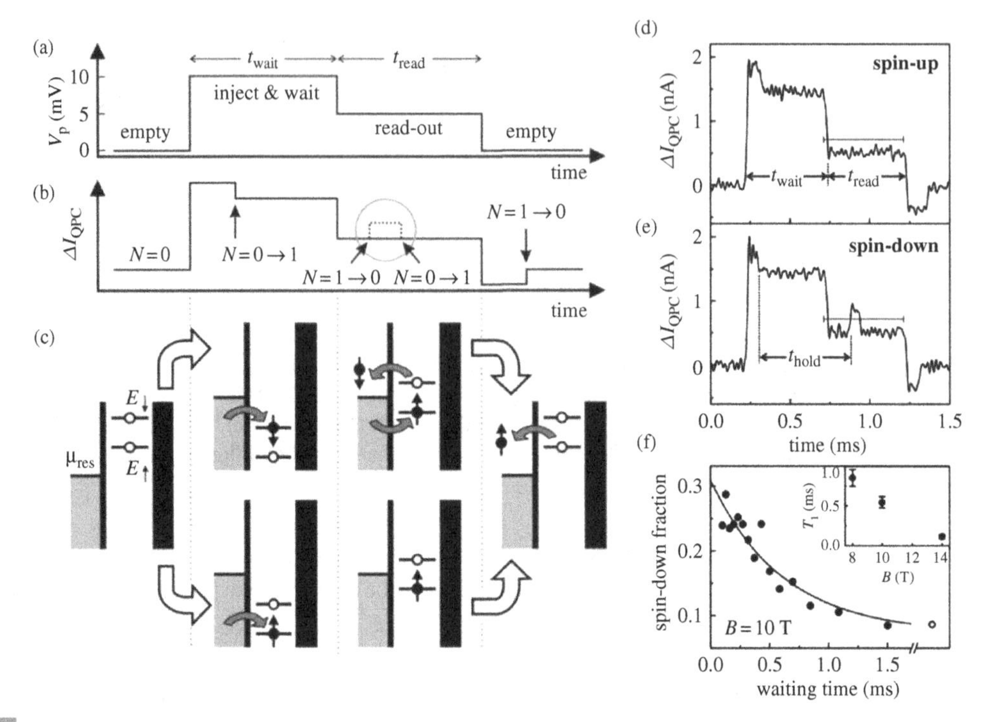

Fig. 5.20. Energy-selective read-out of a single-electron spin [139]. (a) The working sequence. (b) Expected response of the QPC current to the voltage pulses. (c) Energy diagrams depicting the levels during the different stages of the sequence. Measurement outcomes for (d) a spin-up and (e) a spin-down electron. (f) The fraction of spin-down electrons versus the waiting time shows an exponential decay with the spin relaxation time T_1. Inset: T_1 for different magnetic fields. Copyright Wiley-VCH Verlag GmbH & Co. KGaA. Reproduced with permission.

a rate strongly increasing with magnetic field $\propto B^5$. Such strong dependence was revealed experimentally. We note that the experiment requires unusually low tunneling rates Γ: to detect the charge pulse, they should not exceed 10^4 Hz. Nor should they be smaller than the spin relaxation rate.

Exercise 5.16. Quantify possible non-idealities in the work of the described detector. Assume $(t_w)^{-1} \simeq \Gamma \simeq \Gamma_s$. Evaluate:

(i) the probability that the dot is empty after the waiting time;
(ii) the probabilities of finding either a spin-up or a spin-down electron in the dot after the waiting time.
(iii) There is a chance that after the waiting time the dot is in the $|\downarrow\rangle$ state, but this is not detected since the spin relaxes before the electron would tunnel out. Evaluate the probability of this process.

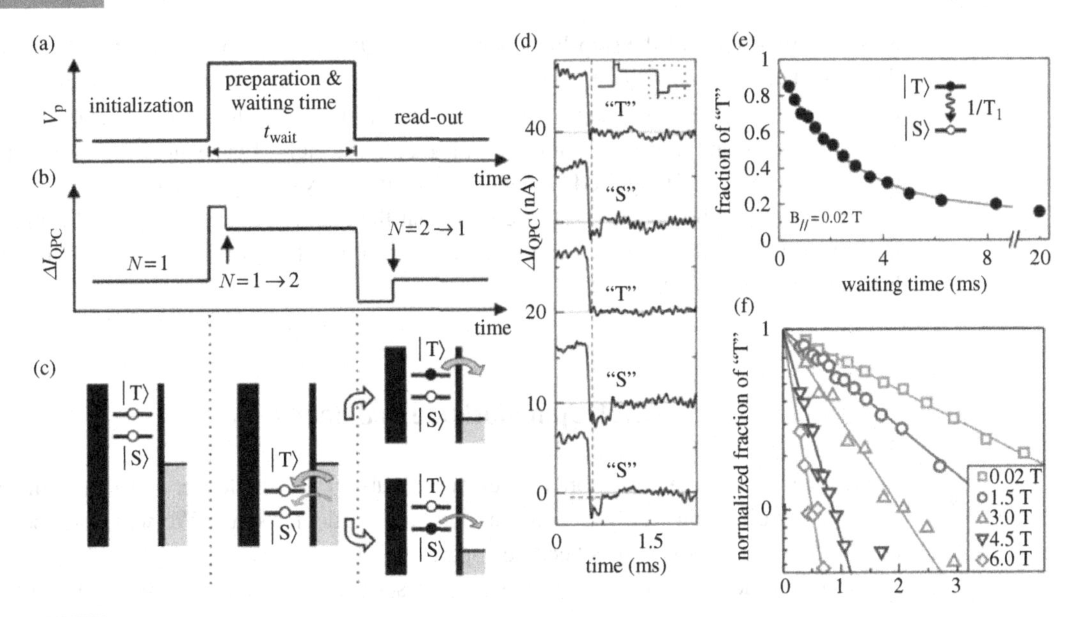

Fig. 5.21. Single-shot rate-selective read-out of two-electron spin states [139]. (a) Working sequence. (b) Expected response of the QPC current. (c) Energy diagrams indicating the positions of the levels during the stages of the sequence. (d) Real-time traces of QPC during the last stage for $t_w = 0.8$ ms. The step-like negative change right after the pulse indicates the singlet. (e) The relaxation of the triplet states measured gives $T_1 = 2.58$ ms. (f) Magnetic-field dependence of the relaxation.

Another type of read-out has been implemented to resolve spin states in a two-electron quantum dot. The ground state in the two-electron configuration is a singlet (see Section 5.4) with two electrons in the same level. The first excited state is a triplet with two electrons in different levels: first and second. The triplet is three-fold-degenerate in the weak magnetic fields used in the experiment, so the triplet components cannot be distinguished. The read-out makes use of the fact that the tunneling rates from singlet (Γ_S) and triplet (Γ_T) differ significantly (by a factor of 20 in the experiment). This is because the second electron level is close to the top of the tunnel barrier; this increases its tunneling amplitude.

The working sequence is as follows. By applying the gate voltage, the dot is put first to the state with a single electron. Then the gate voltage pulse pulls the two-electron levels down and thus enables the tunneling of the second electron to the dot. The resulting state can be either singlet or triplet. The probability of a certain state is determined by the competition of the tunneling rates $\Gamma_{S,T}$. Since there are three degenerate triplet states, the singlet is realized with the probability $\Gamma_S/(\Gamma_S + 3\Gamma_T)$. After the waiting time t_w, during which the relaxation from the triplet to singlet state may take place, the spin state is detected. The read-out pulse pulls the levels up so that tunneling from both states is enabled. If the system is in a triplet state, the tunneling takes place with the faster rate Γ_T. The QPC charge read-out does not show the charge $2e$ between the beginning of the read-out pulse and the moment of tunneling since it is too slow to resolve this. If the system is in the singlet state, the tunneling rate is much slower, so the corresponding signal is resolved (Fig. 5.21). The presence or absence of the step-like change of the QPC current thus serves to differentiate the singlet from either triplet state. Evidently, this read-out can be used to measure the relaxation time of a triplet state. This was found to be rather long ($T_1 = 2.5\,\mathrm{ms}$) in low magnetic fields and quickly decreased with increasing field. Let us note that this time is at least four orders of magnitude longer than the typical working sequences of the charge, phase, and flux qubits. This raises the hope that the manipulation of a spin qubit may be performed with better accuracy and using more complicated working sequences: one just has more time to do this.

5.7.1 Spin blockade in a double-dot

In practice, such a manipulation still remains a substantial challenge. Most experimental work in this direction has been concentrated on a particular setup. We will describe it in detail below, so its advantages become apparent.

We have a double-dot between two leads described in Section 5.5 (Fig. 5.12). Let us concentrate first on two charge configurations: $(1, 1)$ and $(0, 2)$, and assume that the gating of the dots brings these to the same energy. The ground state in the $(0, 2)$ configuration is a spin singlet $|\mathrm{S_g}\rangle$. The first excited state is a triplet. It lies at sufficiently higher energy to be disregarded in our consideration. If there is no tunneling between the dots and no magnetic field, the charge configuration $(1, 1)$ is four-fold-degenerate: an electron in each dot may come with either spin-up or spin-down. The tunneling through the potential barrier separating the dots mixes the states of $(1, 1)$ with those of $(0, 2)$. It is essential to note that the spin conservation imposes strict selection rules on possible matrix elements describing such mixing: singlets are only coupled to singlets, and triplets only to triplets. This implies that only the singlet $|\mathrm{S}\rangle$ of the $(1, 1)$ configuration is coupled to $|\mathrm{S_g}\rangle$ while the triplet states are not affected by tunneling. The magnetic field applied in the z direction splits up these three states according to the projection of the full spin $S_z = -1, 0, 1$. We denote the states as $|\mathrm{T_-}\rangle$, $|\mathrm{T_0}\rangle$, $|\mathrm{T_+}\rangle$, respectively. The Hamiltonian in the basis of the five states involved is given by

$$\hat{H} = \varepsilon|\mathrm{S_g}\rangle\langle\mathrm{S_g}| + \mathcal{T}\left(|\mathrm{S}\rangle\langle\mathrm{S_g}| + |\mathrm{S_g}\rangle\langle S|\right) + 2\mu_\mathrm{e}B(|\mathrm{T_-}\rangle\langle\mathrm{T_-}| - |\mathrm{T_+}\rangle\langle\mathrm{T_+}|), \qquad (5.45)$$

where ε is the energy shift of the $(0, 2)$ configuration with respect to $(1, 1)$. The Hamiltonian is diagonal in triplets so that their energies are $2\mu_\mathrm{e}B$, 0, and $-2\mu_\mathrm{e}B$, respectively. The singlet states are mixed: the diagonalization of the 2×2 matrix gives (see Fig. 5.22)

$$E_{+,-} = \frac{\varepsilon}{2} + \sqrt{\frac{\varepsilon^2}{4} + \mathcal{T}^2}\,. \qquad (5.46)$$

So far we have not considered tunneling to the leads. The voltages of the leads are tuned in such a way that the state $|\mathrm{S_g}\rangle$ quickly decays – an electron tunnels to the right-hand lead. This brings the system to the charge configuration $(0, 1)$, with a random direction of the spin of the remaining electron. The next process is electron tunneling from the left lead

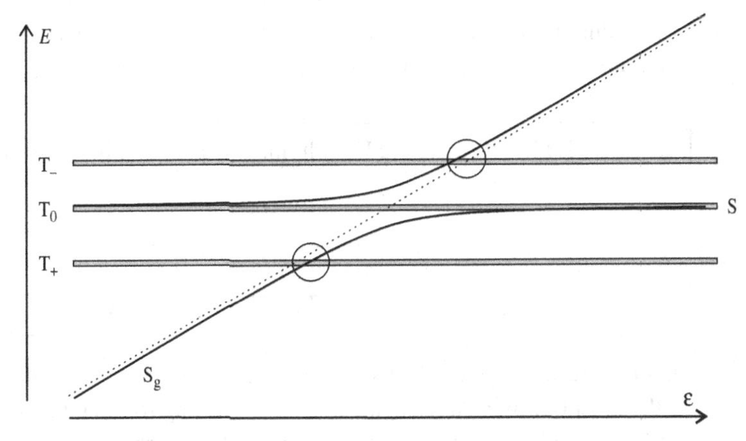

Fig. 5.22. The levels in a double quantum dot in the spin-blockade regime. The crossings of singlet and triplet states are encircled.

to the left dot, so that the system arrives at one of the states corresponding to the charge configuration $(1, 1)$.

Let us note a remarkable feature of the setup. If the resulting state is a triplet state, the charge transport stops: owing to spin conservation, an electron cannot tunnel from the left to the right dot. In other words, the matrix elements between the triplet states and $|S_g\rangle$ are absent. This phenomenon is known as *spin blockade*. In contrast to Coulomb blockade, the tunneling is not forbidden by energy limitations, but rather by the fundamental law of spin conservation.

We are now ready to describe the idealized work of the device at hand. Two qubits, one in each dot, are realized in the basis of the four states of the charge configuration $(1, 1)$. It is expedient to keep the energy shift large, $|\varepsilon| \gg \mathcal{T}$, during preparation and manipulation of the qubits. This is to prevent the mixing of the states $|S\rangle$ and $|S_g\rangle$, so that $|S\rangle$ lives longer than the manipulation time. For manipulation, it is convenient to have two sources of magnetic field, $\boldsymbol{B}_1$ and $\boldsymbol{B}_2$, affecting two dots separately. The Hamiltonian at the manipulation stage can be written as

$$\hat{H} = -\mu_e \boldsymbol{B}_1 \hat{\sigma}^{(1)} - -\mu_e \boldsymbol{B}_2 \hat{\sigma}^{(2)} - J(t)\hat{\sigma}^{(1)} \cdot \hat{\sigma}^{(2)}, \tag{5.47}$$

where the exchange interaction between spins, $J(t) \approx \mathcal{T}^2/\varepsilon$, equals the energy splitting of spin singlet and triplet states and is obtained by the expansion of E^- in Eq. (5.46) in $\mathcal{T}/\varepsilon$. This enables all possible quantum operations with two qubits. Separate qubit rotations are implemented with the pulses of $\boldsymbol{B}_1$ and $\boldsymbol{B}_2$, whereas the modulation of $J(t)$ allows us to implement two-qubit gates. In particular, to implement the CNOT operation, one uses Eq. (5.44).

For the read-out, one reduces the energy shift ε so that the states $|S\rangle$ and $|S_g\rangle$ mix significantly. If a two-qubit gate after the manipulation sequence ends up in state $|S\rangle$, this mixing partially transfers the probability to state $|S_g\rangle$. The electron quickly tunnels to the right lead. In contrast, triplet states are spin-blocked, and the electron does not tunnel out. Detecting the current averaged over many working cycles thus gives the probability of it ending up in state $|S\rangle$. As discussed above, in combination with arbitrary two-qubit operations, this suffices to measure an arbitrary element of the density matrix.

Exercise 5.17. Rewrite the Hamiltonian, Eq. (5.47), in terms of the singlet and triplet states.

5.7.2 Nuclear spins

Of all the factors that could disturb the work of the device, a rather exotic one turns out to be the most limiting. At least, it is the most limiting for quantum dots made in GaAs-based semiconductor heterostructures. The proper functioning is affected most by the magnetic fields originating from nuclear spins. In almost all materials, each nucleus bears a spin. The interaction of the nuclear spins and electrons in solids can often be disregarded: the nuclear spins have a "life on their own" that hardly affects the electron transport under normal circumstances. Nuclear spin dynamics is very slow at electron or even human scales; to give the typical numbers for GaAs, the relaxation time of nuclear spins can be as large as $\tau_{\mathrm{ns}} \sim 100\,\mathrm{s}$. The inhomogeneous distribution of nuclear spin also diffuses, characterized by the diffusion coefficient $D_{\mathrm{N}} \simeq 10^{-13}\,\mathrm{cm}^2/\mathrm{s}$. Nuclear spins are usually randomly oriented. Since they are in fact small magnets, one could wonder why this is: the magnets should be oriented by either mutual interaction or by the external magnetic field if present. The point is that the energy scale of these magnetic interactions is unusually small: their magnetic moments are about 10^{-3} of the electron magnetic moment, and the interaction between the neighboring nuclear spins is $\simeq 10^4$ Hz. The estimate of the ordering temperature is therefore 10^{-7} K, far below any reasonable temperature to be reached in a quantum transport experiment.

The specific feature of GaAs is a sizeable interaction of electron spin with nuclear polarization. If all nuclear spins are aligned in the same direction, the Zeeman splitting of the electrons is $E_{\mathrm{n}} = 0.135\,\mathrm{mV}$, corresponding to a magnetic fields of 5 T. Such energies are certainly noticeable at the quantum transport scale. Since the nuclear spins are disordered, one may think that the average over this disorder cancels the effect. Let us estimate if this is really so. In a typical one-electron quantum dot in GaAs, the electron wave function extends over $N_{\mathrm{QD}} \sim 10^5$–$10^6$ atoms; this yields an estimate of a typical number of nuclei interacting with the electron spin. The energy E_{n} is composed of independent contributions of N_{QD} nuclear spins. If the nuclear spins are disordered, we still expect the fluctuation to be of the order of $\Omega_{\mathrm{ns}} \equiv E_{\mathrm{n}}/\hbar\sqrt{N_{\mathrm{QD}}} \sim 10^8$ Hz. This corresponds to a temperature 10^{-3} K and is therefore negligible at the energy scale of electrons in the quantum dot. On the other hand, it is significant in the context of coherent spin manipulation. It represents a random shift of the operational frequency. This random shift varies in time over the typical scale τ_{ns}.

In particular, for a double-dot device considered above, the interaction of electrons with nuclear spins adds the following term to the qubit Hamiltonian given by Eq. (5.47):

$$\hat{H}_{\mathrm{n}} = -\hbar\boldsymbol{\Omega}^{(1)}\hat{\boldsymbol{\sigma}}^{(1)} - \hbar\boldsymbol{\Omega}^{(2)}\hat{\boldsymbol{\sigma}}^{(2)}, \tag{5.48}$$

where $\mathbf{\Omega}^{(1),(2)}$ are randomly oriented vectors with a typical magnitude of the order of Ω_{ns}, acting in each of the dots.

Moreover, the interaction between electron spin and nuclei can be further enhanced by the polarization of the nuclei induced by the manipulated spin. The point is that the relaxation of electron spin sometimes (rarely) involves the flip of a spin of a certain nucleus in the dot. Note that, depending on the manipulation protocol, there may be a preferential direction for this spin flip. Since the relaxation time of a nuclear spin is enormously slow, even very rare flips can lead to a substantial nuclear polarization. The effective magnetic field created by this polarization again shifts the operational frequency and may create a considerable feedback on the manipulation scheme.

5.7.3 Coherent manipulation of electron spin: example

We described the scene for which experiments on coherent manipulation of electron spins are performed. We concentrate on one particular experiment [140] that gives an example of how the difficulties described above can be circumvented by a creative choice of the manipulation protocols. The authors were not able to achieve the full two-qubit functionality of the device: they work with two quantum states $|\text{S}\rangle$, $|\text{T}_0\rangle$ representing a single qubit thereby. The higher triplet states $|\text{T}_\pm\rangle$ are separated by applying a sufficiently high magnetic field B_z such that the Zeeman energy by far exceeds the random contribution of the nuclear spins. Furthermore, only one "handle," the time-dependent exchange interaction $J(t)$, has been employed for the manipulation. As discussed above (Eq. (5.47)), the exchange $J(t)$ is determined by the energy shift ε, and this allows for an easy gate voltage manipulation. Three working sequences have been realized.

Nuclear spin effect

To initialize the system, the gate voltages have been set to assure the stable $(0, 2)$ charge configuration, so the quantum state is an almost pure $|\text{S}_\text{g}\rangle$. A sweep of the gate voltage transfers this state deep into charge configuration $(1, 1)$. The sweep is fast on the scale of the random nuclear field Ω_{ns} but slow at the scale of the tunneling mixing $\mathcal{T}/\hbar$. Under these conditions, the resulting quantum state is $|\text{S}\rangle$. The effective Hamiltonian in the basis of $|\text{S}\rangle$, $|\text{T}_0\rangle$ is given by

$$\hat{H} = \begin{pmatrix} J(t) & \hbar\Omega \\ \hbar\Omega & 0 \end{pmatrix}, \tag{5.49}$$

with $\Omega = \Omega_z^{(1)} - \Omega_z^{(2)} \simeq \Omega_{\text{ns}}$.

Control question 5.20. Why does the external field B_z not appear in Eq. (5.49)? Why are the nuclear magnetic fields in two dots subtracted? Why do their x, y components not enter the effective Hamiltonian?

Deep in the $(1,1)$ configuration, $J \approx 0$, and the free evolution of the qubit is the rotation about x axis. The qubit is kept rotating for the time τ_S, so that after the evolution it is in the coherent superposition $\cos(\Omega\tau_S)|S\rangle + \sin(\Omega\tau_S)|T\rangle$. Another sweep of the gate voltage brings the system back to the initialization conditions: to the gate voltages corresponding to the stable charge configuration $(0,2)$. There are two possible outcomes. With probability $\cos^2(\Omega\tau_S)$, the state is $|S\rangle$ and is adiabatically transferred to $|S_g\rangle$. In this case, the adjacent QPC measures the $(0,2)$ charge configuration immediately after the sweep. With probability $\sin^2(\Omega\tau_S)$, the state is $|T_0\rangle$. After the sweep, the charge configuration remains as $(1,1)$. This is an excited state with respect to $(0,2)$. However, the spin blockade guarantees that the triplet state persists for a long time ($>10\,\mu$s in the experiment) sufficient to measure the charge configuration $(1,1)$. We note that the read-out exploits the same idea as the two-electron read-out described previously (Fig. 5.21).

If Ω were a fixed frequency, the measurement would yield the singlet probability $P_S = \cos^2(\Omega\tau_S)$. However, it takes a long time and many working cycles to accumulate the data with sufficient precision. Although Ω does not change within a working cycle, it does change during the data accumulation time. Therefore, the probability measured is an average over the distribution of Ω. Since Ω arises from the independent contributions of a large number of nuclei, it is normally distributed with zero average. The average is evaluated in the following way (see also Section 6.7);

$$\langle\cos^2\Omega\tau_S\rangle = \frac{1}{2} + \frac{1}{2}\mathrm{Re}\langle e^{2i\Omega\tau_S}\rangle = \frac{1}{2}\left(1 + e^{-2\langle\Omega^2\rangle\tau_S^2}\right).$$

The probability P_S decreases exponentially with the time, saturating at $1/2$. This exponential decay observed experimentally (Fig. 5.23) allowed a measurement of $\langle\Omega^2\rangle = (10\,\mathrm{ns})^{-2}$, in accordance with preceding estimations. The decay seems to indicate that the duration of the manipulation sequence must be shorter than 10 ns.

Rabi oscillations

The next working sequence we describe attempts to achieve the manipulation at this time scale of 10 ns, and is as follows. The system is again initialized in the charge configuration $(0,2)$, and brought deep into the charge configuration $(1,1)$. In distinction from the previous working sequence the transfer is slow in comparison with the nuclear mixing time $\Omega_{\mathrm{ns}}^{-1}$. The Hamiltonian at this point in the basis of $|S\rangle$ and $|T_0\rangle$ states has the form given by Eq. (5.49) with $J=0$. The slow transfer brings the qubit to the ground state of this Hamiltonian, which is $|\uparrow\downarrow\rangle = (|S\rangle + |T_0\rangle)/\sqrt{2}$ (assuming negative Ω). Subsequently, the qubit is brought rapidly to the vicinity of the point $\varepsilon = 0$ where $J(t)$ is finite, and is rapidly brought back after the time τ_E. This is the manipulation: we give a pulse to the exchange interaction J. The read-out is the reverse of the initialization: one slowly moves the qubit out to the point $\varepsilon \approx 0$, whereby the ground state $|\uparrow\downarrow\rangle$ and the excited state $|\downarrow\uparrow\rangle = (-|S\rangle + |T_0\rangle)/\sqrt{2}$ are adiabatically transformed into $|S\rangle$ and $|T_0\rangle$, respectively. Further rapid increase of ε transfers the qubit in the charge configuration $(0,2)$, provided it is in the $|S\rangle$ state, and leaves it in the configuration $(1,1)$ due to the spin blockade, provided

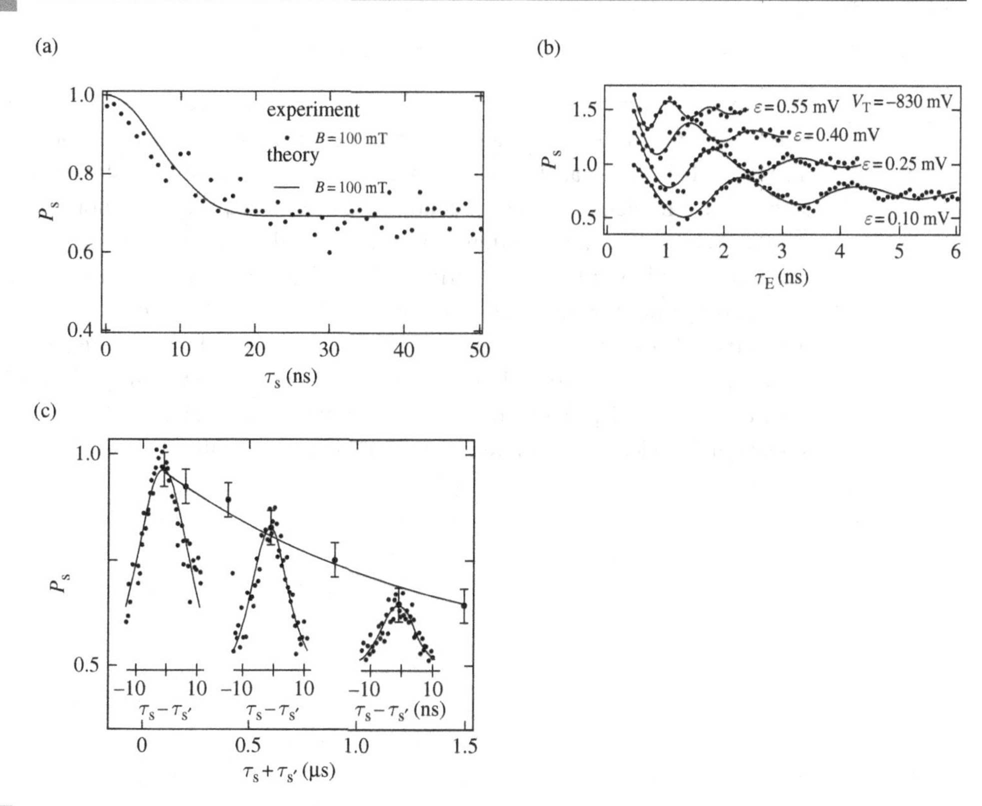

Fig. 5.23. Example of coherent manipulation of electron spin. (a) Nuclear spin effect causes apparent spin decoherence at the nanosecond scale seen in P_S versus the pulse time τ_S. The deviations of P_S from the theoretical saturation value 1/2 are due to non-ideality of the read-out. The theoretical Gaussian fit (solid curve) is adjusted for this. (b) Rabi oscillations in P_S versus the exchange coupling pulse duration τ_E. Different curves correspond to different values of $J(\varepsilon)$ and manifest different periods of the oscillations. The curves are shifted by 0.3 for clarity; the solid line is a fit with damped cosine function. (c) Spin-echo experiment reveals the true coherence time of microsecond scale. The slowly decaying curve corresponds to the precisely equal free evolution intervals $\tau_{S,S'}$: In this case, the effect of the random nuclear field is canceled by the spin-echo sequence. The insets show the decay at $\tau_S - \tau_{S'}$ in the nanosecond range. Taken from Ref. [140]; reprinted with permission from AAAS.

it is in the $|T_0\rangle$ state. Thus, measuring the charge of one of the dots yields the probability of the qubit being in the$|\uparrow\downarrow\rangle$ state after the manipulation.

If we neglect Ω in comparison with $J/\hbar$, the manipulation is the rotation about the z axis by the angle $J\tau_E/2\hbar$. The qubit would perform Rabi oscillations with frequency $J/2\hbar$, the probability $P_{\uparrow\downarrow}$ after the manipulation being given by $\cos^2(J\tau_E/2\hbar)$. A small but finite Ω results in a random shift of the Rabi frequency $\Omega^2\hbar/J$. One expects the apparent decay of the Rabi oscillations at the time scale of the inverse frequency shift, $J/\hbar\Omega^2 \gg \Omega^{-1}$, that is, at a larger time scale than in the preceding nuclear spin effect experiment. The experimental results are shown in Fig. 5.23 and do not display such an improvement. This has been attributed to the extra dephasing brought by the charges: during the pulse, the qubit is sensitive to the charge noise that affects J.

Spin echo

These two experiments, however, do not show the real potential of the device. In fact, the free evolution of the qubit remains coherent for much longer times than the decay constants measured. It is just that the random nuclear fields essentially mask this time scale. The authors have implemented a spin-echo manipulation sequence (Section 5.3) to cancel the random field Ω. The initialization and read-out are the same as in the previous experiment. The free evolution of the qubit during the time interval τ_S in the presence of the random field Ω is followed by the exchange interaction pulse that provides the π-rotation around the z axis. If after the pulse the qubit is left to evolve during precisely the same time $\tau_{S'} = \tau_S$, the rotations around the x axis cancel each other precisely. This is how the true coherence time (1.2 μs), two orders of magnitude better than Ω^{-1} has been revealed. The signal quickly dies if the difference, $\tau_S - \tau_S'$, becomes of the order of Ω^{-1} (see Fig. 5.23).

6 Interaction, relaxation, and decoherence

In Chapter 3, we discussed charging effects in nanostructures: the most important manifestation of electron–electron interaction in quantum transport. In this chapter, we concentrate on another aspect of interactions which we have so far mentioned very briefly. It concerns interaction with *slow modes*. In most cases these slow modes are electromagnetic excitations in the nanostructure and nearby circuit that form an *electromagnetic environment* of the nanostructure. The effect of this interaction is threefold. First, it may affect and alter transport properties of the nanostructure. Secondly, it provides energy *relaxation*: transporting electrons and qubits may exchange their energy with the electromagnetic environment. Thirdly, the environment provides *decoherence*, inducing time-dependent phase shifts to wave functions of propagating electrons and qubits, thereby destroying the quantum coherence of corresponding states.

The physics discussed in this chapter is sometimes involved and various. It requires effort to see a "common denominator" in all effects mentioned. We choose to present material starting from the ideas of *dissipative quantum mechanics*: a branch of quantum mechanics developed in the 1970s and 1980s. For several concrete phenomena this presentation manner deviates from that commonly accepted in the literature. Although this may be inconvenient for the reader, we did this for the sake of the "big picture," which allows us to see links and analogies between formally different phenomena.

The structure of the chapter is as follows. Sections 6.1 and 6.2 are introductory. In Section 6.1, we discuss electromagnetic excitations in linear circuits and the way to treat them quantum-mechanically. In Section 6.2 we review general ideas of dissipative quantum mechanics, which are not specific for quantum transport: the orthogonality catastrophe, shake-up, classification of environments.

These ideas receive the most direct implementation in the context of tunneling of electrons and Cooper pairs in the presence of an electromagnetic environment, which we consider in detail in Section 6.3. Section 6.4 is devoted to situations where it is important that electrons move in the electromagnetic environment before and after tunneling. Altshuler–Aronov corrections and Luttinger liquids fall into this class. Section 6.5 extends our ideas to nanostructures that include more transparent channels. Coulomb blockade at arbitrary transparency is also presented.

The electrons in nanostructures may provide slow modes that are not electromagnetic excitations. This *fermionic* environment is presented in Section 6.6, in which we consider the Kondo effect and Fermi-edge singularity. Sections 6.7 and 6.8 focus on decoherence and relaxation. These phenomena are manifested differently for qubits and electrons, which is why we separate the material into two different sections.

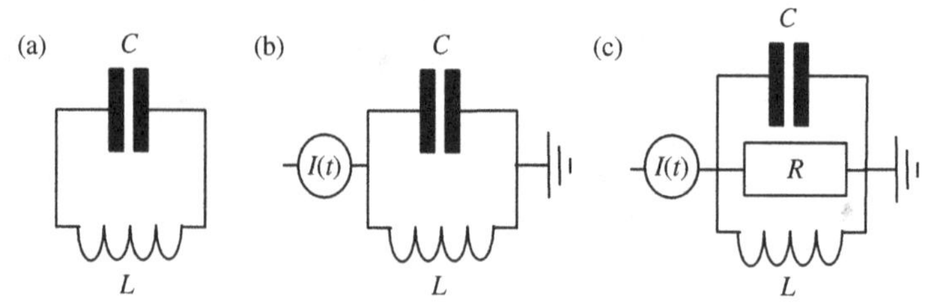

Fig. 6.1. Stages of quantization in electric circuits. (a) The simplest electrical oscillator is quantized as a generic quantum oscillator giving rise to a single boson mode. (b) The response to the time-dependent current source $I(t)$ defines the susceptibility $\chi(\omega)$. (c) To describe dissipative circuit elements (resistor R), infinitely many boson modes are introduced to simulate the susceptibility given by the circuit theory rules.

6.1 Quantization of electric excitations

The purpose of this section is to provide a quantum description of electromagnetic excitations. We start with a simple circuit consisting of a capacitor with capacitance C and an inductor with inductance L (see Fig. 6.1(a)). This circuit is known as the simplest electric oscillator: its excitations are persisting oscillations with frequency $\omega_0 = 1/\sqrt{LC}$. In classical terms, the circuit is described with two equations for the flux in the inductor Φ and the charge in the capacitor Q:

$$\dot{Q} = \Phi/L; \tag{6.1a}$$

$$\dot{\Phi} = -Q/C. \tag{6.1b}$$

Equation (6.1a) is the first Kirchhoff rule: the currents in the capacitor and the inductor are the same. Equation (6.1b) is the second Kirchhoff rule: the sum of voltage drops over a closed loop in the circuit is zero; the voltage drops are therefore opposite.

Let us construct the quantum description of the circuit. In fact, we did most of the work in Section 3.5.2 when discussing the Josephson junction. The point is that a Josephson junction at a small phase difference can be approximated by an inductor. We have established that, in this case, the quantum mechanics of the system is governed by the following Hamiltonian:

$$\hat{H} = E_{\mathrm{J}}\frac{\hat{\phi}^2}{2} + E_{\mathrm{C}}\frac{\hat{N}^2}{2},$$

where the operators $\hat{N}$ and $\hat{\phi}$ are proportional to the charge of the capacitor and the flux in the inductor, and satisfy canonical commutation relations $\left[\hat{N}, \hat{\phi}\right] = -2\mathrm{i}$. Since $\hat{N} = \hat{Q}/e$, $\phi = 2\pi\Phi/\Phi_0$, the Hamiltonian can be written in terms corresponding to classical energies of the capacitor and the inductor as follows:

$$\hat{H} = \frac{\hat{\Phi}^2}{2L} + \frac{\hat{Q}^2}{2C}.$$

The commutation relation is rescaled to $[\hat{Q}, \hat{\Phi}] = -\mathrm{i}\hbar$.

It turns out that all harmonic oscillators – physical systems exhibiting oscillations that obey linear equations of motion – are described by equivalent Hamiltonians. Such a Hamiltonian is a quadratic form of two operators whose commutator is a constant. The most used example is the Hamiltonian of a particle with a mass m moving in a parabolic potential $U(x) = U''x^2/2$. The operators of particle momentum $\hat{p}$ and coordinate $\hat{x}$ satisfy $[\hat{p}, \hat{x}] = -\mathrm{i}\hbar$ and $\hat{H} = \hat{p}^2/2m + U''\hat{x}^2/2$. The Hamiltonians of all oscillators are the same upon linear transformation of the operators involved.

There is a standard form of the oscillator Hamiltonian valid for all imaginable harmonic oscillators. It is given in terms of the boson creation and annihilation operators $\hat{b}^\dagger, \hat{b}$, which satisfy the standard commutation relation $[\hat{b}^\dagger, \hat{b}] = 1$. The Hamiltonian is given by

$$\hat{H} = \hbar\omega_0 \hat{b}^\dagger \hat{b} + \text{const.},$$

irrespective of the nature of the oscillator. The eigenstates of this Hamiltonian differ by the number of excited quanta $n = \hat{b}^\dagger \hat{b}$, the corresponding energies being $E_n = \hbar\omega_0 n$.

Exercise 6.1. Given the Hamiltonian $\hat{H} = A\hat{X}^2/2 + B\hat{Y}^2/2$ and the commutation relation $[\hat{X}, \hat{Y}] = \mathrm{i}C$, bring it to the above standard form. Find ω_0 and express $\hat{X}, \hat{Y}$ in terms of $\hat{b}^\dagger, \hat{b}$.

To specify the nature of the oscillator, one has to specify the forces moving and exciting it. For electric circuits, these forces are voltage and current sources. For example, let us add the current source $I(t)$ to our oscillator circuit. The source increases the energy of the system by $-I(t)\hat{\Phi}$, and this is to be added to the Hamiltonian:

$$\hat{H} = \hbar\omega_0 \hat{b}^\dagger \hat{b} - \hat{\Phi} I(t).$$

The flux operator Φ is linear in $\hat{b}^\dagger, \hat{b}$, and the coefficients of the linear combination are given by the answer to the above exercise,

$$\hat{\Phi} = -\mathrm{i}\left(\frac{\hbar^2 L}{4C}\right)^{1/4} \hat{b}^\dagger + \text{h.c.}$$

Next, we relate the expectation value of the flux operator to the current source. As discussed, in the context of Josephson junctions we can do this solving either quantum Heisenberg equations $\dot{\hat{\Phi}} = \mathrm{i}[\hat{H}, \hat{\Phi}]$ or classical equations (Eqs. (6.1)) with the current source added:

$$\dot{Q} = \Phi/L - I(t); \tag{6.2a}$$

$$\dot{\Phi} = -Q/C. \tag{6.2b}$$

The latter is simpler. We rewrite the above equations in terms of the Fourier components $\Phi(\omega)$, $I(\omega)$ to establish that

$$\langle\Phi(\omega)\rangle = \chi(\omega) I(\omega); \ \chi(\omega) = \frac{L}{1 - (\omega + \mathrm{i}0)^2/\omega_0^2}. \tag{6.3}$$

The so-defined function $\chi(\omega)$ is an example of a generalized *susceptibility*. It is a response function: it gives the flux response on the action of a time-dependent current source

$I(t)$. For any circuit or system, the response function satisfies the causality property: the response depends on the values of the action in the past and not those in the future. In terms of Fourier components, this means that $\chi(\omega)$ is an analytical function of ω in the upper half-plane of complex ω. We add an infinitesimally small imaginary part i0 to the denominator in Eqs. (6.3) to assure this property. The imaginary part of the susceptibility is obtained using the Cauchy formula $\mathrm{Im}\,(x+\mathrm{i}0)^{-1} = -\pi\delta(x)$ and is given by

$$\mathrm{Im}\,\chi(\omega) = \frac{\pi L}{2\omega_0}\left(\delta(\omega-\omega_0) - \delta(\omega+\omega_0)\right).$$

This produces δ-peaks at resonant frequency $\omega = \pm\omega_0$. The imaginary part of the susceptibility is related to dissipation in the circuit. For an oscillator, the dissipation can only occur at a resonant frequency corresponding to the excitation of the bosons of the oscillator mode.

The circuit studied so far is rather unrealistic: it does not contain the dissipative elements, for example, resistors. It is straightforward to introduce such elements at classical level: let us add a resistor in parallel with the capacitor and the inductor. The current in the resistor, $\dot{\Phi}/R$, is proportional to the voltage drop $\dot{\Phi}$ on it, so the classical equations become

$$\dot{Q} = \Phi/L + \dot{\Phi}/R - I(t);\ \dot{\Phi} = -Q/C.$$

The susceptibility is now given by

$$\chi(\omega) = \frac{L}{1-(\omega^2/\omega_0)^2 - \mathrm{i}\omega L/R};$$

$$\mathrm{Im}\,\chi(\omega) = \frac{\omega L^2/R}{\left(1-(\omega/\omega_0)^2\right)^2 + (\omega L/R)^2}.$$

We see that the imaginary part of the susceptibility is not zero at any ω, although at weak dissipation, $R \gg \sqrt{L/C}$, it gives two narrow peaks approximating the δ-peaks of an ideal oscillator. The boson modes are thus present at *any* frequency.

This implies the following quantum description. A circuit with dissipative elements is described by infinitely many boson modes, numbered k, with frequencies ω_k. This is instead of a single mode with frequency ω_0. The frequencies of the modes form a continuous spectrum, filling the whole frequency interval. The Hamiltonian becomes

$$\hat{H} = \sum_k \hbar\omega_k \hat{b}_k^\dagger \hat{b}_k - I(t)\hat{\Phi}; \tag{6.4a}$$

$$\hat{\Phi} = \sum_k C_k \hat{b}_k^\dagger + \mathrm{h.c.} \tag{6.4b}$$

The quantum model is specified with the introduced weight coefficients C_k. Their values also determine the dynamics of the flux Φ. To see how this works, let us relate these coefficients to the susceptibility. We look at the Heisenberg equations of motion for boson operators corresponding to the Hamiltonian, Eq. (6.4a)

$$\dot{\hat{b}}_k = \mathrm{i}[\hat{H}, \hat{b}_k]/\hbar = -\mathrm{i}\omega_k \hat{b}_k + \mathrm{i}C_k I(t)/\hbar.$$

Solving the equations using the Fourier transform method, we find

$$\langle b_k(\omega)\rangle = \frac{C_k/\hbar}{\omega_k - \omega - \mathrm{i}0} I(\omega).$$

Summing up these averages to $\langle\hat{\Phi}\rangle$, we find the susceptibility in terms of C_k as follows:

$$\chi(\omega) = \sum_k \frac{|C_k|^2}{\hbar}\left(\frac{1}{\omega_k - \omega - \mathrm{i}0} + \frac{1}{\omega_k + \omega + \mathrm{i}0}\right).$$

Taking the imaginary part of this equation and using the Cauchy formula again we show that

$$\sum_k \omega |C_k|^2 \delta(\omega - \omega_k) = \frac{\hbar}{\pi}\,\mathrm{Im}\,\chi(\omega) \qquad \text{at } \omega \geq 0. \tag{6.5}$$

This relation in fact provides the quantum description of electric circuits. It relates the parameters of the quantum model – the weight coefficients C_k – with the classical description of the circuit in terms of the susceptibility. Obviously, the relation does not fix all C_k unambiguously. However, this is not necessary. The point is that the quantum model involves infinitely many boson degrees of freedom to describe the dissipative dynamics of a single degree of freedom Φ. The model is therefore highly redundant: all possible choices of C_k satisfying Eq. (6.5) bring about identical results.

Let us employ the quantum description obtained to evaluate the (quantum) fluctuations in the circuit. We define the frequency-dependent noise of flux similar to in Section 1.7.3 as follows:[1]

$$S_\Phi(\omega) = \int dt\, \mathrm{e}^{-\mathrm{i}\omega t}\langle\hat{\Phi}(0)\hat{\Phi}(t)\rangle = \int dt\, \mathrm{e}^{-\mathrm{i}\omega t} S_\Phi(t). \tag{6.6}$$

The time-dependent boson operators become $\hat{b}_k(t) = \mathrm{e}^{\mathrm{i}\omega_k t}\hat{b}_k$. Expressing the operator $\hat{\Phi}$ in terms of the weight coefficients, we see that

$$S_\Phi(\omega) = 2\pi \sum_k |C_k|^2 \left(N_k \delta(\omega - \omega_k) + (1 + N_k)\delta(\omega + \omega_k)\right), \tag{6.7}$$

where $N_k \equiv \langle\hat{b}_k^\dagger\hat{b}_k\rangle$ is the average occupation number of the boson mode. Using Eq. (6.5), we prove that the asymmetric part of the noise does not depend on the occupation and is immediately given by the imaginary part of the susceptibility, that is

$$S_\Phi(-\omega) - S_\Phi(\omega) = 2\hbar\,\mathrm{Im}\,\chi(\omega). \tag{6.8}$$

If the circuit is in thermal equilibrium, the occupation numbers of the boson modes are given by a Bose distribution, $N_k = N_\mathrm{B}(\omega_k) \equiv 1/(\mathrm{e}^{\hbar\omega_k/k_\mathrm{B}T} - 1)$ and depend on frequency only. Substituting this into Eq. (6.7), we find $S(-\omega) = S(\omega)\exp(\hbar\omega/k_\mathrm{B}T)$ and

$$S(\omega) = 2\hbar N_\mathrm{B}(\omega)\,\mathrm{Im}\,\chi(\omega) \tag{6.9}$$

at any frequency. A special limit is given by $\omega \ll k_\mathrm{B}T/\hbar$. In this case, $N_\mathrm{B} \approx k_\mathrm{B}T/\omega$ and

$$S(\omega) \approx S(-\omega) \approx k_\mathrm{B}T\,\mathrm{Im}\,\chi(\omega)/\omega. \tag{6.10}$$

This does not contain $\hbar$ and presents therefore the classical thermal fluctuations. In the opposite limit, $\omega \gg k_\mathrm{B}T/\hbar$, the noise is purely quantum:

$$S(\omega) = 2\hbar\,\mathrm{Im}\,\chi(|\omega|)\Theta(-\omega).$$

[1] There is a factor of 2 difference between the definitions.

We note the similarity with the relations for the equilibrium *current* noise discussed in Section 1.7.3. In that case, the susceptibility is defined as the charge response on the action of the voltage source and is given by $\mathrm{i}\omega G(\omega)$, $G(\omega) \equiv I(\omega)/V(\omega)$ being the frequency-dependent conductance (admittance) of the nanostructure. This is in contrast to the case of flux or voltage noise where the susceptibility $\chi(\omega) = \mathrm{i}\omega Z(\omega)$ is, apart from the factor, the impedance of the circuit. The fluctuations in linear circuits, either quantum or classical, are *Gaussian* and have a well known property that any higher-order correlator of electric variables can be expressed in terms of the noise S_Φ. In the present quantization approach, this is guaranteed by the properties of boson annihilation and creation operators. Later in the chapter, we will need a relation for the correlation of the exponents of electric variables. The Gaussian property means that

$$\langle \mathrm{e}^{\hat{A}} \rangle = \mathrm{e}^{\langle \hat{A}^2 \rangle/2}; \ \langle \mathrm{e}^{\hat{A}} \mathrm{e}^{\hat{B}} \rangle = \mathrm{e}^{\langle \hat{A}^2 \rangle/2 + \langle \hat{A}\hat{B} \rangle + \langle \hat{B}^2 \rangle/2} \tag{6.11}$$

for any operators $\hat{A}, \hat{B}$ that are linear in boson operators. Using this, we prove, for example, that

$$\langle \mathrm{e}^{-\mathrm{i}a\hat{\Phi}(0)} \mathrm{e}^{\mathrm{i}a\hat{\Phi}(t)} \rangle = \mathrm{e}^{-a^2(S(0)-S(t))}. \tag{6.12}$$

The above relations present the quantization of the circuit where only one degree of freedom – either flux or charge – is important. The quantization requires the susceptibility – the response of this degree of freedom on either current or voltage source, respectively. What do we do for a general linear circuit where several degrees of freedom are of importance? The Hamiltonian in this case generally includes the corresponding sources: current ones for flux variables (labeled i) and voltage ones for charge variables (labeled j):

$$\hat{H} = \sum_k \hbar\omega_k \hat{b}_k^\dagger \hat{b}_k - \sum_j I_j \hat{\Phi}_j - \sum_i V_i \hat{Q}_i.$$

The quantum dynamics of the circuit is determined by the corresponding susceptibilities – the responses to the sources:

$$-\mathrm{i}\omega\langle \hat{\Phi}_j(\omega) \rangle = \sum_{j'} Z_{jj'}(\omega) I_{j'}(\omega) + \sum_i K_{ij}(\omega) V_j(\omega);$$
$$-i\omega\langle \hat{Q}_i \rangle = \sum_{i'} G_{ii'}(\omega) V_{i'}(\omega) + \sum_j \tilde{K}_{ij} I_j(\omega),$$

which we choose to present as the matrices of impedances ($Z_{jj'}$), admittances ($G_{ii'}$), and gains ($K_{ji}, \tilde{K}_{ij}$). The operators of the variables are linear in boson variables:

$$\hat{\Phi}_j = \sum_k f_k^{(j)} \hat{b}_k^\dagger + \text{h.c.}; \ \ \hat{Q}_i = \sum_k q_k^{(i)} \hat{b}_k^\dagger + \text{h.c.} \tag{6.13}$$

The correspondence between the weight coefficients and response functions is given by ($\omega > 0$)

$$\begin{aligned} \textstyle\sum_k f_k^{(j)} (f_k^{(j')})^* \, \delta(\omega - \omega_k) &= \frac{\hbar}{\omega\pi} \left(Z_{jj'}(\omega) - Z^*_{j'j}(\omega) \right); \\ \textstyle\sum_k q_k^{(i)} (q_k^{(i')})^* \, \delta(\omega - \omega_k) &= \frac{\hbar}{\omega\pi} \left(G_{ii'}(\omega) - G^*_{i'i}(\omega) \right); \\ \textstyle\sum_k f_k^{(j)} (q_k^{(i)})^* \, \delta(\omega - \omega_k) &= \frac{\hbar}{\omega\pi} \left(K_{ji}(\omega) - \tilde{K}^*_{ij}(\omega) \right). \end{aligned} \tag{6.14}$$

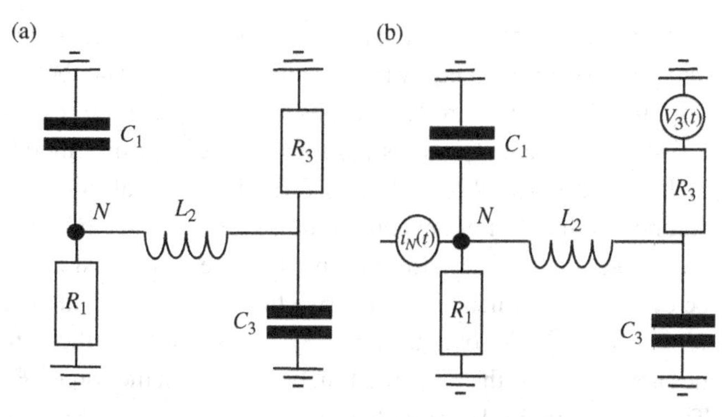

Fig. 6.2. (a) In this example circuit, the variables of interest are the current I_3 in resistor R_3 and the voltage V_N in node N. For quantization, one needs the corresponding susceptibilities. (b) The susceptibilities are obtained by as responses of V_N and I_3 on virtual current (i_N) and voltage sources ($V_3(t)$).

The response functions in an equilibrium circuit satisfy the Onsager symmetry relations, $Z_{jj'} = Z_{j'j}, G_{ii'} = G_{i'i}, K_{ij} = -\tilde{K}_{ji}$. In this case, the weight coefficients q_k can be chosen to be real while f_k are purely imaginary.

To avoid any misunderstanding, we note that no physical detail of the circuit design should represent the sources introduced; one does not have to provide the actual means for applying voltage or injecting electrons. The sources are needed to represent the response functions of the variables of interest and can be completely virtual. Let us illustrate this with a concrete example of the circuit depicted in Fig. 6.2. Suppose the variables of interest are current in resistor R_3 and voltage in node N. To characterize the weight coefficients, we need the response functions. Those are readily obtained if we insert the voltage source V_3 in series with the resistor 3 and inject the current i_N into the node N. The response functions, however, are calculated with the circuit theory rules and are known provided the parameters of the circuit are known.

The noises in this general circuit are defined in terms of the product of the time-dependent operators. For example, the current noises form a matrix given by

$$S_{ij}(\omega) = \int \mathrm{d}t\, \mathrm{e}^{-\mathrm{i}\omega t} \langle \hat{I}_i(0)\hat{I}_j(t)\rangle. \tag{6.15}$$

For $i \neq j$, this yields a *correlator* of the corresponding fluctuating currents.

Exercise 6.2. Evaluate the correlator of the current and voltage specified, $S_{IV}(\omega) = \int \mathrm{d}t\, \mathrm{e}^{-\mathrm{i}\omega t} \langle \hat{I}_3(0)\hat{V}_N(t)\rangle$. For this, present the operators in the form given in Eqs. (6.13), express the correlator in terms of weight coefficients, and use Eqs. (6.14) to find the correspondence between the correlator and response functions. Find the required response functions from the circuit theory analysis.

We have just quantized the electric circuits: we found their quantum description starting from a classical one. Sometimes we come across the reverse problem. Start with a Hamiltonian of a quantum system, for example that of electrons in a metal, and find the description of the system in terms of a handful of variables, for example in terms of voltages and currents appearing in an equivalent electrical circuit. This belongs to the subject of quantum response theory. If there are reasons to assume that the resulting description boils down to a system of linear equations (as for the electrons in a metal), this belongs to the subject of *linear response* theory. Let us formulate several general properties of susceptibilities and noises that do not depend on the concrete choice of variables in use. Let the variables be denoted as $\hat{x}_i$: they do not have to have a dimension of length, nor do they have to be of the same dimension. To derive the relations, we add extra terms to the Hamiltonian, $\hat{H} = \hat{H}_0 - \sum_i F_i(t)\hat{x}_i$, introducing sources that drive the dynamics of $\hat{x}_i$. The susceptibilities are defined as the response functions of x_i on the forces F_i, $\langle \hat{x}_i(\omega)\rangle = \sum_j \chi_{ij}(\omega)F_i(\omega)$. They are given by the *Kubo formula*:

$$\chi_{ij}(\omega) = \frac{\mathrm{i}}{\hbar}\int_{-\infty}^{0} \mathrm{d}t\ \mathrm{e}^{-\mathrm{i}\omega t}[\hat{x}_i(0), \hat{x}_j(t)] \equiv \int_{-\infty}^{\infty} \mathrm{d}t\ \chi_{ij}(\tau). \tag{6.16}$$

Exercise 6.3. Derive the Kubo formula from the Heisenberg equations for operators $\hat{x}_j(t)$.

If we compare this to our definition of noises given in Eq. (6.6), we see that $\chi_{ij}(t) = \mathrm{i}\Theta(-t)(S_{ij}(t) - S_{ji}(-t))/\hbar$, where the factor $\Theta(-t)$ guards the causality: the response depends on the force in the past only. Due to this factor, we cannot immediately relate the frequency components of the χ and S values. However, if we take two mutually reversed susceptibilities, χ_{ij} and χ_{ji}, we can prove that

$$\chi_{ij}(\omega) - \chi_{ji}(-\omega) = \frac{\mathrm{i}}{\hbar}(S_{ij}(\omega) - S_{ji}(-\omega)). \tag{6.17}$$

Since $\chi^*_{ij}(\omega) = \chi_{ij}(\omega)$, we recover Eq. (6.8) for $i = j$. We also note that the definition of the noises imply that $S^*_{ij}(\omega) = S_{ji}(\omega)$. This is all we can say about the system under general circumstances.

We can say much more if we assume that the system is in thermal equilibrium. A convenient basis to work in is that of the Hamiltonian. In this case, the density matrix is diagonal in this basis and $\rho_n \propto \mathrm{e}^{-E_n/k_\mathrm{B}T}$. The noise definition is given by

$$S_{ij} = 2\pi \sum_{n,m} x^{(i)}_{nm} x^{(j)}_{mn} \delta(\hbar\omega + E_m - E_n)\rho_n. \tag{6.18}$$

Comparing this with the expression for S_{ji}, we find that

$$S_{ij}(\omega) = S_{ji}(-\omega)\exp(-\hbar\omega/k_\mathrm{B}T). \tag{6.19}$$

Combining this with Eq. (6.17), we can unambiguously relate the noises and susceptibilities. Traditionally, this is written in the form of the *fluctuation-dissipation* theorem, which relates the symmetrized noises (measured with a classical detector; see Section 1.7.3), $S^{(s)}_{ij}(\omega) = (S_{ij}(\omega) + S_{ji}(-\omega))/2$, and the susceptibilities,

$$S^{(s)}_{ij}(\omega) = \coth\left(\frac{\hbar\omega}{2k_\mathrm{B}T}\right)\frac{1}{2\mathrm{i}}\left(\chi_{ij}(\omega) - \chi^*_{ji}(\omega)\right). \tag{6.20}$$

Yet more relations hold under time-reversibility conditions. One can choose the variables x_i to be either symmetric or antisymmetric with respect to time reversal. In our general circuit example, the charges Q_i are symmetric while the fluxes Φ_i are antisymmetric. The matrix elements of the operators entering Eq. (6.18) are real for symmetric and purely imaginary for the asymmetric ones. Suppose all x_i are symmetric. This implies that $S_{ji}(\omega)$ are real, and $S_{ji} = S_{ij}$. It then follows that the susceptibilities must be symmetric, $\chi_{ij} = \chi_{ji}$. This is the Onsager relation.

Control question 6.1. Suppose some of the variables x_i are antisymmetric. What do the Onsager relations look like?

6.1.1 Electricity in extended conductors

Several times in this chapter we will need to depart from the circuit description and to know how electricity is quantized in extended conductors, where it is characterized by the spatial distribution of voltage, $V(t, \boldsymbol{r})$. As in circuits, the quantization is based on the response functions. The difference is that the response functions depend now on two coordinate arguments $\boldsymbol{r}, \boldsymbol{r}'$, giving, for instance, the voltage response in the point $\boldsymbol{r}$ on the current $I(t)\mathrm{d}\boldsymbol{r}'$ injected in the point $\boldsymbol{r}'$:

$$V(\omega, \boldsymbol{r}) = \int \mathrm{d}\boldsymbol{r}'\, Z(\omega; \boldsymbol{r}, \boldsymbol{r}') I(\boldsymbol{r}'). \tag{6.21}$$

This defined impedance determines all quantum properties of the voltage distribution. In this subsection, we evaluate this impedance for several various geometries addressed in this chapter.[2]

A geometry represents a concrete structure at a certain space scale, a typical distance between $\boldsymbol{r}$ and $\boldsymbol{r}'$. For the sake of a concrete example, let us consider a uniform conductor in the form of a slab with dimensions $L_x \gg L_y \gg L_z$ placed above an ideally conducting plane (gate electrode) (see Fig. 6.3).

At space scales smaller than L_z, the system can be regarded as a three-dimensional infinite homogeneous conductor. In a homogeneous system, the impedance conveniently depends only on the difference of the coordinates $\boldsymbol{r}, \boldsymbol{r}'$. So we work with Fourier transformations with respect to the coordinate difference, $V(\omega, \boldsymbol{q}) = Z(\omega, \boldsymbol{q}) I(\omega, \boldsymbol{q})$. To find the voltage response on the current injected in the point $\boldsymbol{r}'$, we use charge conservation:

$$\mathrm{div}\, \boldsymbol{j} + \frac{\partial \rho}{\partial t} = I(t)\delta(\boldsymbol{r} - \boldsymbol{r}')\,.$$

[2] Strictly speaking, the exact description of a spatially dependent electromagnetic field, taking relativistic retardation into account, requires a vector potential $\mathbf{A}(\boldsymbol{r})$ along with a scalar potential $V(\boldsymbol{r})$ to be considered. The corresponding response functions are matrices representing, for example, current responses on the action of vector potentials. However, in most model problems – and in all problems in this chapter – one can live with scalar potential only.

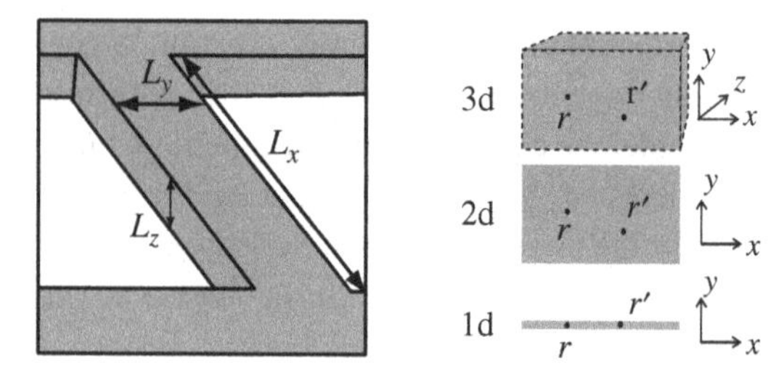

Fig. 6.3. Geometries and separation of scales. The coordinate-dependent impedance of the conducting nanostructure shown on the left ($L_z \ll L_y \ll L_x$), $Z(\omega, r, r')$, can be evaluated in different homogeneous geometries depending on the distance between r and r'.

In metals we have the electroneutrality condition, $\rho(\boldsymbol{r}, t) = 0$. Using this and the relation between current density and potential gradient, $\boldsymbol{j} = -\sigma \nabla V$ (σ being the specific conductivity of the conductor material), we obtain

$$Z(\omega, \boldsymbol{q}) = 1/(\sigma q^2). \tag{6.22}$$

The impedance does not depend on frequency, indicating that the response is *instantaneous* in this approximation.

Let us turn to larger space scales. At $L_z \ll |\boldsymbol{r} - \boldsymbol{r}'| \ll L_y$, the geometry is effectively two-dimensional. The voltage profile in the z direction is fixed and the voltage can only change in two directions x, y, i.e. $V(x, y, z) = V(x, y) f(z)$; $f(z)$ is a unity constant in the film and drops linearly to zero in the space between the film and the gate electrode. Implementing the charge conservation, we should be aware of the capacitance between the film and the gate. This leads to charge accumulation at the surface of the film, $\rho = \tilde{C} V$, $\tilde{C}$ being the capacitance to the gate per unit area, and this charge takes part in overall charge balance. To create the balance, we integrate the current density over the thickness of the film. In this form, it becomes $\boldsymbol{j} = -\nabla V / \tilde{R}$, where $\boldsymbol{j}, \nabla$ have x, y components only. The proportionality coefficient is called the "resistance per square," $\tilde{R} = 1/\sigma L_z$. Substituting all this into the charge conservation law, we obtain

$$-\nabla^2 V / \tilde{R} + \tilde{C} \dot{V} = I(t) \delta(\boldsymbol{r} - \boldsymbol{r}') \Rightarrow Z(\omega, \boldsymbol{q}) = \frac{1}{-\mathrm{i}\omega \tilde{C} + q^2/\tilde{R}}, \tag{6.23}$$

where the wave vector $\boldsymbol{q}$ is two-dimensional as well.

At yet larger scales, $L_z \ll |\boldsymbol{r} - \boldsymbol{r}'| \ll L_y$, the profile of the voltage distribution is fixed in two directions, $V(x, y, z) = V(x) f(y, z)$, so we encounter one-dimensional geometry. The relations are valid for 1d geometry if we redefine $\tilde{C}, \tilde{R}$ as capacitance and resistance per unit length, respectively. Such a 1d wire is commonly called an RC-line. By inspecting the denominator of Eq. (6.23), we find that the propagation of the field is no longer instantaneous; rather, it is a diffusion, with diffusion coefficient $D^* = 1/(\tilde{R}\tilde{C})$.

Until now, we have assumed that the current response on the voltage gradient is resistive. More generally, the response contains resistive and inductive parts, so that $-\nabla V = \tilde{R} \boldsymbol{j} + \tilde{L}\, \mathrm{d}\boldsymbol{j}/\mathrm{d}t$. The induction can come from changing magnetic fields produced by the current, but can also come from electrons accelerated by the electric field (kinetic induction). For 2d geometry, $\tilde{L}$ is the film inductance per square, while for 1d geometry it is the inductance per unit length. We see that at sufficiently large frequencies $\omega \gg \tilde{R}/\tilde{L}$, the inductive response dominates and the *time derivative* of the current is proportional to the voltage gradient, $\mathrm{d}\boldsymbol{j}/\mathrm{d}t = -\nabla V/\tilde{L}$. A plain example of such a system is a coaxial TV cable: at very low frequencies, it is an RC-line; at higher frequencies of TV signal, its response is almost entirely inductive and it becomes an LC-line. For both 1d and 2d geometries, the impedance is given by

$$Z(\omega, \boldsymbol{q}) = \frac{\mathrm{i}\omega}{(\omega + \mathrm{i}0)^2 \tilde{C} - q^2/\tilde{L}}. \tag{6.24}$$

This changes the way in which the electricity propagates. Since the inductive response does not provide the dissipation by itself, the electrical excitations move, conserving their energy. The elementary excitations are called plasmons. The dispersion law for the plasmons is found by inspecting the denominator of Eq. (6.24): $\omega = v_{\mathrm{p}} q$; this implies that the plasmons propagate with velocity $v_{\mathrm{p}} = 1/\sqrt{\tilde{L}\tilde{C}}$, which does not depend on $\boldsymbol{q}$.

As an application of the quantization procedure, let us evaluate the correlator of the voltages in the extended conductor:

$$S_V(\omega, \boldsymbol{q}) = \int \mathrm{d}t\ \mathrm{d}\boldsymbol{r} e^{\mathrm{i}\omega t + \mathrm{i}\boldsymbol{q}\cdot\boldsymbol{r}} \langle \hat{V}(\boldsymbol{r}, 0) \hat{V}(\boldsymbol{r}', t) \rangle.$$

As we see from Eqs. (6.14) and (6.9), this correlator is directly related to the impedance,

$$S_V(\omega, \boldsymbol{q}) = \omega N_{\mathrm{B}}(\omega)\ \mathrm{Re}\ Z(\omega, \boldsymbol{q}). \tag{6.25}$$

For instance, for an RC-line and $\omega \ll k_{\mathrm{B}}T/\hbar$, we find

$$S_V(\omega, \boldsymbol{q}) = \frac{2k_{\mathrm{B}}T\tilde{R}}{(\omega \tilde{C}\tilde{R}/q^2)^2 + 1}. \tag{6.26}$$

Control question 6.2. Estimate the correlation length of the electrical fluctuations in the RC-line at a given frequency ω.

6.1.2 Electricity and electrons

Electricity in conductors is due to electrons. In Section 6.1.1, we evaluated the impedance in various geometries under some simplifying assumptions about the electrons. We have assumed that the electrons ideally screen any stationary charge in the conductor, so there is no net charge except in the conductor-bound areas. We also disregard the effects of electron motion – either diffusive or ballistic – on the impedance. In this subsection, we will evaluate the impedance in a more microscopic way, explicitly taking into account the electron response on the electric field. This response will contain the disregarded effects. We will see that the assumptions are good, and that in most cases the corrections can be disregarded. They are, however, important for the interaction effects to be considered in Section 6.4.

We define the impedance as the frequency-dependent voltage response at point $\boldsymbol{r}$ on the external current injected to the conductor at point $\boldsymbol{r}'$. We stress that this current is *not* a stream of electrons (otherwise, we should be able to provide a physical terminal to each point of the conductor.) Generally, the impedance is evaluated as follows. The external current induces a charge $\rho_{\text{ext}}(\boldsymbol{r},\omega) = I(\boldsymbol{r},\omega)/(-\text{i}\omega)$. This charge induces the electrostatic potential, even in the absence of electrons, $V_{\text{ext}}(\boldsymbol{r},\omega) = \int \text{d}\boldsymbol{r}'\ U(\boldsymbol{r},\boldsymbol{r}')\rho_{\text{ext}}(\boldsymbol{r}',\omega)$. The electrons are present, however, and react on this potential. Generally, the charge response of the electrons on the potential may be presented as follows:

$$\rho_{\text{el}}(\boldsymbol{r}) = \int \text{d}\boldsymbol{r}'\ C_{\text{el}}(\boldsymbol{r},\boldsymbol{r}')V(\boldsymbol{r}',\omega). \tag{6.27}$$

Since the electron charge produces a potential on its own, the actual voltage distribution is found from Eq. (6.27) and the following self-consistency equation:

$$\begin{aligned} V(\boldsymbol{r},\omega) &= \int \text{d}\boldsymbol{r}'\ U(\boldsymbol{r},\boldsymbol{r}')(\rho_{\text{ext}}(\boldsymbol{r}',\omega) + \rho_{\text{el}}(\boldsymbol{r}',\omega)) \\ &= \int \text{d}\boldsymbol{r}' U(\boldsymbol{r},\boldsymbol{r}')(I(\boldsymbol{r}',\omega)/(-\text{i}\omega) + \rho_{\text{el}}(\mathbf{r}',\omega)). \end{aligned}$$

If we turn to a homogeneous geometry and change to Fourier transforms with respect to the coordinate difference, the resulting impedance is given by

$$Z(\omega,\boldsymbol{k}) = \frac{U(\boldsymbol{k})}{1 - C_{\text{el}}(\omega,\boldsymbol{k})U(\boldsymbol{k})}. \tag{6.28}$$

Let us specify the electron response. If the coordinate difference is of the order of Fermi wavelength, there is no other way to evaluate $C_{\text{el}}(\boldsymbol{r},\boldsymbol{r}')$ except full quantum-mechanical perturbation theory involving exact electron wave functions. However, if $|\boldsymbol{r}-\boldsymbol{r}'|$ exceeds the Fermi wavelength, the electron motion is semiclassical. It can be characterized by the probability $\mathcal{P}(t-t',\boldsymbol{r},\boldsymbol{r}')$ of finding an electron at point $\boldsymbol{r}$ at time moment t provided it was at point $\boldsymbol{r}'$ at time moment t'. The kernel C_{el} is related to this probability as follows:

$$\begin{aligned} \rho(\omega,r) &= e^2\nu\left(-V(\omega,\boldsymbol{r}) - \text{i}\omega \int \text{d}\boldsymbol{r}'\ \mathcal{P}(\omega,\boldsymbol{r},\boldsymbol{r}')V(\omega,\boldsymbol{r}')\right) \\ &\Rightarrow C_{\text{el}}(\omega,\boldsymbol{r},\boldsymbol{r}') = -e^2\nu(\delta(\boldsymbol{r}-\boldsymbol{r}') + \text{i}\omega\mathcal{P}(\omega,\boldsymbol{r},\boldsymbol{r}')). \end{aligned} \tag{6.29}$$

Depending on geometry, the electron density of states ν here may be per unit volume (3d), area (2d), or length (1d). An heuristic derivation of this relation is as follows. The stationary voltage ($\omega = 0$) concentrated near a point reduces the electron density by $-eV(r)\nu$ around this point. This is given by the first term in the brackets. If voltage increases in time, the electrons are released from the point and propagate through the nanostructure, increasing the electron density wherever they arrive.

We are ready to consider concrete geometries. Let us start with three dimensions. In this case, $U(\boldsymbol{r}-\boldsymbol{r}') = 1/|\boldsymbol{r}-\boldsymbol{r}'|$ and $U(\boldsymbol{k}) = 4\pi/k^2$. In the stationary case, $C_{\text{el}} = -e^2/\nu$, and we obtain

$$V(\boldsymbol{k}) = \frac{4\pi}{k^2+\kappa^2};\ \kappa \equiv \sqrt{e^2\nu/4\pi}. \tag{6.30}$$

This yields a potential produced by an external charge, and we see that it is not screened completely: the potential persists at distances of the order of κ^{-1}, the screening length, and falls off exponentially at larger distances. Usually the screening length is very small, even smaller than the Fermi wavelength. Let us consider $\omega \neq 0$ and assume diffusive propagation of electrons, so that

$$\mathcal{P}(\omega, \boldsymbol{k}) = \frac{1}{-\mathrm{i}\omega + Dk^2}; \; C_{\mathrm{el}} = -\frac{e^2\nu}{1 - \mathrm{i}\omega/Dk^2}. \tag{6.31}$$

Since we have assumed diffusion, we work at space scales exceeding the mean free path and therefore exceeding the screening length, $k \ll \kappa$. So the impedance is given by

$$Z_{3d} = \frac{1}{\sigma k^2}\left(1 - \frac{Dk^2}{\mathrm{i}\omega}\right) = \frac{1}{\sigma k^2} + \frac{1}{e^2\nu(-\mathrm{i}\omega)}, \tag{6.32}$$

where we have made use of $\sigma = e^2\nu D$. The first term coincides with the simpler form of the impedance given in Eq. (6.22). The second term presents the effect of incomplete screening and can be seen as a capacitance between the source of external charge and the conductor. Since this part of the response is purely imaginary, it does not contribute to the real part of the impedance and thus does not affect the voltage fluctuations from Eq. (6.25).

Dealing with 2d and 1d geometry, we make use of the local electrostatics model expressing the electron interaction in terms of capacitance $\tilde{C}$, $U(\boldsymbol{r} - \boldsymbol{r}') = \tilde{C}^{-1}\delta(\boldsymbol{r} - \boldsymbol{r}')$. The expression for C_{el} remains the same, and the impedance is cast into the following form:

$$Z(\omega, \boldsymbol{k}) = \frac{(\tilde{C}'/\tilde{C})^2}{-\mathrm{i}\omega\tilde{C}' + k^2/\tilde{R}} + \frac{1}{-\mathrm{i}\omega\tilde{C}}\left(1 - \frac{\tilde{C}'}{\tilde{C}}\right), \tag{6.33}$$

where we introduce the effective capacitance $\tilde{C}' = e^2\nu\tilde{C}/(\tilde{C} + e^2\nu)$. Comparing this with Eq. (6.23) we see that also in 2d and 1d cases the impedance acquires a purely capacitive correction (the second term in Eq. (6.33)). The first term is very close to the simplified impedance, Eq. (6.23); the difference is the renormalized capacitance in the denominator and an overall factor.

The capacitance renormalization is usually small, $\tilde{C} \approx \tilde{C}'$, since $\tilde{C} \ll e^2\nu$. Owing to this, the diffusion coefficient of electricity, $D^* = 1/\tilde{R}\tilde{C}$, by far exceeds the diffusion coefficient of electrons $D = 1/\tilde{R}e^2\nu = (1 - \tilde{C}'/\tilde{C})D^*$. To illustrate the smallness, let us consider 1d geometry with a wire of radius r_0 and the distance to the gate $a \gg r_0$. The geometric capacitance $\tilde{C} \approx 2/\ln(r_0/a) \simeq 1$, while $e^2\nu = \pi e^2\nu_{3d}r_0^2 \simeq \kappa r_0^2$. We see that the renormalization is small provided r_0 exceeds the screening length κ^{-1}, that is for any wire whose geometric cross-section allows for many transport channels.

We finish this subsection by introducing a model of a one-channel ballistic wire. It is frequently called the *Luttinger model*. The model is characterized by the Fermi velocity of electrons in the channel v_{F} and the geometric capacitance $\tilde{C}$ per unit length. To evaluate the impedance, we start with $\mathcal{P}(x, x'; t - t')$. Since in the initial point the electron can have either velocity, $\mathcal{P}(x, x'; t - t')$ consists of two contributions corresponding to the electron motion in two opposite directions:

$$\begin{aligned} &\mathcal{P}(x, x'; t - t') \\ &= (\delta(x - x' - v_{\mathrm{F}}(t - t')) + \delta(x - x' + v_{\mathrm{F}}(t - t')))\frac{\Theta(t - t')}{2}. \end{aligned} \tag{6.34}$$

In Fourier components, $\mathcal{P}(\omega, q) = \mathrm{i}\omega/((\omega+\mathrm{i}0)^2 - v_\mathrm{F}^2 q^2)$. Using this and $\nu = 2_s/(\pi v_\mathrm{F})$, we obtain, for the impedance,

$$Z(\omega, k) = \frac{1}{-\mathrm{i}\omega\tilde{C}} \frac{(\omega+\mathrm{i}0)^2 - v_\mathrm{F}^2 k^2}{(\omega+\mathrm{i}0)^2 - v_\mathrm{p}^2 k^2} \tag{6.35}$$

$$= \frac{\tilde{C}'}{\tilde{C}^2} \frac{\mathrm{i}\omega}{(\omega+\mathrm{i}0)^2 - v_\mathrm{p}^2 k^2} + \frac{1}{-\mathrm{i}\omega\tilde{C}}\left(1 - \frac{\tilde{C}'}{\tilde{C}}\right), \tag{6.36}$$

where the effective capacitance is defined as in Eq. (6.33) and the plasmon velocity is given by $v_\mathrm{p} = v_\mathrm{F}\sqrt{\tilde{C}/(\tilde{C} - \tilde{C}')}$. The first term in the impedance is similar to that of the LC-line (see Eq. (6.24)). Elementary excitations of the Luttinger liquid are therefore plasmons. The second term presents a capacitive contribution identical to Eq. (6.33). The specifics of the Luttinger model is that the capacitance renormalization can be significant. As we will see in Section 6.4, this provides significant interaction effects.

Common features of all concrete examples analyzed in this subsection are the capacitive correction and the absence of peculiarities in the denominators of Z that would reflect the electron motion. Naively, one could think of an opportunity to measure the electron motion directly by impedance measurement, hoping that at least part of the electric signal propagates with the electrons moving. The evaluated impedances show that this is not so: the measured impedance detects only the propagation of electricity. We will see in Section 6.4 that tunneling of electrons in a medium is affected by an effective impedance that incorporates both the effects of electricity and electron propagation: measuring the tunnel current, one can thus indirectly observe the electron motion in an electrical measurement.

6.2 Dissipative quantum mechanics

The quantization in electric circuits considered in the previous section can be straightforwardly generalized to any linear dissipative system. Although the quantum description involves many degrees of freedom (boson modes), only one or several variables are important, and the linear susceptibilities corresponding to these variables are sufficient to find all the quantum features. Dissipative quantum mechanics investigates more complicated nonlinear systems. An example problem is of a particle with mass M in a potential $U(x)$ that experiences a friction force $F_\mathrm{f} = -\gamma\dot{x}$ and an optionally external force F_ext. The classical equation of motion is as follows:

$$M\ddot{x} = -\frac{\partial U(x)}{\partial x} - \gamma\dot{x} + F_\mathrm{ext}.$$

If the potential $U(x)$ has a single minimum, one can proceed as follows. The qualitative result can be readily guessed: the classical particle subject to friction would freeze in the minimum, and quantum fluctuations spread the particle around the minimum. A reasonable first step is to expand the potential around the minimum, $U(x) \approx ax^2/2 + \text{const}$. The resulting equations are linear, and we can apply the quantization procedure described in Section 6.1. We need to know the frequency-dependent susceptibility $\chi(\omega)$ which is determined it from the classical equation as the reaction of x on the external force. The corresponding quantum Hamiltonian is given by

$$\hat{H} = H_{\text{env}} - \hat{x} F_{\text{ext}}(t), \tag{6.37}$$

where H_{env} denotes the collection of boson modes and the operator $\hat{x}$ is a linear combination of creation and annihilation boson operators of these modes with the weight coefficients that reproduce $\chi(\omega)$. To describe non-linear effects, we add $H_{nl} = U(x) - ax^2/2$. This term can be taken into account in the framework of perturbation theory, but it is not obvious why one should introduce such complications: since the potential has only a single minimum, the result is not supposed to change qualitatively.

The potential $U(x)$ with *two* minima of (almost) the same depth presents a much more interesting problem. The corresponding classical particle in the presence of friction is bistable: it may freeze in either minimum. Once frozen in a given minimum, it remains there forever, and the system functions like a binary memory cell. This is in contrast to a quantum particle without friction. It may be delocalized in such a two-minima potential owing to the coherent tunneling via the potential barrier between the minima. If the quantum states corresponding to the particle localized in a certain minimum are $|+\rangle$, $|-\rangle$, the stationary states may be the superpositions of the two, for example $(|+\rangle \pm |-\rangle)/\sqrt{2}$, where the particle is equally distributed between both minima.

The most interesting part of the problem is thus how a particle *tunnels between two states in the presence of friction*. So we can concentrate on yet simpler problem: a two-level system, a qubit coupled to a dissipative environment. Although the complete solution of this problem is still not presented, a qualitative understanding was achieved about 20 years ago [141, 142] and was one of the major developments of quantum mechanics. A very short formulation of this understanding is as follows: the system is *either* quantum (delocalized between minima) *or* classical (bistable memory cell) depending on the properties of the environment. It appears that the same scenario holds for more complicated situations, for example for a particle in a periodic potential.

To start with, we truncate the problem. From all the quantum states of the system without dissipation, we concentrate on two and consider the effect of the dissipation on transitions between them.

6.2.1 Spin-boson model

The corresponding framework model for a two-level system in the presence of the environment is called the *spin-boson* model. "Spin" appears here because a qubit is commonly mapped onto two states of a spin-1/2 particle. "Boson" denotes the boson model of the environment that affects the qubit. The total Hamiltonian of the spin-boson model consists of the Hamiltonians of the qubit, the environment, and the coupling between the two:

$$\hat{H} = \hat{H}_{\text{q}} + \hat{H}_{\text{env}} + \hat{H}_{\text{coupling}}. \tag{6.38}$$

The qubit Hamiltonian is given in its usual 2×2 matrix form (see Section 5.3). Let us fix the basis states and the stationary Hamiltonian, having in mind a two-minimum potential with an energy difference ε between the minima. The basis states $|+\rangle$, $|-\rangle$ are localized in minima with energies $\pm\varepsilon/2$. The tunneling matrix element between the states $|\pm\rangle$ is $\mathcal{T}$. The Hamiltonian is given by

$$\hat{H}_{\mathrm{q}} = \frac{\varepsilon}{2}\hat{\sigma}_z + \mathcal{T}\hat{\sigma}_x = \begin{bmatrix} \varepsilon/2 & \mathcal{T} \\ \mathcal{T} & -\varepsilon/2 \end{bmatrix}. \tag{6.39}$$

There are two stationary states of this Hamiltonian, "up" and "down," corresponding to the energies $E_{\mathrm{u,d}} = \pm\sqrt{(\varepsilon/2)^2 + \mathcal{T}^2}$. Their wave functions are given by ($\tan\theta \equiv 2\mathcal{T}/\varepsilon$)

$$|\mathrm{u}\rangle, |\mathrm{d}\rangle = \frac{-\sin\theta\,|+\rangle + (\cos\theta \pm 1)\,|-\rangle}{\sqrt{2(1 \pm \cos\theta)}}.$$

"Down" is the ground state. An important parameter that characterizes the delocalization is the probability of finding the qubit in the upper minimum provided the system is in the ground state, $p_{\mathrm{dec}} = |\langle \mathrm{d}|+\rangle|^2$ at $\epsilon > 0$. From the "down" eigenvector we derive that

$$p_{\mathrm{dec}} = \sin^2(\theta/2); \;\; p_{\mathrm{dec}} = \frac{\mathcal{T}^2}{\varepsilon^2} + \cdots \qquad \text{at } \mathcal{T} \to 0. \tag{6.40}$$

We see that the delocalization probability is in fact small if the tunneling is weak, $\mathcal{T} \ll \varepsilon$. However, this probability at a given $\mathcal{T}$ quickly increases on decreasing the energy bias ε. This signals the breakdown localization, and we use this signal to identify the delocalization in the context of dissipative quantum mechanics.

As in Section 6.1, the environment consists of the non-interacting boson modes, which we number here by m:

$$\hat{H}_{\mathrm{env}} = \sum_m \hbar\omega_m \hat{b}_m^\dagger \hat{b}_m.$$

As to the coupling Hamiltonian, it is assumed that only a single spin component, σ_z, is coupled to the environment. The strength of this coupling is controlled by a parameter q_0:

$$\hat{H}_{\mathrm{coupling}} = -q_0\hat{\sigma}_z\hat{F}; \;\; \hat{F} \equiv \sum_m (C_m\hat{b}_m^\dagger + C_m^*\hat{b}_m).$$

The operator $\hat{F}$ represents the fluctuations of the environment that affect the qubit. We should make sense of the spectral weight coefficients C_m. How should we do this? The best way is to consider a continuous variable q which is, for the spin-boson model, restricted to two values $\pm q_0$. If our model stems from a particle in a two-minimum potential, q is just the coordinate of the particle and $\hat{F}$ is then the friction force the particle may experience. Let us provide an action: move the particle with our hands, so that this coordinate is a fixed $q(t)$. We get the reaction of the environment, which is this friction force. We introduce a dynamical susceptibility $\chi(\omega)$ that relates the action and the reaction:

$$F(\omega) = \chi(\omega)q(\omega).$$

As in Section 6.1 (see Eq. (6.5)), this fixes the weight coefficients such that

$$\operatorname{Im}\chi(\omega) = \frac{\pi}{\hbar}\sum_m |C_m|^2\,\delta(\omega - \omega_m). \tag{6.41}$$

6.2.2 Shifted oscillators and shake-up

The spin-boson model cannot be solved exactly for general parameters. However, if we set $\mathcal{T} = 0$, we can do this, and we find some very instructive solutions. We will use these solutions to proceed with the perturbation expansion in terms of small $\mathcal{T}$. If $\mathcal{T} = 0$, no tunneling may occur and the system is fixed in one of the two classical states: $+$ or $-$. The Hamiltonians for *either* of the states depend on the boson variables only and are given by

$$\hat{H}_{+,-} = \sum_m \hbar\omega_m \hat{b}_m^\dagger \hat{b}_m \pm q_0(C_m \hat{b}_m^\dagger + C_m^* \hat{b}_m).$$

We can handle very well the Hamiltonian that does *not* contain the linear terms $\propto q_0$. The stationary wave functions of this Hamiltonian are those with a fixed number of bosons n_m in each mode, $|\{n_m\}\rangle$, the corresponding energies being $\sum_m \hbar\omega_m N_m$. We are going to perform a trick that kills these linear terms. One can guess the trick from a very simple analogy with a single oscillator. Let us "spoil" the parabolic potential with the term linear in x:

$$U(x) = \alpha\frac{x^2}{2} + \lambda x.$$

Eventually, it stays parabolic with a *shifted* variable x:

$$U(x) = \alpha\frac{x^2}{2} + \lambda x = \frac{\alpha}{2}\left(x - \frac{\lambda}{\alpha}\right)^2 - \left(\frac{\lambda}{\alpha}\right)^2.$$

Thus inspired, let us shift the boson operators. The shift is opposite for the states $+$, $-$ so new operators are different in these situations, i.e.

$$\hat{b}_m^{(\pm)} = \hat{b}_m \pm \lambda_m/2.$$

One checks that linear terms cancel out if $\lambda_m = -2q_0C_m/(\hbar\omega_m)$. The Hamiltonians in terms of new boson operators in this case simply become

$$\hat{H}_{+,-} = \sum_m \hbar\omega_m \hat{b}_m^{(\pm)\dagger}\hat{b}_m^{(\pm)} + \text{const.}$$

The shifted operators satisfy the same boson commutation relations as original operators. Thus we can handle each of the Hamiltonians as described above. The eigenstates of $\hat{H}_+$ are $|\{n_m\}_+\rangle$ and those of $\hat{H}_-$ are $|\{n_m\}_-\rangle$. The important point is that these states are not the same since the annihilation/creation operators for $+$ and $-$ have been shifted with respect to one another. Let us find the relation between the vacuums $|0_+\rangle$ and $|0_-\rangle$. By definition of a vacuum,

$$b_m^{(-)}|0_-\rangle = 0,$$

one can extract no boson from the vacuum. We rewrite this in terms of the annihilation operators of another vacuum:

$$b_m^{(+)}|0_-\rangle = \lambda_m|0_-\rangle.$$

This is recognized as the definition of *coherent state*. A coherent state $|\psi\rangle$ of a single boson mode is defined as an eigenstate of the annihilation operator of this mode:

$$\hat{b}\,|\psi\rangle = \lambda\,|\psi\rangle\,.$$

The coherent states are heavily used in quantum optics. They can be expressed as a superposition of the states with a different certain number of bosons n:

$$|\psi\rangle = \exp\left(-\frac{|\lambda|^2}{2}\right)\sum_n \frac{\lambda^n}{\sqrt{n!}}\,|n\rangle.$$

This formula is proven from the explicit form of the annihilation operator $\hat{b} = \sum_n \sqrt{n}|n-1\rangle\langle n|$. The average number of bosons in the coherent state is $\bar{N} = |\lambda|^2$. We thus come to the following important understanding: *one of the vacuums is a coherent state with respect to the other one, and the shifts λ determine the parameter of this coherent state*:

$$|0_-\rangle = \exp\left(-\frac{|\lambda|^2}{2}\right)\sum_n \frac{\lambda^n}{\sqrt{n!}}\,|n_+\rangle. \tag{6.42}$$

Now we are ready to discuss the *shake-up*. Let the system relax in state $-$, so it gets to the ground state $|0_-\rangle$ (assuming $\epsilon > 0$). Let us suppose we have a knob that allows us to change q. We suddenly turn the knob, shifting the value of q from $-q_0$ to q_0. If we manage to shift very quickly, the wave function does not change and is still $|0_-\rangle$. However, since the q has changed, it is no longer a ground state. Rather, it is a superposition of the ground and all excited states with coefficients given by Eq. (6.42). We thus supply energy to the system: we shake it up. Let us first consider a single mode and estimate the probability of supplying energy E while shaking it up in such a way. The probability $P(E)$ comprises the partial probabilities p_n to excite n bosons and, for a single mode, consists of a series of δ-peaks:

$$P(E) = \sum_n p_n\delta(E - n\hbar\omega).$$

The partial probabilities are given by

$$p_n = |\langle 0_- \mid n_+\rangle|^2 = \mathrm{e}^{-\bar{N}}\frac{\bar{N}^n}{n!}, \tag{6.43}$$

with $\bar{N} = |\lambda_m|^2 = |(2q_0C_m/\hbar\omega_m)|^2$, depending on the mode index n. We see that these probabilities obey the *Poisson distribution*; this is typical for coherent states. The Poisson distribution indicates that the acts of boson excitations are statistically independent.

Most of the important issues in dissipative quantum mechanics can be understood if we look at this one-mode $P(E)$ in two different limits (see Fig. 6.4(b)). If $\bar{N} \ll 1$, the highest probability is that of no energy. The shake-up in this limit most likely occurs without any dissipation. The highest probability of getting some energy is the one-boson one, as the probabilities of exciting a larger number of bosons decreases quickly as this number increases. The opposite limit is $\bar{N} \gg 1$. In this case, $P(E)$ is centered around $\bar{N}$, and it is improbable that more than $\bar{N}$ bosons are excited. Importantly, the probability of exciting none or a few bosons is also strongly suppressed.

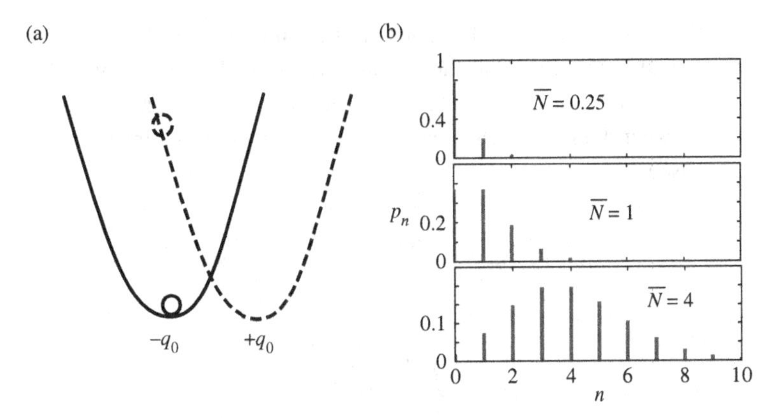

Fig. 6.4. (a) Artistic representation of an oscillator shake-up. Solid (dashed) lines give the potential and the position of the oscillating "particle" before (immediately after) the shake-up. (b) The probabilities of getting n bosons in the course of the shake-up plotted at various $\bar{N}$.

Let us recall now that the environment consists of infinitely many modes that together form a continuous spectrum. There is an oscillator shift and a shake-up in each mode. The energy excitations in different modes are independent, so the total $P(E)$ is thus contributed by all modes and all possible boson numbers n_m in each mode in the following way:

$$P(E) = \sum_{n_m} \exp\left(-\sum_m \bar{N}_m\right) \prod_m \frac{\bar{N}^{n_m}}{n_m!} \delta\left(E - \sum_m n_m \hbar\omega_m\right). \tag{6.44}$$

This cumbersome expression may be reduced to a much more comprehensive form. However, we pause before doing this since it is now a convenient moment to introduce the important concept of the *orthogonality catastrophe*.

6.2.3 Orthogonality catastrophe

Let us figure out what the probability is of emitting strictly *no* bosons during the shake-up of all modes. This probability is a product of the probabilities of emitting no bosons in each mode:

$$p_0 = \exp\left(-\sum_m \bar{N}_m\right), \tag{6.45}$$

and thus has a "quiet" shake-up, without exciting any environmental modes, and, importantly, without any energy dissipation. For any finite number of boson modes, this probability stays finite. This may be different if we go to the continuous limit. The integral presenting the sum over modes in Eq. (6.45) may diverge and the probability vanishes. This situation is called the *orthogonality catastrophe* (first described by P. W. Anderson in 1951). The point is that p_0 is simply the overlap between the two vacuums, $p_0 = |\langle 0_+|0_-\rangle|^2$. If $p_0 = 0$, the two vacuums are orthogonal. This takes place if the average number of bosons emitted in all modes diverges. Orthogonality of two quantum states implies that these states, so to say, belong to two different worlds: they neither know nor care about each other, they do not mix, and no (elastic) tunneling is possible between them. We can readily use Eq. (6.41) to express the probability in terms of the susceptibility $\chi(\omega)$:

$$\sum_m \bar{N}_m = \sum_m \left| \frac{2q_0 C_m}{\hbar\omega_m} \right|^2 = \int_0^\infty d\omega \sum_m \delta(\omega - \omega_m) |C_m|^2 \frac{4q_0^2}{\hbar^2\omega^2}$$
$$= \int_0^\infty d\omega \, \frac{4q_0^2 \operatorname{Im} \chi(\omega)}{\pi \hbar \omega^2}.$$

The integral over frequencies in this expression may diverge either at the upper or the lower limit. The upper ("ultra-violet") divergency is usually irrelevant since it can be removed by the upper cut-off frequency that limits the applicability of a given environmental model at very short time scales. The divergence at low frequencies – long time scales – is more important. We see that the integral diverges at low frequencies if $\operatorname{Im} \chi$ approaches zero at $\omega \to 0$ faster than ω.

6.2.4 P(E) and classification of environments

Let us write Eq. (6.44) for $P(E)$ in a more convenient form. The idea is to introduce an extra integration to present the δ-function of energy as an integral over an exponent. The exponent can be straightforwardly summed up over n_m:

$$P(E) = \sum_{n_m} \exp\left(-\sum_m \bar{N}_m\right) \prod_m \frac{\bar{N}^{n_m}}{n_m!} \delta\left(E - \sum_m n_m \hbar\omega_m\right)$$
$$= \int \frac{dt}{2\pi\hbar} \sum_{n_m} \exp\left(i\left(E - \sum_m n_m \hbar\omega_m\right) t/\hbar\right) \exp\left(-\sum_m \bar{N}_m\right) \prod_m \frac{\bar{N}^{n_m}}{n_m!}$$
$$= \int \frac{dt}{2\pi\hbar} e^{iEt/\hbar} \prod_m \sum_{n_m} \frac{\bar{N}_m^{n_m}}{n_m!} \exp(-in_m\omega_m t - \bar{N}_m)$$
$$= \int \frac{dt}{2\pi\hbar} e^{iEt/\hbar} \prod_m \exp(\bar{N}_m(e^{-i\omega_m t} - 1))$$
$$= \int \frac{dt}{2\pi\hbar} e^{iEt/\hbar} \exp\left(\sum_m \bar{N}_m(e^{-i\omega_m t} - 1)\right)$$
$$= \int \frac{dt}{2\pi\hbar} e^{iEt/\hbar} e^{J(t)-J(0)};$$
$$J(t) = \sum_m e^{-i\omega_m t} \bar{N}_m = \int_0^\infty d\omega \, e^{-i\omega t} \frac{4q_0^2 \operatorname{Im} \chi(\omega)}{\pi\hbar\omega^2}.$$

Finally, let us express $P(E)$ in terms of the Fourier component of $J(t)$, $P_1(E)$. This is the probability of emitting one boson with energy E in any mode in the limit of small q_0. It can be readily expressed in terms of $\chi(\omega)$ as follows:

$$P_1(E) = \frac{4q_0^2 \operatorname{Im} \chi(E/\hbar)}{E^2} \quad \text{at } E > 0, \tag{6.46}$$

where $P(E) = 0$ at negative energies. Finally, $P(E)$ is given by

$$P(E) = \int \frac{dt}{2\pi\hbar} e^{-iEt/\hbar} \exp\left(\int dE' \, P_1(E')(e^{iE't/\hbar} - 1)\right). \tag{6.47}$$

From this one concludes that, in the limit of small q_0, $p_0 \approx 1 - \int dE \, P_1(E) + \cdots$ and

$$P(E) = p_0\delta(E) + P_1(E) + \cdots, \tag{6.48}$$

where the first term corresponds to a "quiet" shake-up with no boson emitted, while the second term describes one-boson excitation.

So far, we have considered the shake-up and evaluated the probability $P(E)$ of giving energy E to the environment. In view of the fact that we are not usually able to shake the system in this way, the quantity does not seem to be particularly useful. The use of $P(E)$ becomes apparent if we recognize that the shake-up takes place by itself in the course of a spontaneous transition between the upper and lower states of the qubit. We concentrate on two states, "+" and "−," separated by an energy bias ε, and consider now a small but finite tunneling matrix element $\mathcal{T}$. The transition rate can be evaluated in the lowest non-vanishing (second) order from the Fermi Golden Rule, which in general is given by

$$\Gamma = \frac{2\pi}{\hbar} \sum_{\text{final states}} |M_{if}|^2 \delta(E_i - E_f).$$

In our case, the matrix element M is just the tunnel matrix element $\mathcal{T}$. In the initial state, the qubit is in the "+" position accumulating the energy ε – the energy splitting between the qubit levels. The environment is in the ground state $|0+\rangle$ corresponding to this position. After the transition, the qubit is in position "−," while the environment is shaken up with respect to the new ground state $|0-\rangle$. Owing to energy conservation, the energy that goes to the environment is ε. Therefore, the transition rate is directly proportional to $P(\varepsilon)$, the probability of disposing this energy in the course of the shake-up:

$$\Gamma(\varepsilon) = \frac{2\pi\mathcal{T}^2}{\hbar} P(\varepsilon). \tag{6.49}$$

If one-boson processes dominate, this probability is just the one-boson probability P_1, given by

$$\Gamma(\varepsilon) = \frac{2\pi\mathcal{T}^2}{\hbar} P_1(\varepsilon). \tag{6.50}$$

Another characteristic quantity is the delocalization probability p_{dec} defined by Eq. (6.40). It is evaluated in second order in $\mathcal{T}$ by perturbations, the resulting expression

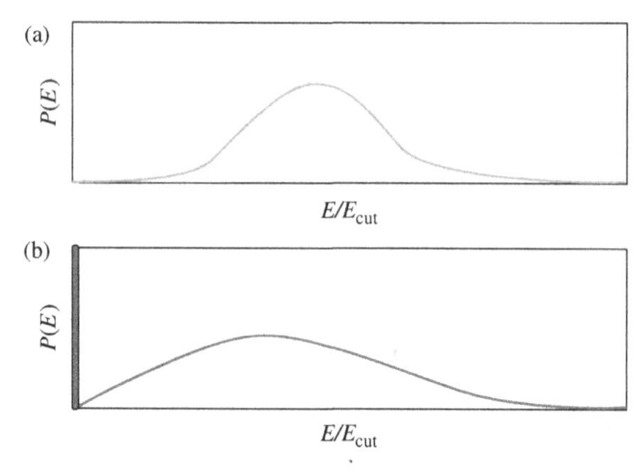

Fig. 6.5. **Types of $P(E)$. (a) First type: orthogonality catastrophe; $P(E)$ strongly suppressed at small energies. (b) Second type: no catastrophe, δ-peak (shown with a bar) at zero energy.**

also involving $P(E)$. In first order in $\mathcal{T}$, the ground state $|0_-\rangle$ acquires corrections proportional to all possible states $|n_m+\rangle$, where the qubit is in the "wrong" position:

$$|g\rangle = |0_+\rangle + \sum_{\{n_m\}} \psi_{\{n_m\}} \left|\{n_m\}_-\right\rangle; \ \psi_{\{n_m\}} = \frac{\Delta \left\langle 0_+ \mid \{n_m\}_- \right\rangle}{\varepsilon + \sum_m \hbar n_m \omega_m}.$$

The delocalization probability, i.e. the probability of finding the system in any of these states, is therefore given by $p_{\text{dec}} = \sum_{\{n_m\}} \left|\psi_{\{n_m\}}\right|^2$. Expressing the squares of matrix elements in terms of $P(E)$, we prove that

$$p_{\text{dec}} = \mathcal{T}^2 \int_0^\infty \mathrm{d}E \, \frac{P(E)}{(\varepsilon + E)^2}. \tag{6.51}$$

If there is no dissipation, $P(E) = \delta(E)$, and Eq. (6.51) reduces to the known result for a qubit not affected by the environment: $p_{\text{dec}} = \mathcal{T}^2/\varepsilon^2$.

There are two main types of $P(E)$; the difference between them is best seen from the plots in Fig. 6.5. For the first type, the orthogonality catastrophe is in place. This implies that the shake-up cannot occur "quietly," that is, without disposing of energy. This also implies that the probability of disposing finite but small amounts of energy is strongly suppressed, although it remains finite. The $P(E)$ of the first type looks very much like a continuous version of the one-mode probability (Fig. 6.4) in the classical limit $\bar{N} \gg 1$, having a peak at large energies. This peak becomes relatively narrow in the classical limit, corresponding to well defined energy transfer in the course of classical shake-up.

For the second type, there is no orthogonality catastrophe and p_0 remains finite. This gives rise to a characteristic δ-peak at zero energy, indicating the possibility of "quiet" shake-up and elastic tunneling. A continuous tail beyond the peak is mainly due to the one-boson excitations; the processes involving more bosons are increasingly suppressed. This resembles the one-mode $P(E)$ at $\bar{N} \ll 1$.

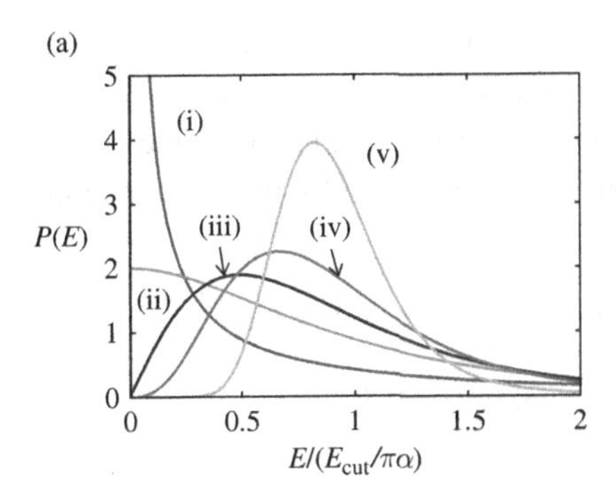

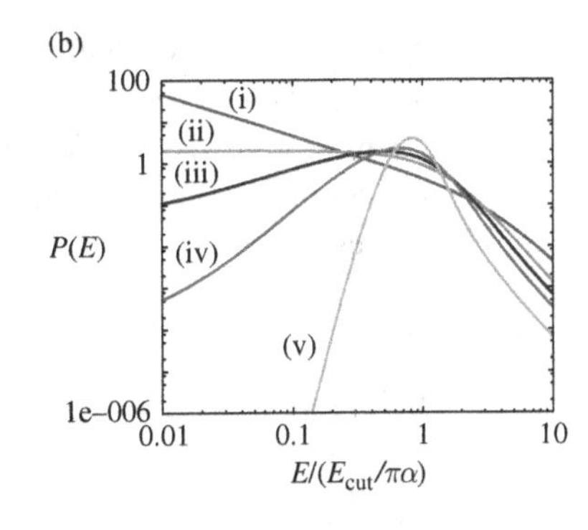

Fig. 6.6. **Ohmic environment: $P(E)$. (a) Different curves correspond to different coupling strengths. (i) $\alpha = 0.1$; (ii) $\alpha = 0.5$; (iii) $\alpha = 1$; (iv) $\alpha = 2$; (v) $\alpha = 10$. (b) The same plot with a log-log scale. The power-law dependence of $P(E)$ becomes apparent.**

> **Control question 6.3.** Which type is "more quantum"?

By continuously changing the parameters of the environment, one can move between the first and second types by crossing a border line. At the border line, the situation is rather unusual. The $P(E)$ does not display the δ-peak, so the orthogonality catastrophe takes place. However, the $P(E)$ is more intense at small energies than in the first type: it assumes a power-law dependence. If the power is negative, it even increases with decreasing energy (see Fig. 6.6).

Let us outline the classification of all possible environments proposed by Leggett in 1981 [143]. The starting point of this classification is the energy dependence of one-boson probability (we have seen that it is directly related to dynamical susceptibility by Eq. (6.46)). The attention is paid to small energies, and the assumption is that there is a cut-off E_{cut} that removes possible problems at high energies. There is a power-law dependence of $P(E)$ at $E \to 0$:

$$P_1(E) \propto E^s (E_{\text{cut}})^{-1-s} \exp(-E/E_{\text{cut}}). \tag{6.52}$$

If $s < -1$, the environment is called *subohmic*, and the $P(E)$ is of the first type mentioned. If $s > -1$, the environment is *superohmic*, and the $P(E)$ is of the second type.

Special attention is paid to the so-called *ohmic* environment ($s = -1$) that corresponds to the borderline between the first and second types. This is because this type of environment is rather widespread: in this case, Im $\chi(\omega) \propto \omega$ at small frequencies, so that at any time $\chi(\omega)$ is analytical and therefore expandable at $\omega = 0$.[3] It is convenient to choose the high-energy cut-off for the ohmic environment in the following form:

$$P_1(E) = \frac{2\alpha}{E} \frac{1}{(E/E_{\text{cut}})^2 + 1}, \tag{6.53}$$

[3] This is always true for the susceptibilities in finite-element electrical circuits. This gave rise to the term "ohmic" since the dissipation in such circuits is due to ohmic resistors.

which corresponds better to a typical form of $\chi(\omega)$ than the exponential cut-off. Here, $\alpha \propto q_0^2$ characterizes the strength of coupling to the environment. The corresponding $P(E)$ assumes a power-law dependence at small energies, with the exponent that depends on the coupling strength to the environment (Fig. 6.6), that is, on the intensity of one-boson process,

$$P_1 = \frac{2\alpha}{E} \Rightarrow P(E) \propto E^{2\alpha-1}.$$

Below we will briefly address each type of environment.

Subohmic

The subohmic environment suppresses the transitions between "+" and "−" in the most efficient way, at least at low energies. The transition rate is exponentially small at low energies, and is given by

$$\Gamma(\varepsilon) \propto \exp\left(-\frac{1}{(\varepsilon/E_{\text{cut}})^{(1/(s+2)-1)}}\right). \tag{6.54}$$

It seems that the + and − states are separated by an almost impenetrable barrier. They are also localized, as we see by examining the delocalization probability. Indeed, the integral in Eq. (6.51) is mostly contributed by $E \simeq E_{\text{cut}}$. The resulting p_{dec} is small, $\simeq(\Delta/E_{\text{cut}})^2$, and does not depend on the energy bias ε. Therefore, it does not increase with $\varepsilon \to 0$ as p_{dec} of the delocalized qubit does.

Ohmic

For an ohmic environment, $P(E)$ is not exponentially small and may even increase with decreasing E. However, the transitions with a given energy loss E generally involve many phonons. One sees this from the fact that $P(E)$ is suppressed in comparison with $P_1(E)$:

$$s = -1;\ P_1 = \frac{2\alpha}{E} \Rightarrow P(E) \propto E^{2\alpha-1} \ll P_1(E) \text{ at } E \to 0.$$

Let us look at the transition rate given by Eq. (6.49) as a function of energy bias ε:

$$\Gamma(\varepsilon) = \frac{2\pi}{\hbar}\mathcal{T}^2\left(\frac{\varepsilon}{E_{\text{cut}}}\right)^{2\alpha-1}. \tag{6.55}$$

The transition rate brings about an uncertainty, $\simeq\hbar\Gamma$, to the energy position of the decaying level +. We compare this uncertainty to the energy itself. We label a quantum state "good" provided the relative energy uncertainty $\hbar\Gamma(\varepsilon)/\varepsilon$ is much smaller than unity. Evaluating this relative uncertainty from Eq. (6.55), we see that $\alpha > 1$ implies that the excited qubit state is "good," at least in the limit of low energies ε. As for $\alpha < 1$, the state becomes "bad" at low energy and decays too fast. This signals that some kind of transition takes place precisely at $\alpha = 1$. What kind of transition is this?

The answer to this question was given by Albert Schmid in 1983 [144]. He proved that the system is localized at $\alpha > 1$ and delocalized otherwise. The transition at $\alpha = 1$ was named after him.

In fact, this seems rather counter-intuitive. A truly quantum system – of which a qubit without dissipation would be an example – displays both (infinitely) slow decay, that is a "good" excited state and delocalization. Intuitively, it would be clear that the "good" state at $\alpha > 1$ is also delocalized. However, it is precisely the other way around. To understand the point, let us look at the integral in Eq. (6.51). We cut it off at E_{cut}, so it becomes

$$p_{\text{dec}} \propto \mathcal{T}^2 \int_0^{E_{\text{cut}}} \text{d}E \, \frac{1}{(\varepsilon + E)^2} (E/E_{\text{cut}})^{2\alpha - 1}.$$

If $\alpha < 1$, the integral converges at energies $\simeq \varepsilon$ and the probability,

$$p_{\text{dec}} \propto \frac{\mathcal{T}^2}{\varepsilon^2} \left(\frac{\varepsilon}{E_{\text{cut}}} \right)^{2\alpha},$$

decreases with decreasing ε, reaching values $\simeq 1$ at sufficiently small ε. This is the delocalization. We see that such behavior requires that $P(E)$ either increases or at least grows slower than E at $E \to 0$. For the ohmic case, this is only possible if the excited state is "bad." Otherwise, the integral converges at $E \simeq E_{\text{cut}}$ and the delocalization probability remains small and not sensitive to ε, $p_{\text{dec}} = (\mathcal{T}/E_{\text{cut}})^2$. The system is localized although the excited state is "good." Actually, this is consistent with the fact that the excited state is also "good" in the subohmic regime owing to the vanishing transition rate (see Eq. (6.54)).

Superohmic

For any superohmic environment ($s > -1$) the transition rate is dominated by a one-boson process. The transition rate is therefore a power-law at low energies:

$$\Gamma(\varepsilon) = \frac{2\pi \Delta^2}{\hbar} P_1(\varepsilon) = \text{const.} \cdot \varepsilon^s.$$

Repeating the analysis carried out for the ohmic regime, we see that the excited state is "bad" if $s < 1$ and "good" if $s > 1$.

Does this imply a localization transition at $s = 1$ similar to the Schmid transition in the ohmic regime? The answer is no; in fact, the state is delocalized in both cases. The orthogonality catastrophe is absent for a superohmic environment. The $P(E)$ is of the second type with the δ-peak at zero energy, which is absent for ohmic and subohmic environments. The integral in Eq. (6.51) for the delocalization probability is mainly contributed by this δ-peak, so that $p_{\text{dec}} = p_0(\mathcal{T}/\varepsilon)^2$. The classification of all possible environments is summarized in Table 6.1.

6.2.5 Finite temperature and renormalization

For a deeper discussion of the spin-boson model, it is constructive to present it in an alternative form. Let us look again at the Hamiltonian of the model, given by

$$\hat{H} = \frac{\varepsilon}{2} \sigma_z + \mathcal{T} \sigma_x + \hat{H}_{\text{env}} - q_0 \sigma_z \hat{F},$$

Table 6.1 Classification of environments

Environment type	Delocalization	"Good" state	Orthogonality catastrophe
Superohmic, $s > 1$	yes	yes	no
Superohmic, $-1 < s < 1$	yes	no	no
Ohmic, $s = -1; \alpha < 1$	yes	no	yes
Ohmic, $s = -1; \alpha > 1$	no	yes	yes
Subohmic, $-2 < s < -1$	no	yes	yes

The behavior changes (top row to bottom row) from completely quantum qubit-like to a completely classical "memory cell."

and let us regard, for the moment, the operator F as a classical external force, $\hat{F} \Rightarrow F(t)$. We note that this force contributes to a time-dependent phase shift between the states $|+\rangle$, $|-\rangle$ of the qubit, this contribution being given by

$$\phi(t) = \frac{2q_0}{\hbar} \int_{-\infty}^{t} \mathrm{d}\tau \; F(\tau).$$

We can define new basis states corresponding to $|+\rangle$, $|-\rangle$ that incorporate this part of the phase shift as follows:

$$|+\rangle_{\text{new}}, |-\rangle_{\text{new}} = \exp \pm \mathrm{i}\phi(t)/2 \, (|+\rangle, |-\rangle) \, .$$

The qubit Hamiltonian in this new basis is given by

$$\hat{H} = \frac{\varepsilon}{2}\sigma_z + \mathcal{T}\mathrm{e}^{\mathrm{i}\phi(t)}\sigma^- + \mathcal{T}\mathrm{e}^{-\mathrm{i}\phi(t)}\sigma^+; \quad 2\sigma^{\pm} \equiv \sigma_x \pm \mathrm{i}\sigma_y,$$

and does not contain the term linear in F. This transformation is similar to the "moving frame" transformation studied in Section 5.3. The same transformation can be done using the operator $\hat{F}$. The resulting Hamiltonian is given by

$$\hat{H} = -\frac{\varepsilon}{2}\sigma_z + \mathcal{T}\mathrm{e}^{\mathrm{i}\hat{\phi}(t)}\sigma^+ + \mathcal{T}\mathrm{e}^{-\mathrm{i}\hat{\phi}}\sigma^- + H_{\text{env}}, \tag{6.56}$$

where the phase shift operator $\hat{\phi}$ is, as usual, linear in boson operators, $\hat{\phi} = \sum_m (f_m^* \hat{b}_m + f_m \hat{b}_m^\dagger)$. Since $\dot{\phi} = q_0 \hat{F}/\hbar$, $f_m = \mathrm{i}q_0 C_m/(\hbar\omega_m)$, and the correspondence between the weight coefficients and the susceptibility is now given by

$$q_0^2 \omega^2 \, \mathrm{Im}\, \chi(\omega) = \pi\hbar \sum_m |f_m|^2 \, \delta(\omega - \omega_m).$$

The transition rate from the upper to the lower level is obtained using second-order perturbations in $\mathcal{T}$ and is given by

$$\Gamma_{+\to-}(\varepsilon) = \mathcal{T}^2 \int_{-\infty}^{\infty} \mathrm{d}t \, \langle \mathrm{e}^{\mathrm{i}\hat{\phi}(0)} \mathrm{e}^{-\mathrm{i}\hat{\phi}(t)} \rangle \mathrm{e}^{\mathrm{i}\varepsilon t}. \tag{6.57}$$

Recalling Eq. (6.49), we understand that this expresses $P(E)$ in terms of the correlator of the exponents of ϕ. We have already evaluated $P(E)$ at zero temperature and in equilibrium, and this is given by Eq. (6.47). We use the present relation to find it for finite temperatures. Using the results of Section 6.1 (Eqs. (6.11) and (6.12)), we see that

$$P(E) = \int \frac{dt}{2\pi\hbar} e^{iEt} e^{S_\phi(t) - S_\phi(t=0)}. \tag{6.58}$$

Using the susceptibility (see Eq. (6.9)) and the fact that $\dot{\phi} = 2q_0 F/\hbar$, we find

$$\begin{aligned} S_\phi(t) - S_\phi(t=0) &= \int \frac{d\omega}{2\pi} \left(e^{-i\omega t} - 1 \right) S_\phi(\omega) \\ &= \int \frac{d\omega}{2\pi} \frac{1}{e^{\hbar\omega/k_B T} - 1} \left(e^{-i\omega t} - 1 \right) \frac{8q_0^2}{\hbar\omega^2} \operatorname{Im} \chi(\omega). \end{aligned} \tag{6.59}$$

These two formulas summarize the environmental effects on tunneling in the most general way and will be used intensively below. Comparing this with Eq. (6.47), we see that the finite temperature effects are taken into account using the following modification:

$$P_1(T, E) = P_1(T = 0, |E|)/(1 - e^{-E/k_B T}). \tag{6.60}$$

Let us review some properties of $P(E)$ at finite temperature. At finite temperature, the qubit may absorb energy from the environment, so that $P(E)$ has some weight at negative energies too. We note that the energy change for the transition rate in the opposite direction, $\Gamma_{-\to+}$, is opposite. Therefore $\Gamma_{-\to+} = (2\pi T^2/\hbar)P(-\varepsilon)$. Since the system is in thermal equilibrium, the rates must satisfy the detailed balance condition $\Gamma_{+\to-}/\Gamma_{-\to+} = \exp(\varepsilon/k_B T)$. This implies

$$P(-E) = P(E)\exp(-E/k_B T)\,. \tag{6.61}$$

At zero temperature, there is always a singularity in $P(E)$ at $E \to 0$. At finite temperature, the singularity is expected to be smoothed, so that $P(E)$ is continuous with all derivatives.

It could seem natural to assume that the temperature modifies $P(E)$ only at the energy scale $|E| \simeq k_B T$ so that it vanishes at negative energies and retains its zero-temperature value at positive energies provided $|E| \gg k_B T$. This is indeed the case for an ohmic environment at $\alpha \simeq 1$ (see Fig. 6.7(a)), and, as seen from Eq. (6.60), for one-boson processes. Generally, the thermal effect on $P(E)$ is more involved. For example, for subohmic environments a transition is typically a many-boson process. In this case, E is associated with the energy of all bosons, while $k_B T$ is a typical energy scale for one boson. Moreover, at finite temperature many-boson processes may also become important for superohmic environments. To see this, let us consider the average number of bosons emitted at finite temperature, given by

$$\bar{N} = \int_{-\infty}^{\infty} dE\ P_1(T, E) = \int_0^\infty dE \coth\left(\frac{E}{2k_B T}\right) P_1(T = 0, E).$$

The integrand is extra divergent at $E \ll T$, $\propto T P_1(E)/E \propto T E^{s-1}$. This is why the average number of bosons also diverges at low energies for superohmic environments, provided $s < 0$.

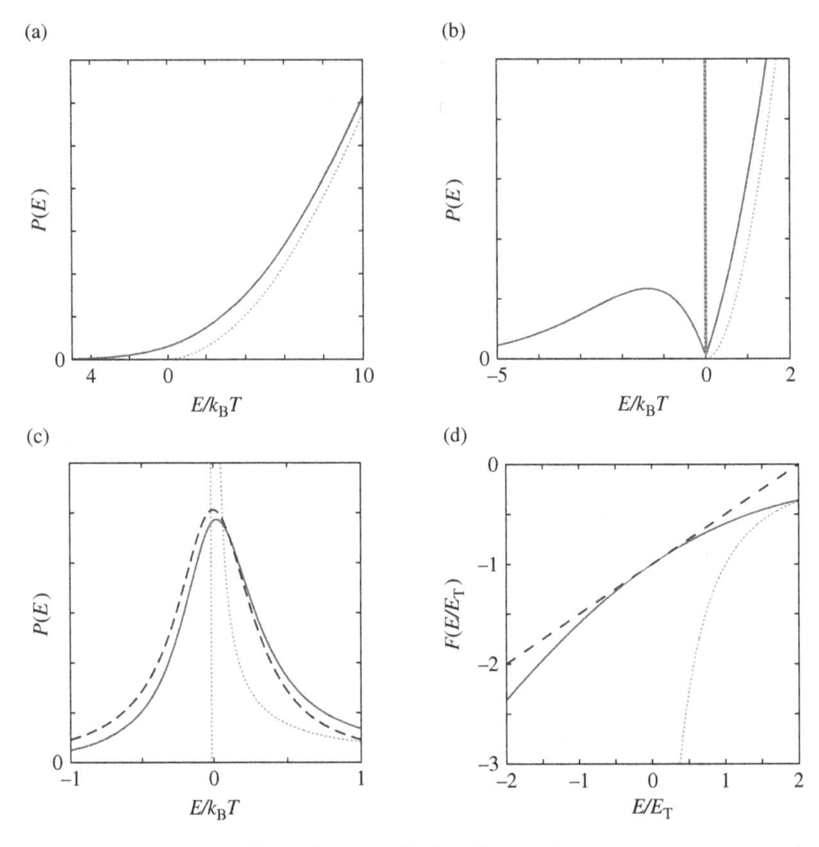

Fig. 6.7. $P(E)$ at finite temperature. The dotted lines in all plots denote the zero-temperature result. (a) Ohmic environment ($\alpha = 1.5$). The $P(E)$ is smoothed and modified at $E \simeq k_B T$. (b) Superohmic environment ($s = 1.5$). The δ-peak is not smoothed, and $P(E)$ is modified at $E \simeq k_B T$. (c) Ohmic environment at small ($\alpha = 0.05$). $P(E)$ is concentrated in a peak that is much narrower than $k_B T$. Dashed line: Lorentzian shape given by Eq. (6.62). (d) Subohmic environment ($s = -3/2$). $P(E)$ is modified at energy scale $E_T \propto T^{1/2} \gg k_B T$. The scaling function $F_{-3/2}(E/E_T) = \ln(P(E))/(E_T/k_B T)$ is presented (see the text). Dashed line: $1 - x/2$.

For superohmic environments with $s = 0$, the picture remains simple: the δ-peak survives while the one-boson contribution is smoothed by temperature at $|E| \simeq k_B T$ (Fig. 6.7(b)). For a superohmic environment with $s > 0$, the δ-peak at zero energy is replaced by a smooth symmetric peak spread over an energy scale much *smaller* than $k_B T$. The same situation pertains to the practically important case of an ohmic environment with $\alpha \ll 1/2\pi$. At $E \ll k_B T$ the noise S_ϕ is essentially classical (see Eq. (6.10)), and the phase is driven by a white noise. This leads to *phase diffusion*: the correlator of phases increases linearly in time, $\langle \phi(0)(\phi(t) - \phi(0)) \rangle = 2D_\phi |t|$, $D_\phi \equiv \alpha\pi(k_B T/\hbar)$. Substituting this into Eq. (6.58), we find a Lorentzian-shaped peak of energy-width w, as follows:

$$P(E) = \int \frac{dt}{2\pi\hbar} e^{iEt/\hbar} e^{-2D_\phi |t|} = \frac{w/\pi}{w^2 + E^2}; \tag{6.62}$$

$$w \equiv 2\pi\alpha k_B T \ll k_B T. \tag{6.63}$$

The energy scale $w \ll k_B T$ is in fact related to the *decoherence* of qubit states, so we recall this scale in Section 6.7.

For subohmic environments, the thermal effect results in an energy scale *greater* than temperature. To estimate this scale, let us recall Eq. (6.54). The typical number of bosons involved can thus be estimated as $N(E) \simeq -\ln P(E) \simeq (E_{\text{cut}}/E)^{1/(s+2)-1}$. The thermal scale E_T is determined from the condition $E_T/N(E_T) \simeq k_B T$, that is, the energy of each boson is of the order of $k_B T$. This gives the temperature-dependent $E_T \simeq k_B T (k_B T/E_{\text{cut}})^{s+1}$. At zero energy, this defines the temperature dependence of $P(E)$, $P(0) = \exp(E_T/k_B T)$. At the thermal scale, the $P(E)$ is given by s-dependent scaling function $F_s(x)$:

$$\ln P(E) = \frac{E_T}{k_B T} F_s\left(\frac{E}{E_T}\right).$$

At small parameter values, $F_s(x) = -1 + x/2$, so that, at $E \ll E_T$, $P(E) = P(0)\exp(E/(2k_B T))$. At $x \to \infty$, $F_s(x) \propto x^{1-1/(s+2)}$; this guarantees that at $E \gg E_T$ the $P(E)$ does not depend on temperature and coincides with Eq. (6.54).

To conclude our review of dissipative quantum mechanics, let us consider *renormalization* techniques in application to the spin-boson model. Generally speaking, a renormalization technique attempts to approximate a change of dynamics of a system upon changing a scale of time (or energy) by a change of parameters characterizing the system Hamiltonian. Therefore, a renormalization technique is not an exact method: the accuracy of its results is usually similar to that of the dimension analysis method in elementary physics. Similar to dimension analysis, renormalization works where no other method does. This is why the renormalization technique is in fact *not* technical: it formulates the essence of the dynamics in the simplest terms possible and thereby refines the understanding of a system, clearing up unnecessary technical details.

To illustrate, let us reconsider the transition rate for an ohmic environment. At low energies, it is given by (see Eq. (6.55))

$$\Gamma(\varepsilon) = \frac{2\pi}{\hbar}\mathcal{T}^2\left(\frac{\varepsilon}{E_{\text{cut}}}\right)^{2\alpha-1}.$$

Since we work at low energies, higher energies $E \simeq E_{\text{cut}}$ are not accessible for us and we cannot know what E_{cut} is. Moreover, we should expect that the transition rate does not depend on the details of the cut-off: we do not have to know what E_{cut} is. The same applies to the "bare" tunneling matrix element $\mathcal{T}$. Those parameters cannot be measured at low energies and therefore present unnecessary technicalities. An equivalent and adequate form of the presentation of the same relation is the scaling law: for any low-energy $\varepsilon_1, \varepsilon_2$,

$$\frac{\Gamma(\varepsilon_1)}{\Gamma(\varepsilon_2)} = \left(\frac{\varepsilon_1}{\varepsilon_2}\right)^{2\alpha-1}.$$

This does not contain any "unphysical" parameters $\mathcal{T}$, E_{cut}.

The idea of renormalization of the spin-boson model is to make the tunneling amplitude dependent on the energy scale. Let us work with a conveniently dimensionless energy scale, $\zeta = \ln(E_{\text{cut}}/E)$, which changes from zero to infinity while E changes from E_{cut} to zero. The renormalized $\mathcal{T}(\zeta)$ is estimated as follows. The phase operator $\hat{\phi}$ is linear in boson operators. Let us separate it into "slow" and "fast" parts, $\hat{\phi} = \hat{\phi}_s + \hat{\phi}_f$, where the slow (fast) part is contributed by boson modes m with energies $\hbar\omega$ smaller (bigger) than the energy $E^*_{\rm cut} = E_{\rm cut}\,{\rm e}^{-\zeta}$. The next step is to average the Hamiltonian, Eq. (6.38), over "fast" modes of higher energy. The only terms which are modified by the averaging are those proportional to $\mathcal{T}$:

$$\mathcal{T}\langle {\rm e}^{{\rm i}\hat{\phi}}\rangle_{\rm fast}\sigma^- = \mathcal{T}\exp\left(-\frac{1}{2}\langle\hat{\phi}_{\rm f}^2\rangle_{\rm fast}\right){\rm e}^{{\rm i}\hat{\phi}_s}\sigma^-, \tag{6.64}$$

where we use Eq. (6.11) and the fact that $\hat{\phi}_s, \hat{\phi}_f$ commute since they originate from separate boson modes. If we now define

$$\mathcal{T}(\zeta) = \mathcal{T}\exp\left(-\frac{1}{2}\langle\hat{\phi}_{\rm f}^2\rangle_{\rm fast}\right),$$

the resulting averaged Hamiltonian is equivalent to the original Hamiltonian upon replacing $E_{\rm cut} \to E^*_{\rm cut}$ and $\mathcal{T} \to \mathcal{T}(\zeta)$. The new cut-off energy $E^*_{\rm cut}$ is called the *running cut-off*. We will bring the renormalization equations to the differential form by inspecting the difference between $\mathcal{T}(\zeta + {\rm d}\zeta)$ and $\mathcal{T}(\zeta)$, in the limit of ${\rm d}\zeta \to 0$. We find

$$\frac{\mathcal{T}(\zeta + {\rm d}\zeta)}{\mathcal{T}(\zeta)} = \exp\left(-\frac{1}{2}\langle\hat{\phi}_{\rm f}^2\rangle_{(\zeta,\,\zeta+{\rm d}\zeta)}\right) \approx 1 - \frac{1}{2}\langle\hat{\phi}_{\rm f}^2\rangle_{(\zeta,\,\zeta+{\rm d}\zeta)}, \tag{6.65}$$

where the averaging is over boson modes with energies in a slice between $E^*_{\rm cut}(\zeta)$ and $E^*_{\rm cut}(\zeta + {\rm d}\zeta)$. Using Eqs. (6.59) and (6.47) we obtain

$$\frac{{\rm d}\mathcal{T}}{{\rm d}\zeta} = -\frac{1}{2}P_1(E^*_{\rm cut})E^*_{\rm cut}\mathcal{T}.$$

For an ohmic environment, $P_1(E) = 2\alpha/E$, and we immediately obtain

$$\frac{{\rm d}\mathcal{T}}{{\rm d}\zeta} = -\alpha\mathcal{T}. \tag{6.66}$$

For super/subohmic dissipation, we cannot proceed in such a direct fashion, since $P_1(E)$, given by Eq. (6.52), still explicitly depends on $E_{\rm cut}$. To get rid of it, we note that $({\rm d}P_1/{\rm d}E)/P_1 = s$ and introduce an auxilliary scale-dependent variable $X(\zeta) \equiv P_1(E^*_{\rm cut})E^*_{\rm cut}$. We obtain two scaling equations that do not depend on cut-offs:

$$\frac{{\rm d}\mathcal{T}}{{\rm d}\zeta} = -X\mathcal{T}, \tag{6.67}$$

$$\frac{{\rm d}X}{{\rm d}\zeta} = -(s+1)X, \tag{6.68}$$

thereby accomplishing the renormalization. While the equations do not depend on the cut-off, the solutions certainly do. The analysis of the solutions reproduces all the qualitative results of this section, including those previously obtained from the explicit solution for $P(E)$ given by Eqs. (6.47), (6.58), and (6.59). To avoid repetition, we do not outline this analysis here: we spare this for the models that defy explicit solutions given in Sections 6.5 and 6.6.

6.3 Tunneling in an electromagnetic environment

In this section, we consider inelastic tunneling in nanostructures. The tunneling is accompanied by energy loss (or possibly energy gain from thermal fluctuations). From a very formal point of view, this is usually caused by electron–electron interaction. However, this point is hardly practical since this is not usually the interaction between nearby electrons. The electric perturbation caused by a tunneling electron quickly propagates through the whole setup and may affect other electrons that are, say, meters apart from the point of tunneling. So in the field of quantum transport it is natural to assume that the interaction takes place between an electron and electromagnetic excitations rather than between two electrons. These excitations have been considered in Section 6.1 in detail, and they form an environment for the tunneling electron.

From a theoretical point of view, we apply the spin-boson model of the previous section to concrete setups [145]. As in the spin-boson model, the essential physics will be incorporated into $P(E)$, the probability of disposing of energy E into the environment.

This section is closely related to Chapter 3, where Coulomb blockade was considered. Basically, we refine the simple theory of interaction described there to encompass inelastic effects. The effect of environment on electron tunneling is sometimes nicknamed *dynamical Coulomb blockade*. First of all, we consider single-electron tunneling, where we will mostly elaborate on a single-junction setup. Next, we concentrate on Josephson circuits, where *Cooper-pair tunneling* comes into play. Another tunneling process in Josephson circuits is *phase tunneling*, where the system jumps between the minima of the Josephson potential separated by the phase difference $\approx 2\pi$. We note a remarkable *duality* between these two types of tunneling.

6.3.1 Single-electron tunneling

A tunnel junction is usually a capacitor whose plates are metallic leads separated by (an almost) insulating tunnel barrier. Let us consider two one-particle states r, l, one on each side of the barrier with energies E_r, E_l (see Section 3.2.7). Let us then consider two states of the whole electron system: state "+" corresponds to r filled with an electron while l is empty, and in state "−" r is empty. The tunneling matrix element T_{rl} mixes these two states. If we forget for the moment about all other states in the system, "+", "−" form a qubit with the following Hamiltonian:

$$\hat{H}_{\mathrm{q}} = \frac{E_r - E_l}{2}\sigma_z + T_{rl}\sigma_x .$$

The transition between "+" and "−" transfers an electron between the leads. Let us now switch to the electric circuit theory description. The capacitor in question is embedded into a circuit that provides a fluctuating voltage difference $\hat{V}$ between the plates of the capacitor. Since the transferred charge is the electron charge e, this contributes to the energy difference between "+" and "−" as $e\hat{V}$, and the full Hamiltonian becomes

$$\hat{H} = \hat{H}_{\mathrm{q}} + \frac{e}{2}\hat{V}\sigma_z + H_{\mathrm{env}},$$

where, as discussed, H_{env} is the collection of boson modes presenting the electrical excitations in the circuit. The Hamiltonian is the same as for the spin-boson model, Eq. (6.38), if we make the associations $(E_r - E_l) \to \varepsilon, T_{rl} \to \mathcal{T}, \hat{V} \to F, e/2 \to q_0$. The transition rate between "+" and "−" is therefore given by Eq. (6.49). The total rate of single-electron tunneling is contributed by all possible pairs of r, l. Since the $P(E)$ is the same for all such pairs, we may use the tunneling Hamiltonian trick described in Eq. (3.32) to arrive at

$$\Gamma_{\mathrm{RL}} = \frac{G_{\mathrm{T}}}{e^2} \int \mathrm{d}E_{\mathrm{R}}\, \mathrm{d}E_{\mathrm{L}}\; f_{\mathrm{R}}(E_{\mathrm{R}})(1 - f(E_{\mathrm{L}}))P(E_{\mathrm{R}} - E_{\mathrm{L}}). \tag{6.69}$$

Let us note a similarity with single-electron tunneling under Coulomb-blockade conditions. An electron present in the right-hand lead (filling factor f_{R}) tunnels to an empty state in the left lead (filling factor $(1 - f_{\mathrm{L}})$) with energy change so that $E_{\mathrm{R}} \neq E_{\mathrm{L}}$. Under Coulomb-blockade conditions, this energy change is the same for all tunnel processes and equals the difference of electrostatic energies $-\Delta E$. The environment provides a random contribution E to this energy change, its probability distribution being given by $P(E)$. Integration over the energies of incoming electrons thus yields

$$\Gamma_{\mathrm{RL}}(\Delta E) = \int \mathrm{d}E\; \Gamma^{(0)}_{\mathrm{RL}}(\Delta E + E)P(E), \tag{6.70}$$

where $\Gamma^{(0)}_{\mathrm{RL}}(\Delta E)$ is the single-electron energy-dependent rate in the absence of coupling to the environment, given by Eq. (3.27):

$$\Gamma^{(0)}_{\mathrm{RL}}(\Delta E) = \frac{G_{\mathrm{T}}}{e^2} \frac{\Delta E}{\exp(\Delta E/k_{\mathrm{B}}T) - 1}.$$

At vanishing temperature, the energy given to the environment is always positive so that the tunneling rate is suppressed.

The time has come to characterize the $P(E)$ for an electromagnetic environment. The susceptibility for the voltage variable $\hat{V}$ is, apart from a factor, the impedance of the circuit seen by the tunneling electron: $\chi(\omega) = \mathrm{i}\omega Z(\omega)$. From the general relation given in Eq. (6.46), with $q_0 = e/2$, we see that

$$P_1(E) = \frac{\Theta(E)}{E} G_{\mathrm{Q}}\, \mathrm{Re}\, Z(E/\hbar), \tag{6.71}$$

at vanishing temperature. At finite temperature,

$$P_1(T, E) = P_1(T = 0, |E|)/(1 - \mathrm{e}^{-E/k_{\mathrm{B}}T}),$$

as mentioned in Section 6.2.5. $P(E)$ is related to $P_1(E)$ in the usual way (Eq. (6.47)).

The environment modifies the single-electron rate according to Eq. (6.69) in any tunnel junction, for example one of the two junctions of a SET transistor. Here we concentrate on a single junction biased by voltage V. There is no Coulomb blockade in a single junction, so the electrostatic energy change is given by $\Delta E = -eV$.

To find the I–V curve of a single tunnel junction, we note that the rate in the opposite direction is that at opposite voltage, $\Gamma_{LR}(V) = -\Gamma_{RL}(-V)$. The current is thus odd in voltage:

$$I = G_T \int dE\, P(E) \left(\frac{E - eV}{\exp\left(\frac{-eV+E}{k_B T}\right) - 1} - \frac{E + eV}{\exp\left(\frac{eV+E}{k_B T}\right) - 1} \right). \tag{6.72}$$

This rather terrifying equation simplifies for vanishing temperatures $k_B T \ll eV$ to the following:

$$eI(V) = G_T \int_0^{eV} dE(eV - E)P(E) \Rightarrow \frac{\partial^2 I}{\partial V^2} = eG_T P(eV). \tag{6.73}$$

To model the environment for any realistic nanostructure, one has to understand at which time and space scale is this environment felt by tunneling electrons. It is clear from the preceding equations that the characteristic scale of energy change is $E_{ch} \simeq \max(eV, k_B T)$. This gives a time scale $\hbar/E_{ch}$. The time scale sets the space scale: the typical distance at which the electricity has propagated given the time scale. One has to remember that the time scale is quite short (shorter than one nanosecond) for any realistic voltage/temperature, so there is *high-frequency* impedance involved in all such models. One could set up a conveniently tunable electromagnetic environment for a tunnel junction by mounting a variable resistor on a desk in a lab and connecting the resistor to the small tunnel junction in the dilution fridge, typically several meters apart. This would not work. The parasitic capacitance of the connecting wires would efficiently shunt this resistor at time scales shorter than one-tenth of a second.

Exercise 6.4. An overlap tunnel junction has been formed by evaporation of metallic films through the mask such that long wires of width 100 nm have been formed. The film resistance per square is 1 Ω, and the capacitance per unit length characterizing the wires is 10 F/cm. The junction is biased at 1 mV. Estimate the time and space scales characterizing the electromagnetic environment. From this, estimate the effective impedance felt by the tunneling electrons.

Having said that, we turn to three most interesting and simple models of electromagnetic environment.

Single-mode environment

Let the electron tunnel between the plates of a capacitor that is a part of the electric oscillator we have considered at the beginning of this chapter (see Fig. 6.8). The frequency-dependent impedance is that of the capacitor and inductor connected in parallel:

$$Z(\omega) = \frac{1}{-i(\omega + i0)C + i/\omega L},$$
$$\mathrm{Re}\, Z(\omega) = \frac{\pi}{2C}\delta(\omega - \omega_0) \qquad \text{at } \omega > 0,$$

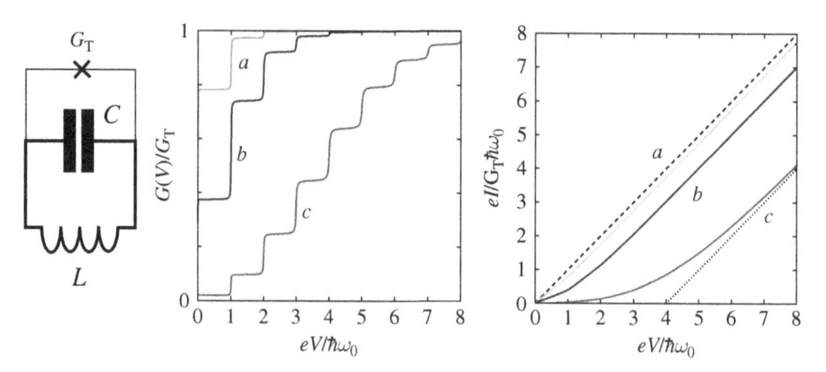

Fig. 6.8. Electron tunneling in a single-mode environment. Curves a, b, and c correspond to $\bar{N} = 0.25, 1$, and 4. The differential conductance shows steps at $eV = n\hbar\omega_0$. The I-V curves show the offset $\bar{N}\hbar\omega_0 \equiv E_c$ at higher voltages. Dashed line: linear I-V curve, $I = G_T V$. Dotted line: the same line off-set for $\bar{N} = 4$.

where $\omega_0 = \sqrt{LC}$. The real part gives a δ-peak at $\omega = \omega_0$, indicating the emission/adsorption from/to a single mode. To make Re Z smoother, we can add a resistor, giving a small width to the δ-peak as was done to produce the plots of Fig. 6.8. The corresponding $P(E)$ consists of a series of δ-peaks at $E_n = n\hbar\omega_0$, their magnitudes being given by the Poisson distribution, Eq. (6.43), with $\bar{N} = \hbar\omega_0/E_C$. Each peak in $P(E)$ gives rise to a step in differential conductance at $eV = n\hbar\omega_0$. At the step, the conductance increases with increasing voltage. This manifests the onset of a new tunneling process: the electron has enough energy to emit n quanta in the course of tunneling. However, the increase is relative. In fact, the conductance at low voltages is suppressed by a factor of $\exp(-\bar{N})$. It takes (infinitely) many step-like increases to reach, at high voltages, the unsuppressed conductance G_T.

The steps in the differential conductance produce cusps in the I–V curve. Visually, these cusps are not very prominent. What is visible on the I–V curves is the voltage *offset* V_{off}: at high voltages, the asymptote $I(V) \approx G_T(V - V_{\text{off}}\,\text{sgn}(V))$ is shifted by V_{off} with respect to the Ohm's law. The value of the offset is naturally given by the average energy loss of a high-energy electron traversing the junction:

$$eV_{\text{off}} = \int \mathrm{d}E\; P(E) = \int \mathrm{d}E\; P_1(E). \tag{6.74}$$

For the single-mode environment, $eV_{\text{off}} = \bar{N}\hbar\omega_0$.

Resistor model

The most straightforward and widely used model of the environment takes into account the capacitance of the junction but disregards the frequency dependence of the parallel impedance: it is represented by a frequency-independent resistor R. The effective environment is of the ohmic type (Eq. (6.53)) with $\alpha = RG_Q$ and cut-off energy $E_{\text{cut}} = \hbar/RC$. The G–V and I–V curves at vanishing temperature $k_B T \ll eV$ (see Figs. 6.9 (a) and (b)) are not unlike the smoothed versions for the single-mode environment with $\bar{N} \simeq \alpha$ (they do not display steps or cusps). The differential conductance is suppressed at low voltages, the suppression being weak/strong for α much smaller/larger than unity. The I–V curves show the voltage offset at sufficiently large voltages. Yet there are differences. At low voltages, the I–V dependence is an anomalous power law, $I \simeq V^{1+\alpha}$. Therefore, the zero-voltage tunneling conductance is zero at any α, even a very small one. For the latter case, the substantial suppression of the conductance is restricted to low voltages and temperatures, $\max(eV, k_B T)/E_C \simeq \exp(-1/\alpha)$. This complicates the experimental observation of the suppression: the tunnel conductors in the *low-resistive* environment $RG_Q \ll 1$ look linear, although in principle they are not.

For I–V curves, the value of the voltage offset is always $V_{off} = E_C/e$, irrespective of R. This is the charging energy of the capacitor charged by the instantaneous electron tunneling. We know that the capacitor will not keep this energy/charge for long: it will discharge via the resistor R at the time scale $\simeq RC$. From the Heisenberg uncertainty relation, we expect that the electrons will feel this energy scale at $E \gg \hbar/RC$. Indeed, for low resistances, the offset achieves its asymptotic value only at $eV \simeq \hbar/RC \gg E_C$. For large resistances, the energies of the order of E_C are already much bigger than $\hbar/RC$, so that the offset is visible at the energies of the order of itself. In the limit $R \to \infty$, the I–V curve is piecewise linear: $I = 0$ for voltages not exceeding e/C and $I = G_T(V - \mathrm{sgn}(V)V_{off})$. This suggests that in this limit (almost) every electron transferring the tunnel junction has to charge the capacitor, very much like it has to do in Coulomb-blockade systems like SETs. This is in agreement with the peak-shaped $P(E)$ at large α.

Let us discuss the temperature dependence of the differential conductance. At large and moderate α, the typical scale of this dependence is $k_B T \simeq E_C$. At $eV < E_C$, the conductance increases with temperature, saturating at G_T. At $eV > E_C$, the conductance exhibits non-monotonous temperature dependence: first it decreases, deviating from its zero-temperature value G_T. This is followed by an increase back to G_T at $k_B T \gg eV$ (Fig. 6.9(c)). At low voltages, $eV \ll V_{off}$, the strongly suppressed conductance shows a dependence at $k_B T \simeq eV$. At small α, the temperature dependence mostly occurs at the scale $k_B T \simeq eV$. Very roughly, the conductance retains its zero-temperature value at $eV > k_B T/3$ and has minimal dependence on voltage at $|eV| < k_B T/3$ (Fig. 6.9(d)).

RC line

An interesting type of environment is felt by a tunnel junction connected to one end of a semi-infinite RC-line made of capacitors and resistors (Fig. 6.10). We evaluate the effective impedance as follows: we know from Eq. (6.23) the Fourier component $Z(q)$ of the RC-line impedance. The junction is connected in a certain point r, so it feels the impedance $Z(r, r' = r) = \int Z(q)\mathrm{d}q/2\pi$. Since the line is semi-infinite, this has to be corrected by a factor of 2 (a semi-infinite line provides one-half of the conductance of the infinite line):

$$Z_{RC}(\omega) = 2\int \frac{\mathrm{d}q}{2\pi} Z(q) = \sqrt{\frac{\tilde{R}}{\mathrm{i}\omega\tilde{C}}}, \tag{6.75}$$

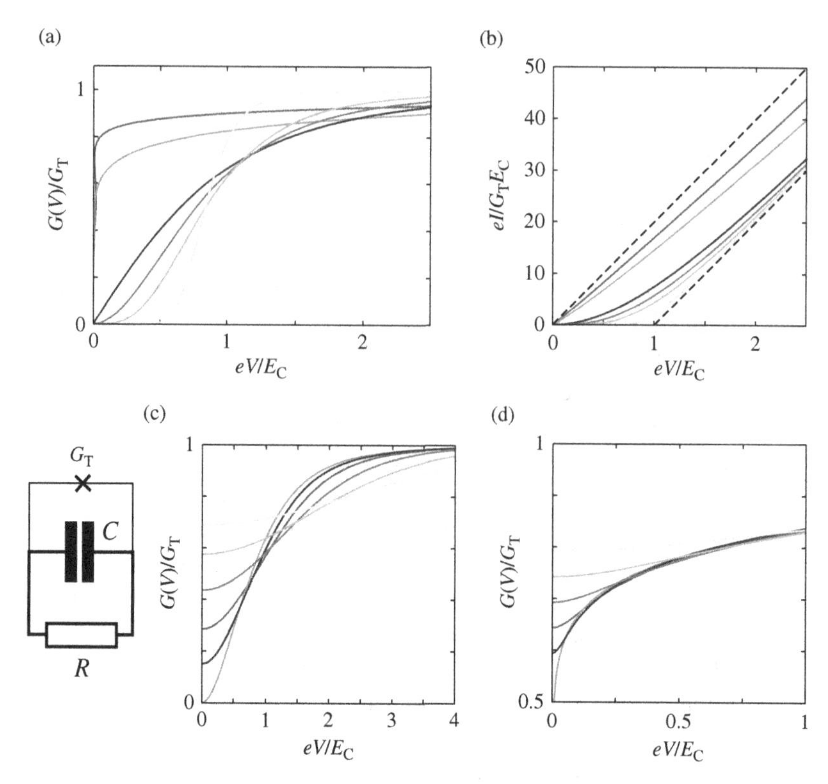

Fig. 6.9. Electron tunneling in an ohmic environment. (a) Differential conductance at vanishing temperature. Upper to lower curves correspond to $\alpha = 0.02, 0.05, 0.5, 1, 2, 20$, at $eV < E_C$. (b) I–V curves for the same set of α values. (c) Temperature dependence of the differential conductance for $\alpha = 1$ and $k_B T/E_C = 0.01, 0.1875, 0.375, 0.75, 1.5, 3$ (from lower to upper curve at $eV < E_C$). (d) The same as in (c) for $\alpha = 0.05$.

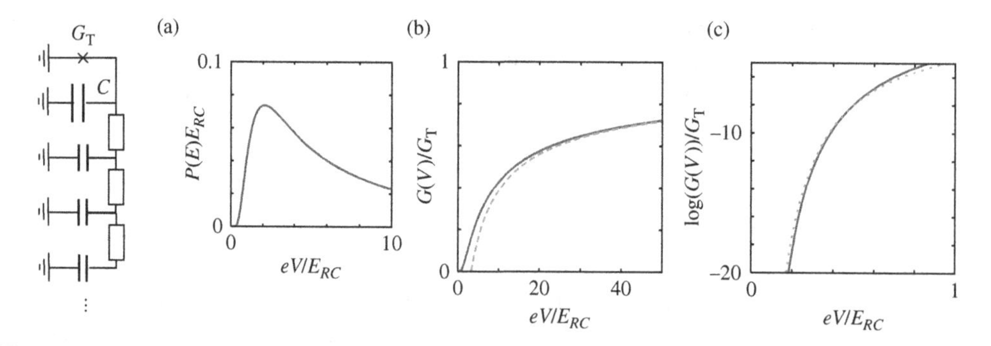

Fig. 6.10. Electron tunneling into RC-line. (a) Energy loss distribution $P(E)$. (b) Differential conductance. Dashed line represents the high-voltage asymptotics. (c) Differential conductance in the region of exponential suppression. Dashed line represents the exponential asymptotics.

$\tilde{R}, \tilde{C}$ being the resistance and capacitance per unit length, respectively. Qualitatively, the electric field penetrates into the line at the length $L(\omega) \simeq \sqrt{\omega \tilde{C} \tilde{R}}$ feeling the impedance $\simeq \tilde{R} L(\omega)$. Using the general considerations of Section 6.2.4, we determine that the RC-line environment is of superohmic type with $s = -3/2$. A natural cut-off energy for $P_1(E)$ is the charging energy E_C associated with the self-capacitance of the wire. We note, however, that we do not need any cut-off: all integrals in Eq. (6.47) converge giving a one-parametric $P(E)$ (Fig. 6.10) as follows:

$$P(E) = \sqrt{\frac{E_{RC}}{E^3}} \exp(-E_{RC}/E), \tag{6.76}$$

with the characteristic energy scale $E_{RC} = (e^2/4\pi\tilde{C})\tilde{R}G_Q$, E_{RC} being the charging energy of the piece of the wire with resistance $2\pi G_Q^{-1}$ of the order of inverse conductance quantum. This form of $P(E)$ is valid, provided the natural cut-off does not matter, $\max(E, E_{RC}) \ll E_C$. It combines a relatively narrow region of exponential suppression at low energies with a long slow-falling tail at higher energies. This leads to a difference of scales that is highly unusual for one-parametric behavior. For instance, from the plots in Figs. 6.10 (b) and (c), we see that the differential conductance is exponentially suppressed for $E < E_{RC}$. However, it still noticeably deviates from G_T at energies that are two orders of magnitude higher.

Small corrections

In all the examples given above, the effect of environment may be strong, and this requires the high-frequency impedance to be at least of the order of G_Q^{-1}. As mentioned, this is by no means a typical situation. If one does not make any special attempts to create a high-impedance environment, one ends up at the scale of so-called vacuum impedance, or wave resistance of a vacuum, characterizing emission of electromagnetic irradiation to outer space, $Z_0 = 1/(\epsilon_0 c) \approx 377\ \Omega$, c being the speed of light. Geometric factors usually suppress the impedance to a fraction of Z_0: for instance, the wave impedance of a TV cable is 50 Ω. This scale is much smaller than G_Q: $Z_0 G_Q/2 \approx 0.015$. Therefore, in a typical situation, the interaction with the electromagnetic environment provides only small (10^{-2}) corrections to linear I–V curves. Expanding $P(E)$ as in Eq. (6.48) and using Eq. (6.72), we evaluate the correction to the differential conductance up to the first power in the environmental impedance:

$$\delta G_T(V)/G_T = \int_0^\infty \frac{d\omega}{\omega} W\left(\frac{\hbar\omega}{k_B T}, \frac{eV}{k_B T}\right) \text{Re } Z(\omega) G_Q;$$
$$W(v, w) \equiv \sum_\pm \pm \frac{1}{e^{-v\pm w} - 1}\left(1 - \frac{v \mp w}{1 - e^{-v\pm w}}\right). \tag{6.77}$$

The correction is additive in $Z(\omega)$. The modes at a given frequency ω contribute to the correction with a weight function W (see Fig. 6.11). At vanishing temperature $k_B \ll \hbar\omega$, this weight function is simply a step function, given by

$$W = -\Theta(|eV| - \hbar\omega),$$

with the correction vanishing at $|eV| > \hbar\omega$. It is instructive to present the correction as a sum of elastic and inelastic parts. The elastic part can be attributed to the renormalization

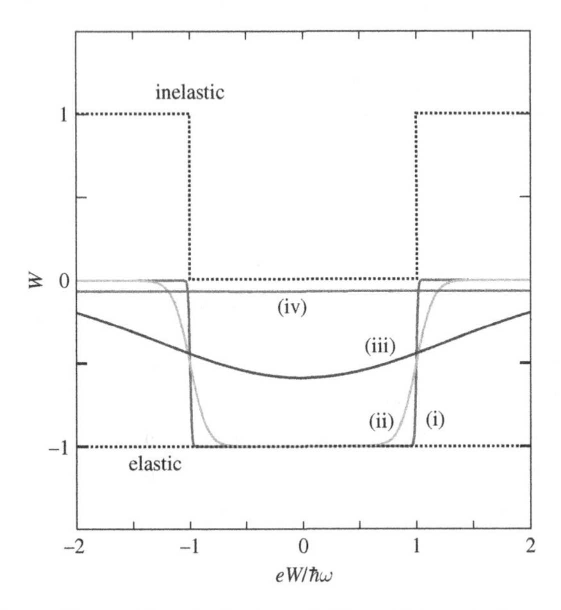

Fig. 6.11. **Weight function for small corrections to the tunnel differential conductance. Curves (i)–(iv) correspond to $k_BT/\hbar\omega = 0.005, 0.05, 0.5, 5$, respectively. Dotted lines represent inelastic and elastic parts of the correction at vanishing temperature.**

of the tunneling amplitudes and therefore does not depend on V. As discussed, the elastic conductance is suppressed and the elastic part is therefore negative. The inelastic part corresponds to the opening of a new transport channel: electron tunneling with emission of the quantum $\hbar\omega$, so it can only take place at $|eV| > \hbar\omega$ and is positive there. As we see, in the case considered the two parts cancel each other at $eV > \hbar\omega$ (Fig. 6.11). Correspondingly, at vanishing temperature the correction to the conductance is given by

$$\frac{\delta G_{\mathrm{T}}(V)}{G_{\mathrm{T}}} = -G_{\mathrm{Q}} \int_{eV/\hbar}^{\infty} \mathrm{Re}\, Z(\omega) \frac{\mathrm{d}\omega}{\omega}; \tag{6.78}$$

$$\frac{\partial(\delta G_{\mathrm{T}})}{\partial V G_{\mathrm{T}}} = G_{\mathrm{Q}} \frac{\mathrm{Re}\, Z(eV/\hbar)}{V}. \tag{6.79}$$

6.3.2 Cooper-pair tunneling

Let us turn to a single Josephson junction embedded into an electromagnetic environment. As discussed in previous chapters, the coherent flow of Cooper pairs in the junction is not accompanied by energy dissipation and cannot be separated into elementary tunneling events. Here we concentrate on the opposite situation: the Cooper pairs traverse the junction one by one, and each separate tunneling event dissipates some energy into the environment.

To see how and when it is possible, we consider a Josephson junction biased by voltage V and we inspect its Hamiltonian in charge representation, N being the number of Cooper pairs transferred:

$$\hat{H} = \frac{E_J}{2}\sum_N (|N+1\rangle\langle N| + |N\rangle\langle N+1|) + 2e(V+\tilde{V})\sum_N N|N\rangle\langle N| + \hat{H}_{\text{env}}. \tag{6.80}$$

Here, we have also added the terms describing the environment (H_{env}) and the coupling to it. The latter term comes with an environment-induced fluctuating voltage drop over the junction, $\tilde{V}$, that is linear in the environment boson operators. This form makes it evident that the junction can be found in the states of the set $|N\rangle$. If all $|N\rangle$ have the same energy, the wave function is spread uniformly over the whole set of N. The dissipationless supercurrent takes place. The applied voltage V shifts the neighboring states by $2eV$. The N states now form a ladder in energy space. Owing to the energy mismatch between the neighboring states, the junction localized in a state with a well defined N so that no net supercurrent is flowing through it. The environment helps the current: the junction may tunnel from $|N\rangle$ to $|N-1\rangle$, emitting $2eV$ to the environment (assuming $V > 0$). A Cooper pair is thereby transferred through the junction in the direction given by the voltage (down the ladder). In the presence of thermal fluctuations, the energy may also be absorbed from the environment. This results in a Cooper-pair transfer in the opposite direction (up the ladder).

If we concentrate on two neighboring states, we immediately recognize the qubit shake-up setup of Section 6.2.2 with the tunneling amplitude given by $E_J/2$. The rates of *Cooper-pair tunneling* are thus given by Eq. (6.49):

$$\Gamma_{\pm} = \frac{\pi E_J^2}{2\hbar} P(\pm 2eV), \tag{6.81}$$

and the current is therefore given by

$$I = 2e(\Gamma_+ - \Gamma_-) = \frac{eE_J^2}{\hbar}(P(2eV) - P(-2eV)). \tag{6.82}$$

$P(E)$ is determined by the impedance of the environment. Since the Cooper pairs bear the charge $2e$, the corresponding one-boson probability, given by

$$P_1(E, T=0) = \frac{\Theta(E)}{E} 4G_Q \,\text{Re}\, Z(E/\hbar), \tag{6.83}$$

is four times bigger than that for single-electron tunneling, Eq. (6.71). In distinction from the single-electron tunneling, the tunneling of Cooper pairs is immediately related to the energy-loss function $P(E)$ of an environment. The plots showing $P(E)$ for various environments (Figs. 6.4, 6.5, 6.6, and 6.7) give us, upon rescaling, the I–V curves of the voltage-biased Josephson junction.

Let us now understand the parameter region where the Cooper-pair tunneling takes place. We start with ohmic dissipation. In this case, $\hbar I/e \simeq (E_J^2/eV)(eV/E_{\text{cut}})^{2\alpha}$ ($\alpha = 2G_Q R$ for Cooper-pair tunneling). We can see the anomalous power law involved as the result of the energy-dependent renormalization of the tunneling amplitude, $E_J^* \simeq E_J(E/E_{cut})^\alpha$. The tunneling rate in this notation reads $\Gamma \simeq E_J^{*2}/\hbar eV$. The rate estimated should be compared with the renormalized amplitude E_J^*. The latter determines the maximum coherent current, $I_c \simeq eE_J/\hbar$, through the junction. The Cooper-pair tunneling rate cannot exceed this quantity since, in this case, the tunnel events would occur too quickly to be regarded as independent ones. So, if our estimate for Γ exceeds E_J^*, the tunneling picture fails; this signals the coherent regime. If $\Gamma \ll E_J^*$, we are in the Cooper-pair tunneling regime. This implies $E_J^* \ll eV$.

If the environment is sufficiently resistive, $\alpha > 1$, $E_J^*(E = 2eV)/eV$ decreases with decreasing voltage. Therefore, the junction is always in the Cooper-pair tunneling regime. The coherent supercurrent in the junction is destroyed by the voltage fluctuations coming from the environment. The junction is localized in charge space. If $\alpha < 1$, the situation is reversed: E_J^*/eV increases with decreasing voltage. At sufficiently small voltage $eV_c \simeq E_J^*(eV_c)$, the junction enters the coherent regime characterized by a supercurrent I_c^*. The junction is delocalized in charge space. In terms of E_{cut} and E_J, $eV_c \simeq I_c^*\hbar/e \simeq E_J(E_J/E_{cut})^{\alpha/(1-\alpha)}$. We will see below how this supercurrent is disrupted by the *phase* tunneling. Therefore, at $\alpha = 1$, the Josephson junction undergoes a transition between the localized and delocalized state: this is the Schmid transition (see Section 6.2.4).

Control question 6.4. Explain the difference between the Schmid transition for a qubit and for a Josephson junction.

Even for a typical, low-impedance, environment ($\alpha \ll 1$), at sufficiently high temperature the voltage fluctuations may become strong enough to destroy the supercurrent and provide Cooper-pair tunneling. The relevant energy scale in this case is much less than $k_B T$, so that $P(E)$ assumes the Lorentzian shape given by Eq. (6.62). The corresponding rates are the same in both directions:

$$\Gamma_+ = \Gamma_- = \frac{E_J^2}{2\hbar} \frac{w}{w^2 + (2eV)^2}; \quad w \equiv 2\pi\alpha k_B T \ll k_B T.$$

As mentioned, the Cooper-pair tunneling regime takes place only if $\Gamma \ll E_J/\hbar$, so we imply that $E_J \ll w$. Since the rates are the same in both directions, there is no net current in this approximation. To get the current, let us note that, by virtue of the detailed balance condition in Eq. (6.61), $P(E) - P(-E) \approx (P(E) + P(-E))(E/k_B T)$ at $E \ll k_B T$. Therefore, the current is reduced by a small factor $\simeq eV/k_B T$ in comparison with $2e\Gamma_\pm$:

$$I = 2e\frac{4eV}{k_B T}\Gamma_\pm = \frac{4E_J^2}{\hbar} \frac{2\pi\alpha V}{w^2 + (2eV)^2}.$$

In the opposite case of a high-impedance environment, one has to go to higher voltages to reach the significant current. The current peaks near the resonant voltage $V_0 \equiv e/C$, where the energy gain from voltage, $2eV$, matches the charging energy of the junction, $(2e)^2/2C = 4E_C$. The shape of the peak is Gaussian with voltage width V_f,

$$I(V) = \frac{E_J^2}{2\sqrt{2\pi} V_f \hbar} \exp\left(-\frac{(V - V_0)^2}{2V_f^2}\right),$$

conforming to the shape of P(E). This shape of the current resonance indicates that the position of the resonance is simply shifted by the quasi-stationary voltage fluctuations. Their normal distribution gives the Gaussian shape, and the square of the width is just the variance of the fluctuations. Indeed, the time scale of the fluctuation is RC-time, $\simeq\alpha\hbar/E_C$, which is much larger than the time scale $\hbar/E_C$ corresponding to the resonant frequency. At low temperatures, $k_B T \ll \hbar RC$, the voltage fluctuations are quantum. They are thermal in the opposite limit. As discussed, the applicability of the Cooper-pair tunneling picture requires $E_J \ll eV_f$.

Exercise 6.5. Compute V_f in both limits, making use of Eq. (6.9).

The name "Cooper-pair tunneling" suggests the quantum nature of the effect. It is not always so: it is worth noting that Cooper-pair tunneling has a simple classical analog in the limit of small impedances, Re $ZG_Q \ll 1$ and voltages exceeding w/e. If in the first approximation we disregard the effect of the environment, the voltage bias of a Josephson junction gives rise to a linear sweep of the phase difference, $\varphi = \varphi_0 + 2eVt/\hbar$. This results in an ac current through the junction, $I(t) = I_c \sin(2eVt/\hbar + \varphi_0)$, which oscillates at the Josephson frequency $2eV/\hbar$. The ac current dissipates in the environment, the dissipation rate $dE/dt = I_c^2 \text{Re}(Z(2eV/\hbar))/2$ being proportional to the real part of the impedance at the frequency. Since the dissipated energy must be supplied by the circuit, we conclude that a small dc current must be running such that $I_{dc}V = dE/dt$. Therefore we obtain

$$I_{dc}(V) = \frac{I_c^2}{2V}\text{Re}(Z(2eV/\hbar)),$$

which coincides with the "quantum" relation in Eq. (6.82) in the relevant limit $P_1(E) \approx P(E)$.

6.3.3 Phase tunneling

Let us consider the Josephson junction from a complementary viewpoint and recall the washboard potential discussed in Section 3.5. If $E_J \gg E_C$, the quantum state of the junction is expected to be well localized in phase in one of the equidistant minima of the washboard potential. As discussed in Section 3.5.5, the system can coherently tunnel between the neighboring minima. This tunneling is characterized by the (exponentially small) amplitude Δ_0, $\Delta_0 = 8\sqrt{E_J\hbar\omega_p/\pi}\exp(-8E_J/\hbar\omega_p)$. Here, $\omega_p = \sqrt{8E_J E_C}/\hbar$ is the so-called plasmon frequency – the oscillation frequency near the bottom of each well, such that $\hbar\omega_p$ gives the first excited state in the well. The bias current I tilts the potential so that the neighboring minima are separated by the energy difference $\pi\hbar I/e$. Now the tunneling can only proceed by emitting/absorbing this energy from the environment. In distinction from Section 3.5, we concentrate here on small tilts ($I/e \ll \omega_p$). The Hamiltonian describing the situation reads as follows:

$$\hat{H} = \frac{\Delta_0}{2} \sum_N (|N+1\rangle\langle N| + |N\rangle\langle N+1|$$
$$+ \frac{\pi\hbar}{e}(I + \tilde{I}) \sum_N N|N\rangle\langle N| + \hat{H}_{\text{env}}. \tag{6.84}$$

Here, N numbers the minima of the washboard potential and $\tilde{I}$ represents the current fluctuations due to the environment.

The best way to proceed is to note the formal equivalence of the above Hamiltonian and that of the Cooper-pair tunneling (Eq. (6.80)). One switches between the Hamiltonians just by switching notation:

$$E_J \Leftrightarrow \Delta_0; \ \frac{\pi\hbar}{e}(I, \tilde{I}) \Leftrightarrow 2e(V, \tilde{V}).$$

This is the manifestation of the *charge–phase* duality: we have encountered this in Section 3.5 when we compared two coherent regimes where charge or phase were delocalized. Now we observe the same duality for incoherent tunneling of charge or phase in the presence of the environment. While the voltage fluctuations $\tilde{V}$ are proportional to the environment impedance, the current fluctuations are proportional to the admittance $Y(\omega) \equiv Z^{-1}(\omega)$. The duality relation is given by

$$2G_Q Z(\omega) \Leftrightarrow Y(\omega)/2G_Q. \tag{6.85}$$

The duality thus relates the phase tunneling in a low (high) impedance environment to the Cooper-pair tunneling in a high- (low-) impedance environment. Using the above duality relations, we immediately obtain the V–I curve of the Josephson junction in the phase tunneling regime: we just rewrite Eqs. (6.82) and (6.83) to obtain the voltage over the junction,

$$V = \frac{\pi\Delta_0^2}{2e}\left(P(\pi\hbar I/e) - P(-\pi\hbar I/e)\right), \tag{6.86}$$

where $P(E)$ is determined by the impedance of the environment with the zero-temperature one-boson probability, given by

$$P_1(E, T=0) = \frac{\Theta(E)}{E} \frac{\text{Re}\, Y(E/\hbar)}{G_Q}. \tag{6.87}$$

Control question 6.5. How many replacements have been made to obtain Eqs. (6.86) and (6.87)?

So we do not have to discuss the V–I curves in the phase tunneling regime in much detail: they are similar to the V–I curves in the Cooper-pair tunneling regime in the corresponding environment. We will only consider ohmic environment in the limit of vanishing temperature and small voltages/currents. We have seen that in this limit for $\alpha > 1$, Cooper-pair tunneling takes place and $I \simeq V^{2\alpha-1}$. What happens at $\alpha < 1$? By virtue of the duality relation, Eq. (6.85), $\alpha > 1$ is mapped onto $\alpha < 1$ if charge and phase are exchanged. We conclude that at $\alpha < 1$ the transport is dominated by the phase tunneling. The $P(E)$ for phase tunneling is proportional to $E^{2/\alpha-1}$; this gives $V \simeq I^{2/\alpha-1}$. Reverting this relation, we obtain $I \simeq V^{\alpha/(2-\alpha)}$. The current is thus increasing with increasing voltage, with the power exponent <1. At the Schmid transition, both Cooper-pair and phase tunneling result in $I \simeq V$. These two kinds of environment-assisted incoherent tunneling take place in the limit of small and large E_J, respectively.

The duality of the Josephson junction, though a powerful tool to analyze the limiting cases, is not exact. The point is that the region of validity of the Hamiltonian, Eq. (6.84), is limited by low energies $\ll\hbar\omega_p$: indeed, it disregards excited states in each washboard potential well. Its dual counterpart, Eq. (6.80), does not suffer from such limitations. There is another system that is described by the Hamiltonian in Eq. (6.84) in a much wider energy region. This is a thin superconducting wire. The discrete variable N in this case yields the number of windings of the superconducting phase along the wire. The thin wire can be approximated by or realized as a chain of a large number of Josephson junctions considered in Section 3.6.2, the discrete variable being the number of vortices enclosed by the chain. The transitions between neighboring N are commonly called *phase slips*. Therefore, the above duality substitutions may be used to relate the physics of the superconducting wire and a single Josephson junction in a wider energy interval [146].

6.4 Electrons moving in an environment

In Section 6.3, we have given a simple theory of electron tunneling in the presence of an electromagnetic environment. The basic assumption was that the tunneling transitions between all electron states eventually feel the same environment and are characterized with the same $P(E)$ function. This is certainly true for a circuit-like environment model where all electrons tunneling to a certain electrode feel the same (fluctuating) potential: that of the electrode. The physical foundation for such a model is (as discussed in Section 6.1.2) that the electrons in a metal usually propagate much slower than the electricity. This is why the electrons near a tunnel junction all feel the potential that is constant in space.

In this section, we will concentrate on the setups where electrons can take over the electromagnetic field propagating in a medium and/or move with comparable speed [147]. In this case, the interaction between environment bosons and electrons assumes the following general form:

$$\hat{H}_{\text{int}} = \sum_q U^{(q)}_{k,k'} \hat{a}^\dagger_k \hat{a}_{k'} \hat{b}^\dagger_q + \text{h.c.}, \tag{6.88}$$

where q labels the electromagnetic modes and k, k' label the electron states. The previously used environment model corresponds to the diagonal $U^{(q)}_{k,k'} = \delta_{kk'} U^{(q)}_k$, with $U^{(q)}_k$ being equal for all states k of the same lead.

Control question 6.6. How is this specific form of interaction related to the assumption of the constant-in-space potential?

The presence of non-diagonal terms in Eq. (6.88) indicates that the electromagnetic environment may cause transitions between the states k, k' of the same lead that do not involve any tunneling. These terms are responsible for electron energy relaxation and decoherence, which we consider in Section 6.8. Here, we concentrate on the effect of the environment on the electron tunneling and demonstrate that in these setups it can still be described by an *effective* impedance and the associated energy loss function $P(E)$. The effective impedance incorporates the distributed impedance $Z(\omega; \boldsymbol{r}, \boldsymbol{r}')$ in the medium (see Section 6.1.2) and the electron propagation, and in principle depends on the electron states before and after tunneling. The effective "energy loss" function $P(E)$ does not have to be positive as previously: it appears to describe both inelastic and elastic tunneling and cannot be directly associated with a positive probability of an inelastic process.

The concept of effective impedance glues together the dissipative quantum mechanics and several important topics of quantum transport, such as Luttinger liquids, Altshuler–Aronov corrections, and escaping the Coulomb blockade, traditionally treated as separate subjects. We address these topics in the following subsections.

To understand the concept of effective impedance, let us restrict ourselves to the first non-vanishing (second) order in perturbation theory in $U^{(q)}$. Second-order processes involve a (virtual) emission/absorption of an environmental boson in the course of tunneling and give the corrections to the elastic tunneling rate in the "absence" of the environment. The corrections are independently contributed to by all boson modes, so it suffices to look at a single mode with energy $\hbar\omega_0$. For simplicity we assume vanishing temperature and that the interaction with the environment takes place in the left lead only. An electron tunnels from a certain state filled state r in the right-hand lead to states k in the left; this tunneling is described by the tunneling Hamiltonian $\hat{H}_{\text{T}} = \sum_k T_k \hat{a}_k^\dagger \hat{a}_r$ (see Section 3.2.7). Let us first concentrate on the *inelastic* correction. The state after the tunneling comprises an emitted boson and an electron on the left in the state k, $E_r = E_k + \hbar\omega_0$. The intermediate virtual state always involves another electron state k_1. If k_1 is empty in the initial state, it is filled in the virtual state reached by the electron transfer from r to k'. If k_1 is filled in the initial state, the virtual state is reached by the emission of the boson, which also promotes an electron from k_1 to k (see Fig. 6.12). The resulting amplitude is contributed to by both virtual states and does not depend on the filling of k_1:

$$Am = \sum_{k_1} T_{k_1} \frac{1}{E_k + \hbar\omega_0} U_{k,k_1}. \tag{6.89}$$

Squaring the amplitude and summing up over k, we produce the correction to the tunneling rate from state r involving the sum over three states k, k_1, and k_2 as follows:

$$(\delta\Gamma_r)_{\text{in}} = \frac{2\pi}{\hbar} \sum_{k,k_{1,2}} \frac{T_{k_1} T_{k_2}^* U_{k,k_1} U_{k,k_2}^* \delta(E_r - E_k - \hbar\omega_0)\Theta(E_k)}{(E_k + \hbar\omega_0 - E_{k_1})(E_k + \hbar\omega_0 - E_{k_2})}. \tag{6.90}$$

In our search for the effective impedance, we look at the following function:

$$\pi X(\epsilon) = \frac{1}{E_k + \epsilon + \text{i}0 - E_{k_2}} \frac{U_{k,k_1} U_{k,k_2}^*}{\epsilon - \hbar\omega_0 + \text{i}0} \frac{1}{E_k + \epsilon + \text{i}0 - E_{k_1}}.$$

The imaginary part is related to dissipation and consists of three δ-functions concentrated at $\epsilon = \hbar\omega_0$, $\epsilon = E_{k_2} - E_k$, and $\epsilon = E_{k_1} - E_k$. The imaginary part defines an effective one-boson energy "loss" function $P_1(\epsilon)$, given by

$$P_1(\epsilon) = \text{Im } X(\epsilon) = P_1^{(\text{in})} + P_1^{(\text{el})},$$
$$P_1^{(\text{in})} \propto \delta(\epsilon - \hbar\omega);\ P_1^{(\text{el})} \propto \delta(E_k + \epsilon - E_{k_{1,2}}),$$

that, similarly to Eqs. (6.48) and (6.50), Eq. (6.77) determines the overall correction to the rate given by

$$\delta\Gamma_r = \frac{2\pi}{\hbar}\sum_k T_{k_1}T_{k_2}^* \left(\int_0^\infty \delta(E_k + \epsilon - E_r)\mathrm{d}\epsilon\, P_1(\epsilon) - \delta(E_k - E_r)\int_0^\infty \mathrm{d}\epsilon\, P_1(\epsilon) \right).$$

The term $P_1^{(\text{in})}$ in P_1 reproduces the derived inelastic correction given by Eq. (6.90) and can be readily associated with the probability of energy loss $\hbar\omega_0$ as before. However, there are other terms, $P^{(\text{el})}$, that are specific for the present situation. The energy loss for these terms correspond to the transitions between the electron states k and $k_{1,2}$. However, this energy loss is compensated by the equal energy mismatch between states r and k. Therefore, the terms give the *elastic* transitions between states r and k_1. This mixing between boson and electron transitions, and correspondingly between the inelastic and elastic processes, is typical for the situation considered and is related to the mixing of two excitations: bosons and electron–hole pairs. Therefore, the effective $P_1(\epsilon)$ is not immediately related to a positive probability of the energy loss, and in fact does not have to be positive, as is clear from Eq. (6.91). The effective impedance $Z_{\text{eff}}(\omega)$ is related to $P_1(E)$ in the usual way (see Eq. (6.71)). In general, the impedance depends on too many details, such as the states $k_{1,2}$ involved, to be of practical value.

The effective impedance does become useful in the semi-classical picture of tunneling outlined in the following. Let us consider electrons propagating in a medium (generally with scattering) that consists of two leads separated by a tunnel barrier (see Fig. 6.12(a)). The overall tunneling rate is composed of the products of individual amplitudes. Each individual amplitude corresponds to tunneling at a given point $\boldsymbol{R}$ at the barrier. This tunneling is between the electron states *coming* into the vicinity of point $\boldsymbol{R}$ and *leaving* it at certain velocity directions in the left and in the right lead, $\boldsymbol{v}_\text{c}^{(\text{L,R})}$, $\boldsymbol{v}_1^{(\text{L,R})}$. For brevity, let us mark an individual amplitude of this kind with a composite index j comprising $\boldsymbol{R}$, $\boldsymbol{v}_{\text{c},1}^{(\text{L,R})}$. Let us note that fixing the initial (final) velocity at a given point determines an electron trajectory going from (coming to) the point. This is why there are two trajectories associated with each index j, one in the left lead $\boldsymbol{r}_j^{(\text{L})}(t)$, and one in the right one, $\boldsymbol{r}_j^{(\text{R})}(t)$. The trajectories hit the barrier at the same point at $t = 0$, $r_j^{(\text{R,L})}(t = 0) = \boldsymbol{R}$. If an electron tunnels from the left to the right lead at the time moment τ, its coordinate at the time moment t is given by either $\boldsymbol{r}_j^{(\text{L})}(t - \tau)$ $(t < \tau$, before the tunneling) or $\boldsymbol{r}_j^{(\text{R})}(t - \tau)$ $(t > \tau$, after the tunneling).

Generally, the total tunneling rate is contributed to by products of all individual amplitudes at different time moments:

$$\Gamma \propto \sum_{j,j'} \int \mathrm{d}t_1\, dt_2(\cdots)\, T_j(t_1)T_{j'}^*(t_2).$$

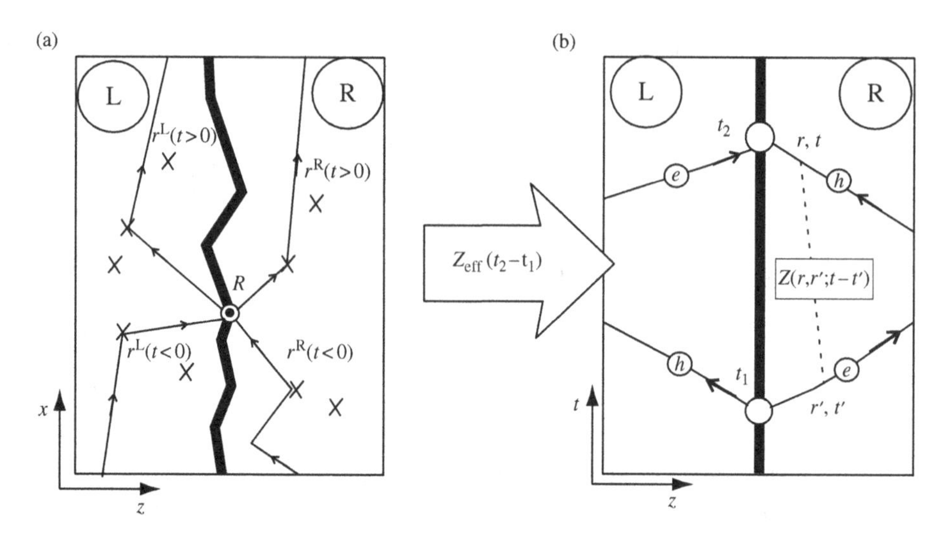

Fig. 6.12. **Effective impedance incorporates information about the electron motion and the extended impedance in the medium. (a) Electron trajectories in the left and right leads for the tunneling event taking place at position *R*. Thick black line denotes the tunnel barrier. (b) Time–space diagram for the effective impedance $Z_{\rm eff}(t_2 - t_1)$.**

In the semi-classical scheme, the interference between different j is disregarded: the semi-classical approximation requires that the electron crosses the barrier at a certain point $\boldsymbol{R}$ with certain velocity directions before and after tunneling. The total rate is thus a sum of partial rates, Γ_j. Each partial rate is characterized by the corresponding amplitude T_j, and can be considered separately from all other partial rates.

The semi-classical approximation does not require that the time moments t_1, t_2 at which the barrier is crossed are the same. The resulting partial rate thus comes about from the interference of two tunneling amplitudes T_j at the time moments $t_{1,2}$ integrated over all possible $t_{1,2}$. This allows us to incorporate the time-dependent phase shifts of the tunneling amplitudes, $T_j \to T_j e^{i\phi_j(\tau)}$, and define the effective energy loss function $P_j(E) = \int dt e^{iEt/\hbar} \langle e^{i\phi_j(\tau)} e^{i\phi_j(\tau+t)} \rangle$, in very much the same manner as we did in Section 6.2.5 (see Eqs. (6.57) and (6.58)). The $P_j(E)$ depends now on the index j, and can, in principle, be different for different tunneling points $\boldsymbol{R}_j$ and/or velocities. The correlator of ϕ_j is determined by the response function which (apart from the frequency factors, see Eq. (6.59)) coincides with the effective impedance sought.

To obtain this function, we concentrate on electron tunneling from the left to the right lead. We note that such tunneling at point $\boldsymbol{R}_j$ at time moment t can be described in two equivalent ways. (i) The creation of an electron on the right and a hole on the left. In this case, we have no excitation before the moment t. At a later time, both quasiparticles follow the outgoing trajectories $\boldsymbol{r}_j^{(\rm L,R)}$ in the corresponding leads. (ii) The annihilation of an electron coming from the left and a hole coming from the right. In this case, both quasiparticles follow incoming trajectories and there are no excitations after the moment t. We choose to present the tunneling at the later time moment t_2 as an annihilation and the tunneling at the earlier time moment $t_1 < t_2$ as a creation.

Now we can evaluate the phase shift at the later time moment as a reaction to the tunneling at the earlier time moment. The created electron and hole produce a time- and coordinate-dependent potential $\Phi(t, \boldsymbol{r})$. Since the electron and hole move along trajectories $\boldsymbol{r}_j^{(\mathrm{R,L})}(t)$, their total charge is given by $q_1(t, \boldsymbol{r}) = e(\delta(\boldsymbol{r} - \boldsymbol{r}_j^{(\mathrm{R})}(t - t_1)) - \delta(\boldsymbol{r} - \boldsymbol{r}_j^{(\mathrm{L})}(t - t_1)))\Theta(t - t_1)$ and gives rise to the following potential:

$$
\begin{aligned}
\Phi(t, \boldsymbol{r}) &= \int \mathrm{d}t' \mathrm{d}\boldsymbol{r}' Z(t - t'; \boldsymbol{r}, \boldsymbol{r}') q_1(t', \boldsymbol{r}') \\
&= e(Z(t - t'; \boldsymbol{r}, \boldsymbol{r}_j^{(\mathrm{R})}(t' - t_1)) - Z(t - t'; \boldsymbol{r}, \boldsymbol{r}_j^{(\mathrm{L})}(t' - t_1))). \qquad (6.91)
\end{aligned}
$$

The phase shift for the tunneling at time moment t_2 is accumulated by the corresponding electron and hole that move along the incoming trajectories in the potential created. It is given by

$$
\begin{aligned}
\phi_j(t) &= \frac{e}{\hbar} \int_{-\infty}^{t_2} \mathrm{d}t (\Phi(t, \boldsymbol{r}_j^{(\mathrm{L})}(t - t_2) - \Phi(t, \boldsymbol{r}_j^{(\mathrm{R})}(t - t_2)) \\
&= \frac{e}{\hbar} \int_{-\infty}^{t_2} \mathrm{d}t \; \mathrm{d}\boldsymbol{r} \; \Phi(t, \boldsymbol{r}) q_2(\boldsymbol{r}, t), \qquad (6.92)
\end{aligned}
$$

with $q_2 = e(\delta(\boldsymbol{r} - \boldsymbol{r}_j^{(\mathrm{L})}(t - t_2)) - \delta(\boldsymbol{r} - \boldsymbol{r}_j^{(\mathrm{R})}(t - t_2)))$. Combining Eqs. (6.91) and (6.92), we obtain the following response function:

$$
\chi(t_2 - t_1) = \int_{t_1}^{t_2} \mathrm{d}t \int_{t_1}^{t_2} \mathrm{d}t' \int \mathrm{d}\boldsymbol{r} \; \mathrm{d}\boldsymbol{r}' q_2(t, \boldsymbol{r}) q_1(t', \boldsymbol{r}') Z(t - t', \boldsymbol{r}, \boldsymbol{r}').
$$

Switching to the frequency representation, we arrive at a compact formula to be used further:

$$
Z_{\mathrm{eff}}(\omega) = \omega^2 \int \mathrm{d}\boldsymbol{r} \; \mathrm{d}\boldsymbol{r}' \mathcal{P}_{\mathrm{A}}(-\omega, \boldsymbol{r}) Z(\omega, \boldsymbol{r}, \boldsymbol{r}') \mathcal{P}_{\mathrm{R}}(\omega, \boldsymbol{r}'). \qquad (6.93)
$$

Here, the *retarded* and *advanced* particle propagators, $\mathcal{P}_{\mathrm{R,A}}$ are defined as follows:

$$
\mathcal{P}_{\mathrm{R,A}} = \int \mathrm{d}t \; \mathrm{e}^{\mathrm{i}\omega t} \Theta(\pm t)(\delta(\boldsymbol{r} - \boldsymbol{r}_j^{(\mathrm{R})}(t)) - \delta(\boldsymbol{r} - \boldsymbol{r}_j^{(\mathrm{L})}(t))), \qquad (6.94)
$$

incorporating the electron motion along the trajectories $r_j^{(\mathrm{L,R})}$. We stress the correspondence of $\mathcal{P}_{\mathrm{R,A}}$ with the electron propagator $\mathcal{P}(t; \boldsymbol{r}, \boldsymbol{r}')$ that has been heavily used in Section 6.1.2. As a reminder, the propagator yields the probability of finding an electron at point $\boldsymbol{r}$ provided it was at point $\boldsymbol{r}'$ at time 0. Since we are considering the rate in the lowest order in tunneling amplitudes, we should neglect the tunneling in the propagator: the electron stays in the left/right lead if it is initially in the left/right lead. Then the correspondence is given by

$$
\mathcal{P}_{\mathrm{R}}(\omega, \boldsymbol{r}) = \mathcal{P}^{(\mathrm{R})}(\omega; \boldsymbol{r}, \boldsymbol{R}_j) - \mathcal{P}^{(\mathrm{L})}(\omega; \boldsymbol{r}, \boldsymbol{R}_j), \qquad (6.95)
$$

$$
\mathcal{P}_{\mathrm{A}}(\omega, \boldsymbol{r}) = \mathcal{P}^{(\mathrm{R})}(-\omega; \boldsymbol{R}_j, \boldsymbol{r}) - \mathcal{P}^{(\mathrm{L})}(-\omega; \boldsymbol{R}_j, \boldsymbol{r}). \qquad (6.96)
$$

Control question 6.7. Can you prove the above correspondence?

Equations (6.95) and (6.96) allow us to evaluate the effective impedance and the corresponding modification of the tunneling rate if we know the electron motion ($\mathcal{P}^{(\mathrm{R,L})}$) and the electrical response ($Z(\omega, \boldsymbol{r}, \boldsymbol{r}')$) in the medium. They work not only for a ballistic electron motion where it moves along a certain trajectory, but also for diffusive motion. In this case, the expressions for $\mathcal{P}_{\mathrm{R,A}}$ are averaged over the trajectories so that the propagator $\mathcal{P}$ satisfies the diffusion equation (see Eq. (6.31)).

Before we go to concrete examples, let us explain why we chose to present the tunnelings at $t_{1,2}$ as creation/annihilation of electron/hole pairs. It seems that a more natural alternative for, say, tunneling at time moment t_2 would be to look at an electron that is on the left at $t < t_2$ and on the right at $t > t_2$. This does not work, leading as it does to an incorrect and divergent expression for the effective impedance. The point is that an electron moving in a medium loses energy. Realistically, it will stop doing so if its energy approaches E_{F}. However, this is not incorporated in our approach, and an electron that exists all the time would suffer an infinite energy loss. One would have to correct this, introducing extra terms that cancel divergencies and eventually reproduce Eq. (6.93). The choice we made is the simplest way to avoid such corrections.

6.4.1 Escaping the Coulomb blockade

Let us consider tunneling through a tunnel junction of conductance G_{T} into an island (Fig. 6.13(a)) separated from a lead by a constriction of resistance R ($G_{\mathrm{T}}R \ll 1$). If we disregard the electron motion through the island to the lead, the tunneling electron feels the impedance $Z_{\mathrm{c}}(\omega) = R/(1 - \mathrm{i}\omega\tau_{RC})$, $\tau_{RC} \equiv RC$. As we have learned in Section 6.3, in this case the dynamical Coulomb blockade strongly suppresses the tunneling at low energies and completely suppresses it at zero energy: the low-voltage zero-temperature conductance

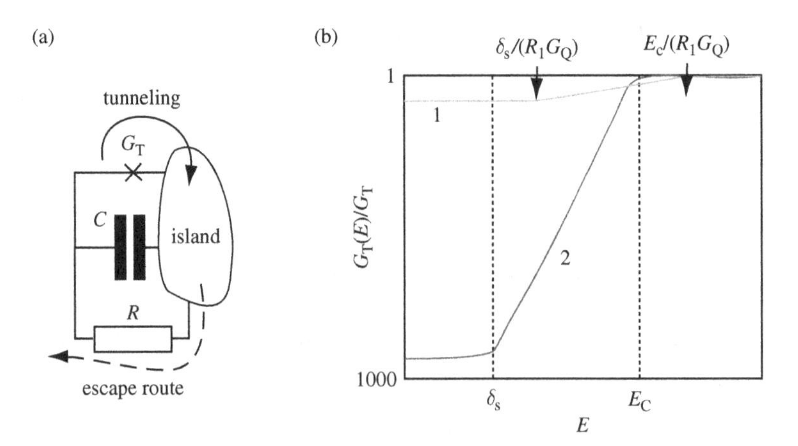

Fig. 6.13. Escaping Coulomb blockade. (a) Electrons can cross the constriction (R) that provides a resistive environment, thereby escaping the Coulomb blockade in the island. (b) Log-log sketch of the energy-dependent conductance. The curves correspond to different environment resistances $R_1 \ll G_{\mathrm{Q}}^{-1} \ll R_2$. Due to the escape, the conductances are finite at vanishing energy.

is given by $G_T(V) \simeq G_T(eV\tau_{RC}/\hbar)RG_Q$, $G_T(0) = 0$. It is intuitively clear that the electron motion can change this. At a sufficiently long time scale τ_d, the electron leaves the island and the resistive constriction behind. Therefore it *escapes* the Coulomb blockade set by the impedance R. Since long time scales correspond to low energies, we expect $G_T(0)$ to remain finite.

The effective impedance allows us to quantify this escape. To evaluate this, let us note that both electric potential and probability are conveniently constant over the island. Thus we can average both $\mathcal{P}_{A,R}(\boldsymbol{r})$ and $Z(\boldsymbol{r},\boldsymbol{r}')$ over the island. The averaged $\mathcal{P}_R$ gives the probability of finding an electron in the island at time moment t if it tunnels there at time moment 0. This probability falls off exponentially with time, $\exp(-t/\tau_d)$, therefore $\mathcal{P}_A = (-i\omega + 1/\tau_d)^{-1} = \mathcal{P}_R(-\omega)$. The time constant τ_d has been already evaluated in Subsection 3.2.7: it is the inverse of the rate Γ_{given} given by 3.35, $\tau_d = Re^2/\delta_S$. It might seem that now we can plug in Z_c and $\mathcal{P}$ to Eq. (6.93). However, the extended impedance at such low frequencies is also affected by the electron motion, as discussed in Section 6.1.2. Adapting Eq. (6.28) to the present situation, we obtain the true impedance $Z(\omega) = 1/(-i\omega C - i\omega e^2/\delta_S(1 + i\omega\mathcal{P}_R))$. Let us note now that electrons propagate slower than electricity, $\tau_d \gg \tau_{RC}$, or, in other terms, $E_C \gg \delta_S$. Under this condition,

$$Z(\omega)\mathcal{P}_R \approx Z_c(\omega)\frac{1}{-i\omega}. \tag{6.97}$$

One can say that the electron motion corrections cancel each other in Z and $\mathcal{P}_R$, a fact that we will use several times in this section.

Control question 6.8. Can you explicitly prove Eq. (6.97)?

We thus end up with $Z_{eff} = Z_c(-i\omega)\mathcal{P}_R(-\omega)$. This finally brings us to

$$Z_{eff}(\omega) = \frac{i\omega R}{(i\omega + \tau_d)(1 + i\omega\tau_{RC})}.$$

The effective impedance vanishes at zero frequency. According to the general classification of environments outlined in Section 6.2, this brings us to the superohmic regime, where the Anderson catastrophe is gone so we expect a finite probability of elastic tunneling. Accordingly, the ratio of zero-voltage and high-voltage tunnel conductances is finite, $G_T(0)/G_T = \exp(-\int_0^\infty G_Q \mathrm{Re}\, Z(\omega)d\omega/\omega)$. Explicitly,

$$\frac{G_T(0)}{G_T} = \left(\frac{\delta_S}{2E_C}\right)^{RG_Q}. \tag{6.98}$$

In a low-resistance environment, the conductance renormalization is not only finite but small, $G_T(0)/G_T = 1 - RG_Q \ln(2E_C/\delta_S)$ (Fig. 6.13(b)). Comparing this with the usual expression for the environment-suppressed tunneling, we see that the voltage-dependent conductance saturates at $eV \simeq \hbar\tau_d \simeq (G/G_Q)\delta_S$. In the opposite case, $RG_Q \gg 1$, the suppression is large and the saturation takes place at $eV \simeq \delta_S$.

We have assumed here that the constriction of resistance R can be treated as an ohmic resistor. While it is always a correct assumption at $R \ll R_Q$, it is not compatible with the coherent transport through the constriction in the opposite limit. Indeed, if the transport is coherent, the constriction is simply another tunnel junction of resistance R. We have addressed the zero-voltage conductance of such a setup in Chapter 3 in the context of elastic co-tunneling to obtain (Eq. (3.57)) $G_T(0)/G_T \simeq (RG_Q)(\delta_S/E_C)$. The suppression is in place; it is expressed in the same parameters as, but eventually is much smaller than, that for our escape model, Eq. (6.98), with an ohmic resistor. Indeed, the finite low-voltage conductance corresponds to the elastic scattering through the whole setup: the tunnel junction, the island, and the constriction. Since the ohmic resistor at $RG_Q \gg 1$ is essentially incoherent, this suppresses the elastic scattering by a factor that is large in comparison with that for a coherent tunnel constriction.

6.4.2 Luttinger liquids

Here we consider tunneling to and in Luttinger liquids. The Luttinger model describes an infinite one-dimensional gas of interacting electrons without scattering. In the context of quantum transport, such a gas can be realized in a sufficiently long waveguide supporting a single transport channel. This is why long one- or few-channel ballistic wires, where interaction effects are important, are commonly termed Luttinger liquids. "Liquid" here stresses the role of interaction as opposed to an almost non-interacting "gas." In Section 6.1.2 we have evaluated both the electron motion and the extended impedance in a Luttinger liquid, expressing them in terms of the parameters of the model: the Fermi velocity v_F the and geometric capacitance per unit length, $\tilde{C}$. We are thus ready to evaluate the effective impedance and its effect on the tunneling.

Tunneling in one-channel wires may be arranged according to three distinct setups (see Figs. 6.14(a), (b), and (c), (c′)). Let us start with setup A, in which an electron tunnels from an external electrode (black in the figure) to the point $x = 0$ somewhere in a (formally infinite) Luttinger liquid. We disregard the interaction in the electrode, which allows us to concentrate on the impedance and the electron motion in the liquid. Owing to the homogeneity of the liquid, the effective impedance can be represented by Fourier components of $\mathcal{P}_{R,A}$ and Z:

$$Z_{\text{eff}} = -\omega^2 \int \frac{dk}{2\pi} \mathcal{P}_A(-\omega, -k) Z(\omega, k) \mathcal{P}_R(\omega, k). \tag{6.99}$$

After tunneling from the external electrode, the electron moves with equal probability in one of the two opposite directions. Let us consider the one with positive velocity. The trajectory is given by $x(t) = v_F t$ in the liquid (right-hand electrode). Equation (6.94) then yields $\mathcal{P}_{R,A}(\omega, k) = (\mp i\omega + 0 \mp ikv_F)^{-1}$. Combining this with Eq. (6.35) for the impedance, and integrating over k, we obtain a frequency-independent effective impedance Z_A, given by

$$G_Q Z_A = \frac{(g + 1/g - 2)}{2_s 2}, \tag{6.100}$$

where we introduce a dimensionless parameter

$$g \equiv \frac{v_F}{v_p} \equiv \sqrt{(\tilde{C} - \tilde{C}')/\tilde{C}},$$

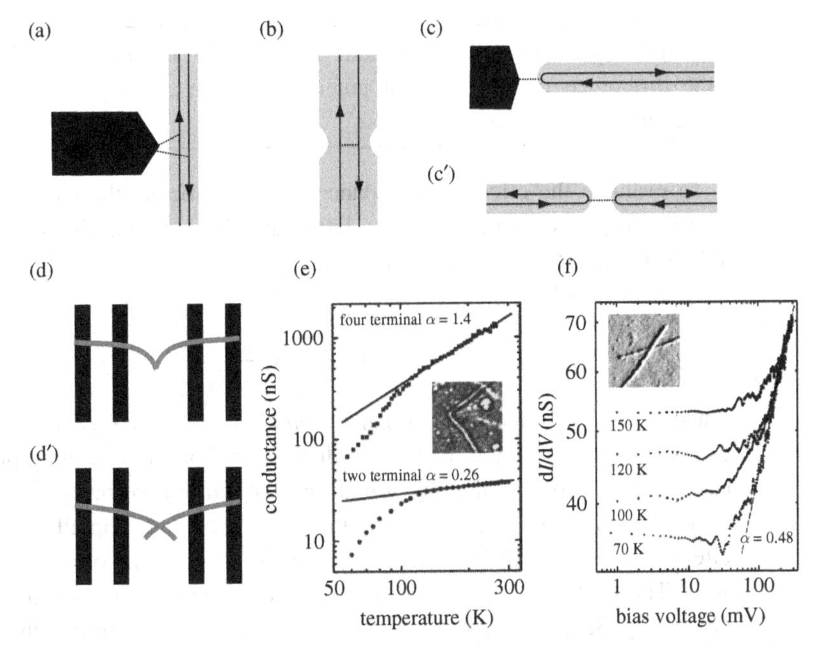

Fig. 6.14. **Tunneling in Luttinger liquids. Three distinct setups are shown. External metal electrodes are shown in black. (a) Setup A, to the middle. (b) Setup B, backscattering. Strictly, the third setup comprises two almost identical setups: (c) setup C, to the ends, and (c′) setup C′, between the ends. Parts (d), (d′), (e), and (f), taken from Ref. [148]: illustrate the experiment and are explained in the text.**

standard in the theory of Luttinger liquids. For repulsive interaction between electrons, $0 < g < 1$. A small g value signals large interactions, while $g = 1$ corresponds to a non-interacting gas. Indeed, $Z_A = 0$ at $g = 1$, and tunneling is purely elastic, as it should be in the absence of interactions. The constant 2_s gives (spin) degeneracy of the electron modes: if, for example, a large magnetic field completely polarizes electrons in the wire, $2_s = 1$. The modification of tunneling for sufficiently low energies is the same as for the resistor model described in Section 6.3 (see, for example, Fig. 6.9). The voltage-dependent tunneling conductance at vanishing temperature and the temperature-dependent conductance at zero voltage obey a power law, as follows:

$$\frac{G_T(V_1)}{G_T(V_2)} = \left(\frac{V_1}{V_2}\right)^{G_Q Z_A}, \qquad \frac{G_T(T_1)}{G_T(T_2)} = \left(\frac{T_1}{T_2}\right)^{G_Q Z_A}.$$

The tunneling is suppressed as the energy decreases.

Setup B in Fig. 6.14(b) is the tunneling between electrons propagating in opposite directions. Since an electron turns back in the course of such process, it is in fact *backscattering*. Such backscattering can be induced by any weak defect in or close to the wire. It can be seen as a (small) reduction of the two-terminal conductance of the wire due to the defect, given by

$$\frac{\mathrm{d}I}{\mathrm{d}V} = \frac{2_s}{2} G_Q - G_B(V), \tag{6.101}$$

where the first term presents the conductance of an ideal wire.

Control question 6.9. Explain Eq. (6.101) making use of Fig. 1.8: what is the difference in the chemical potentials of the electrons propagating in different directions?

To evaluate the effective impedance, let us associate the electrons propagating in a positive (negative) direction with a left (right) lead, so that the trajectories are given by $x^{(\mathrm{R,L})}(t) = \pm v_{\mathrm{F}}t$. This yields $P_{\mathrm{A}}(\omega, k) = (-\mathrm{i}\omega + 0 + kv_{\mathrm{F}})^{-1} - (-\mathrm{i}\omega + 0 + kv_{\mathrm{F}})^{-1}$, and integration over k in (6.99) yields

$$G_{\mathrm{Q}}Z_{\mathrm{B}} = \frac{2(g-1)}{2_s}. \tag{6.102}$$

Since $g < 1$, this effective impedance is *negative*. The backscattering is thus *increased* at decreasing energy, $G_{\mathrm{B}} \simeq V^{G_{\mathrm{Q}}Z_{\mathrm{B}}}$, in distinction from all tunneling processes considered earlier in this book. This may sound unusual, before we recognize that the increase in backscattering implies a decrease in the wire conductance: in all cases, an environmental effect tends to quench low-energy conductance, either that of a tunnel junction or of the wire. The backscattering can be regarded as tunneling provided it is weak enough, i.e. $G_{\mathrm{B}} \ll G_{\mathrm{Q}}$. Since the backscattering grows with decreasing energy, this condition is broken at sufficiently low energy. Starting with this energy, a weak defect would provide a strong effect: it effectively breaks the free electron propagation in the wire, separating it into two weakly connected pieces.

This brings us to setup C, or, to put it in strict terms, to two almost equivalent setups C and C′ (shown in Figs. 6.14(c) and (c′)). For setup C, an electron tunnels from the external electrode to the end of the wire rather than to the middle of it. We cannot immediately make use of Eq. (6.99) since the wire is not homogeneous: near the end, both Z and $\mathcal{P}_{\mathrm{R,A}}$ depend on both coordinates $\boldsymbol{r}, \boldsymbol{r}'$ rather than on their difference only. The effective impedance is evaluated after performing a symmetry trick; the calculation is given as Exercise 6.6, and yields

$$G_{\mathrm{Q}}Z_{\mathrm{C}} = \frac{(1/g-1)}{2_s}. \tag{6.103}$$

This impedance is positive and gives the power-law suppression of the tunneling at low energies. Let us consider the situation when the tunneling commences between the ends of two Luttinger liquids (setup C′ shown in Fig. 6.14(c′)). We note that in the framework of the Luttinger model the extended impedance is local in space: charge in one of the wires does not induce any voltage in another wire. This is a consequence of both the local capacitance model in use and the fact that electrons do not go between the wires. Looking at the general relations in Eqs. (6.93) and (6.95), we recognize that the effective impedance is contributed to separately by two pieces, each arising from integration over the coordinates in the left or right wire. Each piece is the same as in setup C. We conclude that the effective impedance doubles in comparison with setup C, $Z_{\mathrm{C}'} = 2Z_{\mathrm{C}}$.

Control question 6.10. Suppose the wires separated by the tunnel barrier are not identical so that Fermi velocities in the wires differ. What is $Z_{\mathrm{C}'}$ in this case?

Exercise 6.6. The extended impedance in setup C is a function of both coordinates, $x, x' > 0$, which is obtained by the solution of differential equations given in Section 6.1.2 with boundary condition $j(0) = 0$, since both electric and particle current must vanish at the end of the wire. Use symmetry arguments to prove that

$$Z(x, x') = Z_{\mathrm{h}}(x - x') - Z_{\mathrm{h}}(x + x'),$$

Z_{h} being the extended impedance of the homogeneous Luttinger liquid that depends on the difference of the coordinates only. Draw an analogy with the image charge method in electrostatics. Use the above relation to derive Z_{C}.

What happens to all the tunneling effects mentioned if we increase the number of propagating channels in the wire while keeping the same geometric capacitance? An estimate can be obtained from Eqs. (6.100), (6.102), and (6.103) by formally setting 2_s to the (large) number of channels. It may seem that interaction effects are increased: $g \approx 0$ at $2_s \to \infty$. However, all impedances given here come with 2_s in the denominators so $Z_{\mathrm{A,B,C}} \ll G_{\mathrm{Q}}$ in the limit of a large number of transport channels. We conclude that interaction is weak in this limit.

Finite size, low energies

So far we have assumed the wire to be infinite. In practice, it must end somewhere, and the range at which electrons propagate without scattering would hardly exceed a few microns. It is important that the finite size gives rise to a new low energy scale at which either electrons or electricity propagate along the full length L of the wire, $E_{\mathrm{low}} \simeq \hbar \max(v_{\mathrm{p}}, v_{\mathrm{F}})/L$. This energy scale limits the applicability of the power laws at low energy, while at higher energies they are valid until the energy reaches E_{F}. The fate of the wire at energies $\lesssim E_{\mathrm{low}}$ is governed by the general laws of quantum transport and the contacts to the leads. If the contacts are good, resembling the QPC situation, the wire can be regarded as a single scatterer with no appreciable energy dependence of the conductance. If the increased backscattering is still small, the wire is close to an ideal QPC. If contacts are bad, the piece of the wire becomes a quantum dot with the level spacing $\simeq \hbar v_{\mathrm{F}}/L$ and Coulomb energy $E_{\mathrm{C}} = e^2/\tilde{C}L$.

Carbon nanotubes

It might seem that a Luttinger liquid can be easily made by electrostatic shaping of a 2DEG: one would just use two wide top electrodes separated by a narrow gap. Many attempts to make long ballistic wires indeed concerned superconducting heterostructures. However, it appears to be very difficult to provide no backscattering at sufficient length. Since interaction effects tend to increase the backscattering at low energies, a long wire is cut into a collection of weakly connected pieces by occasional weak defects.

Nature and technology provide a better material for quantum wires: carbon nanotubes. Carbon comes in the form of graphite: a collection of weakly coupled atom-thick layers of carbon. (One easy draws a line with a graphite pencil because the cohesion between the layers is small.) Single-wall carbon nanotubes are thin (≈1 nm in diameter) cylinders obtained by "wrapping" of a single layer. A tight, chemically pure carbon lattice makes the nanotubes effectively defectless, this allows for the observation of the Luttinger liquid exponents. Owing to a peculiar band structure of a single graphite layers, the electrons propagating in the carbon nanotubes possess extra degeneracy, so that $2_s = 4$.

Experiment

We illustrate Luttinger liquids where three power-law exponents corresponding to Z_{A}, Z_{C}, $Z_{\mathrm{C'}}$ have been cleverly measured with the same device [148]. A carbon nanotube was placed on a substrate to touch four metal electrodes. An AFM (atomic force microscope) tip was used to affect the nanotube. At the first stage of the experiment, a buckle was made in the middle of it (see Fig. 6.14(d)). Later, the nanotube was pushed further in the same direction by the tip applied to the buckling point. Thereby the nanotube was torn into two pieces that then bounced back to cross in the middle of the structure (see Fig. 6.14(d′)).

The contacts between the nanotube and the metals are effectively of tunneling nature. This is why below the low energy scale E_{low} the nanotube is a double quantum dot subjected to Coulomb blockade. Since the device is short, this low-energy scale is not at all small in absolute units: $E_{\mathrm{C}}/k_{\mathrm{B}} \simeq 70\,\mathrm{K}$, a temperature too high for a typical quantum transport experiment. However, this leaves a wide energy range from E_{low} to E_{F} in which to observe the Luttinger exponents. In practice, the measurements stopped at room temperature and the corresponding voltages.

In the first stage, the two-terminal conductance of the device was dominated by the tunnel barriers between the tube and the metal electrodes: this corresponds to setup A. The temperature dependence (lower curve in Fig. 6.14(e)) of the conductance gave the power $G_{\mathrm{Q}}Z_{\mathrm{A}} = 0.26$ (referred to as α in the figure); this yields $g = 0.26$. One can exclude the contact resistance by making a four-terminal measurement that concentrates on the resistance of the middle of the nanotube. This resistance is at least 30 times smaller, but still lower than G_{Q}^{-1}. This proves that the buckle is eventually quite a strong defect: it must work as a tunnel barrier, so we have setup C′. The corresponding effective impedance was measured to be $G_{\mathrm{Q}}Z_{\mathrm{C'}} = 1.4$; this is in excellent agreement with $g = 0.26$ extracted from the two-terminal measurement.

Tearing the nanotube decreased the resistance by factor of 4 at room temperature. It was thus dominated by the tunnel junction at the crossing. This is a junction between two nanotube pieces similar to the buckle junction. However, the exponent obtained from the differential conductance measured differs by a factor of 3, $G_{\mathrm{Q}}Z_{\mathrm{eff}} = 0.50$. The point is that now the junction is at a significant distance from the ends of the torn pieces. The electrons tunneled between the "bulk" parts of the nanotube pieces. This suggests that $Z_{\mathrm{eff}} = 2Z_{\mathrm{A}}$, in agreement with the experiment.

Bosonization

The fundamental difference between bosons and fermions is somehow smeared in a strict one-dimensional geometry since the particles in these situations cannot get around each other. This enabled the so-called bosonization technique that works in one-dimensional setups and allows for an efficient and compact description of Luttinger liquids and evaluation of the exponents. The first step of bosonization is similar to the quantization of electric circuits: electrical excitations in the Luttinger liquid are quantized and represented as a set of bosons. During the second step, the electron creation/annihilation operators are represented as exponents of the boson fields: bosonization makes the electrons redundant. In our opinion, at this point the level of the theoretical abstraction gets too high. Although we appreciate the elegance of bosonization, we review Luttinger liquids with earthly techniques that easily reveal the links to other quantum transport setups. An interested reader can find an excellent compact review of bosonization aimed at Luttinger liquids and adjacent phenomena of quantum transport in Ref. [149].

6.4.3 Altshuler–Aronov corrections

Let us turn to tunneling into extended metals where electron transport is diffusive. Such tunneling should also be affected by a dynamical Coulomb blockade. From what we know about the phenomenon, it is safe to assume that it is governed by an effective impedance. This impedance can be estimated as the resistance R of a metal piece adjacent to the point of tunneling. If the metal is "good," so the electron transport in there is coherent, such a resistance must be small, not exceeding G_Q^{-1}. In this case, the effect of a dynamical Coulomb blockade is small, being a *correction* $\simeq RG_Q$ to the tunneling rate for non-interacting electrons.

But what does "adjacent" mean? In classical transport, "adjacent" would imply a close proximity: that at atomic scale or perhaps at the scale of the mean free path. In 1979, Altshuler and Aronov [150] addressed the interaction correction to tunneling rate to find that this is not so: the scale defining "adjacent" is typically much bigger than the mean free path and can depend on the energy of the tunneling electron. This work became one of the pillars of mesoscopics and the reference point for quantum transport.

As discussed, an energy scale defines a time scale that gives rise to two space scales: L_{diff} and L_{el} characterizing the propagation of electrons and electricity, respectively. Thus, we can proceed with the effective impedance method described since it can account for phenomena taking place at both scales. We will make full use of the relations from Section 6.1.2. In a common metal, $L_{\mathrm{diff}} \ll L_{\mathrm{el}}$. This scale separation allows for simplification of Eq. (6.93); as we note in Section 6.4.1, we can replace the extended impedance $Z(\omega; \boldsymbol{r}, \boldsymbol{r}')$ by its electric-circuit approximation $Z_{\mathrm{c}}(\omega; \boldsymbol{r}, \boldsymbol{r}')$, given by

$$\int \mathrm{d}\boldsymbol{r}' \, Z(\omega; \boldsymbol{r}, \boldsymbol{r}') \mathcal{P}(\boldsymbol{r}', \boldsymbol{R}) \approx Z_{\mathrm{c}}(\omega; \boldsymbol{r}, \boldsymbol{R}),$$

$\boldsymbol{R}$ being the point where the tunneling takes place. Expressions for Z_{c} in different geometries can be found in Section 6.1.1. The effective impedance becomes

$$Z_{\mathrm{eff}} = \int \mathrm{d}\boldsymbol{r} \, \mathcal{P}_{\mathrm{A}}(-\omega; \boldsymbol{R}, \boldsymbol{r}) Z_{\mathrm{c}}(\omega; \boldsymbol{r}, \boldsymbol{R}).$$

For homogeneous media, one may use the representation in terms of Fourier components.

We start with three-dimensional geometry: tunneling takes place at the surface of a semi-infinite metal with conductivity σ. Combining Eqs. (6.22) and (6.31), we get

$$Z_{\text{eff}} = 2\int \frac{\mathrm{d}\mathbf{k}}{(2\pi)^3} \frac{-\mathrm{i}\omega}{\sigma(-\mathrm{i}\omega + Dk^2)k^2} \rightarrow \text{Re } Z_{\text{eff}} = \frac{1}{\pi\sigma}\sqrt{\frac{\omega}{2D}}. \tag{6.104}$$

The factor of 2 accounts for the fact that the metal is semi-infinite. According to the general classification of environments, we are in a superohmic regime. Employing Eq. (6.78) yields $\delta G_{\text{T}}(V)/G_{\text{T}}(0) = 2(G_{\text{Q}}/\sigma\pi)\sqrt{eV/2\hbar D}$. This prediction[4] made in Ref. [150] gave an explanation of the experiments [151], 152].

To interpret this effective impedance, let us first perform some elementary electricity theory. Let us imagine an ideally conducting sphere of radius r embedded into the metal such that its center coincides with the point of tunneling. If we apply a voltage to the sphere and ground the metal, the resistance equals $1/2\pi\sigma d$.

Control question 6.11. Can you reproduce the above formula for the resistance?

The effective impedance thus equals the effective resistance of the metal with a piece of size $\sim L_{\text{diff}}$ *removed.* This is consistent with the escaping Coulomb blockade considered earlier in this section: the resistance covered by the propagating electrons does not contribute to the effective impedance.

The three-dimensional geometry is slightly confusing since, in this case, the electricity propagates instantly, $L_{\text{el}} = \infty$. In two-dimensional geometry – tunneling into a film – the correction is obtained using the impedance expression in Eq. (6.23) instead of Eq. (6.22). The electric field in two-dimensional geometry diffuses with diffusion coefficient $D^* = 1/\tilde{R}\tilde{C}$, so the electrical scale is estimated as $L_{\text{el}} \simeq \sqrt{D^*/\omega}$. The effective impedance does not depend on frequency and is given by

$$Z_{\text{eff}} = \frac{\tilde{R}}{2\pi} \ln\left(\frac{D^*}{D}\right).$$

This is the impedance of a circle of the film that has a radius L_{el} and a hole in the center, L_{diff} being the radius of the hole.

This finally defines "adjacent" as far as the effective impedance is concerned: within L_{el}, beyond L_{diff}.

If we go to one-dimensional geometry – an RC-line – the resistance is proportional to the length scale, and since $L_{\text{el}} \gg L_{\text{diff}}$, the resistance of a piece "removed" by the electron diffusion is irrelevant. The effect of the dynamical Coulomb blockade is not affected by the diffusion and is given by Eq. (6.76).

We start this paragraph with an explicit warning: it contains plausible, but wrong, statements to illustrate the subtleties of interaction corrections. The resistance of a diffusive sample is due to scattering at impurities. At least in the Born approximation, the

[4] In addition to the contribution of electrical excitations, Ref. [150] also gives the contribution of slow modes of another kind – spin excitations, not considered in this book. The spin-excitation contribution has a similar structure, but is usually numerically small and vanishes at scales exceeding the spin-flip length.

scattering at each impurity can be regarded as a kind of tunneling between electrons crossing the impurity position in different directions. We have already used this analogy when describing backscattering in a Luttinger liquid. It is plausible that the rate of this tunneling, $1/\tau_p$, is also affected by the dynamical Coulomb blockade, perhaps at diffusive length. Since the conductivity is inversely proportional to the rate, we expect an interaction correction to the conductance, $\delta\sigma/\sigma \simeq -RG_Q$, R being the resistance at the scale L_{diff}. Indeed, such interaction corrections are present, have this order of magnitude, and have been derived by Altshuler and Aronov [153]. What is wrong is our reasoning: the conductivity corrections are of a more subtle nature. Indeed, if we run an actual calculation with Eq. (6.93) at the diffusive scale, the effective impedance turns out to be zero: the trajectories of the electron and the hole describing scattering on the impurity are spread over the same volume, resulting in zero total charges $q_{1,2}$. The conductance corrections cannot be caught with the trajectory method in use: they arise from the trajectory branching mentioned in Section 2.6.2. We will come back to this question at the end of Section 6.5.

6.5 Weak interaction

Having given so much attention to tunneling in the previous three sections, we would like to depart from it now and discuss interaction effects in nanostructures with more transmissive channels. As opposed to tunneling, we cannot now analyze strong and weak interactions on an equal footing: we stick to the simpler case of weak interaction. This, however, is a quite natural case in transmissive nanostructures. As we have seen in many examples in Section 6.4, the interaction strength is characterized by an effective impedance, where $ZG_Q \ll 1$ corresponds to a weak interaction. Multi-channel transmissive nanostructures exhibit large conductance; we are decisively in the $G \gg G_Q$ limit. If we associate the effective impedance with the nanostructure conductance, we understand that Coulomb interaction is always weak in this case. We could have called this section "Weak Coulomb blockade," as some other authors have. However, for us the word "blockade" does not obviously imply the degree of strength involved; much like the words "dead" or "pregnant."

In this section, we mostly employ a rather heuristic renormalization analysis introduced in Section 6.2.5. Although it may miss some of the details, it enables us to deliver the main message: weak interaction can lead to large consequences; its effect does not have to be limited to small corrections to the non-interacting picture. The large consequences, however, will take place at a very low energy scale only. We start by studying a transmission renormalization by an ohmic environment and describe the related evolution of transmission distribution. If part of a nanostructure provides such an environment for another part, Coulomb blockade is given a chance: the transport can be suppressed completely at sufficiently low energies, and Section 6.5.2 tells us when this happens. Coulomb blockade is elucidated for a single-node setup: analysis of more complicated setups leads us to a study of Altshuler–Aronov corrections to conductivity.

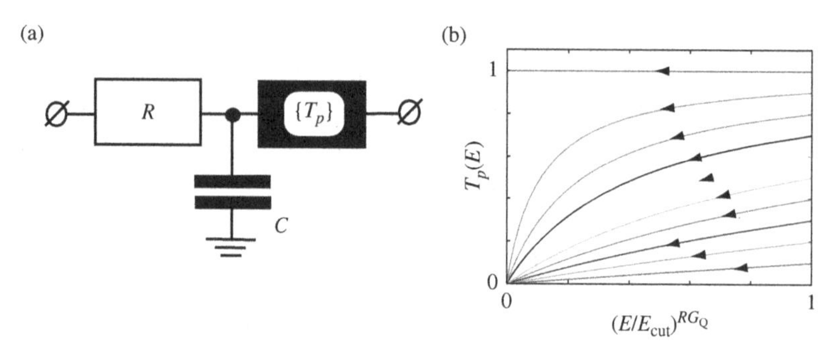

Fig. 6.15. (a) Ohmic resistor R provides an electromagnetic environment for a junction with arbitrary transmission eigenvalues. (b) The interaction causes the flow of transmission eigenvalues towards $T = 0$ at decreasing energy. The curves, from bottom to top, correspond to $T_p(E_{cut})$ ranging from 0.1 to 1 with step 0.1.

6.5.1 Arbitrary transparency

Let us consider a setup similar to that studied in Section 6.3: a junction in series with an ohmic resistor representing the effect of an electromagnetic environment (Fig. 6.15(a)). It is convenient to assume that the resistance of this ohmic resistor R is much smaller than the resistance G^{-1} of the junction: in this case, the electron transfers through the junction are conveniently separated from the dynamics of the electric field fluctuations, which are determined by the smaller resistor. In other words, the junction does not make an environment for itself. We are interested in a sufficiently transmissive junction: it is described by a scattering matrix $\hat{s}$, and there is at least one transmission eigenvalue $T_p \simeq 1$. This implies $G \gtrsim G_Q$, so that the effect of Coulomb blockade is automatically weak: $RG_Q \ll 1$.

We would like to employ the renormalization procedure similar to that used in Section 6.2.5. The idea of the renormalization is to incorporate the interaction effects into the energy dependence of the scattering matrix of the junction. At each step of the renormalization we thus conveniently deal with a nanostructure to be described in the framework of the scattering approach of Chapter 1: the only effect of the interactions is the energy-dependence of the scattering matrix. This convenience is only allowed if the interaction is weak.

We start at sufficiently high energy E_{cut}, at which the interaction effects may be disregarded. For our setup, this energy is set by the RC-time of the circuit, $E_{cut} \simeq \hbar/RC$. Let us work with a conveniently dimensionless energy scale $\zeta = \ln(E_{cut}/E)$ that changes from 0 to ∞ while E changes from E_{cut} to 0. The renormalized scattering matrix $\hat{s}(\zeta)$ is evaluated by small steps; each step reduces the energy by $\mathrm{d}E$. At each step, we start with $\hat{s}(E)$ and consider its change due to the boson modes with energies in the interval $(E - \mathrm{d}E, E)$. This gives us $\hat{s}(E - \mathrm{d}E)$: the renormalization procedure amounts to a differential equation – the flow equation.

To start with, let us look at a single channel and the corresponding 2×2 scattering matrix:

$$\hat{s} = \begin{pmatrix} r & te^{\mathrm{i}\phi} \\ t'e^{-\mathrm{i}\phi} & r \end{pmatrix}. \tag{6.105}$$

Here, we have incorporated the electrical fluctuations induced by the resistor R into the phase factor of the transmission amplitudes; the phase operator $\hat{\phi}$ is linear in boson operators. We proceed as in the Section 6.2.5, separating ϕ into "slow" and "fast" parts, the latter being contributed by the boson modes in the energy interval $(E - \mathrm{d}E, E)$. Adapting the work flow of Section 6.2.5 for our case (Eq. (6.64)), we obtain the renormalization of the phase factor $F \equiv \exp(\mathrm{i}\phi)$ as follows:

$$F \to F\langle \mathrm{e}^{\mathrm{i}\phi_\mathrm{f}} \rangle = F\mathrm{e}^{-\langle \phi_\mathrm{f}^2 \rangle/2} \approx F\left(1 - \frac{\langle \phi_\mathrm{f}^2 \rangle}{2}\right);$$

$$\frac{\mathrm{d}F}{F} = -\frac{G_\mathrm{Q} R}{2} \int_{E-\mathrm{d}E}^{E} \frac{\mathrm{d}E}{E} = -\frac{G_\mathrm{Q} R}{2}\,\mathrm{d}\zeta.$$

The change of the scattering matrix $\delta\tilde{\hat{s}}$ is thus given by

$$\delta\tilde{\hat{s}} = -\frac{G_\mathrm{Q} R}{2}\,\mathrm{d}\zeta \begin{pmatrix} 0 & t\mathrm{e}^{\mathrm{i}\phi} \\ t'\mathrm{e}^{-\mathrm{i}\phi} & 0 \end{pmatrix} = (\tau_z \hat{s} \tau_z - \hat{s})\frac{G_\mathrm{Q} R}{4}.$$

The problem in hand is that the renormalized $\hat{s} + \delta\tilde{\hat{s}}$ is not a scattering matrix since it does not satisfy the unitarity condition. Indeed, the correction represents the average over some scattering matrices, and such an average is generally not a unitary matrix. We circumvent this problem by taking instead of $\hat{s} + \delta\tilde{\hat{s}}$ the closest matrix $\hat{s} + \delta\hat{s}$ that obeys the unitary condition. For any $\delta\tilde{\hat{s}}$,

$$\delta\hat{s} = \delta\tilde{\hat{s}} - \hat{s}(\delta\tilde{\hat{s}})^\dagger \hat{s}. \qquad (6.106)$$

Exercise 6.7. Prove Eq. (6.106) using the following definition of distance between arbitrary matrices $\hat{A}$, $\hat{B}$ in the matrix space $\mathrm{dist}(\hat{A}, \hat{B}) = \mathrm{Tr}((\hat{A} - \hat{B})(\hat{A} - \hat{B})^\dagger)$.

Bringing everything together, we obtain the following flow equation for the scattering matrix:

$$\frac{\mathrm{d}\hat{s}}{\mathrm{d}\zeta} = \frac{G_\mathrm{Q} R}{8}\left(\tau_z \hat{s} \tau_z - \hat{s}\tau_z \hat{s}^\dagger \tau_z \hat{s}\right).$$

This equation can be simplified if we are just interested in the flow of the transmission eigenvalue T_p. Squaring the non-diagonal elements of the matrix, we obtain

$$\frac{\mathrm{d}T_p}{\mathrm{d}\zeta} = -G_\mathrm{Q} R\, T_p(1 - T_p). \qquad (6.107)$$

We can now generalize to a multi-channel junction by stating that Eq. (6.107) holds for any transmission eigenvalue of the junction.

If all transmission eigenvalues $T_p \ll 1$, this reproduces the earlier results concerning tunneling, $T_p \propto E^{G_\mathrm{Q} R}$. We see that for higher transmission the effect of interaction is slowed down by a factor of $(1 - T_p)$. In particular, ideal transmission, $T_p = 1$, is not renormalized at all. However, for any transmission less than unity the interaction effect is the same: at sufficiently low energies, the transmission vanishes. The flow of different values of T_p is plotted in Fig. 6.15(b). For the first time, Eq. (6.107) has been derived for an arbitrary one-channel scatterer in a weakly interacting ($g \approx 1$) Luttinger liquid [154].

Exercise 6.8. Check that in this limit Eq. (6.107) is consistent with the exponents in Eqs. (6.102) and (6.103) for tunneling and backscattering in a Luttinger liquid.

The flow relation in Eq. (6.107) gives rise to a set of interesting and somehow unexpected results. Since the full counting statistics in the scattering approach is defined by the set of transmission eigenvalues, the flow relation defines the interaction effect on the statistics. If we recall the definitions of Landauer conductance ($G = G_Q \sum_p T_p$) and Fano factor ($F = \sum_p T_p(1-T_p)/\sum_p T_p$), we discover the link between the interaction effect on the conductance and the shot noise as follows:

$$\frac{\mathrm{d}G(E)}{\mathrm{d}\ln E}\frac{1}{G(E)} = G_Q R F. \tag{6.108}$$

To check this experimentally, one measures voltage (or temperature) dependence of conductance and shot noise of the same junction in a controlled environment. Such experiments have been recently performed [155].

Control question 6.12. Can you describe the interaction correction to the Fano factor?

The transmission distribution is an important property of a $G \gg G_Q$ conductor. It is worth studying how it changes under the effect of interaction [156]. Simply from the fact that all transmission eigenvalues approach zero, one would conjecture that, at a sufficiently developed interaction effect, any junction would eventually become a tunnel junction, with conductance decreasing as $G \propto (E/E_{\mathrm{cut}})^{G_Q R}$. It turns out that this is one of the two possible scenarios. The alternative scenario is that a junction becomes a *double* tunnel junction, with conductance decreasing with a doubly smaller exponent, $G \propto (E/E_{\mathrm{cut}})^{G_Q R/2}$.

To comprehend this result, we note that the flow equation can be explicitly integrated as follows:

$$T_p(E) = \frac{T_p(E_{\mathrm{cut}})\xi}{1 - T_p(E_{\mathrm{cut}})(1-\xi)};\ \xi \equiv \left(\frac{E}{E_{\mathrm{cut}}}\right)^{G_Q R}.$$

The transmission distribution at a given energy is thus given by

$$\rho_E(T) = \frac{\xi}{[\xi + T(1-\xi)]^2}\rho_{E_{\mathrm{cut}}}\left(\frac{T}{\xi + T(1-\xi)}\right). \tag{6.109}$$

Let us take a distribution that has an inverse square-root singularity at $T \to 1$, $\rho_{E_{\mathrm{cut}}}(T) = a/\sqrt{1-T}$. Substituting this into Eq. (6.106), we obtain in the low-energy limit:

$$\rho_E(T) = a\sqrt{\frac{\xi}{T^3(1-T)}}, \tag{6.110}$$

which is the transmission distribution of a double tunnel junction with identical conductances of the constituting tunnel conductors (see Section 2.6.1). The conductance is indeed proportional to $\sqrt{\xi} = (E/E_{\mathrm{cut}})^{G_Q R/2}$. One also checks that the transmission distribution given by Eq. (6.110) is not affected by the flow apart from the overall coefficient.

We have learned in Section 2.6 that the transmission distributions of various conductors can be subdivided into two large classes. The conductors of the first class have no transmissions at T close to unity. In the low-energy limit they thus become tunnel junctions. The conductors of the second class, diffusive conductors among them, are characterized by an inverse square-root singularity in a transmission distribution at $T \to 1$. They become tunnel junctions in the low-energy limit.

To understand the result heuristically, let us note that the "normal" exponent $G_Q R$ corresponds to tunneling through the whole nanostructure in a single leap. An alternative is to do take two leaps: get to the "middle" of the nanostructure first, and then tunnel out to another lead. Such two-leap tunneling occurs, for example, in the course of resonant tunneling in quantum dots. Each leap transfers charge $e/2$ through the whole circuit, so the corresponding exponent is halved. This explanation suggests that the conductors of the second class always have a "middle": some states (similar to Fabry–Perot resonances in a double junction) provide an intermediate stop for electrons tunneling in two leaps. The scattering approach does not explicitly imply that a nanostructure has a "middle": its only manifestation is the inverse square-root singularity already mentioned.

We also note that the second scenario only holds while the renormalized conductance remains large, $G(E) \gg G_Q$. When it approaches G_Q, there are only a few effective transport channels left, so the quasicontinuous transmission distribution no longer makes sense. As the energy scale decreases further, the conductor behaves as a tunnel junction.

Exercise 6.9. Find the conductance of a diffusive conductor subject to interaction as a function of the scale ξ. (Hint: use Eq. (6.109)). Estimate the energy scale at which $G(E) \simeq G_Q$.

6.5.2 Coulomb blockade at large conductances

Let us note that in Section 6.5.1 we concentrated on the renormalization of the junction, while the shunting resistor R was assumed to be ohmic, not changing its resistance with the energy scale. Although this is instructive and relevant in a number of cases, this does not seem to conform to equality principles. The resistor could itself be a coherent conductor; in that case, it will behave in a way similar to the junction already considered, and therefore should be subject to equal treatment.

To restore the equality, we now turn to a setup where a single node is connected to a number of leads by arbitrary connectors labeled k, each of which is characterized by its own set of transmission eigenvalues $T_p^{[k]}$ (see Fig. 6.16(a)). Let us count a channel per spin direction, so that the conductance of each connector is given by $G^{[k]} = (G_Q/2)\sum_p T_p^{[k]}$. For two leads, we find a direct correspondence to the previous setup: one of the connectors can be chosen to represent the junction and the other to represent the resistor R. We thus see that in the present setup one part of the nanostructure presents an electromagnetic environment for another part.

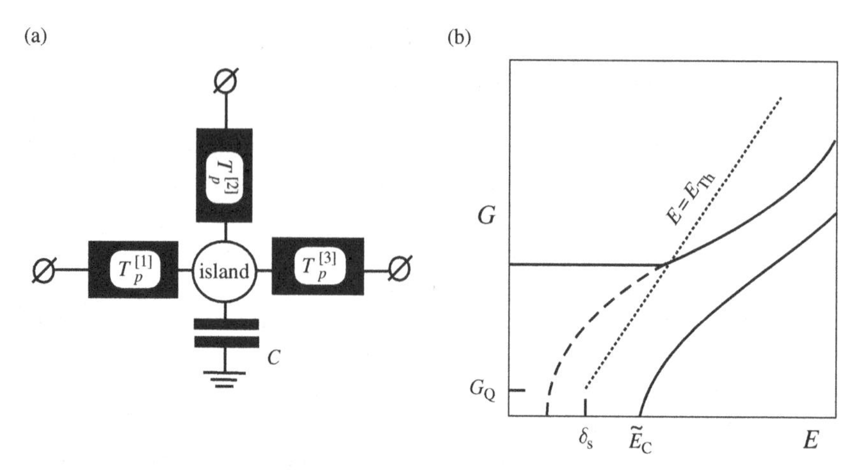

Fig. 6.16. Coulomb blockade at large conductances. (a) Node connected to the leads by arbitrary connectors. Section 6.5.2 addresses whether this island is in the Coulomb blockade regime. (b) The flow of the total conductance of the island versus energy. There are two possible scenarios: (i) Coulomb blockade takes place at $\tilde{E}_C$ (lower solid curve). (ii) Finite conductance renormalization at $E = 0$ (upper solid curve). $E = E_{Th}$ at the dotted line. Dashed line represents the would-be renormalization flow of the upper curve at $\delta_S = 0$.

Let us define the total conductance of the node as $G = \sum_k G^{[k]} = (G_Q/2)\sum_p T_p$, where the sum is over all transmission eigenvalues in all connectors. If $G \ll G_Q$, we recognize that we are dealing with a Coulomb island from Chapter 3. The transport through the island is suppressed at an energy scale $\lesssim E_C$. We also indicate that $G \ll G_Q$ – the requirement of good isolation – is vital for a well developed Coulomb blockade, and something bad happens to it otherwise. In this subsection, we thus investigate the Coulomb island setup in the opposite limit $G \gg G_Q$ to find out the exact fate of this phenomenon.

We proceed by assuming that all transport channels form a common environment with an effective impedance G^{-1} felt by each channel. Adapting Eq. (6.107) to the present situation, we immediately obtain

$$\frac{dT_p}{d\zeta} = -\frac{G_Q}{G} T_p(1 - T_p) = -\frac{2T_p(1 - T_p)}{\sum_p T_p}. \tag{6.111}$$

The difference from Section 6.5.1 is that the effective impedance is also subject to the renormalization. It increases with decreasing energy, so we expect the flow to "accelerate."

The equation is simple enough to solve. As in Section 6.5.1, each transmission eigenvalue evolves with scale ξ as follows:

$$T_p(E) = \frac{T_p(E_{\rm cut})\xi}{1 - T_p(E_{\rm cut})(1 - \xi)}; \; 0 > \xi(E) > 1. \tag{6.112}$$

Unlike in Section 6.5.1, the scale ξ becomes a complicated function of energy, given by

$$\frac{d\xi}{d\zeta} = -\frac{G_Q}{G(\zeta)} = -\frac{2}{\sum_p T_p(E)}. \tag{6.113}$$

Substituting Eq. (6.112) into Eq. (6.113) and integrating the resulting differential equation yields an explicit relation for $\xi(E)$:

$$\ln\left(\frac{E_{\mathrm{cut}}}{E}\right) = -\frac{1}{2}\sum_p \ln(1 - T_p(1-\xi)), \tag{6.114}$$

where T_p are taken at high energy E_{cut}. The transmissions vanish at $\xi \to 0$. Equation (6.114) implies that this takes place at a *finite* energy given by

$$\tilde{E}_{\mathrm{C}} = E_{\mathrm{cut}} \sum_p \sqrt{1 - T_p}.$$

This is the manifestation of flow acceleration already mentioned: the transport is blocked at finite energy $\tilde{E}_{\mathrm{C}}$ rather than in the limit of vanishing energy.

Since the transport is blocked, we recover the Coulomb blockade. The energy scale obtained is simply the effective charging energy of the island renormalized by a large conductance. Since the number of effective channels is of the order of G/G_{Q}, the effective charging energy is exponentially small in comparison with the charging energy E_{C} in the limit of good isolation; thus,

$$\frac{\tilde{E}_{\mathrm{C}}}{E_{\mathrm{C}}} \simeq G \exp(-\alpha G/G_{\mathrm{Q}}), \tag{6.115}$$

$\alpha \simeq 1$ being a dimensionless coefficient characterizing the type of connector.

Exercise 6.10. Assume that all connectors are of the same type and are characterized by the same transmission distribution, or, alternatively, by the function $\mathcal{I}(\phi)$ (see the first part of Eq. (2.94)). Demonstrate that

$$\alpha = \frac{\int_0^\pi \mathrm{d}\phi \mathcal{I}(\phi)}{2\mathcal{I}'(0)}.$$

In particular, $\alpha = 1$ if all connectors are of tunnel nature and $\alpha = \pi^2/4$ if all connectors are diffusive. We learned in Chapter 4 that the large conductance is subject to interference fluctuations of magnitude $\simeq G_{\mathrm{Q}}$: these are termed the universal conductance fluctuations (UCFs). Owing to the exponential dependence on conductance, the UCFs of the effective charging energy are large, of the order of $\tilde{E}_{\mathrm{C}}$ itself.

The renormalization method provides only a crude estimate of $\tilde{E}_{\mathrm{C}}$ and says little about the details of Coulomb blockade at large conductances. For example, at $G \ll G_{\mathrm{Q}}$ the energy of the Coulomb island depends on the charge q induced by a gate electrode: $E(q) = E_{\mathrm{C}}(q/e)^2$ at $|q| < e/2$, periodic in q with period E. The renormalization method does not pick up on this dependence. In the tunnel limit, the exact solution of the problem was sought for almost 20 years, and it was finally obtained in 2007 [157]. According to to Ref. [157], the main contribution to q-dependent energy is cos-shaped; this resembles the ground state of the Cooper-pair box (Section 3.5.1) in the limit of large Josephson energy:

$$E(q) = 4E^*(1 - \cos(q/e)); \quad E^* = \tilde{E}_{\mathrm{C}}(G/\pi G_{\mathrm{Q}})^2.$$

Our considerations of the Coulomb setup are not yet complete: we have to take into account the effects of electron motion. Most importantly, the electrons can escape the island and are no longer subject to interactions. For our setup, this implies that the renormalizations should stop at the energy scale $E_{\text{Th}} = (G(E)/G_{\text{Q}})\delta_{\text{S}}$ related to the escape time, $G(E)$ being the renormalized total conductance of the node.

If the energy E_{Th} is reached before $\tilde{E}_{\text{C}}$, the Coulomb blockade does not take place. Instead, the renormalizations of the total conductance stop at $G(E \simeq E_{\text{Th}})$; this value remains unchanged at lower energies [158]. This is an alternative scenario of low-energy behavior: the transport is not blocked (Fig. 6.16(b)). The transition between the two scenarios takes place at $G(E) \simeq G_{\text{Q}}$; this yields

$$\tilde{E}_{\text{C}} \simeq \delta_{\text{S}}.$$

This also agrees with our picture of quantum dots: for a good quantum dot, the (effective) charging energy should exceed the level spacing. The same criterion can be expressed in terms of high-energy total conductance G (see Eq. (6.115)): the Coulomb blockade only develops if

$$\frac{G}{G_{\text{Q}}} < \frac{1}{\alpha} \ln\left(\frac{E_{\text{C}}}{\delta_{\text{S}}}\right);$$

the energy should exceed mean level spacing. Since the logarithm is never large (say, <10), the existence domain of the Coulomb blockade does not exceed several conductance quanta. Far beyond the domain, the interaction correction to the high-energy conductance is finite and may be small:

$$G(0) - G(E_{\text{cut}}) = -F G_{\text{Q}} \ln\left(\frac{E_{\text{C}}}{\delta_{\text{S}}}\right),$$

where F is the Fano factor of the connectors.

6.5.3 Altshuler–Aronov corrections to the conductivity

Up to now, we have successfully dealt with the interaction corrections in simple circuits: those with one node and no nodes. To extend this reasoning to a larger network, one ascribes a fluctuating voltage to each node and evaluates the effect of the fluctuations on transport properties of the network, eventually using the renormalization technique. The correlation function of the voltage fluctuations is readily obtained from an elementary network analysis in which all connectors are replaced by the corresponding conductances. The fluctuations in different nodes correlate, and this causes a difficulty. Namely, in this case, the effect of the fluctuations cannot be ascribed to a scattering matrix of a single connector and the corresponding transmission coefficient. The correlation of voltages in different nodes leads to the correlation of transmissions in different connectors and therefore complicates a purely scattering approach. One can still proceed by quantum circuit theory, ascribing a matrix $\check{G}(t, t')$ to each node. The voltage fluctuations set up extra correlations between $\check{G}$ in different nodes. While these correlations can be evaluated, they cannot be understood as a renormalization of the corresponding Landauer connectors.

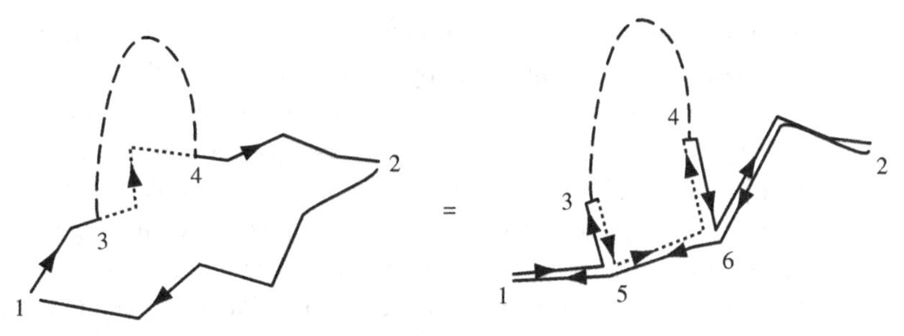

Fig. 6.17. **Aronov–Altshuler corrections to the conductivity as a result of interference (see the text for an explanation).**

This is why the interaction corrections in several-node circuits have not yet been sufficiently investigated. Fortunately the situation is understood in the limit of a very large network containing many nodes. We know that a sufficiently large circuit with approximately equal elements can be regarded as a diffusive conductor. The same applies to a sufficiently large part of it. Therefore, one may concentrate on interaction corrections to the transport taking place inside an extended diffusive conductor. These are called *Altshuler–Aronov corrections to the conductivity* [153, 159].

Why conductivity and not conductance? We have seen in Section 6.5.2 that the interaction corrections cease at an energy scale E_{Th} while persisting at higher energies. The same applies to Althshuler–Aronov corrections. In a diffusive conductor, $E \gg E_{\mathrm{Th}}$ gives rise to a space scale $L_E \simeq \sqrt{\hbar D/E}$ that is much smaller than the conductor dimension. The corrections are formed at the scale L_E and are non-local at this scale. However, if we look at electron propagation at scales $\gg L_E$ but still smaller than the conductor dimension L, the correction amounts to a change of the diffusion coefficient, or conductance. If $k_{\mathrm{B}}T \gg E_{\mathrm{Th}}$, the energy scale of the corrections is defined by temperature. One observes the effect as a temperature-dependent contribution $\delta\sigma$ to the conductivity of the sample. Such a scale separation allows us to see the effect even in bulk samples where the electron propagation over the whole sample is not coherent. Coherence is only required at the L_E scale.

The diffusive conductor can be described in the framework of the Usadel equation for Keldysh 2×2 Green's functions $\check{G}(t, t'; \boldsymbol{r})$ augmented by fluctuating voltages $\check{V}(t, \boldsymbol{r})$ that also retain the Keldysh index to take into account the non-commutativity of the operators $\hat{V}(t, \boldsymbol{r})$. The equation is then averaged over the voltage fluctuations using $S_V(\omega; \boldsymbol{r}, \boldsymbol{r}') = \langle V(t, \boldsymbol{r})V(t', \boldsymbol{r}')\rangle_\omega$. The resulting equation is for the "slow" Green's functions $\check{G}(\epsilon; \boldsymbol{r})$ that change at a scale $\gg L_\epsilon$. If more microscopic details are required, this procedure can be done at the level of the Eilenberger equations [160]. The practical implementation of this scheme requires us to cope with numerous technical difficulties, so we do not outline it here. Rather, we give a qualitative estimation of the effect, and at a later stage we quote Ref. [159] for the final exact result.

The Altshuler–Aronov corrections result from the interference of two electron propagation amplitudes, one of which is affected by interaction with the voltage fluctuations (Fig. 6.17). Both amplitudes go from point 1 to point 2; the unaffected amplitude is reversed since it is complex-conjugated. The affected amplitude is constructed as follows: the electron propagates to point 3 where it emits a virtual boson (dashed line) with energy $\hbar\omega$, and propagates further in a virtual state (dotted line). It picks up the same boson at point 4 and propagates in the usual way to point 2. The resulting probability for it to get from point 1 to 2 is contributed by all possible pairs of the amplitudes. However, the maximum contribution is achieved when the two amplitudes correspond to the propagation over the same path, in a direct and reverse direction, respectively. This path corresponds to a classical trajectory (see also Section 4.4.7). It looks like the amplitudes of opposite directions are attracted to one another, eventually sticking together. If we let them stick, the resulting arrangement includes five classical trajectories: shown as double lines in fig. 6.17. The electron propagates between 1 and 5 and between 6 and 2, following the usual classical trajectories. Its propagation between 5 and 6 is a correction to this common diffusive transport. The correction is thus determined by three trajectories: from 3 to 5, from 4 to 6, and from 6 to 5. These trajectories are less usual since one of the paired amplitudes corresponds to a virtual state shifted in energy by $\hbar\omega$. Because of the uncertainty relation, the time spent on these trajectories is restricted by $\simeq 1/\omega$. Therefore, they persist at the space scale $\sqrt{D/\omega}$. There occurs trajectory branching at points 5 and 6; as discussed in Section 6.4.3, such branching is vital for the correction under consideration. This makes the Althshuler–Aronov corrections to the conductivity distinct from those to the density of states. Another difference is the space scale: while the correction to the density of states is accumulated at scales *exceeding* the diffusion length, the corrections to the conductivity are produced at the space scale *not exceeding* $\sqrt{D/\omega}$.

The energy of the virtual photons emitted/adsorbed can be arbitrarily high. However, it cannot be much smaller than $k_B T$, where such emission–absorption processes are already allowed by energy conservation. If the contribution to the correction increases with decreasing energy, we expect the main contribution to come from $\hbar\omega \simeq k_B T$. The relevant spatial scale is therefore the thermal length $L_T \equiv \sqrt{\hbar D/k_B T}$. Similar to all interaction corrections of this chapter, its relative magnitude is determined by a resistance; in this case, the resistance of the conductor at the L_T scale:

$$\frac{\delta\sigma}{\sigma} \simeq G_Q R(L = L_T).$$

This sets the dependence of the Altshuler–Aronov correction on the effective geometry of the conductor. In one dimension, the resistance increases with increasing L_T (and temperature), $R \propto L_T$. In two dimensions, it does it very slowly, $R \propto \ln(L_T)$. In three dimensions, the resistance decreases, $R \propto 1/L_T$. So we can present the estimation as $\delta\sigma \simeq G_Q(L_T)^{2-d}$, where d is the effective dimension.

Control question 6.13. How is it that $\delta\sigma$ does not depend on the resistance in the estimation?

Control question 6.14. In one dimension, the correction increases with decreasing temperature. Take a finite wire of length L. What is the maximum decrease of the wire conductance, and at which temperature is it achieved?

We recall that the effective dimension is defined by L_T and thus changes with temperature. Consider the geometry of Fig. 6.3. At the highest temperatures, $L_T \ll L_z$, resulting in 3d correction. At lower temperature scales, corresponding to $L_z \ll L_T \ll L_y$ and

$L_y \ll L_{\mathrm{T}} \ll L_x$, the geometry is 2d and 1d, respectively. At lowest temperature, $L_{\mathrm{T}} \simeq L_x$, the correction disappears.

The exact expression, valid for any dimension $d = 1, 2, 3$, is given by [159]

$$\frac{\delta\sigma}{\sigma} = -\frac{2}{d}\int \mathrm{d}\omega\, \omega\frac{\partial}{\partial\omega}\left(\omega \coth\frac{\hbar\omega}{k_{\mathrm{B}}T}\right) \times \mathrm{Re}\int \frac{\mathrm{d}\boldsymbol{q}}{(2\pi)^d} G_{\mathrm{Q}} Z(\omega, \boldsymbol{q}) \frac{Dq^2}{(-\mathrm{i}\omega + Dq^2)^3}. \quad (6.116)$$

The integration here is over ω (the frequency of the boson involved) and $\boldsymbol{q}$ (corresponding to the vector distance between points 3 and 4). There are *three* diffusion-like denominators in Eq. (6.112) corresponding to the three classical trajectories in Fig. 6.17. The impedance takes into account the electron motion (see Section 6.1.2). One can reduce it to the circuit-theory impedance using the trick given in Eq. (6.97). We also note that the impedance $Z(\omega, \boldsymbol{q})$ should be taken at the diffusive length scale $\sqrt{D/\omega}$ rather than at a more natural scale of electricity propagation. This is why we can forget about the latter larger scale, treat the electricity propagation as instantaneous, and use Eq. (6.22) for any dimension. After these substitutions, the integration is readily performed to yield the exact numerical coefficients:

$$\delta\sigma = G_{\mathrm{Q}}(L_{\mathrm{T}})^{2-d} \times \begin{cases} -1.56 & d = 1 \\ -1/\pi \ln(L_{\mathrm{T}}/l) & d = 2 \\ 0.097 & d = 3. \end{cases} \quad (6.117)$$

We disregard interactions with the slow spin fluctuations taken into account in Ref. [159]. The Altshuler–Aronov corrections are routinely observed in the temperature-dependent resistance of diffusive conductors; see Fig. 6.23d for an illustration.

6.6 Fermionic environment

In this section, we consider interaction-related phenomena that cannot be naturally ascribed to a bosonic environment, for example, that of electrical excitations. These phenomena concern a quantum system with discrete states – a quantum dot or a qubit – brought into interaction with electrons of a metal; this is why we talk about a *fermionic* environment. We will discuss the Fermi edge singularity, the Kondo effect, and the concrete implementation of the latter involving a spin-doublet state in a quantum dot. The phenomena of the set have had a long history of research in condensed matter physics before the emergence of the field of quantum transport. They appear in a variety of solids and solid-state transport situations. In addition, models of the phenomena have attracted considerable interest from the purely theoretical community since they provide examples of problems where our understanding of interactions in a many-body system can progress far beyond the perturbative analysis. Eventually, all the models can be solved exactly, in ways both complicated and ingenious. Needless to say, in this book we have to restrict ourselves to (the simplest) quantum transport setups and elementary perturbation theory. We will show that the phenomena are generic for quantum transport, and that the flexibility and controllability of

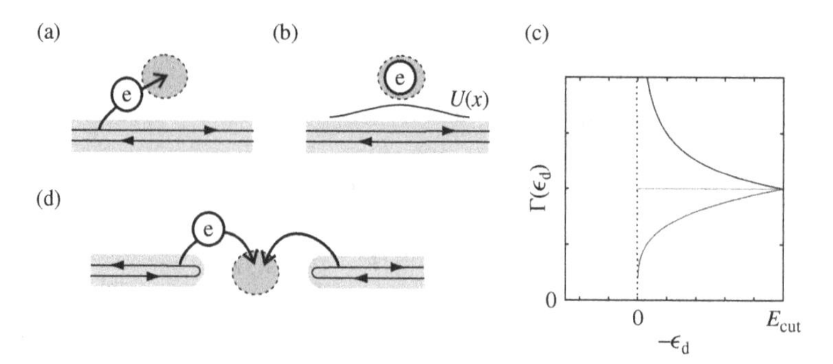

Fig. 6.18. **Fermi edge singularity (FES). (a) A simple setup for FES: electrons from a channel tunnel to a localized state at energy ϵ_d. (b) Interaction in the final state: the electron in the localized state creates potential $U(x)$ affecting the scattering in the channel. (c) FES tunneling rate exhibits a power-law energy dependence. It is either suppressed or enhanced at low energy. (d) The quantum dot setup for Exercise 6.11.**

the setups can bring some new functionalities to the extensively studied phenomena. The main feature of the fermionic environment is its ability to generate low-energy excitations or slow modes – electron–hole pairs, which sometimes can be conveniently regarded as boson modes. This is why at the very theoretical and qualitative level, there is no sharp difference between fermionic and bosonic environments. For example, one can make good use of the general classification of environments derived from the spin-boson model.

6.6.1 Fermi edge singularity

The name "Fermi edge singularity" (FES) is only partially self-explanatory: it only makes clear it has something unusual to do with electrons close to the Fermi surface. The name originates from the apparent singularities in X-ray absorption spectra in metals at frequencies matching the energy difference of a quasilocalized state and the Fermi energy of electrons [161].

Let us consider the FES in a simple quantum transport setup (see Fig. 6.18(a)). We have a localized state – a place for an electron – in the vicinity of a quantum channel. The electrons are allowed to tunnel between the channel and the localized state. Let us set up a charge qubit: the localized state can be either empty (qubit state "0") or occupied (qubit state "1"). To prevent double occupancy of the localized state, we work with a single spin direction and discuss the effect of spin later.

If we disregard interactions, the system is described by the following (and at this stage rather superfluous) Hamiltonian:

$$\hat{H} = \hat{H}_{\text{el}} + \hat{H}_{\text{d}} + \hat{H}_{\text{tun}};$$
$$\hat{H}_{\text{el}} = \sum_k \epsilon_k \hat{a}_k^\dagger \hat{a}_k, \quad \hat{H}_{\text{d}} = \epsilon_{\text{d}} \hat{d}^\dagger \hat{d}, \tag{6.118}$$
$$\hat{H}_{\text{tun}} = \sum_k \left(T_k d^\dagger a_k + T_k^* \hat{a}_k^\dagger \hat{d} \right).$$

Here, $\hat{a}_k^\dagger \hat{a}_k$ are creation/annihilation operators in electron states in the channel, labeled with wavevector k and having energies ϵ_k (counted from the Fermi energy). Operators $\hat{d}^\dagger$ and $\hat{d}$ are creation/annihilation operators in the localized state, and T_k are the tunneling matrix elements (see Eq. (3.30)).

We are interested in the tunneling rate from "0" to "1." The Fermi Golden Rule (see also Eq. (3.33)) readily yields

$$\Gamma(\epsilon_d) = \frac{2\pi}{\hbar} \sum_k |T_k|^2 f(\epsilon_k)\delta(\epsilon_k - \epsilon_d) = \tilde{\Gamma}\Theta(-\epsilon_d),$$

a featureless energy-independent rate we worked with when studying transport in quantum dots (see Section 5.4).

How does the interaction affect this rate? In principle, the system is in an electromagnetic environment: there is Coulomb interaction between electrons of the channel, and perhaps there is also interaction with electrical excitations of the external circuit. We have studied these effects in the previous sections and found that the rate is surpressed. There is one detail that we have missed in the previous sections and that becomes important now. Namely, we have assumed that some time after the tunneling process is accomplished, the electromagnetic excitations have gone and the environment returns to the initial state. This is certainly the case if the tunneling takes place between bulk metals. This is also true if the tunneling takes place at a localized state in the vicinity of an ideally screening metal: the charge of the electron in the occupied state does not influence the electron states in the metal. However, if the screening is not ideal, as in the case of sufficiently narrow constriction, the charge in the occupied state produces a potential $U(x)$ that affects the scattering of the electrons in the channel (Fig. 6.18(b)). The potential persists as long as the localized state is occupied, long after all the relaxation processes in the electromagnetic environment have taken place. This is the point we missed in our previous discussion, and now we are going to concentrate on this effect of *interaction in the final state*.

The effect is straightforward to model: we add the following term to the Hamiltonian:

$$\hat{H}_{\text{int}} = \sum_{k,k'} U_{k,k'} \hat{a}_k^\dagger \hat{a}_{k'} \hat{d}^+ \hat{d},$$

where $U_{k,k'}$ are the matrix elements of the potential $U(x)$. The product of the d-operators makes sure that the potential exists in state "1" only: indeed, $\hat{d}^+\hat{d} = 0$ in state "0" and $\hat{d}^+\hat{d} = 1$ in state "1." As a concrete model, let us assume that the potential is spread over a finite length a that exceeds the electron wavelength $\simeq k_F^{-1}$. In this case, the potential is too smooth to cause scattering between left- and right-going electrons in the channel: its only effect on the scattering is the extra phase shift χ acquired by an electron passing the localized state, given by

$$\chi = -\frac{1}{\hbar v_F} \int_{-\infty}^{\infty} \mathrm{d}x\, U(x).$$

Note that $U_{k,k'}$ is proportional to the Fourier-component of $U(x)$ that falls off at $k - k' \simeq a^{-1}$; we model this with a sharp cut-off at $|k - k'| = a^{-1}$, $U_{k,k'} = \chi\hbar v_F \Theta(|k - k'| - a^{-1})/\mathcal{V}$. Here $\mathcal{V}$ is the "normalization volume" used to convert between discrete and continuous k, $\sum_k = \mathcal{V}\int \mathrm{d}k/2\pi$. The parameter a also defines an upper cut-off energy in our problem, $E_{\mathrm{cut}} = \hbar v_{\mathrm{F}}/a$. We will concentrate on lower energies, assuming $|\epsilon_{\mathrm{d}}| \ll E_{\mathrm{cut}}$.

Let us start with small U, χ and compute the correction to the tunneling matrix element $\propto U$. The uncorrected matrix element is between the initial state, where the channel electrons are in the ground state and the localized state is empty, "0", and the final state, where the localized state is occupied, "1," and there is a hole at k with the energy matching ϵ_{d}, $\epsilon_k = \epsilon_{\mathrm{d}}$. The correction corresponds to a more complicated sequence of events. First, an electron from k' tunnels to the localized state. The energy of this intermediate virtual state with respect to the initial state is thus given by $-\epsilon(k') + \epsilon_{\mathrm{d}}$. Then the interaction term with $U_{k',k}$ transfers the hole from k' to k. The initial and final states are thereby the same, and the correction is obtained by summing up over all k':

$$Am = T_k + \delta Am;\ \ \delta Am = -\sum_{k'} \frac{T_{k'} U_{k',k} \Theta(-\epsilon_{k'})}{-\epsilon_{\mathrm{d}} + \epsilon(k')}$$

$$= T_k \frac{\chi}{\hbar v_{\mathrm{F}}} \int_{k-1/a}^{k+1/a} \frac{\mathrm{d}k}{2\pi} \frac{\Theta(-\epsilon_{k'})}{\hbar v_{\mathrm{F}}(k' - k_{\mathrm{F}}) - \epsilon_{\mathrm{d}}} = T_k \frac{\chi}{2\pi} \int_{\epsilon_{\mathrm{d}} - E_{\mathrm{cut}}}^{0} \frac{\mathrm{d}E}{E - \epsilon_{\mathrm{d}}}$$

$$= -T_k \frac{\chi}{2\pi} \ln\left(\frac{E_{\mathrm{cut}}}{|\epsilon_{\mathrm{d}}|}\right).$$

Some technical comments are necessary when working with the above calculation. The anticommutation of fermionic creation/annihilation operators is important for the sign of correction; to achieve this, we made use of $(\hat{a}^{\dagger}_{k'}\hat{a}_k)(\hat{d}^{\dagger}\hat{a}_{k'}) = -\hat{d}^{\dagger}\hat{a}_k\hat{a}^{\dagger}_{k'}\hat{a}'_k = -\hat{d}^{\dagger}\hat{a}_k$, which is valid if k' is filled. We have neglected the k-dependence of T_k, $T_k \approx T_{k'}$. We switched to integration over energy and performed it with logarithmic accuracy; such low accuracy makes irrelevant the exact form of the cut-off of $U_{k,k'}$ in k-space. We have dealt with the singularity at $E = \epsilon_{\mathrm{d}}$ by taking the principal value of the integral. It is essential that the potential $U_{k,k'}$ is in state "1" only: if it were not, we would see the contribution from an alternative intermediate state. If the potentials were the same in "0" and "1" states, the contributions would cancel each other out, signaling no interaction.

Now we can discuss the result. Remarkably, it displays a logarithmic divergency at low $|\epsilon_{\mathrm{d}}|$ that may make the correction large. Our previous experience suggests that we can implement a renormalization scheme: we compute the tunnel matrix element in small steps, taking into account at each step the contribution of k' in the energy interval $E - \mathrm{d}E < \epsilon(k') < E$. This results in a simple flow equation given by

$$\frac{\mathrm{d}T_k(\zeta)}{\mathrm{d}\zeta} = -\frac{\chi}{2\pi} T_k(\zeta),$$

which brings the power-law energy dependence to the matrix element, $T_k(\epsilon) \propto (|\epsilon|/E_{\mathrm{cut}})^{\chi/2\pi}$. It is important to note that at low energies the matrix element can be suppressed, as well as enhanced, depending on the sign of χ, resulting in a power-law energy dependence of the rate with either positive or negative exponents (Fig. 6.18). Positive U (negative χ) results in enhancement.

There are logarithmic corrections of another type to be taken into account. They manifest as the Anderson orthogonality catastrophe discussed in Section 6.2.3. Let us recognize that the FES also incorporates the concept of two vacuums, ground states of the electron systems, corresponding to states "0" and "1." The interaction in the final state causes the vacuums to differ from one another, so that the overlap of the two is given by $\langle 1_0 | 0_1 \rangle \neq 1$. Let us implement a renormalization scheme to compute this overlap, taking into account the electron states within the energy interval $(E - \mathrm{d}E, E)$. In the first order in $U_{k,k'}$, the "1" vacuum acquires an admixture of the states with a single electron–hole excitation, $|k, k'\rangle \equiv \hat{a}_k^\dagger \hat{a}_{k'} |0_0\rangle$,

$$|0_1\rangle = |0_0\rangle - \sum_{k,k'} \frac{U_{k,k'}}{\epsilon(k) - \epsilon(k')} |k, k'\rangle.$$

The correction to the overlap is the sum of the squares of all admixture coefficients, that is, it is quadratic in $U_{k,k'}$. Concentrating on the electron–hole pairs in the energy interval $(E, E - \mathrm{d}E)$, we arrive at the following flow equation:

$$\frac{\mathrm{d}|\langle 0_0 | 0_1 \rangle|^2}{\mathrm{d}\zeta} = -\alpha_{\mathrm{cl}} |\langle 0_0 | 0_1 \rangle|^2; \; \alpha_{\mathrm{cl}} = \frac{\chi^2}{2\pi^2}.$$

This indeed signals the orthogonality catastrophe: the overlap between two vacuums vanishes in the limit of vanishing energy, $|\langle 0_0 | 0_1 \rangle|^2 = (\epsilon / E_{\mathrm{cut}})^{\alpha_{\mathrm{cl}}}$. The coefficient α_{cl} is equally contributed by left- and right-going electrons since they acquire the same phase shift χ. So far we have disregarded electron spin, now we can easily take it into account: each spin direction contributes to α_{cl} independently, and this doubles α_{cl} for spin-degenerate electrons. We note that, unlike for the tunnel matrix element exponent, the overlap exponent is due to electron states that are not necessarily involved in the tunneling into the localized state. For instance, the localized state may get an electron from another lead or from another state: even in this case, the rate will be affected by the overlap exponent.

Now we can combine both the renormalization of the matrix element and the vacuum overlap into a single expression for the rate:

$$\Gamma(\epsilon_{\mathrm{d}}) = \Theta(-\epsilon_{\mathrm{d}}) \tilde{\Gamma} \left(\frac{|\epsilon_{\mathrm{d}}|}{E_{\mathrm{cut}}} \right)^{(\chi/\pi) + (\chi^2/2\pi^2)}. \tag{6.119}$$

In principle, the whole discussion has so far been restricted to perturbations, that is, to small χ and, correspondingly, to small exponents. It is feasible that at bigger χ the exponents deviate from the values given; for example, there could be $\sin \chi$ instead of χ. An exact solution of the problem [161] shows that this does not happen: the exponent stays as given for any χ. Since the χ are phase shifts, the exponent must be 2π-periodic in χ; this is achieved by bringing them to the interval $(-\pi, \pi)$.

We see that the FES exponent can be of either sign, giving both suppression and enhancement of the tunneling rate at low energies (Fig. 6.18). Let us see how we can reconcile the FES with the general classification of environments. Comparing the rate expressions given by Eqs. (6.55) and (6.119), we understand that, in general, the FES is at the ohmic borderline, the ohmic exponent being given by $\alpha = (1/2)(1 + \chi^2/2\pi^2 + \chi/\pi)$. This establishes equivalence between bosonic and FES fermionic environments, which is sometimes useful in applications. For example, non-interacting fermions correspond to an ohmic bosonic environment with $\alpha = 1/2$. The equivalence also allows for the qualitative analysis of the resulting qubit flipping between "0" and "1." For example, the qubit undergoes a Schmid transition at $\chi \approx 2.3$ corresponding to $\alpha = 1$.

We have understood the FES for a simple setup. It turns out that it can be readily understood for any nanostructure conforming to a scattering matrix [162]. Let the nanostructure be described by a scattering matrix $\hat{s}_0$ in the state "0." When the localized state is occupied, the induced potential changes the scattering in the nanostructure, so the corresponding scattering matrix becomes $\hat{s}_1$. The difference brought by the induced potential – the interaction in the final state – is characterized by the "ratio" of the matrices, a unitary matrix $\hat{R} = \hat{s}_0^{-1}\hat{s}_1$. As for any unitary matrix, it can be diagonalized as follows:

$$R_{\alpha,\beta} \leftarrow \sum_j v_\alpha^{(j)} \exp(\mathrm{i}\chi_j) v_\beta^{(j)*},$$

where α, β label incoming channels and j labels eigenvalues $\exp(\mathrm{i}\chi_j)$ and eigenvectors of the matrix.

The overlap exponent α_{cl} is given in terms of the eigenvalues, $\alpha_{\mathrm{cl}} = \sum_j (\chi_j/2\pi)^2$, and does not depend on the details of tunneling, as expected. As for the details, all incoming electrons can eventually tunnel into the localized state; the probability of this occurring generally depends on the channel. To account for this, one introduces channel-dependent tunnel matrix elements T_α. Within this quite general setup, the tunnel rate is given by

$$\Gamma(\epsilon_{\mathrm{d}}) = \sum_j \Gamma_j \left(\frac{\epsilon_{\mathrm{d}}}{E_{\mathrm{cut}}}\right)^{\chi_j/\pi} \left(\frac{\epsilon_{\mathrm{d}}}{E_{\mathrm{cut}}}\right)^{\alpha_{\mathrm{cl}}}, \tag{6.120}$$

where partial rates $\Gamma_j \propto \sum_\alpha |T_\alpha v_\alpha^{(j)}|$. Since different eigenvalues correspond to different exponents, the above rate is dominated by a single term with the most negative χ_j.

Control question 6.15. What is the correspondence between the general rate given in Eq. (6.120) and the simple setup result in Eq. (6.119)?

Exercise 6.11. Consider a quantum dot housing a localized state (Fig. 6.18(d)). Electrons can tunnel to the dot from one-channel leads on the left and on the right. If the dot is occupied, the reflection amplitude in the left and in the right lead acquires extra phase factors $\mathrm{e}^{i\phi_{\mathrm{L,R}}}$, respectively. Find the energy-dependence of the tunneling rates. Take electron spin into account.

To our knowledge, the FES phenomenon has been observed in the context of quantum transport in Refs. [162] and [163]. It appears to be a promising goal for future experimental research, especially in view of the proposal of tunable FES exponents elaborated in Ref. [164].

6.6.2 Kondo effect

The Kondo effect is usually considered in the context of spin. Indeed, Kondo's discovery resolved a long-standing puzzle concerning enhanced electron scattering by magnetic impurities. However, the Kondo effect is more general and always manifests itself in the presence of *degenerate* discrete quantum states interacting with many degrees of freedom. Unusual and controllable degeneracies can be created in various quantum transport setups. Therefore it is necessary for our purposes to outline the Kondo effect in the most general terms possible.

We have exemplified many times in the book the paradigm of quantum transport: any nanostructure at sufficiently low energies becomes a single scatterer and can be completely characterized by its scattering matrix. The reason is that at low energies the nanostructure is void. The internal degrees of freedom, if present, are "frozen out": it costs energy to excite them, and this energy is not available at sufficiently low temperature/voltage. The ground state of the nanostructure, as well as the whole setup, is the ground state of the electron gas.

This may change if the nanostructure is designed to have a *degenerate* ground state. This is surely possible if we make an isolated nanostructure. It will display discrete quantum states, and will become a kind of qubit. It may be tricky to set several such states at the same energy, but no fundamental law prohibits this. If we bring the qubit into contact with the leads and let electrons scatter at it, the paradigm of quantum transport seems to be violated: the electron scattering may switch the qubit states for no cost, while the qubit retains its internal degree of freedom.

The driving force behind the Kondo effect is the instability of the situation created thereby. The origin of the instability is not obvious and its discovery by Kondo [165] (see also Ref. [166]) came as a big surprise. The instability develops at low energy and essentially changes the properties of the qubit, entangling it with the electrons of the leads. The development of the instability and the fate of the nanostructure may follow several scenarios, but in generic situations at least the nanostructure is void again, and the validity of the paradigm is restored. This can happen in two ways: either the qubit is uncoupled from the nanostructure and would not be switched by scattering, or it is coupled so strongly that its degeneracy is *lifted*.

Let us understand quantitatively the origin of the phenomenon. We start with a generic nanostructure described by a scattering matrix $s_{\alpha\beta}$ in channel space indexed by Greek letters. At the moment, we are not interested in how the channels are distributed over the reservoirs. To this end, we can diagonalize the scattering matrix and neglect the phase shifts obtained. This amounts to setting $s_{\alpha\beta} = \delta_{\alpha\beta}$. Now we connect the qubit. We allow for an arbitrary number of quantum states and label them with Latin letters. If the qubit is connected, the electron can enter channel α while the qubit is in state a and scatter to channel β, switching the qubit to state b. Let the corresponding scattering amplitude be $\mathrm{i}t_{\beta b,\alpha a}$. The factor i is a convenient choice: $t_{\beta b,\alpha a}$ is Hermitian by virtue of unitarity condition. While the coupling is weak, the corresponding amplitudes are small, $|t_{\beta b,\alpha a}|^2 \ll 1$.

This looks very much like the scattering approach for non-interacting electrons, with an extra scattering matrix index corresponding to the qubit state. It is important to note that this is not so: the electrons do interact. Given two electrons and the qubit in state a,

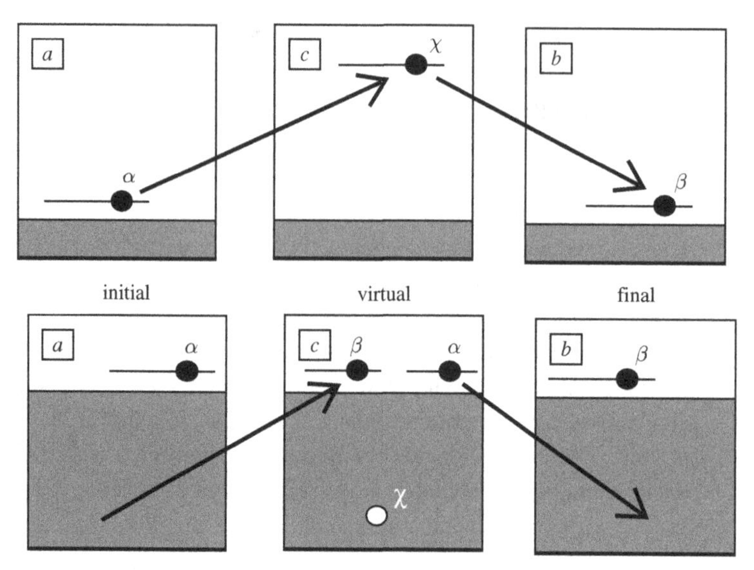

Fig. 6.19. The sequences of the states – electron-like and hole-like – contributing to the renormalization of the scattering amplitude $t_{\beta b,\alpha a}$.

the first electron may switch the qubit to state b and thereby make its presence felt by the second one. We are going to show that this leads to renormalization of the tunneling amplitudes. It is convenient to derive the renormalization using Hamiltonians; let us associate the Hamiltonian (see Eq. (6.118)) given by

$$\hat{H}_{\text{int}} = -\sum_{k,k'} \sum_{\alpha,\beta,a,b} \frac{\hbar}{\sqrt{\mathcal{V} v_{\text{F}}^{\alpha} v_{\text{F}}^{\beta}}} t_{\beta b,\alpha a} |b\rangle\langle a| \hat{a}^{\dagger}_{k,\beta} \hat{a}_{k',\alpha},$$

with the introduced tunneling amplitude, where v_{F}^{α} are the Fermi velocities in the channels.

Let us look at the sequences of states (see Fig. 6.19) that give the second-order correction to the amplitude $t_{\beta b,\alpha a}$. The initial state is an electron in channel α at an energy close to the Fermi surface. There are two types of virtual state: electron-like and hole-like. The electron-like state is achieved by transferring the electron from α to another channel χ and switching the qubit to state c. Since the energies of the virtual state are not limited by energy conservation, the electron can be quite far from the Fermi surface. Then the electron is transferred to channel β and the qubit is switched to b. The hole-like state is obtained by extracting the electron from channel β and switching the qubit to c, while nothing happens to the original electron. Only at the next stage does this electron annihilate the hole in χ, switching the qubit to b. Anticommutation of electron creation/annihilation operators accounts for the negative sign of the hole-like contribution. The final states are the same for both types. To obtain to the flow equation, in full accordance with the renormalization schemes considered previously, we take into account the contribution of the intermediate states in an energy interval $(E - \mathrm{d}E, E)$ at each small renormalization step. This yields

$$\frac{\mathrm{d}t_{\beta b,\alpha a}}{\mathrm{d}\zeta} = \frac{1}{2\pi}\left(t_{\beta b,\chi c}t_{\chi c,\alpha a} - t_{\chi b,\alpha c}t_{\beta c,\chi a}\right). \tag{6.121}$$

One can rewrite Eq. (6.121) in more compact form treating $t_{\beta b,\alpha a}$ as a matrix in the space of both channels and qubit states, $t_{\beta b,\alpha a} \to \hat{t}$. It is helpful to introduce the operation of *partial transpose* that transposes the channel indexes only, $\bar{\hat{t}} \to t_{\alpha b,\beta a}$. Using this notation, the equation becomes

$$\frac{\mathrm{d}\hat{t}}{\mathrm{d}\zeta} = \frac{1}{2\pi}\left(\hat{t}^2 - \overline{\left(\bar{\hat{t}}\right)^2}\right).$$

This flow equation differs from those given earlier in the book: it is not linear and also has a peculiar index structure. This structure means that simple solutions that correspond to separation of the channel and qubit indexes are of no use. For example, we could try $t_{\beta b,\alpha a} \propto \delta_{ab}$ (the electrons do not switch the qubit) or $t_{\beta b,\alpha a} \propto \delta_{\alpha\beta}$ (the electrons do not change channels while scattering). In both cases, the amplitudes do not flow, $\mathrm{d}\hat{t}/\mathrm{d}\zeta = 0$.

Control question 6.16. Can you see this from Eq. (6.121)?

This is why the general solution of the equation is not known.

To find a meaningful partial solution, we recall the discussion of entanglement in Section 5.2.1. We can regard the Hermitian matrix of the amplitudes as a kind of Hamiltonian, and its eigenvectors as a kind of wave function. There is a natural bipartition in this space: that into channels and qubit states, so the concept of entanglement should make sense. Let us try a maximally entangled wave function $|\Psi\rangle$ incorporating N_{ch} channels and N_{q} qubit states. Since it is maximally entangled,

$$\sum_a \Psi^*_{\alpha a}\Psi_{\beta a} = \delta_{\alpha\beta}/N_{\mathrm{ch}}; \; \sum_\alpha \Psi^*_{\alpha a}\Psi_{\alpha b} = \delta_{ab}/N_{\mathrm{q}}.$$

Let us take a traceless matrix that has this $|\Psi\rangle$ as an eigenfunction, with an eigenvalue t, as follows:

$$\hat{t} = K\left(t|\Psi\rangle\langle\Psi| - \frac{t}{N_{\mathrm{ch}}N_{\mathrm{q}}}\hat{1}\right), \tag{6.122}$$

where $K \equiv 1/(1 - 1/N_{\mathrm{ch}}N_{\mathrm{q}})$, and the second term is to make sure that $\mathrm{Tr}(\hat{t}) = 0$. Substituting this into Eq. (6.121), we obtain a simple flow of t, given by

$$\frac{\mathrm{d}t}{\mathrm{d}\zeta} = \frac{K}{2\pi}t^2. \tag{6.123}$$

If high-energy $t(\zeta = 0) = t_0$ is positive, it grows with decreasing energy. The interaction – the Kondo effect – efficiently increases electron scattering till it reaches the strong coupling regime at $t \simeq 1$. Let us note that Eq. (6.123) is solved by $t(\zeta) = (2\pi/K)(\zeta_{\mathrm{K}} - \zeta)^{-1}$. This diverges at $\zeta_{\mathrm{K}} = \zeta$. While the divergence itself is fake, arising from the fact that the validity of renormalizations is restricted to small t, its position gives a good estimation of the energy scale at which t becomes $\simeq 1$. Comparing with $t(\zeta = 0)$, we find this Kondo energy to be

$$T_{\mathrm{K}} \simeq E_{\mathrm{cut}}\mathrm{e}^{-2\pi/Kt_0}. \tag{6.124}$$

The energy dependence can be represented as follows:

$$t(E) = \frac{2\pi}{K}\frac{1}{\ln(E/T_K)} \text{ at } E \gg T_K,$$

which conveniently does not contain the cut-off energy.

If the initial t_0 is negative, strong coupling does not occur and $t(E)$ slowly vanishes at low energies. Its energy dependence is given by the same expression, but with negative ζ_K, and, correspondingly, with T_K that are much larger then E_{cut}, E. The qubit is thus uncoupled from the nanostructure in the low-energy limit. This situation is termed *asymptotic freedom*, and is relatively uninteresting, so we concentrate on the opposite, strong-coupling regime.

One should not take for granted that the partial solution described gives the flow of scattering amplitudes in all possible situations. The solution of the renormalization equation is determined by initial conditions that may be quite complicated, involving many channels and not having to satisfy Eq. (6.122). Only simple models with symmetry properties, like a classic example considered below, immediately result in a flow similar to that exemplified above. Roughly, a general initial condition may be presented by a large linear combination of terms like Eq. (6.122) with different eigenvalues. The terms with larger eigenvalues grow faster and also may suppress the growth of less successful terms. Finally, the strongest (and frequently simplest) survives and determines the relevant scattering amplitudes. This justifies the use of simple models.

To enable the Kondo renormalization, the qubit states do not have to be precisely degenerate: it is only required that their energy splitting is much smaller than the running energy scale E. If the splitting is less than T_K, it does not prevent the transition to the strong-coupling regime. If the splitting is bigger than T_K, the renormalization stops at the splitting energy. It may be that the lowest split levels are still degenerate: the renormalization then proceeds with a "reduced" qubit.

Magnetic impurity

Let us illustrate this with the most usual realization of the Kondo effect in solids: a magnetic impurity with spin $1/2$ in a metal. Near the impurity, the plane waves of the electrons in the metal can be expanded in spherical harmonics: this gives a discrete set of transport channels labeled by discrete values of angular momentum, $L = 0, 1, ...$ A small impurity is felt by s-waves ($L = 0$) only. So, effectively we have two channels corresponding to two spin directions of the electron. In terms of the general model, $N_{ch} = N_q = 2$, $K = 4/3$. The scattering leading to the Kondo effect is due to an exchange interaction between the impurity and electron spins. The rotational symmetry in spin space sets the form of this interaction as follows:

$$\hat{t} = -\frac{t}{3}(\boldsymbol{\sigma}_{el} \cdot \boldsymbol{\sigma}_i),$$

which is luckily of the form given in Eq. (6.122). The eigenfunction of $\hat{t}$ is a maximally entangled state: spin singlet $|\Psi\rangle = (|\uparrow\rangle_{el}|\downarrow\rangle_i - |\downarrow\rangle_{el}|\uparrow\rangle_i)/\sqrt{2}$. Positive t values correspond to antiferromagnetic coupling: it is energetically favorable. Therefore, at a sufficiently small energy scale (temperature) T_K given by Eq. (6.124), with $E_{cut} \simeq E_F$, the impurity crosses to the strong-coupling regime.

What happens to the impurity in this regime? We may give a quick sketch as follows. An electron from the channel with a spin opposite to that of the impurity sits at the impurity, forming a singlet state with bound energy $\simeq T_K$. The degeneracy is thereby neutralized. Owing to this, at low energy $E \ll T_K$, all other electrons pass the impurity without spin switching – they only acquire phase shifts while traversing it. The phase shifts at Fermi energy can be found from the Friedel sum rule (see Eq. (1.149) and discussion there) and correspond to the fact that the extra electron neutralizing the impurity spin is accommodated below the Fermi surface. The phase shifts are $\chi = \pi$ for both electron directions. The exact solution confirms this [167].

It might seem that these phase shifts cannot be observed. However, the scattering the impurity provides to the electron flow comes about from the interference of all spherical harmonics, and is thereby sensitive to the phase shift χ of the s-wave. The scattering cross-section is given by $\sigma = 4\pi k_F^{-2} \sin^2(\chi/2)$; this gives a contribution to the scattering rate $1/\tau_p = v_F c_i \sigma$, c_i being the impurity concentration. Note that $\chi = \pi$ corresponds to the maximum possible scattering rate $(1/\tau)_u = 4\pi v_F c_i / k_F^2 = c_i/(2\nu\hbar)$. The contribution of magnetic impurities to the resistance of a material thus grows with decreasing temperature – this was the puzzle resolved by Kondo.

We introduce two phenomena (spin-flip and energy exchange) that will become important in Section 6.8 and concern the effect of the impurity on the electrons scattered.

Spin-flip

If the electrons can flip the spin of the impurity, this provides a mechanism of spin relaxation. In distinction from the spin-scattering considered in Chapter 1, this spin-flip is incoherent, involving the change of the impurity state. Above the Kondo energy scale, the contribution to the spin-flip rate is given by

$$\frac{1}{\tau_{sf}} = (1/\tau)_u \frac{t^2(E)}{12}; \; t(E) \ll 1. \tag{6.125}$$

The spin-flip rate thus grows slowly with lowering temperature, achieving values $\simeq(1/\tau)_u$ at $E \simeq T_K$. At lower temperatures, the spin-flip ceases since the impurity spin is neutralized.

Two-electron energy exchange

We have mentioned at the beginning of this subsection that the electrons interact through the switching of the qubit: the electron that enters first may change the qubit state and thereby change the scattering state of the following electron. This leads to an energy exchange of the two scattered electrons. Two electrons coming with energies $E \pm E_i$ will go away with energies $E' \pm E_f$ if $E = E'$ conform to energy conservation. A consideration of a two-electron wave function after the scattering event yields that small energy transfers are strongly preferred: $\Gamma \simeq (E_i - E_f)^{-2}$.

For a spin 1/2 impurity, the rate at $t \ll 1$ is given by [168]

$$\Gamma_\mathrm{r} = \left(\frac{t}{3}\right)^4 \frac{(1/\tau)_u}{2\pi^2} \left[\frac{1}{(E_\mathrm{i} - E_\mathrm{f})^2} + \frac{1}{(E_\mathrm{i} + E_\mathrm{f})^2}\right] \delta(E - E'). \tag{6.126}$$

Similar to the spin-flip rate, the energy exchange rate ceases at $T < T_\mathrm{K}$.

Exotic Kondo

A spin-1/2 impurity provides the single most common scenario of the Kondo effect. We will review its realization in quantum dots in Section 6.6.3. There is a large set of different, more exotic and interesting, Kondo scenarios that find a place in quantum transport. Unfortunately we cannot discuss them here and advise the reader to review Refs. [169]–[173] and references therein.

6.6.3 Quantum dot according to Anderson

A quantum dot with an odd number of electrons must be degenerate with respect to spin and is typically in a spin-doublet state. Bringing it in contact with the leads creates a Kondo setup. Forming an analogy with a spin-1/2 impurity, we expect the spin degeneracy to be lifted at a sufficiently low energy scale $\simeq T_\mathrm{K}$. Remarkably, the lifting of the degeneracy results in an increased transmission of the dot, which can reach an ideal value of a quantum point contact. At least theoretically, the conductance of a quantum dot versus gate voltage at very low temperature and voltage should display an alternating conductance pattern: $G = G_\mathrm{Q}$ in odd diamonds and $G \approx 0$ in even diamonds (see Fig. 6.20).

We will analyze this quantitatively in the framework of the *Anderson model*. P. W. Anderson provided many seminal contributions, moving far ahead of the research frontiers of the time. In particular, he proposed a basic model for the Kondo effect in quantum dots long before quantum dots had ever been thought about. The Anderson model was to provide a minimum microscopic description for a magnetic impurity in a metal. Its adaptation for the dot setup is as follows [175]. There is a single spin-degenerate level in the dot, its position ϵ_d with respect to the Fermi level can be tuned with a gate voltage. The charging energy U makes the double occupancy of the dot energetically unfavorable. The Hamiltonian of the dot is thus given by

$$\hat{H}_\mathrm{d} = \epsilon_\mathrm{d}\hat{n}_\mathrm{d} + \frac{U}{2}\hat{n}_\mathrm{d}(\hat{n}_\mathrm{d} - 1),$$

where the total number of electrons $\hat{n}_\mathrm{d}$ in the dot is contributed to by both spin directions, $\hat{n}_\mathrm{d} = \hat{d}^\dagger_\uparrow \hat{d}_\uparrow + \hat{d}^\dagger_\downarrow \hat{d}_\downarrow$, $\hat{d}_\sigma$ being the electron annihilation operators in the dot. Energy consideration yields $n_\mathrm{d} = 1$ in the interval $0 < \epsilon_\mathrm{d} < -U$, $n_\mathrm{d} = 0$ if $\epsilon_\mathrm{d} > 0$, and $n_\mathrm{d} = 2$ if $\epsilon_\mathrm{d} < -U$ (Fig. 6.20). There is no difficulty in applying the Anderson model for multi-electron dots: for a given odd diamond, one just considers a level being filled and disregards all other levels. The n_d in this case is the number of electrons in the level being filled.

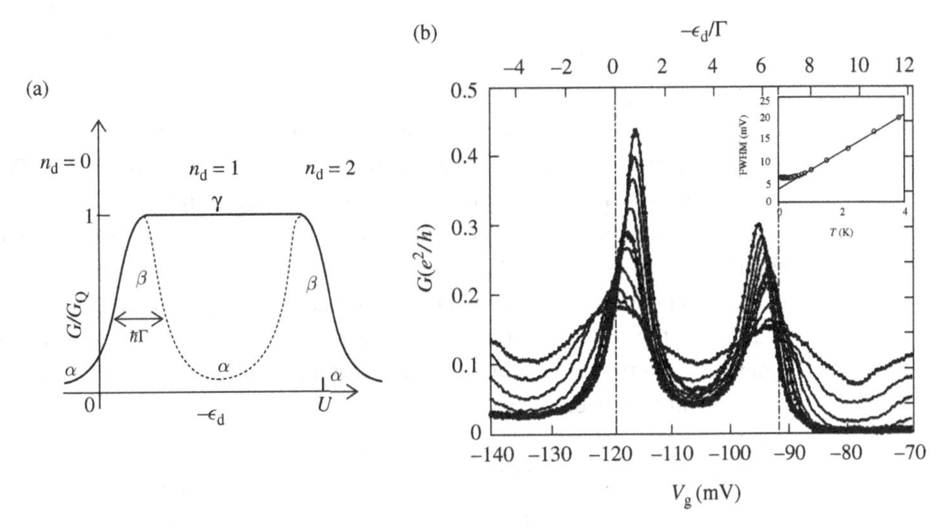

Fig. 6.20. **Kondo effect in quantum dots. (a) Kondo effect may result in an ideal transmission. Sketch of linear conductance of Anderson quantum dot versus gate voltage. Greek letters indicate different transport regimes: α = co-tunneling; β = Coulomb peaks or mixed-valence regime, γ = Kondo regime. (b) Experiment from Ref. [174]. Conductance versus gate voltage at different temperatures. The rise of conductance with temperature manifests the Kondo effect. Curves are for different temperatures, ranging from 100 mK to 3800 mK.**

The dot is coupled to two single-channel leads, the annihilation operators in the left (right) lead being $\hat{a}_{k\sigma}$ ($\hat{b}_{k\sigma}$), $\sigma = \uparrow, \downarrow$. The tunnel matrix elements $T_{a,b}$ couple the electron states to the dot level, so the total Hamiltonian is given by

$$\begin{aligned} \hat{H} &= \hat{H}_{\text{el}} + \hat{H}_{\text{d}} + \hat{H}_{\text{tun}}; \\ \hat{H}_{\text{el}} &= \sum_{k\sigma} \epsilon_k \hat{a}^\dagger_{k\sigma} \hat{a}_{k\sigma} + \sum_{k\sigma} \epsilon_k \hat{b}^\dagger_{k\sigma} \hat{b}_{k\sigma}; \\ \hat{H}_{\text{tun}} &= T_a \sum_{k\sigma} \hat{a}^\dagger_{k\sigma} \hat{d}_{k\sigma} + T_b \sum_{k\sigma} \hat{b}^\dagger_{k\sigma} \hat{d}_{k\sigma} + \text{h.c.} \end{aligned} \tag{6.127}$$

These tunnel matrix elements are related to the transport rates from the level to the corresponding leads: $\Gamma_{a,b} = \pi T_{a,b} \nu / \hbar$. In distinction from the Fermi edge singularity model, the tunnel matrix elements do not depend on the dot occupation. The condition of good isolation requires $\Gamma \ll U$, which we will assume.

Control question 6.17. Can you argue why the good isolation is required and describe the physics of Anderson model in the opposite limit?

The apparent difference to the original Anderson model is the presence of two leads and two tunneling couplings. However, there is a hidden symmetry in the model that eradicates even this difference. We can redefine the channels by making a unitary transformation with a 2×2 matrix $\hat{V}$, as follows:

$$\begin{pmatrix} \hat{f}_{k\sigma} \\ \hat{g}_{k\sigma} \end{pmatrix} = \hat{V} \begin{pmatrix} \hat{a}_{k\sigma} \\ \hat{b}_{k\sigma} \end{pmatrix}; \; \hat{V} \equiv \frac{1}{T_{\mathrm{d}}} \begin{pmatrix} T_a & T_b \\ T_b & -T_a \end{pmatrix}$$

($T_{\mathrm{d}} \equiv \sqrt{T_a^2 + T_b^2}$). Then only channel g will couple the dot level with the tunnel matrix element T_d, while channel f does not even know that the dot is present. This trick allows for a comprehensive evaluation of the transport between the leads.

Let us start by evaluating the co-tunneling amplitude: the correction to the transmission amplitude in the second order in T_{d}. This correction is obviously absent in the f channel. In the g channel, it is contributed by two virtual states: one where the dot is empty (energy denominator E_-) and another where the dot is filled with two electrons in singlet state (energy denominator E_+). For all initial and final configurations of electron and dot spins, the correction is given by

$$\begin{aligned} \hat{t} &= -\mathrm{i}\frac{\Gamma}{2}\left(\frac{1}{E_+} + \frac{1}{E_-}\right)(\boldsymbol{\sigma}_{\mathrm{el}} \cdot \boldsymbol{\sigma}_i) + \mathrm{i}\frac{\Gamma}{2}\left(\frac{1}{E_+} - \frac{1}{E_-}\right) \\ &\equiv -\mathrm{i}\frac{t}{3}(\boldsymbol{\sigma}_{\mathrm{el}} \cdot \boldsymbol{\sigma}_{\mathrm{i}}) + \mathrm{i}t_1. \end{aligned}$$

Since we are interested in transmission from the left to the right, we shall transform back to the channels $\hat{a}, \hat{b}$ with the help of 2×2 matrix $\hat{V}$ as follows:

$$\delta\hat{s} = \hat{V}^{-1} \begin{pmatrix} \hat{t} & 0 \\ 0 & 1 \end{pmatrix} \hat{V} = \frac{\hat{t}}{T_{\mathrm{d}}^2} \begin{pmatrix} T_a^2 & T_a T_b \\ T_a T_b & T_b^2 \end{pmatrix}.$$

The non-diagonal element gives the transmission from the left to the right, with and without spin-flip. Squaring the amplitude elements, we obtain

$$G_{\mathrm{cot}} = G_{\mathrm{Q}} \frac{4\Gamma_a \Gamma_b}{(\Gamma_a + \Gamma_b)^2} \frac{t_0^2/3 + t_1^2}{4} = G_{\mathrm{Q}} \hbar^2 \Gamma_a \Gamma_b \left(\frac{1}{E_+^2} + \frac{1}{E_-^2} + \frac{1}{E_+ E_-}\right). \tag{6.128}$$

Control question 6.18. Can you explain Eq. (6.128)? What is the probability of flipping the electron spin while transmitting?

Now let us take the Kondo effect into account. The phase shift t_1 is not affected by Kondo renormalization, while t grows as described in Section 6.6.2. The strong-coupling regime is achieved at the Kondo energy scale:

$$T_{\mathrm{K}} = \frac{\sqrt{U\hbar\Gamma}}{2} \exp\left(-\frac{\pi E_+ E_+}{(E_+ + E_-)\hbar\Gamma}\right). \tag{6.129}$$

The factor in the exponent is obtained from Eq. (6.124), while the prefactor may be obtained by a renormalization calculation in higher orders or just from the exact solution of the Anderson model [167].

Now we are ready for conductance in the strong-coupling regime T_{K}. We recall that the f channel gets π-shift, while the g channel does not: this sets the scattering matrix in this basis. Transforming back to $\hat{a}, \hat{b}$, we find the scattering matrix to be

$$\hat{s} = \hat{V}^{-1} \begin{pmatrix} -1 & 0 \\ 0 & 1 \end{pmatrix} \hat{V} = \frac{1}{T_{\mathrm{d}}^2} \begin{pmatrix} T_b^2 - T_a^2 & -2T_a T_b \\ -2T_a T_b & T_a^2 - T_b^2 \end{pmatrix}.$$

Squaring the non-diagonal elements gives the conductance in the Kondo regime:

$$G = G_Q \frac{4\Gamma_a \Gamma_b}{(\Gamma_a + \Gamma_b)^2}; \quad G = G_Q \text{ at } \Gamma_a = \Gamma_b.$$

The transmission is thus ideal for symmetric couplings. The asymmetry-related coefficient is the same as for the height of the resonant peaks in the double junction (Section 1.6) and in resonant tunneling through a single level (Section 5.4). This is not surprising since it relates to the same interference trick: neither effect destroys the tunnel barriers, but overcomes them by a fine phase tuning. The difference is that the Kondo effect provides the same phase shift over a wide interval of gate voltages. A cartoon is that the Kondo effect "pins" the resonance to the Fermi level.

An interesting regime in the Anderson model takes place at the cross-overs between the even and odd numbers of electrons in the dot. The cross-over takes place at $\Delta E \simeq \hbar\Gamma$, where Coulomb peaks are observed at temperatures $\simeq\hbar\Gamma$. At lower temperature, the conductance in these regions saturates, and the average number of electrons n_d in the dot continuously changes from 0 to 1 for the left peak or from 1 to 2 for the right one (Figs. 6.20(a) and (b)). This number, in chemical terms, is the valence of the dot, and its non-integer values indicate a *mixed-valence* regime.

If we assume that the phase shift in channel g is given by the Friedel sum rule, we reproduce the zero-voltage conductance in the mixed-valence regime:

$$G_{mv} = G_Q \frac{4\Gamma_a \Gamma_b}{(\Gamma_a + \Gamma_b)^2} \sin^2(\pi n_d/2). \tag{6.130}$$

For symmetric couplings, it changes from 0 deep in even diamonds to G_Q in the middle of the odd diamond.

Experiment

It might seem easy to observe the Kondo effect in quantum dots: just put an odd number of electrons inside and measure the conductance. However, T_K is exponentially small at small conductances, and such small energy scales are beyond reach. One tries to lower the tunnel barriers separating the dot from the leads, but one should be cautious not to raise tunnel conductances to the values of the order of $\simeq G_Q$. In this case, Coulomb peaks are broad, tunneling involves many dot levels, and the discrete states in the dot become questionable: there is no qubit required anymore. This is why the Kondo effect can only be observed in a relatively narrow interval of tunnel conductances, and such observation requires quite some art and diligence. Even in this case, the quantitative comparison with theory frequently remains questionable. The theoretical Kondo effect requires scale separation: $T_K \ll E_{cut}$. In solids, these scales differ by four to six orders of magnitude. In quantum dots, experimental T_K values are comparable with the cut-off scales Γ, E_C, and δ_S.

An experimental criterion for the Kondo regime is the conductance increase with lowering temperature. Let us look at the conductance traces in Fig. 6.20(b) [174]. The quantum dot has been made in a GaAs-based heterostructure by combining lithographic and etching techniques. The size of the dot is about 100 nm so it is expected to contain $\simeq$50 electrons. Nevertheless, there is a clear distinction between even and odd diamonds, and within a single odd diamond the dot can be described by the Anderson model: the authors of Ref. [172] provide a detailed comparison with theory on this basis. The ratio Γ/U was about 1/7, indicating a well isolated dot. The highest temperature still exceeds $\hbar\Gamma$. Two broad Coulomb peaks are seen separating the odd diamonds from even ones. First, with lowering temperature, the conductance decreases everywhere except at the peaks. At further decreasing temperature, the conductance saturates in the even diamonds and at the outer edges of the Coulomb peaks, indicating the mixed-valence regime. Remarkably, the conductance begins to grow in the middle of the odd diamond: this indicates the Kondo effect. Even at the lowest temperature, the conductance traces in Fig. 6.20(b) do not resemble those in Fig. 6.20(a). This is not surprising, given the fact that T_K is determined by Eq. (6.129) and exponentially depends on the gate voltage. It reaches a deep minimum, $\simeq$40 mK, in the middle of the diamond, so the lowest temperature still exceeds T_K there. The scale of the conductance is also affected by the asymmetry of the tunnel couplings.

Control question 6.19. What is the ratio Γ_a/Γ_b expected from the experimental results?

Another popular criterion of the Kondo regime is a narrow peak in differential conductance at zero voltage [174]. The width of the peak gives the experimental estimation of T_K.

6.7 Relaxation and decoherence of qubits

The final two sections of this book are about relaxation and decoherence. We have already mentioned these issues in the book, and we had to: these phenomena actually separate quantum and classical transport. We know already that the environment can strongly influence properties of electrons in nanostructures and qubits since it can absorb energy from them: this is the relaxation. The environment can also affect the behavior of electrons and qubits without changing their energy – it is enough to change their phase: this is called decoherence for qubits and is usually called dephasing for electrons. We start from the discussion of relaxation and dephasing in classical physics. Turning to quantum mechanics, we analyze in detail decoherence and relaxation for a simplest quantum system – a qubit. Both are caused by an environment, so we talk about various types of environments and what decoherence they cause. We reveal the connections between the collection of quantum information and decoherence by addressing continuous weak linear measurement (CWLM) of a qubit.

The most qualitative results of the present section will be used in Section 6.8 in the discussion of the more involved issues of relaxation and dephasing of electrons that propagate in metals.

6.7.1 Classical versus quantum

Importantly, decoherence and relaxation (dissipation) are not specific to quantum mechanics: they occur in classical systems as well. We understand the significance of dissipation in classical physics very well. If we start with a pure classical mechanics, in either Newtonian or Lagrangian formulation, the energy of a body or a system of bodies is always conserved: there is no dissipation in these theories, and this is the point at which they have to be adjusted to describe the real world. If any process is allowed by the laws of mechanics, the time-reversed process is also allowed. In the real world, this is never the case. Some dissipation is always present in the form of friction. The microscopic origin of the friction force lies in the interaction of a body with other degrees of freedom – with the *environment*. The environment can take very different natures: the surface irregularities at atomic scale are responsible for dry friction of a body moving over a table; a particle flowing in air experiences collisions with air molecules; a ship rises on the waves of the sea. The *nature* of an environment is not really important. The only information required is the friction force provided by an environment. This in most cases is proportional to the velocity, $\gamma \dot{x}$.

The effect of this friction force is best illustrated with an example of a classical oscillator. Without friction, a free oscillator performs a periodic motion, with frequency equal to its eigenfrequency ω_0. The amplitude of these oscillations persists in time. In the presence of friction, if it is weak enough, $\gamma \ll m\omega_0$, the motion is still almost periodic, and the amplitude decreases with time as $\exp(-\gamma t/m)$, m being the mass of the oscillator. For strong friction, $\gamma \gg m\omega_0$, the motion is aperiodic – the particle merely relaxes to the equilibrium position. In either case, after some time the oscillator dissipates all energy and stops. This time, τ_d, is determined by the friction coefficient, $\tau_d \sim m/\gamma$. The motion in the presence of friction is not time-reversible – the time-reversed motion would mean that the amplitude is increased with the time to infinity, corresponding to taking energy from the environment.

How does decoherence or dephasing appear in the context of a classical oscillator? If the amplitude is extinguished by dissipation, the phase goes as well. If the friction is strong, we cannot talk at all about the phase since there are no oscillations. Such radical decoherence is easy to understand. But there can be decoherence without an amplitude change and with no relation to dissipation. Let is take a high-quality oscillator and keep the amplitude of the oscillations constant with an external energy source: a clock. Owing to the general imperfectness of this world, the frequency of the oscillator does not remain constant in time but exhibits small fluctuations around the average ω_0: $\omega(t) = \omega_0 + h(t)$, $|h(t)| \ll \omega_0$. While the amplitude remains constant, the frequency fluctuations randomize the phase. On average, the phase equals $\omega_0 t$, as for an ideal oscillator. Importantly, the fluctuations of the phase do not remain small like the frequency fluctuations: they grow with time. Their precise behavior is determined by the statistics of $h(t)$. Let us assume that $h(t)$ obeys Gaussian statistics and that the correlator of $h(t)$ at different time moments is given by $\langle h(t_1)h(t_2)\rangle = h^2 F((t_1 - t_2)/t_h)$. Here, h^2 is the instant variance of $h(t)$ and t_h is a correlation time of the fluctuations, $F(0) = 1$, that vanishes at the values of argument $\simeq 1$. The simplest form of this relation is *white noise*, $\langle h(t_1)h(t_2)\rangle = S_h \delta(t_1 - t_2)$, $S_h = h^2 t_h \int \mathrm{d}x \; F(x)$, which works at time scales $\gg t_h$.

We integrate the equation for the phase $\dot{\varphi} = \omega_0 + h(t)$ to get $\varphi(t) = \omega_0 t + \int_0^t h(t')\mathrm{d}t'$ (we have arbitrarily put $\varphi(0) = 0$). Let us compute the variance of the phase, $\langle \delta\varphi^2(t)\rangle \equiv \langle \varphi(t)^2\rangle - \langle \varphi(t)\rangle^2$, assuming $t \gg t_h$:

$$\begin{aligned}\langle \delta\varphi^2(t)\rangle &= \int_0^t \mathrm{d}t_1 \int_0^t \mathrm{d}t_2 \langle h(t_1)h(t_2)\rangle \\ &\approx \int_0^t \mathrm{d}((t_1+t_2)/2) \int_{-\infty}^{\infty} \mathrm{d}(t_1-t_2) h^2 F((t_1-t_2)/t_h) = S_h t,\end{aligned}$$

where we extend the integration over the time difference to $\pm\infty$ since $t \gg t_h$. We come to an important conclusion that, in the limit of large t, the phase variance is proportional to the time interval. Actually, this is a consequence of the central limit theorem – the sum of a large number of random independent quantities (in our case, $h(t)$ at different time moments) has a Gaussian distribution, no matter what was the distribution of each of the quantities. The fluctuations grow proportionally to the square root of the number of the quantities. Thus, the fluctuations of the phase grow as $\sqrt{t}$. At sufficiently long times they exceed 2π – the phase acquired over a period. This defines the decoherence time τ_φ.

We can quantify this by looking at the correlations of the oscillator displacement $x(t) = x_0 \cos(\varphi(t))$ at different times (assuming $\omega_0 t \gg 1$):

$$\begin{aligned}\langle x(0)x(t)\rangle - \langle x(0)\rangle\langle x(t)\rangle &= (x_0^2/2)\left\langle \mathrm{e}^{\mathrm{i}\varphi(0)}\mathrm{e}^{-\mathrm{i}\varphi(t)}\right\rangle \\ &= (x_0^2/2)\mathrm{e}^{-\langle(\varphi(0)-\varphi(t))^2\rangle/2} \equiv \mathrm{e}^{-t/\tau_\varphi},\ 1/\tau_\varphi = S_h/2. \end{aligned} \tag{6.131}$$

We made use of the fact that the phase fluctuations are Gaussian in handling the average of the exponents: this is the classical analog of the quantum relation given by Eq. (6.11). We see that the decoherence time determines the decay of the correlation function. We stress again that no relaxation occurs in this case: the amplitude of the oscillations x_0 stays the same at all times, as does the energy of the oscillator.

There is another way to arrange decoherence without amplitude decay. It is related to ensemble averaging rather than to fluctuations in time. Let us take an ensemble of many oscillators with constant amplitudes. They have slightly different frequencies ω_i, distributed around ω_0 with a variance h_0^2, $h_0 \ll \omega_0$. The frequencies do *not* fluctuate in time. Let us suppose that the oscillators are small so we cannot measure the signal of each while we can follow the total of their displacements, $X(t) = \sum_i x_i(t)$. The oscillators start in phase, so $X(t) = Nx_0$. Clearly, after time $\simeq 1/h$ the oscillators are no longer in phase and the total signal is therefore reduced. To quantify, we average it over the normal distribution of h, as follows:

$$\langle X(0)X(t)\rangle = Nx_0^2 \int \frac{\mathrm{d}h}{h_0\sqrt{2\pi}} \mathrm{e}^{-h^2/2h_0^2}\mathrm{e}^{\mathrm{i}ht} = Nx_0^2 \mathrm{e}^{-t^2 h_0^2/2}.$$

Similarly to the correlator in Eq. (6.131), the total signal decays exponentially with time owing to the decoherence of different oscillators.

Control question 6.20. Suppose the frequencies of the oscillators of the ensemble also experience fluctuations in time. How should we identify the dominating decoherence mechanism from the measurements of $X(t)$?

We know that both relaxation and decoherence are classical effects. How do they work in quantum systems? Relaxation is accompanied by an energy loss. For quantum systems, this implies the transitions between quantum states of different energy. These states could either belong to a continuous spectrum, like those of electrons tunneling in the environment, or be discrete, like those of qubits. Upon a transition, the wave function of the system changes considerably, so its phase changes as well. Therefore, relaxation is always accompanied by decoherence.

As in classical physics, one can also have decoherence without relaxation. This is a distortion of a phase of the wave function not accompanied by the change of probabilities of being in certain eigenstates. This phase can be, for example, the relative phase of two coefficients in a linear superposition of states with different energies, such as in a qubit. The phase of the electron waves can also be distorted without appreciable change of the electron energy. Such decoherence manifests itself in the destruction of quantum interference of electron waves at the same energy, such as that seen in the Aharonov–Bohm effect. Characteristic times for relaxation and decoherence are denoted by τ_r and τ_φ, respectively. Since dissipation is always accompanied by decoherence, but not vice versa, we always have $\tau_\varphi \le \tau_r/2$.

Although relaxation and decoherence are two distinct phenomena, they can be described in the framework of the same model. We have seen at the beginning of this chapter that dissipation can be included in quantum mechanics as an interaction with an environment represented as a set of external (bosonic) degrees of freedom. A general Hamiltonian would read as follows:

$$\begin{aligned}
\hat{H} &= \hat{H}_{\mathrm{q}} + \hat{H}_{\mathrm{env}} + \hat{H}_{\mathrm{r}} + \hat{H}_{\varphi};\\
\hat{H}_{\mathrm{q}} &= \sum_n E_n |n\rangle\langle n|, \quad \hat{H}_{\mathrm{env}} = \sum_k \hbar\omega_k \hat{b}_k^\dagger \hat{b}_k,\\
\hat{H}_{\mathrm{r}} &= \sum_{n'\neq n,k} V_{nn'k}|n\rangle\langle n'|\hat{b}_k + \text{h.c.},\\
\hat{H}_{\varphi} &= \sum_{nk} V_{nk}|n\rangle\langle n|\hat{b}_k + \text{h.c.}
\end{aligned}$$

Here $\hat{H}_{\mathrm{q}}$ is the Hamiltonian of the quantum system (E_n being the energy levels) and $\hat{H}_{\mathrm{env}}$ represents the environment as a set of harmonic oscillators labeled k with frequencies $\hbar\omega_k$. Two other terms give the coupling between the system and the environment. The contribution $\hat{H}_\varphi$ only has terms diagonal in n; it leaves the system in the same state $|n\rangle$, although it emits/absorbs a boson to/from the environment. Therefore, these couplings are not associated with the energy transfer and are responsible for decoherence. In contrast, $\hat{H}_{\mathrm{r}}$ terms contain off-diagonal matrix elements responsible for the transitions between the different states of the system. The energy is transferred between the system and the environment in the course of such transitions: this models the relaxation.

6.7.2 Relaxation and decoherence in a qubit

Let us specialize to the case of a qubit – a two-level system with states $|+\rangle$ and $|-\rangle$ that have energies $\pm\Delta/2$. We take the simplest model, given by

$$\hat{H} = (\Delta/2 + \hat{\varepsilon}_{\parallel})\,(|+\rangle\langle+| - |-\rangle\langle-|) + \hat{\varepsilon}_{\perp}\,(|+\rangle\langle-| + |-\rangle\langle+|)\,, \tag{6.132}$$

where $\hat{\varepsilon}_{\parallel,\perp}$ are the fields linear in boson operators:

$$\hat{\varepsilon}_{\parallel,\perp} = \sum_m \varepsilon^{(m)}_{\parallel,\perp}\hat{b}^{\dagger}_m + \left(\varepsilon^{(m)}\right)^*_{\parallel,\perp}\hat{b}_m.$$

We did not explicitly write the energy of the environment. The terms with Δ describe the Hamiltonian of the qubit, $\hat{\varepsilon}_{\perp}$ represents the coupling with the diagonal, and $\hat{\varepsilon}_{\perp}$ represents the coupling with the off-diagonal matrix elements. Therefore the fields $\hat{\varepsilon}_{\parallel,\perp}$ are responsible for the decoherence and relaxation, respectively. Note that Δ comes in the combination $\Delta/2 + \hat{\varepsilon}_{\parallel}$: the latter acts as a fluctuating part of the energy splitting in the qubit. Since the working frequency of the qubit is determined by this energy splitting, the field $\varepsilon_{\parallel}$ gives the frequency fluctuations, precisely like in the model of classical decoherence considered in Section 6.7.1. Both fields $\hat{\varepsilon}_{\parallel,\perp}$ have the dimension of energy. A simplification of the model is that we represent the relaxation by a single field: there could be two, corresponding to two possible matrix elements $|+\rangle\langle-|$, $|-\rangle\langle+|$. The model formulated is a straightforward extension of the spin-boson model of Section 6.2. There, we were concentrating on quantum and non-perturbative effects, while here we have a simpler task, solved using the Fermi Golden Rule and some classical reasoning.

Control question 6.21. A spin qubit in a constant magnetic field $\boldsymbol{B}||z$ is subject to an environment that produces a fluctuating magnetic field. Can you express $\varepsilon_{\parallel}$ and $\varepsilon_{\perp}$ in terms of the components of the fluctuating field? How many field components are responsible for the relaxation?

Control question 6.22. The Hamiltonian of the spin-boson model, Eq. (6.38), can be brought to the form given in Eq. (6.132) by diagonalizing the qubit Hamiltonian. What are the fields $\hat{\varepsilon}_{\parallel,\perp}$ in terms of $\hat{F}$?

The coupling to the environment is described by the coefficients $\varepsilon^{(m)}_{\perp,\parallel}$, or, in other words, by the shifts of the corresponding oscillators. We know from Sections 6.1 and 6.2 that, fortunately, we do not need the details of the coefficients. What we need to know is a single function of frequency defining the spectral density of the field correlations: the (quantum) noise of the corresponding field. If the environment is at equilibrium characterized by temperature T, the noise is given by (see Eq. (6.9))

$$S(\omega) = N_B(\omega)\mathrm{sign}(\omega)S_{\mathrm{q}}(|\omega|), \tag{6.133}$$

$$N_B(\omega) \equiv \frac{1}{\exp(\hbar\omega/k_{\mathrm{B}}T) - 1}.$$

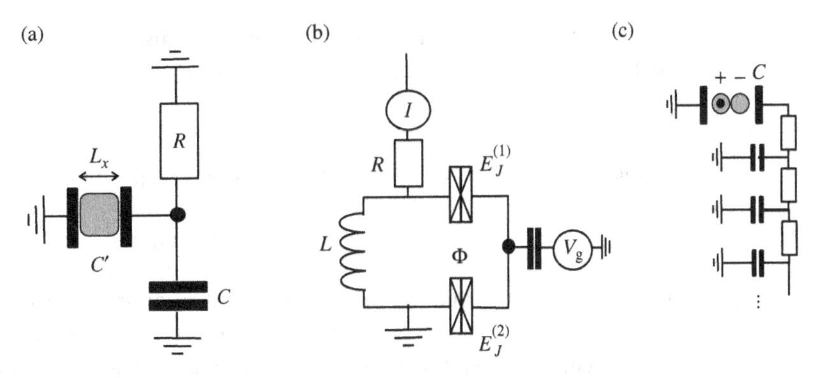

Fig. 6.21. **Relaxation and decoherence in qubit setups. See (a) Exercise 6.12, (b) Exercise 6.13, (c) Exercise 6.14.**

where S_q is temperature-independent and is immediately related to the imaginary part of the corresponding response function, $S_q(\omega) = 2\hbar\chi(\omega)$. At vanishing temperature, $S(\omega) = \Theta(-\omega)S_q(|\omega|)$. All fluctuations at vanishing temperature are quantum; this explains why we use the subscript "q" for S_q.

Let us look at the relaxation of the qubit due to the field $\epsilon_\perp$. At vanishing temperature, the environment cannot emit energy, and thus the only transitions are from state $|+\rangle$ (higher energy) to state $|-\rangle$ (lower energy). The inverse relaxation time is just the transition rate, given by the Fermi Golden Rule:

$$\frac{1}{\tau_r} \equiv \Gamma_\downarrow = \frac{2\pi}{\hbar}\sum_m |\varepsilon_m|^2\delta(\Delta - \hbar\omega_m) = S_\perp(-\Delta/\hbar)/\hbar^2. \tag{6.134}$$

At finite temperature, both absorption and emission are possible. At equilibrium, the up and down transition rates are related as follows:

$$\begin{pmatrix} \Gamma_\downarrow \\ \Gamma_\uparrow \end{pmatrix} = \frac{1}{\hbar^2}\begin{pmatrix} S_\perp(-\Delta/\hbar) \\ S_\perp(\Delta/\hbar) \end{pmatrix} = \begin{pmatrix} N_B(\Delta/\hbar)+1 \\ N_B(\Delta/\hbar) \end{pmatrix}\Gamma_\downarrow(T=0). \tag{6.135}$$

This guarantees that the equilibrium probabilities, $p_\pm$, of finding the qubit in either state obey a Boltzmann distribution given by $p_+/p_- = \exp(-\Delta/k_B T)$.

Exercise 6.12. Let us consider a rectangular quantum dot of dimensions L_x, L_y placed between the plates of a capacitor C' (see Fig. 6.21 (a), gray rectangle). The wave functions of the levels and the energies are given by Eqs. (4.17) and (4.18). A single electron is in the dot. Evaluate the relaxation rate for the transition between the first excited ($n_x = 2, n_y = 1$) and the lowest levels. For this: (i) express the field $\varepsilon_\perp$ in terms of the voltage in the node; (ii) find the susceptibility $\chi(\omega)$ for this voltage, disregarding C' in comparison with C; (iii) find the noise of $\varepsilon_\perp$.

Let us turn to decoherence. The decoherence is determined by diagonal matrix elements of the interaction, and for them $\hat{\varepsilon}_\parallel$ appear to be the fluctuating corrections to the level spacing Δ. Let us start with the assumptions that this correction can be handled like a classical noisy variable and that we have white noise given by

$$\langle \varepsilon_{\parallel}(t)\varepsilon_{\parallel}(t')\rangle = S_{\parallel}(0)\delta(t-t'). \tag{6.136}$$

This white noise induces the change in phase of a superposition of the qubit states:

$$\varphi(0) - \varphi(t) = \frac{2}{\hbar}\int_0^t \varepsilon_{\parallel}(t')\mathrm{d}t' \tag{6.137}$$

(the factor of 2 comes from the fact that the fluctuation in the level spacing is 2ε). Thus, in the long-time limit, the phase fluctuations are given by $\langle \delta\varphi^2\rangle = 4S(0)_{\parallel}t/\hbar^2$ (again, in accordance with the central limit theorem, they grow as t). Recalling Eq. (6.131), we obtain the decoherence time:

$$\frac{1}{\tau_{\varphi}} = \frac{2}{\hbar^2}S_{\parallel}(0). \tag{6.138}$$

Exercise 6.13. A Cooper-pair box (see Section 3.5.1) is biased as shown in Fig. 6.21(b). This provides us with an opportunity to tune the effective Josephson energy (see Eq. (3.75)). However, the resistor R in the bias line provides flux fluctuations that result in decoherence. Quantify the decoherence rate assuming the limit of small Josephson energy, so that the Hamiltonian of the CPB is given by Eq. (3.84).

In fact, noise produced by an environment is not always white. Its spectral density depends on the frequency, and changes at frequencies of order of $\omega \ll k_{\mathrm{B}}T/\hbar$ according to Eq. (6.133). While low-frequency fluctuations ($\hbar\omega \ll k_{\mathrm{B}}T$) are not distinguishable from a classical fluctuating signal, the high-frequency ones ($\hbar\omega \gg k_{\mathrm{B}}T$) are essentially quantum. Which frequencies are relevant for the determination of the decoherence time? Eventually, they are determined by the decoherence time itself: one can digest it, for example, by inspecting the Fourier transform of Eq. (6.131). In particular, if $k_{\mathrm{B}}T \gg \hbar/\tau_{\varphi}$ the noise creating the decoherence is classical at the relevant frequencies; this justifies our starting assumptions. If $1/\tau_{\varphi}$, evaluated using Eq. (6.138) the previous formula, appears to be bigger or comparable with $k_{\mathrm{B}}T/\hbar$, something goes wrong: most likely, the interaction with the environment at quantum level drastically changes the properties of the qubit.

Even at low frequencies, $\ll k_{\mathrm{B}}T/\hbar$, there may be a substantial frequency-dependence of the noise, in which case the noise is *colored* rather than white. In this case, the correlator of the phase factors is no longer a simple exponential function. The time-dependent time variance is given by

$$\langle(\varphi(0)-\varphi(t))^2\rangle = \int \frac{\mathrm{d}\omega}{\pi\hbar^2}\frac{4\sin^2(\omega t/2)}{\omega^2}S(\omega),$$

and may be a complicated function of time. Still, the estimate of τ_{φ} is the time at which the variance becomes of the order of unity, and this time determines the relevant frequencies of the noise. Thus we can estimate the decoherence time using a simple self-consistency relation without making complicated integrals:

$$\frac{1}{\tau_{\varphi}} = \frac{2}{\hbar^2}S_{\parallel}\left(\omega = \frac{1}{\tau_{\varphi}}\right). \tag{6.139}$$

This is derived from Eq. (6.138) by taking the noise at a relevant frequency scale, the scale to be determined from the same equation. The results depend essentially on the form of the function $S_{\parallel}(\omega)$ in the limit of small frequencies, which is the subject of the general classification of environments outlined in Section 6.2.4

Let us go through the classification of environments, specifically looking at the decoherence. Inspecting Eq. (6.46), we understand that $P_1(E) \simeq E^s$ corresponds to $S(\omega)_{\parallel} \simeq (k_B T/\hbar)(\omega/\omega_c)^{s+1}$. Here ω_c is not a cut-off frequency; rather, it is a constant of proper dimension to characterize the intensity of the noise. Solving the above self-consistency relation yields

$$\frac{1}{\tau_\varphi} \simeq (k_B T/\hbar)\left(\frac{\hbar\omega_c}{k_B T}\right)^{1+1/s}.$$

For $s < -1$ (a *sub-ohmic environment*), the decoherence rate $1/\tau_\varphi$ decreases at low temperatures, as one expects from the decreasing noise. However, it decreases slower than the temperature itself. For example, $1/\tau_\varphi \propto T^{2/3}$ for $s = -3/2$. This indicates a problem at low temperatures, $T \ll \hbar\omega_c/k_B$: the estimate of the decoherence rate exceeds the temperature scale, and quantum effects cannot be disregarded. Indeed, we have seen in Section 6.2.4 that at low energies a qubit in a sub-ohmic environment is no longer a qubit. The strong interaction with the environment makes it a classical memory cell.

Exercise 6.14. A charge qubit is placed into a capacitor C connected to an RC-line (Fig. 6.21(c)) with resistance $\tilde{R}$ and capacitance $\tilde{C}$ per unit length. Give an estimation of the decoherence rate. Give the condition of importance of the quantum effects.

For $s = -1$ (an *ohmic environment*), the fluctuations of $\varepsilon_{\parallel}$ are white-noise-like, and $1/\tau_\varphi \propto \alpha(k_B T/\hbar)$. The importance of the quantum effects thus depends on the strength of the coupling α. The decoherence is due to classical noise only for a weak ohmic environment $\alpha \ll 1$. We recall we have studied this situation in the context of the phase diffusion (Eq. (6.62)).

The environments with $s > -1$ are *super-ohmic*. If $-1 < s < 0$, the decoherence rate quickly drops with decreasing temperature. For example, $1/\tau_\varphi \propto T^2$ if $s = -1/2$. For super-ohmic environments with $s > 0$, the self-consistency relation gives the only solution: $1/\tau_\varphi = 0$, no decoherence. In fact, some decoherence must be present since the noise is present. The estimation implies that the decoherence is *incomplete*. The variance of the phase does not grow unrestricted with time, but rather saturates at a finite value. Therefore the memory of the initial phase persists.

How are the relaxation and decoherence reflected in the equations describing the dynamics of the qubit density matrix? The dynamics of diagonal elements $\rho_{++} = p_+$, $\rho_{--} = p_-$ are not affected by decoherence and are governed by simple master equations as follows:

$$\begin{aligned}\frac{dp_+}{dt} &= -\Gamma_\downarrow p_+ + \Gamma_\uparrow p_-,\\ \frac{dp_-}{dt} &= -\Gamma_\uparrow p_- + \Gamma_\downarrow p_+.\end{aligned} \tag{6.140}$$

The decoherence affects the evolution of the non-diagonal element as follows:

$$\frac{d\rho_{-+}}{dt} = \left(i\frac{\Delta}{\hbar} - \frac{\Gamma_{\uparrow} + \Gamma_{\downarrow}}{2} - \frac{1}{\tau_{\varphi}}\right)\rho_{-+}. \tag{6.141}$$

We see that the non-diagonal element is damped under stationary conditions; this damping is due to both the relaxation transitions and the intrinsic decoherence rate $1/\tau_{\varphi}$. One can use the damping time as a definition of decoherence time in accordance with the classical reasoning that the amplitude damping kills the phase as well. A reverse situation can be envisaged: if $1/\tau_{\varphi} \gg \Gamma_{\uparrow,\downarrow}$, the non-diagonal matrix element will vanish long before the relaxation takes place. If an initial condition is a quantum superposition of two qubit states, the coherence of the superposition is lost at the time scale τ_{φ}, without a change in the probabilities of finding the qubit in either state.

6.7.3 Decoherence and measurement

The traditional concept of projective quantum measurement allows us to construct an adequate model for the loss of quantum information and therefore the irreversible aspects of quantum mechanics. However, it cannot immediately explain what happens to a qubit in the process of an actual data acquisition. The imaginable projective measurement instantly sets a qubit in one of the two states: this implies strong coupling with the measuring device. We know that a realistic measurement cannot be instant since the signal fluctuates and has to be averaged over a large time interval to achieve a decent measurement accuracy. Also, it is better if the coupling between the qubit and the measuring device is weak, and the response of the device is usually linear in input signal. If we want a model to incorporate all these features, we come to the concept of *continuous weak linear measurement* (CWLM) (see Refs. [176]–[178]) – a long word combination to make sure that nobody confuses it with the projective measurement.

We consider here the simplest CWLM setup. Since the measuring system – a detector – is linear, it can be described by a large combination of boson modes, like any other environment considered in this section. The output of the detector is represented by the *output variable* $\hat{O}$, which is a field linear in boson operators. For convenience, we make it dimensionless. We understand that this is a matter of choice: the output of a realistic detector may be voltage, current, or any other physical quantity, but we can always express it in dimensionless units. The detector is designed to measure the z component of the qubit pseudospin in the basis of the qubit Hamiltonian. To this end, its coupling to the qubit is given by

$$H_{\text{int}} = \hat{\sigma}_z \hat{\varepsilon}_{\parallel}.$$

The field $\hat{\varepsilon}_{\parallel}$, which is again linear in boson operators, is called the *input variable*.[5] Let us note that it is the same field that is responsible for the decoherence of the qubit (see Eq. (6.132)) so we just re-use the notation.

[5] Not to be confused with the detector input: that is $\hat{\sigma}_z$, and the input variable is coupled to it.

What is the detector response to the qubit pseudospin? To quantify it, we regard first the qubit pseudospin as a classical input signal, $\hat{\sigma}_z \to \sigma$. Then we can compute the response using the Kubo formula (see Eq. 6.16)):

$$\langle \hat{O} \rangle = \frac{\mathrm{i}}{\hbar} \int_{-\infty}^{0} \mathrm{d}t \, [\hat{O}(0), \hat{\varepsilon}_{\parallel}(t)] \equiv A\sigma.$$

The average output signal thus amplifies (with coefficient A) the component of the qubit pseudospin. Let us note that for a general linear detector there is also a reverse susceptibility given by

$$A' = \frac{\mathrm{i}}{\hbar} \int_{-\infty}^{0} \mathrm{d}t \, [\hat{O}(0), \hat{\varepsilon}_{\parallel}(t)].$$

The reverse susceptibility indicates that if we do something with the detector output – for example, connect it to another stage of amplification or to a plotting device – we influence the input, which is not good. In our case, this would create an undesired extra field on the qubit, $h^{(z)} \propto A'$. For these reasons, practical amplifiers have to have a good separation between output and input, $|A| \gg |A'|$. In more philosophical terms, this is a condition to regarding the output signal as classical information. As discussed in Section 5.2, classical information can be read/copied further without distorting the source, that is the input signal. This is also how we see. A signal source produced in our eye on most occasions is not influenced by the fact that we close our eyes, as it would be for the case of an appreciable A'.

So, we require $|A| \gg |A'|$. We note that this is incompatible with the condition of thermal equilibrium frequently assumed in this chapter. Indeed, thermal equilibrium would imply $A = A'$. It is also clear that a system in thermal equilibrium is never a practical amplifier, which would require the dissipation from an external energy source to do their job. Therefore, we cannot use the fluctuation–dissipation theorem to relate the noises and susceptibilities. We can still use Eq. (6.17), since it does not require thermal equilibrium. At zero frequency,

$$A \approx A - A' = (2/\hbar)\mathrm{Im}\, S_{O\parallel}(0).$$

The symmetrized correlator of the noises is given by $S^{(s)}_{O\parallel} = \mathrm{Re}\, S_{O\parallel}(0)$.

Let us understand that we cannot measure this average signal $\langle \hat{O} \rangle = A\sigma$ instantly. The output signal is subject to noise, typically to white noise. To approximate the average, we accumulate the signal. We integrate it over a time interval $(t_0, t_0 + \tau)$ and divide it by the duration τ of the measurement, $\bar{O} = \int_{t_0}^{t_0+\tau} O(t)\mathrm{d}t/\tau$: this makes a single measurement. The measurement result $\bar{O}$ is randomly changing from measurement to measurement and obeys a normal distribution with variance S_O/τ and average $A\sigma$; thus,

$$P(\bar{O}) = \frac{1}{\sqrt{\pi \tau S_O}} \exp\left(-\frac{(\bar{O} - A\sigma)^2 \tau}{2S_O} \right).$$

The accuracy thus improves with increasing duration. How long should we measure in order to resolve two qubit states $\sigma_z = \pm 1$? The average outputs are separated by $2A$; this is to be compared with the signal variance. We conclude that the qubit states can only be resolved if the duration of measurement exceeds a typical "measurement time" $\tau_m = S_{\mathcal{O}}/A^2$, which characterizes the detector speed.

The essence of CWLM theory is that this "measurement time" cannot exceed one-half of the decoherence time of the qubit, that being determined by the noise $S_{\|}$. Suppose one tries to design a faster detector, either by boosting its sensitivity A or by reducing its output noise. Unfortunately, a by-product of the design efforts is the increased decoherence of the qubit.

This is quantified by an inequality involving the characteristics of the mentioned detector: the noise of the output variable $S_{\mathcal{O}}$, the noise of the input variable $S^{(s)}_{\mathcal{O}\|}$, the symmetrized noise correlator $S_{\|}$, and the response of the detector:

$$S_{\|}S_{\mathcal{O}} \geq |S_{\mathcal{O}\|}(0)|^2 = \left(S^{(s)}_{\mathcal{O}\|}\right)^2 + \left(\frac{\hbar A}{2}\right)^2. \tag{6.142}$$

Since the decoherence time is given by Eq. (6.138), this proves $\tau_\varphi \leq 2\tau_m$. In other words, an accurate measurement results in full decoherence of the qubit. This makes the connection with the concept of projective measurement: the wave function is always scrambled by an efficient data acquisition.

Exercise 6.15. Prove the inequality given in Eq. (6.142) using the representation of the fields involved in terms of boson creation/annihilation operators. Make use of the Cauchy–Schwartz inequality: if a set of positive numbers P_k defines an inner product of two complex vectors $(A, B) = \sum_k A_k^* B_k P_k$, $|(A, B)|^2 \leq (A, A)(B, B)$.

One can quantify this in more detail by studying the joint statistics of the qubit and output signal. It is convenient to define it in terms of the integrated output $x(t) = \int_0^t \mathrm{d}t_1 \mathcal{O}(t_1)$. Suppose the density matrix of the qubit is $\hat{\rho}(0)$. The joint density matrix of the qubit and integrated output x, $\hat{\rho}(x,t)$, yields the probability of finding value x of the output *and* the qubit in a certain state at time moment t. If the qubit is not coupled to the detector, this matrix obeys a diffusion equation $\dot{p} = -(S_{\mathcal{O}}/2)p''$, conforming to the normal distribution of the output with the variance $S_{\mathcal{O}}t$ increasing in time. Coupling to the qubit makes it a drift-diffusion equation: $\langle x \rangle = \pm At$, depending on the qubit state. Combining this with the qubit dynamics from Eq. (6.140), we come to the following evolution equations:

$$\begin{aligned}
\dot{p}_+ &= -(S_{\mathcal{O}}/2)p''_+ - Ap_+ - \Gamma_\downarrow p_+ + \Gamma_\uparrow p_-;\\
\dot{p}_- &= -(S_{\mathcal{O}}/2)p''_- + Ap_- - \Gamma_\uparrow p_- + \Gamma_\downarrow p_+;\\
\frac{\mathrm{d}\rho_{-+}}{\mathrm{d}t} &= -(S_{\mathcal{O}}/2)p''_- + \left(\mathrm{i}\frac{\Delta}{\hbar} - \frac{\Gamma_\uparrow + \Gamma_\downarrow}{2} - \frac{1}{\tau_\varphi}\right)\rho_{-+},
\end{aligned} \tag{6.143}$$

where we set $S_{\mathcal{O}\|} = 0$ in the final equation. This is to be solved with initial conditions $\hat{\rho}(0, x) = \hat{\rho}(0)\delta(x)$. The solution is of the form of two Gaussian packets that get wider as a result of diffusion and drift in opposite directions with velocities $\pm A$.

Exercise 6.16. Suppose initially the qubit is in the equal-weight superposition of qubit states, $\hat{\rho}(0) = \sigma_x$. Find the solution of the equation for $p_{\pm}(x)$, disregarding the relaxation. From this, find the joint probability $P(x_1, x_2)$ of the results of two successive measurements of the duration τ. Hint: the result is a superposition of two normal distributions.

The result of Exercise 6.16 provides a quantitative illustration of the quantum non-demolition measurement (see Section 5.6.2) with a linear detector. If the duration of each measurement $\tau \ll \tau_{\text{m}}$, the results spread over a wide range and barely correlate: the measurement is too short to be complete. In the opposite limit, the distribution of the results is concentrated near $\pm A$ and they do correlate: the result of the second measurement is the same as the result of the first one with an overwhelming probability.

Exercise 6.17. Consider Eqs. (6.143) at a time scale bigger than the relaxation time, assuming $\Gamma_\uparrow = \Gamma_\downarrow$. What is the average result of the measurement and its variation if the measurement duration $\tau \gg 1/\Gamma$. How does it compare to a projective measurement?

6.8 Relaxation and dephasing of electrons

Let us turn now to the relaxation and decoherence of the delocalized quantum states – electron waves. Mostly we will discuss electrons in extended diffusive conductors where these processes have been thoroughly investigated both theoretically [159] and experimentally. Traditionally, the process called decoherence for qubits is known as dephasing if we talk about electrons. The point is that an apparent "decoherence" of electron waves in $G \ll G_{\text{Q}}$ conductors can be brought about by spin-orbit interaction, magnetic field, and even by the energy difference, as in superconductors. These mechanisms have been studied in Chapters 2 and 4, and have nothing to do with an environment or inelastic processes of any kind. The term "dephasing" is thus reserved for environment-induced effects.

We start this section with a short discussion to elucidate the link between the propagation time of electrons in a nanostructure and the relaxation or dephasing rate; this is important for manifestations of relaxation and dephasing. We list the mechanisms of the phenomena in Section 6.8.1. The most important (and interesting) mechanism is that arising from voltage fluctuations, so we discuss them in detail in Section 6.8.2. Finally, in Section 6.8.3 we present the recent experimental data supporting and illustrating the theory given in the rest of this section.

Relaxation and decoherence for qubits and for electrons are very similar, at least at the quantitative level. However, their common manifestations are different. The relaxation or decoherence of a qubit is usually seen in a time domain: the qubit is manipulated and then measured over a time interval. The relaxation or decoherence becomes visible if the corresponding time becomes of the order of the duration of the time interval.

It might seem that the same should hold for electrons. Consider, for example, the interference between two electron waves of the same energy with amplitudes $A_{1,2}$ that traverse an Aharonov–Bohm interferometer. Part of the current in this device will reflect this interference, $\delta I \propto \mathrm{Re}(A_1 A_2^*)$. Suppose the wave in one of the arms is subject to an environmental effect of some kind such that it picks up a random fluctuating phase shift $\varphi(t)$. This phase shift will enter the current, $\delta I \propto \cos(\varphi(t))$. Let us now average over the fluctuating phase. Since the variation of the phase grows with time (see, for example, Eq. (6.131)), we conclude that the interference current is averaged out at the time scale τ_φ. This would imply that it is possible to observe interference effects only in the course of a sufficiently fast measurement of the duration not exceeding τ_φ. They would average out at a bigger time scale. This is not what we observe in practice or read in this book. What is wrong with the above reasoning?

What is wrong is that we have forgotten that both waves are coherent before entering the interferometer and that the result of the interference, the current, is formed right after they leave the device to enter a lead. The detected phase shift is thus the difference of the phases before entering and after leaving the device, $\Delta\varphi = \varphi(t_l) - \varphi(t_e)$. The relevant time scale is not the observation time. Rather, it is the time $t_l - t_e = t_d$ required to cross the interferometer. The resulting interference current survives the averaging over the fluctuations, though it is suppressed by the following factor:

$$\delta I \propto \langle \mathrm{e}^{\mathrm{i}(\varphi(t_l)-\varphi(t_e))} \rangle = \exp\left(-\frac{t_d}{\tau_\varphi}\right).$$

This only vanishes if the electron loses phase while in the interferometer.

The same applies to relaxation. Let us recall the setup studied in Section 2.4. There, we considered a double junction in two limiting situations. If there is no inelastic scattering, the filling factor in the node was a superposition of the two in the leads, $f_N(E) = (G_1 f_L(E) + G_2 f_R(E))/(G_1 + G_2)$, $G_{1,2}$ being the junction conductances. It displayed a two-step structure as a function of energy, the steps being at the chemical potentials of the leads. If the inelastic scattering dominates, the filling factor is a smooth function of energy, a Fermi distribution at increased temperature (see Eq. (2.65)).

What separates these two limiting situations? Suppose we can characterize the relaxation of electrons inside the node with a single relaxation time τ_r. If the dwell time τ_d of the electrons in the node is short, there is no chance that the electrons experience the relaxation and one can disregard it. The filling factor exhibits two steps. Otherwise, there is no chance of the electrons leaving the island without interacting with other electrons, and the filling factor is smooth. As we remember, the dwell time corresponds to the Thouless energy of the setup, $\hbar/\tau_d = E_{Th} = (G/G_Q)\delta_S$.

To observe the relaxation, one characterizes the energy-dependent filling factor in the node or in a node of a more complicated nanostructure. The most straightforward way to do this is with a non-invasive voltage probe of very low conductance connecting the node to the third lead biased at voltage V_3 (filling factor $f_3(E) = \Theta(eV_3 - E)$ at vanishing temperature). The current through such a probe is given by

$$eI_p(V_3) = \int \mathrm{d}E (f_N(E) - \Theta(eV_3 - E)) G(E - eV_3), \tag{6.144}$$

where $G(E)$ is the energy-dependent conductance of the probe, which should be sufficiently small not to perturb the current balance in the node. If $G(E) = \text{const.}$, as we frequently assume, this does not work: the probe would just measure the voltage in the node, giving no information. If we know the precise shape of $G(E)$, however, we do have a chance to measure I_p. Equation (6.144) can be represented in terms of the *Fourier transforms* of the functions involved, $I_p(t) = G(t) f_N(t)$. So, we find $f_N(t) = I_p(t)/G(t)$ and run the inverse Fourier transform to determine $f_N(E)$.

Exercise 6.18. Prove Eq. (6.144) and explain in terms of Fourier transforms why it gives no information about $f_N(E)$ if $G(E) = \text{const.}$

We conclude that, for a nanostructure of a given size, the importance of dephasing/relaxation is simply determined by the ratio τ_φ/τ_d of the dephasing time to the dwell time in the nanostructure. This is a yes/no issue: the dephasing/relaxation is either important or not. Correspondingly, the transport is either incoherent or coherent.

A more interesting picture emerges in extended conductors – we concentrate on diffusive ones. The dephasing time sets a length scale $L_\varphi = \sqrt{D\tau_\varphi}$. The transport is coherent at scales shorter than L_φ and incoherent at scales larger than L_φ. Thouless proposed a convenient scheme to comprehend this. Let us subdivide a large conductor into smaller sub-conductors, each of size L_φ. This subdivision depends on the effective geometry that in turn is determined by the ratio of L_φ to the dimensions of the conductor. A film of thickness less than L_φ is 2d; a wire with transverse dimensions less than L_φ is 1d. Let us note that each sub-conductor – an elementary nanostructure – is on *mesoscopic border*, that is, the dwell time in a subconductor equals the dephasing time.

Let us look at the weak localization correction to the conductance. For a coherent conductor, $(\delta G)_{WL} \sim G_Q$ (see Section 4.4). A long incoherent conductor is presented as a chain of coherent conductors each of size L_φ. Summing up the resistors in series, we find that the weak localization correction of the entire wire is of the order of $G_Q L_\varphi/L$. It is certainly reduced in comparison with the coherent case, but it survives and can be observed: this is a neat way to see coherent effects in incoherent conductors.

To make it quantitative, let us recall Eq. (4.127), which expresses the weak localization correction in terms of the cooperon propagator:

$$\delta\sigma_{WL}(B) = -G_Q D \int_0^\infty dt \; P_{coop}(\boldsymbol{r}, t' + t; \boldsymbol{r}, t') \left(3\exp(-t/\tau_{SO}) - 1\right).$$

The propagator with delay t accounts for the electron trajectories that come back to the same point after time t. To account for the fluctuating phase picked up along the trajectory, we multiply the propagator $P_{coop}(t' + t, t)$ by the factor $\exp(-t/\tau_\varphi)$. To use such a simple suppression factor is a gross simplification. We have seen in Section 6.7.2 that this generally does not work for colored noise. We will see further why this simplification is sufficient. Performing the integration over time, we find that the decoherence "rate" due to a magnetic field, $1/\tau_H$, and the dephasing rate, $1/\tau_\varphi$, always add up. The correction to

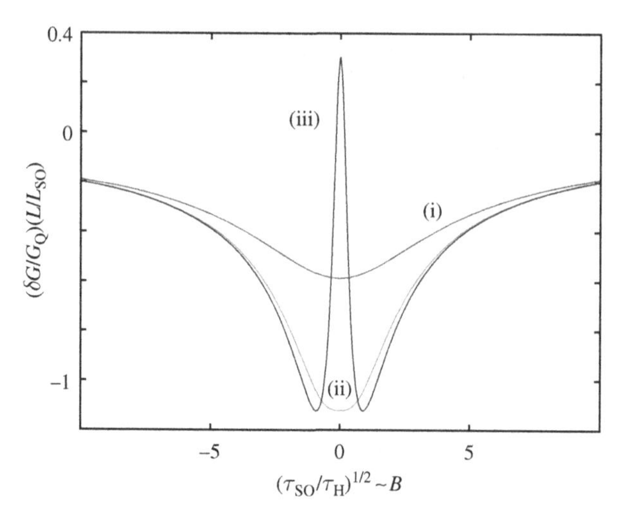

Fig. 6.22. **Weak localization correction to the conductance of a 1d wire versus magnetic field. Curves: (i) $\tau_\varphi/\tau_{SO} = 0.1$; (ii) $\tau_\varphi/\tau_{SO} = 1$; (iii) $\tau_\varphi/\tau_{SO} = 10$.**

the total conductance in 1d geometry is obtained by replacing $1/\tau_H$ by $1/\tau_H + 1/\tau_\varphi$ and is given by

$$\delta G_{\mathrm{WL}} = G_{\mathrm{Q}} \frac{L_\varphi}{L} \frac{1}{2} \left(\frac{1}{\sqrt{1 + (\tau_\varphi/\tau_H)}} - \frac{3}{\sqrt{1 + (\tau_\varphi/\tau_{\mathrm{SO}}) + (\tau_\varphi/\tau_H)}} \right). \tag{6.145}$$

The weak localization correction displays an interesting pattern as a function of magnetic field $B \propto 1/\sqrt{\tau_H}$. If $\tau_\varphi \ll \tau_{\mathrm{SO}}$, it gives a broad dip of the width $\tau_H \simeq \tau_\varphi$. Upon decreasing the dephasing rate, the curve crosses over to a superposition of a narrow peak $\tau_H \simeq \tau_\varphi$ and a broad ($\tau_H \simeq \tau_{\mathrm{SO}}$) dip (see Fig. 6.22).

Exercise 6.19. Obtain a similar relation for 2d geometry, concentrating on the magnetoconductance $\delta G_{\mathrm{WL}}(0) - \delta G_{\mathrm{WL}}(B)$. Make use of Eq. (4.127). Assume that the effect of magnetic field can be incorporated into τ_H (this is not correct if the magnetic field is perpendicular to the plane of the film, but suits for a parallel field).

6.8.1 Mechanisms of relaxation and decoherence

Let us list the relevant dissipation mechanisms in metallic solids that cause both dephasing and relaxation.

In any solid, the electron relaxation can be mediated by the *electron–phonon interaction*: the energy is transferred from electrons to the vibrations of the crystalline lattice. We have not discussed this interaction, although it is very important in solid state physics. The point is that the electron–phonon scattering rate strongly depends on energy,

$$\frac{1}{\tau_{\mathrm{d}}} \simeq \lambda \frac{E}{\hbar} \left(\frac{E}{\hbar \omega_{\mathrm{D}}} \right)^2, \tag{6.146}$$

where λ is a dimensionless electron–phonon coupling constant (typical value $\lambda \sim 0.2$) and ω_D is the Debye frequency. The maximum phonon energy is estimated as 100 K. Typical electron energies are of order of temperature. At 100 K the relaxation rate is therefore of the order of energy, and the electrons are barely coherent. At room temperature, the phonons are not only a powerful source of relaxation, but also the scattering on phonons dominates the resistance. However, at moderately low temperatures of 1 K the phonon relaxation rate is already small, 10^7–10^8 Hz; usually it cannot compete with other inelastic processes and may be safely disregarded. Phonons do not provide any special dephasing mechanism that is more intensive than the relaxation: the dephasing and relaxation rates are of the same order of magnitude.

At low temperatures, the electron–electron scattering thus becomes more important. In this case, the energy is not transferred from the conductor, but is merely redistributed between the electrons. At low energy scales, the electrons do not significantly change their momentum in the course of such scattering. In this case, the scattering is reduced to the interaction of an electron, with *voltage fluctuations* produced by all other electrons. The fluctuations vary slowly both in space (at scales $\gg k_F^{-1}$) and in time. For the sake of a crude estimation, we can approximate the voltage noise inside the conductor by that of an effective resistance R,

$$\langle \hat{V}(t)\hat{V}(t')\rangle_\omega \simeq \max(k_B T, \hbar\omega)R.$$

Let us recall from Section 6.7.2 that, for both decoherence and relaxation, the rate is determined by the spectral density of energy fluctuations. Since for charged electrons the energy fluctuations are just given by $\delta E = eV$, we can estimate the rates as follows:

$$\frac{1}{\tau_{r,\varphi}} \simeq \frac{E}{\hbar} G_Q R, \qquad (6.147)$$

where E denotes a scale of electron energy corresponding to either temperature or voltage difference across the nanostructure. We see from this that the voltage fluctuations are an intrinsic mechanism of the dissipation in quantum transport. Any resistance eventually brings about dephasing and relaxation. If the electron propagation is coherent throughout the whole nanostructure, or the nanostructure is at the border between coherent and incoherent regimes, we estimate R as the resistance of the whole nanostructure. At the border, the rates should be of the order of the dwell time in the nanostructure, that is, $\tau_{r,\varphi} E_{Th} \simeq \hbar$. Combining this with Eq. (6.147), we find that the incoherent regime takes precedence if the electron energy scale exceeds E_{in}, that is

$$E_{in} \simeq E_{Th} \frac{G}{G_Q} \simeq \left(\frac{G}{G_Q}\right)^2 \delta_S,$$

G being the conductance of the nanostructure. This estimation is applicable at $G \gg G_Q$, and in this case the resulting energy scale $E_{in} \gg E_{Th}$. This scale defines the mesoscopic border (see Fig. 1 in the Introduction to this book). For extended conductors, the effective resistance is a resistance of a small part of the conductor: we will analyze this in detail in Section 6.8.2.

Finally, there are various spin-dependent effects. The elastic part of the spin-dependent scattering is provided by spin-orbit interaction and does not lead to any dissipation. As we have seen, it leads to apparent decoherence, changing the interference corrections at the length scale of L_{SO}. This is to be contrasted with scattering on stray magnetic impurities, which are commonly present in all but very clean samples. We mentioned in Section 6.6.2 two things the impurities do to electrons: spin-flips and two-electron energy exchange. Those are sources of dephasing and relaxation, respectively. As seen from Eqs. (6.125) and (6.126), the upper boundary of these rates is given by the unitary rate, $1/\tau_u = c_i/(2\nu\hbar)$, c_i being the impurity concentration.

Control question 6.23. The density of states in a metal is given by $\nu = 0.1\,\text{eV}$ per atom. There is one magnetic impurity per 10^6 atoms of the metal. What is the upper boundary of the dephasing rate?

At temperatures $T \gg T_K$, the spin flip rate is stronger than the relaxation rate so the impurities produce more dephasing than relaxation. It is expected that both rates cease below the Kondo temperature, but for many magnetic atoms the Kondo temperature is in the millikelvin range, so they remain active in spoiling the electron coherence. Although magnetic impurities are certainly an extrinsic source of dissipation, and might seem easy to excise, in practice they remain the most important factor determining dephasing and relaxation in metals (see Section 6.8.3).

6.8.2 Voltage fluctuations

Let us discuss the voltage fluctuations as a source of relaxation and dephasing in extended diffusive conductors. We give here a set of qualitative estimations and quote Ref. [159] for exact results.

Let us estimate the relaxation time first. The electron that experiences an energy loss $\hbar\omega$ does it at a time scale ω^{-1}. During this time interval, it propagates a distance of the order of $\sqrt{D\omega}$. The effective resistance defining the voltage fluctuations is thus the resistance of a piece of conductor of this size. Let us, in addition, assume that the typical energy loss is of the order of the energy itself, and the typical energy is of the order of temperature. This brings us to the following estimation:

$$\frac{1}{\tau_r} \simeq \frac{k_B T}{\hbar} G_Q R(L_T), \tag{6.148}$$

where $R(L_T)$ is the resistance of a piece of conductor of size $L_T = \sqrt{\hbar D/k_B T}$ (called the thermal length). Naturally, the rate should be smaller than the energy loss involved, and this holds as long as $R(L_T) \ll G_Q^{-1}$. Let us now recall that the resistance depends on the length scale differently in different effective geometries: $R(L) \propto L$ in one dimension; $R(L) \propto$ const. in two dimensions, and $R(L) \propto L^{-1}$ in three dimensions. This implies that the temperature dependence of the relaxation rate is also different for different dimensions, $1/\tau_r \propto T^{d/2}$, $d = 1, 2, 3$, In all dimensions, the rate decreases with decreasing temperature.

However, in one dimension the rate decreases slower than temperature and becomes of the order of $k_B T$ at a certain temperature $\tilde{T}$. At this temperature, the effective resistance $R(L_T)$ becomes of the order of G_Q^{-1}. This indicates a problem and signals a cross-over to a different transport regime: that of strong localization (see Section 4.5). The cross-over is difficult to quantify, as are the details of the transport in this regime. However, the big picture is clear: the extended conductor is subdivided into the subconductors of $R \simeq G_Q$. The Coulomb blockade catches up, so each subconductor becomes a Coulomb island. Thus, at lower temperatures we are dealing with a strongly disordered Coulomb array, and we expect an insulator in the limit of vanishing temperature.

Control question 6.24. Can you express $\tilde{T}$ for one dimension in terms of the resistance $\tilde{R}$ per unit length? And in terms of the total resistance of the wire and the average level spacing δ_S in the wire?

It may seem that the same problem with the rate getting too high would take place in three dimensions upon *increasing* temperature. However, it is clear that the estimation for the 3d rate is obtained in diffusive limit. This restricts its validity to energy scales $k_B T \lesssim 1/\tau_p$ and corresponding space scales much longer than l. At the scale of the mean free path, the effective resistance of a 3d metal is still smaller than G_Q^{-1}, so no problems arise.

Let us turn to dephasing. In this case, we do not care about the energy loss. Along the lines of Section 6.7.2, we regard the voltage noise as classical Nyquist noise:

$$\langle V(t)V(t')\rangle = k_B T R \delta(t - t'),$$

where R is the effective resistance. This leads to a familiar estimation, $1/\tau_\varphi \simeq (k_B T/\hbar)(G_Q R)$, but what shall we take now for the resistance? If the length of the nanostructure were shorter than the dephasing length, a good decision may be to take the full resistance of the structure. For extended conductors, the electrons can only explore a small part of it before they lose the information on the initial phase. The distance covered in the course of the diffusive motion during the time τ_φ is $L_\varphi = (D\tau_\varphi)^{1/2}$. Since the electrons can only see this part of the conductor, the effective resistance should be taken at the scale L_φ. We thus make the following estimation:

$$\frac{1}{\tau_\varphi} \simeq \frac{k_B T}{\hbar} R\left(L_\varphi = \sqrt{D\tau_\varphi}\right) G_Q, \tag{6.149}$$

which is a self-consistent equation for τ_φ similar to Eq. (6.139). Let us look at its solutions in various effective geometries.

In one dimension (a wire), $R \propto L$, and thus $1/\tau_\varphi \propto T^{2/3}$. Following the general classification of environments, we conclude that the voltage fluctuations in diffusive wires provide a sub-ohmic environment with $s = -3/2$. Let us contrast the relaxation and dephasing rates in one dimension as follows:

$$\frac{1}{\tau_\varphi} = \frac{k_B T}{\hbar}\left(\frac{T}{\tilde{T}}\right)^{-1/3}; \quad \frac{1}{\tau_r} = \frac{k_B T}{\hbar}\left(\frac{T}{\tilde{T}}\right)^{-1/2}. \tag{6.150}$$

We see that the dephasing rate is faster: the electrons lose the phase first and only later equilibrate their energy. We also see a problem at sufficiently low temperatures; this is expected for sub-ohmic environments. The dephasing rate becomes of the order of temperature at the same temperature as the relaxation rate. We recall that this indicates transition to the strong localization regime.

In two dimensions, the estimation yields

$$\frac{\hbar}{\tau_\varphi} \simeq \frac{k_B T}{\hbar} G_Q R_\Box \ln(L_\varphi/L_T) \simeq \frac{k_B T}{\hbar} G_Q R_\Box \ln\left(\frac{1}{G_Q R_\Box}\right), \tag{6.151}$$

$R_\Box$ being the sheet resistance of the film. This signals an Ohmic environment and white noise. The log factor accounts for the fact that the dephasing is contributed to by all the frequencies: from $1/\tau_\varphi$ up to $k_B T/\hbar$. The dephasing rate prevails the relaxation in two dimensions by this factor.

In three dimensions, the self-consistency equation formally has the solutions $1/\tau_\varphi \propto T^{-2}$, which does not make any sense: this would imply infinitely strong dephasing at zero temperature. Indeed, we formally obtain $k_B T \ll 1/\tau_\varphi$, which means that the scale for the resistance has been chosen incorrectly. As a matter of fact, the dephasing is contributed to by frequencies not exceeding $k_B T/\hbar$. So we have to take $R(L_T)$ for the effective resistance. The estimation for the dephasing rate is the same as for the relaxation rate, $1/\tau_\varphi \simeq 1/\tau_r \propto T^{3/2}$. This implies that in three dimensions the dephasing is always accompanied by relaxation.

Let us provide some quantitative details of relaxation using the voltage fluctuations. The inelastic scattering can be described by a master equation for the energy-dependent electron filling factor $f(E)$ $(\tilde{f} \equiv 1 - f)$ as follows:

$$\left(\frac{\partial f(E)}{\partial t}\right)_{\text{in}} = \int d\omega \left(-S_r(\omega) f(E)\tilde{f}(E+\hbar\omega) + S_r(-\omega) f(E+\hbar\omega)\tilde{f}(E)\right), \tag{6.152}$$

where $S_r(\omega)$ gives the spectral intensity of the voltage fluctuations. The first term gives the rate of transitions from the electron states with energy E to the states with energy $E + \hbar\omega$ and comes with the corresponding filling factors. The second term describes the reverse transitions from $E + \hbar\omega$ to E, corresponding to the energy change $-\hbar\omega$.

In turn, the voltage fluctuations originate from the (virtual) electron transitions. We know that in thermal equilibrium conditions a noise and a susceptibility are related by Eq. (6.9). Since we are considering the relaxation, we must be ready for an electron distribution that differs greatly from the Fermi one. In this case, the situation can be characterized with an analog of the boson non-equilibrium distribution function that is contributed by the electron transitions with given energy transfer $\hbar\omega$ and all possible energies E:

$$N_{\text{nq}}(\omega) \equiv \frac{1}{\hbar\omega} \int dE\ f(E)\tilde{f}(E - \hbar\omega) \neq N_B(\omega) \text{ if } f(E) \neq f_F(E).$$

We choose $N_{\text{nq}}(\omega)$ in such a way that the spectral density is related to a susceptibility by a non-equilibrium analog of Eq. (6.9):

$$S_r(\omega) = 2\hbar\chi_r(\omega) N_{\text{nq}}(\omega).$$

Exercise 6.20. Consider a wire connected to two reservoirs biased at voltage difference V ($eV \gg k_B T$). The length of the wire is small in comparison with $L_r \equiv \sqrt{D\tau_r}$, so the electron distribution is a two-step function (see Eq. (2.14)). Compute $N_{nq}(\omega)$ in all points of the wire. Compare $N_{nq}(\omega)$ with $N_B(\omega)$ at small ω, and give the effective temperature at all points of the wire.

The susceptibility $\chi_r(\omega)$ depends on frequency only. It is given in terms of the extended impedance $Z(\omega, \boldsymbol{q})$ and the diffuson propagators that account for the electron motion:

$$2\hbar\chi_r(\omega) = G_Q \int \frac{d\boldsymbol{q}}{(2\pi)^d} \,\mathrm{Re}\, Z(\omega, \boldsymbol{q}) \frac{\omega D q^2}{|-\omega + Dq^2|^2} \simeq G_Q R(L = \sqrt{D/\omega}).$$

Similar to the case of Altshuler–Aronov corrections to the conductivity (see Section 6.5.3), the diffusion guarantees that the relevant space scale is $\sqrt{D/\omega}$. Therefore we can regard the electricity propagation as instant and use Eq. (6.97) to replace the impedance with its circuit-theory value.

When applying the inelastic collision term, it is usual that the energy-dependent electron filling factor varies in space. Indeed, the non-equilibrium conditions can only be maintained with a constant flux of electrons from different reservoirs with different filling factors. The filling factor near the entrance to a reservoir matches that in the reservoir, so it has to change across the sample. To account for this, one plugs the inelastic collision term given in Eq. (6.152) into the diffusion equation:

$$\frac{\partial f(E, \boldsymbol{r})}{\partial t} = \nabla(D \nabla f(E, \boldsymbol{r})) + \left(\frac{\partial f(E, \boldsymbol{r})}{\partial t}\right)_{\mathrm{in}}.$$

The solution varies at a typical space scale $L_r = \sqrt{D\tau_r}$, provided it is longer than the sample size. This is a much longer scale than $\sqrt{D/\omega}$ entering the susceptibility, so the filling factor does not change at distances $\simeq \sqrt{D/\omega}$. This justifies the above local approximation: the spectral intensity at a point $\boldsymbol{r}$ is determined by N_{nq} at the same point.

The inelastic collision term given by Eq. (6.152) is generally too complicated to be characterized with a single rate τ_r. So we cannot readily improve the qualitative estimations we did before, except in some special cases. To construct one of these special cases, let us add only a few excited electrons with energy E to the metal in the ground state. Since the hot electrons are few, they do not significantly change the spectral intensity, so it is given by the equilibrium expression $S_r = -2\hbar\chi_r\Theta(-\omega)$. We evaluate the transition rate of an electron with energy E to all other states with energy $E > 0$ as follows:

$$\frac{1}{\tau_{ee}(E)} = \int_0^E d\omega \, 2\hbar |\chi_r(\omega)|.$$

Exercise 6.21. Compute the rate $1/\tau_{ee}(E)$ for $d = 1, 2, 3$. Use the expressions for the extended impedance found in Section 6.1.2. What is the average energy loss during the transition?

Let us discuss the quantitative details of dephasing. It is not a priori obvious that one can characterize all manifestations of the dephasing with a single rate $1/\tau_\varphi$ corresponding to white noise and exponential dephasing. It is necessary to concentrate on a concrete manifestation, and the most natural one is the magnitude of the weak localization correction. This can be expressed in terms of the cooperon propagator. To evaluate the effect of voltage fluctuations, one includes the fluctuating voltage $V(t, \boldsymbol{r})$ in Eq. (4.60) for the cooperon:

$$\left(\frac{\partial}{\partial t} + \mathrm{i}2eV(t,\boldsymbol{r}) - D\left(\nabla_r + \mathrm{i}\frac{2e\boldsymbol{A}}{c}\right)^2\right) P^{BB'}_{\text{coop}} = \delta(\boldsymbol{r}-\boldsymbol{r}')\delta(t-t'). \qquad (6.153)$$

This equation for the cooperon has to be solved and then averaged over the voltage fluctuations, assuming the classical Nyquist noise. This is a rather complicated mathematical problem that allows for an elegant solution [159]. In one dimension, this yields (assuming $\tau_\varphi \ll \tau_{\text{SO}}$)

$$\delta G_{\text{1d}} = -2^{-1/2} G_{\text{Q}} \frac{L_\varphi}{L} \frac{Ai(\tau_\varphi/2\tau_H)}{Ai'(\tau_\varphi/2\tau_H)},$$

where $Ai(x)$ is the Airy function, and $1/\tau_\varphi$ is given by Eq. (6.150) with $k_{\text{B}}\tilde{T} = (\pi/8)G_{\text{Q}}R\delta_{\text{S}}$ (see Control question 6.24).

It turns out that the combination of Airy functions can be approximated by a square root with accuracy 4%. This justifies Eq. (6.145) and also the gross approximation made for the single dephasing rate.

In two dimensions, the method yields the exact coefficient for the estimate given in Eq. (6.151):

$$\left(\frac{1}{\tau_\varphi}\right)_{\text{2d}} = \frac{k_{\text{B}}T}{\hbar}\frac{G_{\text{Q}}}{R_\square} \ln\left(\frac{1}{G_Q R_\square}\right).$$

6.8.3 Experiments

As mentioned in Section 6.8.2, the weak localization correction is sensitive to the dephasing time. The magnitude and the temperature dependence of this correction therefore reveals the dephasing time and its temperature dependence. We illustrate this with a recent experiment conducted by Pierre *et al.* [179]. The authors have painstakingly measured the zero-voltage resistance of a large series of long metal wires made of copper, silver, and gold, while varying magnetic field and temperature. While this does not sound like an extremely novel and promising project, the results were not at all boring.

The wires were really long at nano-scale; their lengths ranged from 100 to 400 μm, with the total resistance in the kilo-ohm range. The magnetic-field dependence of the resistance exhibited a narrow dip. This is a manifestation of the weak localization correction, its sign indicating that the dephasing length L_φ exceeds L_{SO}. Both lengths were extracted from fits with Eq. (6.145) (see Fig. 6.23(a)). The coherence length L_φ never exceeded 20 μm so that the transport over the whole wire was never coherent. The wire consisted of at least 20 coherent subconductors.

The authors reported a striking dependence of the coherence time on the chemical purity of the materials used. The wires made from 99.999% and 99.9999% pure silver were easy

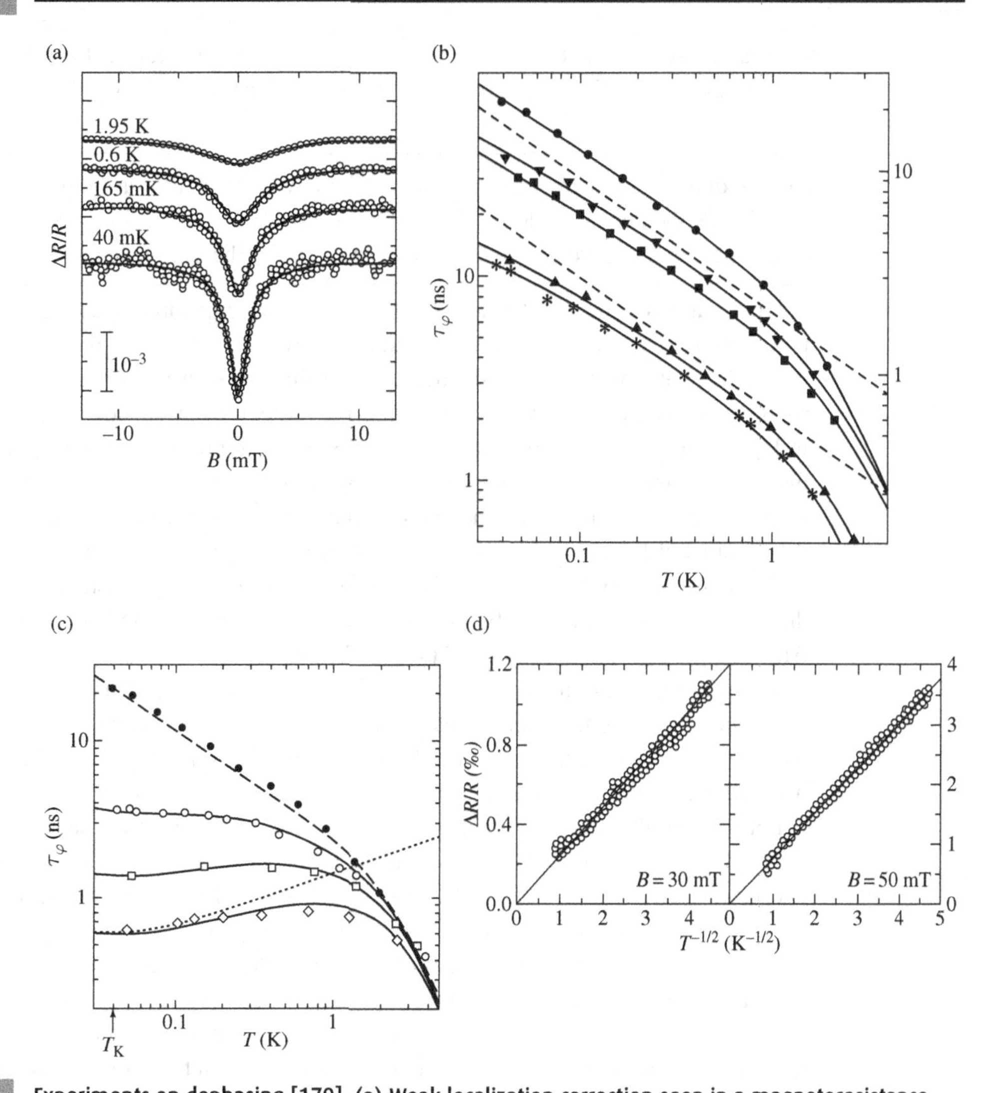

Fig. 6.23. Experiments on dephasing [179]. (a) Weak localization correction seen in a magnetoresistance trace. The dephasing length L_φ is extracted from the fit. (b) In "good" samples, the temperature dependence of the dephasing time mostly follows a $T^{-2/3}$ law (dashed lines in this log–log plot). The deviations above 1 K are due to phonons. (c) In "bad" samples (white symbols), the dephasing time apparently saturates at low temperatures owing to spin-flip at magnetic impurities. (d) Temperature dependence of the resistivity is governed by Altshuler–Aronov corrections, Eq. (6.117).

to distinguish from other measurement results. This degree of purity is very difficult to control, so the results vary significantly for formally identical samples. The samples can be divided into two groups: "good" ones and "bad" ones.

The "good" samples showed an unsaturating increase of τ_φ with decreasing temperature (see Fig. 6.23(b)) down to the lowest temperatures accessible ($\simeq$40 mK). At temperatures <1 K, the power-law dependence $1/\tau_\varphi \propto T^{2/3}$ has been confirmed in accordance with Eq. (6.143). This proves that the dominant dephasing mechanism is indeed related to voltage fluctuations. For temperatures above 1 K, the temperature dependence becomes steeper and eventually crosses over to $1/\tau_\varphi \propto T^3$, signaling electron–phonon scattering.

Control question 6.25. Suppose the $T^{-2/3}$ law holds at arbitrarily low temperatures. At 40 mK the dephasing length in a 200 μm sample was measured to be $L_\varphi = 20$ μm. At which temperature does it become of the order of the sample length?

"Bad" samples showed the saturation of τ_φ at low temperature (see Fig. 6.23(c)). Less pure samples tended to be "bad," indicating the role of impurities, presumably magnetic ones. The main contribution to the dephasing is thus the spin-flip scattering at these impurities. An extra piece of evidence is a non-monotonous dependence of τ_φ (lowest curve in Fig. 6.23(c)). This is consistent with the Kondo amplification of the spin-flip scattering. As an extra check, the samples were doped by manganese (magnetic impurities with $T_K = 40$ mK). This led to an increased dephasing rate as expected. The coherence was thus destroyed by a very low concentration of the magnetic impurities that did not produce a noticeable effect on the resistance of the samples. The temperature dependence of the resistance was dominated by interaction effects and can be accurately described by the Altshuler–Aronov relation, Eq. (6.117), for the 1d case.

The spin-flip scattering is predicted to freeze out below T_K. This was not seen in the experiment presented. Different magnetic impurities have different T_K, some T_K being exponentially small. This may mask the quenching of the spin-flip dephasing rate. Recent work addresses this issue [180].

Let us describe an experiment on the relaxation of the electron distribution function [181]. The setup (Fig. 6.24(a)) consists of a 20 μm long metallic silver wire between two reservoirs biased at voltage U, and a local probe electrode to which a voltage V is applied. The probe was a tunnel junction ($R_T \simeq 100\,\mathrm{k\Omega}$) in series with a rather resistive ($R \simeq 1.5\,\mathrm{k\Omega}$) aluminum wire. The probing can be done in two ways. At small magnetic field, the aluminum wire is in a superconducting state. At equilibrium ($U = 0$), the tunnel differential conductance peaks sharply at $V = \pm\Delta/e$, Δ being the superconducting energy gap. A two-step structure in the non-equilibrium wire results in four peaks: at $\pm\Delta/e$ and at $U \pm \Delta$. The actual electron distribution function is determined by the deconvolution of the tunnel conductance.

This method does not work at higher magnetic fields >0.1 T since the superconductivity in the aluminum wire is destroyed. The alternative method exploits the dynamical Coulomb-blockade effect (see Section 6.3), which is quite noticeable owing to the large normal resistance of the aluminum wire. At equilibrium ($U = 0$), the dynamical Coulomb blockade gives rise to a conductance dip at $V = 0$. For the two-step distribution, it gives rise to two dips of approximately half the magnitude, at $V = 0$ and $V = U$ (Fig. 6.24(b)).

Two samples of different purity (99.9999% and 99.999%) have been measured. The pure sample (sample 2) showed a sharp two-step distribution function (Fig. 6.24(c)). This indicated that the relaxation time was bigger than the dwell time $t_d = L^2/D \simeq 20$ ns in the wire. The fit with the 1d relaxation rate, Eq. (6.145), was satisfactory. The impure sample (sample 1) had the same resistance and dwell time. The contribution of voltage fluctuations to the relaxation rate is therefore expected to be the same. However, the distribution

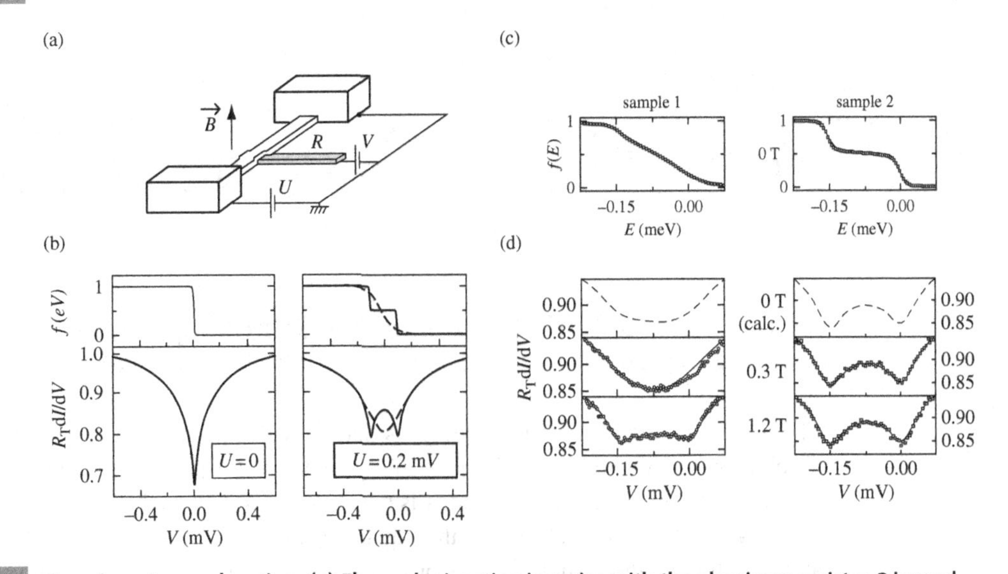

Fig. 6.24. **Experiments on relaxation. (a) The probe junction in series with the aluminum resistor R is used to measure the electron distribution in the silver wire biased with U. (b) The principle of the measurement of the electron distribution function. The Coulomb-blockade singularity splits if a sharp two-step structure is present. (c) The distributions measured in impure and pure samples. The relaxation rate is enhanced by magnetic impurities. (d) Magnetic field has no visible effect on the distribution in the pure sample, but sharpens it in the impure sample.**

function in the impure sample was much more smooth: no obvious two-step structure was observed. The attempt to fit the curve with Eq. (6.145) indicated a 20 times faster relaxation rate.

A plausible source for the enhanced relaxation are the magnetic impurities present in the impure sample. To check this hypothesis, the authors took measurements in a high magnetic field. The traces of the differential conductance of the tunnel junction are presented in Fig. 6.24(d). The distribution function in the pure sample did not depend on the magnetic field as expected for the relaxation due to voltage fluctuations. In contrast, the distribution function has definitely sharpened in the impure sample. Indeed, the relaxation rate due to magnetic impurities is expected to be reduced by the applied magnetic field. In this case, there is a preferential direction for impurity spins. To flip the spin, the electrons must give Zeeman energy $E_H = g\mu_B B$ to it ($E_H = 0.18\,\text{meV}$ at $B = 1.5\,\text{T}$). This energy must be provided by the voltage source, so one expects the reduction of the relaxation rate at $eU < E_H$, as seen in the experiment. No significant reduction of the relaxation rate has been observed at larger $U = 0.3\,\text{mV}$. As far as detailed comparison with theory is concerned, good fits require an order of magnitude larger concentration of magnetic impurities than expected from material analysis of the samples and from the measurements of dephasing. So the electron relaxation still presents puzzles and requires more experimental and theoretical research, along with most of the topics of quantum transport covered in the book.

A Appendix A Survival kit for advanced quantum mechanics

A.1 Green's function of the Schrödinger equation

The wave function of a particle, $\Psi(\boldsymbol{r}, t)$ is determined from the Schrödinger equation. Its absolute value squared gives the probability of finding a particle at point $\boldsymbol{r}$ at the time moment t. A more informative quantity is the *Green's function* of the Schrödinger equation, given by $G(\boldsymbol{r}, t; \boldsymbol{r}', t')$. Its absolute value squared gives the *conditional* probability: the probability of finding the particle at $(\boldsymbol{r}, t)$, provided it was at the point $\boldsymbol{r}'$ at time moment t'. The Green's function solves the time-dependent Schrödinger equation with the delta-function source on the right-hand side as follows:

$$\mathrm{i}\hbar \frac{\partial G}{\partial t} - \hat{H}_{\boldsymbol{r}} G = \delta(t - t')\delta(\boldsymbol{r} - \boldsymbol{r}'), \tag{A.1}$$

where $\hat{H}_{\boldsymbol{r}}$ is the Hamiltonian of the free particle (see Eq. (1.4)) at point $\boldsymbol{r}$. Integrating Eq. (A.1) from $t = t' - 0$ to $t = t' + 0$, we find

$$G(t = t' + 0) - G(t = t' - 0) = -\frac{\mathrm{i}}{\hbar}\delta(\boldsymbol{r} - \boldsymbol{r}'). \tag{A.2}$$

The Green's function experiences a jump at $t = t'$. Two of the solutions are particularly useful. The *retarded* Green's function G^{R} describes the evolution forwards in time. It is only non-zero at $t > t'$. Using the completeness of the basis $\psi_n(\boldsymbol{r})$ of the stationary Schrödinger equation, we obtain

$$\begin{aligned} G^{\mathrm{R}}(\boldsymbol{r}, t; \boldsymbol{r}', t') &= \sum_n \int_{-\infty}^{\infty} \frac{\mathrm{d}\omega}{2\pi} \mathrm{e}^{-\mathrm{i}\omega(t-t')} \frac{\psi_n(\boldsymbol{r})\psi_n^*(\boldsymbol{r}')}{\hbar\omega - E_n + \mathrm{i}0} \\ &= -\frac{\mathrm{i}}{\hbar} \sum_n \mathrm{e}^{-\mathrm{i}E_n(t-t')/\hbar} \psi_n(\boldsymbol{r})\psi_n^*(\boldsymbol{r}')\Theta(t - t'), \end{aligned} \tag{A.3}$$

where E_n is the energy of the state described by the wavefunction ψ_n. Note that the integrand of Eq. (A.3) is analytical in the upper half-plane of the complex variable ω; this guarantees that the function is retarded. Similarly, the *advanced* Green's function is the solution of Eq. (A.1) which vanishes at $t > t'$. It can be represented as the following integral:

$$G^{\mathrm{A}}(\boldsymbol{r}, t; \boldsymbol{r}', t') = \sum_n \int_{-\infty}^{\infty} \frac{\mathrm{d}\omega}{2\pi} \mathrm{e}^{-\mathrm{i}\omega(t-t')} \frac{\psi_n(\boldsymbol{r})\psi_n^*(\boldsymbol{r}')}{\hbar\omega - E_n - \mathrm{i}0}, \tag{A.4}$$

and the integrand is analytical in the lower half-plane.

Equations (A.3) and (A.4) are convenient for a finite system with discrete quantum states $|n\rangle$. For an infinite system, the Green's functions must be expressed via scattering states (see, for example, Ref. [182]).

It is convenient to view the Green's function as a matrix in space-time. The unit matrix is the delta-function, and the Green's function is the inverse of the evolution operator $\hat{K}^{\mathrm{R,A}} = \mathrm{i}\hbar\partial/\partial t - \hat{H}_{\boldsymbol{r}} \pm \mathrm{i}0$, where the upper and lower signs correspond to the retarded and advanced Green's functions, respectively. In matrix notation, $\hat{K}^{\mathrm{R,A}}G^{\mathrm{R,A}} = 1$, the operator $\hat{K}$ is a diagonal matrix given by

$$\hat{K}(\boldsymbol{r},t;\boldsymbol{r}',t') \equiv \hat{K}(\boldsymbol{r},t)\delta(\boldsymbol{r}-\boldsymbol{r}')\delta(t-t'),$$

and the matrix product means the integration over the intermediate coordinate and time. This relation facilitates writing down the perturbation series for the Green's function.

If the Hamiltonian contains a weak perturbation $\hat{V}$, $\hat{H} = \hat{H}_0 + \hat{V}$, the retarded Green's function of the perturbed system is given by

$$\begin{aligned} G^{\mathrm{R}} &= (\hat{K}^{\mathrm{R}})^{-1} = (\hat{K}_0^{\mathrm{R}} - \hat{V})^{-1} = \left[\hat{K}_0^{\mathrm{R}}\left(1-(\hat{K}_0^{\mathrm{R}})^{-1}\hat{V}\right)\right]^{-1} \\ &= \left(1-(\hat{K}_0^{\mathrm{R}})^{-1}\hat{V}\right)^{-1}(\hat{K}_0^{\mathrm{R}})^{-1} = \left[1+G_0^{\mathrm{R}}\hat{V}+G_0^{\mathrm{R}}\hat{V}G_0^{\mathrm{R}}\hat{V}+\cdots\right]G_0^{R} \\ &= G_0^{\mathrm{R}} + G_0^{\mathrm{R}}\hat{V}G_0^{\mathrm{R}} + G_0^{\mathrm{R}}\hat{V}G_0^{\mathrm{R}}\hat{V}G_0^{\mathrm{R}} + \cdots, \end{aligned} \tag{A.5}$$

where $\hat{K}_0^{\mathrm{R}} = \mathrm{i}\hbar\partial/\partial t - \hat{H}_{\boldsymbol{r}} - \mathrm{i}0$ and $G_0 = \hat{K}_0^{-1}$. A similar relation holds for the advanced Green's function.

For practical applications of this scheme, one needs some advanced methods. The methods used in this book are explained in Section 2.3 (the Keldysh approach and semi-classical methods) and in Section 4.4 (diffuson–cooperon expansion).

A.2 Second quantization

- Electrons are identical quantum-mechanical particles obeying Fermi statistics. The wave function of several electrons is antisymmetric – it changes sign every time we interchange two electrons.
- The Fock space is the space of all antisymmetric solutions of the Schrödinger equation, containing solutions for N electrons with all possible N (including $N = 0$). An element of the Fock state $|n_1, n_2, \ldots\rangle$ is labeled by the *occupation numbers* of the states $1, 2, \ldots$: 0 if the state is empty and 1 if it is occupied. A special element of the Fock state is a vacuum, $|0, 0, \ldots\rangle$, when there are no particles in the system.
- The states of the Fock space can be obtained from the vacuum by application of a number of *creation operators*. The creation operator $\hat{a}_i^\dagger$ adds an electron into the state $|i\rangle$, described by the wavefunction $\psi_i(\boldsymbol{r})$, as follows:

$$\begin{aligned} &\hat{a}_i^\dagger\,|n_1,\ldots,n_{i-1},0,n_{i+1},\ldots\rangle = \pm\,|n_1,\ldots,n_{i-1},1,n_{i+1},\ldots\rangle, \\ &\hat{a}_i^\dagger\,|n_1,\ldots,n_{i-1},1,n_{i+1},\ldots\rangle = 0, \end{aligned} \tag{A.6}$$

where the sign is positive (negative) for an even (odd) number of particles in all the states k with $k < i$.

- The conjugated *annihilation operator* $\hat{a}_i$ takes an electron out of the same state:

$$\begin{aligned} \hat{a}_i \left|n_1, \ldots, n_{i-1}, 1, n_{i+1}, \ldots\right\rangle &= \pm \left|n_1, \ldots, n_{i-1}, 0, n_{i+1}, \ldots\right\rangle, \\ \hat{a}_i \left|n_1, \ldots, n_{i-1}, 0, n_{i+1}, \ldots\right\rangle &= 0. \end{aligned} \tag{A.7}$$

- Creation and annihilation operators obey the anticommutation rules:

$$\left[\hat{a}_i^\dagger, \hat{a}_j\right]_+ \equiv \hat{a}_i^\dagger \hat{a}_j + \hat{a}_j \hat{a}_i^\dagger = \delta_{ij}; \quad \left[\hat{a}_i^\dagger, \hat{a}_j^\dagger\right]_+ = \left[\hat{a}_i, \hat{a}_j\right]_+ = 0. \tag{A.8}$$

- The operator $\hat{a}_i^\dagger \hat{a}_i$ describes the number of electrons in state $|i\rangle$. The average of this operator gives the average number of particles n_i in the state $|i\rangle$.
- Any operator can be expressed in terms of the creation and annihilation operators. For example, the Hamiltonian of non-interacting electrons has the form $\hat{H} = \sum_i E_i \hat{a}_i^\dagger \hat{a}_i$.
- Field operators create/annihilate a particle at a certain space point:

$$\hat{\psi}(\boldsymbol{r}) = \sum_i \psi_i(\boldsymbol{r}) \hat{a}_i; \quad \hat{\psi}^\dagger(\boldsymbol{r}) = \sum_i \psi_i^*(\boldsymbol{r}) \hat{a}_i^\dagger. \tag{A.9}$$

The anticommutation rules for the field operators are given by

$$\left[\hat{\psi}^\dagger(\boldsymbol{r}), \hat{\psi}(\boldsymbol{r}')\right]_+ = \delta(\boldsymbol{r} - \boldsymbol{r}'), \quad \left[\hat{\psi}^\dagger(\boldsymbol{r}), \hat{\psi}^\dagger(\boldsymbol{r}')\right]_+ = \left[\hat{\psi}(\boldsymbol{r}), \hat{\psi}(\boldsymbol{r}')\right]_+ = 0. \tag{A.10}$$

- The Hamiltonian is expressed in terms of the field operators as follows:

$$\begin{aligned} \hat{H} = &\int \mathrm{d}\boldsymbol{r}\, \hat{\psi}^\dagger(\boldsymbol{r}) \left[-\frac{\hbar^2}{2m} \nabla^2 + U(\boldsymbol{r}) \right] \hat{\psi}(\boldsymbol{r}) \\ &+ \frac{1}{2} \int \mathrm{d}\boldsymbol{r}\, \mathrm{d}\boldsymbol{r}'\, \hat{\psi}^\dagger(\boldsymbol{r}) \hat{\psi}^\dagger(\boldsymbol{r}') V(\boldsymbol{r} - \boldsymbol{r}') \hat{\psi}(\boldsymbol{r}') \hat{\psi}(\boldsymbol{r}), \end{aligned}$$

where V is the electron–electron interaction.

- For bosons (phonons or photons), the many-body wavefunction is symmetric. The creation and annihilation operators for bosons, defined as

$$\begin{aligned} \hat{b}_i^\dagger \left|n_1, \ldots, n_i, \ldots\right\rangle &= \sqrt{n_i + 1} \left|n_1, \ldots, n_i + 1, \ldots\right\rangle, \\ \hat{b}_i \left|n_1, \ldots, n_i, \ldots\right\rangle &= \sqrt{n_i} \left|n_1, \ldots, n_i - 1, \ldots\right\rangle, \end{aligned} \tag{A.11}$$

obey the following commutation rules:

$$\left[\hat{b}_i, \hat{b}_j^\dagger\right]_- \equiv \hat{b}_i \hat{b}_j^\dagger - \hat{b}_j^\dagger \hat{b}_i = \delta_{ij}; \quad \left[\hat{b}_i^\dagger, \hat{b}_j^\dagger\right]_- = \left[\hat{b}_i, \hat{b}_j\right]_- = 0. \tag{A.12}$$

- The field operators are defined in the same way as for fermions, and obey the commutation (rather than anticommutation) relations. All the operators are expressed via the field operators in the same way as for the case of fermions.

A.3 WKB approximation

- The WKB (Wentzel–Kramers–Brillouin), or semiclassical, approximation applies when typical scales of the potential greatly exceed the wavelength of a particle.
- The wavefunction of an electron for one-dimensional motion in a smooth potential $U(x)$ is given by

$$\psi(x) = \frac{C}{\sqrt{p}} \exp\left(\pm \frac{\mathrm{i}}{\hbar} \int p \, \mathrm{d}x \right).$$

Here $p(x) = \sqrt{2m(E - U(x))}$ is the classical momentum of the electron. The $\pm$ signs in the exponent must be chosen from the boundary conditions (wavefunctions do not grow at infinity).

- If the electron performs a finite motion, the position of its energy levels are determined by the Bohr–Sommerfeld quantization condition,

$$\oint p \, \mathrm{d}x = 2\pi \hbar (n + 1/2), \quad n = 0, 1, 2, \ldots, \tag{A.13}$$

where the integral is calculated over the period of the classical (finite) motion of the electron.

- The semiclassical probability of transmission through a potential barrier is given by

$$T = \exp\left(-\frac{2}{\hbar} \int |p| \mathrm{d}x \right), \tag{A.14}$$

where the integral is taken over the classically forbidden region – where the momentum p is imaginary.

Appendix B Survival kit for superconductivity

This survival kit is intended to provide basic knowledge of superconductivity necessary for understanding the material of the book. We recommend Refs. [41], [56], and [183] to the reader who wishes to acquire a deeper understanding of superconductivity concepts.

B.1 Basic facts

- Below a certain temperature T_c (the *superconducting transition temperature*), the electrical resistance of some metals vanishes. In particular, the most commonly used superconductors are aluminum and niobium; alkali, noble, and magnetic metals never become superconducting. The highest transition temperature found among pure metals is about 9 K for niobium; among "usual" superconducting compounds it is 39 K for magnesium diboride. There are compounds with even higher transition temperatures, of over 100 K, known as high-temperature superconductors. They possess unusual symmetries, which lead to very uncommon physical properties. We do not consider them in this book.
- Superconductors are ideal diamagnets: weak magnetic fields do not penetrate the bulk of superconductors (*Meissner effect*). A high magnetic field destroys superconductivity. This *critical field*, H_c, can vary from 1 G (approximately 10^{-4} T) for tungsten to 1980 G for niobium.
- There is a narrow layer at the boundary of the superconductor where an external magnetic field decreases exponentially to zero value in the bulk. The characteristic length of this decay δ_p, known as the *penetration depth*, is temperature-dependent and diverges at the transition temperature proportionally to $(T_c - T)^{-1/2}$.
- The transition to the superconducting state is a second-order phase transition in zero magnetic field. It may become a first-order phase transition in finite magnetic fields.
- In the superconducting state, the specific heat depends exponentially on the temperature, $C \propto \exp(-\Delta/k_B T)$, $k_B T \ll \Delta$. At the transition point, the specific heat experiences a jump.

B.2 Microscopic theory of superconductivity

The microscopic theory of superconductivity was independently proposed by Bogoliubov and Bardeen, Cooper, and Schrieffer; it is commonly known as BSC theory, referring to the latter three authors.

- Conceptually, superconductivity is similar to superfluidity in helium-4: at low temperatures, liquid helium flows along capillaries without friction. Superconductivity can be interpreted as superfluidity of *Cooper pairs* – pairs consisting of two electrons.
- To form a pair, or indeed a bound state, one needs a mechanism that provides attraction between electrons – this attraction has to overcome the Coulomb repulsion. In metals, this attraction comes from the electron–phonon interaction. The phonon-mediated attraction between electrons has the form of a local interaction, $V_{\text{e-ph}} = -(2\lambda/\nu)\delta(\boldsymbol{r}_1 - \boldsymbol{r}_2)$, where λ is the electron–phonon interaction constant. The attraction is effective only at frequencies lower than the typical phonon frequency ω_D. If this attraction compensates for the Coulomb repulsion, the net interaction between the electrons is attractive, and the metal undergoes a transition into a superconducting state at low temperatures.
- The ground state of two electrons with attraction is a bound state with energy $E = -2\Delta$, with

$$\Delta = \hbar\omega_\text{D} \exp(-1/\lambda). \tag{B.1}$$

  This quantity has an extremely important role in the theory of superconductivity and is known as the *superconducting energy gap.*
- The superconducting gap is related to the temperature T_c of the superconducting transition. The microscopic theory yields that, at zero temperature, $\Delta = 1.76k_\text{B}T_\text{c}$. It is temperature-dependent and vanishes at the transition temperature as $(T_\text{c} - T)^{1/2}$.
- The order parameter characterizing the superconducting phase transition is a *complex* number with the absolute value equal to the gap Δ and the phase φ. This is why the superconducting phase transition can be regarded as spontaneous breaking of the gauge invariance.
- In the ground state, a superconductor can support current – a *supercurrent.* Supercurrent is non-dissipative and is related to the gradient of the superconducting phase, $\boldsymbol{j}_\text{s} = en_\text{s}(\boldsymbol{v}_\text{s} - (e/mc)\boldsymbol{A})$, where n_s and $\boldsymbol{v}_\text{s} = (\hbar/2m)\nabla\varphi$ are the *superfluid density* and *superfluid velocity*, respectively, and $\boldsymbol{A}$ is the vector potential due to the external magnetic field. In the bulk of the superconductor, the supercurrent vanishes. If there is a phase difference between two pieces of superconductor, there is a supercurrent flowing between them – this is the *Josephson effect*; see Section 1.8.
- The excitation spectrum of a superconductor has the following form:

$$\epsilon_{\boldsymbol{p}} = \sqrt{\Delta^2 + \xi_{\boldsymbol{p}}^2}, \tag{B.2}$$

  where $\xi_{\boldsymbol{p}}$ are the excitation energies for quasiparticles (electrons and holes) in a normal metal. There are no excitations with energy below Δ. In particular, if a superconductor is brought into contact with a normal metal, electrons from the normal metal with energies below Δ cannot penetrate the superconductor.
- The BCS density of states for the excitations has the following form:

$$\nu_\text{BCS}(E) = \sum_{\boldsymbol{p}} \delta\left(E - \sqrt{\Delta^2 + \xi_{\boldsymbol{p}}^2}\right) \propto \frac{\theta(E - \Delta)}{\sqrt{E - \Delta}}. \tag{B.3}$$

  It vanishes inside the gap and diverges at $E = \Delta$.

- The size of a Cooper pair, estimated from the uncertainty relation as $\xi \equiv \delta r \sim \hbar/\delta p \sim \hbar v_F/\Delta$, for $v_F \sim 10^6$ m/s and $\Delta/k_B \sim T_c \sim 10$ K, becomes $\xi \sim 1\,\mu$m, which is much bigger than the mean distance between electrons. Thus, one cannot imagine Cooper pairs as rigid small ball-like structures. Instead, one may think of superconducting correlations between electrons at the distance of the order of ξ – the *superconducting correlation length*. For example, if a superconductor is brought into contact with a normal metal, electrons at a distance ξ from the superconductor would still feel the correlations – this is the *proximity effect*; see Section 2.8.
- The dimensionless ratio of penetration depth to correlation length $\kappa = \delta_p/\xi$ is responsible for the behavior of the superconductor in a magnetic field. For $\kappa < 1/\sqrt{2}$ (Type I superconductors), the energy of the interface between the superconductor and the normal metal is positive. In the opposite case, $\kappa > 1/\sqrt{2}$ (Type II superconductors), the interface energy is negative.
- In Type II superconductors, the proliferation of normal metal pieces in a magnetic field becomes energetically favorable. An external magnetic field in a huge range penetrates the superconductor as narrow filaments – known as *Abrikosov vortices*. The superfluid velocity circulates around the vortex core (of the size of $\xi \ll \delta_p$). Each vortex carries precisely one flux quantum Φ_0, this being a consequence of the fact that the phase shift around a closed contour (in this case, drawn around a vortex) is a multiple integer of 2π. The vortices usually form a triangular lattice.

Appendix C Unit conversion

For nanostructures, appropriate units of energy are millivolts (meV) and appropriate units for length are nanometers (nm). Below, we express other units and important scales in terms of these two.

- **Conversion between energy, frequency, and temperature**.
  We use the expressions $eV = \hbar\omega = k_B T$.

	meV	THz	K
1 meV	1	1.52	11.6
1 THz	0.66	1	7.61
1 K	0.086	0.13	1

- **Electron charge and conductance quantum**:
  $e^2 = 1.44 \times 10^3$ meV nm;
  $G_Q \equiv e^2/\pi\hbar = 6.97 \times 10^5$ m/s $= 7.75 \times 10^{-5}$ S;
  $R_Q \equiv G_Q^{-1} = 12.9\,\mathrm{k\Omega}$.
- **Fermi velocity**, typical values:
  for aluminum, $v_F = 2 \times 10^6$ m/s; $\hbar v_F = 1.3 \times 10^3$ meV nm;
  for a carbon nanotube, $v_F = 8 \times 10^5$ m/s, $\hbar v_F = 5.3 \times 10^2$ meV nm.
- **Capacitance**: 1 aF $= 10^{-18}$ F $= 23.04$ nm.
- **Speed of light as conductance**: $c = 3.33 \times 10^{-2}$ S; vacuum impedance $4\pi/c = 377\,\Omega$.
- **Electron mass**:
  free electrons, $\hbar^2/m = 755$ meV/nm^2;
  Ga As, $\hbar^2/m = 1.13 \times 10^4$ meV/nm^2.
- **Flux quantum**: $\Phi_0 \equiv \pi\hbar c/e = 2.07 \times 10^3$ T nm^2.
- **Bohr magneton** $e\hbar/2mc$:
  free electrons, 5.78×10^{-2} meV/T;
  GaAs, 0.86 meV/T.

References

[1] Y. Imry, *Introduction to Mesoscopic Physics* (Oxford: Oxford University Press, 1997).

[2] B. L. Altshuler, P. A. Lee, and R. A. Webb, (eds.) *Mesoscopic Phenomena in Solids* (New York: North-Holland, 1991).

[3] L. I. Glazman, G. B. Lesovik, D. E. Khmel'nitskii, and R. I. Shekhter, *Pis'ma Zh. Éksp. Teor. Fiz.* **48**, (1988), 218 [*JETP Lett.* **48**, (1988), 238].

[4] B. J. van Wees, H. van Houten, C. W. J. Beenakker, J. G. Williamson, L. P. Kouwenhoven, D. van der Marel, and C. T. Foxon, *Phys. Rev. Lett.* **60**, (1988), 848.

[5] D. A. Wharam, T. J. Thornton, R. Newbury, *et al.*, *J. Phys. C* **21**, (1988), L209.

[6] I. A. Larkin and V. B. Shikin, *Phys. Lett. A* **151**, (1990), 335.

[7] L. I. Glazman and I. A. Larkin, *Semicond. Sci. Technol.* **6**, (1991), 32.

[8] R. Landauer, *Phil. Mag.* **21**, (1970), 863.

[9] M. Büttiker, *Phys. Rev. B* **46**, (1992), 12485.

[10] L. S. Levitov, H. W. Lee, and G. B. Lesovik, *J. Math. Phys.* **37**, (1996), 4845.

[11] M. Henny, S. Oberholzer, C. Strunk, and C. Schönenberger, *Phys. Rev. B* **59** (1999), 2871.

[12] G. B. Lesovik, *Pis'ma Zh. Éksp. Teor. Fiz.* **49**, (1989), 513 [*JETP Lett.* **49**, (1989), 592].

[13] M. Büttiker, *Phys. Rev. Lett.* **65**, (1990), 2901.

[14] Ya. M. Blanter and M. Büttiker, *Phys. Rep.* **336**, (2000), 1.

[15] C. W. J. Beenakker and M. Büttiker, *Phys. Rev. B* **46**, (1992), 1889.

[16] B. Reulet, J. Senzier, and D. E. Prober, *Phys. Rev. Lett.* **91**, (2003), 196601.

[17] M. Büttiker, *Phys. Rev. Lett.* **57**, (1986), 1761.

[18] A. D. Benoit, S. Washburn, C. P. Umbach, R. B. Laibowitz, and R. A. Webb, *Phys. Rev. Lett.* **57**, (1986), 1765.

[19] Y. Gefen, Y. Imry, and M. Azbel, *Phys. Rev. Lett.* **52**, (1984), 129.

[20] D. Yu. Sharvin and Yu. V. Sharvin, *Pis'ma Zh. Teor. Éksp. Fiz.* **34**, (1981), 285 [*JETP Lett.* **34**, (1981), 272].

[21] R. A. Webb, S. Washburn, C. P. Umbach, and R. B. Laibowitz, *Phys. Rev. Lett.* **54**, (1985), 2696.

[22] S. Washburn, H. Schmid, D. Kern, and R. A. Webb, *Phys. Rev. Lett.* **59**, (1987), 1791.

[23] M. Büttiker, *J. Phys.: Condensed Matter* **5**, (1993), 9361.

[24] P. K. Tien and J. P. Gordon, *Phys. Rev.* **129**, (1963), 647.

[25] G. B. Lesovik and R. Loosen, *Pis'ma Zh. Teor. Éksp. Fiz.* **65**, (1997), 280 [*JETP Lett.* **65**, (1997), 295].

[26] U. Gavish, Y. Levinson, and Y. Imry, *Phys. Rev. B* **62**, (2000), R10637.

[27] P. W. Brouwer, *Phys. Rev. B* **58**, (1998), R10135.

[28] L. S. Levitov. In *Quantum Noise in Mesoscopic Physics*, ed. Yu. V. Nazarov (Dordrecht: Kluwer Academic Publishing, 2003), p. 373; also available as cond-mat/0103617.

[29] A. F. Andreev, *Zh. Teor. Éksp. Fiz.* **46**, (1964), 1823 [*Sov. Phys. JETP* **19**, (1964), 1228].

[30] C. J. Lambert and R. Raimondi, *J. Phys.: Condensed Matter* **10**, (1998), 901.

[31] G. E. Blonder, M. Tinkham, and T. M. Klapwijk, *Phys. Rev. B* **25**, (1982), 4515.

[32] C. W. J. Beenakker and H. van Houten, *Phys. Rev. Lett.* **66**, (1991), 3056.

[33] E. N. Bratus, V. S. Shumeiko, and G. Wendin, *Phys. Rev. Lett.* **74**, (1995), 2110.

[34] D. Averin and A. Bardas, *Phys. Rev. Lett.* **75**, (1995), 1831.

[35] J. C. Cuevas and W. Belzig, *Phys. Rev. B* **70**, (2004), 214512.

[36] R. Cron, M. F. Goffman, D. Esteve, and C. Urbina, *Phys. Rev. Lett.* **86**, (2001), 4104.

[37] E. Scheer, P. Joyez, D. Esteve, C. Urbina, and M. H. Devoret, *Phys. Rev. Lett.* **78**, (1997), 3535.

[38] N. Agrait, A. Levy Yeyati, and J. M. van Ruitenbeek, *Phys. Rep.* **377**, (2003), 81.

[39] S. M. Cronenwett, H. J. Lynch, D. Goldhaber-Gordon, L. P. Kouwenhoven, C. M. Marcus, K. Hirose, N. S. Wingreen, and V. Umansky, Phys. Rev. Lett. **88**, 226805 (2002).

[40] J. M. Ziman, *Principles of the Theory of Solids* (Cambridge: Cambridge University Press, 1972).

[41] A. A. Abrikosov, *Fundamentals of the Theory of Metals* (New York: North-Holland, 1988).

[42] A. I. Larkin and Yu. N. Ovchinnikov, *Zh. Teor. Éksp. Fiz.* **55**, (1968), 2262 [*Sov. Phys. JETP* **28**, (1969), 1200].

[43] G. Eilenberger, *Z. Phys.* **214**, (1968), 195.

[44] W. Belzig, F. K. Wilhelm, C. Bruder, G. Schön, and A. D. Zaikin, *Superlatt. Miscrostr.* **25**, (1999), 1251.

[45] Yu. V. Nazarov, *Superlatt. Miscrostr.* **25**, (1999), 1221.

[46] P. M. Ostrovsky, M. A. Skvortsov, and M. V. Feigel'man, *Phys. Rev. Lett.* **87**, (2001), 027002.

[47] K. M. Schep and G. E. W. Bauer, *Phys. Rev. B* **56**, (1997), 15860.

[48] Yu. V. Nazarov. In *Quantum Dynamics of Submicron Structures*, eds. H. A. Cerdeira, B. Kramer, and G. Schön (Dordrecht: Kluwer Academic Publishing, 1995), p. 687.

[49] O. Agam, I. Aleiner, and A. Larkin, *Phys. Rev. Lett.* **85**, (2000), 3153.

[50] S. Oberholzer, E. V. Sukhorukov, and C. Schönenberger, *Nature* **415**, (2002), 765.

[51] O. N. Dorokhov, *Solid State Commun.* **51**, (1984), 381.

[52] M. N. Baibich, J. M. Broto, A. Fert, *et al.* Phys. Rev. Lett. **61**, (1998), 2472.

[53] I. Žutić, J. Fabian, and S. Das Sarma, *Rev. Mod. Phys.* **76**, (2004), 323.

[54] A. Brataas, Yu. V. Nazarov, and G. E. W. Bauer, *Phys. Rev. Lett.* **84**, (2000), 2481.
[55] C. W. J. Beenakker, *Rev. Mod. Phys.* **69**, (1997), 731.
[56] M. Tinkham, *Introduction to Superconductivity* (New York: McGraw-Hill, 1995).
[57] J. Rammer and H. Smith, *Rev. Mod. Phys.* **58**, (1986), 323.
[58] A. L. Shelankov, *J. Low Temp. Phys.* **60**, (1985), 29.
[59] Yu. V. Nazarov, *Phys. Rev. Lett.* **73**, (1994), 1420.
[60] T. H. Stoof and Yu. V. Nazarov, *Phys. Rev. B* **54**, (1996), R772.
[61] H. Grabert and M. H. Devoret, eds. *Single Charge Tunneling: Coulomb Blockade Phenomena in Nanostructures* (New York: Plenum, 1992).
[62] P. Delsing. In *Single Charge Tunneling: Coulomb Blockade Phenomena in Nanostructures*, eds. H. Grabert and M. H. Devoret (New York: Plenum, 1992), p. 249.
[63] T. A. Fulton and G. J. Dolan, *Phys. Rev. Lett.* **59**, (1987), 109.
[64] L. J. Geerligs, V. F. Anderegg, P. A. M. Holweg, J. E. Mooij, H. Pothier, D. Esteve, C. Urbina, and M. H. Devoret, *Phys. Rev. Lett.* **64**, (1990), 2691.
[65] H. Pothier, P. Lafarge, C. Urbina, D. Esteve, and M. H. Devoret, *Europhys. Lett.* **17**, (1992), 249.
[66] D. V. Averin and Yu. V. Nazarov, *Phys. Rev. Lett.* **65**, (1990), 2446.
[67] M. Matters, J. J. Versluys, and J. E. Mooij, *Phys. Rev. Lett.* **78**, (1997), 2469.
[68] L. S. Kuzmin and D. B. Haviland, *Phys. Rev. Lett.* **67**, (1991), 2890.
[69] R. F. Voss and R. A. Webb, *Phys. Rev. Lett.* **47**, (1981), 268.
[70] M. H. Devoret, J. M. Martinis, and J. Clarke, *Phys. Rev. Lett.* **55**, (1985), 1908.
[71] R. Fazio and H. S. J. van der Zant, *Phys. Rep.* **355**, (2001), 235.
[72] K. B. Efetov, *Zh. Teor. Éksp. Fiz.* **78**, (2000), 2017 [*Sov. Phys. JETP* **51**, (1980), 1015].
[73] Y. Aharonov and A. Casher, *Phys. Rev. Lett.* **53**, (1984), 319.
[74] V. L. Berezinskii, *Zh. Teor. Éksp. Fiz.* **59**, (1970), 907 [*Sov. Phys. JETP* **32**, (1971), 493].
[75] J. M. Kosterlitz and D. J. Thouless, *J. Phys. C* **6**, (1973), 1181.
[76] Ch. Leemann, Ph. Lerch, G.-A. Racine, and P. Martinoli, *Phys. Rev. Lett.* **56**, (1986), 1291.
[77] J. E. Mooij, B. J. van Wees, L. J. Geerligs, M. Peters, R. Fazio, and G. Schön, *Phys. Rev. Lett.* **65**, (1990), 645.
[78] M. T. Tuominen, J. M. Hergenrother, T. S. Tighe, and M. Tinkham, *Phys. Rev. Lett.* **69**, (1992), 1997.
[79] J. M. Hergenrother, J. G. Lu, and M. Tinkham, *IEEE Trans. Appl. Superconductivity,* **5**, (1995), 2604.
[80] T. M. Eiles, J. M. Martinis, and M. H. Devoret, *Phys. Rev. Lett.* **70**, (1993), 1862.
[81] P. Hadley, E. Delvigne, E. H. Visscher, S. Lähteenmäki, and J. E. Mooij, *Phys. Rev. B* **58**, (1998), 15317.
[82] D. V. Averin, A. N. Korotkov, A. J. Manninen, and J. P. Pekola, *Phys. Rev. Lett.* **78**, (1997), 4821.
[83] M. L. Mehta, *Random Matrices* (New York: Academic Press, 2004).
[84] T. Guhr, A. Müller-Groeling, and H. A. Weidenmüller, *Phys. Rep.* **299**, (1998), 189.
[85] B. D. Simons and B. L. Altshuler, *Phys. Rev. B* **48**, (1993), 5422.

[86] B. L. Altshuler and B. I. Shklovskii, *Zh. Teor. Éksp. Fiz.* **91**, (1986), 220 [*Sov. Phys. JETP* **64**, (1986), 127].
[87] A. D. Mirlin, *Phys. Rep.* **326**, (2000), 259.
[88] V. E. Kravtsov and I. V. Lerner, *Phys. Rev. Lett.* **74**, (1995), 2563.
[89] D. Mailly, C. Chapelier, and A. Benoit, *Phys. Rev. Lett.* **70**, (1993), 2020.
[90] B. L. Altshuler, Y. Gefen, and Y. Imry, *Phys. Rev. Lett.* **66**, (1991), 88.
[91] A. Schmid, *Phys. Rev. Lett.* **66**, (1991), 80.
[92] V. Chandrasekhar, R. A. Webb, M. J. Brady, M. B. Ketchen, W. J. Gallagher, and A. Kleinsasser, *Phys. Rev. Lett.* **67**, (1991), 3578.
[93] L. P. Lévy, G. Dolan, J. Dunsmuir, and H. Bouchiat, *Phys. Rev. Lett.* **64**, (1990), 2074.
[94] Ya. M. Blanter and A. D. Mirlin, *Phys. Rev. B* **63**, (2001), 113315.
[95] P. W. Brouwer, *Phys. Rev. B* **51**, (1995), 16878.
[96] C. M. Marcus, A. J. Rimberg, R. M. Westervelt, P. F. Hopkins, and A. C. Gossard, *Phys. Rev. Lett.* **69**, (1992), 506.
[97] S. Hikami, A. I. Larkin, and Y. Nagaoka, *Progr. Theor. Phys.* **63**, (1980), 707.
[98] G. Campagnano and Yu. V. Nazarov, *Phys. Rev. B* **74**, (2006), 125307.
[99] Yu. V. Nazarov, *Ann. Phys.* **8**, (1999), SI193.
[100] P. A. Lee, A. D. Stone, and H. Fukuyama, *Phys. Rev. B* **35**, (1987), 1039.
[101] D. J. Thouless, *Phys. Rep. Lett.* **39**, (1977), 1167.
[102] T. F. Rosenbaum, R. F. Milligan, M. A. Paalanen, G. A. Thomas, R. N. Bhatt, and W. Lin, *Phys. Rev. B* **27**, (1983), 7509.
[103] S. Katsumoto, F. Komori, N. Sano, and S. Kobayashi, *J. Phys. Soc. Japan*, **56**, (1987), 2259.
[104] B. I. Shklovskii and A. L. Efros, *Electronic Properties of Doped Semiconductors* (Berlin: Springer-Verlag, 1984).
[105] M. E. Raikh and I. M. Ruzin. In *Mesoscopic Phenomena in Solids*, eds. B. L. Altshuler, P. A. Lee, and R. A. Webb (New York: North-Holland, 1991), p. 303.
[106] A. A. Mogilyanskii and M. E. Raikh, *Zh. Teor. Éksp. Fiz.* **95**, (1989), 1870 [*Sov. Phys. JETP* **68**, (1990), 1081].
[107] Ya. M. Blanter and M. E. Raikh, *Phys. Rev. B* **63**, (2001), 075304.
[108] M. A. Nielsen and I. L. Chuang, *Quantum Computation and Quantum Information* (Cambridge: Cambridge University Press, 2000).
[109] P. W. Shor, *SIAM J. Comp.* **26**, (1997), 1484; also available as arXiv:quant-ph/9508027.
[110] A. Kitaev, arXiv:quant-phys/9511026.
[111] L. K. Grover, *Phys. Rev. Lett.* **79**, (1997), 325.
[112] A. Einstein, B. Podolsky, and N. Rosen, *Phys. Rev.* **47**, (1935), 777.
[113] J. S. Bell, *Physics* **1**, (1964), 195.
[114] J. F. Clauser, M. A. Horne, A. Shimony, and R. A. Holt, *Phys. Rev. Lett.* **23**, (1969), 880.
[115] N. Gisin and R. Thew, *Nature Photonics* **1**, (2007), 165.

[116] J. C. Egues, P. Recher, D. S. Saraga, V. N. Golovach, G. Burkard, E. V. Sukhorukov, and D. Loss. In *Quantum Noise in Mesoscopic Physics*, ed. Yu. V. Nazarov (Dordrecht: Kluwer Academic Publishing, 2003), p. 241.

[117] G. Burkard, *J. Phys. Condensed Matter* **19**, (2007), 233202.

[118] C. W. J. Beenakker. In *International School of Physics Enrico Fermi, Vol. 162, Quantum Computers, Algorithms and Chaos*, eds. G. Casati, D. L. Shepelyansky, P. Zoller, and G. Benenti (Amsterdam: IOS Press, 2006), p. 307; also available as arXiv:cond-mat/0508488.

[119] Y. Alhassid, *Rev. Mod. Phys.* **72**, (2000), 895.

[120] I. L. Aleiner, P. W. Brouwer, and L. I. Glazman, *Phys. Rep.* **358**, (2002), 309.

[121] B. L. Altshuler, Y. Gefen, A. Kamenev, and L. S. Levitov, *Phys. Rev. Lett.* **78**, (1997), 2803.

[122] A. V. Khaetskii and Yu. V. Nazarov, *Phys. Rev. B* **61**, (2000), 12639.

[123] S. I. Erlingsson, Yu. V. Nazarov, and V. I. Fal'ko, *Phys. Rev. B* **64**, (2001), 195306.

[124] A. T. Johnson, L. P. Kouwenhoven, W. de Jong, N. C. van der Waart, C. P. J. M. Harmans, and C. T. Foxon, *Phys. Rev. Lett.* **69**, (1992), 1592.

[125] H. W. C. Postma, *Carbon Nanotube Junctions and Devices* (The Netherlands: Delft University Press, 2001).

[126] T. H. Oosterkamp, T. Fujisawa, W. G. van der Wiel, K. Ishibashi, R. V. Hijman, S. Tarucha, and L. P. Kouwenhoven, *Nature* **395**, (1998), 873.

[127] Y. Nakamura, Yu. A. Pashkin, and J. S. Tsai, *Nature* **398**, (1999), 786.

[128] D. Vion, A. Aassime, A. Cottet, P. Joyez, H. Pothier, C. Urbina, D. Esteve, and M. H. Devoret, *Science* **296**, (2002), 886.

[129] Y. Makhlin, G. Schön, and A. Schnirman, *Rev. Mod. Phys.* **73**, (2001), 357.

[130] Yu. A. Pashkin, T. Yamamoto, O. Astafiev, Y. Nakamura, D. V. Averin, and J. S. Tsai, *Nature* **421**, (2003), 823.

[131] J. M. Martinis, S. Nam, J. Aumentado, and C. Urbina, *Phys. Rev. Lett.* **89**, (2002), 117901.

[132] N. Katz, M. Ansmann, R. C. Bialczak, *et al.*, *Science* **312**, (2006), 1498.

[133] M. Steffen, M. Ansmann, R. C. Bialczak, N. Katz, M. Neeley, E. M. Weig, A. N. Cleland, and J. M. Martinis, *Science* **313**, (2006), 1423.

[134] I. Chiorescu, Y. Nakamura, C. J. P. M. Harmans, and J. E. Mooij, *Science* **299**, (2003), 1869.

[135] A. Lupascu, S. Saito, T. Picot, P. C. de Groot, C. J. P. M. Harmans, and J. E. Mooij, *Nature Physics* **3**, (2007), 119.

[136] A. O. Niskanen, K. Harrabi, F. Yoshihara, Y. Nakamura, S. Lloyd, and J. S. Tsai, *Science* **316**, (2007), 723.

[137] R. Hanson, L. P. Kouwenhoven, J. R. Petta, S. Tarucha, and L. M. K. Vandersypen, *Rev. Mod. Phys.* **79**, (2007), 1217.

[138] D. Loss and D. P. Di Vincenzo, *Phys. Rev. A* **57**, (1998), 120.

[139] L. P. Kouwenhoven, J. M. Elzerman, R. Hanson, L. H. Willems van Beveren, and L. M. K. Vandersypen, *Physica Stat. Sol. (b)* **243**, (2006), 3682.

[140] J. R. Petta, A. C. Johnson, J. M. Taylor, *et al.*, *Science* **309**, (2005), 2180.

[141] A. J. Leggett, S. Chakravarty, A. T. Dorsey, M. P. A. Fisher, A. Garg, and W. Zwerger, *Rev. Mod. Phys.* 1 (1985), **59**.
[142] U. Weiss, *Quantum Dissipative Systems* (Singapore: World Scientific, 1999).
[143] A. O. Calderia and A. J. Leggett, *Ann. Phys.* **149**, (1983), 374.
[144] A. Schmid, *Phys. Rev. Lett.* **51**, (1983), 1506.
[145] G.-L. Ingold and Yu. V. Nazarov. In *Single Charge Tunneling: Coulomb Blockade Phenomena in Nanostructures*, eds. H. Grabert and M. H. Devoret (New York: Plenum, 1992), p. 21; also available as arXiv:cond-mat/0508728.
[146] J. E. Mooij and Y. V. Nazarov, *Nature Physics* **2**, (2006), 169.
[147] Y. V. Nazarov, *Zh. Teor. Éksp. Fiz.* **95**, (1989), 975. [*Sov. Phys. JETP* **68**, (1989), 561].
[148] H. W. Ch. Postma, M. de Jonge, Z. Yao, and C. Dekker, *Phys. Rev. B* **62**, (2000), R10653.
[149] M. P. A. Fisher and L. I. Glazman. In *Mesoscopic Electron Transport*, eds. L. P. Kouwenhoven, G. Schön, and L. L. Sohn (Dordrecht: Kluwer Academic Publishing, 1997), p. 331; also available as arXiv:cond-mat/9610037.
[150] B. L. Altshuler and A. A. Aronov, *Solid State Commun.* **30**, (1979), 115.
[151] S. Bermon and S. K. So, *Solid State Commun.* **27**, (1978), 727.
[152] R. C. Dynes and J. P. Garno, *Phys. Rev. Lett.* **46**, (1981), 137.
[153] B. L. Altshuler and A. G. Aronov, *Zh. Teor. Éksp. Fiz.* **77**, (1979), 2028. [*Sov. Phys. JETP* **50**, (1979), 968].
[154] K. A. Matveev, D. X. Yue, and L. I. Glazman, *Phys. Rev. Lett.* **71**, (1993), 3351.
[155] C. Altimiras, U. Genneser, A. Cavanna, D. Mailly, and F. Pierre, *Phys. Rev. Lett.* **99**, (2007), 256805.
[156] M. Kindermann and Y. V. Nazarov, *Phys. Rev. Lett* **91**, (2003), 136802.
[157] S. L. Lukyanov, *Nucl. Phys. B* **784**, (2007), 151.
[158] D. A. Bagrets and Y. V. Nazarov, *Phys. Rev. Lett.* **94**, (2005), 056801.
[159] B. L. Altshuler and A. G. Aronov. In *Electron–Electron Interactions in Disordered Systems*, eds. A. L. Efros and M. Pollak (Amsterdam: North-Holland, 1985), p. 27.
[160] G. Zala, B. N. Narozhny, and I. L. Aleiner, *Phys. Rev. B* **64**, (2001), 214204.
[161] P. Nozieres and C. T. De Dominicis, *Phys. Rev.* **178**, (1969), 1097.
[162] A. K. Geim, P. C. Main, N. La Scala, Jr. *et al., Phys. Rev. Lett.* **72**, (1994), 2061.
[163] H. Frahm, C. von Zobeltitz, N. Maire, and R. J. Haug, *Phys. Rev. B* **74**, (2006), 035329.
[164] D. A. Abanin and L. S. Levitov, *Phys. Rev. Lett.* **93**, (2004), 126802.
[165] J. Kondo, *Prog. Theor. Phys.* **32**, (1964), 37.
[166] P. Nozieres, *J. Low Temp. Phys.* **17**, (1974), 31.
[167] A. M. Tsvelick and P. B. Wiegmann, *Adv. Phys.* **32**, (1983), 453.
[168] A. Kaminski and L. I. Glazman, *Phys. Rev. Lett.* **86**, (2001), 2400.
[169] R. M. Potok, I. G. Rau, H. Shtrikman, Y. Oreg and D. Goldhaber-Gordon, *Nature* **446**, (2007), 167.
[170] P. Jarillo-Herrero, J. Kong, H. S. J. van der Zant, C. Dekker, L. P. Kouwenhoven, and S. De Franceschi, *Nature* **434**, (2005), 484.

[171] S. Sasaki, S. De Franceschi, J. M. Elzerman, W. G. van der Wiel, M. Eto, S. Tarucha, and L. P. Kouwenhoven, *Nature* **405**, (2000), 764.

[172] L. Borda, G. Zaránd, W. Hofstetter, B. I. Halperin, and J. von Delft, *Phys. Rev. Lett.* **90**, (2003), 026602.

[173] M. Pustilnik, L. Borda, L. I. Glazman, and J. von Delft, *Phys. Rev. B* **69**, (2004), 115316.

[174] D. Goldhaber-Gordon, J. Göres, M. A. Kastner, H. Shtrikman, D. Mahalu, and U. Meirav, *Phys. Rev. Lett.* **81**, (1998), 5225.

[175] L. I. Glazman and M. E. Raikh, *Pis'ma Zh. Teor. Éksp. Fiz.* **47**, (1988), 378. [*JETP Lett.* **47**, (1988), 452].

[176] D V. Averin. In *Quantum Noise in Mesoscopic Physics*, ed. Yu. V. Nazarov (Dordrecht: Kluwer Academic Publishing, 2003), p. 229.

[177] A. A. Clerk, S. M. Girvin, and A. D. Stone, *Phys. Rev. B* **67**, (2003), 165324.

[178] A. N. Korotkov, *Phys. Rev. B* **60**, (1999), 5737.

[179] F. Pierre, A. B. Gougam, A. Anthore, H. Pothier, D. Esteve, and N. O. Birge, *Phys. Rev. B* **68**, (2003), 085413.

[180] G. M. Alzoubi and N. O. Birge, *Phys. Rev. Lett.* **97**, (2006), 226803.

[181] A. Anthore, F. Pierre, H. Pothier, and D. Esteve, *Phys. Rev. Lett.* **90**, (2003), 076806.

[182] H. U. Baranger and A. D. Stone, *Phys. Rev. B* **40**, (1989), 8169.

[183] P.-G. de Gennes, *Superconductivity of Metals and Alloys* (New York: Perseus Books, 1999).

Index

addition energy, 213, 412
adiabatic manipulation, 403
adiabatic pumping, 93, 94, 97, 247
advanced algorithms, 381, 382, 384
  order-finding, 384
  quantum Fourier transform, 381, 383
Aharonov–Bohm effect
  ballistic, 69
  diffusive, 318, 320, 347, 361, 362, 541
Aharonov–Casher effect, 285
Altshuler–Aronov corrections, 457, 500
  conductivity, 520, 521, 523, 557, 559
  density of states, 511, 513
Anderson localization, 366
Anderson model, 423, 534–538
Andreev conductance, 101, 174, 200, 201, 204, 292
  ballistic, 103
  diffusive, 200
  double junction, 180
  tunnel junction, 103
  two ballistic contacts, 181
Andreev reflection, 98–104
  multiple, 109–112, 293
  single connector, 204
  two-electron tunneling, 292
arrays
  Josephson arrays, 278–287
  tunnel junction arrays, 220–223, 228, 248, 251, 253, 372
avoided crossing, 306, 307, 319

background charge, 217
BCS density of states, 197, 289, 290
beam splitter, 56, 57, 59, 61
Bell inequalities, 397
Bell parameter, 392, 393
billiard, 311, 312, 331
binomial distribution, 45, 46
BKT transition, 285–287
Bloch frequency, 276
Bloch sphere, 376, 377, 399
Boltzmann equation, 130–133, 137, 142, 143, 373
bosonization, 511
break junction, 112, 113
Breit–Wigner formula, 423
Brownian motion, 307, 332

canonical ensemble, 320–323
capacitance matrix, 217, 218, 278
capacitive response, 84
carbon nanotubes, 509, 510
chaotic cavity, 182, 325, 327
charge quantization, 211, 212, 215–217, 238, 270
charge qubits, 404, 427, 432, 433, 436
charge soliton, 221
charge-vortex duality, 212
charging energy, 211, 221
  charge qubits, 432, 433
  charging energy matrix, 217, 218, 221
  Cooper-pair box, 266–268, 270, 271, 274
  environment, 491, 493, 496, 519
  flux qubits, 439
  insulators, 372
  isolated island, 213
  Josephson arrays, 278, 279, 282, 283, 296
  Josephson junction, 268
  quantum dots, 411, 412, 415, 520, 534
  RC-time, 216
  SET, 224, 237, 243, 249, 253
  single-electron box, 215
  spin qubits, 445
  tunnel junction arrays, 223
  two islands, 219
circular ensembles, 308
  COE, 308, 309, 325, 326, 329, 330
  CSE, 308, 309, 325, 326
  CUE, 308, 325, 326, 329, 330
CNOT gate, 378, 380, 404, 405, 445
co-tunneling current, 256–258, 293, 427
co-tunneling rate, 255, 258, 424, 427
coherent state, 474
coherent tunneling, 274, 275, 471
collision integral, 132
commutation relations, 268
  charge and phase, 267, 268, 458
  shifted oscillators, 473
conductance quantization, 22, 23, 115
conductance quantum, 21, 36, 38, 73, 114, 188, 216, 315, 365, 366, 493
connector, 155, 156, 171, 172, 184, 189, 517, 519
  action, 165, 167, 172–175, 209, 348, 355
  diffusive, 185

finite-element separation, 166
full counting statistics, 207–210
leakage current, 167
matrix current, 164–168, 172–174, 176, 189, 348
quantum corrections, 348, 355, 357, 358
spin currents, 189–193
superconductivity, 196–200, 202–204
transmission distribution, 176–178, 184
continuous weak linear measurement, 538, 546
Cooper-pair box, 212, 265–267, 269–271, 295, 519
charge qubit, 432, 433
Cooper-pair tunneling, 494, 495, 497
cooperon, 316, 341–343, 345, 346, 352, 354–356, 358, 361–363, 551, 558
Coulomb blockade, 181, 211, 214, 231, 234, 237, 249, 264, 287, 293, 519, 555
carbon nanotubes, 510
double quantum dots, 428
environment, 488
escaping, 504, 505, 512
FCS, 233, 244
Josephson effect, 264, 265, 436
large conductances, 217, 513, 514, 517, 519, 520
many islands, 217, 220, 221, 223, 228, 230
memory cell, 245
parity effect, 289
quantum dots, 310, 407, 412, 414, 422
quasiparticles, 289
SET, 224–226, 237, 240, 241, 249, 250, 252, 258
single-electron box, 214, 215
turnstile, 246
Coulomb diamonds, 211
Coulomb shards, 228
NSN SET, 292
parity effect, 288
quantum dots, 412, 423, 424
SET, 226, 227, 238, 239
Coulomb gap, 372, 373
Coulomb oscillations, 211, 240
Coulomb shards, 228, 230, 246, 413
Coulomb staircase, 211, 240–242
counting statistics, *see* full counting statistics
current conservation
circuit theory, 155, 160, 162, 177, 178, 184, 195, 197, 200, 201
scattering theory, 51, 52, 58, 121, 129
superconductivity, 155
current copier, 253
CWLM, *see* continuous weak linear measurement

decoherence, 541
classical physics, 539, 540
diffusive conductors, 345–347, 350, 351, 358, 359, 361, 362, 500, 549–552, 554
quasiparticle leakage, 167, 194, 197
qubit, 381, 398, 433, 435, 438, 442, 446, 457, 485, 538, 542–546, 548
density matrix, 138
charge qubit, 413, 430, 431
entanglement, 389, 394, 396
environment, 464, 545, 548
equation of motion, 212, 296–298
extended, 139
phase qubit, 438
quantum state tomography, 405, 406
reduced, 394, 395
spin qubit, 452
density of states, 12, 17, 134, 136, 311
BCS, 197, 289, 290
Coulomb gap, 372, 373
insulators, 371
proximity effect, 196–198
dephasing, 538, 549, 551–556, 558, 560
classical, 539
depletion layer, 15, 16
diabatic manipulation, 403
diagram technique, 139, 337
diffusion coefficient, 134, 312, 332, 340, 452, 466, 469, 512
diffusive wires, 47, 48, 331, 555
diffuson, 316, 339–343, 345, 352, 354–356, 358, 362, 363, 557
discrete levels, 309, 310, 319, 406, 409, 415, 439
disorder, 29, 30, 35–37, 127, 128, 171
background charges, 223
diagram technique, 338, 339
interference, 360, 363
localization, 365, 366, 368, 369, 371
persistent currents, 319–321
quantum dots, 312, 313, 315
spin qubit, 452
2DEG, 15
white noise, 132, 143
distribution function
Boltzmann equation, 130–133
DMPK equation, 331–335
double junction, 157, 170
Eilenberger equation, 147
equilibrium, 7
FCS, 42, 58
level spacings, 301, 303
one-dimensional, 135, 136, 560, 561
Poisson kernel, 325, 327
Porter–Thomas, 305
reservoirs, 19, 33, 51, 168, 415, 488
spin currents, 190
distribution of transmission eigenvalues, 35, 36, 124, 172, 174, 175, 180, 326, 327, 336
DMPK equation, 331, 333–335, 358, 363, 364, 367, 368
doping, 14, 279
double quantum dots, 428
drift-diffusion equation, 133–135, 137, 144, 548

dynamical Coulomb blockade, 487, 504, 511–513, 560
dynamical phase, 66, 67, 71, 73, 75, 76

effective temperature, 159, 302, 307, 557
Eilenberger equation, 142–144
  supercurrent, 147, 148, 152
Einstein relation, 134
elastic co-tunneling
  quantum dots, 423–425
  SET, 251, 252, 263, 291, 506
electron–electron interaction, 48, 213, 407, 410, 415, 457, 487
electron–phonon interaction, 552
entanglement, 387, 393–396, 433, 531
environment, 457, 471
  classification, 479
  decoherence, 381
  fermionic, 523, 524
  Josephson junction, 277, 293, 494–499
  ohmic, 479, 480, 483–486, 498, 528, 545, 556
  quantum dots, 411, 415
  qubit relaxation, 447, 471, 483, 538, 542–544
  single-mode, 489
  spin-boson, 471, 472, 475
  subohmic, 480, 545, 555
  superohmic, 481, 484, 492
  transition rate, 477, 478
  tunneling, 487–489, 499, 500, 508, 513
EPR quantum key distribution, 390
exchange field, 114, 116, 118, 119, 122, 151
exchange interaction, 116, 445, 451, 453, 454, 456, 532
excitons, 220, 252, 253

Fano factor, 47, 93, 174, 180, 187, 208, 209, 331, 516, 520
FCS, *see* full counting statistics
Fermi edge singularity, 523, 524, 535
ferromagnet, 116, 121, 187–189
filling factor, *see* distribution function
finite-element approach, 157
fluctuation-dissipation theorem, 47, 61, 92, 547
Fokker–Planck equation, 307, 332
friction, 470–472, 539
Friedel sum rule, 94, 533, 537
full counting statistics, 41, 58, 89, 97, 168, 170, 173, 174, 516
  Coulomb blockade, 233, 243
  density matrix, 297
  Keldysh formalism, 139, 143
  multi-terminal, 205
  third cumulant, 46, 48, 49, 244

gauge-invariant phase, 66, 282
Gaussian ensembles, 300, 307–309
  GOE, 300, 301, 303, 314, 316, 323
  GSE, 300, 302, 303, 314, 316
  GUE, 300–304, 307, 314, 316, 323
giant magnetoresistance, 187
Green's function
  circuit theory, 163–165, 167–171, 191, 195, 197, 198, 201, 205, 206, 209
  diffuson, 339, 343
  Keldysh, 138, 140, 141
  non-crossing approximation, 337–339
  semiclassical, 137, 141, 142, 144–153, 161, 163, 168

Hadamard gate, 382, 388, 390
hopping conduction, 368

impedance, 465, 468
  Bloch oscillations, 276
  effective impedance, 500, 502, 504–506, 508, 510–513, 518
  large frequencies, 467
  Luttinger model, 469, 470, 509
  metals, 466
  ohmic, 490
  RC-line, 491, 492
  single-mode, 489
  tunneling, 489
  voltage correlations, 467
incoherent tunneling, 274, 499
inductive response, 84, 467
inelastic co-tunneling
  NSN SET, 290
  quantum dots, 413, 426
  SET, 251
  SSS SET, 293

Josephson effect, 105, 107, 108, 211, 212, 264, 265, 269
Josephson frequency, 108, 112, 276, 293, 497
Josephson junction, 106, 212, 265, 267, 268
  charge qubits, 432
  current-biased, 266, 270, 274
  environment, 494–499
  flux qubits, 439, 441
  phase qubits, 436

Keldysh approach, 138, 139
Keldysh contour, 138, 139
Kondo effect, 457, 523, 529, 531, 532, 534–538
Kubo formula, 464, 547

Landauer formula, 21, 24, 34, 38, 41, 45, 46, 83, 94, 121, 125, 126, 173, 315, 328, 423
  multi-terminal, 50, 51, 55
Laplace equations, 134, 155, 156
leakage currents, 159, 162, 167, 168, 172, 175, 198, 199
level correlation function, 314, 315, 317, 322

level statistics, 309, 311, 313–315, 320, 322, 323
Levitov formula, 43, 44, 46, 58, 59, 103, 168, 173, 174, 205, 209, 244
localization length, 364, 365, 367, 370, 371
Luttinger liquid, 470, 506, 509–511, 513, 515
Luttinger model, 469, 470, 506, 508
Lyapunov exponent, 312

macroscopic quantum tunneling, 212, 274, 277, 436
magnetic impurity, 532, 534, 554
magnetopolarizability, 321–323
master equation, 231, 232, 297
  FCS, 233
  quantum dots, 413, 416, 418, 419, 427
  qubit, 545
  two states, 238, 261
matrix current, 144, 145, 151, 159, 161–164, 166, 175, 195, 196, 201
matrix voltage, *see* Green's function
maximally entangled state, 389–395, 532
mean free path, 84, 133, 320, 334, 339, 363
  diffusion, 133, 135, 144, 145, 147, 151, 167, 193, 313, 340, 341, 343, 357, 469
  localization, 363, 367
  relaxation, 555
mean level spacing, 213, 311, 313–315, 344, 366, 367, 428, 520
memory cell, 237, 245, 246, 248, 250, 471, 482
mini-gap, 198
Mott's law, 371, 373

noise
  colored, 551
  Nyquist–Johnson, 47
  quantum, 90–93
  shot, 47, 174, 326, 331, 516
  voltage fluctuations, 496–498, 520, 521, 549, 553–556, 558, 560, 561
  white, 484, 539, 544, 556, 558
NSN SET, 290, 291, 294
nuclear spin, 452, 453, 455

offset charge, 218, 248
optimal fluctuation, 368
orthogonality catastrophe, 457, 475, 478, 479, 527

pair correlation function, 303
parametric statistics, 305–307
parity effect, 212, 287–289
parity tunneling, 291, 292, 294
percolation, 371, 372
Perron–Frobenius operator, 312, 315, 317
persistent current, 106, 318–323
phase diffusion, 484, 545
phase qubits, 436, 439
phase shifts
  Aharonov–Bohm, 65, 66, 71, 114, 282
  averaging, 126–128, 176, 205
  combining scattering matrices, 76
  distribution, 179
  double junction, 126
  dynamical, 64, 65, 78
  environment, 457
  interference, 71, 264, 360, 533
  qubit operations, 383, 403, 404
phase slips, 499
photon-assisted tunneling, 88, 93, 401
plasmon, 272, 274, 276, 277, 467, 470, 497
Poisson distribution, 44, 46–48, 311, 314, 474, 490
Poisson kernel, 325
Porter–Thomas distribution, 305
proximity effect, 196, 197, 202
pseudorate, 297, 298

QPC, *see* quantum point contact
quantronium, 433–435, 441
quantum computer, 374, 375, 378, 380, 381, 406, 407, 445, 446
quantum dots
  chaotic, 312, 313, 315
  discrete levels, 214, 309, 310, 313, 406
  disordered, 312, 317
  elastic co-tunneling, 423, 425, 426
  interactions, 409, 412
  Kondo effect, 534, 537
  large conductances, 520
  single-electron tunneling, 412, 420
  single-photon source, 397
  spin qubit, 445, 446, 452
quantum information, 381, 386–388, 390, 396, 397, 546
quantum non-demolition measurement, 441, 549
quantum phase transition, 278, 279, 284
quantum point contact, 17, 21, 23, 55
  connector, 174, 180
  detector, 429, 446
  kinetic inductance, 84
  non-ideal, 357
  spin filter, 115
  supercurrent, 106
quantum search, 381, 384–386
quantum state tomography, 405, 406
quantum teleportation, 387–390, 396, 397
quasicharge, 270, 275, 276
quasiparticle poisoning, 292

Rabi oscillations, 401, 435, 454, 455
Ramsey sequence, 402
random matrix, 194, 299
  Hermitian, 300–302, 305, 306, 313, 314, 316
  unitary, 308, 324, 325, 351

read-out, 380–382, 384, 387, 388, 430
charge qubit, 431, 432, 434, 435
flux qubit, 440, 441, 443
phase qubit, 436–438
spin qubit, 446, 447, 449–451, 456
reduced density matrix, 389
reflection coefficient, 10, 35, 93, 103, 332
reflection matrix, 32, 119, 395
relaxation, 457, 541, 561
classical physics, 539, 540
electrons, 549, 550, 552–556, 558, 560, 561
quantum dots, 414–416, 418–420, 427
qubit, 441, 442, 538, 542, 543, 545, 546
renormalization technique, 485, 486, 496, 514–520, 526, 527, 530, 532, 536
resonant manipulation, 399, 400, 402, 403, 435, 444
resonant pair tunneling, 294
resonant tunneling, 68, 69, 181, 244, 407, 413, 422, 425, 426, 517, 537
rotating wave approximation, 399
rotation by pulse, 399

scaling, 365, 367, 484–486
scattering matrix
combining scattering matrices, 76, 78
scatterers connected to a node, 80
spin, 118
scattering rate, 132, 533, 552
Schmid transition, 481, 499, 528
self-averaging, 126–128, 154, 171, 176, 365, 368
shake-up, 457, 473–475, 477, 478, 495
side bands, 86–88
single-electron pump, 247, 248
single-electron tunneling
environment, 487, 488
quantum dots, 413, 427
SET, 212, 246, 250, 258, 260–262
spectral rigidity, 305
spin accumulation, 120–122, 169, 189–193
spin blockade, 450, 451, 454
spin current, 120–122, 189, 192
spin echo, 402, 403, 456
spin filter, 115
spin qubit, 450
spin relaxation, 188, 190, 192, 193, 446–450, 452, 453, 533
spin singlet, 396, 410, 450, 451, 532
spin triplet, 410
spin-boson model, 471–473, 481, 485, 487, 488, 524, 542
spin-flip, 120, 453, 536, 554
spin-orbit interaction, 114, 117, 346, 447
cooperon, 315, 316, 340, 341
decoherence, 123, 190, 345, 359, 549, 553
random matrix, 300, 314, 316
scattering matrix, 118, 119
SQUID, 107, 108, 320, 437, 444
SSS SET, 293, 294
strong localization, 299, 363, 555, 556
subgap structure, 110
supercurrent, 106–108, 146, 147, 150, 151, 153–155, 204, 293, 294, 434, 495, 496
SWAP gate, 382, 383, 445

thermometer, 240
Thouless energy, 197, 315–317, 321, 347, 350, 362, 367, 550
Tien–Gordon effect, 85, 88, 89
Toffoli gate, 379, 380
transfer matrix, 78–80, 331
transmission coefficient, 10, 11, 18, 23, 24, 35, 53, 67, 68, 111, 136, 178, 446, 520
transmission matrix, 32, 37, 78, 79
transport spectroscopy, 413, 421, 422, 426
tunneling Hamiltonian, 235, 236, 253, 255, 295, 500
tunneling rates
connector, 87
quantum dots, 414–416, 419, 448, 449
SET, 224, 233–235, 249, 261
turnstile, 237, 246–248, 251
two-dimensional electron gas, 14, 15, 117
two-electron tunneling, 292

unitary mixer, 127, 128, 171
universal conductance fluctuations, 70, 71, 74, 330, 337
Usadel equation, 144–146, 172, 174, 183, 198
discretization, 166–168
leakage currents, 161, 162, 197
supercurrent, 151, 154

variable range hopping, 371
variational principle, 164, 165, 175, 209
voltage probe, 52–54, 199, 550
von Neumann entropy, 394
vortex, 212, 282–287
vortex–antivortex pairs, 286

weak antilocalization, 123
weak localization
Aharonov–Bohm effect, 71, 74
chaotic cavities, 326, 327, 329
dephasing, 551, 552, 558
diffusive conductors, 336, 337, 342, 345, 347–351, 354–362
DMPK equation, 331, 335
Wiedemann–Franz law, 136
Wigner lattice, 222
Wigner representation, 141
Wigner–Dyson distribution, 303

Zeeman splitting, 114–117, 151, 314, 315, 319, 410, 452
Zener tunneling, 404